Lecture Notes in Computer Science 4818

Commenced Publication in 1973
Founding and Former Series Editors:
Gerhard Goos, Juris Hartmanis, and Jan van Leeuwen

Lecture Notes in Computer Science 4915

Ivan Lirkov   Svetozar Margenov
Jerzy Waśniewski (Eds.)

Large-Scale
Scientific Computing

6th International Conference, LSSC 2007
Sozopol, Bulgaria, June 5-9, 2007
Revised Papers

 Springer

Volume Editors

Ivan Lirkov
Bulgarian Academy of Sciences
Institute for Parallel Processing
1113 Sofia, Bulgaria
E-mail: ivan@parallel.bas.bg

Svetozar Margenov
Bulgarian Academy of Sciences
Institute for Parallel Processing
1113 Sofia, Bulgaria
E-mail: margenov@parallel.bas.bg

Jerzy Waśniewski
Technikal University of Denmark
Department of Informatics and Mathematical Modelling
2800 Kongens Lyngby, Denmark
E-mail: jw@imm.dtu.dk

Library of Congress Control Number: Applied for

CR Subject Classification (1998): G.1, D.1, D.4, F.2, I.6, J.2, J.6

LNCS Sublibrary: SL 1 – Theoretical Computer Science and General Issues

ISSN   0302-9743
ISBN-10   3-540-78825-5 Springer Berlin Heidelberg New York
ISBN-13   978-3-540-78825-6 Springer Berlin Heidelberg New York

Springer is a part of Springer Science+Business Media

springer.com

© Springer-Verlag Berlin Heidelberg 2008
Printed in Germany

Typesetting: Camera-ready by author, data conversion by Scientific Publishing Services, Chennai, India
Printed on acid-free paper   SPIN: 12246786   06/3180   5 4 3 2 1 0

Preface

The 6th International Conference on Large-Scale Scientific Computations (LSSC 2007) was held in Sozopol, Bulgaria, June 5–9, 2007. The conference was organized by the Institute for Parallel Processing at the Bulgarian Academy of Sciences in cooperation with SIAM (Society for Industrial and Applied Mathematics). Partial support was also provided from project BIS-21++ funded by the European Commission in FP6 INCO via grant 016639/2005.

The conference was devoted to the 60th anniversary of Richard E. Ewing. Professor Ewing was awarded the medal of the Bulgarian Academy of Sciences for his contributions to the Bulgarian mathematical community and to the Academy of Sciences. His career spanned 33 years, primarily in academia, but also included industry. Since 1992 he worked at Texas A&M University being Dean of Science and Vice President of Research, as well as director of the Institute for Scientific Computation (ISC), which he founded in 1992. Professor Ewing is internationally well known with his contributions in applied mathematics, mathematical modeling, and large-scale scientific computations. He inspired a generation of researchers with creative enthusiasm for doing science on scientific computations. The preparatory work on this volume was almost done when the sad news came to us: Richard E. Ewing passed away on December 5, 2007 of an apparent heart attack while driving home from the office.

Plenary Invited Speakers and Lectures:

- O. Axelsson, Mesh-Independent Superlinear PCG Rates for Elliptic Problems
- R. Ewing, Mathematical Modeling and Scientific Computation in Energy and Environmental Applications
- L. Grüne, Numerical Optimization-Based Stabilization: From Hamilton-Jacobi-Bellman PDEs to Receding Horizon Control
- M. Gunzburger, Bridging Methods for Coupling Atomistic and Continuum Models
- B. Philippe, Domain Decomposition and Convergence of GMRES
- P. Vassilevski, Exact de Rham Sequences of Finite Element Spaces on Agglomerated Elements
- Z. Zlatev, Parallelization of Data Assimilation Modules

The success of the conference and the present volume in particular are the outcome of the joint efforts of many colleagues from various institutions and organizations. First, thanks to all the members of the Scientific Committee for their valuable contribution forming the scientific face of the conference, as well as for their help in reviewing contributed papers. We especially thank the organizers of the special sessions. We are also grateful to the staff involved in the local organization.

Traditionally, the purpose of the conference is to bring together scientists working with large-scale computational models of environmental and industrial problems and specialists in the field of numerical methods and algorithms for modern high-speed computers. The key lectures reviewed some of the advanced achievements in the field of numerical methods and their efficient applications. The conference lectures were presented by the university researchers and practical industry engineers including applied mathematicians, numerical analysts and computer experts. The general theme for LSSC 2007 was "Large-Scale Scientific Computing" with a particular focus on the organized special sessions.

Special Sessions and Organizers:

- Robust Multilevel and Hierarchical Preconditioning Methods — J. Kraus, S. Margenov, M. Neytcheva
- Domain Decomposition Methods — U. Langer
- Monte Carlo: Tools, Applications, Distributed Computing — I. Dimov, H. Kosina, M. Nedjalkov
- Operator Splittings, Their Application and Realization — I. Farago
- Large-Scale Computations in Coupled Engineering Phenomena with Multiple Scales — R. Ewing, O. Iliev, R. Lazarov
- Advances in Optimization, Control and Reduced Order Modeling — P. Bochev, M. Gunzburger
- Control Systems — M. Krastanov, V. Veliov
- Environmental Modelling — A. Ebel, K. Georgiev, Z. Zlatev
- Computational Grid and Large-Scale Problems — T. Gurov, A. Karaivanova, K. Skala
- Application of Metaheuristics to Large-Scale Problems — E. Alba, S. Fidanova

More than 150 participants from all over the world attended the conference representing some of the strongest research groups in the field of advanced large-scale scientific computing. This volume contains 86 papers submitted by authors from over 20 countries.

The 7th International Conference LSSC 2009 will be organized in June 2009.

December 2007

Ivan Lirkov
Svetozar Margenov
Jerzy Waśniewski

Table of Contents

III   Monte Carlo: Tools, Applications, Distributed Computing

IV   Operator Splittings, Their Application and Realization

V   Recent Advances in Methods and Applications for Large Scale Computations and Optimization of Coupled Engineering Problems

VI   Control Systems

VII   Environmental Modelling

VIII   Computational Grid and Large-Scale Problems

IX   Application of Metaheuristics to Large-Scale Problems

X   Contributed Talks

Part I

Plenary and Invited Papers

Mesh Independent Convergence Rates Via Differential Operator Pairs

Owe Axelsson[1] and János Karátson[2]

[1] Department of Information Technology, Uppsala University,
Sweden & Institute of Geonics AS CR, Ostrava, Czech Republic
[2] Department of Applied Analysis, ELTE University, Budapest, Hungary

Abstract. In solving large linear systems arising from the discretization of elliptic problems by iteration, it is essential to use efficient preconditioners. The preconditioners should result in a mesh independent linear or, possibly even superlinear, convergence rate. It is shown that a general way to construct such preconditioners is via equivalent pairs or compact-equivalent pairs of elliptic operators.

1 Introduction

Preconditioning is an essential part of iterative solution methods, such as conjugate gradient methods. For (symmetric or unsymmetric) elliptic problems, a primary goal is then to achieve a mesh independent convergence rate, which can enable the solution of extremely large scale problems. An efficient way to construct such a preconditioner is to base it on an, in some way, simplified differential operator. The given and the preconditioning operators should then form an equivalent pair, based on some inner product. Then the finite element discretization of these operators form the given matrix and its preconditioner, that is, if the given elliptic boundary value problem

$$Lu = f$$

is suitably discretized to an algebraic system $L_h u_h = f_h$, then another, equivalent operator S considerably simpler than L, is discretized in the same FEM subspace to form a preconditioner S_h, and the system which is actually solved is

$$S_h^{-1} L_h u_h = S_h^{-1} f_h.$$

By use of equivalent pairs of operators, one can achieve a mesh independent linear convergence rate. If, in addition, the operator pairs are compact-equivalent, then one can achieve a mesh independent superlinear convergence rate. The purpose of this presentation is to give a comprehensive background to the above, and to illustrate its applications for some important classes of elliptic problems. Mesh independence and equivalent operator pairs have been rigorously dealt with previously in [12,15], while superlinear rate of convergence and compact-equivalent pairs have been treated in [6,8] (see also the references

I. Lirkov, S. Margenov, and J. Waśniewski (Eds.): LSSC 2007, LNCS 4818, pp. 3–15, 2008.
© Springer-Verlag Berlin Heidelberg 2008

therein). Since in general the problems dealt with will be nonsymmetric, we first recall some basic results on generalized conjugate gradient methods, which will be used here. Equivalent and compact-equivalent pairs of operators are then discussed. Then some applications are shown, including a superlinear convergence result for problems with variable diffusion coefficients.

2   Conjugate Gradient Algorithms and Their Rate of Convergence

Let us consider a linear system

$$Au = b \tag{1}$$

with a given nonsingular matrix $A \in \mathbf{R}^{n \times n}$, $f \in \mathbf{R}^n$ and solution u. Letting $\langle ., . \rangle$ be a given inner product on $\mathbf{R}^n$ and denoting by A^* the adjoint of A w.r.t. this inner product, in what follows we assume that

$$A + A^* > 0,$$

i.e., A is positive definite w.r.t. $\langle ., . \rangle$. We define the following quantities, to be used frequently in the study of convergence:

$$\lambda_0 := \lambda_0(A) := \inf\{\langle Ax, x \rangle : \|x\| = 1\} > 0, \qquad \Lambda := \Lambda(A) := \|A\|, \tag{2}$$

where $\|.\|$ denotes the norm induced by the inner product $\langle ., . \rangle$.

2.1   Self-adjoint Problems: The Standard CG Method

If A is self-adjoint, then the standard CG method reads as follows [3]: let $u_0 \in \mathbf{R}^n$ be arbitrary, $d_0 := -r_0$; for given u_k and d_k, with residuals $r_k := Au_k - b$, we let

$$u_{k+1} = u_k + \alpha_k d_k, \quad d_{k+1} = -r_{k+1} + \beta_k d_k, \tag{3}$$

where $\alpha_k = -\dfrac{\langle r_k, d_k \rangle}{\langle Ad_k, d_k \rangle}$, $\beta_k = \dfrac{\|r_{k+1}\|^2}{\|r_k\|^2}$.

To save computational time, normally the residual vectors are also formed by recursion:

$$r_{k+1} = r_k + \alpha_k Ad_k, \tag{4}$$

further we use $\langle r_k, d_k \rangle = -\|r_k\|^2$ for α_k, i.e, $\alpha_k = \|r_k\|^2 / \langle Ad_k, d_k \rangle$. In the study of convergence, one considers the error vector $e_k = u - u_k$ and is generally interested in its energy norm

$$\|e_k\|_A = \langle Ae_k, e_k \rangle^{1/2}. \tag{5}$$

Now we briefly summarize the minimax property of the CG method and two convergence estimates, based on [3]. We first note that the construction of the algorithm implies $e_k = P_k(A)e_0$ with some $P_k \in \pi_k^1$, where π_k^1 denotes the set of polynomials of degree k, normalized at the origin. Moreover, we have the optimality property

$$\|e_k\|_A = \min_{P_k \in \pi_k^1} \|P_k(A)e_0\|_A. \tag{6}$$

If $0 < \lambda_1 \leq \ldots \leq \lambda_n$ are the eigenvalues of A, then (6) implies

$$\frac{\|e_k\|_A}{\|e_0\|_A} \leq \min_{P_k \in \pi_k^1} \max_{\lambda \in \sigma(A)\}} |P_k(\lambda)|, \tag{7}$$

which is a basis for the convergence estimates of the CG method.

Using elementary estimates via Chebyshev polynomials, we obtain from (7) the linear convergence estimate

$$\left(\frac{\|e_k\|_A}{\|e_0\|_A}\right)^{1/k} \leq 2^{1/k} \frac{\sqrt{\lambda_n} - \sqrt{\lambda_1}}{\sqrt{\lambda_n} + \sqrt{\lambda_1}} = 2^{1/k} \frac{\sqrt{\kappa(A)} - 1}{\sqrt{\kappa(A)} + 1} \qquad (k = 1, 2, ..., n), \tag{8}$$

where $\kappa(A) = \lambda_n/\lambda_1$ is the standard condition number.

To show a superlinear convergence rate, another useful estimate is derived if we consider the decomposition

$$A = I + E \tag{9}$$

and choose $P_k(\lambda) := \prod_{j=1}^{k} \left(1 - \frac{\lambda}{\lambda_j}\right)$ in (7), where $\lambda_j := \lambda_j(A)$ are ordered according to $|\lambda_1 - 1| \geq |\lambda_2 - 1| \geq \ldots$ Then a calculation [3] yields

$$\left(\frac{\|e_k\|_A}{\|e_0\|_A}\right)^{1/k} \leq \frac{2}{k\,\lambda_0} \sum_{j=1}^{k} |\lambda_j(E)| \qquad (k = 1, 2, ..., n). \tag{10}$$

Here by assumption $|\lambda_1(E)| \geq |\lambda_2(E)| \geq \ldots$ If these eigenvalues accumulate in zero then the convergence factor is less than 1 for k sufficiently large and moreover, the upper bound decreases, i.e. we obtain a superlinear convergence rate.

2.2 Nonsymmetric Systems

For nonsymmetric matrices A, several CG algorithms exist (see e.g. [1,3,11]). First we discuss the approach that generalizes the minimization property (6) for nonsymmetric A and avoids the use of the normal equation, see (18) below. A general form of the algorithm, which uses least-square residual minimization, is the generalized conjugate gradient–least square method (GCG-LS method) [2,3]. Its full version uses all previous search directions when updating the new approximation, whose construction also involves an integer $t \in \mathbf{N}$, further, we let $t_k = \min\{k,t\}$ $(k \geq 0)$. Then the algorithm is as follows: let $u_0 \in \mathbf{R}^n$ be arbitrary, $d_0 := Au_0 - b$; for given u_k and d_k, with $r_k := Au_k - b$, we let

$$\begin{cases} u_{k+1} = u_k + \sum_{j=0}^{k} \alpha_{k-j}^{(k)} d_{k-j} \text{ and } d_{k+1} = r_{k+1} + \sum_{j=0}^{t_k} \beta_{k-j}^{(k)} d_{k-j}, \\ \text{where } \beta_{k-j}^{(k)} = -\langle Ar_{k+1}, Ad_{k-j}\rangle / \|Ad_{k-j}\|^2 \quad (j = 0, \ldots, s_k) \\ \text{and the numbers} \quad \alpha_{k-j}^{(k)} \quad (j = 0, \ldots, k) \quad \text{are the solution of} \\ \sum_{j=0}^{k} \alpha_{k-j}^{(k)} \langle Ad_{k-j}, Ad_{k-l}\rangle = -\langle r_k, Ad_{k-l}\rangle \quad (0 \leq l \leq k). \end{cases} \tag{11}$$

There exist various truncated versions of the GCG-LS method that use only a bounded number of search directions, such as GCG-LS(k), Orthomin(k), and GCR(k) (see e.g. [3,11]). Of special interest is the GCG-LS(0) method, which requires only a single, namely the current search direction such that (11) is replaced by

$$\begin{cases} u_{k+1} = u_k + \alpha_k d_k, \text{ where } \alpha_k = -\langle r_k, Ad_k\rangle / \|Ad_k\|^2; \\ d_{k+1} = r_{k+1} + \beta_k d_k, \text{ where } \beta_k = -\langle Ar_{k+1}, Ad_k\rangle / \|Ad_k\|^2. \end{cases} \tag{12}$$

Proposition 1. *(see, e.g., [2]). If there exist constants $c_1, c_2 \in \mathbf{R}$ such that $A^* = c_1 A + c_2 I$, then the truncated GCG-LS(0) method (12) coincides with the full version (11).*

The convergence estimates in the nonsymmetric case often involve the residual

$$r_k = Ae_k = Au_k - b. \tag{13}$$

as this is readily available. It follows from [2] that

$$\|r_{k+1}\| \leq \left(1 - \left(\frac{\lambda_0}{\Lambda}\right)^2\right)^{k/2} \|r_0\| \qquad (k = 1, 2, \ldots, n). \tag{14}$$

The same estimate holds for the GCR and Orthomin methods together with their truncated versions, see [11].

An important occurrence of the truncated GCG-LS(0) algorithm (12) arises when the decomposition

$$A = I + E \tag{15}$$

holds for some antisymmetric matrix E, which often comes from symmetric part preconditioning. In this case $A^* = 2I - A$ and hence Proposition 1 is valid [2]. The convergence of this iteration is then determined by E, and using that E has imaginary eigenvalues, one can easily verify as in [7] that $1 - (\lambda_0/\Lambda)^2 = \|E\|^2/(1+\|E\|^2)$. Hence (14) yields that the GCG-LS(0) algorithm (12) converges as

$$\left(\frac{\|r_k\|}{\|r_0\|}\right)^{1/k} \leq \frac{\|E\|}{\sqrt{1 + \|E\|^2}} \qquad (k = 1, 2, \ldots, n). \tag{16}$$

On the other hand, if A is normal and we have the decomposition (9), then the residual errors satisfy a similar estimate to (10) obtained in the symmetric case, see [3]:

$$\left(\frac{\|r_k\|}{\|r_0\|}\right)^{1/k} \leq \frac{2}{k\lambda_0} \sum_{j=1}^{k} |\lambda_j(C)| \qquad (k = 1, 2, ..., n). \qquad (17)$$

Again, this shows superlinear convergence if the eigenvalues $\lambda_j(C)$ accumulate in zero. If A is non-normal then, as shown in [3], the superlinear estimate remains uniform in a family of problems if the order of the largest Jordan block is bounded as $n \to \infty$.

Another common way to solve (1) with nonsymmetric A is the CGN method, where we consider the normal equation

$$A^*Au = A^*b \qquad (18)$$

and apply the symmetric CG algorithm (3) for the latter [13]. In order to preserve the notation r_k for the residual $Au_k - b$, we replace r_k in (3) by s_k and let $r_k = A^{-*}s_k$, i.e., we have $s_k = A^*r_k$. Further, A and b are replaced by A^*A and A^*b, respectively. From this we obtain the following algorithmic form: let $u_0 \in \mathbf{R}^n$ be arbitrary, $r_0 := Au_0 - b$, $s_0 := d_0 := A^*r_0$; for given d_k, u_k, r_k and s_k, we let

$$\begin{cases} z_k = Ad_k, \\[2mm] \alpha_k = -\dfrac{\langle r_k, z_k \rangle}{\|z_k\|^2}, \quad u_{k+1} = u_k + \alpha_k d_k, \quad r_{k+1} = r_k + \alpha_k z_k; \\[2mm] s_{k+1} = A^*r_{k+1}, \\[2mm] \beta_k = \dfrac{\|s_{k+1}\|^2}{\|s_k\|^2}, \quad d_{k+1} = s_{k+1} + \beta_k d_k. \end{cases} \qquad (19)$$

The convergence estimates for this algorithm follow directly from the symmetric case. Using $\|e_k\|_{A^*A} = \|Ae_k\| = \|r_k\|$ and that (2) implies $\kappa(A^*A)^{1/2} = \kappa(A) \leq \Lambda/\lambda_0$, we obtain

$$\left(\frac{\|r_k\|}{\|r_0\|}\right)^{1/k} \leq 2^{1/k} \frac{\Lambda - \lambda_0}{\Lambda + \lambda_0} \qquad (k = 1, 2, ..., n). \qquad (20)$$

On the other hand, having the decomposition (9), using the relation $\|(A^*A)^{-1}\| = \|A^{-1}\|^2 \leq \lambda_0^{-2}$ and $A^*A = I + (C^* + C + C^*C)$, the analogue of the superlinear estimate (10) for equation $A^*Au = A^*b$ becomes

$$\left(\frac{\|r_k\|}{\|r_0\|}\right)^{1/k} \leq \frac{2}{k\lambda_0^2} \sum_{i=1}^{k} \left(|\lambda_i(C^* + C)| + \lambda_i(C^*C)\right) \qquad (k = 1, 2, ..., n). \qquad (21)$$

3   Equivalent Operators and Linear Convergence

We now give a comprehensive presentation of the equivalence property between pairs of operators, followed by a basic example for elliptic operators. First a brief outline of some theory from [12] is given.

Let $B : W \to V$ and $A : W \to V$ be linear operators between the Hilbert spaces W and V. Let B and A be invertible and let $D := D(A) \cap D(B)$ be dense, where $D(A)$ denotes the domain of an operator A. The operator A is said to be equivalent in V-norm to B on D if there exist constants $K \geq k > 0$ such that

$$k \leq \frac{\|Au\|_V}{\|Bu\|_V} \leq K \qquad (u \in D \setminus \{0\}). \tag{22}$$

The condition number of AB^{-1} in V is then bounded by K/k. Similarly, the W-norm equivalence of B^{-1} and A^{-1} implies this bound for $B^{-1}A$. If A_h and B_h are finite element approximations (orthogonal projections) of A and B, respectively, then the families (A_h) and (B_h) are V-norm uniformly equivalent with the same bounds as A and B.

In practice for elliptic operators, it is convenient to use H^1-norm equivalence, since this avoids unrealistic regularity requirements (such as $u \in H^2(\Omega)$). We then use the weak form satisfying

$$\langle A_w u, v \rangle_{H_D^1} = \langle Au, v \rangle_{L^2} \qquad (u, v \in D(A)), \tag{23}$$

where $H_D^1(\Omega)$ is defined in (26). The fundamental result on H^1-norm equivalence in [15] reads as follows: if A and B are invertible uniformly elliptic operators, then A_w^{-1} and B_w^{-1} are H^1-norm equivalent if and only if A and B have homogeneous Dirichlet boundary conditions on the same portion of the boundary.

In what follows, we use a simpler Hilbert space setting of equivalent operators from [8] that suffices to treat most practical problems. We recall that for a symmetric coercive operator, the energy space H_S is the completion of $D(S)$ under the inner product $\langle u, v \rangle_S = \langle Su, v \rangle$, and the coercivity of S implies $H_S \subset H$. The corresponding S-norm is denoted by $\|u\|_S$, and the space of bounded linear operators on H_S by $B(H_S)$.

Definition 1. Let S be a linear symmetric coercive operator in H. A linear operator L in H is said to be S-bounded and S-coercive, and we write $L \in BC_S(H)$, if the following properties hold:

(i) $D(L) \subset H_S$ and $D(L)$ is dense in H_S in the S-norm;
(ii) there exists $M > 0$ such that

$$|\langle Lu, v \rangle| \leq M \|u\|_S \|v\|_S \qquad (u, v \in D(L));$$

(iii) there exists $m > 0$ such that

$$\langle Lu, u \rangle \geq m \|u\|_S^2 \qquad (u \in D(L)).$$

The weak form of such operators L is defined analogously to (23), and produces a variationally defined symmetrically preconditioned operator:

Definition 2. For any $L \in BC_S(H)$, let $L_S \in B(H_S)$ be defined by

$$\langle L_S u, v \rangle_S = \langle Lu, v \rangle \qquad (u, v \in D(L)).$$

Remark 1. (i) Owing to Riesz representation theorem the above definition makes sense. (ii) L_S is coercive on H_S. (iii) If $R(L) \subset R(S)$ (where $R(.)$ denotes the range), then $L_S|_{D(L)} = S^{-1}L$.

The above setting leads to a special case of equivalent operators:

Proposition 2. *[9] Let N and L be S-bounded and S-coercive operators for the same S. Then*

    (a) N_S and L_S are H_S-norm equivalent,
    (b) N_S^{-1} and L_S^{-1} are H_S-norm equivalent.

Definition 3. For given $L \in BC_S(H)$, we call $u \in H_S$ the *weak solution* of equation $Lu = g$ if $\langle L_S u, v \rangle_S = \langle g, v \rangle$ $(v \in H_S)$. (Note that if $u \in D(L)$ then u is a strong solution.)

Example. A basic example of equivalent elliptic operators in the S-bounded and S-coercive setting is as follows. Let us define the operator

$$Lu \equiv -\mathrm{div}\,(A\,\nabla u) + \mathbf{b} \cdot \nabla u + cu \qquad \text{for } u_{|\Gamma_D} = 0,\ \frac{\partial u}{\partial \nu_A} + \alpha u_{|\Gamma_N} = 0, \quad (24)$$

where $\dfrac{\partial u}{\partial \nu_A} = A\nu \cdot \nabla u$ and ν denotes the outer normal derivative, with the following properties:

Assumptions 3.1

  (i) $\Omega \subset \mathbf{R}^d$ is a bounded piecewise C^1 domain; Γ_D, Γ_N are disjoint open measurable subsets of $\partial\Omega$ such that $\partial\Omega = \overline{\Gamma}_D \cup \overline{\Gamma}_N$;

  (ii) $A \in C^1(\overline{\Omega}, \mathbf{R}^{d \times d})$ and for all $x \in \overline{\Omega}$ the matrix $A(x)$ is symmetric; $\mathbf{b} \in C^1(\overline{\Omega})^d$, $c \in L^\infty(\Omega)$, $\alpha \in L^\infty(\Gamma_N)$;

  (iii) there exists $p > 0$ such that $A(x)\xi \cdot \xi \geq p\,|\xi|^2$ for all $x \in \overline{\Omega}$ and $\xi \in \mathbf{R}^d$; $\hat{c} := c - \frac{1}{2}\,\mathrm{div}\,\mathbf{b} \geq 0$ in Ω and $\hat{\alpha} := \alpha + \frac{1}{2}\,(\mathbf{b} \cdot \nu) \geq 0$ on Γ_N;

  (iv) either $\Gamma_D \neq \emptyset$, or $\hat{c}$ or $\hat{\alpha}$ has a positive lower bound.

Let S be a symmetric elliptic operator on the same domain Ω:

$$Su \equiv -\mathrm{div}\,(G\,\nabla u) + \sigma u \qquad \text{for } u_{|\Gamma_D} = 0,\ \tfrac{\partial u}{\partial \nu_G} + \beta u_{|\Gamma_N} = 0, \quad (25)$$

with analogous assumptions on G, σ, β. Let

$$H_D^1(\Omega) = \{u \in H^1(\Omega),\ u_{|\Gamma_D} = 0\}, \quad \langle u, v \rangle_S = \int_\Omega (G\,\nabla u \cdot \nabla v + \sigma uv) + \int_{\Gamma_N} \beta uv\, d\sigma \quad (26)$$

which is the energy space H_S of S. Then the following result can be proved:

Proposition 3. *[9]. The operator L is S-bounded and S-coercive in $L^2(\Omega)$.*

The major results in this section are mesh independent convergence bounds corresponding to some preconditioning concepts. Let us return to a general Hilbert space H. To solve $Lu = g$, we use a Galerkin discretization in $V_h = \text{span}\{\varphi_1, \ldots, \varphi_n\} \subset H_S$, where φ_i are linearly independent. Let

$$\mathbf{L}_h := \left\{ \langle L_S \varphi_i, \varphi_j \rangle_S \right\}_{i,j=1}^n$$

and, for the discrete solution, solve

$$\mathbf{L}_h\, \mathbf{c} = \mathbf{b}_h \tag{27}$$

with $\mathbf{b}_h = \{\langle g, \varphi_j \rangle\}_{j=1}^n$. Since $L \in BC_S(H)$, the symmetric part of $\mathbf{L}_h$ is positive definite.

First, let L be symmetric itself. Then its S-coercivity and S-boundedness turns into the spectral equivalence relation

$$m\|u\|_S^2 \leq \langle L_S u, u \rangle_S \leq M\|u\|_S^2 \qquad (u \in H_S). \tag{28}$$

Then $\mathbf{L}_h$ is symmetric too. Let

$$\mathbf{S}_h = \left\{ \langle \varphi_i, \varphi_j \rangle_S \right\}_{i,j=1}^n \tag{29}$$

be the stiffness matrix of S, to be used as preconditioner for $\mathbf{L}_h$. This yields the preconditioned system

$$\mathbf{S}_h^{-1}\mathbf{L}_h\, \mathbf{c} = \mathbf{S}_h^{-1}\mathbf{b}_h. \tag{30}$$

Now $\mathbf{S}_h^{-1}\mathbf{L}_h$ is self-adjoint w.r.t. the inner product $\langle \mathbf{c}, \mathbf{d} \rangle_{\mathbf{S}_h} := \mathbf{S}_h\, \mathbf{c} \cdot \mathbf{d}$.

Proposition 4. *(see, e.g., [10]). For any subspace $V_h \subset H_S$,*

$$\kappa(\mathbf{S}_h^{-1}\mathbf{L}_h) \leq \frac{M}{m} \tag{31}$$

independently of V_h.

Consider now nonsymmetric problems with symmetric equivalent preconditioners. With $\mathbf{S}_h$ from (29) as preconditioner, we use the bounds (2) for the GCG-LS and CGN methods:

$$\lambda_0 = \lambda_0(\mathbf{S}_h^{-1}\mathbf{L}_h) := \inf\{\mathbf{L}_h\, \mathbf{c} \cdot \mathbf{c} : \ \mathbf{S}_h\, \mathbf{c} \cdot \mathbf{c} = 1\}, \quad \Lambda = \Lambda(\mathbf{S}_h^{-1}\mathbf{L}_h) := \|\mathbf{S}_h^{-1}\mathbf{L}_h\|_{\mathbf{S}_h}.$$

These bounds can be estimated using the S-coercivity and S-boundedness

$$m\|u\|_S^2 \leq \langle L_S u, u \rangle_S, \quad |\langle L_S u, v \rangle_S| \leq M\|u\|_S\|v\|_S \qquad (u, v \in H_S). \tag{32}$$

Proposition 5. *[9]. For any subspace $V_h \subset H_S$,*

$$\frac{\Lambda(\mathbf{S}_h^{-1}\mathbf{L}_h)}{\lambda_0(\mathbf{S}_h^{-1}\mathbf{L}_h)} \leq \frac{M}{m} \tag{33}$$

independently of V_h.

Consequently, by (14), the GCG-LS algorithm (11) for system (30) satisfies

$$\left(\frac{\|r_k\|_{\mathbf{S}_h}}{\|r_0\|_{\mathbf{S}_h}}\right)^{1/k} \leq \left(1 - \left(\frac{m}{M}\right)^2\right)^{1/2} \qquad (k = 1, 2, ..., n), \tag{34}$$

which holds as well for the GCR and Orthomin methods together with their truncated versions; further, by (20), the CGN algorithm (19) for system (30) satisfies

$$\left(\frac{\|r_k\|_{\mathbf{S}_h}}{\|r_0\|_{\mathbf{S}_h}}\right)^{1/k} \leq 2^{1/k}\frac{M - m}{M + m} \qquad (k = 1, 2, ..., n). \tag{35}$$

Finally, let now $\mathbf{S}_h := (\mathbf{L}_h + \mathbf{L}_h^T)/2$ be the symmetric part of $\mathbf{L}_h$. Here $\mathbf{L}_h = \mathbf{S}_h + \mathbf{Q}_h$ with $\mathbf{Q}_h := (\mathbf{L}_h - \mathbf{L}_h^T)/2$, and $L_S = I + Q_S$ where Q_S is antisymmetric in H_S, further, $\mathbf{S}_h^{-1}\mathbf{L}_h = \mathbf{I}_h + \mathbf{S}_h^{-1}\mathbf{Q}_h$ where $\mathbf{S}_h^{-1}\mathbf{Q}_h$ is antisymmetric w.r.t. the inner product $\langle .,.\rangle_{\mathbf{S}_h}$. Then the full GCG algorithm reduces to the simple truncated version (12), further, we obtain the mesh independent estimate $\|\mathbf{S}_h^{-1}\mathbf{Q}_h\|_{\mathbf{S}_h} \leq \|Q_S\|$, whence by (16),

$$\left(\frac{\|r_k\|_{\mathbf{S}_h}}{\|r_0\|_{\mathbf{S}_h}}\right)^{1/k} \leq \frac{\|Q_S\|}{\sqrt{1 + \|Q_S\|^2}} \qquad (k = 1, 2, ..., n). \tag{36}$$

4   Compact-Equivalent Operators and Superlinear Convergence

We now present the property of compact-equivalence between operator pairs, based on [8], which is a refinement of the equivalence property and provides mesh independent superlinear convergence. We use the Hilbert space setting of Definition 1 and include a main example (which, moreover, is a characterization) for elliptic operators.

Definition 4. Let L and N be S-bounded and S-coercive operators in H. We call L and N *compact-equivalent in H_S* if

$$L_S = \mu N_S + Q_S \tag{37}$$

for some constant $\mu > 0$ and compact operator $Q_S \in B(H_S)$.

Remark 2. If $R(L) \subset R(N)$, then compact-equivalence of L and N means that $N^{-1}L$ is a compact perturbation E of constant times the identity in the space H_S, i.e., $N^{-1}L = \mu I + E$.

One can characterize compact-equivalence for elliptic operators. Let us take two operators as in (24):

$$L_1 u \equiv -\text{div}\,(A_1 \nabla u) + \mathbf{b}_1 \cdot \nabla u + c_1 u \qquad \text{for } u_{|\Gamma_D} = 0,\ \frac{\partial u}{\partial \nu_{A_1}} + \alpha_1 u_{|\Gamma_N} = 0,$$

$$L_2 u \equiv -\text{div}\,(A_2 \nabla u) + \mathbf{b}_2 \cdot \nabla u + c_2 u \qquad \text{for } u_{|\Gamma_D} = 0,\ \frac{\partial u}{\partial \nu_{A_2}} + \alpha_2 u_{|\Gamma_N} = 0$$

where we assume that L_1 and L_2 satisfy Assumptions 3.1. Then the following fundamental result holds:

Proposition 6. *[8].   The elliptic operators L_1 and L_2 are compact-equivalent in $H_D^1(\Omega)$ if and only if their principal parts coincide up to some constant $\mu > 0$, i.e. $A_1 = \mu A_2$.*

Now we discuss preconditioned CG methods and corresponding mesh independent superlinear convergence rates. Let us consider an operator equation $Lu = g$ in a Hilbert space H for some S-bounded and S-coercive operator L, and its Galerkin discretization as in (27). Let us first introduce the stiffness matrix $\mathbf{S}_h$ as in (29) as preconditioner.

Proposition 7. *[8]. If L and S are compact-equivalent with $\mu = 1$, then the CGN algorithm (19) for system (30) yields*

$$\left(\frac{\|r_k\|_{\mathbf{S}_h}}{\|r_0\|_{\mathbf{S}_h}} \right)^{1/k} \le \varepsilon_k \qquad (k = 1, 2, ..., n), \tag{38}$$

where $\varepsilon_k \to 0$ is a sequence independent of V_h.

A similar result holds for the GCG-LS method, provided however that Q_S is a normal compact operator in H_S and the matrix $\mathbf{S}_h^{-1}\mathbf{Q}_h$ is $\mathbf{S}_h$-normal [6]. These properties hold, in particular, for symmetric part preconditioning. The sequence ε_k contains similar expressions of eigenvalues as (17) or (21) related to Q_S, which we omit for brevity.

For elliptic operators, we can derive a corresponding result. Let L be the elliptic operator in (24) and S be the symmetric operator in (25). If the principal parts of L and S coincide, i.e., $A = G$, then L and S are compact-equivalent by Proposition 6, and we have $\mu = 1$. Hence Proposition 7 yields a mesh independent superlinear convergence rate. Further, by [8], an explicit order of magnitude in which $\varepsilon_k \to 0$ can be determined in some cases. Namely, when the asymptotics for symmetric eigenvalue problems $Su = \mu u$, $u_{|\Gamma_D} = 0$, $r\left(\frac{\partial u}{\partial \nu_A} + \beta u\right)_{|\Gamma_N} = \mu u$ satisfies $\mu_i = O(i^{2/d})$, as is the case for Dirichlet problems, then

$$\varepsilon_k \le O\left(\frac{\log k}{k}\right) \quad \text{if } d = 2 \quad \text{and} \quad \varepsilon_k \le O\left(\frac{1}{k^{2/d}}\right) \quad \text{if } d \ge 3. \tag{39}$$

5 Applications of Symmetric Equivalent Preconditioners

We consider now symmetric preconditioning for elliptic systems defined on a domain $\Omega \subset R^N$. Let

$$L_i \equiv -div(A_i \nabla u_i) + \mathbf{b}_i \cdot \nabla u_i + \sum_{j=1}^{l} V_{ij} u_j = g_i$$

$$u_i = 0 \text{ on } \partial\Omega_D, \quad \frac{\partial u_i}{\partial \nu_A} + \alpha_i u_i = 0 \text{ on } \partial\Omega_N, i = 1, 2, \cdots l.$$

$$(40)$$

Here it is assumed that $\mathbf{b}_i \in C^1(\Omega)^N$, $g_i \in L^2(\Omega)$ and $V_{ij} \in L^\infty(\Omega)$, and the matrix $V = \{V_{ij}\}_{i,j=1}^{l}$ satisfies the coercivity property pointwise in Ω, $\lambda_{min}(V+V^T) - max\, div\mathbf{b}_i \geq 0$, pointwise in Ω, where λ_{min} denotes the smallest eigenvalue.

Then system (40) has a unique solution $u \in H_D^1(\Omega)^l$.

As preconditioning operator we use the l-tuple $S = (S_1, \cdots, S_l)$ of independent operators, $S_i u_i \equiv -div(A_i \nabla u_i) + h_i u$, where $u_i = 0$ on $\partial\Omega_D$, $\dfrac{\partial u_i}{\partial \nu_A} + \beta_i u_i = 0$ on $\partial\Omega_N$ and $\beta_i \geq 0, i = 1, 2, \cdots l$.

Now we choose a FEM subspace $V_h \subset H_D^1(\Omega)^l$ and look for the solution u_h of the corresponding system $L_h \mathbf{c} = \mathbf{b}$ using a preconditioner S_h being the stiffness matrix of S.

One can readily verify that there occurs a superlinear convergence of the preconditioned CGM which, furthermore, is mesh independent.

An application where such systems arise is in meteorology, where the chemical reaction terms have been linearized in a Newton nonlinear iteration method ([16]). Another important application of equivalent pairs of elliptic operators arises for the separable displacement preconditioning method for elasticity systems, formulated in displacement variables. There, the equivalence of the given and the separable displacement operators can be proven using Korn's inequality, see [4,5] for details and further references.

We have shown that a superlinear convergence takes place for operator pairs (i.e., the given and its preconditioner) which are compact-equivalent. The main theorem states that the principal, i.e., the dominating (second order) parts of the operators must be identical, apart from a constant factor. This seems to exclude an application for variable coefficient problems, where for reasons of efficiency we choose a preconditioner which has constant, or piecewise constant coefficients, assuming we want to use a simple operator such as the Laplacian as preconditioner.

However, we show now how to apply some method of scaling or transformation to reduce the problem to one with constant coefficients in the dominating part. We use then first a direct transformation of the equation. Let

$$Lu \equiv -div(a\nabla u) + \mathbf{b} \cdot \nabla u + cu = g,$$

$$(41)$$

where $a \in C^1(\Omega), a(x) \geq p > 0$.

Here a straightforward computation shows that

$$\frac{1}{a} Lu = -div(\nabla u) + \frac{1}{a}(\mathbf{b} - \nabla a)\nabla u + \frac{c}{a} u = \frac{g}{a},$$

i.e.,the principal part consists simply of the Laplacian operator, $-\Delta$.

A case of special importance occurs when a is written in the form $a = e^{-\phi}, \phi \in C^1(\Omega)$ and $\mathbf{b} = 0$. Then $-\nabla a = e^{-\phi}\nabla\phi = a\nabla\phi$ and (41) takes the form

$$\frac{1}{a}Lu = -\Delta u + \nabla\phi\nabla u + e^{\phi}cu = e^{\phi}g.$$

This is a convection-diffusion equation with a so called potential vector field, $\mathbf{v} = \nabla\phi$. Such problems occur frequently in practice, e.g. in modeling of semiconductors.

When the coefficient a varies much over the domain Ω one can apply transformations of both the equation and the variable, to reduce variations of gradients $(\|\nabla u\|)$ of $O(\max(a)/\min(a))$ to $O(\max\sqrt{a}/\min(\sqrt{a}))$. Let then $u = a^{1/2}v$ and assume that $a \in C^2(\Omega)$. Then a computation shows that

$$-a^{-1/2}\frac{\partial}{\partial x_i}(a\frac{\partial u}{\partial x_i}) = -a^{-1/2}\frac{\partial}{\partial x_i}(a^{-1/2}\frac{\partial v}{\partial x_i} - \frac{1}{2}a^{-1/2}\frac{\partial a}{\partial x_i}v$$
$$= -\frac{\partial^2 v}{\partial x_i^2} + a^{-1/2}\frac{\partial^2(a^{-1/2})}{\partial x_i^2},$$

and $a^{-1/2}\mathbf{b}\nabla u = a^{-1}\mathbf{b}\nabla v - \frac{1}{2}(\mathbf{b}\cdot\nabla u/a^2)v$. Hence

$$a^{-1/2}(Lu - g) = \Delta v + \widehat{\mathbf{b}}\cdot\nabla v + \widehat{c}v - \widehat{g}, \qquad (42)$$

where $\widehat{\mathbf{b}} = a^{-1}\mathbf{b}, \widehat{c} = a^{-1}c - \frac{1}{2}\mathbf{b}\cdot\nabla u/a^2 + a^{-1/2}\Delta(a^{-1/2})$ and $\widehat{g} = a^{-1/2}g$.

Remark 3. It is seen that when $\mathbf{b} = 0$ both the untransformed (41) and transformed (42) operators are selfadjoint.

The relation $Nv \equiv a^{-1/2}Lu$ shows that

$$\langle Nv, v\rangle_{L^2(\Omega} = \langle a^{-1/2}Lu, a^{1/2}u\rangle_{L^2(\Omega} = \langle Lu, u\rangle_{L^2(\Omega}$$

holds for all $u \in D(L)$. The positivity of the coefficient a shows hence that $\|u\|_{H^1}$ and $\|v\|_{H^1}$ are equivalent, and N inherits the H^1-coercivity of L, i.e., the relation $\langle Lu, u\rangle_{L^2(\Omega} \geq m\|u\|_{H^1(\Omega)}^2$ is replaced by $\langle Nv, v\rangle_{L^2(\Omega} \geq \widehat{m}\|v\|_{H^1(\Omega)}^2$ for a certain constant $\widehat{m} > 0$. This shows the we may apply e.g. the Laplacian operator $(-\Delta)$ as preconditioner to N which, being a compact equivalent pair, implies a superlinear and meshindependent rate of convergence of CGM.

Acknowledgement

The second author was supported by the Hungarian Research Office NKTH under Öveges Program and by the Hungarian Research Grant OTKA No.K 67819.

References

1. Ashby, S.F., Manteuffel, T.A., Saylor, P.E.: A taxonomy for conjugate gradient methods. SIAM J. Numer. Anal. 27(6), 1542–1568 (1990)
2. Axelsson, O.: A generalized conjugate gradient least square method. Numer. Math. 51, 209–227 (1987)
3. Axelsson, O.: Iterative Solution Methods. Cambridge University Press, Cambridge (1994)
4. Axelsson, O.: On iterative solvers in structural mechanics; separate displacement orderings and mixed variable methods. Math. Comput. Simulation 50(1-4), 11–30 (1999)
5. Axelsson, O., Karátson, J.: Conditioning analysis of separate displacement preconditioners for some nonlinear elasticity systems. Math. Comput. Simul. 64(6), 649–668 (2004)
6. Axelsson, O., Karátson, J.: Superlinearly convergent CG methods via equivalent preconditioning for nonsymmetric elliptic operators. Numer. Math. 99(2), 197–223 (2004)
7. Axelsson, O., Karátson J.: Symmetric part preconditioning of the CGM for Stokes type saddle-point systems. Numer. Funct. Anal. Optim. (to appear)
8. Axelsson, O., Karátson J.: Mesh independent superlinear PCG rates via compact-equivalent operators. SIAM J. Numer. Anal. (to appear)
9. Axelsson, O., Karátson J.: Equivalent operator preconditioning for linear elliptic problems (in preparation)
10. D'yakonov, E.G.: The construction of iterative methods based on the use of spectrally equivalent operators. USSR Comput. Math. and Math. Phys. 6, 14–46 (1965)
11. Eisenstat, S.C., Elman, H.C., Schultz., M.H.: Variational iterative methods for nonsymmetric systems of linear equations. SIAM J. Numer. Anal. 20(2), 345–357 (1983)
12. Faber, V., Manteuffel, T., Parter, S.V.: On the theory of equivalent operators and application to the numerical solution of uniformly elliptic partial differential equations. Adv. in Appl. Math. 11, 109–163 (1990)
13. Hestenes, M.R., Stiefel, E.: Methods of conjugate gradients for solving linear systems. J. Res. Nat. Bur. Standards, Sect. B 49(6), 409–436 (1952)
14. Karátson J., Kurics T.: Superlinearly convergent PCG algorithms for some nonsymmetric elliptic systems. J. Comp. Appl. Math. (to appear)
15. Manteuffel, T., Parter, S.V.: Preconditioning and boundary conditions. SIAM J. Numer. Anal. 27(3), 656–694 (1990)
16. Zlatev, Z.: Computer treatment of large air pollution models. Kluwer Academic Publishers, Dordrecht (1995)

Bridging Methods for Coupling Atomistic and Continuum Models

Santiago Badia[1,2], Pavel Bochev[1], Max Gunzburger[3], Richard Lehoucq[1], and Michael Parks[1]

[1] Sandia National Laboratories, Computational Mathematics and Algorithms, P.O. Box 5800, MS 1320, Albuquerque NM 87185, USA
{sibadia,pbboche,rblehou,mlparks}@sandia.gov
[2] CIMNE, Universitat Politècnica de Catalunya
Jordi Girona 1-3, Edifici C1, 08034 Barcelona, Spain
[3] School of Computational Science, Florida State University,
Tallahassee FL 32306-4120, USA
gunzburg@scs.fsu.edu

Abstract. We review some recent developments in the coupling of atomistic and continuum models based on the blending of the two models in a bridge region connecting the other two regions in which the models are separately applied. We define four such models and subject them to patch and consistency tests. We also discuss important implementation issues such as: the enforcement of displacement continuity constraints in the bridge region; and how one defines, in two and three dimensions, the blending function that is a basic ingredient in the methods.

Keywords: Atomistic to continuum coupling, blended coupling, molecular statics.

1 Coupling Atomistic and Continuum Models

For us, continuum models are PDE models that are derived by invoking a (physical) continuum hypothesis. In most situations, these models are *local* in nature, e.g., forces at any point and time depend only on the state at that point. Atomistic models are discrete models. In particular, we consider molecular statics models; these are particle models in which the position of the particles are determined through the minimization of an energy, or, equivalently, by Newton's laws expressing force balances. These models are, in general, *nonlocal* in nature, e.g., particles other than its nearest neighbors exert a force on a particle.

There are two types of situations in which the coupling of atomistic and continuum models arise. In the *concurrent domain* setting, the atomistic model is used to determine information, e.g., parameters such as diffusion coefficients, viscosities, conductivities, equations of state, etc., or stress fields, etc., that are needed by the continuum model. Both models are assumed to hold over the same domain. Typically, these parameters are determined by taking statistical averages of the atomistic solution at points in the domain and, in this setting,

I. Lirkov, S. Margenov, and J. Waśniewski (Eds.): LSSC 2007, LNCS 4818, pp. 16–27, 2008.

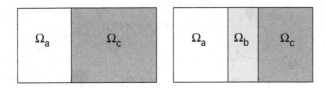

Fig. 1. Non-overlapping (left) and overlapping (right) coupling of atomistic and continuum models

usually, only one-way coupling is needed, e.g., the atomistic model is used to determine the continuum parameters. However, in more realistic situations, two-way coupling is needed, e.g., the atomistic model may need the macroscopic temperature as an input.

In the *domain decomposition* setting (which is the one we consider in this paper), the atomistic and continuum models are applied in different subdomains. The atomistic model is valid everywhere but is computationally expensive to use everywhere. So, it is applied only in regions where "singularities" occur, e.g., cracks, dislocations, plastic behavior, etc., and a continuum model is applied in regions where, e.g., ordinary elastic behavior occurs. There remains the question of how one couples the atomistic to the continuum model; there are two approaches to effect this coupling. For *non-overlapping* coupling, the atomistic and continuum models are posed on disjoint domains that share a common interface. For *overlapping* coupling, the regions in which the atomistic and continuum model are applied are connected by a *bridge* region in which both models are applied. See the sketches in Fig. 1.

Atomistic-to-continuum (AtC) coupling is distinct from most continuum-to-continuum couplings due to the *non-local nature of atomistic models.* Although the are no "active" particles in the region in which only the continuum model is applied, in a setting in which particles interact nonlocally, the forces exerted by the missing particles on the active particles are not accounted for; this discrepancy gives rise to what is known as *ghost force* phenomena.

In this paper, we consider AtC coupling methods that use overlapping regions because, in that case, it is easier to mitigate the ghost force effect. Note that one should not simply superimpose the two models in the bridge region since this leads to a non-physical "doubling" of the energy in Ω_b. Instead, the two models must be properly *blended* in this region. Such models are considered in [1,2,3,4,6]; here, we review the results of [1,2,6].

2 Blended AtC Coupled Models

We assume that the atomistic model is valid in the *atomistic* and *bridge* regions, Ω_a and Ω_b, respectively; see Fig. 1. The continuum model is valid in the *continuum region* Ω_c and the bridge region Ω_b but is not valid in the atomistic region Ω_a. We want to "seamlessly" blend the two models together using the bridge region Ω_b according to the following principles: the atomistic model

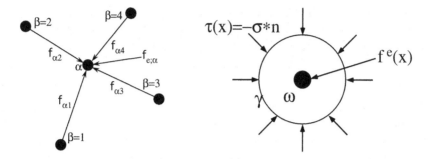

Fig. 2. Left: forces acting on the particle located at $\mathbf{x}_\alpha$. Right: forces acting on a point $\mathbf{x}$ in the continuum region.

"dominates" the continuum model near the interface surface between the atomistic and bridge regions and the continuum model "dominates" the atomistic model near the interface surface between the continuum and bridge regions.

In the atomistic region Ω_a, we assume that the force on the particle α located at the position $\mathbf{x}_\alpha$ is due to *externally applied force* $\mathbf{f}_{e;\alpha}$ and the *forces exerted by other particles* $\mathbf{f}_{\alpha,\beta}$ within the ball $\mathcal{B}_\alpha = \{\mathbf{x} \in \Omega : |\mathbf{x} - \mathbf{x}_\alpha| \leq \delta\}$ for some given δ. See the sketch in Fig. 2.

The inter-particle forces are determined from a potential function, e.g., if $\mathbf{x}_\alpha$ and $\mathbf{x}_\beta$ denote the positions of the particles α and β, then $\mathbf{f}_{\alpha,\beta} = -\nabla \Phi(|\mathbf{x}_\alpha - \mathbf{x}_\beta|)$, where $\Phi(\cdot)$ is a prescribed potential function. Instead of using the particle positions $\mathbf{x}_\alpha$, one instead often uses the displacements $\mathbf{u}_\alpha$ from a reference configuration.

Let $\mathcal{N}_\alpha = \{\beta \mid \mathbf{x}_\beta \in \mathcal{B}_\alpha, \beta \neq \alpha\}$, i.e., $\mathcal{N}_\alpha$ is the set of the indices of the particles[1] located within $\mathcal{B}_\alpha$, other than the particle located at $\mathbf{x}_\alpha$ itself. Then, for any particle α, force equilibrium gives

$$\mathbf{f}_\alpha + \mathbf{f}_{e;\alpha} = \mathbf{0},$$

where $\mathbf{f}_\alpha = \sum_{\beta \in \mathcal{N}_\alpha} \mathbf{f}_{\alpha,\beta}$. We assume that in Ω_a there are two kinds of particles: particles whose positions are specified in advance and particles whose positions are determined by the force balance equations. The set of indices of the second kind of particles is denoted by $\mathcal{N}_a$. It is convenient to recast the force balance equation for the remaining particles in an equivalent variational form

$$\sum_{\alpha \in \mathcal{N}_a} \mathbf{v}_\alpha \cdot \mathbf{f}_\alpha = -\sum_{\alpha \in \mathcal{N}_a} \mathbf{v}_\alpha \cdot \mathbf{f}_{e;\alpha} \qquad \forall \mathbf{v}_\alpha \in \mathbb{R}^d, \alpha \in \mathcal{N}_a.$$

In the continuum region Ω_c, the Cauchy hypothesis implies that the forces acting on any continuum volume ω enclosing the point $\mathbf{x}$ are given by the *externally applied volumetric force* $\mathbf{f}^e$ and the *force exerted by the surrounding*

---

[1] Note that for some α, the set $\mathcal{N}_\alpha$ may include the indices of some particles whose positions are specified.

material $\mathbf{f}_c = -\int_\gamma \sigma \cdot \mathbf{n}\, d\gamma$, where γ denotes the boundary of ω and σ denotes the stress tensor. See the sketch in Fig. 2. Note that $-\sigma \cdot \mathbf{n}$ is the stress force acting on a point on γ.

We assume that $\sigma(\mathbf{x}) = \sigma(\mathbf{x}, \nabla \mathbf{u}(\mathbf{x}))$ and is possibly nonlinear in both its arguments. Here, $\mathbf{u}(\mathbf{x})$ denotes the continuous displacement at the point $\mathbf{x}$. For a homogeneous material, $\sigma(\mathbf{x}) = \sigma(\nabla \mathbf{u}(\mathbf{x}))$, i.e., the stress does not explicitly depend on position. In the equilibrium state, we have that

$$-\int_\gamma \sigma \cdot \mathbf{n}\, d\gamma + \int_\omega \mathbf{f}^e\, d\omega = \mathbf{0}, \qquad \text{so that} \qquad \int_\omega \left(\nabla \cdot \sigma + \mathbf{f}^e \right) d\omega = \mathbf{0}.$$

Then, since ω is arbitrary, we conclude that at any point $\mathbf{x}$ in the continuum region, we have the force balance

$$\nabla \cdot \sigma + \mathbf{f}^e = \mathbf{0}.$$

For simplicity, we assume that we only have displacement boundary conditions.

Again, it will be convenient if we recast the continuum force balance equations in the equivalent variational form

$$\int_{\Omega_c} \sigma(\mathbf{u}) : \varepsilon(\mathbf{v})\, d\Omega = \int_{\Omega_c} \mathbf{f}^e \cdot \mathbf{v}\, d\Omega \qquad \forall \mathbf{v} \in \mathbf{H}_0^1(\Omega_c),$$

where we have the strain tensor $\varepsilon(\mathbf{v}) = \frac{1}{2}(\nabla \mathbf{v} + \nabla \mathbf{v}^T)$ and the homogeneous displacement test space $\mathbf{H}_0^1(\Omega_c)$.

2.1   Blended Models in the Bridge Region

We introduce the blending functions $\theta_a(\mathbf{x})$ and $\theta_c(\mathbf{x})$ satisfying $\theta_a + \theta_c = 1$ in Ω with $0 \le \theta_a, \theta_c \le 1$, $\theta_c = 1$ in Ω_c and $\theta_a = 1$ in Ω_a. Let $\theta_\alpha = \theta_a(\mathbf{x}_\alpha)$ and $\theta_{\alpha,\beta} = \theta_a \left(\frac{\mathbf{x}_\alpha + \mathbf{x}_\beta}{2} \right)$ or $\theta_{\alpha,\beta} = \frac{\theta_\alpha + \theta_\beta}{2}$.

We introduce four ways to blend the atomistic and continuum models. Let $\mathcal{N}_b$ denote the set of indices of the particles in Ω_b whose positions are not fixed by the boundary conditions.

Blended model I

$$-\int_{\Omega_b} \theta_c \sigma(\mathbf{u}) : \varepsilon(\mathbf{v})\, d\Omega + \sum_{\alpha \in \mathcal{N}_b} \theta_\alpha \mathbf{v}_\alpha \cdot \mathbf{f}_\alpha = -\int_{\Omega_b} \theta_c \mathbf{f}^e \cdot \mathbf{v}\, d\Omega - \sum_{\alpha \in \mathcal{N}_b} \theta_\alpha \mathbf{v}_\alpha \cdot \mathbf{f}_{e;\alpha}$$

$$\forall \mathbf{v} \in \mathbf{H}_0^1(\Omega_c) \text{ and } \mathbf{v}_\alpha \in \mathbb{R}^d, \alpha \in \mathcal{N}_b.$$

Blended model II

$$-\int_{\Omega_b} \theta_c \sigma(\mathbf{u}) : \varepsilon(\mathbf{v})\, d\Omega + \sum_{\alpha \in \mathcal{N}_b} \mathbf{v}_\alpha \cdot \sum_{\beta \in \mathcal{N}_\alpha} \theta_{\alpha,\beta} \mathbf{f}_{\alpha,\beta} = -\int_{\Omega_b} \theta_c \mathbf{f}^e \cdot \mathbf{v}\, d\Omega$$

$$- \sum_{\alpha \in \mathcal{N}_b} \theta_\alpha \mathbf{v}_\alpha \cdot \mathbf{f}_{e;\alpha} \qquad \forall \mathbf{v} \in \mathbf{H}_0^1(\Omega_c), \mathbf{v}_\alpha \in \mathbb{R}^d, \alpha \in \mathcal{N}_b.$$

Blended model III

$$-\int_{\Omega_b} \sigma(\mathbf{u}) : \varepsilon(\theta_c \mathbf{v}) \, d\Omega + \sum_{\alpha \in \mathcal{N}_b} \theta_\alpha \mathbf{v}_\alpha \cdot \mathbf{f}_\alpha = -\int_{\Omega_b} \theta_c \mathbf{f}^e \cdot \mathbf{v} \, d\Omega$$

$$-\sum_{\alpha \in \mathcal{N}_b} \theta_\alpha \mathbf{v}_\alpha \cdot \mathbf{f}_{e;\alpha} \qquad \forall \mathbf{v} \in \mathbf{H}_0^1(\Omega_c) \text{ and } \mathbf{v}_\alpha \in \mathbb{R}^d, \alpha \in \mathcal{N}_b.$$

Blended model IV

$$-\int_{\Omega_b} \sigma(\mathbf{u}) : \varepsilon(\theta_c \mathbf{v}) \, d\Omega + \sum_{\alpha \in \mathcal{N}_b} \mathbf{v}_\alpha \cdot \sum_{\beta \in \mathcal{N}_\alpha} \theta_{\alpha,\beta} \mathbf{f}_{\alpha,\beta} = -\int_{\Omega_b} \theta_c \mathbf{f}^e \cdot \mathbf{v} \, d\Omega$$

$$-\sum_{\alpha \in \mathcal{N}_b} \theta_\alpha \mathbf{v}_\alpha \cdot \mathbf{f}_{e;\alpha} \qquad \forall \mathbf{v} \in \mathbf{H}_0^1(\Omega_c) \text{ and } \mathbf{v}_\alpha \in \mathbb{R}^d, \alpha \in \mathcal{N}_b.$$

Methods I and II were introduced in [1,6] while Method III and IV were introduced in [2]. An important observation is that in the bridge region Ω_b, near the continuum region Ω_c, we have that θ_a is small so that $\theta_{\alpha,\beta}$ and θ_α are small as well. Thus, blended models of the type discussed here automatically mitigate any ghost force effects, i.e., any ghost force will be multiplied by a small quantity such as $\theta_{\alpha,\beta}$ or θ_α.

2.2   Displacement Matching Conditions in the Bridge Region

In order to complete the definition of the blended model, one must impose constraints that tie the atomistic displacements $\mathbf{u}_\alpha$ and the continuum displacements $\mathbf{u}(\mathbf{x})$ in the bridge region Ω_b. These take the form of

$$\mathcal{C}\big(\mathbf{u}_\alpha, \mathbf{u}(\mathbf{x})\big) = 0 \qquad \text{for } \alpha \in \mathcal{N}_b \text{ and } \mathbf{x} \in \Omega_b$$

for some specified constraint operator $\mathcal{C}(\cdot, \cdot)$.

One could slave all the atomistic displacements in the bridge region to the continuum displacements, i.e., set

$$\mathbf{u}_\alpha = \mathbf{u}(\mathbf{x}_\alpha) \qquad \forall \alpha \in \mathcal{N}_b.$$

We refer to such constraints as *strong constraints*. Alternatively, the atomistic and continuum displacements can be matched in an average sense to define *loose constraints*. For example, one can define a triangulation $\mathcal{T}^H = \{\Delta_t\}_{t=1}^{T_b}$ of the bridge region Ω_b; this triangulation need not be the same as that used to effect a finite element discretization of the continuum model. Let $\mathcal{N}_t \neq \emptyset$ denote indices of the particles in Δ_t. One can then match the atomistic and continuum displacements in an average sense over each triangle Δ_t:

$$\sum_{\alpha \in \mathcal{N}_t} \mathbf{u}(\mathbf{x}_\alpha) = \sum_{\alpha \in \mathcal{N}_t} \mathbf{u}_\alpha \qquad \text{for } t = 1, \dots, T_b.$$

Once a set of constraints has been chosen, one also has to choose a means for enforcing them. One possibility is to enforce them *weakly* through the use of the

Lagrange multiplier rule. In this case, the test functions $\mathbf{v}_\alpha$ and $\mathbf{v}(\mathbf{x})$ and trial functions $\mathbf{u}_\alpha$ and $\mathbf{u}(\mathbf{x})$ in the variational formulations are not constrained; one ends up with saddle-point type discrete systems

$$
\begin{pmatrix} \mathcal{A}_{a,\theta_a} & 0 & \mathcal{C}_a \\ 0 & \mathcal{A}_{c,\theta_c} & \mathcal{C}_c \\ \mathcal{C}_a^* & \mathcal{C}_c^* & 0 \end{pmatrix} \cdot \begin{pmatrix} \text{atomistic unknowns} \\ \text{continuum unknowns} \\ \text{Lagrange multipliers} \end{pmatrix} = \text{RHS.}
$$

Note that the coupling of the atomistic and continuum variables is effected only through the Lagrange multipliers.

A second possibility is to enforce the constraints *strongly*, i.e., require that all candidate atomistic and continuum displacements satisfy the constraints. In this case, the test functions $\mathbf{v}_\alpha$ and $\mathbf{v}(\mathbf{x})$ in the variational formulations should be similarly constrained. One ends up with simpler discrete systems of the form

$$
\widetilde{\mathcal{A}}_{a,\theta_a,\theta_c} \cdot (\text{atomistic unknowns}) + \widetilde{\mathcal{A}}_{c,\theta_c,\theta_a} \cdot (\text{continuum unknowns}) = \text{RHS.}
$$

Note that the atomistic and continuum variables are now tightly coupled. The second approach involves fewer degrees of freedom and results in better behaved discrete systems but may be more cumbersome to apply in some settings.

2.3 Consistency and Patch Tests

To define an AtC coupled problem, one must specify the following data sets:

- $F = \left\{ \mathbf{f}_\alpha^e \right\}_{\alpha \in \mathcal{N}_a \cup \mathcal{N}_b}$ (external forces applied to the particles);
- $P = \left\{ \mathbf{u}_\alpha \right\}_{\alpha \notin \mathcal{N}_a \cup \mathcal{N}_b}$ (displacements of the particles whose positions are fixed)
- $B = \left\{ \mathbf{f}^e(\mathbf{x}) \right\}_{\mathbf{x} \in \Omega_c \cup \Omega_b}$ (external forces applied in the continuum region);
- $D = \left\{ \mathbf{u}(\mathbf{x}) \right\}_{\mathbf{x} \in \partial(\Omega_c \cup \Omega_b)}$ (continuum displacements on the boundary).

We subject the AtC blending methods we have defined to two tests whose passage is crucial to their mathematical and physical well posedness. To this end, we define two types of test problems. The set $\{F, P, B, D\}$ defines a *consistency test problem* if the pure atomistic solution $\mathbf{u}_\alpha$ and the pure continuum solution $\mathbf{u}(\mathbf{x})$ are such that the constraint equations, i.e, $C\big(\mathbf{u}_\alpha, \mathbf{u}(\mathbf{x})\big) = 0$, are satisfied on Ω. Further, a consistency test problem defines a *patch test problem* if the pure continuum solution $\mathbf{u}(\mathbf{x})$ is such that $\varepsilon(\mathbf{u}) = \text{constant}$, i.e., it is a solution with constant strain.

If we assume that $\{F, P, B, D\}$ defines a patch test problem with atomistic solution $\mathbf{u}_\alpha$ and continuum solution $\mathbf{u}(\mathbf{x})$, then, an AtC coupling method *passes the patch test* if $\{\mathbf{u}_\alpha, \mathbf{u}(\mathbf{x})\}$ satisfies the AtC model equations. Similarly, an AtC coupling method *passes the consistency test* if $\{\mathbf{u}_\alpha, \mathbf{u}(\mathbf{x})\}$ satisfies the AtC model equations for any consistency test problem. Note that passing the consistency test implies passage of the patch test, but not conversely.

Our analyses of the four blending methods (see [2]) have shown that Methods I and IV are not consistent and do not pass patch test problems; Method III is consistent and thus also passes any patch test problem; and Method II is conditionally consistent: it is consistent if, for a pair of atomistic and continuum solutions $\mathbf{u}_\alpha$ and $\mathbf{u}$, respectively

$$-\int_\Omega \theta_c \sigma(\mathbf{u}) : \varepsilon(\mathbf{v}) \, d\Omega + \int_\Omega \sigma(\mathbf{u}) : \varepsilon(\theta_c \mathbf{v}) \, d\Omega$$

$$+ \sum_{\alpha \in \mathcal{N}_a \cup \mathcal{N}_b} \mathbf{v}_\alpha \cdot \sum_{\beta \in \mathcal{N}_\alpha} \theta_{\alpha,\beta} \mathbf{f}_{\alpha,\beta} - \sum_{\alpha \in \mathcal{N}_a \cup \mathcal{N}_b} \theta_\alpha \mathbf{v}_\alpha \cdot \sum_{\beta \in \mathcal{N}_\alpha} \mathbf{f}_{\alpha,\beta} = 0$$

and passes patch tests if this condition is met for patch test solutions.

From these results, we can forget about Methods I and IV and it seems that Method III is better than Method II. The first conclusion is valid but there are additional considerations that enter into the relative merits of Methods II and III. Most notably, Method II is the only one of the four blended models that satisfies[2] Newton's third law. In addition, the violation of patch and consistency tests for Method II is tolerable, i.e., the error introduced can be made smaller by proper choices for the model parameters, e.g., in a 1D setting, we have shown (see [1]) that the patch test error is proportional to $\frac{s^2}{L_b h}$, where s = particle lattice spacing (a material property), h = finite element grid size, and L_b = width of the bridge region Ω_b. While we cannot control the size of s, it is clear that in realistic models this parameter is small. Also, the patch test error for Method II can be made smaller by making L_b larger (widening the bridge region) and/or making h larger (having more particles in each finite element).

2.4 Fully Discrete Systems in Higher Dimensions

We now discuss how the fully discrete system can be defined in 2D; we only consider Method II for which we have

$$-\int_{\Omega_b \cup \Omega_c} \theta_c \sigma(\mathbf{u}) : \varepsilon(\mathbf{v}) \, d\Omega + \sum_{\alpha \in \mathcal{N}_b \cup \mathcal{N}_a} \mathbf{v}_\alpha \cdot \sum_{\beta \in \mathcal{N}_\alpha} \theta_{\alpha,\beta} \mathbf{f}_{\alpha,\beta} =$$

$$-\int_{\Omega_b \cup \Omega_c} \theta_c \mathbf{f}^e \cdot \mathbf{v} \, d\Omega - \sum_{\alpha \in \mathcal{N}_b \cup \mathcal{N}_a} \theta_\alpha \mathbf{v}_\alpha \cdot \mathbf{f}_{e;\alpha}$$

$$\forall \mathbf{v} \in \mathbf{H}_0^1(\Omega_b \cup \Omega_c) \text{ and } \mathbf{v}_\alpha \in \mathbb{R}^d, \alpha \in \mathcal{N}_b \cup \mathcal{N}_a.$$

We also consider the case of the strong enforcement of the hard constraints

$$\mathbf{u}_\alpha = \mathbf{u}(\mathbf{x}_\alpha) \qquad \forall \alpha \in \mathcal{N}_b.$$

The other methods and looser constraints handled using the Lagrange multiplier rule can be handled in a similar manner.

---

[2] Related to this observation is the fact that Method II is the only blended method that has symmetric weak form provided the weak forms of the pure atomistic and continuum problems are also symmetric.

To discretize the continuum contributions to the blended model, we use a finite element method. Let $W^h \subset \mathbf{H}_0^1(\Omega_b \cup \Omega_c)$ be a nodal finite element space and let $\{\mathbf{w}_j^h(\mathbf{x})\}_{j=1}^J$ denote a basis for W^h. Then, the continuum displacement $\mathbf{u}(\mathbf{x})$ is approximated by

$$\mathbf{u}(\mathbf{x}) \approx \mathbf{u}^h(\mathbf{x}) = \sum_{j=1}^J c_j \mathbf{w}_j^h(\mathbf{x}).$$

Let $\mathcal{S}_c = \Big\{ j \in \{1, \ldots, J\} \mid \mathbf{x}_j \in \Omega_c \Big\}$ and $\mathcal{S}_b = \Big\{ j \in \{1, \ldots, J\} \mid \mathbf{x}_j \in \Omega_b \Big\}$ denote the set of indices of the nodes in Ω_c and Ω_b, respectively. We use continuous, piecewise linear finite element spaces with respect to partition of $\Omega_b \cup \Omega_c$ into a set of T triangles $\mathcal{T}^h = \{\Delta_t\}_{t=1}^T$; higher-order finite element spaces can also be used.

For $j = 1, \ldots, J$, we let $\mathcal{T}_j^h = \{t : \Delta_t \in \mathrm{supp}(\mathbf{w}_j)\}$, i.e., $\mathcal{T}_j^h$ is the set of indices of the triangles sharing the finite element node $\mathbf{x}_j$ as a vertex. Thus, we have that

$$\int_{\mathrm{supp}(\mathbf{w}_j^h)} F(\mathbf{x}) \, d\Omega = \sum_{t \in \mathcal{T}_j^h} \int_{\Delta_t} F(\mathbf{x}) \, d\Omega.$$

The standard choice for the quadrature rule, since we are using piecewise linear finite element functions, is the mid-side rule for triangles. Thus, if $\widehat{\mathbf{x}}_{\Delta;k}$, $k = 1, \ldots, 3$, are the vertices of a triangle Δ, we have the quadrature rule $\int_\Delta F(\mathbf{x}) \, d\Omega \approx \frac{V_\Delta}{3} \sum_{q=1}^3 F(\mathbf{x}_{\Delta;q})$, where V_Δ denotes the volume of the triangle Δ, $\mathbf{x}_{\Delta;1} = \frac{\widehat{\mathbf{x}}_{\Delta;1} + \widehat{\mathbf{x}}_{\Delta;2}}{2}$, $\mathbf{x}_{\Delta;2} = \frac{\widehat{\mathbf{x}}_{\Delta;2} + \widehat{\mathbf{x}}_{\Delta;3}}{2}$, and $\mathbf{x}_{\Delta;3} = \frac{\widehat{\mathbf{x}}_{\Delta;3} + \widehat{\mathbf{x}}_{\Delta;1}}{2}$.

In the continuum region Ω_c, we have the discretized continuum model

$$- \sum_{t \in \mathcal{T}_j^h} \frac{V_{\Delta_t}}{3} \sum_{q=1}^3 \sigma\Big((\mathbf{x}_{\Delta_t;q}), \nabla \mathbf{u}^h(\mathbf{x}_{\Delta_t;q}) \Big) : \nabla \mathbf{w}_j^h(\mathbf{x}_{\Delta_t;q})$$

$$= - \sum_{t \in \mathcal{T}_j^h} \frac{V_{\Delta_t}}{3} \sum_{q=1}^3 \mathbf{f}^e(\mathbf{x}_{\Delta_t;q}) \cdot \mathbf{w}_j^h(\mathbf{x}_{\Delta_t;q}) \qquad \text{for } j \in \mathcal{S}_c.$$

Of course, in the atomistic region Ω_a, we have that

$$\sum_{\beta \in \mathcal{N}_\alpha} \mathbf{f}_{\alpha,\beta} = -\mathbf{f}_\alpha^e \qquad \text{for } \alpha \in \mathcal{N}_a.$$

Due to the way we are handling the constraints, we have that the atomistic test and trial functions in the bridge region Ω_b are slaved to the continuum test and trial functions, i.e., $\mathbf{u}_\alpha = \mathbf{u}^h(\mathbf{x}_\alpha)$ and $\mathbf{v}_\alpha = \mathbf{w}_j^h(\mathbf{x}_\alpha)$.

For $j \in \mathcal{S}_b$, let $\mathcal{N}_j = \{\alpha \mid \mathbf{x}_\alpha \in \mathrm{supp}(\mathbf{w}_j^h)\}$ denote the set of particle indices such that the particles are located within the support of the finite element basis function $\mathbf{w}_j^h$. Then, in the bridge region Ω_b, we have that

$$-\sum_{t\in T_j^h}\frac{V_{\Delta_t}}{3}\sum_{q=1}^{3}\theta_c(\mathbf{x}_{\Delta_t;q})\sigma\Big((\mathbf{x}_{\Delta_t;q}),\nabla\mathbf{u}^h(\mathbf{x}_{\Delta_t;q})\Big):\nabla\mathbf{w}_j^h(\mathbf{x}_{\Delta_t;q})$$

$$+\sum_{\alpha\in\mathcal{N}_j}\sum_{\beta\in\mathcal{N}_\alpha}\theta_{\alpha,\beta}\mathbf{f}_{\alpha,\beta}\cdot\mathbf{w}_j^h(\mathbf{x}_\alpha)$$

$$=-\sum_{t\in T_j^h}\frac{V_{\Delta_t}}{3}\sum_{q=1}^{3}\theta_c(\mathbf{x}_{\Delta_t;q})\mathbf{f}^e(\mathbf{x}_{\Delta_t;q})\cdot\mathbf{w}_j^h(\mathbf{x}_{\Delta_t;q})$$

$$-\sum_{\alpha\in\mathcal{N}_j}\theta_\alpha\mathbf{f}_\alpha^e\cdot\mathbf{w}_j^h(\mathbf{x}_\alpha)\qquad\text{for }j\in\mathcal{S}_b.$$

In 3D, one cannot use mid-face or mid-edge quadrature rules as one can in 1D and 2D, even for uncoupled continuum problems. Instead, one must use rules for which at least some of the quadrature points are in the interior of tetrahedra. Other than this, the development of a fully discretized method follows the same process as in the 2D case.

2.5   Choosing the Blending Function in 2D and 3D

For the blending function $\theta_c(\mathbf{x})$ for $\mathbf{x}\in\Omega_b$, we of course have that $\theta_a(\mathbf{x})=1-\theta_c(\mathbf{x})$, $\theta_a(\mathbf{x})=1$, $\theta_c(\mathbf{x})=0$ in Ω_a, $\theta_a(\mathbf{x})=0$. and $\theta_c(\mathbf{x})=1$ in Ω_c.

In many practical settings, the domain Ω_b is a rectangle in 2D or is a rectangular parallelepiped in 3D. In such cases, one may simply choose $\theta_c(\mathbf{x})$ in the bridge region to be the tensor product of global 1D polynomials connecting the atomistic and continuum regions across the bridge region. One could choose linear polynomials in each direction such that their values are zero at the bridge/atomistic region interface and one at the bridge/continuum region interface. If one wishes to have a smoother transition from the atomistic to the bridge to the continuum regions, one can choose cubic polynomials in each direction such that they have zero value and zero derivative at the bridge/atomisitic region interface and value one and zero derivative at the bridge/continuum region interface.

For the general case in 2D, we triangulate the bridge region Ω_b into the set of triangles having vertices $\{\mathbf{x}_{b;i}\}_{i=1}^{l}$. In practice, this triangulation is the same as that used for the finite element approximation of the continuum model in the bridge region but, in general, it may be different. For the two triangulations to be the same, we must have that the finite element triangulation is conforming with the interfaces between the bridge region and the atomistic and continuum regions, i.e., those interfaces have to be made up of edges of triangles of the finite element triangulation. The simplest blending function is then determined by setting $\theta_c(\mathbf{x})=\xi^h(\mathbf{x})$, where $\xi^h(\mathbf{x})$ is a continuous, piecewise linear function with respect to this triangulation. The nodal values of $\xi^h(\mathbf{x})$ are chosen as follows. Set $\xi^h(\mathbf{x}_{b;i})=0$ at all nodes $\mathbf{x}_{b;i}\in\overline{\Omega}_a\cap\overline{\Omega}_b$, i.e., on the interface between the atomistic and bridge regions. Then, set $\xi^h(\mathbf{x}_{b;i})=1$ at all nodes $\mathbf{x}_{b;i}\in\overline{\Omega}_b\cap\overline{\Omega}_c$, i.e., on the interface between the continuum and bridge regions. For the remaining nodes $\mathbf{x}_{b;i} \in \Omega_b$, there are several ways to choose the values of ξ^h. One way is to choose them according to the relative distances to the interfaces. A more convenient way is to let ξ^h be a finite element approximation, with respect to the grid, of the solution of Laplace's equation in Ω_b that satisfies the specified values at the interfaces. Once $\xi^h(\mathbf{x})$ is chosen, we set $\theta_a(\mathbf{x}) = 1 - \xi^h(\mathbf{x})$ for all $\mathbf{x} \in \Omega_b$ and choose $\theta_\alpha = \theta_a(\mathbf{x}_\alpha) = 1 - \xi^h(\mathbf{x}_\alpha)$ so that

$$\theta_{\alpha,\beta} = 1 - \frac{\xi^h(\mathbf{x}_\alpha) + \xi^h(\mathbf{x}_\beta)}{2} \quad \text{or} \quad \theta_{\alpha,\beta} = 1 - \xi^h\left(\frac{\mathbf{x}_\alpha + \mathbf{x}_\beta}{2}\right).$$

This recipe can be extended to 3D settings.

One may want $\theta_c(\mathbf{x})$ to have a smoother transition from the atomistic to the bridge to the continuum regions. To this end, one can choose $\xi^h(\mathbf{x})$ to not only be continuous, but to be continuously differentiable in the bridge region and across the interfaces. Note that in 2D, this requires the use of the fifth-degree piecewise polynomial Argyris element or the cubic Clough-Tocher macro-element; see [5]. Such elements present difficulties in a finite element approximation setting, but are less problematical in an interpolatory setting.

3   Simple Computational Examples in 1D

For $0 < a < c < 1$, let $\Omega = (0,1)$, $\Omega_a = (0,a)$, $\Omega_b = (a,c)$, and $\Omega_c = (c,1)$. In $\overline{\Omega}_c \cup \overline{\Omega}_b = [a,1]$, we construct the uniform partition $x_j = a + (j-1)h$ for $j = 1, \ldots, J$ having grid size h. We then choose W^h to be the continuous, piecewise linear finite element space with respect to this partition. Without loss of generality, we define the bridge region Ω_b using the finite element grid, i.e., we assume that there are finite element nodes at $x = a$ and $x = c$; this arrangement leads to a more convenient implementation of blending methods in 2D and 3D. In $\overline{\Omega}_a \cup \overline{\Omega}_b = [0,c]$, we have a uniform particle lattice with lattice spacing s given by $x_\alpha = (\alpha - 1)s$, $\alpha = 1, \ldots, N$. Note that the lattice spacing s is a fixed material property so that there is no notion of $s \to 0$. One would think that one can still let $h \to 0$; however, it makes no sense to have $h < s$.

We consider the atomistic model to be a one-dimensional linear mass-spring system with two-nearest neighbor interactions and with elastic moduli K_{a1} and K_{a2} for the nearest-neighbor and second nearest-neighbor interactions; only the two particles to the immediate left and right of a particle exert a force on that particle. The continuum model is one-dimensional linear elasticity with elastic modulus K_c. We set $K_{a1} = 50$, $K_{a2} = 25$, $K_c = K_{a1} + 2K_{a2} = 100$. A unit point force is applied at the finite element node at the end point $x = 1$ and the displacement of the particle located at the end point $x = 0$ is set to zero. Using either the atomistic or finite element models, the resulting solutions are ones having uniform strain 0.01; thus, we want a blended model solution to also recover the uniform strain solution.

We choose $h = 1.5s$ and $s = 1/30$ so that we have $a = 0.3$, $c = 0.6$, 20 particles, and 16 finite element nodes; there are no particles located at either $x = a$ or $x = c$, the end points of the bridge region Ω_b. For the right-most

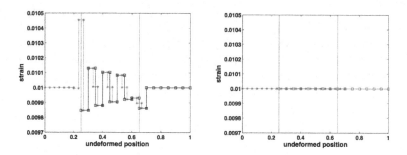

Fig. 3. Strain for Method II (left) and Method III(right)

particle $x_{20} < c$, we have that $\theta_a(x_{20}) \neq 0$. To avoid the ghost forces associated with the missing bond to the right of the 20th particle, a 21st particle is added to the right of $x = c$. Since $x_{21} \in \Omega_c$, we have that $\theta_a(x_{21}) = 0$ so that we need not be concerned with its missing bond to the right; this is a way that blending methods mitigate the ghost force effect. We see from Fig. 3 that Method III passes the patch test but Method II does not. However, the degree of failure for Method II is "small." From Fig. 4, we see that Method I fails the patch test; the figure for that method is for the even simpler case of nearest-neighbor interactions. Similarly, Method IV fails the patch test.

In Fig. 5, we compare the blended atomistic solutions in the bridge region, obtained using Method III with both strong and loose constraints, with that obtained using the fully atomistic solution. We consider a problem with a uniform load and zero displacements at the two ends; we only consider nearest-neighbor interactions. The loose constraint allows the atomistic solution to be free to reproduce the curvature of the fully atomistic solution, leading to better results. The strong constraint is too restrictive, forcing the atomistic solution to follow the finite element solution; it results in a substantial reduction in the accuracy in the bridge region.

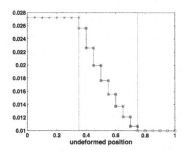

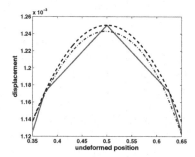

Fig. 4. Strain for Method I

Fig. 5. Fully atomistic solution (dashed line); loose constraints (dash-dotted line); and strong constraints (solid line)

Acknowledgments

The authors acknowledge the contributions to part of this work by Mohan Nugge-hally, Catalin Picu, and Mark Shephard and especially Jacob Fish of the Rens-selaer Polytechnic Institute. SB was supported by the European Community through the Marie Curie contract NanoSim (MOIF-CT-2006-039522). For PB, RL, and MP, Sandia is a multiprogram laboratory operated by Sandia Cor-poration, a Lockheed Martin Company, for the United States Department of Energy's National Nuclear Security Administration under Contract DE-AC04-94-AL85000. MG was supported in part by the Department of Energy grant number DE-FG02-05ER25698.

References

1. Badia, S., et al.: A Force-Based Blending Model for Atomistic-to-Continuum Cou-pling. Inter. J. Multiscale Comput. Engrg. (to appear)
2. Badia, S., et al.: On Atomistic-to-Continuum (AtC) Coupling by Blending. SIAM J. Multiscale Modeling and Simulation (submitted)
3. Belytschko, T., Xiao, S.: Coupling Methods for Continuum Model with Molecular Model. Inter. J. Multiscale Comput. Engrg. 1, 115–126 (2003)
4. Belytschko, T., Xiao, S.: A Bridging Domain Method for Coupling Continua with Molecular Dynamics. Comp. Meth. Appl. Mech. Engrg. 193, 1645–1669 (2004)
5. Ciarlet, P.: The finite element method for elliptic problems. SIAM, Philadelphia (2002)
6. Fish, J., et al.: Concurrent AtC Coupling Based on a Blend of the Continuum Stress and the Atomistic Force. Comp. Meth. Appl. Mech. Engrg. (to appear)

Parallelization of Advection-Diffusion-Chemistry Modules

István Faragó[1], Krassimir Georgiev[2], and Zahari Zlatev[3]

[1] Eötvös Loránd University, Pázmány P. s. 1/c, H-1117 Budapest, Hungary
faragois@cs.elte.hu
[2] Institute for Parallel Processing, Bulgarian Academy of Sciences
Acad. G. Bonchev, Bl. 25A, 1113 Sofia, Bulgaria
georgiev@parallel.bas.bg
[3] National Environmental Research Institute, Aarhus University
Frederiksborgvej 399, P.O. Box 358, DK-4000 Roskilde, Denmark
zz@dmu.dk

Abstract. An advection-diffusion-chemistry module of a large-scale air pollution model is split into two parts: (a) advection-diffusion part and (b) chemistry part. A simple sequential splitting is used. This means that at each time-step first the advection-diffusion part is treated and after that the chemical part is handled. A discretization technique based on central differences followed by Crank-Nicolson time-stepping is used in the advection-diffusion part. The non-linear chemical reactions are treated by the robust Backward Euler Formula. The performance of the combined numerical method (splitting procedure + numerical algorithms used in the advection-diffusion part and in the chemical part) is studied in connection with six test-problems. We are interested in both the accuracy of the results and the efficiency of the parallel computations.

1 Statement of the Problem

Large-scale air pollution models are usually described mathematically by systems of PDEs (partial differential equations):

$$
\frac{\partial c_s}{\partial t} = -\frac{\partial(u c_s)}{\partial x} - \frac{\partial(v c_s)}{\partial y} - \frac{\partial(w c_s)}{\partial z}
$$

$$
+\frac{\partial}{\partial x}\left(K_x \frac{\partial c_s}{\partial x}\right) + \frac{\partial}{\partial y}\left(K_y \frac{\partial c_s}{\partial y}\right) + \frac{\partial}{\partial z}\left(K_z \frac{\partial c_s}{\partial z}\right)
$$

$$
+E_s - (\kappa_{1s} + \kappa_{2s})c_s + Q_s(t, c_1, c_2, \ldots, c_q), \quad s = 1, 2, \ldots, q.
$$

(1)

The number q of equations in (1) is equal to the number of chemical species. The other quantities in (1) are (i) the concentrations c_s of the chemical species, (ii) the wind velocities u, v and w, (iii) the diffusion coefficients K_x, K_y, and K_z, (iv) the emission sources E_s, (v) the deposition coefficients κ_{1s} and κ_{2s}, and (vi) the non-linear terms $Q_s(t, c_1, c_2, \ldots, c_q)$ describing the chemical reactions. More

I. Lirkov, S. Margenov, and J. Waśniewski (Eds.): LSSC 2007, LNCS 4818, pp. 28–39, 2008.
© Springer-Verlag Berlin Heidelberg 2008

details about large-scale air pollution models can be found in Zlatev [13] and Zlatev and Dimov [14] as well as in the references given in these two monographs.

In order to check better the accuracy of the numerical methods used, the following simplifications in (1) were made: (i) the three-dimensional model was reduced to a two-dimensional model, (ii) the deposition terms were removed, (iii) a constant horizontal diffusion was introduced, and (iv) a special wind velocity wind was designed. The simplified model is given below:

$$\frac{\partial c_s^*}{\partial \tau} = -(\eta - 1)\frac{\partial c_s^*}{\partial \xi} - (1 - \xi)\frac{\partial c_s^*}{\partial \eta} + K\left(\frac{\partial^2 c_s^*}{\partial \xi^2} + \frac{\partial^2 c_s^*}{\partial \eta^2}\right) \tag{2}$$

$$+ E_s^*(\xi, \eta, \tau) + Q_s^*(\tau, c_1^*, c_2^*, \ldots, c_q^*),$$

where the independent variables ξ, η and τ vary in the intervals:

$$\xi \in [0, 2], \quad \eta \in [0, 2], \quad \tau \in [0, M(2\pi)], \quad M \geq 1. \tag{3}$$

The system of PDEs (2) must be considered together with some initial and boundary conditions. It will be assumed here that some appropriate initial and boundary conditions are given. Some further discussion related to the initial and boundary conditions will be presented in the remaining part of this paper.

The replacement of the general wind velocity terms $u = u(t, x, y)$ and $v = v(t, x, y)$ from (1) with the special expressions $\eta - 1$ and $1 - \xi$ in (2) defines a rotational wind velocity field (i.e., the trajectories of the wind are concentric circles with centres in the mid of the space domain and particles are rotated with a constant angular velocity).

If only the first two terms in the right-hand-side of (2) are kept, i.e., if pure advection is considered, then the classical rotation test will be obtained. This test has been introduced in 1968 simultaneously by Crowley and Molenkampf ([3] and [10]). In this case, the centre of the domain is in the point $(\xi_1, \eta_1) = (1.0, 1.0)$. High concentrations, which are forming a cone (see the upper left-hand-side plot in Fig. 1) are located in a circle with centre at $(\xi_0, \eta_0) = (0.5, 1.0)$ and with radius $r = 0.25$. If $\tilde{x} = \sqrt{(\xi - \xi_0)^2 + (\eta - \eta_0)^2}$, then the initial values for the original Crowley-Molenkampf test are given by $c_s^*(\xi, \eta, 0) = 100(1 - \tilde{x}/r)$ for $r < \tilde{x}$ and $c_s^*(\xi, \eta, 0) = 0$ for $r \geq \tilde{x}$. It can be proved that $c_s^*(\xi, \eta, k\,2\pi) = c_s^*(\xi, \eta, 0)$ for $k = 1, 2, \ldots, M$, i.e., the solution is a periodic function with period 2π. It can also be proved that the cone defined as above will accomplish a full rotation around the centre (ξ_1, η_1) of the domain when the integration is carried out from $\tau = k\,2\pi$ to $\tau = (k+1)\,2\pi$, where $k = 0, 1, \ldots, M-1$ (which explains why Crowley-Molenkampf test is often called the rotation test).

The advection process is dominating over the diffusion process, which is defined mathematically by the next term in the right-hand-side of (2). This means in practice that the constant K is very small, which is a very typical situation in air pollution modelling.

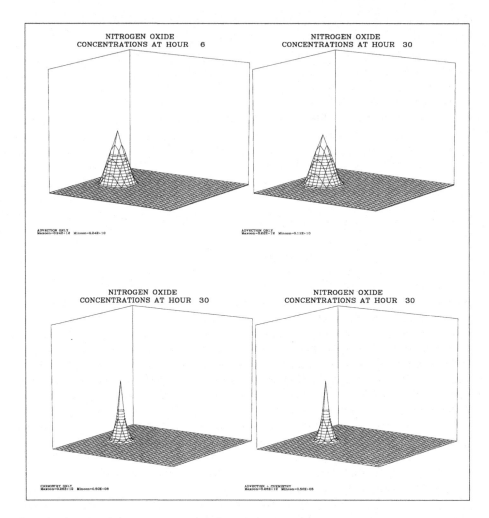

Fig. 1. Graphical representation of the solution of (5) when there are no emission sources (puff); (a) the initial solution is shown in the upper left-hand-side plot, (b) the solution at $t = 30$ when only the advection-diffusion module is run is shown in the upper right-hand-side plot, (c) the solution at $t = 30$ when only the chemistry module is run is shown in the lower left-hand-side plot, and (d) the solution at $t = 30$ when both the advection-diffusion module and the chemistry module is run is shown in the lower right-hand-side plot

If the first two terms and the last term in the right-hand-side of (2) are kept, i.e., if both advection and chemistry are treated, then an extension of the classical Crowley-Molenkampf rotation test is obtained, which was proposed and studied, twenty years later, in Hov et al. [9].

The short discussion given above shows that (2) is a further extension of the test proposed in Hov et al. [9].

The intervals for the independent variables ξ, η and τ, which are used in (3) are very convenient when the original Crowley-Molenkampf test is considered. For the extended test, where diffusion and chemistry terms are added, it is more convenient to use the intervals $[a_1, b_1]$, $[a_2, b_2]$ and $[a, b]$, which can be obtained by using the following substitutions in (2):

$$x = a_1 + \frac{b_1 - a_1}{2}\,\xi, \qquad y = a_2 + \frac{b_2 - a_2}{2}\,\eta, \qquad t = a + \frac{b - a}{2\pi}\,\tau. \tag{4}$$

The result is the following system of partial differential equations, which will be used in the following part of this paper:

$$\frac{\partial c_s}{\partial t} = -\mu(y - y_1)\frac{\partial c_s}{\partial x} - \mu(x_1 - x)\frac{\partial c_s}{\partial y} + K\left(\frac{\partial^2 c_s}{\partial x^2} + \frac{\partial^2 c_s}{\partial y^2}\right) \tag{5}$$

$$+ E_s(x, y, t) + Q_s(t, c_1, c_2, \ldots, c_q),$$

where

$$x_1 = \frac{a_1 + b_1}{2}, \qquad y_1 = \frac{a_2 + b_2}{2}, \qquad \mu = \frac{2\pi}{b - a}. \tag{6}$$

The actual intervals used in Section 4 are obtained by setting $a_1 = a_2 = 0.0$, $a = 6.0$, $b_1 = b_2 = 500.0$ and $b = 30.0$. The units for a_1, a_2, b_1, and b_2 are kilometres, while hours are used for a and b. Note that the length of the time-interval, which is actually used, is 24 hours, which is important, because this allows us to study the diurnal variation of some chemical species.

The numerical algorithms which might be used in the solution of (5), the accuracy of the solutions obtained when different discretizations are applied, and the possibility to design efficient parallel devices will be studied in this paper.

2 · Sequential Splitting

The advection-diffusion-chemistry module (5) is divided into two parts: advection-diffusion part and chemistry part. The two parts are combined in a sequential splitting procedure (i.e., after the discretization the treatment of the first part is followed by the treatment of the second part at every time-step). The advection-diffusion part is represented by the following system of PDEs (the equations of this system are independent, i.e., there is no coupling):

$$\frac{\partial g_s}{\partial t} = -\mu(y - y_1)\frac{\partial g_s}{\partial x} - \mu(x_1 - x)\frac{\partial g_s}{\partial y} + K\left(\frac{\partial^2 g_s}{\partial x^2} + \frac{\partial^2 g_s}{\partial y^2}\right). \tag{7}$$

The last two terms in (5) form the second part of the splitting procedure, the chemistry part:

$$\frac{\partial h_s}{\partial t} = E_s + Q_s(t, h_1, h_2, \ldots, h_q). \tag{8}$$

It is necessary to introduce some discretization in order to explain how the two parts (7) and (8) can be coupled. Assume that $t \in [a, b]$, $x \in [a_1, b_1]$ and

$y \in [a_2, b_2]$ and consider the equidistant increments $\triangle t$, $\triangle x$ and $\triangle y$. It will furthermore be assumed that $\triangle x = \triangle y$. Introduce the three equidistant grids:

$$\mathbf{G_t} = \{t_n \,|\, t_0 = a, \; t_n = t_{n-1} + \triangle t, \; n = 1, 2, \ldots, N_t, \; t_{N_t} = b\}, \tag{9}$$

$$\mathbf{G_x} = \{x_i \,|\, x_0 = a_1, \; x_i = x_{i-1} + \triangle x, \; i = 1, 2, \ldots, N_x, \; x_{N_x} = b_1\}, \tag{10}$$

$$\mathbf{G_y} = \{y_j \,|\, y_0 = a_2, \; y_j = y_{j-1} + \triangle y, \; j = 1, 2, \ldots, N_y, \; y_{N_y} = b_2\}. \tag{11}$$

It will be assumed that $N_x = N_y$, which together with $\triangle x = \triangle y$ implies that the space domain is a square. The assumptions $N_x = N_y$ and $\triangle x = \triangle y$ are in fact not needed, they are only made in order to simplify the further explanations. The approximations $c_{s,n,i,j} \approx c_s(t_n, x_i, y_j)$, $g_{s,n,i,j} \approx g_s(t_n, x_i, y_j)$, and $h_{s,n,i,j} \approx h_s(t_n, x_i, y_j)$ are obtained by some numerical algorithms (numerical algorithms will be discussed in the next section). It is convenient to consider the vectors: $\tilde{c}_n$, $\tilde{g}_n$ and $\tilde{h}_n$. Vector $\tilde{c}_n$ contains all values of $c_{s,n,i,j}$ at time t_n, this means that index n is fixed, while the remaining three indices are varying ($s = 1, 2, \ldots, q$, $i = 0, 1, 2, \ldots, N_x$, and $j = 0, 1, 2, \ldots, N_y$). The other two vectors, $\tilde{g}_n$ and $\tilde{h}_n$, are formed in a similar way.

Assume now that all computations up to some time-point t_{n-1} have been completed. Then we set $\tilde{g}_{n-1} = \tilde{c}_{n-1}$ and calculate an approximation $\tilde{g}_n$ by solving (7) using the selected numerical algorithm with $\tilde{g}_{n-1}$ as initial value. We continue by setting $\tilde{h}_{n-1} = \tilde{g}_n$ and solving (8) by using the selected for the second part numerical algorithm (which is as a rule different from that used in the solution of the first part) with $\tilde{h}_{n-1}$ as initial value. After the solution of the (8) we set $\tilde{c}_n = \tilde{h}_n$ and everything is prepared for the next time-step (for calculating $\tilde{c}_{n+1}$). It remains to explain how to start the computational process (how to calculate $\tilde{c}_1$). There is no problem with this because the initial value $\tilde{c}_0$ of (5) must be given.

One of the great problems related to the application of splitting techniques is the problem with the calculation of boundary conditions for the different sub-problems. This problem does not exist for the splitting technique discussed in this section, because if good boundary conditions can be obtained for the original problem (5), then these boundary conditions can be used in the treatment of the first part (7), while it is obvious that no boundary conditions are needed when the chemistry part (8) is handled. Thus, the sequential splitting procedure is in some sense optimal with regard to the boundary conditions when (5) is solved.

The use of splitting procedure is very useful in the case where parallel computations are to be carried out. It is easily seen that the system of PDEs (7) is in fact consisting of q independent equations (one equation per each chemical compound). This means that we have q parallel tasks when the advection-diffusion part is treated. It is also quite obvious that the second system of PDEs, (8), will contain $(N_x + 1)(N_y + 1)$ independent equations (one per each grid-point in the space domain) when some suitably chosen discretization is applied, which leads to $(N_x + 1)(N_y + 1)$ parallel tasks. Thus, parallel tasks appear in a very natural way when splitting procedures are used in connection with large-scale scientific models.

3   Selection of Numerical Algorithms

The operators which appear in the advection-diffusion part (7) and in the chemistry part (8) have different properties. Therefore, it is appropriate to apply different numerical algorithms during the treatment of (7) and (8) in order to exploit better the properties of the two operators.

3.1   Numerical Algorithms for the Advection-Diffusion Part

It was pointed out in the previous section that the system of PDEs (7) consists of q independent systems. Therefore, it is quite sufficient to fix some s, $s \in \{1, 2, \ldots, q\}$ and to consider the solution of the system obtained in this way. The discretization of the space derivatives in (7) by central differences followed by Crank-Nicolson time-stepping (see, for example, [11]) lead to the solution of a system of linear differential equations:

$$(I - A)\, \bar{g}_{n+1} = (I + A)\, \bar{g}_n + \bar{\omega}_n. \tag{12}$$

$I - A$ and $I + A$ are five-diagonal matrices of order $(N_x+1)(N_y+1)$, which depend on both the wind velocity and the diffusivity coefficient. For the special choice of the latter two parameters made in Section 1, these matrices are constant in time. Vector $\bar{\omega}_n$ is induced by the boundary conditions. Dirichlet boundary conditions have been used in this study. Vector $\bar{g}_n$ is a calculated, in the previous time-step, approximation of the concentrations of chemical species s at the $(N_x+1)(N_y+1)$ grid-points of the space domain. Similarly, vector $\bar{g}_{n+1}$ is the approximation that is to be calculated at the current time-step.

LAPACK subroutines for banded matrices, [2], can successfully be used when N_x and N_y are not large, while iterative methods (the simple Gauss-Seidel method, [8], was actually used) are more efficient when N_x and N_y are large.

3.2   Numerical Algorithms for the Chemistry Part

It was pointed out in the previous section that the system of PDEs (8) consists of $(N_x + 1)(N_y + 1)$ independent systems after the application of discretization based on (10) and (11). Therefore, it is quite sufficient to fix some pair (i, j) with $i \in \{0, 1, 2, \ldots, N_x\}$ and $j \in \{0, 1, 2, \ldots, N_y\}$ and to consider the solution of the system with q equations obtained in this way. There are no spatial derivatives in (8) which means that this system of PDEs will directly be reduced to a system of ordinary differential equations when the discretization by the grids (9)–(11) is performed:

$$\frac{d\bar{h}}{dt} = \bar{E} + \bar{Q}(t, \bar{h}), \tag{13}$$

where (a) the indices i and j are omitted and (b) $\bar{h}$, $\bar{E}$ and $\bar{Q}$ are vectors with q components which are concentrations and emissions and right-hand sides calculated at the grid-point defined by the pair (i, j).

The simple Backward Euler Formula is applied to solve (13). This algorithm results in the following non-linear system of algebraic equations:

$$\bar{h}_{n+1} = \bar{h}_n + \triangle t \left[\bar{E}_{n+1} + \bar{Q}(t_{n+1}, \bar{h}_{n+1}) \right], \tag{14}$$

where (a) $\bar{E}_{n+1}$ contains the emissions of all chemical species that are emitted at the grid-point under consideration and at time t_{n+1}, (b) $\bar{Q}$ is a non-linear fiction depending on the concentrations of all chemical species at the grid-point under consideration and calculated at time t_{n+1}, (c) $\bar{h}_n$ is a vector containing concentrations of all species at the grid-point under consideration that have been calculated in the previous time-step, and (d) $\bar{h}_{n+1}$ is similar to $\bar{h}_n$, but its components are the concentrations that are to be computed at the current time-step.

The non-linear system of algebraic equations (14) is solved by using the Newton iterative method [7]. This leads to the solution, at each iteration, of a system of linear algebraic equations with a coefficient matrix $B = I - \triangle t \, \partial \bar{Q} / \partial \bar{h}$ containing the Jacobian matrix of function $\bar{Q}$. It is important to solve the system of linear algebraic equations efficiently (in many large-scale air pollution codes the treatment of the chemical part takes about 90% of the CPU time). It is possible to apply some of the LAPACK subroutines for dense matrices [2], because matrix B is small (of order 56 in our case). Matrix B is sparse, but not very sparse and, therefore, the attempts to exploit directly the sparsity by using the codes from [4,5,6,12] were not very successful. A special matrix technique, similar to that described in [14] has been applied. This technique is based on a further sparsifying the matrix by removing small non-zero elements and using some of the special rules discussed in [14]. The numerical results indicate that this technique works very well.

It must be emphasized here that although the system (14) is small, it has to be solved extremely many times. For example, if $N_x = N_y = 512$ it is necessary to solve 263169 systems of order 56 at every time-step.

4   Numerical Tests

Several numerical test were performed in order to illustrate (a) the accuracy of the algorithms chosen and (b) the ability of the code to run in parallel. Some of the results will be presented and discussed in this section.

4.1   Organization of the Experiments

The advection-diffusion-chemistry module (5) is run over the interval $[6, 30]$, which means that the computations are started at 6:00, the length of the time-interval is 24 hours and the end of the computations is at 6:00 on the next day.

Two types of experiments were run. In the first of them there are no emissions (this is a simulation of a puff). High concentrations forming a cone are located in a sub-domain. Now if only advection-diffusion terms are kept, then the cone is simply rotated around the centre of the space domain. If only chemistry terms are kept, then there are changes of the size of the concentrations, but without any transport. If both advection-diffusion and chemistry terms are kept, then

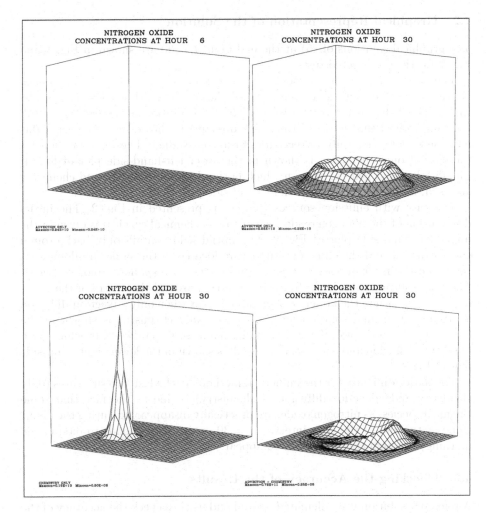

Fig. 2. Graphical representation of the solution of (5) when there are emission sources (plume); (a) the initial solution is shown in the upper left-hand-side plot, (b) the solution at $t = 30$ when only the advection-diffusion module is run is shown in the upper right-hand-side plot, (c) the solution at $t = 30$ when only the chemistry module is run is shown in the lower left-hand-side plot, and (d) the solution at $t = 30$ when both the advection-diffusion module and the chemistry module is run is shown in the lower right-hand-side plot

the cone is rotated and the size of the concentrations are varied. The solutions at the end of the time-interval are presented graphically in Fig. 1.

In the second type of experiments, emission sources are inserted at the same sub-domain where the cone was located in the previous case (a plume is simulated in these experiments). Again three sub-cases were considered: (a) only advection-diffusion, (b) only chemistry and (c) both advection diffusion and chemistry. The solutions at the end of the time-interval are presented graphically in Fig. 2.

4.2   Graphical Representation of the Solution

The graphical solutions of (5) at the end-point $t = 30$ are shown in Fig. 1 and Fig. 2 for the six selected tests.

The case without emission sources (puff) is presented in Fig. 1. The initial distribution of the nitrogen oxide (one of the 56 chemical species) is given in the upper left-hand-side plot of Fig. 1. The final solution (obtained after a full rotation around the centre of the space domain) is shown in the upper right-hand-side plot when only advection-diffusion is studied. The final solution for the case of only chemistry is shown in the lower left-hand-side plot of Fig. 1. Finally, the most general case, in which both advection-diffusion and chemistry are used, is shown in the lower right-hand-side plot of Fig. 1.

The case with emission sources (plume) is presented in Fig. 2. The initial distribution of the nitrogen oxide (one of the 56 chemical species) is given in the upper left-hand-side plot of Fig. 2. The initial field contain only back-ground concentrations and the emission sources are located in the same sub-domain as the sub-domain where the cone is located in the corresponding plot of Fig. 1. The final solution (obtained after a full rotation around the centre of the space domain) is shown in the upper right-hand-side plot when only advection-diffusion is studied. The final solution for the case of only chemistry is shown in the lower left-hand-side plot of Fig. 2. Finally, the most general case, in which both advection-diffusion and chemistry are used, is shown in the lower right-hand-side plot of Fig. 2.

The major difficulty for the numerical methods in the last two sub-cases (only chemistry and advection-diffusion + chemistry) is due to the fact that some chemical species, as nitrogen oxide, are practically disappearing during the night, while at the end of the time-interval ($t = 30$, which corresponds to 6:00 in the morning) the high concentrations re-appear.

4.3   Checking the Accuracy of the Results

A reference solution was calculated, stored and used to check the accuracy of the results. The reference solution was calculated by using a discretization defined by $N_x^{ref} = N_y^{ref} = 1024$ and $N_t^{ref} = 30720$. It was not possible to store all values of the reference solution. Values of the reference solution were stored for each of the chemical compounds at the end of every hour on a 33×33 field.

Let vector $\hat{c}_{k,s}^{ref}$ contain the reference solution for the sth chemical compound, $s = 1, 2, \ldots, q$, $q = 56$, calculated at the end of hour k, $k = 1, 2, \ldots, 24$. Vector $\hat{c}_{k,s}^{ref}$ contains $33 \times 33 = 1089$ components.

Let vector $\hat{c}_{k,s}$ be the corresponding vector calculated by another discretization with some $N_x < N_x^{ref}$, $N_y < N_y^{ref}$ and $N_t < N_t^{ref}$.

The error of the calculated solution for the chemical species s can be obtained by using the following formula:

$$ERROR_s = \max_{k=0,1,\ldots,24} \left(\frac{\|\hat{c}_{k,s} - \hat{c}_{k,s}^{ref}\|}{\max(\|\hat{c}_{k,s}^{ref}\|, 1.0)} \right). \tag{15}$$

The global error of the calculated solution (the error over all chemical compounds) can be obtained by using a slight modification of (15):

$$ERROR = \max_{k=0,1,\dots,24,\ s=1,2,\dots,56} \left(\frac{\|\hat{c}_{k,s} - \hat{c}_{k,s}^{ref}\|}{\max(\|\hat{c}_{k,s}^{ref}\|, 1.0)} \right). \tag{16}$$

It is clear that if $N_x = N_y = 16$, then the reference solution 33×33 fields must be projected into 17×17 fields when the error is to be computed.

Some accuracy results are given in Table 1 for the case where there are no emission sources (puff) and in Table 2 for the case where there are some emission sources. It is immediately seen that the results obtained when there are emission sources (plume) are much more accurate than the results obtained without emissions (puff). Also the rate of convergence is higher in the case of plume.

Table 1. Results obtained when (5) is run by using different discretizations in the case where there are no emissions (puff). RATE is showing the rate by which the accuracy is improved (related to the previous run).

N_x	N_y	N_t	ERROR	RATE
16	16	480	$1.99 * 10^{+0}$	-
32	32	960	$8.72 * 10^{-1}$	2.285
64	64	1920	$7.84 * 10^{-1}$	1.112
128	128	3840	$3.43 * 10^{-1}$	2.288
256	256	7680	$2.25 * 10^{-1}$	1.522
512	512	15360	$5.35 * 10^{-2}$	4.208

Table 2. Results obtained when (5) is run by using different discretizations in the case where there are emissions sources (plume). RATE is showing the rate by which the accuracy is improved (related to the previous run).

N_x	N_y	N_t	ERROR	RATE
16	16	480	$3.73 * 10^{-1}$	-
32	32	960	$7.68 * 10^{-2}$	4.860
64	64	1920	$2.18 * 10^{-2}$	3.527
128	128	3840	$8.73 * 10^{-3}$	2.497
256	256	7680	$2.86 * 10^{-3}$	3.055
512	512	15360	$1.33 * 10^{-3}$	2.140

4.4   Parallel Computations

Parallel computations were carried out on the SUN computers of the Danish Centre for Scientific Computing [15]. These are shared memory computers. Therefore OpenMP tools were used (see also [1] and [16]). Up to 32 processors were used.

In the advection-diffusion part there are $q = 56$ parallel tasks (see the end of Section 2). The number of parallel tasks in the chemistry part is $(N_x+1)(N_y+1)$.

Table 3. Results obtained when (5) is run on different numbers of processors in the case where there are no emissions sources (puff). The discretization parameters are: $N_x = N_y = 512$ and $N_t = 15360$. The computing times are measured in CPU hours, the speed-ups are given in brackets.

Processors	Advection	Chemistry	Total	Overhead
1	28.67	33.78	66.14	5.6%
2	15.15 (1.9)	19.69 (1.7)	37.28 (1.8)	6.5%
4	8.33 (3.4)	10.21 (3.3)	20.29 (3.3)	8.6%
8	5.24 (5.5)	5.53 (6.1)	12.08 (5.5)	10.8%
16	2.57 (11.2)	2.49 (13.6)	6.16 (10.8)	17.9%
32	2.19 (13.1)	1.26 (26.8)	4.55 (14.5)	24.1%

Table 4. Results obtained when (5) is run on different numbers of processors in the case where there are emissions sources (plume). The discretization parameters are: $N_x = N_y = 512$ and $N_t = 15360$. The computing times are measured in CPU hours, the speed-ups are given in brackets.

Processors	Advection	Chemistry	Total	Overhead
1	26.74	35.53	65.97	5.6%
2	14.05 (1.9)	20.36 (1.7)	37.00 (1.8)	7.0%
4	7.59 (3.5)	10.44 (3.4)	19.75 (3.3)	8.7%
8	4.96 (5.4)	5.63 (6.3)	11.90 (5.5)	11.0%
16	2.99 (8.9)	2.70 (13.2)	6.92 (9.6)	17.8%
32	2.10 (12.7)	1.30 (27.3)	4.54 (14.5)	25.1%

Some results are given in Table 3 and Table 4. "Advection", "Chemistry", and "Total" refer to the times spent in the advection-diffusion part, the chemical part, and the total computing time. The computing times are measured in CPU hours. The speed-ups are given in brackets. "Overhead" is calculated by using the formula: Overhead = 100 (Total - Advection - Chemistry) / Total. The discretization was performed by using: $N_x = N_y = 512$ and $N_t = 15360$.

Acknowledgements

This research was partly supported by a Collaborative Linkage Grant "Monte Carlo Sensitivity Studies of Environmental Security" within the NATO Programme Security through Science.

References

1. Alexandrov, V., et al.: Parallel runs of a large air pollution model on a grid of Sun computers. Mathematics and Computers in Simulation 65, 557–577 (2004)
2. Anderson, E., et al.: LAPACK: Users' Guide. SIAM, Philadelphia (1992)
3. Crowley, W.P.: Numerical advection experiments. Monthly Weather Review 96, 1–11 (1968)

4. Demmel, J.W., et al.: A supernodal approach to sparse partial pivoting. SIAM Journal of Matrix Analysis and Applications 20, 720–755 (1999)
5. Demmel, J.W., Gilbert, J.R., Li, X.S.: An asynchronous parallel supernodal algorithm for sparse Gaussian elimination. SIAM Journal of Matrix Analysis and Applications 20, 915–952 (1999)
6. Duff, I.S., Erisman, A.M., Reid, J.K.: Direct Methods for Sparse Matrices. Oxford University Press, Oxford-London (1986)
7. Golub, G.H., Ortega, J.M.: Scientific Computing and Differential Equations. Academic Press, Boston-San Diego-New York-London-Sydney-Tokyo-Toronto (1992)
8. Golub, G.H., Van Loan, C.F.: Matrix Computations, Maryland. Johns Hopkins University Press, Baltimore (1983)
9. Hov, Ø., et al.: Comparison of numerical techniques for use in air pollution models with non-linear chemical reaction. Atmospheric Environment 23, 967–983 (1988)
10. Molenkampf, C.R.: Accuracy of finite-difference methods applied to the advection equation. Journal of Applied Meteorology 7, 160–167 (1968)
11. Strikwerda, J.C.: Finite difference schemes and partial differential equations. Second Edition. SIAM, Philadelphia (2004)
12. Zlatev, Z.: Computational methods for general sparse matrices. Kluwer Academic Publishers, Dordrecht-Boston-London (1991)
13. Zlatev, Z.: Computer Treatment of Large Air Pollution Models. In: Environmental Science and Technology Science, vol. 2, Kluwer Academic Publishers, Dordrecht-Boston-London (1995)
14. Zlatev, Z., Dimov, I.: Computational and Environmental Challenges in Environmental Modelling. In: Studies in Computational Mathematics, Amsterdam-Boston-Heidelberg-London-New York-Oxford-Paris-San Diego-San Francisco-Singapore-Sydney-Tokyo, vol. 13, Elsevier, Amsterdam (2006)
15. Danish Centre for Scientific Computing at the Technical University of Denmark, Sun High Performance Computing Systems (2002), http://www.hpc.dtu.dk
16. Open MP tools (1999), http://www.openmp.org

Comments on the GMRES Convergence for Preconditioned Systems

Nabil Gmati[1] and Bernard Philippe[2]

[1] ENIT — LAMSIN. B.P. 37, 1002 Tunis Belvédère, Tunisie
[2] INRIA/IRISA, Campus de Beaulieu, 35042 RENNES Cedex, France

Abstract. The purpose of this paper is to comment a frequent observation by the engineers studying acoustic scattering. It is related to the convergence of the GMRES method when solving systems $Ax = b$ with $A = I - B$. The paper includes a theorem which expresses the convergence rate when some eigenvalues of B have modulus larger than one; that rate depends on the rate measured when solving the system obtained by spectral projection onto the invariant subspace corresponding to the other eigenvalues. The conclusion of the theorem is illustrated on the Helmholtz equation.

1 Introduction

The purpose of this paper is to comment a frequent observation by the engineers studying acoustic scattering. It is related to the convergence of the GMRES method when solving systems $Ax = b$ with $A = I - B \in \mathbb{R}^{n \times n}$. If the spectral radius $\rho(B)$ is smaller than one, it is easy to see that GMRES converges better than the convergence obtained by the Neumann series $A^{-1} = \sum_{k \geq 0} B^k$; engineers usually claim that when some eigenvalues of B lie out the unit disk, GMRES still converges. Our attempt is to explain that effect.

First, let us define the context precisely. For any matrix $A \in \mathbb{R}^{n \times n}$ and from an initial guess x_0, the GMRES method [12] iteratively builds a sequence of approximations x_k $(k \geq 1)$ of the solution x such that the residual $r_k = b - Ax_k$ satisfies

$$\|r_k\| = \min_{\substack{q \in \mathcal{P}_k \\ q(0) = 1}} \|q(A)r_0\| , \tag{1}$$

where $\mathcal{P}_k$ is the set of polynomials of degree k or less. Throughout this paper, the considered norms are the 2-norms.

Since the implementation of the method involves the construction by induction of an orthonormal system V_k which makes its storage mandatory, the method needs to be restarted at some point $k = m$ to be tractable. The corresponding method is denoted $GMRES(m)$.

The studied decomposition, $A = I - B$, arises from the introduction of a preconditioner as usually done to improve the convergence. This is a non singular

I. Lirkov, S. Margenov, and J. Waśniewski (Eds.): LSSC 2007, LNCS 4818, pp. 40–51, 2008.

matrix M that is easy to invert (i.e. solving systems $My = c$ is easy) and such that M^{-1} is some approximation of A^{-1}. It can be applied on the left side of the system by solving $M^{-1}Ax = M^{-1}b$ or on the right side by solving $AM^{-1}y = b$ with $x = M^{-1}y$. By considering the corresponding splitting $A = M - N$, we shall investigate the behavior of GMRES on the system

$$(I - B)x = b , \tag{2}$$

where B is respectively $B = M^{-1}N$ (left side) or $B = NM^{-1}$ (right side). By rewriting condition (1) in that context, the sequence of residuals satisfies (see the proof of Theorem 1) :

$$\|r_k\| = \min_{\substack{q \in \mathcal{P}_k \\ q(1) = 1}} \|q(B)r_0\| . \tag{3}$$

In this paper, after some comments on various interpretations of the convergence of GMRES, we first consider the situation where $\rho(B) < 1$. We demonstrate that it is hard to have bounds that illustrate the usual nice behavior of GMRES in such situations. Therefore, for the situation where some eigenvalues of B lie outside the unit disk, we propose to link the residual evolution to the residual evolution of a projected system, which discards all the outer eigenvalues.

2   About the Convergence of GMRES

In exact arithmetic and in finite dimension n, GMRES should be considered as a direct method since for every x_0, there exists a step $k_{end} \leq n$ such that $x_{k_{end}} = x$. Unfortunately, computer arithmetic is not exact and in most situations $k_{end} = n$ is much too big to be reached. However, when B is rank deficient, $k_{end} < n$. For instance, when the preconditioner M is defined by a step of the Multiplicative Schwarz method obtained from an algebraic overlapping 1-D domain decomposition, it can be proved that k_{end} is no bigger than the sum of the orders of the overlaps [1].

When the residuals does not vanish at some point $k \leq m$ where m is the maximum feasible size for the basis V_k, restarting becomes necessary. In that situation, the method GMRES(m) builds a sequence of approximations $x_0^{(K)}$ corresponding to all the initial guesses of the outer iterations. It is known that, unless the symmetric part of $I - B$ is positive or negative definite, stagnation may occur which prevents convergence.

In this paper, we limit the study of the convergence to the observation of the residual decrease during one outer iteration.

2.1   Existing Bounds

In [5], Embree illustrates the behavior of several known bounds. He proves that, for non normal matrices, any of the bounds can be overly pessimistic in some situations. Here, we only consider two bounds in the case where $\rho(B) < 1$.

By selecting the special polynomial $q(Z) = z^k$ in (3), we get the bound

$$\|r_k\| \leq \|B^k\| \|r_0\| . \tag{4}$$

The bound indicates that asymptotically there should be at least a linear convergence rate when $\rho = \rho(B) < 1$. For instance, if B can be diagonalized by $B = XDX^{-1}$, then the following bound holds :

$$\|r_k\| \leq cond(X)\rho^k \|r_0\| , \tag{5}$$

where $cond(X)$ is the condition number of the matrix of eigenvectors. It is clear that the bound might be very poor for large values of the condition number. For instance, it might happen that $cond(X)\rho^n > 1$, in which case no information can be drawn from (5).

The bound can be improved by an expression involving the condition numbers of the eigenvalues [5]. These bounds can be generalized when matrix B involves some Jordan blocks [13]. However, as mentioned earlier, there are always situations in which the bounds blow up.

In order to get a bound $\|r_k\| \leq K\gamma^k$ with a constant K smaller than the condition numbers of the eigenvalues, other expressions have been obtained from the field $W(A)$ of values of A ($W(A) = \{u^H Au | u \in \mathbb{C}^n, \|u\| = 1\}$). That set is convex and includes the eigenvalues (and therefore their convex hull). It is contained in the rectangle limited by the extremal eigenvalues of the symmetric and skew-symmetric parts of A. It is easy to see that $W(A) = 1 - W(B)$. When A is positive definite (or negative definite), which is equivalent to assuming that $1 \notin W(B)$, Beckermann [2] proved that

$$\|r_k\| \leq (2 + \gamma)\gamma^k \|r_0\| \tag{6}$$

where $\gamma = 2\sin\left(\frac{\beta}{4-2\beta/\pi}\right) < \sin\beta$ with β defined by : $\cos\beta = \frac{dist(0,W(A))}{\max |W(A)|}$. That result slightly improved earlier results [4].

The behaviors of the two bounds (5) and (6) are illustrated on a special test matrix. The chosen matrix is built by the following instruction (in MATLAB syntax): A=0.5*eye(100)+0.25*gallery('smoke',100). The parameters which are involved in the two bounds are displayed in Table 1. In Figure 1, the field of values of A is displayed (inner domain of the shaded region) as well as a zoom of the set at the origin. Below, the explicit bounds are plotted and compared to the residuals of GMRES applied to $Ax = b$ with $b = (1, \cdots, 1)^T/\sqrt{n}$. Although that matrix may be considered rather special, the behavior is common to the situation of non-normal matrices. The expressions of the bounds are of

Table 1. Characteristics of the matrix used in Figure 1

Spectral radius of $B = I - A$	: 0.752
Condition number of the eigenvectors	: 6.8×10^{13}
Parameter γ (from the field of values)	: 0.996

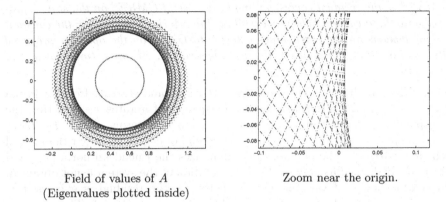

Field of values of A Zoom near the origin.
(Eigenvalues plotted inside)

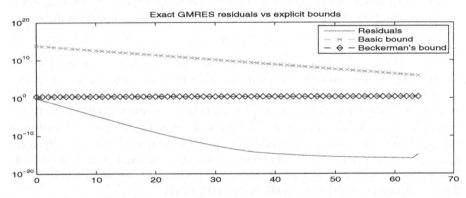

Fig. 1. Classical behavior of explicit bounds

the shape $K\gamma^k$, and the non-normality will impact either the constant K which blows up or the rate γ which will be very close to 1.

From that experience, it is clear that nothing can be expected from explicit bounds to illustrate the claim of the engineers mentioned in the introduction. The obstacle is twofold: (i) even for $\rho(B) < 1$ the bounds are often useless; (ii) the bounds have a linear behavior. Several authors have attempted to exhibit superlinear bounds (see for instance [9,10]) but the advantage is more theoretical than practical.

In that perspective, it is better to show how some eigenvalues of B outside the unit disk may deteriorate the convergence with respect to the situation $\rho(B) < 1$.

2.2   When Some Eigenvalues of B Lie Outside the Unit Disk

Let us denote by $D = \{z \in \mathbb{C}, \ |z| < 1\}$ the open unit disk and by $(\lambda_i)_{i=1,n}$ the eigenvalues of B.

Theorem 1. *If $A = I - B \in \mathbb{R}^{n \times n}$ is non singular, with p eigenvalues of B outside the open unit disk D, then for GMRES and for $k \geq p$:*

$$\|r_k\| \leq K\|r_{k-p}^{red}\|$$

where r_k^{red} is the residual corresponding to the use of GMRES on the system projected onto the invariant subspace excluding the exterior eigenvalues; the constant K only depends on the eigenvalues exterior to D and on the operator projected onto the invariant subspace of all other eigenvalues. All projections considered are spectral.

Proof. When applied to the system $(I - B)x = b$ and starting from an initial guess x_0, the GMRES iteration builds a sequence of iterates x_k such that the corresponding residual can be expressed in polynomial terms: $r_k = \pi(I - B) r_0$ where the polynomial $\pi \in P_k(\mathbb{C})$ is a polynomial of degree no larger than k and such that $\pi(0) = 1$. The polynomial π minimizes the Euclidean norm $\|r_k\|$ and it is uniquely defined as long as the exact solution of the system is not obtained. Through a change of variable $\tau(z) = \pi(1 - z)$, the residual is expressed as $r_k = \tau(B) r_0$ where the normalizing condition becomes $\tau(1) = 1$. Therefore, for any polynomial $\tau \in \mathcal{P}_k(\mathbb{C})$ such that $\tau(1) = 1$, the following bound stands:

$$\|r_k\| \leq \|\tau(B) r_0\|. \tag{7}$$

We shall now build a special residual polynomial. For that purpose, let us decompose the problem onto the two supplementary invariant subspaces $span(X_1) \oplus span(X_2)$ where X_1 and X_2 are bases of the invariant subspaces corresponding to respectively the p largest eigenvalues of B and the $n - p$ other eigenvalues. By denoting $X = [X_1, X_2]$, $Y = X^{-T} = [Y_1, Y_2]$ and $P_i = X_i Y_i^T$ the spectral projector onto $span(X_i)$ for $i = 1, 2$, the matrix B can be decomposed into $B = B_1 + B_2$ where $B_i = P_i B P_i$ for $i = 1, 2$. Therefore the nonzero eigenvalues of B_1 lie outside D whereas the spectrum of B_2 is included in D. Moreover, for any polynomial π, the decomposition $\pi(B) = \pi(B_1)P_1 + \pi(B_2)P_2$ holds.

Let π_1 the polynomial of degree p that vanishes at each eigenvalue $\lambda_i \notin D$ and such that $\pi_1(1) = 1$. The polynomial π_1 is uniquely defined by

$$\pi_1(z) = \prod_{\lambda_i \notin D} \frac{z - \lambda_i}{1 - \lambda_i}. \tag{8}$$

By construction, that polynomial satisfies the following property

$$\pi_1(B_1)P_1 = 0. \tag{9}$$

Let τ_2 be the residual polynomial corresponding to the iteration $k - p$ when solving the system projected onto the invariant subspace $span(X_2)$ (spectral projection):

$$\begin{cases} (P_2 - B_2)x = P_2 b, \\ P_1 x = 0. \end{cases}$$

Therefore the residual of the reduced system is $r_{k-p}^{red} = \tau_2(B_2)P_2 r_0$. By considering the polynomial $\tau = \pi_1 \tau_2$, since $\tau(B_1)P_1 = 0$ we get

$$\begin{aligned} \tau(B) r_0 &= \tau(B_1) P_1 r_0 + \tau(B_2) P_2 r_0, \\ &= \pi_1(B_2) \tau_2(B_2) P_2 r_0, \\ &= \pi_1(B_2) r_{k-p}^{red}, \\ &= \pi_1(B_2) P_2 r_{k-p}^{red}. \end{aligned}$$

The last transformation is not mandatory but highlights that the operator which defines the constant K is zero on $span(X_1)$. By inequality (7), the conclusion of the theorem holds with $K = \|\pi_1(B_2)P_2\|$.

The result of the theorem may be interpreted informally as saying that p eigenvalues of B outside the unit disk only introduce a delay of p steps in the convergence of GMRES. This fact will be illustrated in the experiments of the next section.

3   Illustration

3.1   Solving a Helmholtz Problem

Let $\Omega_i \subset \mathbb{R}^3$ be a bounded obstacle with a regular boundary Γ and Ω_e be its unbounded complementary. The Helmholtz problem models a scattered acoustic wave propagating through Ω_e; it consists in determining u such that

$$\begin{cases} \Delta u + \kappa^2 u = 0 \text{ in } \Omega_e, \\ \partial_n u = f \text{ on } \Gamma, \\ (\frac{x}{|x|}.\nabla - i\kappa)u = e^{i\kappa|x|}O(\frac{1}{|x|^2}) \quad x \in V_\infty, \end{cases} \tag{10}$$

where κ is the wave number and where V_∞ is a neighborhood of infinity. The last condition represents the Sommerfeld radiation condition. To solve the boundary value problem (10), we may consider an integral equation method (eventually coupled to finite element method for non-constant coefficients). The efficiency of this approach has been investigated by several authors, e.g. [8,11]. An alternative approach consists in using a coupled method which combines finite elements and integral representations [6]. This approach avoids the singularities of the Green function. The idea simply amounts to introducing a fictitious boundary Σ surrounding the obstacle. The Helmholtz problem is posed in the truncated domain Ω_c (delimited by Γ and Σ) with a non-standard outgoing condition using the integral formula which is specified on Σ,

$$\Delta u + \kappa^2 u = 0 \text{ in } \Omega_c, \partial_n u = f \text{ on } \Gamma, \tag{11}$$

$$(\partial_n - i\kappa)u(x) = \int_\Gamma (u(y)\partial_n K(x-y) - f(y)K(x-y))\,\mathrm{d}\gamma(y), \forall x \in \Sigma, \tag{12}$$

where $K(x) = (\frac{x}{|x|}.\nabla - i\kappa)G_\kappa(x)$ and G_κ is the Green function. Observe that the integral representation is used only on Σ which avoids occurrences of singularities. We suppose that this problem is discretized by a Lagrange finite element method. Let N_{Ω_c} be the total number of degrees of freedom on Ω_c and N_Σ (resp. N_Γ) be the number of degrees of freedom on Σ (resp. Γ). The shape function associated with node x_α is denoted w_α. Let u_α be the approximation of the solution u at x_α and $v = (u_\alpha)_\alpha$.

The linear system can be formulated as follows:

$$(A - C)v = b \tag{13}$$

where

$$A_{\alpha,\beta} = \int_{\Omega_c} \left(\nabla w_\alpha(x) \nabla \overline{w}_\beta(x) - \kappa^2 w_\alpha(x) \overline{w}_\beta(x) \right) dx - i\kappa \int_\Sigma w_\alpha(x) \overline{w}_\beta(x) d\sigma(x)$$

$$C_{\alpha,\beta} = \int_\Sigma \int_\Gamma w_\alpha(x) \partial_n K(x - y) \, d\gamma(y) \overline{w}_\beta(x) \, d\sigma(x)$$

The matrix of the system (13) is complex, non Hermitian and ill-conditioned. Matrix A is a sparse matrix but matrix C, which represents non-local coupling terms enforced by the integral term, is dense. Since, solving a system in A is easier than solving system (13), the matrix A is chosen as a preconditioner. In other words, the linear system is formulated as follow:

$$(I_{N_{\Omega_c}} - B)v = c, \tag{14}$$

where $B = A^{-1}C$ and $c = A^{-1}b$. Direct methods adapted to sparse matrices are good candidates for solving the preconditioning step.

3.2   Definition of the Test Problems

The numerical results deal with examples of acoustic scattering with an incident plane wave. Four tests are defined (see Figure 2).

The scatterer is considered to be a cavity. In the first test, the computational domain is rectangularly shaped while in the second and third tests, the domains are nonconvex. The cavity of Test 2 is the same as in Test 1. Cavities and domains are the same for Test 3 and Test 4. The meshing strategy implies two rows of finite elements in the non convex cases (Tests 2, 3 and 4). The characteristic parameters for every test are listed in Table 2.

Table 2. Characteristics of the tests

	N_{el}	N_{Ω_c}	N_Σ	h	κ
Test # 1	1251	734	88	0.06	1
Test # 2	543	403	134	0.06	1
Test # 3	2630	1963	656	0.005	1
Test # 4	2630	1963	656	0.005	62

N_{el} : number of finite elements;
N_{Ω_c} : number of degrees of freedom in Ω_c;
N_Σ : number of degrees of freedom in Σ;
h : mean diameter of the mesh;
κ : wave number.

3.3   Numerical Results

For the four tests, system (13) is solved by GMRES preconditioned by A which corresponds to solving system (14). In Figures 3 and 4, for every test, the plot

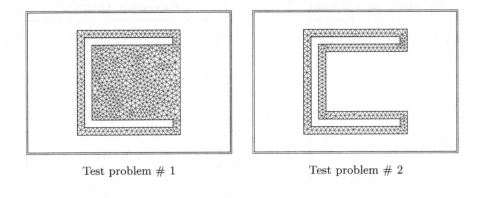

Test problem # 1   Test problem # 2

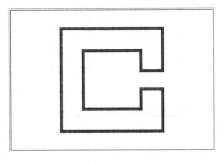

Test problems # 3 & # 4

Fig. 2. Cavities, domains and characteristics for the test problems

of the eigenvalues of B is reported, with respect to the unit disk (left figure), as well as the sequence of the GMRES residuals (right figure). Parameter p denotes the number of eigenvalues outside the unit disk. In order to illustrate Theorem 1, when $p > 0$ the residuals of the projected system are plotted with a shift (delay) p on the iteration numbers (dashed line). The same curve is plotted (solid line) with the residuals multiplied by the constant K in the bound given by the theorem.

The study of the spectrum of $B = A^{-1}C$ in Test 1 and Test 2 shows that when the fictitious boundary is located far from the cavity (Test 1), the whole spectrum is included in the unit disk. In that situation, GMRES converges in a few iterations. In Test 2, three eigenvalues exit the unit disk which deteriorates the convergence. This is foreseeable by reminding the connection between the method coupling Finite Element Method with an integral representation and the Schwarz method: the larger computational domain (which corresponds to the overlapping in the Schwarz method) — the faster the convergence [3,7].

In Test 3 and Test 4, the same cavity with the same domain and the same discretization are considered. Only the wave numbers differ. The mesh size corresponds to twenty finite elements by wave length when $\kappa = 62$. When κ increases, the number of eigenvalues outside the unit disk decreases. However, it can be

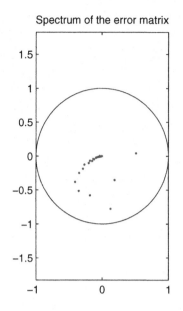
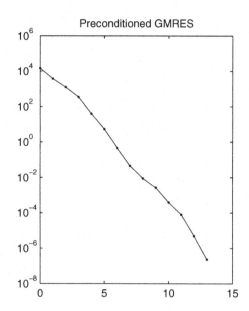

Helmholtz Test $1 - p = 0$.

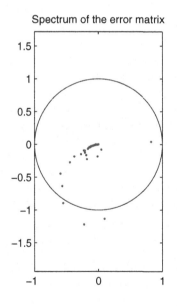
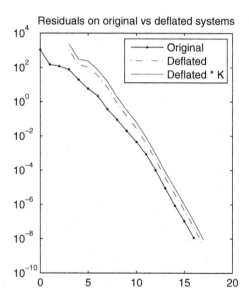

Helmholtz Test $2 - p = 3$.

Fig. 3. Spectra of B and GMRES residual evolutions

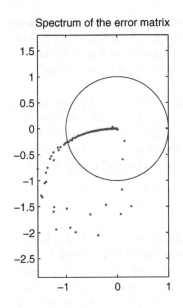
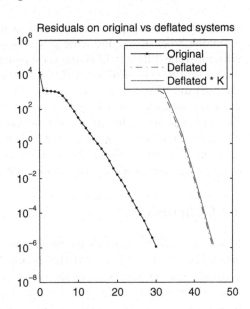

Helmholtz Test 3 − p = 29.

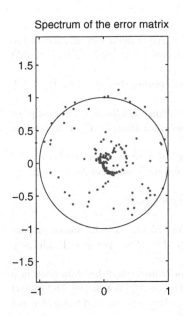
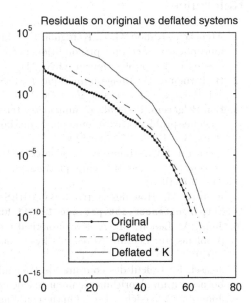

Helmholtz Test 4 − p = 11.

Fig. 4. Spectra of B and GMRES residual evolutions

seen that the convergence is not necessarily improved: the number of iterations is smaller when $\kappa = 1$ inducing $p = 29$ compared to $p = 11$ when $\kappa = 62$. This illustrates the difficulty of characterizing the convergence by only considering the modulus of the eigenvalues of B. Clustering of eigenvalues and non-normality of the operator play an important role.

In the situations where $p > 0$ (Test 2, 3 and 4), the bound given by the theorem appears to be often quite accurate. The worst case arises in Test 3 in which the number of outer eigenvalues is high ($p = 29$); in that situation the delay introduced by the shift p in the iteration number is too high because the convergence occurs at the same time that all the outer eigenvalues are killed.

4   Conclusion

Exhibiting realistic bounds for the successive residuals obtained when solving a linear system $Ax = b$ by GMRES is hopeless except in special situations. The knowledge of the spectrum of the operator is not even sufficient to fully understand the behavior of the convergence. These comments were already discussed by several authors but in this paper we have tried to explain the favorable situation where only a few eigenvalues of B ($A = I - B$) are of modulus larger than 1.

References

1. Atenekeng-Kahou, G.A., Kamgnia, E., Philippe, B.: An explicit formulation of the multiplicative Schwarz preconditioner. Research Report RR-5685, INRIA (2005) (Accepted for publication in APNUM)
2. Beckermann, B.: Image numérique, GMRES et polynômes de Faber. C. R. Acad. Sci. Paris, Ser. I 340, 855–860 (2005)
3. Ben Belgacem, F., et al.: Comment traiter des conditions aux limites à l'infini pour quelques problèmes extérieurs par la méthode de Schwarz alternée. C. R. Acad. Sci. Paris, Ser. I 336, 277–282 (2003)
4. Eisenstat, S.C., Elman, H.C., Schultz, M.H.: Variational iterative methods for non-symmetric systems of linear equations. SIAM Journal on Numerical Analysis 20(2), 345–357 (1983)
5. Embree, M.: How descriptive are GMRES convergence bounds?, http://citeseer.ist.psu.edu/196611.html
6. Jami, A., Lenoir, M.: A new numerical method for solving exterior linear elliptic problems. In: Lecture Notes in Phys., vol. 90, pp. 292–298. Springer, Heidelberg (1979)
7. Jelassi, F.: Calcul des courants de Foucault harmoniques dans des domaines non bornés par un algorithme de point fixe de cauchy. Revue ARIMA 5, 168–182 (2006)
8. Johnson, C., Nédélec, J.-C.: On the coupling of boundary integral and finite element methods. Math. Comp. 35(152), 1063–1079 (1980)
9. Kerhoven, T., Saad, Y.: On acceleration methods for coupled nonlinear elliptic systems. Numer. Maths 60(1), 525–548 (1991)

10. Moret, I.: A note on the superlinear convergence of GMRES. SIAM Journal on Numerical Analysis 34(2), 513–516 (1997)
11. Nédélec, J.-C.: Acoustic and electromagnetic equations Integral Representations for Harmonic Problems. In: Applied Mathematical Sciences, vol. 144, Springer, NewYork (2001)
12. Saad, Y., Schultz, M.H.: GMRES: a generalized minimal residual algorithm for solving nonsymmetric linear systems. SIAM J. Sci. Stat. Comput. 7(3), 856–869 (1986)
13. Tichý, P., Liesen, J., Faber, V.: On worst-case GMRES, ideal GMRES, and the polynomial numerical hull of a Jordan block. Electronic Transactions on Numerical Analysis (submitted, June 2007)

Optimization Based Stabilization of Nonlinear Control Systems

Lars Grüne

Mathematical Institute, University of Bayreuth, 95440 Bayreuth, Germany
lars.gruene@uni-bayreuth.de
http://www.math.uni-bayreuth.de/~lgruene/

Abstract. We present a general framework for analysis and design of optimization based numerical feedback stabilization schemes utilizing ideas from relaxed dynamic programming. The application of the framework is illustrated for a set valued and graph theoretic offline optimization algorithm and for receding horizon online optimization.

1 Introduction

The design of feedback controls for nonlinear systems is one of the basic problem classes in mathematical control theory. Among the variety of different design objectives within this class, the design of stabilizing feedbacks is an important subproblem, on the one hand because it captures the essential difficulties and on the other hand because it often occurs in engineering practice. The rapid development of numerical methods for optimization and optimal control which has let to highly efficient algorithms which are applicable even to large scale nonlinear systems, naturally leads to the idea of using such algorithms in feedback stabilization. While the basic principles underlying the relation between optimal control and stabilization have been understood since the 1960s in the context of the linear quadratic regulator design, the application to nonlinear systems poses new problems and challenges. These are caused, for instance, due to the complicated dynamical behavior which even low dimensional nonlinear systems can exhibit, due to hybrid or switching structures incorporated in the dynamics or simply due to the size of the problems to be solved, e.g. when discretized PDEs are to be controlled.

In this paper we investigate the foundation of optimization based feedback stabilization for nonlinear systems and develop a framework which allows to design and analyze different numerical approaches. The main goal of our framework is to give rigorous mathematical stability proofs even in the case when the numerical approximation is inaccurate, in the sense that the optimal control problem which should be solved theoretically is only very coarsely approximated by our numerical algorithm. Our approach was motivated and inspired by two sources: on the one hand by classical Lyapunov function stability theory, and in this context our condition can be understood as a preservation of the Lyapunov function property under (not necessarily small) numerical approximation

I. Lirkov, S. Margenov, and J. Waśniewski (Eds.): LSSC 2007, LNCS 4818, pp. 52–65, 2008.

errors. On the other hand we make use of relaxed dynamic programming methods, which essentially ensure that even in the presence of errors we can still give precise bounds for the performance of the feedback controller derived from the numerical information.

The organization of this paper is as follows. After describing the setup and the (theoretical) relation between optimization and stabilization in Section 2, we develop our conditions for optimization based stabilization with coarse numerical approximations in Section 3. In Section 4 we apply these conditions to a graph theoretic offline optimization approach while in Section 5 we show how these conditions can contribute to the analysis and design of unconstrained receding horizon control schemes.

2   Setup and Preliminaries

We consider a nonlinear discrete time system given by

$$x(n+1) = f(x(n), u(n)), \quad x(0) = x_0 \tag{1}$$

with $x(n) \in X$ and $u(n) \in U$ for $n \in \mathbb{N}_0$. We denote the space of control sequences $u : \mathbb{N}_0 \to U$ by $\mathcal{U}$ and the solution trajectory for some $u \in \mathcal{U}$ by $x_u(n)$. Here the state space X is an arbitrary metric space, i.e., it can range from a finite set to an infinite dimensional space.

A typical class of systems we consider are sampled-data systems governed by a controlled differential equation $\dot{x}(t) = g(x(t), \tilde{u}(t))$ with solution $\varphi(t, x_0, \tilde{u})$ for initial value x_0. These are obtained by fixing a sampling period $T > 0$ and setting

$$f(x, u) := \varphi(T, x, \tilde{u}) \quad \text{with} \quad \tilde{u}(t) \equiv u. \tag{2}$$

Then, for any discrete time control function $u \in \mathcal{U}$ the solutions x_u of (1),(2) satisfy $x_u(n) = \varphi(nT, x_0, \tilde{u})$ for the piecewise constant continuous time control function $\tilde{u} : \mathbb{R} \to U$ with $\tilde{u}|_{[nT,(n+1)T)} \equiv u(n)$. Note that with this construction the discrete time n corresponds to the continuous time $t = nT$.

2.1   Infinite Horizon Optimal Control

Our goal is to find a feedback control law minimizing the infinite horizon cost

$$J_\infty(x_0, u) = \sum_{n=0}^{\infty} l(x_u(n), u(n)), \tag{3}$$

with running cost $l : X \times U \to \mathbb{R}_0^+$. We denote the optimal value function for this problem by

$$V_\infty(x_0) = \inf_{u \in \mathcal{U}} J_\infty(x_0, u).$$

Here a (static state) feedback law is a control law $F : X \to U$ which assigns a control value u to each state x and which is applied to the system according to the rule

$$x_F(n+1) = f(x_F(n), F(x_F(n))), \quad x_F(0) = x_0. \tag{4}$$

From dynamic programming theory (cf. e.g. [2]) it is well known that the optimal value function satisfies Bellman's optimality principle, i.e.,

$$V_\infty(x) = \min_{u \in U} \{l(x, u) + V_\infty(f(x, u))\} \tag{5}$$

and that the optimal feedback law F is given by

$$F(x) := \operatorname*{argmin}_{u \in U} \{l(x, u) + V_\infty(f(x, u))\}. \tag{6}$$

Remark 1. In order to simplify and streamline the presentation, throughout this paper it is assumed that in all relevant expressions the minimum with respect to $u \in U^m$ is attained. Alternatively, modified statements using approximate minimizers could be used which would, however, considerably increase the amount of technicalities needed in order to formulate our assumptions and results.

2.2   Asymptotic Feedback Stabilization

Our main motivation for considering infinite horizon optimal control problems is the fact that these problems yield asymptotically stabilizing feedback laws. In order to make this statement precise, we first define what we mean by an asymptotically stabilizing feedback law.

Let us assume that the control system under consideration has an equilibrium $x^* \in X$ for some control $u^* \in U$, i.e.,

$$f(x^*, u^*) = x^*.$$

Asymptotic stability can be elegantly formulated using the concept of comparison functions. To this end, as usual in nonlinear stability theory, we define the class $\mathcal{K}$ of continuous functions $\delta : \mathbb{R}_0^+ \to \mathbb{R}_0^+$ which are strictly increasing and satisfy $\delta(0) = 0$ and the class $\mathcal{K}_\infty$ of functions $\delta \in \mathcal{K}$ which are unbounded and hence invertible with $\delta^{-1} \in \mathcal{K}_\infty$. We also define the (discrete time) class $\mathcal{KL}$ of continuous functions $\beta : \mathbb{R}_0^+ \times \mathbb{N}_0 \to \mathbb{R}_0^+$ which are of class $\mathcal{K}$ in the first argument and strictly decreasing to 0 in the second argument. Examples for $\beta \in \mathcal{KL}$ are, for instance,

$$\beta(r, n) = Ce^{-\sigma n}r \quad \text{or} \quad \beta(r, n) = \frac{C\sqrt{r}}{1 + n}.$$

Then we say that a feedback law $F : X \to U$ asymptotically stabilizes the equilibrium x^*, if there exists $\beta \in \mathcal{KL}$ such that for all initial values $x_0 \in X$ the solution of (4) satisfies

$$\|x_F(n)\|_{x^*} \le \beta(\|x_0\|_{x^*}, n) \tag{7}$$

using the brief notation $\|x\|_{x^*} = d(x, x^*)$ for the distance of a point $x \in X$ to the equilibrium x^*, where $d(\cdot, \cdot)$ is a metric on X. Note that $\|\cdot\|_{x^*}$ does not need to be a norm. In less formal words, Condition (7) demands that, by virtue of β

being of class $\mathcal{K}$ in its first argument, any solution starting close to x^* remains close to x^* for all future times and that, since β is decreasing to 0 in its second argument, any solution converges to x^* as $n \to \infty$. This $\mathcal{KL}$ characterization of asymptotic stability is actually equivalent to the ε–δ formulation often found in the literature.

In order to obtain an asymptotically stabilizing optimal feedback F from (6) we proceed as follows: For the running cost l we define

$$l^*(x) := \inf_{u \in U} l(x, u)$$

and choose l in such a way that there exist $\gamma_1 \in \mathcal{K}_\infty$ satisfying

$$\gamma_1(\|x\|_{x^*}) \le l^*(x). \tag{8}$$

Then, if an asymptotically stabilizing feedback law exists, under suitable boundedness conditions on l (for details see [11, Theorem 5.4]; see also [3] for a treatment in continuous time) there exist $\delta_1, \delta_2 \in \mathcal{K}_\infty$ such that the inequality

$$\delta_1(\|x\|_{x^*}) \le V_\infty(x) \le \delta_2(\|x\|_{x^*}) \tag{9}$$

holds. Furthermore, from the optimality principle we can deduce the inequality

$$V_\infty(f(x, F(x))) \le V_\infty(x) - l(x, F(x)) \le V_\infty(x) - l^*(x) \le V_\infty(x) - \gamma_1(\|x\|_{x^*}).$$

Since $V_\infty(x) \le \delta_2(\|x\|_{x^*})$ implies $\|x\|_{x^*} \ge \delta_2^{-1}(V_\infty(x))$ we obtain

$$\gamma_1(\|x\|_{x^*}) \ge \gamma_1(\delta_2^{-1}(V_\infty(x)))$$

and thus

$$V_\infty(f(x, F(x))) \le V_\infty(x) - \gamma_1(\delta_2^{-1}(V_\infty(x))).$$

For the solution $x_F(n)$ from (4) this implies

$$V_\infty(x_F(n)) \le \sigma(V_\infty(x_0), n)$$

for some suitable $\sigma \in \mathcal{KL}$. Thus, using (9), we eventually obtain

$$\|x_F(n)\|_{x^*} \le \delta_1^{-1}(\sigma(\delta_2(\|x_0\|_{x^*}, n))) =: \beta(\|x_0\|_{x^*}, n).$$

Using the monotonicity of the involved $\mathcal{K}_\infty$ functions it is an easy exercise to show that $\beta \in \mathcal{KL}$. This proves that the infinite horizon optimal feedback indeed asymptotically stabilizes the equilibrium x^*.

Essentially, this proof uses that V_∞ is a Lyapunov function for the closed loop system (4) controlled by the infinite horizon optimal feedback law F.

3   The Relaxed Optimality Principle

The relation between asymptotic feedback stabilization and infinite horizon optimal control paves the way for applying powerful numerical algorithms from the area of optimal control to the feedback stabilization problem. Thus, it is no surprise that in the literature one can find numerous approaches which attempt to proceed this way, i.e., find a numerical approximation $\widetilde{V} \approx V_\infty$ and compute a numerical approximation $\widetilde{F}$ to the optimal feedback law using

$$\widetilde{F}(x) := \operatorname*{argmin}_{u \in U} \left\{ l(x, u) + \widetilde{V}(f(x, u)) \right\}. \tag{10}$$

Examples for such approaches can be found, e.g., in [17,15,8] for general nonlinear systems and in [6,21] for homogeneous systems; the approach is also closely related to semi–Lagrangian finite element discretization schemes for Hamilton-Jacobi PDEs, see, e.g., [1, Appendix 1], [7,4]. All these schemes rely on the fact that the numerically computed function $\widetilde{V}$ closely approximates the true optimal value function V_∞ and typically fail in case of larger numerical errors.

Unfortunately, however, except for certain special situations like, e.g., unconstrained linear quadratic problems, even sophisticated numerical techniques can only yield good approximations $\widetilde{V} \approx V_\infty$ in low dimensional state spaces X. Hence, in general it seems too demanding to expect a highly accurate numerical approximation to the optimal value function and thus we have to develop concepts which allow to prove stability of numerically generated feedback laws even for rather coarse numerical approximations $\widetilde{V}$.

The main tool we are going to use for this purpose is a rather straightforward and easily proved "relaxed" version of the dynamic programming principle. This fact, which we are going to formalize in the following theorem, has been used implicitly in many papers on dynamic programming techniques during the last decades. Recently, it has been extensively studied and used by Lincoln and Rantzer in [18,20].

In order to formulate the theorem we need to define the *infinite horizon value function* $V_\infty^{\widetilde{F}}$ of a feedback law $\widetilde{F} : X \to U$, which is given by

$$V_\infty^{\widetilde{F}}(x_0) := \sum_{n=0}^{\infty} l(x_{\widetilde{F}}(n), \widetilde{F}(x_{\widetilde{F}}(n))),$$

where $x_{\widetilde{F}}$ is the solution from (4) with $\widetilde{F}$ instead of F.

Theorem 1. *(i) Consider a feedback law $\widetilde{F} : X \to U$ and a function $\widetilde{V} : X \to \mathbb{R}_0^+$ satisfying the inequality*

$$\widetilde{V}(x) \geq \alpha l(x, \widetilde{F}(x)) + \widetilde{V}(f(x, \widetilde{F}(x))) \tag{11}$$

for some $\alpha \in (0, 1]$ and all $x \in X$. Then for all $x \in X$ the estimate

$$\alpha V_\infty(x) \leq \alpha V_\infty^{\widetilde{F}}(x) \leq \widetilde{V}(x) \tag{12}$$

holds.

(ii) If, in addition, the inequalities (8) and (9) with $\widetilde{V}$ instead of V_∞ hold, then the feedback law $\widetilde{F}$ asymptotically stabilizes the system for all $x_0 \in X$.

Proof. (i) Using (11) for $x = x_{\widetilde{F}}(n)$ and all $n \in \mathbb{N}_0$ we obtain

$$\alpha l(x_{\widetilde{F}}(n), \widetilde{F}(x_{\widetilde{F}}(n))) \leq \widetilde{V}(x_{\widetilde{F}}(n)) - \widetilde{V}(x_{\widetilde{F}}(n+1)).$$

Summing over n yields

$$\alpha \sum_{n=0}^{m} l(x_{\widetilde{F}}(n), \widetilde{F}(x_{\widetilde{F}}(n))) \leq \widetilde{V}(x_{\widetilde{F}}(0)) - \widetilde{V}(x_{\widetilde{F}}(m+1)) \leq \widetilde{V}(x_{\widetilde{F}}(0)).$$

For $m \to \infty$ this yields that $\widetilde{V}$ is an upper bound for $\alpha V_\infty^{\widetilde{F}}$ and hence (12), since the first inequality in (12) is obvious.

(ii) From (11) we immediately obtain

$$\widetilde{V}(f(x, \widetilde{F}(x))) \leq \widetilde{V}(x) - \alpha l(x, \widetilde{F}(x)).$$

Now we can proceed exactly as in Section 2.2 using $\alpha\gamma_1$ instead of γ_1 in order to conclude asymptotic stability. □

The contribution of Theorem 1 is twofold: On the one hand, in (i) it gives an estimate for the infinite horizon value $V_\infty^{\widetilde{F}}$ based on $\widetilde{V}$ and α, on the other hand, in (ii) it ensures that the corresponding feedback $\widetilde{F}$ is indeed asymptotically stabilizing.

We emphasize the fact that no relation between $\widetilde{V}$ and V_∞ is needed in order to obtain these results. Hence, we can use this theorem even if $\widetilde{V}$ is only a very rough approximation to V_∞, provided, of course, that our numerical scheme is such that $\alpha \in (0, 1]$ satisfying (11) can be found.

In the following two sections we present two numerical approaches for which this is indeed possible. In Section 4 we discuss an offline optimization method particularly suitable for low dimensional systems, whose main advantages are the cheap online evaluation of the feedback and its capability to be easily extended to hybrid systems, i.e., systems with additional discrete states and switching rules. In this context we will see that a suitable extension of the basic algorithm results in a method for which the assumptions of Theorem 1 can be verified.

In Section 5 we investigate receding horizon control, an online optimization technique particularly suitable for smooth dynamics for which fast online optimization is possible even for large scale systems. For these schemes we will see that Theorem 1 induces conditions on the infinite horizon running cost l which can be used in order to considerably reduce the complexity of the online optimization problem.

4   A Set Oriented and Graph Theoretic Approach

In this section we describe an offline optimization method which is based on a set oriented discretization method followed by graph theoretic optimization methods. Since the method is described in detail in [9], here we only sketch the main ideas and in particular the relevance of Theorem 1 in this context.

In order to apply the method, we assume that the state space X is a compact subset of a finite dimensional Euclidean space and consider a partition $\mathcal{P}$ of X consisting of finitely many disjoint sets P, called cells or boxes. For each cell $P \in \mathcal{P}$ we define a map $F : \mathcal{P} \times U \to 2^{\mathcal{P}}$ by setting

$$F(P, u) := \{Q \in \mathcal{P} \mid f(x, u) \in Q \text{ for some } x \in P\}.$$

Furthermore, we fix a target region T consisting of cells in $\mathcal{P}$ and containing the desired equilibrium x^*.

The basic idea of the approach as presented in [16] is to define a weighted directed graph $G = (\mathcal{P}, E, w)$ consisting of *nodes* $\mathcal{P}$, *edges* $E \subset \mathcal{P} \times \mathcal{P}$ and *weights* $w : E \to \mathbb{R}_0^+$ capturing the dynamics of F and the running cost l. This is accomplished by setting

$$E := \{(P, Q) \subset \mathcal{P} \times \mathcal{P} \mid Q \in F(P, U)\}$$

and

$$w((P, Q)) := \inf\{l(x, u) \mid x \in P, \ u \in U : f(x, u) \in Q\}.$$

For $P, Q \in \mathcal{P}$ a *path* $p(P, Q)$ joining P and Q is a sequence of edges $e_1 = (P_0, P_1), e_2 = (P_1, P_2), \ldots, e_l = (P_{l-1}, P_l)$ with $P_0 = P$ and $P_l = Q$. Its *length* is defined as

$$L(p(P, Q)) := \sum_{k=1}^{l} w(e_k).$$

The *shortest path problem* then consists of computing the value

$$V_{\mathcal{P}}(P) := \inf\{L(p(P, Q)) \mid Q \subset T\}$$

with the convention $\inf \emptyset = \infty$, i.e., it computes the length of the shortest path in the graph joining P and some cell Q in the target T. Such a shortest path problem can be efficiently solved using, e.g., Dijkstra's algorithm, see [16].

This shortest path problem assigns a value to each node of the graph G and thus to each cell P of the partition $\mathcal{P}$. If for each $x \in X$ we denote by $\rho(x) \in \mathcal{P}$ the cell containing x, then we can define the numerical value function on X as $\widetilde{V}_{\mathcal{P}}(x) = V_{\mathcal{P}}(\rho(x))$. For this function it turns out that

$$\widetilde{V}_{\mathcal{P}}(x) \leq V_{\infty}(x) \quad \text{and} \quad \widetilde{V}_{\mathcal{P}_i}(x) \to V_{\infty}(x),$$

where the convergence holds for subsequently finer partitions $\mathcal{P}_i$ with target sets $T_i \to \{x^*\}$, see [16]. Furthermore, as shown in [8], under suitable conditions this convergence is even uniform and the feedback defined using (10) asymptotically stabilizes the system — due to the use of the target set T not necessarily exactly at x^* but at least at a neighborhood of x^*, a property called *practical stabilization*.

The main limitation of this approach is that typically rather fine partitions $\mathcal{P}$ are needed in order to make the resulting feedback work, thus the method often exceeds the computer's memory capability. In other words, we are exactly in the situation described at the beginning of Section 3.

A remedy for this problem can be obtained by analyzing the construction of $\widetilde{V}_{\mathcal{P}}$ in more detail. In fact, the shortest path problem leads to the optimality principle

$$V_{\mathcal{P}}(P) = \inf_{(P,Q)\in E}\{w((P,Q)) + V_{\mathcal{P}}(Q)\}$$

which by construction of the graph is equivalent to

$$\widetilde{V}_{\mathcal{P}}(x) = \inf_{u\in U, x'\in\rho(x)}\{l(x',u) + \widetilde{V}_{\mathcal{P}}(f(x',u))\}.$$

If we could change this last equation to

$$\widetilde{V}_{\mathcal{P}}(x) = \inf_{u\in U}\sup_{x'\in\rho(x)}\{l(x',u) + \widetilde{V}_{\mathcal{P}}(f(x',u))\}, \tag{13}$$

and if $\widetilde{F}(x) \in U$ denotes the minimizer of (13), then we immediately obtain

$$\widetilde{V}_{\mathcal{P}}(x) \geq l(x,\widetilde{F}(x)) + \widetilde{V}_{\mathcal{P}}(f(x,\widetilde{F}(x))),$$

i.e., (11) with $\alpha = 1$ — independently of how fine the partition $\mathcal{P}$ is chosen.

This is indeed possible by modifying the shortest path problem defining $\widetilde{V}_{\mathcal{P}}$:

Instead of a set E of edges $e = (P,Q)$ we now define a set H of *hyperedges*, i.e., pairs $h = (P,\mathcal{N})$ with $\mathcal{N} \in 2^P$, by

$$H := \{(P,\mathcal{N}) \subset \mathcal{P} \times 2^{\mathcal{P}} \mid \mathcal{N} = F(P,u) \text{ for some } u \in U\}$$

with weights

$$w((P,\mathcal{N})) := \inf\{\sup_{x\in P} l(x,u) \mid u \in U : \mathcal{N} = F(P,u)\}$$

leading to a *directed hypergraph*. It turns out (see [10,19]) that Dijkstra's algorithm can be efficiently extended to hypergraphs, leading to values $V_{\mathcal{P}}(P)$ of the nodes satisfying the optimality principle

$$V_{\mathcal{P}}(P) = \inf_{(P,\mathcal{N})\in H}\sup_{Q\in\mathcal{N}}\{w((P,Q)) + V_{\mathcal{P}}(Q)\}$$

which is equivalent to the desired equation (13).

In order to illustrate the benefit of this hypergraph approach whose solution "automatically" satisfies (11) with $\alpha = 1$, we consider the classical inverted pendulum on a cart given by

$$\left(\frac{4}{3} - m_r\cos^2\varphi\right)\ddot{\varphi} + \frac{1}{2}m_r\dot{\varphi}^2\sin 2\varphi - \frac{g}{\ell}\sin\varphi = -u\frac{m_r}{m\ell}\cos\varphi,$$

where we have used the parameters $m = 2$ for the pendulum mass, $m_r = m/(m+M)$ for the mass ratio with cart mass $M = 8$, $\ell = 0.5$ as the length of the pendulum and $g = 9.8$ for the gravitational constant. We use the corresponding sampled-data system (2) with $T = 0.1$ and the running cost

$$l(\varphi_0,\dot{\varphi}_0,u) = \frac{1}{2}\int_0^T 0.1\varphi(t,\varphi_0,\dot{\varphi}_0,u)^2 + 0.05\dot{\varphi}(t,\varphi_0,\dot{\varphi}_0,u)^2 + 0.01u^2 dt \tag{14}$$

and choose $X = [-8,8] \times [-10,10]$ as the region of interest.

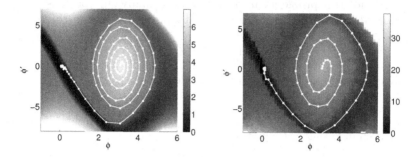

Fig. 1. Approximate optimal value function and resulting feedback trajectory for the inverted pendulum on a 2^{18} box partition using the graph approach (left) and on a 2^{14} box partition using the hypergraph approach (right)

Figure 1 compares the two approaches on the respective coarsest possible partitions on which stabilization was achieved. It is clearly visible that the hypergraph approach (right) leads to both considerably fewer partition cells and to a much faster convergence of the controlled trajectory.

5   A Receding Horizon Approach

Receding horizon control — also known as model predictive control — is probably the most successful class of optimization based control methods and is widely used in industrial applications.

In its simplest form, receding horizon control consists in truncating the infinite horizon functional, i.e., for $N \in \mathbb{N}$ we consider the functional

$$J_N(x_0, u) = \sum_{n=0}^{N-1} l(x_u(n), u(n)), \tag{15}$$

with optimal value function $V_N(x_0) := \inf_{u \in \mathcal{U}} J_N(x_0, u)$.

This problem can be solved by various numerical techniques, e.g. by converting the problem into a static optimization problem with the dynamics as constraints, which can be solved by the SQP method, cf., e.g., [12].

In order to get a feedback law F_N from this finite horizon problem, at each time instant we measure the current state x_n and (online) minimize (15) with $x_0 = x_n$. This yields an optimal control sequence $u^*(0), \ldots, u^*(N-1)$ from which we obtain the feedback by setting

$$F_N(x_n) := u^*(0),$$

i.e., by taking the first element of the optimal control sequence.

The questions we want to investigate now is whether this scheme does yield a stabilizing feedback control F_N and, if yes, what is the performance of this controller, i.e, what is $V_\infty^{F_N}$. Our main motivation for this analysis is to derive conditions on the running cost l under which we can ensure stability and good performance even for short optimization horizons N, leading to low complexity and thus short computational time for solving the online optimization problem.

In the literature, the majority of papers dealing with stability issues of receding horizon control uses additional terminal constraints and costs, typically requiring $x(N)$ to lie in a neighborhood of the equilibrium to be stabilized. This modification is known to enhance stability both in theory and in practice, however, its main disadvantage is that the operating region of the resulting controller is restricted to the feasible set, i.e., to the set of initial conditions for which the terminal constraints are feasible. This set, in turn, depends on the optimization horizon N and, thus, in order to obtain large operating regions, typically large optimization horizons N are needed leading to complex optimization problems and high computational effort.

Stability results for receding horizon problems without terminal costs have been presented in [14,5] using convergence $V_N \to V_\infty$. Another way to obtain such results has been pursued in [13] based on the convergence $|V_N - V_{N+1}| \to 0$ as $N \to \infty$. The next proposition is one of the main results from [13], which will allow us to apply Theorem 1 to $\widetilde{V} = V_N$ and $\widetilde{F} = F_N$.

Proposition 1. *Consider $\gamma > 0$ and $N \geq 2$ and assume that the inequalities*

$$V_2(x) \leq (\gamma + 1)l(x, F_1(x)) \quad and \quad V_k(x) \leq (\gamma + 1)l(x, F_k(x)), \ k = 3, \ldots, N$$

hold for all $x \in X$. Then the inequality

$$\frac{(\gamma + 1)^{N-2}}{(\gamma + 1)^{N-2} + \gamma^{N-1}} V_N(x) \leq V_{N-1}(x)$$

holds for $x \in X$.

The proof can be found in [13] and relies on a inductive application of the optimality principle for V_k, $k = 1, \ldots, N$.

Combining Proposition 1 and Theorem 1 we immediately arrive at the following theorem, which was first proved in [13] and whose proof we repeat for convenience of the reader.

Theorem 2. *Consider $\gamma > 0$, let $N \in \mathbb{N}$ be so large that $(\gamma + 1)^{N-2} > \gamma^N$ holds and let the assumption of Proposition 1 holds. Then the inequality*

$$\frac{(\gamma + 1)^{N-2} - \gamma^N}{(\gamma + 1)^{N-2}} V_\infty(x) \leq \frac{(\gamma + 1)^{N-2} - \gamma^N}{(\gamma + 1)^{N-2}} V_\infty^{F_N}(x) \leq V_N(x) \leq V_\infty(x)$$

holds. If, in addition, the inequalities (8) and (9) hold, then the feedback law F_N asymptotically stabilizes the system.

Proof. First note that the first and the last inequality in the assertion are obvious. In order to derive the middle inequality, from Proposition 1 we obtain

$$V_N(f(x, F_N(x))) - V_{N-1}(f(x, F_N(x))) \leq \frac{\gamma^{N-1}}{(\gamma + 1)^{N-2}} V_{N-1}(f(x, F_N(x))).$$

Using the optimality principle for V_N

$$V_N(x) = l(x, F_N(x)) + V_{N-1}(f(x, F_N(x)))$$

and the assumption of Proposition 1 for $k = N$, we obtain the inequality $V_{N-1}(f(x, F_N(x))) \leq \gamma l(x, F_N(x))$ and can conclude

$$V_N(f(x, F_N(x))) - V_{N-1}(f(x, F_N(x))) \leq \frac{\gamma^N}{(\gamma + 1)^{N-2}} l(x, F_N(x)).$$

Thus, using the optimality principle for V_N once again yields

$$V_N(x) \geq l(x, F_N(x)) + V_N(f(x, F_N(x))) - \frac{\gamma^N}{(\gamma + 1)^{N-2}} l(x, F_N(x))$$

implying the assumption of Theorem 1(i) for $\widetilde{V} = V_N$ and $\widetilde{F} = F_N$ with

$$\alpha = 1 - \frac{\gamma^N}{(\gamma + 1)^{N-2}} = \frac{(\gamma + 1)^{N-2} - \gamma^N}{(\gamma + 1)^{N-2}}$$

and thus the asserted inequalities.

Asymptotic stability now follows immediately from Theorem 1(ii) since the proved inequality together with (8) and (9) implies (9) for $\widetilde{V} = V_N$.   □

We emphasize that the decisive condition for stability is $(\gamma + 1)^{N-2} > \gamma^N$. In particular, the larger γ is, the larger the optimization horizon N must be in order to meet this condition. Hence, in order to ensure stability for small N, we need to ensure that γ is small.

An estimate for γ can, e.g., be obtained if a null–controlling control sequence is known, i.e., if for each x_0 we can find a sequence $u \in \mathcal{U}$ such that $l(x_u(n), u(n))$ converges to 0 sufficiently fast. In this case, for each $k \in \mathbb{N}$ we can estimate

$$V_k(x_0) \leq V_\infty(x_0) \leq J_\infty(x_0, u) \quad \text{and} \quad l^*(x_0) \leq l(x_0, F_k(x_0))$$

and an estimate for γ can then be computed comparing $J_\infty(x_0, u)$ and $l^*(x_0)$. In particular, such an analysis can be used for the design of running costs l which lead to small values of γ and thus to stability for small optimization horizons N.

We illustrate this procedure for a control system governed by a reaction-advection-diffusion PDE with distributed control given by

$$y_t = y_x + \nu y_{xx} + \mu y(y + 1)(1 - y) + u \tag{16}$$

with solutions $y = y(t, x)^1$ for $x \in \Omega = (0, 1)$, boundary conditions $y(t, 0) = y(t, 1) = 0$, initial condition $y(0, x) = y_0(x)$ and distributed control $u(t, \cdot) \in L^2(\Omega)$. The corresponding discrete time system (1), whose solutions and control functions we denote by $y(n, x)$ and $u(n, x)$, respectively, is the sampled-data system obtained according to (2) with sampling period $T = 0.025$.

---

[1] Note the change in the notation: x is the independent state variable while $y(t, \cdot)$ is the new state, i.e., X is now an infinite dimensional space.

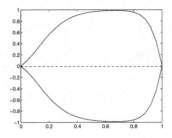

Fig. 2. Equilibria for $u \equiv 0$; solid=asymptotically stable, dashed=unstable

For our numerical computations we discretized the equation in space by finite differences on a grid with nodes $x_i = i/M$, $i = 0, \ldots, M$, using backward (i.e., upwind) differences for the advection part y_x. Figure 2 shows the equilibria of the discretized system for $u \equiv 0$, $\nu = 0.1$, $\mu = 10$ and $M = 25$.

Our goal is to stabilize the unstable equilibrium $y^* \equiv 0$, which is possible because with the additive distributed control we can compensate the whole dynamics of the system. In order to achieve this task, a natural choice for a running cost l is the tracking type functional

$$l(y(n, \cdot), u(n, \cdot)) = \|y(n, \cdot)\|_{L^2(\Omega)}^2 + \lambda \|u(n, \cdot)\|_{L^2(\Omega)}^2 \qquad (17)$$

which we implemented with $\lambda = 0.1$ for the discretized model in MATLAB using the lsqnonlin solver for the resulting optimization problem.

The simulations shown in Figure 3 reveal that the performance of this controller is not completely satisfactory: for $N = 11$ the solution remains close to $y^* = 0$ but does not converge while for $N = 3$ the solution even grows.

The reason for this behavior lies in the fact that in order to control the system to $y^* = 0$, in (16) the control needs to compensate for y_x, i.e., any stabilizing control must satisfy $\|u(n, \cdot)\|_{L^2(\Omega)}^2 \gtrsim \|y_x(n, \cdot)\|_{L^2(\Omega)}^2$. Thus, for any stabilizing control sequence u we obtain $J_\infty(y_0, u) \gtrsim \lambda \|y_x(n, \cdot)\|_{L^2(\Omega)}^2$ which — even for small values of λ — may be considerably larger than $l^*(y) = \|y\|_{L^2(\Omega)}^2$, resulting

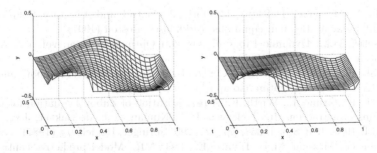

Fig. 3. Receding horizon with l from (17), $N = 3$ (left) and $N = 11$ (right)

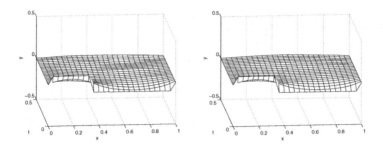

Fig. 4. Receding horizon with l from (18), $N = 2$ (left) and $N = 3$ (right)

in a large γ and thus the need for a large optimization horizon N in order to achieve stability.

This effect can be avoided by changing l in such a way that $l^*(y)$ includes $\|y_x\|^2_{L^2(\Omega)}$, e.g., by setting

$$l(y(n,\cdot), u(n,\cdot)) = \|y(n,\cdot)\|^2_{L^2(\Omega)} + \|y_x(n,\cdot)\|^2_{L^2(\Omega)} + \lambda\|u(n,\cdot)\|^2_{L^2(\Omega)}. \quad (18)$$

For this l the control effort needed in order to control (16) to $y^* = 0$ is proportional to $l^*(y)$. Thus, γ is essentially proportional to λ and thus, in particular, small for our choice of $\lambda = 0.1$ which implies stability even for small optimization horizon N. The simulations using the corresponding discretized running cost illustrated in Figure 4 show that this is indeed the case: we obtain asymptotic stability even for the very small optimization horizons $N = 2$ and $N = 3$.

Acknowledgment

I would like to thank Oliver Junge and Karl Worthmann for computing the examples in Sections 4 and 5, respectively.

References

1. Bardi, M., Capuzzo Dolcetta, I.: Optimal Control and Viscosity Solutions of Hamilton-Jacobi-Bellman equations. Birkhäuser, Boston (1997)
2. Bertsekas, D.P.: Dynamic Programming and Optimal Control, vol. 1,2. Athena Scientific, Belmont, MA (1995)
3. Camilli, F., Grüne, L., Wirth, F.: Control Lyapunov functions and Zubov's method. In: SIAM J. Control Optim. (to appear, 2008)
4. Camilli, F., Grüne, L., Wirth, F.: A regularization of Zubov's equation for robust domains of attraction. In: Treleaven, P.C., Nijman, A.J., de Bakker, J.W. (eds.) PARLE 1987. LNCS, vol. 258, pp. 277–290. Springer, Heidelberg (1987)
5. Grimm, G., Messina, M.J., Tuna, S.E., Teel, A.R.: Model predictive control: for want of a local control Lyapunov function, all is not lost. IEEE Trans. Automat. Control 50(5), 546–558 (2005)

6. Grüne, L.: Homogeneous state feedback stabilization of homogeneous systems. SIAM J. Control Optim. 38, 1288–1314 (2000)
7. Grüne, L.: An adaptive grid scheme for the discrete Hamilton–Jacobi–Bellman equation. Numer. Math. 75(3), 319–337 (1997)
8. Grüne, L., Junge, O.: A set oriented approach to optimal feedback stabilization. Syst. Control Lett. 54(2), 169–180 (2005)
9. Grüne, L., Junge, O.: Approximately optimal nonlinear stabilization with preservation of the Lyapunov function property. In: Proceedings of the 46th IEEE Conference on Decision and Control, New Orleans, Louisiana (2007)
10. Grüne, L., Junge, O.: Global optimal control of perturbed systems. J. Optim. Theory Appl. 136 (to appear, 2008)
11. Grüne, L., Nešić, D.: Optimization based stabilization of sampled–data nonlinear systems via their approximate discrete–time models. SIAM J. Control Optim. 42, 98–122 (2003)
12. Grüne, L., Nešić, D., Pannek, J.: Model predictive control for nonlinear sampled–data systems. In: Gianni, P. (ed.) ISSAC 1988. LNCS, vol. 358, Springer, Heidelberg (1989)
13. Grüne, L., Rantzer, A.: On the infinite horizon performance of receding horizon controllers. In: Preprint, Universitat Bayreuth, IEEE Trans. Automat. Control (2006) (to appear, 2008),
www.math.uni-bayreuth.de/Igruene/publ/infhorrhc.html
14. Jadbabaie, A., Hauser, J.: On the stability of receding horizon control with a general terminal cost. IEEE Trans. Automat. Control 50(5), 674–678 (2005)
15. Johansen, T.A.: Approximate explicit receding horizon control of constrained nonlinear systems. Automatica 40(2), 293–300 (2004)
16. Junge, O., Osinga, H.M.: A set oriented approach to global optimal control. ESAIM Control Optim. Calc. Var. 10(2), 259–270 (2004)
17. Kreisselmeier, G., Birkhölzer, T.: Numerical nonlinear regulator design. IEEE Trans. Autom. Control 39(1), 33–46 (1994)
18. Lincoln, B., Rantzer, A.: Relaxing dynamic programming. IEEE Trans. Autom. Control 51, 1249–1260 (2006)
19. von Lossow, M.: A min-max version of Dijkstra's algorithm with application to perturbed optimal control problems. In: Proceedings of the GAMM Annual meeting, Zürich, Switzerland (to appear, 2007)
20. Rantzer, A.: Relaxed dynamic programming in switching systems. IEE Proceedings — Control Theory and Applications 153, 567–574 (2006)
21. Tuna, E.S.: Optimal regulation of homogeneous systems. Automatica 41(11), 1879–1890 (2005)

Part II

Robust Multilevel and Hierarchical Preconditioning Methods

On Smoothing Surfaces in Voxel Based Finite Element Analysis of Trabecular Bone

Peter Arbenz and Cyril Flaig

Institute of Computational Science, ETH Zürich, CH-8092 Zürich

Abstract. The (micro-)finite element analysis based on three-dimensional computed tomography (CT) data of human bone takes place on complicated domains composed of often hundreds of millions of voxel elements. The finite element analysis is used to determine stresses and strains at the trabecular level of bone. It is even used to predict fracture of osteoporotic bone. However, the computed stresses can deteriorate at the jagged surface of the voxel model.

There are algorithms known to smooth surfaces of voxel models. Smoothing however can distort the element geometries. In this study we investigate the effects of smoothing on the accuracy of the finite element solution, on the condition of the resulting system matrix, and on the effectiveness of the smoothed aggregation multigrid preconditioned conjugate gradient method.

1 Introduction

In view of the growing importance of osteoporosis due to the obsolescence of the population in industrialized countries an accurate analysis of individual bone strength is in dire need. In fact according to the WHO, lifetime risk for osteoporotic fractures in women is estimated close to 40%; in men risk is 13% [7]. With the advent of fast and powerful computers, simulation techniques are becoming popular for investigating the mechanical properties of bones and predicting the strength of a given patient's bones. In order to gain an improved comprehension of structure and strength of bone, large scale computer simulations are executed based on the theory of (non)linear elasticity and the finite element method.

Today's approach is based on three-dimensional computed tomography (CT) whereby bones are scanned with a resolution of 50-100μm. Using a direct voxel-conversion technique the three-dimensional computer reconstructions of bone can be converted to a finite element mesh, that can be used to perform a 'virtual experiment', i.e., to simulate a mechanical test in great detail and with high precision. The resulting procedure is called microstructural finite element (μFE) analysis.

The approach based on the FE analysis leads to linear systems of equations

$$K\mathbf{u} = \mathbf{f}, \tag{1}$$

where the stiffness matrix K is symmetric positive-definite, the components of the vector $\mathbf{u}$ are the displacements at the nodes of the voxel mesh. $\mathbf{f}$ contains external loads or prescribed displacements.

I. Lirkov, S. Margenov, and J. Waśniewski (Eds.): LSSC 2007, LNCS 4818, pp. 69–77, 2008.

The system of equations (1) can be solved very efficiently by the conjugate gradient algorithm preconditioned by smoothed aggregation multigrid. Systems of up to hundreds of millions of degrees of freedom have been solved on large scale computers within a couple of minutes [1, 2, 3].

The voxel approach has deficiencies though. In particular, the jagged domains leading to exceeding stresses at the nodes of the mesh corresponding to corners of the domain. A straightforward procedure is to smooth the surface of the computational domain. Taubin [9, 10] suggested a surface fairing algorithm that does not shrink the body it embraces. Boyd and Müller [4] have applied this algorithm to voxel based models. In this note we investigate this latter algorithm in a parallel environment. In section 2 we discuss how smoothing can be done with piecewise trilinear isoparametric hexahedral elements. In section 3 we discuss the effects of the flexible elements on visualization, stresses, and condition of the stiffness matrix. We also mention how the computational work can be reduced by splitting distorted hexahedra in piecewise linear tetrahedral elements.

2   Smoothing

In bone structure analysis the computational domain is composed of a multitude of tiny cubes, so-called voxels, that are generated directly from the output of the CT scanner. Surface patches and edges are always aligned with the coordinate directions. In contrast to the originally smooth object, the voxel model has a jagged surface. The stresses induced by the computed displacements can have singularities at edges and corners of the surface but also of interfaces between different materials. A straightforward approach to get rid of the singular stresses is to smooth the surface and material interfaces of the computational domain.

In computer graphics there is a well-known procedure to smooth polygonal surfaces called mesh fairing [9, 10]. The coordinates $\mathbf{x}$ of the mesh vertices are moved according to the diffusion equation

$$\frac{\partial \mathbf{x}}{\partial t} = D \cdot \Delta \mathbf{x}, \qquad D > 0, \tag{2}$$

where the Laplacian at $\mathbf{x}$ is approximated by

$$\Delta \mathbf{x}_i = \sum_{j \in \mathfrak{N}(i)} w_{ij}(\mathbf{x}_j - \mathbf{x}_i), \qquad \sum_{j \in \mathfrak{N}(i)} w_{ij} = 1. \tag{3}$$

Here, $\mathfrak{N}(i)$ denotes the neighbor nodes of node i. The choice of the weights evidently affects the quality of the smoothing [10]. The most effective scheme is due to Fujiwara, where w_{ij} is proportional to $1/\|\mathbf{x}_j - \mathbf{x}_i\|$.

Applying Euler's method with time step Δt to (2) we get

$$\mathbf{x}_i^{\text{new}} = \mathbf{x}_i^{\text{old}} + \lambda \Delta \mathbf{x}_i^{\text{old}}, \qquad \lambda = D\Delta t. \tag{4}$$

Boyd and Müller [4] applied this technique to the voxel model by classifying nodes as fixed nodes, surface nodes, interface nodes, inner nodes and near-surface nodes. Fixed and inner nodes must not move; this is accomplished by leaving empty their set of neighbors. Surface and interface nodes produce a surface mesh which is smoothed by the algorithm. Since only surface and interface nodes can qualify as neighbors of surface and interface nodes, these nodes are subjected to Taubin's original 2D algorithm. Near-surface nodes should compensate for the movements of the surface and interface nodes, such that hexahedra near the surface are not seriously distorted.

The smoothing procedure (4) shrinks the volume of the 3D object it is applied to. To compensate this effect, Taubin suggests to replace λ in every second step by $-\mu$ where $\mu = \lambda$ or slightly bigger [9]. The 'negative diffusion' has the effect of (approximately) restoring the volume of the 3D object.

This smoothing procedure was incorporated into our fully-parallel μ-finite element code PARFE [8] that is based on the Trilinos framework [11], see [6] for details. A piecewise trilinear finite element space was implemented on the distorted mesh based on isoparametric hexahedral elements. The matrix elements have been computed approximately by the 64-point tensor product Gaussian quadrature rule.

3 Results

In this section we discuss four effects of smoothing, (1) the visual quality, (2) the condition number of the stiffness matrix, (3) the scalability of the smoothing procedure, and (4) the cost of the assembling of the matrix. The computations have been done on the Cray XT3 at the Swiss National Supercomputer Center.

To show the *visual quality* of the smoothing algorithm a sphere was smoothed with λ and μ as proposed in [10, 4]. A sphere has an absolutely smooth surface to which the surface coordinates should converge. The smoothing procedure in fact generates quite a smooth surface, see Fig. 1. The implemented procedure

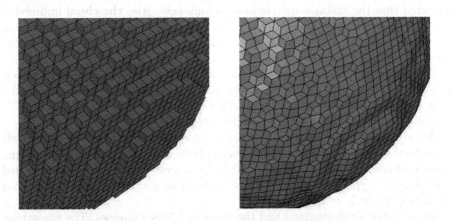

Fig. 1. A sample sphere. On the left is the original; on the right is a smoothed sphere subject to 32 smoothing steps with $\lambda = 0.4$ and $1/\mu + 1/\lambda = 0.1$.

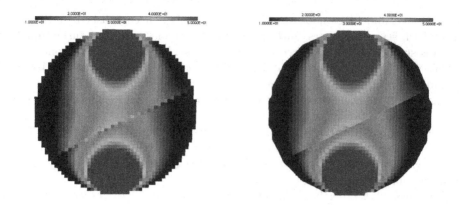

Fig. 2. A clip through a sphere consisting of two materials. The color indicates the stress. The original jagged sphere is on the left; the smoothed sphere is on the right after 32 smoothing steps with the parameters as in Fig. 1. The stresses do not oscillate so much in the smoothed version as in the unsmoothed version.

Table 1. Impact of the number of smoothing steps on maximal and mean stress

	Sphere			cube1		
Smoothing steps	0	8	16	0	8	16
Max stress (MPa)	367.7	365.3	365.6	231.6	235.2	237.0
Mean stress (MPa)	20.8	20.73	20.4	19.6	20.1	20.3

not only smoothes surfaces, but also interfaces between differing materials, as can be seen from Fig. 2. To investigate how the visual impression changes as the number of smoothing steps increases we consider a bone specimen consisting of 98'381 voxels. From Fig. 3 we see that already very few smoothing steps lead to dramatically improved surfaces. After 28 steps, however, some voxels get so much distorted that the stiffness matrix loses definiteness. Also the visual impression does not improve much beyond this point. It is possible but to time consuming to check individual elements for strong distortion and detach them from the smoothing process.

The deformed hexahedra not only have an effect on the visual impression, but also change the distribution of the stresses. Boyd and Müller [4] report that the peak stresses are lowered by a factor 4 for a sphere model. Camacho et al. [5] describe a similar effect for the von Mises stresses.

We used two models to measure the stress: a sphere consisting of two materials and a bone sample (cube1 in [2]) consisting of one material. We applied a varying number of smoothing steps with $\lambda = 0.4$. As suggested in [10, 4] we determined μ from $1/\lambda + 1/\mu = 0.1$. Both models are fixed at $z = 0$ and a load is exerted on the top plane ($z = z_{\max}$).

Table 1 shows the maximal and the mean von Mises stresses. The smoothing procedure has a minimal effect on these. The maximal stress is observed at the verge of the loading. Fig. 2 shows that the stresses propagate through the

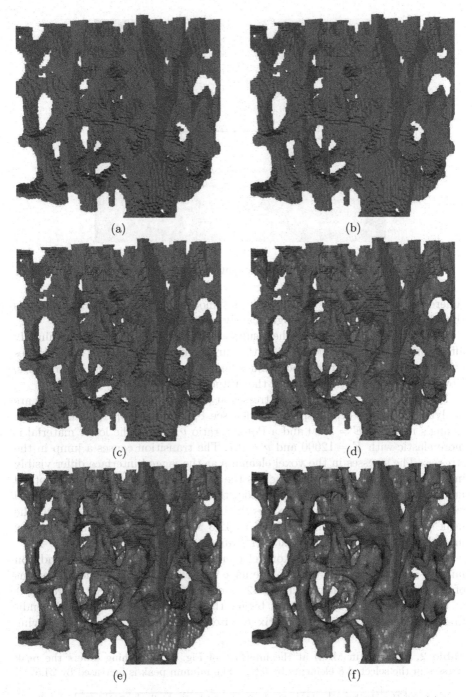

(a)

(b)

(c)

(d)

(e)

(f)

Fig. 3. A sample mesh representing trabecular bone. The unsmoothed mesh (a) was smoothed with 4 (b), 8 (c), 16 (d), 28 (e), and 64 (f) steps, respectively. The step size was 0.4. With more than 28 smoothing steps of the deformations became too severe.

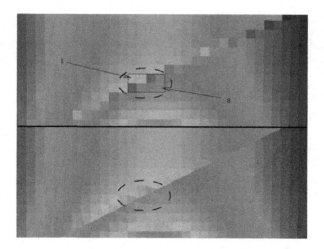

Fig. 4. Section of Figure 2 at the material interface

sphere and not along the surface where the smoothing takes place. Hence, in this situation, smoothing does not significantly affect the stresses.

In the artificial bone sample the results are similar. Smoothing only minimally affects the stresses. The elements on the surface often have unconstrained nodes or edges that are often displaced much more by smoothing than fixed nodes. However, they are not relevant for the stiffness of the model.

The changes of the von Mises stresses at the transition of the materials are analyzed by means of the sphere model, see Fig. 4. The upper material has a Young's modulus $E = 5000$ and a Poisson ratio $\nu = 0.3$. The lower material is more elastic with $E = 12000$ and $\nu = 0.1$. The transition causes a jump in the stresses. The stresses in the voxel elements at the jagged interface differ visibly in the same material as well as across the material interface. The unsmoothed model generates high peak stresses, too, that are not observed in the physical experiment. After 32 smoothing steps the interface becomes nearly a plane. The peak stress decreases from 46.8 MPa to 42.3 MPa. The resulting stresses vary much less at the material interface, cf. Table 2.

The smoothing procedure distorts elements which in turn affects the condition number of the stiffness matrix K. We have investigated the condition by means of two models. The first model, cube2, is obtained from cube1 by mirroring it at three faces. Thus it is 8 times bigger than cube1. The (estimated) condition numbers of the stiffness matrix K after 16 smoothing steps with varying

Table 2. Stresses (in MPa) at the interface of Fig. 4. Smoothing lowers the peak stresses in the selected 8 elements by 9.6%. The minium peak is increased by 24.5%.

Hexahedra	1	2	3	4	5	6	7	8
Unsmoothed	38.8	39.6	46.8	25.3	47.0	25.7	31.0	32.0
Smoothed	40.1	41.2	42.3	31.5	42.5	31.8	32.9	33.6

Table 3. Condition numbers of the stiffness matrix depending on the Euler step size λ. 16 smoothing steps were applied to the model `cube2`.

smoothing step size λ	0.0	0.3	0.4	0.475	0.5	0.51
without preconditioning	$3.4 \cdot 10^5$	$3.4 \cdot 10^5$	$3.4 \cdot 10^5$	$3.4 \cdot 10^5$	$4.5 \cdot 10^5$	—
with ml preconditioner	245.2	237.4	239.0	246.1	248.1	—

Table 4. Condition numbers of the stiffness matrix depending on the Euler step size λ. 16 smoothing steps were applied to the 2-materials sphere.

smoothing step size λ	0.0	0.5	0.6	0.67	0.685
without preconditioning	$1.75 \cdot 10^6$	$1.65 \cdot 10^6$	$2.31 \cdot 10^6$	$1.51 \cdot 10^7$	—
with ml preconditioner	507.1	447.0	486.2	664.5	—

Euler step size λ are given in Table 3. For either the preconditioned or the un-preconditioned system, the condition numbers are not affected much as long as $\lambda \leq 0.5$.

For the two-materials sphere corresponding numbers are found in Table 4. Here the condition numbers vary more. Little smoothing improves the condition; too large step sizes lead to indefinite systems.

In a third test we fixed the step size $\lambda = 0.4$ and varied the number of smoothing steps, cf. Table 5. Here we observe a slow but gradual increase of the condition number up to 28 smoothing steps. Beyond this point some of the voxel elements seem to flip over causing indefinite matrices.

To investigate *weak scalability* we chose the artificial bone displayed in Fig. 3 that is inclosed in a cube and can be mirrored at all faces to generate arbitrarily large bones, see [2]. Not surprisingly, the computations, in particular the new smoothing and assembling procedures, show perfect weak scalability up to $216 = 6^3$ processors. Notice that all voxels are considered flexible if smoothing is applied at all. This is justified by our application, trabecular bone, where usually 3/4 of the nodes are near the surface. Surprisingly, the assembling time only increased by a factor 8; apparently most of the time in this phase is due to memory accesses. Nevertheless, the assembling has become as time consuming as the solution phase. To decrease the cost of assembly we split the (distorted) hexahedra in six tetrahedra with linear basis functions. By this we regained a factor of 5 in the assembling time, however at the cost of stiffer structures, see [6].

We have tested *strong scalability* by means of a bone model of a fixed fracture of the distal radius with $38'335'350$ degrees of freedom. Because the patient's arm

Table 5. `cube2` model: condition numbers of the stiffness matrix depending on the number of smoothing steps with fixed $\lambda = 0.4$

# of smoothing steps	8	16	24	26	28	30
without preconditioner	$3.41 \cdot 10^5$	$3.40 \cdot 10^5$	$3.40 \cdot 10^5$	$3.40 \cdot 10^5$	$7.12 \cdot 10^5$	—
with ml preconditioner	234.9	239.0	246.9	248.4	253.1	—

Table 6. Fixed fracture model with 38'335'350 degrees of freedom

#CPU	Time			Speedup		
	Smoothing	Assembling	Solving	Smoothing	Assembling	Solving
160	3.922	76.79	59.29	1	1	1
240	2.712	52.44	38.76	1.45	1.46	1.53
320	2.002	39.63	32.01	1.96	1.94	1.85
480	1.382	26.69	21.22	2.84	2.87	2.79

could not be fixed perfectly the mesh has no trabecular structure. The model consists of a full mesh. So, each node has the maximal number of neighbors and communication volume between compute nodes is relatively high. Table 6 shows the execution times and speedups for smoothing, assembling, and solving the preconditioned system. The speedups are almost linear.

4   Conclusions

We have parallelized a smoothing procedure originally proposed by Taubin [10] that has been adapted by Boyd and Müller [4] for application to trabecular bones. We have observed that (1) the smoothing procedure results in a large subjective improvement of the visualization, that (2) the condition of the stiffness matrix is not increased too much as long as the elements are not distorted too severely, and that (3) the smoothing procedure applied to the model shows a reduced variation of the stresses at material transitions. However, drastically lower stresses on the surface were not obtained. Smoothing entails that the local stiffness matrices must be computed for each element which results in increased simulation times.

Acknowledgments

We acknowledge helpful discussions with Profs. Steve Boyd and Ralph Müller. The computations on the Cray XT3 have been performed in the framework of a Large User Project grant of the Swiss National Supercomputing Centre (CSCS).

References

1. Adams, M.F., et al.: Ultrascalable implicit finite element analyses in solid mechanics with over a half a billion degrees of freedom. In: ACM/IEEE Proceedings of SC(2004)
2. Arbenz, P., et al.: A scalable multi-level preconditioner for matrix-free μ-finite element analysis of human bone structures. Internat. J. Numer. Methods Engrg (to appear)
3. Arbenz, P., et al.: Multi-level μ-finite element analysis for human bone structures. In: Kågström, B., et al. (eds.) PARA 2006. LNCS, vol. 4699, Springer, Heidelberg (2007)

4. Boyd, S.K., Müller, R.: Smooth surface meshing for automated finite element model generation from 3D image data. J. Biomech. 39(7), 1287–1295 (2006)
5. Camacho, D.L.A., et al.: An improved method for finite element mesh generation of geometrically complex structures with application to the skullbase. Journal of Biomechanics 30, 1067–1070 (1997)
6. Flaig, C.: Smoothing surfaces in voxel based finite element analysis of trabecular bones. Master thesis, ETH Zurich, Inst. of Computational Science (March 2007)
7. Melton III, L.J., et al.: How many women have osteoporosis? J. Bone Miner. Res. 20(5), 886–892 (2005) Reprinted in J. Bone Miner. Res. 20(5), 886–892 (2005)
8. The PARFE Project Home Page, http://parfe.sourceforge.net/
9. Taubin, G.: A signal processing approach to fair surface design. In: SIGGRAPH 1995. Proceedings of the 22nd Annual Conference on Computer Graphics and Interactive Techniques, pp. 351–358. ACM Press, New York (1995)
10. Taubin, G.: Geometric signal processing on polygonal meshes, Eurographics State of the Art Report. 11 pages (September 2000), available from http://mesh.brown.edu/taubin/pdfs/taubin-eg00star.pdf
11. The Trilinos Project Home Page, http://software.sandia.gov/trilinos/

Application of Hierarchical Decomposition: Preconditioners and Error Estimates for Conforming and Nonconforming FEM

Radim Blaheta

Department of Applied Mathematics, Institute of Geonics AS CR
Studentská 1768, 70800 Ostrava-Poruba, Czech Republic
blaheta@ugn.cas.cz

Abstract. A successive refinement of a finite element grid provides a sequence of nested grids and hierarchy of nested finite element spaces as well as a natural hierarchical decomposition of these spaces. In the case of numerical solution of elliptic boundary value problems by the conforming FEM, this sequence can be used for building both multilevel preconditioners and error estimates. For a nonconforming FEM, multilevel preconditioners and error estimates can be introduced by means of a hierarchy, which is constructed algebraically starting from the finest discretization.

1 Introduction

Let us consider a model elliptic boundary value problem in $\Omega \subset R^2$,

$$\text{find} \quad u \in V \; : \; a(u, v) = b(v) \qquad \forall v \in V, \tag{1}$$

where $V = H_0^1(\Omega)$, $b(v) = \int_\Omega f v dx$ for $f \in L_2(\Omega)$ and

$$a(u, v) = \int_\Omega \sum_{ij}^{2} k_{ij} \frac{\partial u}{\partial x_i} \frac{\partial v}{\partial x_j} \, dx \,. \tag{2}$$

Above $K = (k_{ij})$ is a symmetric and uniformly bounded positive definite matrix.

This type of boundary value problems are most frequently solved by the finite element method (FEM). A successive refinement of a finite element grid provides a sequence of nested grids and hierarchy of nested finite element spaces as well as a natural hierarchical decomposition of these spaces. This sequence can be used for building both multilevel preconditioners and error estimates. In Section 2, we describe such hierarchy for conforming Courant type finite elements. We also mention the strengthened Cauchy-Bunyakowski-Schwarz (CBS) inequality, which is important for characterization of the hierarchical decomposition. In Section 3, we show that the hierarchical decomposition allows to construct preconditioners and error estimates. Section 4 is devoted to hierarchical decompositions constructed algebraically for nonconforming Crouzeix-Raviart FEM. We show that this decomposition allows again to introduce both preconditioners and error estimates.

I. Lirkov, S. Margenov, and J. Waśniewski (Eds.): LSSC 2007, LNCS 4818, pp. 78–85, 2008.

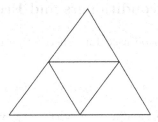

Fig. 1. A regular decomposition of a triangle

2 Hierarchical Decomposition for Conforming FEM

Let us consider a coarse triangular finite element grid $\mathcal{T}_H$ in Ω and a fine grid $\mathcal{T}_h$, which arises by a refinement of the coarse elements, see Fig. 1 for the most typical example. We assume that $\Omega = \bigcup\{E: E \in \mathcal{T}_H\}$.

By $\overline{\mathcal{N}}_H$ and $\overline{\mathcal{N}}_h$, we denote the sets of nodes corresponding to $\mathcal{T}_H$ and $\mathcal{T}_h$, respectively. Further, $\mathcal{N}_H = \{x \in \overline{\mathcal{N}}_H,\ x \notin \partial\Omega\}$, $\mathcal{N}_h = \{x \in \overline{\mathcal{N}}_h,\ x \notin \partial\Omega\}$. Naturally, $\mathcal{N}_h = \mathcal{N}_H \cup \mathcal{N}_H^+$, where $\mathcal{N}_H^+$ is the complement of $\mathcal{N}_H$ in $\mathcal{N}_h$.

Now, we can introduce the finite element spaces V_H and V_h ($V_H \subset V_h$) of functions which are continuous and linear on the elements of the triangulation $\mathcal{T}_H$ and $\mathcal{T}_h$, respectively.

The space V_h allows a natural hierarchical decomposition. Let $\{\phi_i^H\}$ and $\{\phi_i^h\}$ be the standard nodal finite element bases of V_H and V_h, i.e. $\phi_i^H(x_j) = \delta_{ij}$ for all $x_j \in \mathcal{N}_H$, $\phi_i^h(x_j) = \delta_{ij}$ for all $x_j \in \mathcal{N}_h$. Then V_h can be also equipped with a *hierarchical basis* $\{\bar{\phi}_i^h\}$, where

$$\bar{\phi}_i^h = \begin{cases} \phi_i^h & \text{if } x_i \in \mathcal{N}_H^+, \\ \phi_i^H & \text{if } x_i \in \mathcal{N}_H. \end{cases}$$

It gives a *natural hierarchical decomposition* of the space V_h,

$$V_h = V_H \oplus V_H^+, \quad V_H^+ = \text{span } \{\phi_i^h,\ x_i \in \mathcal{N}_H^+\}. \tag{3}$$

The decomposition (3) is characterized by the strengthened CBS inequality with the constant $\gamma = \cos(V_H, V_H^+)$, which is defined as follows:

$$\gamma = \cos(V_H, V_H^+)$$
$$= \sup\left\{ \frac{|a(u,v)|}{\|u\|_a \|v\|_a} : u \in V_H,\ u \neq 0,\ v \in V_H^+,\ v \neq 0 \right\}. \tag{4}$$

Above $\|u\|_a = \sqrt{a(u,u)}$ is the energy norm. If $\mathcal{T}_h$ arises from $\mathcal{T}_H$ by a regular division of the coarse grid triangles into 4 congruent triangles (see Fig. 1) and if the coefficients $K = (k_{ij})$ are constant on the coarse grid elements then $\gamma < \sqrt{3/4}$ for general anisotropic coefficients and arbitrary shape of the coarse grid elements. For more details, see [1] and the references therein.

3   Hierarchical Preconditioners and Error Estimates

The decomposition (3) can be used for construction of preconditioners for the
FE matrices A_h and $\bar{A}_h$,

$$\langle A_h \mathbf{u}, \mathbf{v} \rangle = \langle \bar{A}_h \bar{\mathbf{u}}, \bar{\mathbf{v}} \rangle = a(u, v) \tag{5}$$

for $u = \sum \mathbf{u}_i \phi_i^h = \sum \bar{\mathbf{u}}_i \bar{\phi}_i^h$ and $v = \sum \mathbf{v}_i \phi_i^h = \sum \bar{\mathbf{v}}_i \bar{\phi}_i^h$. Both nodal and hierarchical basis FE matrices A_h and $\bar{A}_h$ then have a hierarchic decomposition

$$A_h = \begin{bmatrix} A_{11} & A_{12} \\ A_{21} & A_{22} \end{bmatrix} \begin{matrix} \mathcal{N}_H^+ \\ \mathcal{N}_H \end{matrix} \quad \text{and} \quad \bar{A}_h = \begin{bmatrix} \bar{A}_{11} & \bar{A}_{12} \\ \bar{A}_{21} & \bar{A}_{22} \end{bmatrix} \begin{matrix} \mathcal{N}_H^+ \\ \mathcal{N}_H \end{matrix}. \tag{6}$$

Note that the diagonal blocks A_{11}, A_{22} of A_h carry only the local information.
On the opposite, the diagonal blocks $\bar{A}_{11} = A_{11}$ and $\bar{A}_{22} = A_H$ of $\bar{A}_h$ carry both
local and global information on the discretized problem.

Note also that the relation between A_h and $\bar{A}_h$ implies the identity between
the Schur complements,

$$S_h = \bar{S}_h, \quad S_h = A_{22} - A_{21} A_{11}^{-1} A_{12}, \quad \bar{S}_h = \bar{A}_{22} - \bar{A}_{21} \bar{A}_{11}^{-1} \bar{A}_{12}.$$

The standard *hierarchic multiplicative preconditioner* then follows from an
approximate factorization of A_h with Schur complement S_h replaced by $\bar{A}_{22}$,

$$B_h = \begin{bmatrix} I & 0 \\ A_{21} A_{11}^{-1} & I \end{bmatrix} \begin{bmatrix} A_{11} & \\ & \bar{A}_{22} \end{bmatrix} \begin{bmatrix} I & A_{11}^{-1} A_{12} \\ 0 & I \end{bmatrix}. \tag{7}$$

Note that getting efficient preconditioners assumes that

- A_{11} is approximated for a cheaper computation. The simplest approximation
   is the diagonal of A_{11}, see [2], more accurate approximation can use incomplete factorization or a locally tridiagonal element-by-element approximation
   of A_{11}, see [3],
- $\bar{A}_{22} = A_H$ is also approximated. A natural way how to do it is to use
   hierarchical decomposition recursively and to solve the system with $\bar{A}_{22}$ by a
   few inner iterations with a proper hierarchical preconditioner. In a multilevel
   setting, we can get an optimal preconditioner, see [4,5,6].

Another application of the hierarchical decomposition is in error estimation,
see [7,10] and the references therein. If $u \in V$ is the exact solution, $u_H \in V_H$
and $u_h \in V_h$ are the finite element solutions of the problem (1) in V_H and V_h,
respectively, and if there is a constant $\beta < 1$ such that

$$\| u - u_h \|_a \le \beta \| u - u_H \|_a, \tag{8}$$

(saturation condition) then the Galerkin orthogonality allows to show that

$$\| w_h \|_a \le \| u - u_H \|_a \le \frac{1}{1 - \beta^2} \| w_h \|_a, \tag{9}$$

where $w_h = u_h - u_H$, see [7]. Thus $\eta = \parallel w_h \parallel_a$ can serve as an *efficient and reliable error estimator*.

A cheaper error estimator $\bar{\eta}$ can be computed via the hierarchical decomposition (3). Let $\bar{\eta} = \parallel \bar{w}_h \parallel_a$, where

$$\bar{w}_h \in V_H^+ : \quad a(\bar{w}_h, v_h) = b(v_h) - a(u_h, v_h) \qquad \forall v_h \in V_H^+ \tag{10}$$

then

$$\parallel \bar{w}_h \parallel_a \leq \parallel u - u_H \parallel_a \leq \frac{1}{(1 - \beta^2)(1 - \gamma^2)} \parallel \bar{w}_h \parallel_a \tag{11}$$

where γ is the CBS constant from (4).

Algebraically,

$$\bar{\eta} = \langle A_{11} \mathbf{w}_1, \mathbf{w}_1 \rangle^{1/2}, \tag{12}$$

where

$$\mathbf{w}_1 : A_{11} \mathbf{w}_1 = \mathbf{b}_1 - \bar{A}_{12} \mathbf{w}_2, \tag{13}$$

$$\mathbf{w}_2 : \bar{A}_{22} \mathbf{w}_2 = \mathbf{b}_2. \tag{14}$$

A still cheaper estimators can be computed by using the approximations of A_{11}. In this respect, the locally tridiagonal approximation introduced by Axelsson and Padiy [3], which is robust with respect to anisotropy and element shape, is a good candidate for obtaining a cheap reliable and efficient hierarchic error estimator. The multiplicative preconditioner of A_{11} has more than two times better κ.

4   Nonconforming Finite Elements

Let $\mathcal{T}_h$ be a triangulation of Ω, $\mathcal{M}_h$ be the set of midpoints of the sides of triangles from $\mathcal{T}_h$, $\mathcal{M}_h^0$ and $\mathcal{M}_h^1$ consist of those midpoints from $\mathcal{M}_h$, which lie inside Ω and on the boundary $\partial \Omega$. Then the Crouzeix-Raviart finite element space V_h is defined as follows

$$V_h = \{ v \in U_h : \quad v(x) = 0 \quad \forall x \in \mathcal{M}_h^1 \}, \tag{15}$$

$$U_h = \{ v \in L_2(\Omega) : \quad v|_e \in P_1 \quad \forall e \in \mathcal{T}_h, \quad [v](x) = 0 \quad \forall x \in \mathcal{M}_h^0 \}, \tag{16}$$

where $[v](x)$ denotes the jump in $x \in \mathcal{M}_h^0$.

The finite element solution $u_h \in V_h$ of (1) is now defined as

$$u_h \in V_h : \quad a_h(u_h, v_h) = b(v_h) \qquad \forall v_h \in V_h \tag{17}$$

where a_h is the broken bilinear form,

$$a_h(u_h, v_h) = \sum_{T \in \mathcal{T}_h} \int_T \sum_{ij} k_{ij} \frac{\partial u_h}{\partial x_i} \frac{\partial v_h}{\partial x_j} dx. \tag{18}$$

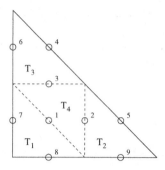

Fig. 2. A macroelement E with 9 midpoint nodes m_i

If $\mathcal{T}_H$ and $\mathcal{T}_h$ are two nested triangulations and V_H and V_h are the corresponding Crouzeix-Raviart spaces then $V_H \not\subseteq V_h$ and it is impossible to repeat the constructions of Section 2. But still there is a possibility to introduce a hierarchical basis in V_h algebraically, one such possibility, the DA splitting, is described in [8]. The construction is associated with the coarse triangles $E \in \mathcal{T}_H$ considered as macroelements composed from four congruent triangles $T \in \mathcal{T}_h$, see Fig. 2.

Let $\phi_1^h, \ldots, \phi_9^h$ be the nodal basis functions of the macroelement E, i.e. $\phi_i^h(m_j) = \delta_{ij}$. Then a hierarchical basis on E can be created from the following basis functions,

$$
\begin{aligned}
\bar{\phi}_l^h &= \phi_i^h \quad \text{for} \quad li = 11,\,22,\,33 \\
\bar{\phi}_l^h &= \phi_i^h - \phi_j^h \quad \text{for} \quad lij = 445, 567, 689 \\
\bar{\phi}_l^h &= \phi_i^h + \phi_j^h + \phi_k^h \quad \text{for} \quad lijk = 7145, 8267, 9389
\end{aligned}
\tag{19}
$$

The last triple will be called aggregated basis functions.

The hierarchical basis on a macroelement can be extended to a hierarchical basis in the whole space V_h. Using this hierarchical basis, the space V_h can be decomposed as follows

$$V_h = V_A \oplus V_A^+,$$

where V_A is spanned on the aggregated basis functions and V_A^+ is spanned on the remaining basis functions. For this decomposition, $\gamma = \cos(V_A, V_A^+) = \sqrt{3/4}$, see [8].

In [8,9], it is shown that the decomposition can be used for defining optimal order hierarchical preconditioners.

Now, we shall investigate the use of the DA hierarchical decomposition for the hierarchical error estimation in the case of nonconforming Crouzeix-Raviart FEM.

Let u_H, u_h be the nonconforming finite element solutions of (1) in V_H and V_h, respectively, and let us define

$$
u_A \in V_A : \quad a_h(u_A, v_h) = b(v_h) \quad \forall v_h \in V_A,
\tag{20}
$$

$$
w_A \in V_A^+ : \quad a_h(w_A, v_h) = b(v_h) - a(u_A, v_h) \quad \forall v_h \in V_A^+.
\tag{21}
$$

Algebraically, let $\bar{A}_h u_h = \bar{b}_h$ be the algebraic version of (17) in the introduced hierarchical basis and the hierarchical decomposition of this system gives the following block form,

$$\begin{bmatrix} \bar{A}_{11} & \bar{A}_{12} \\ \bar{A}_{21} & \bar{A}_{22} \end{bmatrix} \begin{bmatrix} \mathbf{u}_1 \\ \mathbf{u}_2 \end{bmatrix} = \begin{bmatrix} \mathbf{b}_1 \\ \mathbf{b}_2 \end{bmatrix} \tag{22}$$

with the first and second block corresponding to V_A^+ and V_A, respectively. Then

$$u_A \sim \mathbf{w}_2 = \bar{A}_{22}^{-1} \mathbf{b}_2, \tag{23}$$

$$w_A \sim \mathbf{w}_1 = \bar{A}_{11}^{-1} (\mathbf{b}_1 - \bar{A}_{12} \mathbf{w}_2). \tag{24}$$

We shall also consider the algebraic system

$$A_H \mathbf{u}_H = \mathbf{b}_H \tag{25}$$

corresponding to V_H and the broken energy norms

$$\| v_H \|_H = \sqrt{a_H(v_H, v_H)}, \qquad \| v_h \|_h = \sqrt{a_h(v_h, v_h)}$$

for $v_H \in V_H$ and $v_h \in V_h$, respectively.

Now, our aim is to investigate if

$$\eta = \| w_A \|_h = \sqrt{\langle \bar{A}_{11} \mathbf{w}_1, \mathbf{w}_1 \rangle} \tag{26}$$

is again a possible error estimator. We shal do it in three steps.

1. First, under the assumption that the saturation condition is valid, i.e. there is a $\beta < 1$,

$$\| u - u_h \|_h \leq \beta \| u - u_H \|_H,$$

it is possible to use $\| u_h - u_H \|_h$ as an error estimator for $\| u - u_H \|_H$, because

$$\frac{1}{1+\beta} \| u_h - u_H \|_h \leq \| u - u_H \|_H \leq \frac{1}{1-\beta} \| u_h - u_H \|_h. \tag{27}$$

Note that (27) follows from the triangle inequality. It is not possible to use the Galerkin orthogonality as it was done for (9).

2. Second, we shall investigate a relation between $\| u_h - u_H \|_h$ and $\| u_h - u_A \|_h$. For example, if f is constant on the elements $E \in \mathcal{T}_H$, then the vector $\mathbf{b}_2$ from (23) and $\mathbf{b}_H$ from (25) are equal. From [8], we have $\bar{A}_{22} = 4A_H$, thus $\mathbf{w}_2 = \frac{1}{4} \mathbf{u}_H$. Then

$$\| u_h - u_A \|_h^2 = \| u_h \|_h^2 - \| u_A \|_h^2 = \| u_h \|_h^2 - \frac{1}{4} \| u_H \|_h^2$$

$$\| u_h - u_H \|_h^2 = \| u_h \|_h^2 - \| u_H \|_h^2 + c_h$$

where c_h is the consistency term, $c_h = b(u_H) - a_h(u_h, u_H)$. It can be proved [11] that $c_h \to 0$ for $h \to 0$. Thus, for h sufficiently small

$$\| u_h - u_H \|_h \leq \| u_h - u_A \|_h \quad (\text{but not } \| u_h - u_H \|_h \sim \| u_h - u_A \|_h).$$

3. Third, the norm $\| u_h - u_A \|_h$ can be estimated by $\| w_A \|_h$. It holds, that

$$\| w_A \|_h^2 \leq \| u_h - u_A \|_h^2 \leq (1 - \gamma^2)^{-1} \| w_A \|_h^2,$$

where $\gamma = \sqrt{3/4}$ is the strengthened CBS constant for the DA splitting.
   The proof is simple. First,

$$\| w_A \|_h^2 = a_h(u_h - u_A, w_A) \leq \| u_h - u_A \|_h \| w_A \|_h.$$

Next, let $u_h = \hat{u}_A + \hat{w}_A$, $\hat{u}_A \in V_A$, $\hat{w}_A \in V_A^+$. Then

$$\| u_h - u_A \|_h^2 = a_h(u_h - u_A, u_h - u_A) = a_h(u_h - u_A, \hat{u}_A - u_A + \hat{w}_A)$$
$$= a_h(u_h - u_A, \hat{w}_A) = a_h(w_A, \hat{w}_A) \leq \| w_A \|_h \| \hat{w}_A \|_h$$
$$\| u_h - u_A \|_h^2 = \| \hat{u}_A - u_A + \hat{w}_A \|_h^2$$
$$\geq \| \hat{u}_A - u_A \|_h^2 + \| \hat{w}_A \|_h^2 - 2 \mid a_h(\hat{u}_A - u_A, \hat{w}_A) \mid$$
$$\geq (1 - \gamma^2) \| \hat{w}_A \|_h^2$$

Consequently,

$$(1 - \gamma^2) \| \hat{w}_A \|_h^2 \leq \| u_h - u_A \|_h^2 = a_h(w_A, \hat{w}_A) \leq \| w_A \|_h \| \hat{w}_A \|_h$$
$$\text{i.e.} \quad (1 - \gamma^2) \| \hat{w}_A \|_h \leq \| w_A \|_h$$

and

$$\| u_h - u_A \|_h^2 \leq (1 - \gamma^2)^{-1} \| w_A \|_h^2.$$

5   Conclusions

The first aim of the paper is to show that the progress in construction and analysis of the hierarchical multilevel preconditioners, as e.g. the mentioned locally tridiagonal approximation [3] to the pivot block can be exploited also for development of hierarchical error estimates.

The second aim is to extend the hierarchical error estimate concept to nonconforming finite elements with the aid of an auxiliary algebraic subspace V_A, $V_H \sim V_A$, $V_A \subset V_h$. This extension could provide an evaluation of the error and its distribution on an early stage of multilevel iterations and gives a chance to improve the discretization in the case of insufficient accuracy.

We have shown that a crucial point for this extension will be the approximation property of the algebraic space V_A. For the DA construction, the approximation property is not sufficient and we can get error estimator which is reliable but not efficient. A possible remedy could be in the use of the generalized DA decompositions, see [12,13]. In this respect, a further investigation is required.

Acknowledgements

The support given by the grant 1ET400300415 of the Academy of Sciences of the Czech Republic is greatly acknowledged. The author also thanks to the referees for a careful reading.

References

1. Axelsson, O., Blaheta, R.: Two simple derivations of universal bounds for the C.B.S. inequality constant. Applications of Mathematics 49, 57–72 (2004)
2. Axelsson, O., Gustafsson, I.: Preconditioning and two-level multigrid methods of arbitrary degree of approximations. Mathematics of Computation 40, 219–242 (1983)
3. Axelsson, O., Padiy, A.: On the additive version of the algebraic multilevel iteration method for anisotropic elliptic problems. SIAM Journal on Scientific Computing 20(5), 1807–1830 (1999)
4. Axelsson, O., Vassilevski, P.: Algebraic Multilevel Preconditioning Methods I. Numerische Mathematik 56, 157–177 (1989)
5. Axelsson, O., Vassilevski, P.: Algebraic Multilevel Preconditioning Methods II. SIAM Journal on Numerical Analysis 27, 1569–1590 (1990)
6. Axelsson, O., Vassilevski, P.: A black box generalized conjugate gradient solver with inner iterations and variable-step preconditioning. SIAM Journal on Matrix Analysis and Applications 12, 625–644 (1991)
7. Bank, R.: Hierarchical bases and the finite element method. Acta Numerica 5, 1–43 (1996)
8. Blaheta, R., Margenov, S., Neytcheva, M.: Uniform estimate of the constant in the strengthened CBS inequality for anisotropic non-conforming FEM systems. Numerical Linear Algebra with Applications 11, 309–326 (2004)
9. Blaheta, R., Margenov, S., Neytcheva, M.: Robust optimal multilevel preconditioners for nonconforming FEM systems. Numerical Linear Algebra with Applications 12, 495–514 (2005)
10. Brenner, S., Carstensen, C.: Finite element methods. In: Stein, E., de Borst, R., Hughes, T.J.R. (eds.) Encyclopedia of Computational Mechanics, pp. 73–118. J. Wiley, Chichester (2004)
11. Brenner, S., Scott, L.: The Mathematical Theory of Finite Element Methods. Springer, Heidelberg (2002)
12. Kraus, J., Margenov, S., Synka, J.: On the multilevel preconditioning of Crouzeix-Raviart elliptic problems. Numerical Linear Algebra with Applications (to appear)
13. Margenov, S., Synka, J.: Generalized aggregation-based multilevel preconditioning of Crouzeix-Raviart FE elliptic problems. In: Boyanov, T., et al. (eds.) NMA 2006. LNCS, vol. 4310, pp. 91–99. Springer, Heidelberg (2007)

Multilevel Preconditioning of Rotated Trilinear Non-conforming Finite Element Problems[*]

Ivan Georgiev[1,3], Johannes Kraus[2], and Svetozar Margenov[3]

[1] Institute of Mathematics and Informatics, Bulgarian Academy of Sciences
Acad. G. Bonchev, Bl. 8, 1113 Sofia, Bulgaria
john@parallel.bas.bg
[2] Johann Radon Institute for Computational and Applied Mathematics,
Altenbergerstraße 69, A-4040 Linz, Austria
johannes.kraus@oeaw.ac.at
[3] Institute for Parallel Processing, Bulgarian Academy of Sciences
Acad. G. Bonchev, Bl. 25A, 1113 Sofia, Bulgaria
margenov@parallel.bas.bg

Abstract. In this paper algebraic two-level and multilevel preconditioning algorithms for second order elliptic boundary value problems are constructed, where the discretization is done using Rannacher-Turek non-conforming rotated trilinear finite elements. An important point to make is that in this case the finite element spaces corresponding to two successive levels of mesh refinement are not nested in general. To handle this, a proper two-level basis is required to enable us to fit the general framework for the construction of two-level preconditioners for conforming finite elements and to generalize the method to the multilevel case.

The proposed variants of hierarchical two-level basis are first introduced in a rather general setting. Then, the involved parameters are studied and optimized. The major contribution of the paper is the derived estimates of the constant γ in the strengthened CBS inequality which is shown to allow the efficient multilevel extension of the related two-level preconditioners. Representative numerical tests well illustrate the optimal complexity of the resulting iterative solver.

1 Introduction

In this paper we consider the elliptic boundary value problem

$$
\begin{aligned}
Lu \equiv -\nabla \cdot (\mathbf{a}(\mathbf{x})\nabla u(\mathbf{x})) &= f(\mathbf{x}) \quad \text{in} \quad \Omega, \\
u &= 0 \quad \text{on} \quad \Gamma_D, \\
(\mathbf{a}(\mathbf{x})\nabla u(\mathbf{x})) \cdot \mathbf{n} &= 0 \quad \text{on} \quad \Gamma_N,
\end{aligned}
\tag{1}
$$

where Ω is a polyhedral domain in $\mathbb{R}^3$, $f(\mathbf{x})$ is a given function in $L^2(\Omega)$, the coefficient matrix $\mathbf{a}(\mathbf{x})$ is symmetric positive definite and uniformly bounded in

[*] The authors gratefully acknowledge the support by the Austrian Academy of Sciences, and by EC INCO Grant BIS-21++ 016639/2005. The first and third authors were also partially supported by Bulgarian NSF Grant VU-MI-202/2006.

I. Lirkov, S. Margenov, and J. Waśniewski (Eds.): LSSC 2007, LNCS 4818, pp. 86–95, 2008.
© Springer-Verlag Berlin Heidelberg 2008

Ω, $\mathbf{n}$ is the outward unit vector normal to the boundary $\Gamma = \partial\Omega$, and $\Gamma = \bar{\Gamma}_D \cup \bar{\Gamma}_N$. We assume that the elements of the diffusion coefficient matrix $\mathbf{a(x)}$ are piece-wise smooth functions on $\bar{\Omega}$.

The weak formulation of the above problem reads as follows:
given $f \in L^2(\Omega)$ find $u \in \mathcal{V} \equiv H_D^1(\Omega) = \{v \in H^1(\Omega) : v = 0 \text{ on } \Gamma_D\}$, satisfying

$$\mathcal{A}(u,v) = (f,v) \quad \forall v \in H_D^1(\Omega), \text{ where } \mathcal{A}(u,v) = \int_\Omega \mathbf{a(x)}\nabla u(\mathbf{x}) \cdot \nabla v(\mathbf{x})d\mathbf{x}. \quad (2)$$

We assume that the domain Ω is discretized by the partition $\mathcal{T}_h$ which is obtained by a proper refinement of a given coarser partition $\mathcal{T}_H$. We assume also that $\mathcal{T}_H$ is aligned with the discontinuities of the coefficient matrix $\mathbf{a(x)}$ so that over each element $E \in \mathcal{T}_H$ the coefficients of $\mathbf{a(x)}$ are smooth functions.

The variational problem (2) is discretized using the finite element method, i.e., the continuous space $\mathcal{V}$ is replaced by a finite dimensional subspace $\mathcal{V}_h$. Then the finite element formulation is:
find $u_h \in \mathcal{V}_h$, satisfying $\mathcal{A}_h(u_h, v_h) = (f, v_h) \quad \forall v_h \in \mathcal{V}_h$, where

$$\mathcal{A}_h(u_h, v_h) = \sum_{e \in \mathcal{T}_h} \int_e \mathbf{a}(e)\nabla u_h \cdot \nabla v_h d\mathbf{x}. \quad (3)$$

Here $\mathbf{a}(e)$ is a piece-wise constant symmetric positive definite matrix, defined by the integral averaged values of $\mathbf{a(x)}$ over each element from the coarser triangulation $\mathcal{T}_H$. We note that in this way strong coefficient jumps across the boundaries between adjacent finite elements from $\mathcal{T}_H$ are allowed. The next sections are devoted to the study of two-level and multilevel preconditioners for the case of non-conforming Rannacher-Turek finite elements. A unified hierarchical splitting of the FEM spaces is developed, followed by uniform estimates of the related CBS constants. The numerical results that are presented towards the end of the paper support the theoretical analysis.

2   Rannacher-Turek Finite Elements

Nonconforming finite elements based on *rotated* multilinear shape functions were introduced by Rannacher and Turek [10] as a class of simple elements for the Stokes problem.

The unit cube $[-1, 1]^3$ is used as a reference element $\hat{e}$ to define the isoparametric rotated trilinear element $e \in \mathcal{T}_h$. Let $\psi_e : \hat{e} \to e$ be the trilinear bijective mapping between the reference element $\hat{e}$ and e. The polynomial space of scalar shape functions $\hat{\phi}_i$ on the reference element $\hat{e}$ is defined by

$$\hat{\mathcal{P}} := \{\hat{\phi}_i : 1 \le i \le 6\} = \text{span}\{1, x, y, z, x^2 - y^2, y^2 - z^2\},$$

and the shape functions ϕ_i on e are computed from $\hat{\phi}_i$ via the relations $\{\phi_i\}_{i=1}^6 = \{\hat{\phi}_i \circ \psi_e^{-1}\}_{i=1}^6$. Two different discretization variants, i.e., two different sets of

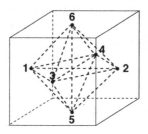

Fig. 1. Node numbering and connectivity pattern of the reference element $\hat{e}$

shape functions $\hat{\phi}_i$ are considered. For the variant MP (mid point), $\{\hat{\phi}_i\}_{i=1}^6$ are found by the point-wise interpolation condition

$$\hat{\phi}_i(b_\Gamma^j) = \delta_{ij},$$

where $b_\Gamma^j, j = 1, 6$ are the centers of the faces of the cube $\hat{e}$. Alternatively, the variant MV (mean value) corresponds to the integral mean-value interpolation condition

$$|\Gamma_{\hat{e}}^j|^{-1} \int_{\Gamma_{\hat{e}}^j} \hat{\phi}_i d\Gamma_{\hat{e}}^j = \delta_{ij},$$

where $\Gamma_{\hat{e}}^j$ are the faces of the reference element $\hat{e}$. For the explicit form of the reference-element shape functions we refer to [6].

3   Hierarchical Two-Level Splittings

Let us consider two consecutive discretizations $\mathcal{T}_H$ and $\mathcal{T}_h$. Note that in this case $\mathcal{V}_H$ and $\mathcal{V}_h$ are not nested.

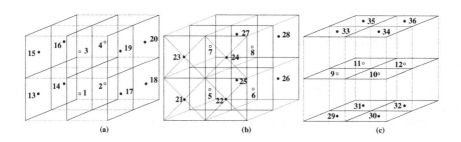

Fig. 2. Node numbering in macro-element

3.1 "First Reduce" (FR) Two-Level Splitting

We follow the idea of [5,8] to define an algebraic two-level preconditioner. For that reason, let $\varphi_E = \{\phi_i(x,y)\}_{i=1}^{36}$ be the macro-element vector of the nodal basis functions and A_E be the macro-element stiffness matrix corresponding to $E \in \mathcal{T}_h$. The global stiffness matrix A_h can be written as $A_h = \sum_{E \in \mathcal{T}_h} A_E$ where the summation is understood as the FEM assembly procedure. Next, we introduce the following macro-element level transformation matrix J_E in the 2×2 block diagonal form $J_E = \frac{1}{4} \begin{bmatrix} I \\ & J_{E,22} \end{bmatrix}$, where I is the 12×12 identity matrix and

$$
J_{E,22} = \begin{bmatrix} P & & & & & \\ & P & & & & \\ & & P & & & \\ & & & P & & \\ & & & & P & \\ & & & & & P \\ E_1 & E_2 & E_3 & E_4 & E_5 & E_6 \end{bmatrix}.
\tag{4}
$$

Each block E_i is a 6×4 zero matrix except for its i-th row which is composed of all ones, and $P = \begin{bmatrix} -1 & 1 & -1 & 1 \\ -1 & -1 & 1 & 1 \\ 1 & -1 & -1 & 1 \end{bmatrix}$. The matrix J_E defines locally a two-level hierarchical basis $\widetilde{\varphi}_E$, namely, $\widetilde{\varphi}_E = J_E \varphi_E$. The hierarchical two-level macro-element stiffness matrix is given by $\widetilde{A}_E = J_E A_E J_E^T$, and the related global stiffness matrix reads as $\widetilde{A}_h = \sum_{E \in \mathcal{T}_h} \widetilde{A}_E$. We split the two-level stiffness matrix $\widetilde{A}_h$ into 2×2 block form $\widetilde{A}_h = \begin{bmatrix} \widetilde{A}_{11} & \widetilde{A}_{12} \\ \widetilde{A}_{21} & \widetilde{A}_{22} \end{bmatrix}$, where $\widetilde{A}_{11}$ corresponds to interior nodal unknowns with respect to the macro-elements $E \in \mathcal{T}_h$. We observe that $\widetilde{A}_{11}$ is a block-diagonal matrix whose diagonal blocks are of size 12×12; Each such block corresponds to the interior points $\{1, 2, \ldots, 12\}$ of one macro element. The first step of the "First Reduce" (FR) algorithm is to eliminate these unknowns exactly, which can be done locally, i.e., separately for each macro element. Therefore the Schur complement $B = \widetilde{A}_{22} - \widetilde{A}_{21}\widetilde{A}_{11}^{-1}\widetilde{A}_{12}$ can be assembled from the local macro-element contributions B_E and the (sub)matrix B_{22} in

$$
B = \sum_{E \in \mathcal{T}_h} B_E = \begin{bmatrix} B_{11} & B_{12} \\ B_{21} & B_{22} \end{bmatrix} = \begin{bmatrix} \sum_E B_{E,11} & \sum_E B_{E,12} \\ \sum_E B_{E,21} & \sum_E B_{E,22} \end{bmatrix},
\tag{5}
$$

which corresponds to the sums of basis functions over each macro-element face, can be associated with the coarse grid. It is important to note that

$$
\ker(B_{E,22}) = \ker(A_e) = \text{span}\{(1,1,1,1,1,1)^T\}
\tag{6}
$$

which allows us to apply a local analysis to estimate the constant γ corresponding to the splitting defined by the block partition (5). We proceed as follows:

Step 1: We compute the local Schur complements arising from static condensation of the "interior degrees of freedom" and obtain the (24×24) matrix B_E.

Step 2: Following the theory, it suffices to compute the minimal eigenvalue of the generalized eigenproblem $S_E \mathbf{v}_E = \lambda_E^{(1)} B_{E,22} \mathbf{v}_E$ where $\mathbf{v}_E \neq (c, c, \ldots, c)^T \forall c \in \mathbb{R}$ and $S_E = B_{E,22} - B_{E,21} B_{E,11}^{-1} B_{E,12}$. Then

$$\gamma^2 \leq \max_{E \in \mathcal{T}_h} \gamma_E^2 = \max_{E \in \mathcal{T}_h} (1 - \lambda_E^{(1)}). \tag{7}$$

3.2   Two-Level Splitting by Differences and Aggregates (DA)

Similarly to the FR case, the DA splitting is easily described for one macro-element. If $\phi_1, \ldots, \phi_{36}$ are the standard nodal basis functions for the macro-element, then we define

$$
\begin{aligned}
\mathcal{V}(E) &= \text{span}\,\{\phi_1, \ldots, \phi_{36}\} = \mathcal{V}_1(E) \oplus \mathcal{V}_2(E)\,, \\
\mathcal{V}_1(E) &= \text{span}\,\{\phi_1, \ldots, \phi_{12}, \phi_{14} + \phi_{16} - (\phi_{13} + \phi_{15}), \phi_{15} + \phi_{16} - (\phi_{13} + \phi_{14}), \\
&\qquad \phi_{13} + \phi_{16} - (\phi_{14} + \phi_{15}), \ldots, \phi_{34} + \phi_{36} - (\phi_{33} + \phi_{35}), \\
&\qquad \phi_{35} + \phi_{36} - (\phi_{33} + \phi_{34}), \phi_{33} + \phi_{36} - (\phi_{34} + \phi_{35})\} \\
\mathcal{V}_2(E) &= \text{span}\,\{\phi_{13} + \phi_{14} + \phi_{15} + \phi_{16} + \sum_{j=1}^{12} \beta_{1j}\phi_j, \ldots, \phi_{33} + \phi_{34} + \phi_{35} + \phi_{36} \\
&\qquad + \sum_{j=1}^{12} \beta_{6j}\phi_j\}\,.
\end{aligned}
$$

The related transformation matrix is $J_E = \frac{1}{4} \begin{bmatrix} I \\ J_{E,21} \ J_{E,22} \end{bmatrix}$, where I is the 12×12 identity matrix, $J_{E,21} = \begin{bmatrix} 0 \\ \mathcal{B} \end{bmatrix}$, where $\mathcal{B} = (\beta_{ij})_{6 \times 12}$ and $J_{E,22}$ is given by (4). The vector of the macro-element basis functions $\varphi_E = \{\phi_i\}_{i=1}^{36}$ is transformed to a new hierarchical basis $\tilde{\varphi}_E = \{\tilde{\phi}_i\}_{i=1}^{36} = J_E \varphi_E$. Moreover, we have

$$\tilde{A}_E = J_E A_E J_E^T = \begin{bmatrix} \tilde{A}_{E,11} \ \tilde{A}_{E,12} \\ \tilde{A}_{E,21} \ \tilde{A}_{E,22} \end{bmatrix} \begin{matrix} \} \ \tilde{\phi}_i \in \mathcal{V}_1(E) \\ \} \ \tilde{\phi}_i \in \mathcal{V}_2(E) \end{matrix}. \tag{8}$$

According to the local definitions, we can similarly construct the new hierarchical basis $\tilde{\varphi} = \{\tilde{\varphi}_h^{(i)}\}_{i=1}^{N_h}$ for the whole finite element space $\mathcal{V}_h$, which is split into the coarse space $\mathcal{V}_2$ and its complement $\mathcal{V}_1$, i.e.,

$$\mathcal{V}_h = \mathcal{V}_1 \oplus \mathcal{V}_2\,. \tag{9}$$

Now, we are in a position to analyze the constant $\gamma = \cos(\mathcal{V}_1, \mathcal{V}_2)$ for the splitting (9). Again, as in the previous section, we would like to perform this analysis locally, by considering the corresponding problems on macro-elements. For this purpose we need to have satisfied the condition

(i) $\ker(\tilde{A}_{E,22}) = \ker(A_e)$,

which is equivalent to $\sum\limits_{i=1}^{6} \beta_{ij} = 1$, $\forall j \in \{1, 2, \ldots, 12\}$.

There are obviously various DA splittings satisfying the condition (i). For more details about aggregation based preconditioners see the review paper [4].

When the two-level algorithm is recursively generalized to the multilevel case, it is useful if

(ii) $\tilde{A}_{E,22}$ is proportional to A_e.

It seems to be rather complicated to find a parameter matrix $\mathcal{B}$, which satisfies the condition (ii) in the general case of Rannacher-Turek trilinear finite elements.

4 Uniform Estimates of the CBS Constant

We study in this section both splitting algorithms, FR and DA, for both discretization variants, MP and MV, of cubic rotated trilinear finite elements and the isotropic model problem.

4.1 FR Algorithm

Following (7) we compute the local CBS constant and derive the following global estimates for the isotropic model problem on a mesh composed of cubic elements. The bounds are uniform with respect to the size of the discrete problem and any possible jumps of the coefficients.

Variant MP: For the FR splitting the obtained result is $\lambda_E^{(1)} = 13/21$, which implies $\gamma_E^2 = 1 - \lambda_E^{(1)} = 8/21$, and therefore $\gamma_{MP}^2 \leq 8/21$.

Variant MV: For the FR splitting we further have $\lambda_E^{(1)} = 1/2$, which implies $\gamma_E^2 = 1 - \lambda_E^{(1)} = 1/2$, and therefore $\gamma_{MV}^2 \leq 1/2$.

Let us remind once again, that the obtained estimates hold theoretically for the two-level algorithm only. This is because the matrix $B_{E,22}$ is only associated with the coarse discretization $e \in \mathcal{T}_H$ and is not proportional to the related element stiffness matrix A_e. The CBS constants, however, show a very stable behavior in the FR multilevel setting, which has been verified numerically, cf. Table 1.

4.2 DA Algorithm

Due to the symmetry of the model problem, the non-zero part $\mathcal{B}$ of the lower-left block $J_{E,21}$ of the transformation matrix J_E can be simplified to the form

$$\mathcal{B} = \begin{bmatrix} a\,a\,a\,a\,b\,c\,b\,c\,b\,c\,b\,c \\ a\,a\,a\,a\,c\,b\,c\,b\,c\,b\,c\,b \\ b\,c\,b\,c\,a\,a\,a\,a\,b\,b\,c\,c \\ c\,b\,c\,b\,a\,a\,a\,a\,c\,c\,b\,b \\ b\,b\,c\,c\,b\,b\,c\,c\,a\,a\,a\,a \\ c\,c\,b\,b\,c\,c\,b\,b\,a\,a\,a\,a \end{bmatrix}.$$

Table 1. Multilevel behavior of γ^2 for "First reduce" algorithm

variant	ℓ	$\ell-1$	$\ell-2$	$\ell-3$	$\ell-4$	$\ell-5$
MP	0.38095	0.39061	0.39211	0.39234	0.39237	0.39238
MV	0.5	0.4	0.39344	0.39253	0.39240	0.39238

Let us write the condition (ii) in the form $\tilde{A}_{E,22} = pA_e$. Then, (ii) is reduced to a system of three nonlinear equations for (a, b, c), with a parameter p. It appears, that the system for (a, b, c) has a solution if $p \in [p_0, \infty)$. In such a case, we can optimize the parameter p, so that the CBS constant is minimal. The obtained results are summarized below.

Variant MP

Lemma 1. *There exists a DA two-level splitting satisfying the condition (ii) if and only if $p \geq 3/14$. Then, the obtained solutions for (a, b, c) are invariant with respect to the local CBS constant $\gamma_E^2 = 1 - 1/(8p)$, and for the related optimal splitting $\gamma_{MP}^2 \leq 5/12$.*

Variant MV: The same approach is applied to get the estimates below.

Lemma 2. *There exists a DA two-level splitting satisfying the condition (ii) if and only if $p \geq 1/4$. Then, the obtained solutions for (a, b, c) are invariant with respect to the local CBS constant $\gamma_E^2 = 1 - 1/(8p)$, and for the related optimal splitting $\gamma_{MV}^2 \leq 1/2$.*

The computed (local) estimates for γ^2 for the FR algorithm are always smaller than the related ones for the DA algorithm. One can also observe a nice convergence to the value of $\theta \approx 0.39238$ for both variants, MP and MV, see Table 1.

5   Computational Complexity

The CBS constant is not only used to analyze the related two-level preconditioners but it is also involved in the construction of the acceleration matrix polynomial P_β in algebraic multilevel iteration (AMLI) methods [1,2]. A main result in [1,2] is that the AMLI method (based on the multiplicative preconditioner) is of optimal order of computational complexity, if

$$1/(\sqrt{1-\gamma^2}) < \beta < \tau, \tag{10}$$

where $\tau \approx N^{(k)}/N^{(k-1)}$ is the reduction factor of the number of degrees of freedom from a given fine to the next coarser mesh, which in our case is approximately 8. The left-hand side inequality in (10) assures that the condition number will be bounded uniformly in the number of levels whereas the right-hand side inequality allows to estimate the computational work $w^{(\ell)}$ that is required for one application of the preconditioner at level ℓ of the finest discretization, i.e., $w^{(\ell)} \leq c\,(N^{(\ell)} + \beta\,N^{(\ell-1)} + \ldots + \beta^\ell\,N^{(0)}) < c\,N/(1 - \frac{\beta}{\tau})$, where $N = N^{(\ell)}$.

The work for the construction of the proposed AMLI preconditioners is also proportional to N. This can easily be seen by observing that

- the matrices $A^{(k)}$, $0 \leq k \leq \ell$, have at most 11 nonzero entries per row,
- every two-level transformation $J^{(k)}$ is the identity for interior unknowns,
- in case of the FR splitting the remaining rows of $J^{(k)}$, which are given according to $J_{E,22}$, see (4), have 4 nonzero entries per row,
- in case of the DA splitting the remaining rows of $J^{(k)}$ are given according to $[J_{E,21}, J_{E,22}]$, which results in 4 or at most 28 nonzeros per row,
- the costs for the elimination of the interior nodal unknowns is $\mathcal{O}(N^{(k)})$,
- the (global) product $\widetilde{A}^{(k)} = J^{(k)} A^{(k)} J^{(k)T}$ requires $\mathcal{O}(N^{(k)})$ operations,
- alternatively, the hierarchical basis matrix $\widetilde{A}^{(k)}$ can be assembled from the local contributions $\widetilde{A}_E^{(k)} = J_E A_E^{(k)} J_E^T$, at total costs of $\mathcal{O}(N^{(k)})$ operations.

Clearly, the storage requirement for the preconditioner is $\mathcal{O}(N)$ as well.

6   Numerical Results

We solved the model problem (1) using the preconditioned conjugate gradient (PCG) method combined with the multiplicative variant of the multilevel preconditioner based on either DA or FR splitting. The computational domain is $\Omega = (0,1)^3$ and both discretization variants, MP and MV, are considered. The mesh size is varied in the range $h = 1/8$ to $h = 1/128$ resulting in 512 to 2 097 157 finite elements with 1 728 to 6 340 608 nodes, respectively. For any element e in $\mathcal{T}_h$ the matrix $\mathbf{a}(e)$ in (3) is defined by $\mathbf{a}(e) := \alpha(e) \cdot I$, where the following situation of a jump in the coefficient $\alpha = \alpha(e)$ is considered:

$$\alpha(e) = \left\{ \begin{array}{l} 1 \text{ in } (I_1 \times I_1 \times I_1) \bigcup (I_2 \times I_2 \times I_1) \bigcup (I_2 \times I_1 \times I_2) \bigcup (I_1 \times I_2 \times I_2) \\ \varepsilon \qquad\qquad\qquad\qquad\qquad\qquad \text{elsewhere} \end{array} \right\},$$

where $I_1 = (0, 0.5]$ and $I_2 = (0.5, 1)$, and $\varepsilon = 10^{-3}$.

Table 2 summarizes the number of PCG iterations that reduce the residual norm by a factor 10^8 when performing the V-cycle AMLI. In Table 3 we list the corresponding results for the linear AMLI W-cycle employing the matrix polynomial $Q_1(t) = (1 - P_2(t))/t = q_0 + q_1 t$ for stabilizing the condition number. We use the coefficients

$$q_0 = \frac{2}{\sqrt{1 - \gamma^2}}, \quad q_1 = -\frac{1}{1 - \gamma^2}, \tag{11}$$

Table 2. Linear AMLI V-cycle: number of PCG iterations

MP: $1/h$	8	16	32	64	128	MV: $1/h$	8	16	32	64	128
DA: $\varepsilon = 1$	9	12	16	20	24	DA: $\varepsilon = 1$	12	17	22	29	38
$\varepsilon = 10^{-3}$	9	12	16	20	25	$\varepsilon = 10^{-3}$	12	17	22	30	39
FR: $\varepsilon = 1$	8	11	14	18	22	FR $\varepsilon = 1$	10	14	17	21	26
$\varepsilon = 10^{-3}$	8	11	14	18	22	$\varepsilon = 10^{-3}$	10	14	17	21	26

Table 3. Linear AMLI W-cycle: number of PCG iterations

MP: $1/h$	8	16	32	64	128
DA: $\varepsilon = 1$	9	10	10	10	10
$\varepsilon = 10^{-3}$	9	10	10	10	10
FR: $\varepsilon = 1$	8	9	9	9	9
$\varepsilon = 10^{-3}$	8	9	9	9	9

MV: $1/h$	8	16	32	64	128
DA: $\varepsilon = 1$	12	15	15	16	16
$\varepsilon = 10^{-3}$	12	15	16	16	16
FR: $\varepsilon = 1$	10	12	12	12	12
$\varepsilon = 10^{-3}$	10	12	12	12	12

Table 4. Non-linear AMLI W-cycle: number of (outer) GCG iterations

MP: $1/h$	8	16	32	64	128
DA: $\varepsilon = 1$	9	9	9	9	9
$\varepsilon = 10^{-3}$	9	10	10	10	10
FR: $\varepsilon = 1$	8	9	9	9	9
$\varepsilon = 10^{-3}$	8	9	9	9	9

MV: $1/h$	8	16	32	64	128
DA: $\varepsilon = 1$	12	12	12	12	12
$\varepsilon = 10^{-3}$	12	12	12	12	12
FR: $\varepsilon = 1$	10	11	11	11	11
$\varepsilon = 10^{-3}$	10	11	11	11	11

which is in accordance with the analysis in [1,2] for exact inversion of the pivot block. The reported numerical experiments, however, indicate that this is a proper choice even when using inexact solves based on an ILU factorization (with no additional fill-in). Finally, Table 4 refers to the (variable-step) non-linear AMLI method stabilized by two inner generalized conjugate gradient iterations at every intermediate level, cf., [3,7,9] (and using a direct solve on the coarsest mesh with mesh size $1/h = 4$, as in the other tests).

In accordance with the theory the preconditioners are perfectly robust with respect to jump discontinuities in the coefficients $\mathbf{a}(e)$ if they do not occur inside any element of the coarsest mesh partition. The results slightly favor the FR approach, and, they illustrate well the optimal complexity of the iterative solvers, using a W-cycle, for both of the splittings and for both discretization variants.

References

1. Axelsson, O., Vassilevski, P.: Algebraic Multilevel Preconditioning Methods I. Numer. Math. 56, 157–177 (1989)
2. Axelsson, O., Vassilevski, P.: Algebraic Multilevel Preconditioning Methods II. SIAM J. Numer. Anal. 27, 1569–1590 (1990)
3. Axelsson, O., Vassilevski, P.: Variable-step multilevel preconditioning methods, I: self-adjoint and positive definite elliptic problems. Num. Lin. Alg. Appl. 1, 75–101 (1994)
4. Blaheta, R.: Algebraic multilevel methods with aggregations: an overview. In: Lirkov, I., Margenov, S., Waśniewski, J. (eds.) LSSC 2005. LNCS, vol. 3743, pp. 3–14. Springer, Heidelberg (2006)
5. Blaheta, R., Margenov, S., Neytcheva, M.: Uniform estimate of the constant in the strengthened CBS inequality for anisotropic non-conforming FEM systems. Numerical Linear Algebra with Applications 11(4), 309–326 (2004)

6. Georgiev, I., Margenov, S.: MIC(0) Preconditioning of Rotated Trilinear FEM Elliptic Systems. In: Margenov, S., Waśniewski, J., Yalamov, P. (eds.) LSSC 2001. LNCS, vol. 2179, pp. 95–103. Springer, Heidelberg (2001)
7. Kraus, J.: An algebraic preconditioning method for M-matrices: linear versus non-linear multilevel iteration. Num. Lin. Alg. Appl. 9, 599–618 (2002)
8. Margenov, S., Vassilevski, P.: Two-level preconditioning of non-conforming FEM systems. In: Griebel, M., et al. (eds.) Large-Scale Scientific Computations of Engineering and Environmental Problems, Vieweg. Notes on Numerical Fluid Mechanics, vol. 62, pp. 239–248 (1998)
9. Notay, Y.: Robust parameter-free algebraic multilevel preconditioning. Num. Lin. Alg. Appl. 9, 409–428 (2002)
10. Rannacher, R., Turek, S.: Simple non-conforming quadrilateral Stokes Element. Numerical Methods for Partial Differential Equations 8(2), 97–112 (1992)

A Fixed-Grid Finite Element Algebraic Multigrid Approach for Interface Shape Optimization Governed by 2-Dimensional Magnetostatics[*]

Dalibor Lukáš[1] and Johannes Kraus[2]

[1] Department of Applied Mathematics, VŠB–Technical University of Ostrava, 17. listopadu 15, 708 33 Ostrava–Poruba, Czech Republic
[2] RICAM, Austrian Academy of Sciences, Altenberger Strasse 69, 4040 Linz, Austria

Abstract. The paper deals with a fast computational method for discretized optimal shape design problems governed by 2–dimensional magnetostatics. We discretize the underlying state problem using linear Lagrange triangular finite elements and in the optimization we eliminate the state problem for each shape design. The shape to be optimized is the interface between the ferromagnetic and air domain. The novelty of our approach is that shape perturbations do not affect grid nodal displacements, which is the case of the traditional moving–grid approach, but they are rather mapped to the coefficient function of the underlying magnetostatic operator. The advantage is that there is no additional restriction for the shape perturbations on fine discretizations. However, this approach often leads to a decay of the finite element convergence rate, which we discuss. The computational efficiency of our method relies on an algebraic multigrid solver for the state problem, which is also described in the paper. At the end we present numerical results.

1 Introduction

Shape optimization covers a class of problems in which one looks for an optimal shape of a part of the boundary or interface of a body subjected to a physical field. The optimality means minimization of a given objective functional among admissible shapes. We will restrict ourselves to the case of interface shape optimization with the physics modelled by a linear partial differential equation (PDE). The abstract setting of the problem reads as follows:

$$\min_{(\alpha,u)\in\mathcal{U}_{\mathrm{ad}}\times V} \mathcal{I}(\alpha, u) \quad \text{s.t.} \quad A(\alpha)u = b \quad \text{on } V', \tag{1}$$

[*] This research has been supported by the Czech Grant Agency under the grant GAČR 201/05/P008, by the Czech Ministry of Education under the project MSM6198910027, by the Czech Academy of Science under the project AVČR 1ET400300415, and by the Austrian Science Fund FWF under the project SFB F013.

I. Lirkov, S. Margenov, and J. Waśniewski (Eds.): LSSC 2007, LNCS 4818, pp. 96–104, 2008.

where $\mathcal{U}_{\mathrm{ad}}$ is a nonempty compact subset of admissible piecewise smooth functions α describing some parts of the interface between $\Omega_0(\alpha)$ and $\Omega_1(\alpha)$ which provide a distinct decomposition of a given domain $\Omega \subset \mathbb{R}^2$. Further, $\mathcal{I}$ denotes an objective continuous functional, V is a Hilbert space of functions over Ω with the dual V', $A(\alpha) \in \mathcal{L}(V, V')$ denotes the PDE operator which continuously depends on α, where $\mathcal{L}(V, V')$ consists of linear continuous operators from V to V', $b \in V'$ denotes a physical field source term, and $u \in V$ is the unique solution to the underlying PDE problem.

There is a number of methods solving the problem (1). Let us classify them regarding how they treat the PDE constraint. The following state elimination (nested, black–box) method, cf. [7], is most traditional in shape optimization:

$$\min_{\alpha \in \mathcal{U}_{\mathrm{ad}}} \mathcal{I}(\alpha, A(\alpha)^{-1}b).$$

On the other hand, we can prescribe the state equation via a Lagrange multiplier and solve the following nonlinear saddle–point problem

$$\min_{(\alpha, u) \in \mathcal{U}_{\mathrm{ad}}} \max_{\lambda \in V} \left\{ \mathcal{I}(\alpha, u) + \langle A(\alpha)u - b, \lambda \rangle_{V' \times V} \right\},$$

where $\langle ., . \rangle_{V' \times V}$ denotes the duality pairing. This so–called one–shot (simultaneous, primal–dual, all–at–once) method is superior in case of topology optimization, smooth dependence of $\mathcal{I}(\alpha, u)$ and $A(\alpha)$ thanks to a sparsity of the Hessian of the Lagrange functional, which allows to use Newton methods, cf. [4,6].

Another classification of solution methods follows when we take into account the structure of $A(\alpha)$. Without loss of generality, let us think about the 2–dimensional linear magnetostatic state problem, the classical formulation of which is as follows:

$$\begin{cases} -\nu_0 \triangle u_0(x) & = J(x) & \text{for } x \in \Omega_0(\alpha), \\ -\nu_1 \triangle u_1(x) & = 0 & \text{for } x \in \Omega_1(\alpha), \\ u_0(x) \quad -u_1(x) & = 0 & \text{for } x \in \Gamma(\alpha), \\ \nu_0 \nabla u_0(x) \cdot \mathbf{n}_0(\alpha)(x) - \nu_1 \nabla u_1(x) \cdot \mathbf{n}_0(\alpha)(x) & = 0 & \text{for } x \in \Gamma(\alpha), \\ u(x) & = 0 & \text{for } x \in \Gamma_{\mathrm{D}}, \\ \frac{\partial u}{\partial n} & = 0 & \text{for } x \in \Gamma_{\mathrm{N}}, \end{cases} \tag{2}$$

where $\nu_0 \gg \nu_1 > 0$ denote the reluctivity of the air and ferromagnetics, respectively, J denotes the electric current density, $\Gamma(\alpha) := \partial\Omega_0(\alpha) \cap \partial\Omega_1(\alpha)$ denotes the interface, $\overline{\Gamma_{\mathrm{D}}} \cup \overline{\Gamma_{\mathrm{N}}}$ is a distinct decomposition of $\partial\Omega$ into the Dirichlet and Neumann part, $\mathbf{n}_0(\alpha)$ denotes the outward unit normal vector to $\Omega_0(\alpha)$ and where u consists of $u|_{\Omega_0(\alpha)} := u_0$ and $u|_{\Omega_1(\alpha)} := u_1$. Then, a straightforward approach is the following weak formulation of (2) in the Sobolev space $V := H^1_{0,\Gamma_{\mathrm{D}}} := \{v \in H^1(\Omega) : v = 0 \text{ on } \Gamma_{\mathrm{D}}\}$:

$$\text{Find } u \in V : \int_{\Omega} \nu(\alpha)(x) \nabla u(x) \nabla v(x) \, dx = \int_{\Omega} J(x)v(x) \, dx \ \forall v \in V, \tag{3}$$

where $\nu(\alpha)$ consists of $\nu(\alpha)|_{\Omega_0(\alpha)} := \nu_0$ and $\nu(\alpha)|_{\Omega_1(\alpha)} := \nu_1$ and where J is extended to Ω by zero. Another formulation of (2) prescribes the third and fourth equations of (2), which are called interface conditions, in a weaker sense using the Lagrange formalism again. This might be viewed as a sort of fictitious domain method, cf. [8], or a domain decomposition method. The formulation is as follows: Find $(u_0, u_1, \lambda_t, \lambda_n) \in V_0 \times V_1 \times H^{1/2}(\Gamma(\alpha)) \times H^{1/2}(\Gamma(\alpha))$ as a solution of

$$
\begin{pmatrix}
A_0 & , & 0 & , \text{sym.} \\
0 & , & A_1 & , \text{sym.} \\
I(\alpha) & , & -I(\alpha) & , & 0 \\
B_0(\alpha) & , & -B_1(\alpha) & , & 0
\end{pmatrix}
\begin{pmatrix}
u_0 \\
u_1 \\
\lambda_t \\
\lambda_n
\end{pmatrix}
=
\begin{pmatrix}
J \\
0 \\
0 \\
0
\end{pmatrix},
\tag{4}
$$

where the saddle–point structure corresponds to the equations 1–4 in (2) such that for $d = 0, 1$ we define $V_d := \{v \in H^1(\Omega) : v = 0 \text{ on } \Gamma_D \cap \partial\Omega_d(\alpha)\}$, $A_d u_d := -\nu_d \triangle u_d$, $I(\alpha)u_d := u_d|_{\Gamma(\alpha)}$ and $B_d(\alpha)u_d := \nu_d \nabla u_d(x) \cdot \mathbf{n}_0(\alpha)$. The advantage of this approach is that the PDE–operators A_0 and A_1 are independent of the evolving shape. Thus one can approximate (4) via a discretization of a fixed domain and efficient saddle–point solvers [5,9,14] can be used. However, the formulation (4) poses a lower–order convergence rate of finite element approximations and the optimization functional is nondifferentiable. Recently, there has been a growing number of applications of discontinuous Galerkin methods, cf. [3,12], which turned out to be another proper framework for the interface shape optimization.

The optimization method proposed in this paper is based on state elimination. Our treatment of the interface conditions is half the way from the weak formulation (3) to the domain decomposition approach (4). We approximate the weak formulation (3) on a finite element grid that does not follow the shape α and we map the shape perturbations to the coefficient function of the underlying magnetostatic operator.

For solution of discretized state problems (3) we use an algebraic multigrid (AMG) method. AMG methods [1,13] are known as efficient and robust linear solvers for elliptic boundary-value problems. Our approach is based on the computation of so-called edge matrices, which provide a good starting point for building efficient AMG components, while keeping the set-up costs low [10]. The resulting AMGm solver we are using, see [11], lies in-between classical AMG [13], i.e., strong and weak edges affect the coarsening and the formation of interpolation molecules, and AMG based on element interpolation–so-called AMGe [2], i.e., small-sized neighborhood matrices serve for the computation of the actual interpolation coefficients.

The rest of the paper is organized as follows: in Section 2 we propose the fixed–grid finite element discretization scheme and we discuss its convergence rate, in Section 3 we describe an algebraic multigrid method under consideration and in Section 4 we provide numerical results.

2   Fixed–Grid Finite Element Method

Let $\Omega \subset \mathbb{R}^2$ be a polygonal domain and let $\overline{\Omega} = \overline{\Omega_0(\alpha)} \cup \overline{\Omega_1(\alpha)}$ be its distinct decomposition that is controlled by a piecewise smooth function α such that graph$(\alpha) \subset \partial\Omega_0(\alpha) \cap \partial\Omega_1(\alpha)$. The smoothness improves regularity of the state solution and consequently convergence rate of the method as we will see later in this section. We consider the problem (3). Denote by $h > 0$ a discretization parameter and let $\mathcal{T}_h := \{T_i : i = 1, 2, \ldots, m_h\}$ be a shape regular triangulation of Ω that does not take care of α. We approximate V by the following linear Lagrange finite element subspace of V:

$$V_h := \left\{ v_h(x) \in C(\overline{\Omega}) \mid \forall T_i \in \mathcal{T}_h : v_h|_{T_i} \in P^1(T_i) \text{ and } \forall x \in \Gamma_D : v_h(x) = 0 \right\},$$

where $C(\overline{\Omega})$ denotes the space of functions continuous on $\overline{\Omega}$ and $P^1(T)$ denotes the space of linear polynomials over a triangle T. Let us further assume that the source term $J(x)$ is element-wise constant and that the discretization $\mathcal{T}_h$ follows the jumps of J, i.e. $J(x) = J_i$ on each T_i. The linear form of (3) is thus approximated in a conforming way as follows:

$$b_h(v_h) \equiv b(v_h) := \int_{\Omega} J(x)v_h(x)\,dx = \sum_{T_i \in \mathcal{T}_h} J_i \int_{T_i} v_h(x)\,dx, \quad v_h \in V_h.$$

However, our discretization does not respect the jumps of the coefficient function $\nu(\alpha)(x)$, which leads to a non–conforming discretization of the bilinear form. Let the triangulation be decomposed as follows, see also Fig. 1 (a):

$$\mathcal{T}_h = \mathcal{T}_{h,0}(\alpha) \cup \mathcal{B}_h(\alpha) \cup \mathcal{T}_{h,1}(\alpha),$$

where for $d = 0, 1$ we define $\mathcal{T}_{h,d}(\alpha) := \{T_i \in \mathcal{T}_h \cap \Omega_d(\alpha) \mid T_i \cap \text{graph}(\alpha) = \emptyset\}$ and where $\mathcal{B}_h(\alpha) := \{T_i \in \mathcal{T}_h \mid T_i \cap \text{graph}(\alpha) \neq \emptyset\}$. Then the discretized bilinear form is evaluated as follows:

$$a_h(\alpha)(u_h, v_h) \equiv a(\alpha)(u_h, v_h) := \int_{\Omega} \nu(\alpha)(x)\nabla u_h(x)\nabla v_h(x)\,dx$$

$$= \sum_{T_i \in \mathcal{T}_{h,0}(\alpha)} \nu_0 \int_{T_i} \nabla u_h(x)\nabla v_h(x)\,dx + \sum_{T_i \in \mathcal{T}_{h,1}(\alpha)} \nu_1 \int_{T_i} \nabla u_h(x)\nabla v_h(x)\,dx$$

$$+ \sum_{T_i \in \mathcal{B}_h(\alpha)} \frac{\nu_0|T_i \cap \Omega_0(\alpha)| + \nu_1|T_i \cap \Omega_1(\alpha)|}{|T_i|} \int_{T_i} \nabla u_h(x)\nabla v_h(x)\,dx, \quad u_h, v_h \in V_h,$$

where $|D|$ denotes the area of D.

2.1 Convergence Rate

The approximation estimate is given by Céa's lemma:

$$\|u - u_h\|_V \leq C \min_{v_h \in V_h} \|u - v_h\|_V,$$

where $C > 0$ is a generic constant (continuity over ellipticity of the bilinear form) which is independent of h in case of shape regular discretizations. Let $\Pi_h : V \rightarrow$

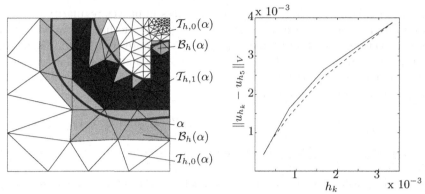

Fig. 1. (a) Decomposition of the discretization controlled by the shape α – only the three inner non-straight curves are controlled, their end points are connected by straight lines; (b) Convergence curve for the fixed–grid approach (solid line) and for the conforming approach (dashed line) computed on 5 levels, where $\|u_{h_k} - u\|_V \approx \|u_{h_k} - u_{h_5}\|_V$ for $k = 0, 1, \ldots, 4$

V_h denote the finite element interpolation operator, e.g. of the Clément–type, and let us choose $v_h := \Pi_h u$. Then, we arrive at the following:

$$\|u - \Pi_h u\|_V^2 \leq \sum_{\substack{T_i \in \mathcal{T}_{h,0}(\alpha) \\ T_i \cap \mathrm{supp}(J) = \emptyset}} \|u - \Pi_h u\|_{H^1(T_i)}^2 + \sum_{\substack{T_i \in \mathcal{T}_{h,0}(\alpha) \\ T_i \subset \mathrm{supp}(J)}} \|u - \Pi_h u\|_{H^1(T_i)}^2$$

$$+ \sum_{T_i \in \mathcal{T}_{h,1}(\alpha)} \|u - \Pi_h u\|_{H^1(T_i)}^2 + \sum_{T_i \in \mathcal{B}_h(\alpha)} \|u - \Pi_h u\|_{H^1(T_i)}^2.$$

Since our discretization respects the jumps of $J(x)$ and does not respect the jumps of $\nu(\alpha)(x)$, the solution u is regular everywhere except for $T_i \in \mathcal{B}_h(\alpha)$. Combining the previous estimates and the standard regularity argument implies that

$$\|u - u_h\|_V \leq C'h + C''h^{-1} \max_{T_i \in \mathcal{B}_h(\alpha)} \|u - \Pi_h u\|_{H^1(T_i)},$$

where the factor h^{-1} is related to the number of elements in $\mathcal{B}_h(\alpha)$. Therefore, the rate of convergence depends on the order of regularity of u across the coefficient jump interface $\Gamma(\alpha)$. Recall that for our shape optimization purposes $\Gamma(\alpha)$ is a smooth curve.

The convergence rate remains an open question. In order to indicate it, we refer to Fig. 1 (b), where we compare the convergence curve for the case of conforming discretization (respecting the coefficient jump) to the fixed–grid case. We can see that both the curves slightly deteriorate from the linear convergence, but the conforming discretization does not improve much. We used uniform refinement of the grid in Fig. 1 (a), where for levels $k = 0, 1, \ldots, 5$ the number of elements/number of nodes are respectively $317/186$, $1268/688$, $5072/2643$, $20288/10357$, $81152/41001$ and $324608/163153$ and where the corresponding discretizatization parameter is $h_k \approx 0.0033/2^k$.

3   Algebraic Multigrid

We are solving the discretized state problem using an algebraic multigrid method that agrees with classical AMG [13], except for the coarse-grid selection and the interpolation component, which are controlled by edge matrices in case of our approach, see [11]. Note that a novelty here is an application to the fixed–grid shape optimization.

One can also view this as involving an auxiliary problem–the one determined by the edge matrices–in the coarsening process. The coarse-grid matrices, however, are still computed via the usual Galerkin triple matrix product, i.e., $A_{k+1} = P_k^T A_k P_k$ at all levels $k = 0, 1, \ldots, \ell - 1$.

The basic idea is to construct suitable small-sized computational molecules from edge matrices and to choose the interpolation coefficients in such a way that they provide a local minimum energy extension with respect to the considered interpolation molecule. Assuming that "weak" and "strong" edges have been identified, the coarse grid has been selected, and a set of edge matrices is available, one defines a so-called interpolation molecule for every f-node i (to which interpolation is desired), cf. [11]:

$$M(i) := \sum_{k \in \mathcal{S}_i^c} E_{ik} + \sum_{j \in \mathcal{N}_i^f : \exists k \in \mathcal{S}_i^c \cap \mathcal{N}_j} E_{ij} + \sum_{k \in \mathcal{S}_i^c \cap \mathcal{N}_j : j \in \mathcal{N}_i^f} E_{jk}, \qquad (5)$$

where the following symbols respectively denote $\mathcal{D}_f$ fine nodes (f-nodes), $\mathcal{D}_c$ coarse nodes (c-nodes), $\mathcal{D} := \mathcal{D}_f \cup \mathcal{D}_c$ all nodes, $\mathcal{N}_i$ direct neighbors of node i, $\mathcal{N}_i^f := \mathcal{N}_i \cap \mathcal{D}_f$ fine direct neighbors, $\mathcal{S}_i$ strongly connected direct neighbors of node i and $\mathcal{S}_i^c := \mathcal{S}_i \cap \mathcal{D}_c$ strongly connected coarse direct neighbors. This molecule arises from assembling all edge matrices E_{pq} associated with three types of edges: The first sum corresponds to the strong edges connecting node i to some coarse direct neighbor k (interpolatory edges). The second sum represents edges connecting the considered f-node i to any of its fine direct neighbors j being directly connected to at least one c-node k that is strongly connected to node i. Finally, the last sum in (5) corresponds to these latter mentioned connections (edges) between fine direct neighbors j and strongly connected coarse direct neighbors k of node i.

The interpolation molecule (5) then serves for the computation of the actual interpolation weights: For a given f-node i let

$$M(i) = M = \begin{pmatrix} M_{ff} & M_{fc} \\ M_{cf} & M_{cc} \end{pmatrix} \qquad (6)$$

be the interpolation molecule where the 2×2 block structure in (6) corresponds to the n_M^f f-nodes and the n_M^c c-nodes the molecule is based on. Consider now the small-sized (local) interpolation matrix

$$P_M = P = \begin{pmatrix} P_{fc} \\ I_{cc} \end{pmatrix}$$

associated with (6). Since M (for the problems under consideration) is symmetric and positive semidefinite (SPSD) we may apply the following concept [2]: For any vector $\mathbf{e}^T = (\mathbf{e}_f^T, \mathbf{e}_c^T) \perp \ker(M)$ we denote by

$$\mathbf{d}_f := \mathbf{e}_f - P_{fc}\mathbf{e}_c \tag{7}$$

the defect of (local) interpolation. With the objective of an energy minimizing coarse basis we choose P_{fc} to be the argument that minimizes

$$\max_{\mathbf{e} \perp \ker(M)} \frac{(\mathbf{e}_f - P_{fc}\mathbf{e}_c)^T(\mathbf{e}_f - P_{fc}\mathbf{e}_c)}{\mathbf{e}^T M \mathbf{e}}.$$

Using (7) and $G := P_{fc}^T M_{ff} P_{fc} + P_{fc}^T M_{fc} + M_{cf} P_{fc} + M_{cc}$ one finds

$$\min_{P_{fc}} \max_{\mathbf{d}_f, \mathbf{e}_c} \frac{\mathbf{d}_f^T \mathbf{d}_f}{\begin{pmatrix} \mathbf{d}_f + P_{fc}\mathbf{e}_c \\ \mathbf{e}_c \end{pmatrix}^T \begin{pmatrix} M_{ff} & M_{fc} \\ M_{cf} & M_{cc} \end{pmatrix} \begin{pmatrix} \mathbf{d}_f + P_{fc}\mathbf{e}_c \\ \mathbf{e}_c \end{pmatrix}}$$

$$= \min_{P_{fc}} \max_{\mathbf{d}_f} \frac{\mathbf{d}_f^T \mathbf{d}_f}{\mathbf{d}_f^T \left[M_{ff} - (M_{ff}P_{fc} + M_{fc})G^{-1}(P_{fc}^T M_{ff} + M_{cf}) \right] \mathbf{d}_f}. \tag{8}$$

Assuming that M_{ff} and G both are SPD the denominator of (8) for an arbitrary vector $\mathbf{d}_f$ is maximized and thus the minimum is attained for

$$P_{fc} := -M_{ff}^{-1} M_{fc}, \tag{9}$$

which results in $1/(\lambda_{\min}(M_{ff}))$. This motivates to choose the interpolation coefficients for node i to equal the entries in the corresponding row of (9).

4   Numerical Results

We consider a problem of optimal shape design of pole heads of a direct current (DC) electromagnet, which is depicted in Fig. 2 (a), while we simplify the geometry so that only two opposite pole heads and coils are present. The goal is to achieve homogeneous magnetic field in a small square Ω_m in the middle among the pole heads, which is evaluated by the following objective functional:

$$\mathcal{I}(u) := \frac{1}{2|\Omega_m|} \int_{\Omega_m} \|\mathrm{curl}(u(x)) - B^{\mathrm{avg}}(u(x))\mathbf{n}_m\|^2 \, dx + \frac{\varepsilon_u}{2|\Omega|} \int_{\Omega} \|\nabla u\|^2 \, dx,$$

where $\mathrm{curl}(u) := (\partial u/\partial x_2, -\partial u/\partial x_1)$ is the magnetic flux density, $\Omega := (-0.2, 0)^2$ (in meters), $\Omega_m := (-0.01, 0)^2$, $\varepsilon_u := 10^{-3}$ introduces a regularization in $H^1(\Omega)$ and where

$$B^{\mathrm{avg}}(u(x)) := \frac{1}{|\Omega_m|} \int_{\Omega_m} \mathrm{curl}(u(x))\mathbf{n}_m \, dx.$$

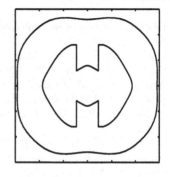

Fig. 2. (a) Original electromagnet; (b) Optimized design

An admissible shape α consists of 3 Bézier curves that are the non–straight curves depicted in Fig. 1 (a). For them we consider 7, 4, and 7 design control parameters (18 in total), respectively, and we further introduce two other shapes α_l and α_u, which are again triples of such Bézier curves, that form box constraints for the set of admissible shapes $\mathcal{U}_{\mathrm{ad}}$. This optimization is subjected to the state equation (3), which we denote by $u(\alpha)$.

The optimization algorithm includes a steepest descent method, a projection to the box constraints and a numerical differentiation for calculation of $\nabla_\alpha \mathcal{I}(u(\alpha)(x))$. The optimized design is depicted in Fig. 2 (b). For the discretization of the state equation we used the finite element fixed–grid approach of Section 2 and the AMG method of Section 3, which was accelerated by the preconditioned conjugate gradients method(PCG). We set up the AMG preconditioner only for the first system at each discretization level and used this setup at the level as a preconditioner for all the other forthcoming systems, which are perturbed by different shapes via the fixed–grid approach. Both the PCG and optimization relative precision were 10^{-8}. The preliminary numerical results are presented in Table 1. The AMG preconditioner certainly deteriorates for perturbed systems, because it is not re–setup, which is a cheap operation that we will use in the next version. However, we could by far not achieve such moderate decay of AMG for example as in the case of geometric multigrid solver, where the iterations grows easily up to hundreds.

Table 1. Numerical results

discretization level	number of elements	number of nodes	number of AMG levels	PCG iterations	optimization iterations
0	317	186	2	6	4
1	9649	4985	4	10–15	5
2	34658	17629	6	15–43	2
3	134292	67721	8	29–61	4

References

1. Brandt, A.: Algebraic multigrid theory: the symmetric case. Appl. Math. Comput 19, 23–56 (1986)
2. Brezina, M., et al.: Algebraic multigrid based on element interpolation (AMGe). SIAM J. Sci. Comput 22, 1570–1592 (2000)
3. Arnold, D., et al.: Unified analysis of discontinuous Galerkin methods for elliptic problems. SIAM J. Numer. Anal 39, 1749–1779 (2002)
4. Borzi, A.: Multigrid Methods for Optimality Systems, habilitation thesis, TU Graz (2003)
5. Bramble, J.H., Pasciak, J.E., Vassilev, A.T.: Analysis of the inexact Uzawa algorithm for saddle point problems. SIAM J. Numer. Anal 34, 1072–1092 (1997)
6. Dreyer, T., Maar, B., Schulz, V.: Multigrid optimization in applications. J. Comp. Appl. Math 120, 67–84 (2000)
7. Haslinger, J., Neittaanmäki, P.: Finite Element Approximation for Optimal Shape, Material and Topology Design. Wiley, Chinchester (1997)
8. Haslinger, J., Klarbring, A.: Fictitious domain/mixed finite element approach for a class of optimal shape design problems, RAIRO. Modélisation Math. Anal. Numér 29, 435–450 (1995)
9. Lukáš, D., Dostál, Z.: Optimal multigrid preconditioned semi–monotonic augmented Lagrangians Applied to the Stokes Problem. Num. Lin. Alg. Appl. (submitted)
10. Kraus, J.K.: On the utilization of edge matrices in algebraic multigrid. In: Lirkov, I., Margenov, S., Waśniewski, J. (eds.) LSSC 2005. LNCS, vol. 3743, pp. 121–129. Springer, Heidelberg (2006)
11. Kraus, J.K., Schicho, J.: Algebraic multigrid based on computational molecules, 1: Scalar elliptic problems. Computing 77, 57–75 (2006)
12. Rivière, B., Bastian, P.: Discontinuous Galerkin methods for two–phase flow in porous media, Research report of Interdisciplinary Center for Scientific Computing, University Heidelberg (2004)
13. Ruge, J.W., Stüben, K.: Efficient solution of finite difference and finite element equations by algebraic multigrid (AMG). In: Paddon, D.J., Holstein, H. (eds.) Multigrid Methods for Integral and Differential Equations. The Institute of Mathematics and Its Applications Conference Series, pp. 169–212. Clarendon Press, Oxford (1985)
14. Schöberl, J., Zulehner, W.: Symmetric indefinite preconditioners for saddle point problems with applications to PDE–constrained optimization problems, SFB Report 2006–19, University Linz (2006)

The Effect of a Minimum Angle Condition on the Preconditioning of the Pivot Block Arising from 2-Level-Splittings of Crouzeix-Raviart FE-Spaces

Josef Synka

Industrial Mathematics Institute, Johannes Kepler University,
Altenberger Str. 69, A-4040 Linz, Austria
josef.synka@jku.at

Abstract. The construction of efficient two- and multilevel preconditioners for linear systems arising from the finite element discretization
of self-adjoint second order elliptic problems is known to be governed by
robust hierarchical splittings of finite element spaces. In this study we
consider such splittings of spaces related to nonconforming discretizations using Crouzeix-Raviart linear elements: We discuss the standard
method based on differences and aggregates, a more general splitting and
the first reduce method which is equivalent to a locally optimal splitting.
All three splittings are shown to fit a general framework of differences
and aggregates. Further, we show that the bounds for the spectral condition numbers related to the additive and multiplicative preconditioners of
the coarse grid complement block of the hierarchical stiffness matrix for
the three splittings can be significantly improved subject to a minimum
angle condition.

Keywords: Multilevel preconditioning, hierarchical splittings, CBS constant, differences and aggregates, first reduce, anisotropy, nonconforming
elements.

1 Introduction

The discrete weak formulation of the self-adjoint elliptic boundary value prob-
lems, as considered in this study, reads as follows: Given $f \in L^2(\Omega)$, find $u_h \in \mathcal{V}_h$
such that $\mathcal{A}_h(u_h, v_h) = (f, v_h) \ \forall v_h \in \mathcal{V}_h$ is satisfied, where

$$\mathcal{A}_h(u_h, v_h) := \sum_{e \in \mathcal{T}_h} \int_e a(e) \, \nabla u_h(\mathbf{x}) \cdot \nabla v_h(\mathbf{x}) \, d\mathbf{x}, \quad (f, v_h) := \int_e f \, v_h \, d\mathbf{x}, \quad (1)$$

and $\mathcal{V}_h := \{v \in L^2(\Omega) : v|_e$ is linear on each $e \in \mathcal{T}_h$, v is continuous at
the midpoints of the edges of triangles from $\mathcal{T}_h$ and $v = 0$ at the
midpoints on $\Gamma_D\}$.

Thereby, $\Omega \subset \mathbb{R}^2$ denotes a polygonal domain and $f(\mathbf{x})$ is a given function in
$L^2(\Omega)$. The matrix $a(\mathbf{x}) := (a_{ij}(\mathbf{x}))_{i,j \in \{1,2\}}$ is assumed to be bounded, symmetric and uniformly positive definite (SPD) on Ω with piecewise smooth functions

I. Lirkov, S. Margenov, and J. Waśniewski (Eds.): LSSC 2007, LNCS 4818, pp. 105–112, 2008.

$a_{ij}(\mathbf{x})$ in $\overline{\Omega} := \Omega \cup \partial\Omega$, and $\mathbf{n}$ represents the outward unit normal vector onto the boundary $\Gamma := \partial\Omega$ with $\Gamma = \overline{\Gamma}_D \cup \overline{\Gamma}_N$. The domain Ω is assumed to be discretized using triangular elements and that the fine-grid partitioning, denoted by $\mathcal{T}_h$, is obtained by a uniform refinement of a given coarser triangulation $\mathcal{T}_H$. If the coefficient functions $a_{ij}(\mathbf{x})$ are discontinuous along some polygonal interfaces, we assume that the partitioning $\mathcal{T}_H$ is aligned with these lines to ensure that $a(\mathbf{x})$ is sufficiently smooth over each element $E \in \mathcal{T}_H$.

The construction of multilevel hierarchical preconditioner is based on a two-level framework for piecewise linear Crouzeix-Raviart finite elements (cf., e.g., [3, 4, 5]). The needed background for the estimates on the constant γ in the CBS inequality can be found in [1]. It is well-known that for hierarchical basis splitting of the conforming finite elements that under certain assumptions γ can be estimated locally by considering a single finite macro-element $E \in \mathcal{T}_H$, which means that $\gamma := \max_{E \in \mathcal{T}_H} \gamma_E$.

In Section 2 we construct a general framework for hierarchical two-level decompositions based on differences and aggregates. In Section 3 we show that the construction of efficient hierarchical basis functions (HB) multilevel preconditioners is independent of the splitting used. The bound of the spectral condition number for the additive and multiplicative preconditioner for the coarse grid complements is generalized in Section 4 under the assumption of a minimum angle condition. The study is concluded in Section 5 with a brief summary of the main results.

2   Hierarchical Two-Level Decompositions

For the derivation of estimates for the CBS-constant, it is known, that it suffices to consider an isotropic (Laplacian) problem in an arbitrarily shaped triangle T. Let us denote the angles in such a triangle, as illustrated in Fig. 1, by θ_1,

θ_2 and $\theta_3 := \pi - (\theta_1 + \theta_2)$. Without loss of generality we assume that these angles satisfy the ordering $\theta_1 \geq \theta_2 \geq \theta_3$., Then, with $a := \cot\theta_1$, $b := \cot\theta_2$ and $c := \cot\theta_3$, the following relations hold in a triangle (cf. [2]):

$$|a| \leq b \leq c \qquad \text{and} \qquad ab + ac + bc = 1. \qquad (2)$$

A simple computation shows that the standard nodal basis element stiffness matrix for a non-conforming Crouzeix-Raviart (CR) linear finite element A_e^{CR} coincides with that for the conforming (c) linear element A_e^c, up to a factor 4 and can be written as

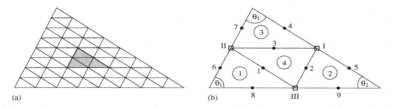

(a)   (b)

Fig. 1. Crouzeix-Raviart finite element (a) Discretization (b) Macro-element in detail

$$A_e^{CR} = 2 \begin{pmatrix} b+c & -c & -b \\ -c & a+c & -a \\ -b & -a & a+b \end{pmatrix}. \tag{3}$$

The hierarchical stiffness matrix at macro-element level is then obtained by assembling four such matrices according to the numbering of the nodal points, as shown in Fig. 1(b). Note that for the non-conforming Crouzeix-Raviart finite element, where the nodal basis functions correspond to the midpoints along the edges of the triangle rather that at its vertices (cf. Fig. 1), the natural vector spaces $\mathcal{V}_H := \text{span}\{\phi_I, \phi_{II}, \phi_{III}\}$ and $\mathcal{V}_h := \text{span}\{\phi_i\}_{i=1}^9$, associated with the basis functions at the coarse and fine mesh, respectively (cf. Fig.1(b)), are no longer nested. Instead of a direct construction with $\mathcal{V}_2 := \mathcal{V}_H$, as feasible for conforming elements, one now has to choose the hierarchical basis functions such that the vector space can be written as a direct sum of the resulting vector subspaces $\mathcal{V}_1$ and $\mathcal{V}_2$. At macro-element level this requires, $\mathcal{V}(E):=\text{span}\{\Phi_E\} = \mathcal{V}_1(E) \oplus \mathcal{V}_2(E)$, where $\Phi_E := \{\phi_E^{(i)}\}_{i=1}^9$ is the set of "midpoint" basis functions of the four congruent elements in the macro-element E (cf. Fig. 1(b)).

In this study we consider three different splittings based on half-difference and half-sum (aggregates) basis functions to ensure the direct sum condition. The splitting of $\mathcal{V}(E)$ can be defined in the general form as

$$\mathcal{V}_1(E) := \text{span}\left\{\phi_1, \phi_2, \phi_3, \phi_1^D + \phi_4 - \phi_5, \phi_2^D + \phi_6 - \phi_7, \phi_3^D + \phi_8 - \phi_9\right\},$$
$$\mathcal{V}_2(E) := \text{span}\left\{\phi_1^C + \phi_4 + \phi_5, \phi_2^C + \phi_6 + \phi_7, \phi_3^C + \phi_8 + \phi_9\right\}, \tag{4}$$

where $\phi_i^D := \sum_k d_{ik}\phi_k$ and $\phi_i^C := \sum_k c_{ik}\phi_k$ with $i, k \in \{1, 2, 3\}$. The transformation matrix corresponding to this general splitting is given by

$$J_E = J_E(C, D) = \begin{bmatrix} I_3 & D & C \\ 0 & J_- & J_+ \end{bmatrix} \quad (\in \mathbb{R}_{9\times9}), \tag{5}$$

where I_3 denotes the 3×3 identity matrix and the matrices involving the half-difference and the half-sum basis functions in the transformation read as

$$J_- := \frac{1}{2} \begin{bmatrix} 1 & -1 & & \\ & 1 & -1 & \\ & & 1 & -1 \end{bmatrix}^T \quad \text{and} \quad J_+ := \frac{1}{2} \begin{bmatrix} 1 & 1 & & \\ & 1 & 1 & \\ & & 1 & 1 \end{bmatrix}^T. \tag{6}$$

D and C are 3×3 matrices whose entries d_{ij} respectively c_{ij} will be specified later. The matrix J_E transforms the vector of the macro-element basis functions $\phi_E := (\phi^{(i)})_{i=1}^9$ to the hierarchical basis vector $\tilde{\phi}_E := (\tilde{\phi}^{(i)})_{i=1}^9 = J_E^T \phi_E$ and the hierarchical stiffness matrix at macro-element level is obtained as

$$\tilde{A}_E = J_E^T A_E J_E = \begin{bmatrix} \tilde{A}_{E,11} & \tilde{A}_{E,12} \\ \tilde{A}_{E,12}^T & \tilde{A}_{E,22} \end{bmatrix} \begin{matrix} \} \in \mathcal{V}_1(E) \\ \} \in \mathcal{V}_2(E) \end{matrix}. \tag{7}$$

The related global stiffness matrix is obtained as $\tilde{A}_h := \sum_{E\in\mathcal{T}_H} \tilde{A}_E$.

Following the local definitions, we similarly construct the hierarchical basis $\tilde{\phi} := \{\tilde{\phi}^{(i)}\}_{i=1}^{N_h}$ for the whole finite element space $\mathcal{V}_h$ with its standard nodal finite element basis given by $\phi := \{\phi^{(i)}\}_{i=1}^{N_h}$. The corresponding splitting for the whole space then reads $\mathcal{V}_h = \mathcal{V}_1 \oplus \mathcal{V}_2$.

The transformation matrix $J = J(C, D)$ such that $\tilde{\phi} = J^T \phi$ is then used for the transformation of the global matrix A_h to its hierarchical form $\tilde{A}_h = J^T A_h J$, and (by a proper permutation of rows and columns) the latter admits the 3×3-block representation

$$
\tilde{A}_h = \begin{bmatrix} \tilde{A}_{11} & \tilde{A}_{12} & \tilde{A}_{13} \\ \tilde{A}_{12}^T & \tilde{A}_{22} & \tilde{A}_{23} \\ \tilde{A}_{13}^T & \tilde{A}_{23}^T & \tilde{A}_{33} \end{bmatrix} \begin{array}{l} \left.\vphantom{\begin{matrix}a\\a\end{matrix}}\right\} \in \mathcal{V}_1 \\[1.2em] \left.\vphantom{a}\right\} \in \mathcal{V}_2 \end{array} \tag{8}
$$

according to the interior, half-difference and half-sum basis functions, which are associated with the locally introduced splitting in (4). The upper-left 2×2 block is thus related to the vector space $\mathcal{V}_1$, while the lower-right block $\tilde{A}_{33}$ relates to $\mathcal{V}_2$. Note that due to the structure of J_E, the relation $\tilde{A}_{11} = A_{11}$ holds.

We consider three different splittings, which can be derived from (5) for different settings of the matrices C and D: Using $D = 0$ and $C = \frac{1}{2} \operatorname{diag}(1, 1, 1) = \frac{1}{2} I$ yields the standard splitting by differences and aggregates (DA) (cf. Blaheta et al. [5]), while keeping $D = 0$ and setting $C = \frac{1}{2} I + \mu(\mathbb{1} - 3I)$ with $\mu \in [0, \frac{1}{4}]$ properly chosen results in the generalized DA-splitting (GDA), as considered in Margenov and Synka, [8][1]. Note that the GDA splitting depends on a minimum angle condition commonly used in the triangular mesh (i.e. all angles must be larger than θ_{min}) and that for $\mu = 0$ the DA-approach is retrieved.

The "First Reduce" (FR) or optimized DA (ODA) splitting is obtained by setting $D = -A_{11}^{-1} \tilde{A}_{12}$ and $C = -A_{11}^{-1} \tilde{A}_{13}$, which is known as the harmonic extension (cf. Kraus, Margenov and Synka [7]). Note that each of the above basis transformations is induced by a local transformation matrix J_E at macroelement level, i.e., $J\big|_E = J_E \ \forall E \in \mathcal{T}_H$ with $J \in \{J_{DA}, J_{GDA}, J_{FR}\}$.

3   Construction of Preconditioners for the Coarse Grid Complements for the Three Splittings Considered

For the construction of multilevel hierarchical preconditioners for the non-conforming finite element spaces arising from a discretization based on Crouzeix-Raviart linear elements, we follow the ideas of Blaheta et al. [6]. The multilevel preconditioner is obtained by recursive application of a two-level preconditioner M for A_h. Thereby, it is crucial that the spectral condition number $\kappa(M^{-1} A_h)$ of the preconditioned matrix $M^{-1} A_h$ is uniformly bounded with respect to the discretization parameter h, the shape of triangular finite elements and arbitrary coefficient anisotropy. To build a hierarchical preconditioner, we need a suitable decomposition of the form $\mathcal{V}_h = \mathcal{V}_1 \oplus \mathcal{V}_2$, which is satisfied by all three splittings based on differences and aggregates, as discussed at the end of Section 2.

---

[1] Here $\mathbb{1}$ denotes the 3×3 matrix of all ones.

We introduce the transformation matrices $J_\pm$ and J_S, defined as

$$J_\pm := \begin{bmatrix} I_3 & 0 & 0 \\ 0 & J_- & J_+ \end{bmatrix} \quad \text{and} \quad J_S = \begin{bmatrix} I_3 & D_S & C_S \\ 0 & I & 0 \\ 0 & 0 & I \end{bmatrix}, \tag{9}$$

respectively, where I_3 designates the 3×3 identity matrix, J_- and J_+ are defined by (6), and the matrices D_S and C_S are to be set to D and C of the corresponding splitting, as indicated by the index $S \in \{DA, GDA, FR\}$ (see Section 2). With these definitions, we can now write all three splittings in the form

$$J_{E;S} = J_\pm J_S. \tag{10}$$

Note that with $\bar{A}_h := J_\pm^T A_h J_\pm$ it can be seen easily that $J_{E;FR}$ also contains the reduction step to eliminate the interior nodes in the first reduce splitting since D_{FR} and C_{FR} simply provide the harmonic extensions. For the DA- and GDA-splitting the elimination of the interior nodes, i.e., of the block $\bar{A}_{11} = A_{11}$, which is block-diagonal, has to be performed separately. This elimination can be done exactly and locally. If the preconditioner is constructed in this way (cf. [6]), then for all three splittings we finally obtain the block-structure

$$\begin{bmatrix} A_{11} & 0 \\ 0 & B \end{bmatrix} \quad \text{with} \quad B = \begin{bmatrix} B_{11} & B_{12} \\ B_{21} & B_{22} \end{bmatrix}. \tag{11}$$

Remark 1. For such a construction of the preconditioner, we then have $C_S = -A_{11}^{-1}\bar{A}_{12}$ in $J_{E;S}$ for any $S \in \{DA, GDA, FR\}$, yielding the same 2×2 left upper block in the hierarchical representation, as obtained for the FR-splitting.

Lemma 1. *Under the above assumptions the Schur complement, denoted by B, is invariant with respect to the matrices C_S and D_S, which appear in the transformation matrix $J_{E;S}$.*

Proof. Follows from direct calculation.

Remark 2. From these observations it becomes evident that it is sufficient to restrict to only one splitting for the construction of preconditioners for the coarse grid complement block based on an exact inversion of the A_{11}-block. The uniform upper bound of the condition number, as reported in [6] for the DA-decomposition for arbitrary meshes, thus equally apply to the GDA- and FR-splitting.

4 Improved Upper Bounds of the Spectral Condition Number for Optimal Preconditioners for the Block B_{11} Based on a Minimum Angle Condition

The matrix B consists of 6×6 macroelement contributions B_E, which can be written in a 2×2-block structure as its global counterpart in (11) by splitting it according to the two-level semi-difference and semi-sum basis functions. At macroelement level the upper left block $B_{11,E}$ is found explicitly as

$$B_{11,E} = \frac{1}{r} \begin{bmatrix} 3 + 2(b^2 + bc + c^2) & 1 + 2c^2 & -(1 + 2b^2) \\ 1 + 2c^2 & 3 + 2(a^2 + ac + c^2) & -(1 + 2a^2) \\ -(1 + 2b^2) & -(1 + 2a^2) & 3 + 2(a^2 + ab + b^2) \end{bmatrix} \quad (12)$$

with $r := 3(a+b+c) - 2abc$. It can be observed easily that $B_{11,E}$ is symmetric and all off-diagonal entries are negative except for the (1,2)-entry. Using the relations in (2) one immediately obtains $|B_{11,E}(2,3)| \leq |B_{11,E}(1,3)| \leq B_{11,E}(1,2)$ and that $|B_{11,E}(1,3)| = B_{11,E}(1,2)$ holds only for $b = c$ (isosceles triangle). Hence the strongest off-diagonal coupling of the macro-element stiffness matrix is given by the (1-2) coupling as it is for the element stiffness matrix (cf. (3) and (2)).

A preconditioner of $B_{11,E}$ of additive type, $C_{11,E}$, is now derived by setting the weakest couplings in $B_{11,E}$ (here negative off-diagonal entries) to zero. C_{11} is then obtained by assembling the modified matrices $C_{11,E}$. In Blaheta et al., [6], it was shown that $B_{11,E}$ and $C_{11,E}$ are spectrally equivalent and that the spectral condition number $\kappa(C_{11,E}^{-1} B_{11,E})$ is uniformly bounded by $(1 + \sqrt{\bar{\mu}})/(1 - \sqrt{\bar{\mu}})$ with $\bar{\mu} = 7/15$. Since C_{11} inherits the properties of $C_{11,E}$ (cf. [6]), this result can be extended to the assembled global preconditioner C_{11}.

Under the assumption of a minimum angle condition, which is commonly used in mesh generators, viz. that all angles in the triangle are greater than or equal to a minimum angle, we will now show that these results can be generalized and that the upper bound of $7/15$ is too pessimistic in practical applications. Without loss of generality we now use the relation

$$\theta_1 \geq \theta_2 \geq \theta_3 \geq \theta_{min}, \quad (13)$$

which renders the relations in (2) to $|a| \leq b \leq c \leq \cot(\theta_{min})$ and $ab + ac + bc = 1$.

The spectral condition number $\kappa(C_{11,E}^{-1} B_{11,E})$ is obtained by considering the generalized eigenvalue problem $B_{E,11}\mathbf{v} = \lambda C_{E,11}\mathbf{v}$ and its corresponding characteristic equation $\det(B_{E,11} - \lambda C_{E,11}) = 0$, where $B_{E,11} - \lambda C_{E,11}$ reads as

$$\begin{bmatrix} \mu[3 + 2(b^2 + bc + c^2)] & \mu(1 + 2c^2) & -(1 + 2b^2) \\ \mu(1 + 2c^2) & \mu[3 + 2(a^2 + ac + c^2)] & -(1 + 2a^2) \\ -(1 + 2b^2) & -(1 + 2a^2) & \mu[3 + 2(a^2 + ab + b^2)] \end{bmatrix} \quad (14)$$

with $\mu = 1 - \lambda$. The shifted eigenvalues $\mu_i := 1 - \lambda_i$, $i = 1, 2, 3$ are easily obtained as $\mu_1 = 0$ and $\mu_{2,3} = \sqrt{\mu(a,b)}$ with

$$\mu(a,b) = \frac{(a+b)^2 [1 + 2(a^2 - ab + b^2)]}{(2 + a^2 + b^2)[3 + 2(a^2 + ab + b^2)]}. \quad (15)$$

Lemma 2. *The function μ, as obtained in the analysis of the additive preconditioner to B_{11}, has the following properties:*

1. *For fixed $a := \cot\theta_1$, $\mu(a,.)$ is strictly monotonically increasing. Its maximum is thus attained at $\mu(\cot\theta_1, \tan\frac{\theta_1}{2}) = \dfrac{1 - 4s^2 + 10s^4}{1 + 8s^2 + 14s^4 - 8s^6} =: \bar{\mu}_{\theta_1}(s)$ with $s := \sin\frac{\theta_1}{2}$.*

2. Let $\theta_1 \in [\frac{\pi}{3}, \frac{\pi}{2}]$, then $\mu(a, b) \leq 1/5$.

3. Let $\theta_1 \in [\frac{\pi}{2}, \pi]$, then $\bar{\mu}_{\theta_1}$ is strictly increasing and bounded by $7/15$, which is thus a uniform bound for μ. Using the minimum angle condition yields $\theta_1 \in (\frac{\pi}{2}, \pi - 2\theta_{min}]$. The upper bound for μ then depends on θ_{min} and generalizes to

$$\bar{\mu}(\theta_{min}) = \frac{11 + 12\cos(2\theta_{min}) + 5\cos(4\theta_{min})}{31 + 29\cos(2\theta_{min}) + \cos(4\theta_{min}) - \cos(6\theta_{min})}, \tag{16}$$

which is less than $7/15$ for $\theta_{min} > 0$, but equal to $7/15$ for $\theta_{min} = 0$.

Proof. 1) With $x_0 := a + b$, $x_1 := 1 + a^2 + b^2 + (a - b)^2$, $x_2 := 3 + a^2 + b^2 + (a + b)^2$, and $x_3 := 2 + a^2 + b^2$, which are all strictly positive for non-degenerated triangles, we can rewrite Eq. (15) as $\mu^2(a, b) = \frac{x_0^2 x_1}{x_2 x_3}$. Since $\mu^2(a, .)$ is continuous and differentiable with regard to b, it is strictly monotonically increasing iff $\frac{\partial \mu^2(a,b)}{\partial b} > 0$: With the denominator being strictly positive this is equivalent to $x_1 x_3 (3 + a^2 - ab) x_0 x_2 [(b - a)(2 + a^2 + ab) + b] > 0$, which holds from (2). Fixing $a := \cot\theta_1$ with $\theta_1 \in [\pi/3, \pi - 2\theta_{min}]$, the angle θ_2 can only be in the range $[\pi/2 - \theta_1/2, \pi - \theta_1 - \theta_{min}]$ according to Relation (13) by taking $\theta_{min} \in [0, \pi/4]$, where the latter is a sufficiently large range for θ_{min} in practice. Consequently, $b \leq \tan\frac{\theta_1}{2}$. Using $s := \sin\frac{\theta_1}{2}$, $c := \cos\frac{\theta_1}{2}$ and the trigonometric relations between sine and cosine, the upper bound for $\mu_{2,3}^2(a, b)$ (with a fixed) is thus given by $\bar{\mu}_{\theta_1}(s)$.

2) Let $\theta_1 \in [\frac{\pi}{3}, \frac{\pi}{2}]$. We show that $\bar{\mu}_{\theta_1}(.)$, as defined in the Lemma, is bounded by $\bar{\mu}_{\theta_1}(1/\sqrt{2}) = 1/5$: Since the denominator of $\bar{\mu}_{\theta_1}$ is strictly positive for the given θ_1-range and $s \in [1/2, 1/\sqrt{2}]$, we have $\bar{\mu}_{\theta_1}(s) \leq 1/5 \Leftrightarrow 4(1 - 7s^2 + 9s^4 + 2s^6) \leq 0 \Leftrightarrow 8(s^2 - 1/2)(s^4 + 5s^2 - 1) \leq 0 \Leftrightarrow p(s) := s^4 + 5s^2 - 1 \geq 0$. Since $p(1/2) = 5/16$ and the polynomial p is clearly strictly increasing for the given s-range, this proves our assertion. (Note that for $\theta_1 = \pi/3$ we have $\bar{\mu}_{\theta_1}(1/2) = 1/6$, but the minimum of $\bar{\mu}_{\theta_1}$ is attained at $\theta_1^* = 68.0726$ degrees with $\bar{\mu}_{\theta_1^*}(\sin(\theta_1/2)) = 0.15716$.)

3) Let $\theta_1 \in [\frac{\pi}{2}, \pi - 2\theta_{min}]$. Then $s := \sin\frac{\theta_1}{2} \in [1/\sqrt{2}, \cos\theta_{min}] \subset [1/\sqrt{2}, 1]$ for all $\theta_{min} \in [0, \pi/4]$. To show that $\bar{\mu}_{\theta_1}$ is strictly increasing we show that $g(t) := \bar{\mu}_{\theta_1}(1/\sqrt{2} + t) > 0$ is strictly increasing for $t \in [0, \cos\theta_{min} - 1/\sqrt{2}]$: A simply computation shows that $g'(t) > 0 \Leftrightarrow 21\sqrt{2} + 330t + 800\sqrt{2}t^2 + 2144t^3 + 2200\sqrt{2}t^4 + 4016t^5 + 2912\sqrt{2}t^6 + 2752t^7 + 720\sqrt{2}t^8 + 160t^9 > 0$. Since all coefficients are positive this condition holds true for all values of t in the given range. Hence, $\bar{\mu}_{\theta_1}$ is strictly increasing and attains its maximum at $\theta_1 = \pi - 2\theta_{min}$ where $s = \cos\theta_{min}$. Inserting this expression for s in $\bar{\mu}_{\theta_1}$ gives the stated formula (16) for the upper bound of $\bar{\mu}_{\theta_1}$ and also of $\mu_{2,3}^2$ depending on the minimum angle in the mesh, as stated in the Lemma. This completes the proof. $\square$

Remark 3. The spectral condition number for the additive and multiplicative preconditioner, where the latter is discussed in full detail in [6], are respectively given by $\kappa_a = \frac{1+\sqrt{\bar{\mu}}}{1-\sqrt{\bar{\mu}}}$ and $\kappa_m = \frac{1}{1-\sqrt{\bar{\mu}}}$, where the upper bound $\bar{\mu} = \bar{\mu}(\theta_{min})$ is

Table 1. Upper bounds for μ, κ_{add} and κ_{mult} for different settings of θ_{min}

θ_{min} (deg.)	0	20	25	30	45
$\bar{\mu}$	$7/15 \simeq 0.4\dot{6}$	0.391	0.355	0.310	0.2
$\kappa_a(\bar{\mu})$	5.312	4.336	3.944	3.560	2.618
$\kappa_m(\bar{\mu})$	$15/8=1.875$	1.642	1.549	1.460	1.25

as defined in Lemma 2. For $\theta_1 \in [\frac{\pi}{3}, \frac{\pi}{2}]$ we obtain $\kappa_a \leq \frac{1+\sqrt{1/5}}{1-\sqrt{1/5}} \simeq 2.618$ and $\kappa_m \leq 5/4$. The effect of the minimum angle condition for $\theta_1 \in [\pi/2, \pi - 2\theta_{min}]$ is summarized in Table 1, where $\bar{\mu}$ is the upper bound of $\mu_{2,3}^2$ depending on θ_{min}. It can be seen easily that the upper bound is significantly improved for values of $\theta_{min} \geq 20$ degrees, as commonly used in mesh generators in practice.

5 Concluding Remarks

In this paper we studied the effect of a minimum angle condition on the spectral condition number for preconditioners of additive and multiplicative type. It was shown that the construction of optimal preconditioners based on an exact elimination of the interior nodes is the same for all three splittings based on aggregates and differences, as discussed in Section 2. The main result in this study is that the upper bound for the (spectral) condition number, as stated in [6], can be improved for practical applications under the assumption of a minimum angle condition.

References

1. Axelsson, O.: Iterative solution methods. Cambridge University Press, Cambridge (1994)
2. Axelsson, O., Margenov, S.: An optimal order multilevel preconditioner with respect to problem and discretization parameters. In: Minev, Wong, Lin (eds.) Advances in Computations, Theory and Practice, vol. 7, pp. 2–18. Nova Science, New York (2001)
3. Axelsson, O., Vassilevski, P.S.: Algebraic multilevel preconditioning methods, I. Numerische Mathematik 56, 157–177 (1989)
4. Axelsson, O., Vassilevski, P.S.: Algebraic multilevel preconditioning methods, II. SIAM Journal on Numerical Analalysis 27, 1569–1590 (1990)
5. Blaheta, R., Margenov, S., Neytcheva, M.: Uniform estimate of the constant in the strengthened CBS inequality for anisotropic non-conforming FEM systems. Numerical Linear Algebra with Applications 11, 309–326 (2004)
6. Blaheta, R., Margenov, S., Neytcheva, M.: Robust optimal multilevel preconditioners for non-conforming finite element systems. Numerical Linear Algebra with Applications 12, 495–514 (2005)
7. Kraus, J., Margenov, S., Synka, J.: On the multilevel precondtioning of Crouzeix-Raviart elliptic problems. Numerical Linear Algebra with Applications (to appear)
8. Margenov, S., Synka, J.: Generalized aggregation-based multilevel preconditioning of Crouzeix-Raviart FEM elliptic problems. In: Boyanov, T., et al. (eds.) NMA 2006. LNCS, vol. 4310, pp. 91–99. Springer, Heidelberg (2007)

Part III

Monte Carlo: Tools, Applications, Distributed Computing

Development of a 3D Parallel Finite Element Monte Carlo Simulator for Nano-MOSFETs

Manuel Aldegunde[1], Antonio J. García-Loureiro[1], and Karol Kalna[2]

[1] Departamento de Electrónica y Computación, Universidad de Santiago de Compostela, Campus Sur, 15782 Santiago de Compostela, Spain
manuelxx@usc.es, elgarcia@usc.es
[2] Device Modelling Group, Department of Electronics & Electrical Engineering, University of Glasgow, Glasgow G12 8LT, United Kingdom

Abstract. A parallel 3D Monte Carlo simulator for the modelling of electron transport in nano-MOSFETs using the Finite Element Method to solve Poisson equation is presented. The solver is parallelised using a domain decomposition strategy, whereas the MC is parallelised using an approach based on the distribution of the particles among processors. We have obtained a very good scalability thanks to the Finite Element solver, the most computationally intensive stage in self-consistent simulations. The parallel simulator has been tested by modelling the electron transport at equilibrium in a 4 nm gate length double gate MOSFET.

1 Introduction

Monte Carlo (MC) methods are widely used to simulate the behaviour of semiconductor devices. They become decisive to obtain proper on-state characteristics of the transistors since they can account for non-equilibrium effects which are neglected in simpler approaches as drift-diffusion. However, they are computationally very intensive and their use in problems such as studies of fluctuation effects is often prohibitive. Therefore, a speed-up of the simulation process in order to save computational time through the use of parallel machines is highly desirable [8,2].

In self-consistent simulations, Poisson equation has to be solved in a 3D mesh at every time step. The size of the time step depends on the material and doping characteristics of the device, but it can be as small as 0.1 fs, resulting in thousands of iterations for each bias point for simulations in the order of ps. Furthermore, it is often necessary to use a very fine mesh, making the simulation expensive not only in terms of processor time, but also in memory requirements.

In this paper, we present a 3D parallel simulator using a Monte Carlo (MC) method to model the electron transport and the Finite Element Method (FEM) on a tetrahedral mesh to solve the Poisson equation. To initialise by achieving a start of the ensemble MC simulation close to equilibrium, the Poisson and current continuity equations are solved using FEM at beginning.

The simulator has been developed for distributed–memory computers [4]. It employs a Multiple Instruction–Multiple Data strategy (MIMD) under the Single Program–Multiple Data paradigm (SPMD) which is achieved by using the

I. Lirkov, S. Margenov, and J. Waśniewski (Eds.): LSSC 2007, LNCS 4818, pp. 115–122, 2008.
© Springer-Verlag Berlin Heidelberg 2008

Message Passing Interface (MPI) standard library [5]. We have chosen the MPI library due to its availability on many computer systems which guarantees the portability of the code [15].

The paper is organised as follows. Section 2 describes the FE solver and its parallelisation. Section 3 is dedicated to the MC module and Section 4 to parallelisation of the global system. Section 5 presents results regarding parallel efficiency and, finally, some concluding remarks are addressed in Section 6.

2   3D Parallel Finite Element Solver

In this section we present the main stages of the FE solver as well as the parallelisation methodology used, based on the domain decomposition strategy.

2.1   Finite Element Solver

The FEM is applied to discretise Poisson and current continuity equations using tetrahedral elements, which allow the simulation of complex domains with great flexibility. At the pre–processing the simulation domain representing the device is triangulated into tetrahedrons using our in-house mesh generator [1]. Then, using the program METIS [7], the solution domain is partitioned into sub–domains and each assigned to an individual processor. The same program is subsequently used to achieve an improved ordering of the nodes of each sub–domain. Finally, the FEM based on tetrahedral element is applied using the Ritz-Galerkin approximation which means that the shape functions are build from piecewise linear functions [14].

2.2   Parallelisation

For the initial solution, Poisson and current continuity equations are decoupled using Gummel methods and linearised using Newton's algorithm. All these algorithms are implemented fully in parallel manner as follows.

The linearised systems obtained from Poisson and current continuity equations are solved using domain decomposition technique [3]. The solution domain Ω is partitioned in p subdomains Ω_i as

$$\Omega = \bigcup_{i=1}^{p} \Omega_i \tag{1}$$

and the domain decomposition methods attempt to solve the problem on the entire domain Ω by concurrent solutions on each subdomain Ω_i.

Each node belonging to a subdomain is an unknown of the whole problem. It is important to distinguish between three types of unknowns: (i) interior nodes, those that are coupled only with local nodes, (ii) local interface nodes, those coupled with external nodes as well as local nodes, and (iii) external interface nodes, which are those nodes in other subdomains coupled with local nodes. We label the nodes according to their subdomains, first the internal nodes and then both the local and external interface nodes. As a result, the linear system associated with the problem has the following structure:

$$
\begin{pmatrix}
B_1 & & & E_1 \\
& B_2 & & E_2 \\
& & \cdot & \\
& & \cdot & \\
& & B_p & E_p \\
F_1 & F_2 & F_p & C
\end{pmatrix}
\begin{pmatrix}
x_1 \\
x_2 \\
\cdot \\
\cdot \\
x_s \\
y
\end{pmatrix}
=
\begin{pmatrix}
f_1 \\
f_2 \\
\cdot \\
\cdot \\
f_s \\
g
\end{pmatrix}
\tag{2}
$$

where, for the subdomain Ω_i, y represents the vector of all interface unknowns, B_i represents the equations of internal nodes, C the equations of interface nodes, E_i the subdomain to the interface coupling seen from the subdomains and F_i represents the interface to the subdomain coupling seen from the interface nodes.

We have used the program METIS to partition the mesh into subdomains. The same program was subsequently used to relabel the nodes in the subdomains in order to obtain a more suitable rearrangement to reduce the bandwidth of the matrix. The PSPARSLIB [12] parallel sparse iterative solvers library modified to take advantage of the reordering of the matrices has been used to solve the linear system (2). A great advantage of this library is its optimisation for various powerful multicomputers. Among many domain decomposition techniques supported within this library the best result was obtained when using the additive Schwarz technique [13]. This technique is similar to a block–Jacobi iteration and consists of updating all the new components from the same residual. The basic additive Schwarz iteration can be described as follows:

1.– Obtain $y_{i,ext}$
2.– Compute a local residual $r_i = (b - Ax)_i$
3.– Solve $A_i \delta_i = r_i$
4.– Update solution $x_i = x_i + \delta_i$

The linear system $A_i \delta_i = r_i$ is solved using standard ILUT preconditioner combined with FGMRES [11].

3   Monte Carlo Simulation of Transport

The starting point of the Monte Carlo program is the definition of the physical system of interest including the material parameters and the values of physical quantities, such as lattice temperature T_0 and electric field $\mathbf{E}$ [6]. At this level, we also define the parameters that control the simulation: the duration of each subhistory, the desired precision of the result, and so on. The next step is a preliminary calculation of each scattering rate as a function of electron energy. The choice of the dispersion relation $\varepsilon(\mathbf{k})$ usually depends on the simulated transport problem. We have decided to have an analytical bandstructure which is a good compromise between accuracy and efficiency if our intention is to model the carrier transport in nanoscaled MOSFETs. A typical supply voltage is expected to be smaller than 1 V and we do not intend to investigate any breakdown device mechanisms which can occur at very large electric fields. Therefore, a nonparabolic dispersion law $\hbar k_\alpha^2/(2m_\alpha) = \varepsilon(1 + \alpha\varepsilon), \alpha = x, y, z$, is used taking into account an anisotropic shape of material valleys.

The subsequent step is the generation of the flight duration. The electron wave vector $\mathbf{k}$ changes continuously during a free flight because of the applied field. Thus, if $\lambda[\mathbf{k}(t)]$ is the scattering probability for an electron in the state $\mathbf{k}$ during the small time interval dt then the probability that the electron, which already suffered a scattering event at time $t = 0$ has not yet suffered further scattering after time t is

$$\exp\left[-\int_0^t dt' \, \lambda[\mathbf{k}(t')]\right]$$

which, generally, gives the probability that the interval $(0, t)$ does not contain a scattering. Consequently, the probability $P(t)$ that the electron will suffer its next collision during time interval dt at a time t is

$$P(t)dt = \lambda[\mathbf{k}(t)] \, \exp\left[-\int_0^t dt' \, \lambda[\mathbf{k}(t')]\right] dt$$

The free-flight time t can be generated from the equation

$$r = \int_0^t dt' \, P(t'),$$

where r is a random number between 0 and 1.

Once the electron free flight is terminated, the scattering mechanism has to be selected. The weight of the i-th scattering mechanism (when n scattering mechanisms are present) is given by a probability

$$P_i(\mathbf{k}) = \frac{\lambda_i(\mathbf{k})}{\lambda(\mathbf{k})}, \qquad \lambda(\mathbf{k}) = \sum_{i=1}^n \lambda_i(\mathbf{k})$$

Generating random number r between 0 and 1 and testing the inequalities

$$\sum_{i=1}^{j-1} \frac{\lambda_i(\mathbf{k})}{\lambda(\mathbf{k})} < r < \sum_{i=1}^j \frac{\lambda_i(\mathbf{k})}{\lambda(\mathbf{k})}, \qquad j = 1, \ldots, n$$

we will accept the j-th mechanism if the j-th inequality is fulfilled. It should be noted that the discussed selection of the free flight time and the scattering channel can be simplified by introducing the self-scattering λ_0 [9,10].

The next step is the choice of the state after scattering. Once the scattering mechanism that caused the end of the electron free flight has been determined, the new state after scattering of the electron, $\mathbf{k}_f$ must be chosen as the final state of the scattering event. If the free flight ended with a self-scattering, $\mathbf{k}_f$ must be taken as equal to $\mathbf{k}_i$, the state before scattering. When, in contrast, a true scattering occurred, then $\mathbf{k}_f$ must be generated stochastically according to the differential cross section of that particular scattering mechanism. The last step of the simulation is the collection of statistical averages.

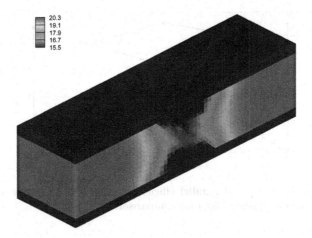

Fig. 1. The electron density (log scale) in a 4 nm gate length DG MOSFET with a body thickness of 3.3 nm at equilibrium obtained from the parallel 3D MC device simulator

4   Parallelisation of the Global System

We have described in Section 2 how Poisson equation is parallelised using a domain decomposition strategy. However, this is not, in general, the best way to share the load of the MC engine. We have to perform a load balancing since the regions with the highest electron density will have the highest number of particles to simulate. Furthermore, they will have more scattering events (ionised impurity scattering in regions of high doping and interface roughness scattering in the inversion layer) and, as a consequence, more computational weight. Although a partitioning algorithm weighted using the electron density profile could help to overcome this difficulty, this is well beyond the scope of this paper. With this in mind, we have chosen a different strategy to parallelise the MC engine in the simulation. The main idea is to replicate the whole mesh and simulate an equal number of particles in each processor. Although this requires the replication of the mesh a number of times, this will not increase the use of the total memory required by the program significantly since even for a relatively big mesh it requires only a few MB of memory. The other drawback of the strategy is the higher number of communications, but we think that this is compensated by their regularity and the balancing of the computations. In this scheme, the free flights are computed completely in parallel, requiring the first communications for the boundary conditions in the contacts. The next communication is to share the local values of the electron density. Taking advantage of the linearity of the charge assignment process this can be done with a single reduction step. The electrostatic potential must be sent to every processor after the resolution of the Poisson equation to update the electric fields of the whole mesh. Finally, the particle flights are carried out in the local tetrahedral mesh under an electric

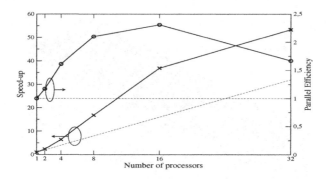

Fig. 2. Speedup (crosses) and parallel efficiency (circles) of one time step. We also show the ideal values (dashed lines) for comparison.

field interpolated using the shape functions of the corresponding tetrahedron to minimise the self-forces.

5   Double Gate MOSFET Test

The developed MC code have been tested by simulating a double gate (DG) MOSFET with a channel length of 4 nm with a body thickness of 3.3 nm and a gate dielectric stack of 0.54 nm and a source/drain doping of 1×10^{20} cm^{-3}. Fig. 1 illustrates the electron density in the DG MOSFET at equilibrium obtained during a time average after 5 ps.

The investigations of parallel performance of the 3D MC device code have been carried out by simulating the DG MOSFET over a small period of 20 fs. All the results presented here were obtained on an HP Superdome cluster with 128 Itanium 2 1.5 GHz processors. The scalability of the program can be characterised by the speedup (S) or the parallel efficiency (PE), defined for p processors as:

$$S(p) = \frac{t_1}{t_p} \qquad PE(p) = \frac{t_1}{t_p p} \qquad (3)$$

where t_1 and t_p are the times employed running the simulation on one and p processors, respectively. Fig. 2 shows their dependence with the number of processors. We can see that their values are over the linear scaling limit. This effect is due to the efficiency of Poisson solver and it starts to vanish for a high number of processors when operations corresponding to non scaling stages (e.g., the electric field update) begin to be of the same order of magnitude than the scaling operations (the linear system solving stage or the particle flights). This limit is intrinsic to our parallelisation strategy and cannot be avoided. Its impact for a given number of processors will mainly depend on the size of the employed mesh.

Table 1 shows the execution times of different parts of the main iteration in the Monte Carlo loop. We obtain a very good scalability for the particle flight times

Table 1. Mean simulation times (s) for one time step ($t_{iteration}$), linear Poisson solver ($t_{Poisson}$), particle flights (t_{flight}), contacts stage ($t_{contacts}$), reduction of electron concentration ($t_{reduction}$), communication of the potentials ($t_{comun.poten.}$) and electric field update (E_{update}) for different number of processors (p)

p	$t_{iteration}$	$t_{Poisson}$	t_{flight}	$t_{contacts}$	$t_{reduction}$	$t_{comun.poten.}$	E_{update}
1	27.50	24.9	0.68	0.35	0.013	0.019	0.207
2	11.74	11.1	0.33	0.20	0.010	0.007	0.204
4	4.27	3.87	0.173	0.146	0.0085	0.0054	0.203
8	1.64	1.08	0.102	0.212	0.0188	0.0050	0.198
16	0.745	0.386	0.0420	0.101	0.005	0.0042	0.200
32	0.5152	0.186	0.0209	0.079	0.009	0.006	0.206

being limited by the contacts stage, where communications and replicated code are required in order to maintain the equilibrium. The other limiting stage is the update of the electric field, which requires a constant time independently of the number of processors. These two stages remain unimportant for a low number of processors, but they limit the scalability for a high number of processors, when these times are of the same order of magnitude or even larger than those of the MC free flights or the solution of Poisson equation.

6 Conclusion

We have presented a parallel 3D device simulator which relies on the ensemble MC method to model the semiconductor transport in the framework of Boltzmann transport equation and on the FE method to solve the Poisson equation on unstructured tetrahedral meshes. The developed parallel 3D MC device simulator has been tested by modelling, at equilibrium, the 4 nm gate length Si DG MOSFET with a body thickness of 3.3 nm and a gate dielectric stack of 0.54 nm.

The parallelisation of the Poisson equation was based on domain decomposition methods whereas the parallelisation of the MC method was based on particle redistribution. This mixture of parallelisation strategies limits the parallel performance of the program, and the maximum optimal number of processors will depend on the size of the mesh. However, before the saturation in the performance occurs, the scalability is superlinear (overtaking the linear limit) making the use of parallelism ideal for multi-core processors and production runs requiring multiple simulations.

Acknowledgement

This work was supported by Spanish Government (MCYT) under the project TIN2004-07797-C02 and by Project HPC-EUROPA (RII3-CT–2003-506079), with the support of European Community-Research Infrastructure Action under FP6 Structuring European Research Area Programme. We are particularly grateful to CESGA (Galician Supercomputing Centre) for providing access to HP Superdome System. M. Aldegunde thanks Ministerio de Educación y Ciencia de España for the awarded fellowship (FPU). K. Kalna would also like to thank UK EPSRC (EP/D070236/1) for its support.

References

1. Aldegunde, M., Pombo, J.J., García-Loureiro, A.: Modified octree mesh generation for Manhattan type structures with narrow layers applied to semiconductor devices. In: Numer, J. (ed.) Model.-Electron. Netw. Device Fields, vol. 19(6), pp. 473–489 (2006)
2. Banse, F., et al.: Implementation of a Bi-Parallel Monte Carlo Device Simulation on Two Architectures. In: Bubak, M., Hertzberger, B., Sloot, P.M.A. (eds.) HPCN-Europe 1998. LNCS, vol. 1401, pp. 193–202. Springer, Heidelberg (1998)
3. Barrett, R., et al.: Templates for the Solution of Linear Systems: Building Blocks for Iterative Methods. SIAM, Philadelphia (2004)
4. García-Loureiro, A.J., et al.: Parallel finite element method to solve the 3D Poisson equation and its application to abrupt heterojunction bipolar transistors. Int. J. Numer. Methods Eng. 49(5), 639–652 (2000)
5. Group, W., Lusk, E., Skjellum, A.: Using MPI. MIT Press, Boston (1996)
6. Jensen, G.U., et al.: Monte Carlo Simulation of Semiconductor Devices. Comput. Phys. Commun. 67(1), 1–61 (1991)
7. Karypis, G., Kumar, V.: METIS: A software package for partitioning unstructured graphs, partitioning meshes, and computing fill–reducing orderings of sparse matrices. Univ. of Minnesota (1997)
8. Ranawake, U.A., et al.: PMC-3D: A Parallel Three-Dimensional Monte Carlo Semiconductor Device Simulator. IEEE Trans. Comput-Aided Des. Integr. Circuits Syst. 13(6), 712–724 (1994)
9. Rees, H.D.: Calculation of distribution functions by exploiting the stability of the steady state. J. Phys. Chem. Solids 30(3), 643–655 (1969)
10. Rees, H.D.: Calculation of steady state distribution functions by exploiting stability. Phys. Lett. A 26(9), 416–417 (1968)
11. Saad, Y.: Iterative Methods for Sparse Linear Systems. PWS Publishing Co. (1996)
12. Saad, Y., Lo, G.-C., Kuznetsov, S.: PSPARSLIB users manual: A portable library of parallel sparse iterative solvers. Technical report, Univ. of Minnesota, Dept. of Computer Science (1997)
13. Seoane, N., García-Loureiro, A.: Study of parallel numerical methods for semiconductor device simulation. Int. J. Numer. Model.-Electron. Netw. Device Fields 19(1), 15–32 (2006)
14. Zienkiewicz, O.C.: The Finite Element Method. McGraw-Hill, New York (1977)
15. The MPI Standard, http://www-unix.mcs.anl.gov/mpi/index.html

Numerical Study of Algebraic Problems Using Stochastic Arithmetic

René Alt[1], Jean-Luc Lamotte[1], and Svetoslav Markov[2]

[1] CNRS, UMR 7606, LIP6, University Pierre et Marie Curie, 4 pl. Jussieu, 75252
Paris cedex 05, France
`Rene.Alt@lip6.fr, Jean-Luc.Lamotte@lip6.fr`
[2] Institute of Mathematics and Informatics, Bulgarian Academy of Sciences,
Acad. G. Bonchev, bl. 8, 1113 Sofia, Bulgaria
`smarkov@bio.bas.bg`

Abstract. A widely used method to estimate the accuracy of the numerical solution of real life problems is the CESTAC Monte Carlo type method. In this method, a real number is considered as an N-tuple of Gaussian random numbers constructed as Gaussian approximations of the original real number. This N-tuple is called a "discrete stochastic number" and all its components are computed synchronously at the level of each operation so that, in the scope of granular computing, a discrete stochastic number is considered as a granule. In this work, which is part of a more general one, discrete stochastic numbers are modeled by Gaussian functions defined by their mean value and standard deviation and operations on them are those on independent Gaussian variables. These Gaussian functions are called in this context *stochastic numbers* and operations on them define *continuous stochastic arithmetic (CSA)*. Thus operations on stochastic numbers are used as a model for operations on imprecise numbers. Here we study some new algebraic structures induced by the operations on stochastic numbers in order to provide a good algebraic understanding of the performance of the CESTAC method and we give numerical examples based on the Least squares method which clearly demonstrate the consistency between the CESTAC method and the theory of stochastic numbers.

Keywords: stochastic numbers, stochastic arithmetic, standard deviations, least squares approximation.

1 Introduction

A widely used method to estimate the accuracy of the numerical solution of real life problems is the CESTAC method, see for example [4,6,8,13,14,15,16,17]. In this method, real numbers are considered as vectors of N Gaussian random numbers constructed to be Gaussian approximations of the same value. This vector is called a "discrete stochastic number". The CESTAC method has been implemented in a software called CADNA in which discrete stochastic numbers are computed one operation after the other. In other words all their components are computed synchronously at the level of each operation so that, in the

I. Lirkov, S. Margenov, and J. Waśniewski (Eds.): LSSC 2007, LNCS 4818, pp. 123–130, 2008.

scope of granular computing [19], a discrete stochastic number is considered as a granule. Moreover in CADNA the components of a the discrete stochastic numbers are randomly rounded up or down with same probability to take into account the rounding of floating point operators in the same way that directed rounding is used in softwares implementing interval arithmetic. In this work, which is part of a more general one, discrete stochastic numbers are modeled by Gaussian functions defined by their mean value and standard deviation and operations on them are those on independent Gaussian variables. These Gaussian functions are called in this context "stochastic numbers" and operations on them define *continuous stochastic arithmetic (CSA)* also called more briefly *stochastic arithmetic*. Operations on stochastic numbers are used as a model for operations on imprecise numbers. Some fundamental properties of stochastic numbers are considered in [5,18]. Here we study numerically the performance of the CESTAC method [1,2,3,10,11] using numerical examples based on the Least squares method. Our experiments clearly demonstrate the consistency between the CESTAC method and the theory of stochastic numbers and present one more justification for both the theory and the computational practice.

The operations addition and multiplication by scalars are well-defined for stochastic numbers and their properties have been studied in some detail. More specifically, it has been shown that the set of stochastic numbers is a commutative monoid with cancelation law in relation to addition. The operator multiplication by -1 (negation) is an automorphism and involution. These properties imply a number of interesting consequences, see, e. g. [10,11].

In what follows we first briefly present some algebraic properties of the system of stochastic numbers with respect to the arithmetic operations addition, negation, multiplication by scalars and the relation inclusion. These theoretical results are the bases for the numerical experiments presented in the paper.

2   Stochastic Arithmetic Theory (SAT) Approach

A *stochastic number a* is written in the form $a = (m, s)$. The first component m is interpreted as *mean value*, and the second component s is the *standard deviation*. A stochastic number of the form $(m; 0)$ has zero standard deviation and represents a (pure) mean value, whereas a stochastic number of the form $(0; s)$ has zero mean value and represents a (pure) standard deviation. In this work we shall always assume $s \geq 0$. Denote by $\mathbb{S}$ the set of all stochastic numbers, $\mathbb{S} = \{(m; s) \mid m \in \mathbb{R}, s \in \mathbb{R}^+\}$. For two stochastic numbers $(m_1; s_1)$, $(m_2; s_2) \in \mathbb{S}$, we define addition by

$$(m_1; s_1) + (m_2; s_2) \stackrel{def}{=} \left(m_1 + m_2; \sqrt{s_1^2 + s_2^2} \right), \tag{1}$$

Multiplication by real scalar $\gamma \in \mathbb{R}$ is defined by:

$$\gamma * (m_1; s_1) \stackrel{def}{=} (\gamma m_1; |\gamma| s_1). \tag{2}$$

In particular multiplication by -1 *(negation)* is

$$- 1 * (m_1; s_1) = (-m_1; \ s_1), \tag{3}$$

and subtraction of $(m_1; s_1)$, $(m_2; s_2)$ is:

$$(m_1; s_1) - (m_2; s_2) \stackrel{def}{=} (m_1; s_1) + (-1) * (m_2; s_2) = \left(m_1 - m_2; \ \sqrt{s_1^2 + s_2^2}\right). \tag{4}$$

Symmetric stochastic numbers. A symmetric (centered) stochastic number has the form $(0; s), s \in \mathbb{R}$. The arithmetic operations (1)–(4) show that mean values subordinate to familiar real arithmetic whereas standard deviations induce a special arithmetic structure that deviates from the rules of a linear space. If we denote addition of standard deviations defined by (1) by "$\oplus$" and multiplication by scalars by "$*$", that is:

$$s_1 \oplus s_2 = \sqrt{s_1^2 + s_2^2}, \quad \gamma * s_1 = |\gamma| s_1,$$

then we can say that the space of standard deviations is an abelian additive monoid with cancellation, such that for any two standard deviations $s, t \in \mathbb{R}^+$, and real $\alpha, \beta \in \mathbb{R}$:

$$\alpha * (s \oplus t) = \alpha * s \oplus \alpha * t,$$
$$\alpha * (\beta * s) = (\alpha\beta) * s,$$
$$1 * s = s,$$
$$(-1) * s = s,$$
$$\sqrt{\alpha^2 + \beta^2} * s = \alpha * s \oplus \beta * s.$$

Examples. Here are some examples for computing with standard deviations:

$$1 \oplus 1 = \sqrt{2}, \quad 1 \oplus 2 = \sqrt{5}, \quad 3 \oplus 4 = 5, \quad 1 \oplus 2 \oplus 3 = \sqrt{14}.$$

Note that $s_1 \oplus s_2 \oplus ... \oplus s_n = t$ is equivalent to $s_1^2 + ... + s_n^2 = t^2$.

Inclusion. Inclusion of stochastic numbers plays important roles in applications. Inclusion relation "$\subseteq_s$" between two stochastic numbers $X_1 = (m_1; s_1)$, $X_2 = (m_2; s_2) \in \mathbb{S}$ is defined by [3]

$$X_1 \subseteq_s X_2 \Longleftrightarrow (m_2 - m_1)^2 \le s_2^2 - s_1^2. \tag{5}$$

Relation (5) is called *stochastic inclusion*, briefly: s-inclusion.

It is easy to prove [3] that addition and multiplication by scalars are (inverse) s-inclusion isotone (invariant with respect to s-inclusion), that is

$$X_1 \subseteq X_2 \Longleftrightarrow X_1 + Y \subseteq X_2 + Y, \quad X_1 \subseteq X_2 \Longleftrightarrow \gamma * X_1 \subseteq \gamma * X_2$$

3   The CESTAC Method

Suppose that some mathematical value r has to be computed with a numerical method implemented on a computer. The initial data are imprecise and the computer uses floating point number representation. In the CESTAC method a real number r, intermediate or final result, is considered as a Gaussian random variable with mean value m and standard deviation σ that have to be approximated. So r is a stochastic number.

In practice a stochastic number is approximated by an N-tuple with components r_j, $j = 1, ..., N$, which are empirical samples representing the same theoretical value. As seen before, this vector is called *discrete stochastic number*. The operations on these samples are those of the computer in use followed by a random rounding. The samples corresponding to imprecise initial values are randomly generated with a Gaussian distribution in a known confidence interval.

Following the classical rules of statistics, the mean value m is the best approximation of the exact value r and the number of significant digits on m is computed by:

$$C_m = log_{10}\left(\frac{\sqrt{N}\,|\bar{r}|}{\sigma\,\tau_\eta}\right), \tag{6}$$

wherein $m = N^{-1}\sum_{j=1}^{N} r_j$, $\sigma^2 = (N-1)^{-1}\sum_{j=1}^{N}(r_j - m)^2$ and τ_η is the value of Student's distribution for $k - 1$ degrees of freedom and a probability level p. Most of the time p is chosen to be $p = 0.95$ so that $\tau_\eta = 4.303$. This type of computation on samples approximating the same value is called *Discrete Stochastic Arithmetic (DSA)*.

It has been shown [5] that if one only wants the accuracy of r, i.e., its number of significant decimal digits, then $N = 3$ suffices. This is what is chosen in the software named CADNA [20] which implements the CESTAC method. But if one wants a good estimation of the error on r then a greater value for N, experimentally at least $N = 5$ must be chosen. The experiments given below use $N = 5$ and $N = 20$ showing that the two series of results are very close and that a large value for N is unnecessary.

The goal of next section 4 is to compare the results obtained with *Continuous Stochastic Arithmetic (CSA)* and the theory developed in this paper with results obtained with the CESTAC method implementing *Discrete Stochastic Arithmetic (DSA)*.

It should be remarked that DSA which is used in the CESTAC method takes into account round-off errors at the level of each floating point operation because of the random rounding done at this level. On the contrary CSA is a theoretical model in which data are imprecise but arithmetic operations are supposed exact. This is why in our experiments relative errors on data are chosen to be of order 10^{-2}–10^{-3} whereas the computations are done using double presision arithmetic. Thus experiments on computer can be considered very close to the theoretical CSA model.

4   Application: Linear Regression

As said before, in the CESTAC method, each stochastic variable is represented by an N-tuple of gaussian random values with known mean value m and standard deviation σ. The method also uses a special arithmetic called *Discrete Stochastic Arithmetic (DSA)*, which acts on the above mentioned N-tuples.

To compare the two models, a specific library has been developed which implements both continuous and discrete stochastic arithmetic. The computations are done separately. The *CSA* implements the mathematical rules defined in Section 2. The comparison has been done on the one-dimensional linear regression method for numeric input data.

4.1   Derivation of a Formula for Regression

Let (x_i, y_i), $i = 1, ..., n$, be a set of n pairs of numbers where all x_i are different, $x_1 < x_2 < ... < x_n$. As well-known the regression line that fits the (numeric) input data (x, y), $x = (x_1, x_2, ..., x_n) \in \mathbb{R}^n$, $y = (y_1, y_2, ..., y_n) \in \mathbb{R}^n$, is

$$l : \eta = (S_{xy}/S_{xx})(\xi - \overline{x}) + \overline{y}, \qquad (7)$$

wherein $\overline{x} = (\sum x_i)/n$, $\overline{y} = (\sum y_i)/n$ (all sums run from 1 to n), and

$$S_{xx} = \sum (x_i - \overline{x})^2 > 0, \qquad S_{xy} = \sum (x_i - \overline{x})(y_i - \overline{y}) = \sum (x_i - \overline{x})y_i.$$

Note that that l passes through the point $(\overline{x}, \overline{y})$.

The expression in the right hand-side of (7) can be rewritten in the form:

$$L : \eta = (S_{xy}/S_{xx})(\xi - \overline{x}) + \overline{y}$$
$$= (1/S_{xx}) \left(\sum (x_i - \overline{x})y_i \right) (\xi - \overline{x}) + \left(\sum y_i \right) /n$$
$$= \sum ((x_i - \overline{x})(\xi - \overline{x})/S_{xx} + 1/n) \, y_i.$$

Thus the line (7) can be represented in the form

$$l : \eta = \sum \gamma_i(\xi) \, y_i, \qquad (8)$$

wherein the functions

$$\gamma_i(\xi) = \gamma_i(x; \xi) = (x_i - \overline{x})(\xi - \overline{x})/S_{xx} + 1/n, \quad i = 1, 2, ..., n, \qquad (9)$$

depend only on x and not on y.

Since γ_i is linear, it may have at most one zero. Denoting by ξ_i the zero of the linear function $\gamma_i(\xi)$, we have

$$\xi_i = \overline{x} + S_{xx}/(n(\overline{x} - x_i)), \quad i = 1, ..., n. \qquad (10)$$

If $x_i = \overline{x}$, then $\gamma_i = 1/n > 0$.

In every interval $[\xi_j, \xi_{j+1}]$ the signs of γ_i do not change and can be easily calculated [9].

Table 1. Results obtained with $\varepsilon = 0$, $\delta = 0.01$

u_i	DSA 5 samples	DSA 20 samples	CSA	CADNA
$(0;\varepsilon)$	$(2.98587;0.000930)$	$(2.99719;0.009134)$	$(3.00000;0.009591)$	$0.298E{+}001$
$(1.5;\varepsilon)$	$(5.99914;0.003732)$	$(6.00011;0.005948)$	$(6.00000;0.006164)$	$0.599E{+}001$
$(2.5;\varepsilon)$	$(8.00291;0.005348)$	$(7.99994;0.006061)$	$(8.00000;0.004690)$	$0.800E{+}001$

4.2 Experiments

As said above, the results obtained with CSA and those obtained with the CES-TAC method with N samples, i.e., with DSA have to be compared. Here the successive values $N = 5$ and $N = 20$ have been chosen to experiment the efficiency of the CESTAC method with different sizes of discrete stochastic numbers. The CSA is based on operations defined on Gaussian random variable $(m; \sigma)$.

The regression method (8) has been implemented with CSA and DSA. We consider the situation when the values of the function y_i are imprecise and abscissas x_i are considered exact.

For all examples presented below, we take the couples of values from the line $v = 2u + 3$. The values chosen for abscissas are $x_1 = 1$, $x_2 = 2$, $x_3 = 2.5$, $x_4 = 4$, $x_5 = 5.5$, and the values y_i considered as imprecise are obtained as follows:

In the case of CSA they are chosen as $y_1 = (5; \delta)$, $y_2 = (7; \delta)$, $y_3 = (8; \delta)$, $y_4 = (11; \delta)$, $y_5 = (14; \delta)$ and δ is chosen as $\delta = 0.01$.

In the case of the CESTAC method (DSA) the data for the y_i are randomly generated with Gaussian distributions whose mean values are the centers of the above stochastic numbers and standard deviation δ.

From formula (8), three values of v_i corresponding to three input values considered as imprecise $u_i = (0; \varepsilon)$; $(1.5; \varepsilon)$; $(2.5; \varepsilon)$ are computed with DSA and with CSA and different values of ε. They are reported in tables 1–3.

The tables show that the mean values obtained with CSA are very close to the mean values obtained with DSA.

Let us now call $(m_v; \sigma_v)$ the values provided by CSA for the above least squares approximation at some point u.

CSA can be considered a good model of DSA if the mean value $\overline{v}$ of the samples obtained at point u with the DSA is in the theoretical confidence interval provided by CSA, in other words if:

$$m_v - 2\sigma_v \leq \overline{v} \leq m_v + 2\sigma_v \qquad (11)$$

with a probability of 0.95. This formula can be rewritten as: $-2\sigma_v \leq \overline{v} - m_v \leq +2\sigma_v$, $|\overline{v} - m_v| \leq +2\sigma_v$.

Table 2. Results obtained with $\varepsilon = 0.01$, $\delta = 0.01$

u_i	DSA 5 samples	DSA 20 samples	CSA	CADNA
$(0;\varepsilon)$	$(2.99372;0.021080)$	$(3.00001;0.018596)$	$(3.00000;0.032542)$	$0.29E{+}001$
$(1.5;\varepsilon)$	$(5.99043;0.015064)$	$(5.99956;0.024394)$	$(6.00000;0.031702)$	$0.59E{+}001$
$(2.5;\varepsilon)$	$(8.01482;0.017713)$	$(7.99296;0.017689)$	$(8.00000;0.031449)$	$0.80E{+}001$

Table 3. Results obtained with $\varepsilon = 0.1$, $\delta = 0.01$

u_i	DSA 5 samples	DSA 20 samples	CSA	CADNA
$(0;\varepsilon)$	(2.76260;0.122948)	(3.03062;0.213195)	(3.00000;0.311120)	Non significant
$(1.5;\varepsilon)$	(5.86934;0.205126)	(6.11816;0.179552)	(6.00000;0.311033)	0.5E+001
$(2.5;\varepsilon)$	(7.97106;0.219142)	(8.07687;0.229607)	(8.00000;0.311008)	0.7E+001

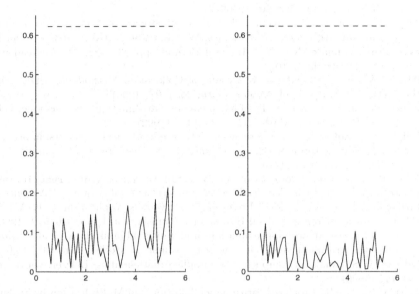

Fig. 1. The dash line represents $2\sigma_v$ and the full line $|\overline{v} - m_v|$, the left figure is computed with $N = 5$ and the right one with $N = 20$

The regression line has been computed with the previous data and $\varepsilon = 0.01$ and from $u = 0.5$ to $u = 5.5$ with a step of 0.1. Figure 1 shows the curves $|\overline{v} - m_v|$ and $2\sigma_{v_i}$ for N=5 and N=20. On our samples formula (11) is always respected.

5   Conclusion

Starting from a minimal set of empirically known facts related to stochastic numbers, we formally deduce a number of properties and relations. We investigate the set of all stochastic numbers and show that this set possesses nice algebraic properties. We point out to the distinct algebraic nature of the spaces of mean-values and standard deviations. Based on the algebraic properties of the stochastic numbers we propose a natural relation for inclusion, called stochastic inclusion. Numerical examples based on Lagrange interpolation demonstrate the consistency between the CESTAC method and the presented theory of stochastic numbers. This is one more justification for the practical use of the CADNA software.

References

1. Alt, R., Lamotte, J.-L., Markov, S.: Numerical Study of Algebraic Solutions to Linear Problems Involving Stochastic Parameters. In: Lirkov, I., Margenov, S., Waśniewski, J. (eds.) LSSC 2005. LNCS, vol. 3743, pp. 273–280. Springer, Heidelberg (2006)
2. Alt, R., Lamotte, J.-L., Markov, S.: Abstract structures in stochastic arithmetic. In: Bouchon-Meunier, B., Yager, R.R. (eds.) Proc. 11th Conference on Information Processing and Management of Uncertainties in Knowledge-based Systems (IPMU 2006), EDK, Paris, pp. 794–801 (2006)
3. Alt, R., Markov, S.: On Algebraic Properties of Stochastic Arithmetic. Comparison to Interval Arithmetic. In: Kraemer, W., Gudenberg, J.W.v. (eds.) Scientific Computing, Validated Numerics, Interval Methods, pp. 331–342. Kluwer Academic Publishers, Dordrecht (2001)
4. Alt, R., Vignes, J.: Validation of Results of Collocation Methods for ODEs with the CADNA Library. Appl. Numer. Math. 20, 1–21 (1996)
5. Chesneaux, J.M., Vignes, J.: Les fondements de l'arithmétique stochastique. C.R. Acad. Sci., Paris, Sér. I, Math. 315, 1435–1440 (1992)
6. Delay, F., Lamotte, J.-L.: Numerical simulations of geological reservoirs: improving their conditioning through the use of entropy. Mathematics and Computers in Simulation 52, 311–320 (2000)
7. NTLAB—INTerval LABoratory V. 5.2., www.ti3.tu-harburg.de/~rump/intlab/
8. Lamotte, J.-L., Epelboin, Y.: Study of the numerical stability of a X-RAY diffraction model. In: Computational Engineering in Systems Applications, CESA 1998 IMACS Multiconference, Nabeul-Hammamet, Tunisia, vol. 1, pp. 916–919 (1998)
9. Markov, S.: Least squares approximations under interval input data, Contributions to Computer Arithmetic and Self-Validating Numerical Methods. In: Ullrich, C.P. (ed.) IMACS Annals on computing and applied mathematics, Baltzer, vol. 7, pp. 133–147 (1990)
10. Markov, S., Alt, R.: Stochastic arithmetic: Addition and Multiplication by Scalars. Appl. Numer. Math. 50, 475–488 (2004)
11. Markov, S., Alt, R., Lamotte, J.-L.: Stochastic Arithmetic: S-spaces and Some Applications. Numer. Algorithms 37(1–4), 275–284 (2004)
12. Rokne, J.G.: Interval arithmetic and interval analysis: An introduction, Granular computing: An emerging paradigm, Physica-Verlag GmbH, 1–22 (2001)
13. Scott, N.S., et al.: Numerical 'health check' for scientific codes: The CADNA approach. Comput. Physics communications 176(8), 507–521 (2007)
14. Toutounian, F.: The use of the CADNA library for validating the numerical results of the hybrid GMRES algorithm. Appl. Numer. Math. 23, 275–289 (1997)
15. Toutounian, F.: The stable $A^T A$-orthogonal s-step Orthomin(k) algorithm with the CADNA library. Numer. Algo. 17, 105–119 (1998)
16. Vignes, J., Alt, R.: An Efficient Stochastic Method for Round-Off Error Analysis. In: Miranker, W.L., Toupin, R.A. (eds.) Accurate Scientific Computations. LNCS, vol. 235, pp. 183–205. Springer, Heidelberg (1986)
17. Vignes, J.: Review on Stochastic Approach to Round-Off Error Analysis and its Applications. Math. and Comp. in Sim. 30(6), 481–491 (1988)
18. Vignes, J.: A Stochastic Arithmetic for Reliable Scientific Computation. Math. and Comp. in Sim. 35, 233–261 (1993)
19. Yao, Y.Y.: Granular Computing: basic issues and possible solutions. In: Wang, P.P. (ed.) Proc. 5th Joint Conference on Information Sciences, Atlantic City, N. J., USA, February 27– March 3, Assoc. for Intelligent Machinery, vol. I, pp. 186–189 (2000)
20. http://www.lip6.fr/cadna

Monte Carlo Simulation of GaN Diode Including Intercarrier Interactions

A. Ashok[1], D. Vasileska[1], O. Hartin[2], and S.M. Goodnick[1]

[1] Department of Electrical Engineering and Center for Solid State Electronics
Research Arizona State University, Tempe, AZ 85287-5706, USA
[2] Frescale Inc, Tempe, AZ, USA
aashwin@asu.edu, vasileska@asu.edu, lee.hartin@freescale.com,
goodnick@asu.edu

Abstract. Gallium Nitride (GaN) is becoming increasingly more attractive for a wide range of applications, such as optoelectronics, wireless communication, automotive and power electronics. Switching GaN diodes are becoming indispensable for power electronics due to their low on-resistance and capacity to withstand high voltages. A great deal of research has been done on GaN diodes over the decades but a major issue with previous studies is the lack of explicit inclusion of electron-electron interaction, which can be quite important for high carrier densities encountered. Here we consider this electron-electron interaction, within a non-parabolic band scheme, as the first attempt at including such effects when modeling nitride devices. Electron-electron scattering is treated using a real space molecular dynamics approach, which exactly models this interaction within a semi-classical framework. It results in strong carrier-carrier scattering on the biased contact of the resistor, where rapid carrier relaxation occurs.

1 Introduction

Even though gallium nitride (GaN) is of primary importance in optoelectronics, for high-density optical storage and solid-state lighting, electronic devices based on this wide-bandgap semiconductor are becoming increasingly attractive for diverse applications, such as wireless communication, automotive or power conversion [12,7]. Besides space and military end-users, GaN offers the utmost perspectives for high-power amplifier manufacturers (3G/4G base station or WiMAX) [14]. The key advantages offered by GaN technology for RF power electronics reside in the combination of higher output power density (even at high frequency), higher output impedance (easier matching), larger bandwidth and better linearity than other existing technologies. These features directly stem from the physical properties of the semiconductor: the large bandgap results in a breakdown electrical field ten times larger than Si. This allows transistor operation at high bias voltage. The high saturation velocity, combined with the high current density in the two-dimensional electron gas, ensures high power handling up to mm-wave frequencies.

I. Lirkov, S. Margenov, and J. Waśniewski (Eds.): LSSC 2007, LNCS 4818, pp. 131–138, 2008.

The most successful GaN device nowadays is the AlGaN/GaN high electron mobility transistor (HEMT). One of the key challenges for large adoption of this technology by the market remains the control of the trap effects (essentially surface traps resulting from piezoelectric character of the devices), on the RF characteristics, often known as the DC/RF dispersion in nitrides [15]. Finally, besides HEMT devices required for RF power applications, switching GaN diodes are very much needed for power electronics regarding their capabilities to withstand high voltages and their low on-resistance [13]. Vertical pin diodes, with low leakage current, have been successfully demonstrated, paving the way to new type of GaN electronic devices, such as permeable base transistors [3].

In this paper we present simulation results for a GaN resistor for the purpose of understanding the role of the electron-electron interactions on the carrier thermalization at the contacts. We find that carrier-carrier interaction is very effective in the carrier thermalization, which suggests that in the contact portion of the device, non-parabolic model for the energy dispersion relation can be used without any loss of accuracy in simulation of HEMT devices at moderately high biases. The paper is organized as follows. In Section 2 we discuss state-of the art in GaN devices modeling. In Section 3 we explain our theoretical model. The resistor simulation and the thermalization of the carriers at the contacts is discussed in Section 4. Conclusions regarding this work are presented in Section 5.

2   Overview of Existing Simulations of Various GaN Device Structures

There are several investigations performed on the theoretical properties of wurtzite phase GaN over the past two decades. However, most of these calculations typically used a modified mobility drift-diffusion model or the Ensemble Monte Carlo (EMC) method using several valleys with analytical bandstructures. These models fail to treat transport properties properly in the high electric field regime where carrier energies are high and the bandstructure gets complicated. There are several reports which calculate the electron transport properties using full-band EMC but most of them used the same parameters as zinc blende materials to calculate electron-phonon scattering rates using the deformation potential.

The first transport simulation using Monte Carlo (MC) methods was reported by Littlejohn, Hauser, and Glisson in 1975 [9]. This simulation included a single valley (Gamma Valley) with both parabolic and non-parabolic bands. Acoustic scattering, polar optical phonon scattering, piezoelectric scattering and ionized impurity scattering were taken into account in these calculations. Velocity saturation and negative differential transconductance in GaN were predicted. In 1993, Gelmont, Kim and Shur [5] pointed out that intervalley electron transfer played a dominant role in GaN in high electric field leading to a strongly inverted electron distribution and to a large negative differential conductance.

They used a non-parabolic, two valley model including Γ and U valleys. Polar optical phonon, piezoelectric, deformation potential and ionized impurity scattering mechanisms were taken into account. The intervalley coupling coefficient of GaAs was utilized in these calculations. Mansour, Kim and Littlejohn also used a two-valley model to simulate the high-temperature dependence of the electron velocity [10]. They included acoustic phonon, polar optical phonon, intervalley phonon and ionized impurity scatterings. Bhapkar and Shur in 1997 came up with an improved multi-valley model that included a second Γ valley in addition to the Γ and U valleys [2]. The energy gap between the two valleys was modified to 2 eV from the earlier 1.5 eV used in all the previous simulations. Scattering mechanisms taken into account were acoustic phonon, polar optical phonon, ionized impurity, piezoelectric and inter valley scattering. This model has been adopted in this work.

All these simulations mentioned above used analytical, non-parabolic band structures. A full-band MC simulation is another approach to get more accurate results at higher electric fields. Full band MC simulations have been reported previously by the Georgia Tech Group. Kolnik et al. reported the first full band MC simulation for both wurtzite and zinc-blende GaN [8]. They considered acoustic, polar optical and intervalley scattering in their calculations. Brennan et al. performed full band MC simulations and compared the results for different III–V materials. He reported a higher electron velocity for wurtzite GaN than the previous simulation data [4]. Both these simulations could not verify their results as no experimental velocity data was reported until then.

Barker et al. reported recently the measurements of the velocity-field characteristics in bulk GaN and AlGaN/GaN test structures using a pulsed I–V measurement technique [1]. These experimental results are comparable to the theoretical models of Kolnik and Brennan and Yu and Brennan. Most other simulations along these lines have reported lower velocity characteristics than that of Kolnik and Brennan. Some groups like Matulionis et al. [11] have suggested that lattice heating could play a very big role in lowering the peak velocity at high electric fields. Some other groups feel that this may be attributed due to the hot phonon effect. Though there seems to be no consensus about this, more work needs to be done in order to better understand the underlying physics. Recently, Yamakawa and co-workers [17] using full-band cellular automata particle-based simulator examined theoretically the RF performance of GAN MESFETs and HEMTs.

Note that none of the above studies included the short-range electron-electron interactions in their theoretical model. From modeling Si MOSFETs, it is well known that the electron-electron (e-e) interactions lead to rapid thermalization of the carriers coming from the channel into the drain end of a MOSFET device [6]. This work has motivated us to examine the role of the carrier-carrier interactions on carrier thermalization at the contacts for devices fabricated in GaN technology. For that purpose we have simulated n^+-n-n^+ diode made of GaN. Details of the theoretical model used are given in the following section.

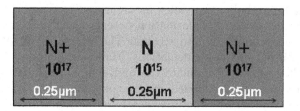

Fig. 1. Simulated n$^+$-n-n$^+$GaN diode

3   Theoretical Model

The GaN n$^+$-n-n$^+$ diode simulated in this work is shown in Fig. 1. The n$^+$ region was doped to 10^{17} cm^{-3} while the n region was doped to 10^{15} cm^{-3}. The bulk transport simulations were performed first to obtain the correct material parameters that fit the theoretical and experimental results. The ensemble Monte Carlo transport kernel incorporated a multi-valley model mentioned in the previous section and various scattering mechanisms such as acoustic, polar optical phonon, ionized impurity, inter valley and piezoelectric scattering were included.

The device simulator comprises of a 1D Poisson solver coupled with the 3D Monte Carlo transport kernel. LU Decomposition scheme is used for the numerical solution of the matrix problem. Carriers are initialized and distributed randomly within the device and are subjected to a free flight and scatter process. The charge distribution is updated after every time step and Poisson solver is called to update the potential. This procedure is repeated until we reach the final time at which point the carriers achieve steady state conditions. This gives us the Hartree potential which does not include the short range carrier-carrier interactions.

To include the short range electron-electron interactions we employed the P3M approach utilized by C. Wordelman *et al.* [16] where the short range interaction is calculated as a direct particle-particle force summation. This short range force is calculated for each carrier by using a lookup force table and the total force on an electron is then computed as the sum of the short range force and the long range mesh force. The P3M — Ensemble Monte Carlo coupled model includes the short range electron-electron interactions which lead to the rapid thermalization of these carriers at the drain end of the device.

4   Resistor Simulations

The drift velocity characteristics for bulk GaN as a function of electric field are shown in Fig. 2. The theoretical results obtained from our model are in close agreement to the experimental results and validates the choice of material parameters used in the simulations. Simulations were run for various applied drain bias and the drain current was calculated. Figure 3 shows the potential

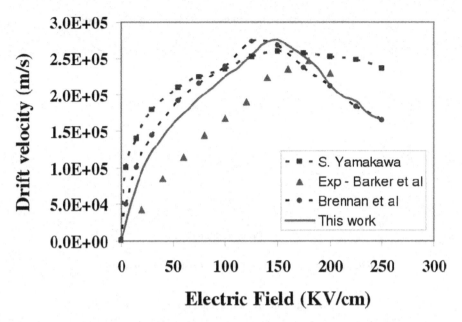

Fig. 2. Velocity — field characteristics for Bulk GaN [12, 13, and 15]

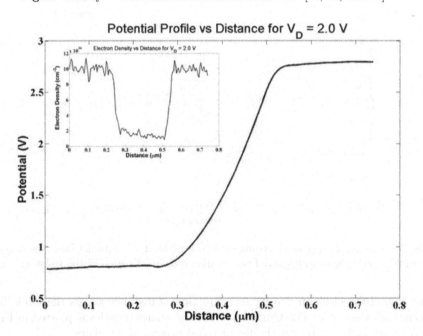

Fig. 3. Potential Profile and Electron Density along the length of the GaN diode for $V_D = 2V$

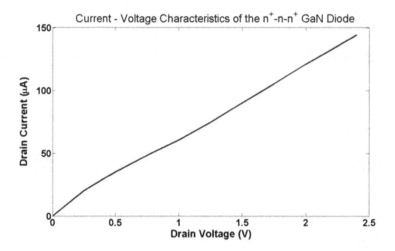

Fig. 4. Current Voltage Characteristics of the diode

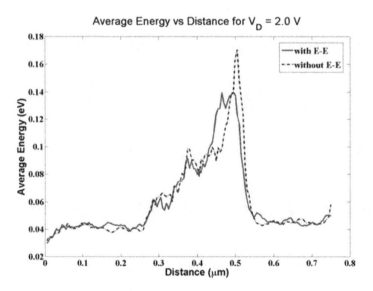

Fig. 5. Average energy of electrons with and without e–e interactions. The n region extends from $0.25\mu m$ to $0.5\mu m$. This is called a channel region of the resistor.

profile and the electron density as a function of distance across the diode for a drain bias $V_D = 2$ V. The Current — Voltage characteristics is plotted in Fig. 4 and we are able to observe the linear trend typical of a resistor.

The inclusion of electron-electron interactions facilitates the rapid thermalization of carriers near the drain contact. This can be clearly seen when we compare the average energy of carriers with and without the inclusion of the electron-electron interaction. Note we only consider the carriers that cross the drain depletion region. This is shown in Fig. 5 and we see that the carriers rapidly lose energy near the drain end of the diode.

5   Conclusions

In summary, we have investigated the role of electron-electron interactions on carrier thermalization at the cathode contact. We find significant carrier thermalization which suggests that modeling HEMT devices does not require use of full-band dispersion at the contacts at moderately high biases.

Acknowledgements

The authors would like to acknowledge the financial support from Freescale Inc., Tempe, AZ, USA.

References

1. Barker, J., et al.: Bulk GaN and AlGaN/GaN heterostructure drift velocity measurements and comparison to theoretical models. Journal of Applied. Physics 97, 63705 (2005)
2. Bhapkar, U.V., Shur, M.S.: Monte Carlo calculation of velocity-field characteristics of wurtzite GaN. Journal of Applied Physics 82(4), 1649–1655 (1997)
3. Camarchia, V., et al.: Theoretical investigation of GaN permeable base transistors for microwave power applications. Semiconductor Science and Technology 21(1), 13–18 (2006)
4. Farahmand, M., Brennan, K.F.: Comparison between wurtzite phase and zincblende phase GaN MESFETs using a full band Monte Carlo simulation. on Electron Devices 47, 493–497 (2000)
5. Gelmont, B., Kim, K., Shur, M.: Monte Carlo simulation of electron transport in gallium nitride. Journal of Applied Physics 74(3), 1818–1821 (1993)
6. Gross, W.J., Vasileska, D., Ferry, D.K.: Discrete impurity effects in nano-scale devices. VLSI Design 13, 75–78 (2001)
7. Hudgins, J.L., et al.: An assessment of wide bandgap semiconductors for power devices. IEEE Trans. on Power Electronics 18(3), 907–914 (2003)
8. Kolnik, J., et al.: Electronic transport studies of bulk zincblende and wurtzite phases of GaN based on an ensemble Monte Carlo including a full zone band structure. Journal of Applied Physics 78, 1033–1038 (1995)
9. Littlejohn, M.A., Hauser, J.R., Glisson, T.H.: Monte Carlo calculation of the velocity-field relationship for gallium nitride. Applied Physics Letters 26, 625–627 (1975)
10. Mansour, N.S., Kim, K.W., Littlejohn, M.A.: Theoretical study of electron transport in gallium nitride. Journal of Applied Physics 77(6), 2834–2836 (1995)
11. Matulionis, A., et al.: Hot-phonon temperature and lifetime in a biased AlGaN/GaN channel estimated from noise analysis. Physical Review B 68, 35338 (2003)

12. Ohno, Y., Kuzuhara, M.: Application of GaN-based heterojunction FETs for advanced wireless communications. IEEE Trans. on Electron Devices 48(3), 517–523 (2001)
13. Sokolov, V.N., et al.: Terahertz generation in submicron GaN diodes within the limited space-charge accumulation regime. Journal of Applied Physics 98(6), 064501–064507 (2005)
14. Trew, R.J., Shin, M.W., Gatto, V.: High power applications for GaN-based devices. Solid State Electronics 41(10), 1561–1567 (1997)
15. Vetury, R., et al.: The impact of surface states on the DC and RF characteristics of AlGaN/GaN HFETs. IEEE Trans. on Electron Devices 48(3), 560–566 (2001)
16. Wordelman, C., Ravaioli, U.: Integration of a particle-particle-particle-mesh algorithm with the ensemble Monte Carlo method for the simulation of ultra-small semiconductor devices. IEEE Trans. on Electron Devices 47(2), 410–416 (2000)
17. Yamakawa, S., et al.: Quantum-corrected full-band cellular Monte Carlo simulation of AlGaN/GaN HEMTs. Journal of Computational Electronics 3(3-4), 299–303 (2004)

Wigner Ensemble Monte Carlo: Challenges of 2D Nano-Device Simulation

M. Nedjalkov[1], H. Kosina[1], and D. Vasileska[2]

[1] Institute for Microelectronics, Technical University of Vienna
Gusshausstrasse 27-29/E360, A-1040 Vienna, Austria
[2] Department of Electrical Engineering,
Arizona State University, Tempe, AZ 85287-5706, USA

Abstract. We announce a two dimensional WIgner ENSemble (WIENS) approach for simulation of carrier transport in nanometer semiconductor devices. The approach is based on a stochastic model, where the quantum character of the carrier transport is taken into account by generation and recombination of positive and negative particles. The first applications of the approach are discussed with an emphasis on the variety of raised computational challenges. The latter are large scale problems, introduced by the temporal and momentum variables involved in the task.

1 Introduction

The Wigner formulation of the quantum statistical mechanics provides a convenient kinetic description of carrier transport processes on the nanometer scale, characteristic of novel nanoelectronic devices. The approach, based on the concept of phase space considers rigorously the spatially-quantum coherence and can account for processes of de-coherence due to phonons and other scattering mechanisms using the models developed for the Boltzmann transport. Almost two decades ago the coherent Wigner equation has been utilized in a deterministic 1D device simulators [3,4,1]. The latter have been refined towards self-consistent schemes which take into account the Poisson equation, and dissipation processes have been included by using the relaxation time approximation. At that time it has been recognized that an extension of the deterministic approaches to two dimensions is prohibited by the enormous increase of the memory requirements, a fact which remains true even for todays computers. Indeed, despite the progress of the deterministic Boltzmann simulators which nowadays can consider even 3D problems, the situation with Wigner model remains unchanged. The reason is that, in contrast to the sparse Boltzmann scattering matrix, the counterpart provided by the Wigner potential operator is dense. A basic property of the stochastic methods is that they turn the memory requirements of the deterministic counterparts into computation time requirements. Recently two Monte Carlo methods for Wigner transport have been proposed [7,5]. The first one has been derived by an operator splitting approach. The Wigner function is presented by an ensemble of particles which are advanced in the phase space and carry

I. Lirkov, S. Margenov, and J. Waśniewski (Eds.): LSSC 2007, LNCS 4818, pp. 139–147, 2008.

the quantum information via a quantity called affinity. The latter is updated at consecutive time steps and actually originates from the Wigner potential, whose values are distributed between particles according their phase space position. This ensemble method has been applied in a self-consistent scheme to resonant-tunneling diodes (RTD's), the scattering with phonons is accounted in a rigorous way [7]. Recently it has been successfully extended to quasi two dimensional simulations of double gate MOSFET's [6]. The second method is based on a formal application of the Monte Carlo theory on the integral form of the Wigner equation. The action of the Wigner potential is interpreted as generation of couples of positive and negative particles. The quantum information is carried by their sign, all other aspects of their evolution including the scattering by phonons are of usual Boltzmann particles. The avalanche of generated particles is controlled by the inverse process: two particles with opposite sign entering given phase space cell annihilate. The approach offers a seamless transition between classical and quantum regions, a property not yet exploited for practical applications.

WIENS is envisaged as an union of theoretical and numerical approaches, algorithms and experimental code for 2D Wigner simulation of nanostructures. In contrast to device simulators which, being tools for investigation of novel structures and materials rely on well established algorithms, WIENS is comprised by mutually related elements which must be developed and tested for relevance and viability. Many open problems need to be addressed such as the choice of the driving force in the Wigner equation, pure quantum versus mixed classical-quantum approaches, the correct formulation of the boundary conditions, appropriate values for the parameters and a variety of possible algorithms. We present the first results in this direction. In the next section a semi-discrete formulation of the Wigner equation for a typical MOSFET structure is derived. An Ensemble particle algorithm is devised in the framework of the second approach. Next, simulation experiments are presented and discussed. It is shown that, despite the nanometer dimensions, the temporal and momentum scales introduce a large scale computational problem.

2   Semi-discrete Wigner Equation

A system of carriers is considered in a typical 2D structure, for example of a MOSFET shown in Fig. 1. The device shape is a perfect rectangle with the highly doped Source and Drain regions at the left and right bottom ends respectively. At the very bottom, to the left and right of the Gate are shown parts of the leads. These supply carriers and thus specify the boundary conditions in the two tinny strips marked in black. It is assumed that the current flows between the device and the leads only, that is, at the rest of the device boundary including the region under the gate carriers are reflected. The state of the carriers is characterized by a wave function which becomes zero at and outside these boundaries. Accordingly the density matrix $\rho(\mathbf{r}_1, \mathbf{r}_2)$, will vanish if some of the arguments $\mathbf{r} = (x, y)$ is outside of the rectangle $(0, \mathbf{L}) = (0, L_x); (0, L_y)$ determined by the device dimensions. This condition holds everywhere but on the leads. We postpone the discussion of the leads and first introduce the Wigner function as obtained from the density matrix by the continuous Wigner-Weyl transform:

$$f_w(\mathbf{r}, \mathbf{k}, t) = \frac{1}{(2\pi)^2} \int_{-\infty}^{\infty} ds e^{-i\mathbf{k}\mathbf{s}} \rho(\mathbf{r} + \frac{\mathbf{s}}{2}, \mathbf{r} - \frac{\mathbf{s}}{2}, t); \tag{1}$$

where $\mathbf{r} = \frac{\mathbf{r}_1 + \mathbf{r}_2}{2}$, $\mathbf{s} = \mathbf{r}_1 - \mathbf{r}_2$.

The condition $0 < \mathbf{r}_1, \mathbf{r}_2 < \mathbf{L}$, (which holds everywhere but on the leads) gives rise to the following condition for $\mathbf{s}$:

$$-\mathbf{L}_c < \mathbf{s} < \mathbf{L}_c \qquad \mathbf{L}_c = 2\min(\mathbf{r}, (\mathbf{L} - \mathbf{r})). \tag{2}$$

The confinement of $\mathbf{s}$ allows to utilize a discrete Fourier transformation in the definition (1). Two details must be adjusted. $\mathbf{L}_c$, which is actually the coherence length, depends on $\mathbf{r}$. Thus the discretization of the wave vector space will change with the position which is inconvenient. Fortunately, since any partially continuous function of $\mathbf{s}$ defined in given interval $(\mathbf{a}, \mathbf{b})$ can be presented by the Fourier states of a larger interval, we can conveniently fix the coherence length to the maximal value of the minimum in (2): $\mathbf{L}_c = \mathbf{L}$. As the values of the function outside $(\mathbf{a}, \mathbf{b})$ are extrapolated to 0, we must formally assign to ρ a domain indicator $\theta_D(\mathbf{s})$, which becomes zero if (2) is violated. However, according the above considerations, θ_D is implicitly included in ρ. From a physical point of view the problem whether to consider the leads or not is still open: Usually leads are included in 1D simulations and the integral in (1) is truncated somewhere deep in them for numerical reasons. In a recent study [2] it is argued that in the leads there are processes which entirely destroy the quantum interference between the point $\mathbf{r} - \mathbf{s}$, Fig. 1, and the corresponding counterpart $\mathbf{r} + \mathbf{s}$. In this way the integral in (1) has to be truncated already at the point where the line along $\mathbf{s}$ enters into the gate (G) region. Alternatively, if leads are considered, the segment of this line lying in the gate region (and the corresponding counterpart in the device) are excluded from the integral. The correlation is enabled after the point of entering into the leads. However, even in this case, $\mathbf{s}$ remains bounded.

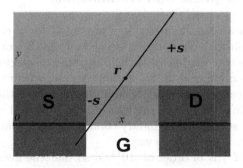

Fig. 1. Typical 2D structure

In this case $\mathbf{L}_c$ must be augmented to $(0, L_x); (-L_y, L_y)$. By changing $\mathbf{s}$ to $2\mathbf{s}$ and applying the discrete Fourier transform:

$$f(\mathbf{r}, \mathbf{n}) = \frac{1}{\mathbf{L}_c} \int_{\mathbf{L}_c/2}^{\mathbf{L}_c/2} ds e^{i2\mathbf{n}\Delta\mathbf{k}\mathbf{s}} \rho(\mathbf{r}, \mathbf{s})$$

$$\rho(\mathbf{r}, \mathbf{s}) = \sum_{\mathbf{n}=-\infty}^{\infty} e^{i2\mathbf{n}\Delta\mathbf{k}\mathbf{s}} f(\mathbf{r}, \mathbf{n})$$

$$\Delta\mathbf{k} = \pi/\mathbf{L}_c$$

to the von-Neumann equation for the density matrix we obtain a Wigner equation which is continuous in space and discrete in momentum:

$$\frac{\partial f(\mathbf{r}, \mathbf{M}, t)}{\partial t} + \frac{\hbar}{m}\mathbf{M}\Delta\mathbf{k}\frac{\partial f}{\partial \mathbf{r}}(\mathbf{r}, \mathbf{M}, t) = \sum_{\mathbf{m}=-\infty}^{\infty} V_w(\mathbf{r}, \mathbf{m}) f_w(\mathbf{r}, (\mathbf{M} - \mathbf{m}), t).$$

Here the Wigner potential is defined as:

$$V_w(\mathbf{r}, \mathbf{M}) = \frac{1}{i\hbar} \frac{1}{\mathbf{L}_c} \int_{-\mathbf{L}_c/2}^{\mathbf{L}_c/2} ds e^{-i2\mathbf{M}\Delta\mathbf{k}\mathbf{s}} (V(\mathbf{r}+\mathbf{s}) - V(\mathbf{r}-\mathbf{s}))\theta_D(\mathbf{s}) \qquad (3)$$

We note the presence of the domain indicator in this definition. According the particle sign approach [5], the Wigner potential generates particles with a frequency given by the Wigner out-scattering rate γ obtained from (3):

$$\gamma(\mathbf{r}) = \sum_{\mathbf{M}=0}^{\infty} |V_w(\mathbf{r}, \mathbf{M})| \qquad (4)$$

The shape and magnitude of γ strongly depend on the treatment of the leads, as it will be shown in what follows.

3   Numerical Aspects and Simulations

The above theoretical considerations are presented for a coherent transport, where the interaction with phonons is switched off. As the corresponding coherent algorithm is the core module of WIENS, it must be carefully developed and tested. We furthermore focus on the stationary case, where the boundary conditions control the carrier transport. The initial picture of the algorithm under development is of an ensemble of particles which is evolved in the phase space at consecutive time steps. The boundary conditions are updated after each step in the usual for device simulations way.

The magnitude of γ is of order of $[10^{15}/s]$ so that the avalanche of particles does not allow an individual treatment of each particle as in the classical Ensemble Monte Carlo algorithm. Particles must be stored on grid points of a mesh in the phase space, where the inverse process of annihilation occurs. Thus along with the wave vector spacing $\Delta\mathbf{k}$, also the real space must be divided into cells of size $\Delta\mathbf{r}$. Actually two arrays are needed to store the particles.

At the beginning of an evolution step the initial one, f_1, is occupied by particles, while the second one, f_2, is empty. Particles are consecutively taken from f_1, initiating from randomly chosen phase space coordinates around the corresponding grid point. A selected particle evolves untill the end of the time step and then is stored in f_2. The stored wave vector corresponds to the initial value since there is no accelerating field. The particle gives rise to secondary, ternary etc. particles which are evolved in the phase space for the rest of the time and then stored in f_2. As they are generated on the same grid in the wave vector space, the assignment in straightforward. As a rule the particles injected by the boundary conditions are slow (low wave vector), while these generated by the Wigner potential are fast. The wave vector ranges over several orders of magnitude so that the task for the position assignment becomes a large scale problem: A straightforward approach is to assign the particle position to the nearest grid point. For fast particles which cross several cells during the time step this introduces small error in the spatial evolution. More dramatic is the situation with the slow particles: if during the time step a particle crosses a distance less than a half of the mesh step it can remain around a grid point for a long time. This is an example for artificial diffusion which is treated in the following way:

(i) Slow particles, e.g., these which belong to the ground cell around the origin of the wave vector are treated in a standard ensemble approach: the phase space evolution is followed continuously throughout the device.

(ii) The grid assignment is chosen stochastically, according a probability proportional to the distance to the neighborhood grid points.

Another large-scale aspect is related to the range of the time constants involved in the problem. The existence of very fast particles imposes an evolution step of few hundreds of femtosecond. The time step t_a between successive assignments to the grid is of order of femtosecond, while the total evolution must be above picosecond in order to reach the stationary conditions. Accordingly large computational times are expected. Thus the first objective of the computations is to investigate the convergence, to optimize where possible the algorithm and to find appropriate values for the parameters leading to stable results. The latter are a necessary condition for solving the physical aspects of the problem: finding a proper normalization, a choice of the boundary conditions and investigation of the quantum phenomena will be focused on a next step.

Two structures, A and B of the type shown on Fig. 1 are considered in the experiments. A Γ valley semiconductor with a small effective mass (0.036) has been chosen to enhance the effects of tunneling. The potential and dimensions of device A are twice as small as compared to device B.

The potential of B is shown in Fig. 2 and Fig. 3 as a contour plot respectively. The potential is obtained from the self-consistent Boltzmann-Poisson solution: In the quantum case the entire potential is used to calculate $\gamma(\mathbf{r})$ and the scattering probability table. The driving force is zero so that the carriers perform a free motion throughout the device until the reflecting boundaries.

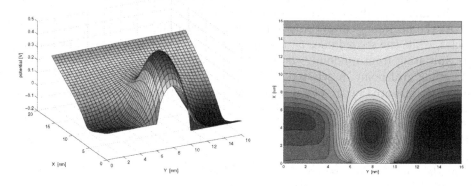

Fig. 2. Device potential   **Fig. 3.** Contour plot of the potential

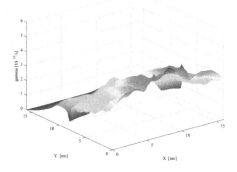

Fig. 4. $\gamma(\mathbf{r})$, device B

Fig. 4 shows $\gamma(\mathbf{r})$ computed for the case including the leads. At the injecting boundaries in the source and drain regions (compare Fig. 1) the generation rate is very high so that an injected particle feels the Wigner potential already at the boundary. On contrary, if leads are excluded the generation rate is zero at this boundary.

Figures 5 and 6 show γ in device A, computed for either of the two cases. There is a profound difference of the shape and magnitude of this quantity on the two pictures. The contour plots of the classical and quantum densities in device A, no leads considered, are compared in Fig. 7 and 8. Both densities have the same shape, however the quantum counterpart is more spread inside the device which can be related to tunneling effects. We note that the quantum density is due to effects of generation and recombination only, so that the basic similarity was the first encouraging result. In particular, since the small generation rate $\simeq 10^{14}/s$ the time t_a between successive assignments is a femtosecond, so that the effect of artificial diffusion is negligible. Here the algorithms treating this effect have

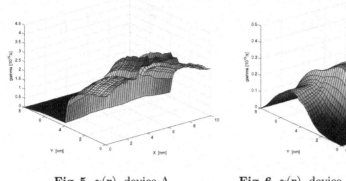

Fig. 5. $\gamma(\mathbf{r})$, device A   **Fig. 6.** $\gamma(\mathbf{r})$, device A, leads are excluded

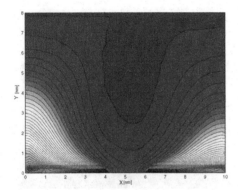   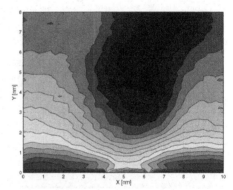

Fig. 7. Boltzmann carrier density   **Fig. 8.** Wigner carrier density

been introduced and tested by reducing t_a. Fig. 9 demonstrates the much slower convergence of the quantum current as a function of the evolution time.

The potential of device B, Fig. 3, has a shape of two valleys separated by a high barrier with a maximum between the valleys. Between this maximum and the high potential at the bottom of the base there is a saddle point. The carrier density is expected to follow this pattern: carriers fill the valleys while their number should decrease with the raise of the potential. This is essentially the behavior of the quantum densities at Fig. 10 and Fig. 11. The former is obtained for a spatial step of 0.2nm giving rise to a Wigner function with 41.10^6 elements. Since the great number of particles the evolution time reached after 30 days of CPU time of a regular 2.5 GHz PC is one picosecond only. Fig. 11 corresponds to a 0.4nm step: the dimension of the Wigner function is around one order of magnitude smaller and the convergence is as much faster. The use of such step became possible due to the algorithms avoiding the artificial diffusion. A difference in the normalization factor along with some nuances in the shape exist, e.g., in the density around the saddle point and the position of the left peak.

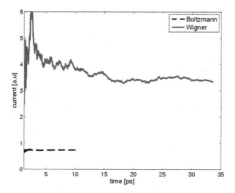

Fig. 9. Convergence of the current

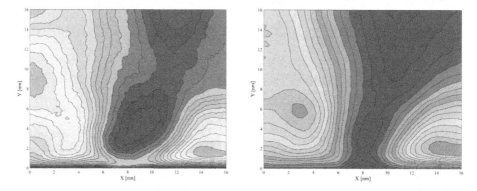

Fig. 10. Carrier density for a 0.2 nm mesh   **Fig. 11.** Carrier density for a 0.4 nm mesh

Several factors can be responsible for this difference: the annihilation mesh is different, the number of the used **k** states, the option that Fig. 10 is not yet in a stationary state. These problems can not be answered without implementation of MPI and GRID technologies which is currently underway.

4   Conclusions

A stochastic approach is developed within a semi-discrete Wigner-Weyl transform, for which the problem of 2D Wigner transport is not an impossible numerical task. The obtained first results are qualitative and mainly demonstrate the convergence, which, furhtermore characterizes a large-scale computational problem. MPI and GRID technologies must be implemented to address the physical aspects such as proper boundary conditions, normalization and resolution of the incorporated quantum phenomena.

Acknowledgment

This work has been partially supported by the Österreichische Forschungsgemeinschaft (ÖFG), Project MOEL239.

References

1. Biegel, B., Plummer, J.: Comparison of self-consistency iteration options for the Wigner function method of quantum device simulation. Phys. Rev. B 54, 8070 (1996)
2. Ferrari, G., Bordone, P., Jacoboni, C.: Electron Dynamics Inside Schort-Coherence Systems. Physics Letters A 356, 371 (2006)
3. Frensley, W.R.: Boundary conditions for open quantum systems driven far from equilibrium. Rev. of Modern Physics 62, 745 (1990)
4. Kluksdahl, N.C., et al.: Self-consistent study of resonant-tunneling diode. Phys. Rev B 39, 7720 (1989)
5. Nedjalkov, M., et al.: Unified particle approach to Wigner-Boltzmann transport in small semiconductor devices. Phys. Rev. B 70, 115319 (2004)
6. Querlioz, D., et al.: Fully Quantum Self-Consistent Study of Ultimate DG-MOSFETs Including Realistic Scattering Using Wigner Monte Carlo Approach, IEDM — Int.Electron Devices Meeting (2006)
7. Shifren, L., Ringhofer, C., Ferry, D.K.: A Wigner Function-Based Quantum Ensemble Monte Carlo Study of a Resonant Tunneling Diode. IEEE Trans.Electron Devices 50, 769 (2003)

Monte Carlo Simulation for Reliability Centered Maintenance Management

Cornel Resteanu[1], Ion Vaduva[2], and Marin Andreica[3]

[1] National Institute for Research and Development in Informatics,
8–10 Averescu Avenue, 011455, Bucharest 1, Romania
resteanu@ici.ro
[2] University of Bucharest, Department of Mathematics and Informatics
14 Academiei St, 010014, Bucharest 1, Romania
vaduva@fmi.unibuc.ro
[3] Academy of Economic Studies,
6 Romana Square, 010572, Bucharest 1, Romania
marinandreica@yahoo.com

Abstract. The paper presents the *RELSYAN* software, which is a tool for reliability analysis of the reparable systems. One focus on the software capability to consider, for a system, a model conceived by a pattern chart (block diagram) plus one or more fault trees. The core of the discourse is the simulation, using the Monte Carlo method, the functioning of the system, taking into account different maintenance strategies, spare parts and manpower policies, mission profiles, skills in operation and external stress factors. The benefits of fast simulations are outlined.

Keywords: Reparable Systems, Pictorial Modeling, Reliability Analysis, Maintenance Management, Monte Carlo Simulation.

1 Introduction

Due to the development of large industrial plants in Romania, in the computer science era, efforts have been made so far as to design information systems meant to contribute to their optimal control. In the paper, advanced solutions in simulating the systems functioning (which is the main yield of the Brite-Euram 95 – 1007 project *"Innovative software tools for reliability centred maintenance management"*) will be presented. As the analytical and Markov chains based methods have a lot of drawbacks, for example they do not allow time variable failure rates [1], *RELSYAN* will implement the Monte-Carlo simulation method to analyse the infrastructure behaviour of a manufacturing/service/benchmarking/other type system [7,3]. In order to create a solid basis for decision-making in maintenance management, the software is able to simulate the life of a system infrastructure taking into account different maintenance strategies, spare parts and manpower policies, mission profiles, skills in operation and external stress factors.

I. Lirkov, S. Margenov, and J. Waśniewski (Eds.): LSSC 2007, LNCS 4818, pp. 148–156, 2008.
© Springer-Verlag Berlin Heidelberg 2008

2 System's Model Building

The system structure and behavior will be specified using graphical editors [6] which produce pattern charts (block diagrams) and fault trees. The pattern chart, whose hierarchical editing is determined by the organizational structure, will ask for data in the basic entities in maintenance database and organize them in a graphical formalism which represents the system. Conceptually, the resulting *drawing* is a directed graph. The graph nodes will be represented by pictograms, either shadowed-outline rectangles if they indicate non-terminal structure elements (blocks), or regular-outline rectangles if they indicate terminal structure elements (components). The graph arcs are connections among the structure elements. They can be viewed in terms of failure effects, or even more, they can acquire information about the nature of the tasks performed by a structure element on behalf of another structure element. In order to build a pattern chart, a user has two types of micro-structures of pictograms which he/she may invoke. The first type, known as *standard micro-structures*, contains series, parallel (k-out-of-n, consecutive or not), stand-by, with fault alarm device, bridge micro-structures of components or blocks. The standard micro-structures are pre-defined. The second type, known as *specific (or user) micro-structures*, includes user's productions. Highly recurrent micro-structures will always direct towards using such a facility. In fact, *RELSYAN* software will in time suppress the differentiation between standard and specific micro-structures.

The fault tree cannot be but an alternative to the pattern chart. It can supplement the system reliability analysis, which must be a very complex analysis. The so-called stress factors (environmental, climate, and ambient conditions, natural disasters such as flood, hurricanes, and earthquakes, conflicts, fires, sabotages, long mental and physical exhaustion, nonobservance of the maintenance schedule, wrong operation of human factors, etc.) will always escape from the pattern charts. A fault tree is a logical diagram describing the interactions between a potential critical event (an accident) of a system and the causal factors of such an event. The *RELSYAN* software only refers to the quality-oriented fault trees, while the quantitative analysis stays with the Monte-Carlo method, which applies in simulating the system functioning. If the pattern chart insists on the structural description, embedding the functional description, the fault tree is mostly concerned with the functional description taking into account even the external stress factors. While making a fault tree description, the user calls for a generic pictogram, a rectangle with a proceeding symbol, the undeveloped symbol, which generates by means of customization any of the graphical symbols used in this case, namely and/or gates, out/in transfers and basic events. To the elements of both types of diagrams, the user can attach attributes referring to their identification and individual time-behaviour, if they are terminal elements. Various failure and repair time models (exponential, normal, lognormal, Weibull, Gamma, Pareto, Birnbaum-Saunders, Gumbel-small extreme, inverse-Gaussian, Rayleigh, Fisk, etc.) are possible. Discrete random events are simulated by using various discrete probability distributions (Bernoulli, binomial, Pascal, Poisson, etc). Pattern charts and fault trees can be validated in an explicit mode so that the user might be able to solve special situations such as: incorrectness, inconsistency, incompleteness.

3   System Function in *RELSYAN* Context

The structure function Φ bridges the components state vector $\underline{x} = (x_1, x_2, \ldots, x_n)$ and the system state, namely $\Phi : \{0,1\}^n \to \{0,1\}$ with $\Phi(\underline{x}) = x_s$, where

$$x_s = \begin{cases} 1, & \text{if the system works} \\ 0, & \text{if the system has a breakdown} \end{cases}$$

is the synthetical form of the system state. For a proper presentation of the system function associated algorithms, one consider:

— $V \neq \Phi, V = (v_1, v_2, \ldots, v_n)$ the vertices (or nodes) set, where $n =$card V;
— $A \subseteq V \times V$ the arcs set, as an ordered set of 2-uples (v_i, v_j), $v_i, v_j \in V$.

$DG = (V, A)$ is the directed graph which, in the condition that it is free of circuits, can be the inner model for pattern charts and with the condition $od(v_i) = 1$, $(\forall)$ $i = \overline{1, n}$ can be the inner model for fault trees. The inner model is stored in a multi-m-tree shaped file, which is dynamically balanced throughout editing the external model. This file can be used both for restoring the drawing, by key-reading, and for rapidly restoring the graph, by sequential reading. Notice that the same mechanism is valuable both for pattern chart and fault tree. An m-tree contains a model page. The pages are linked by a system of double pointers. Each page contains a sub-graph / sub-tree stored in a specific gamma format (all emergent arcs from a node in a single record). In order to obtain a very performing system function, one should next consider the system as a directed graph where two privileged nodes categories are indicated, the input nodes and the output nodes. The rest of nodes are considered to be intermediary nodes. Based on the arc matrix $A = (a_{ij})_{1 \leq i,j \leq n}$, the path matrix $P = (p_{ij})_{1 \leq i,j \leq n}$ is obtained, where:

$$a_{ij} = \begin{cases} 1, & \text{if the } i\text{-th node is linked with the } j\text{-th node,} \\ 0, & \text{otherwise,} \end{cases}$$

and

$$p_{ij} = \begin{cases} 1, & \text{if there is a path from the } i\text{-th node to the } j\text{-th node,} \\ 0, & \text{otherwise.} \end{cases}$$

Theoretically, the path matrix will be calculated by consecutive raisings $A = A^{(1)}$, $A^{(k+1)} = A^{(k)} \otimes A$, for $k = 1, 2, \ldots$, until $A^{(k+1)} = A^k$ is obtained. The operation $\otimes$ is defined as follows: if we denote $A = (a_{ij})_{1 \leq i,j \leq n}$, $B = (b_{ij})_{1 \leq i,j \leq n}$, $C = (c_{ij})_{1 \leq i,j \leq n}$, then

$$A \otimes B = C \quad \text{with} \quad c_{ij} = \max_{i \leq k \leq n} \{\min(a_{ik}, b_{kj})\}.$$

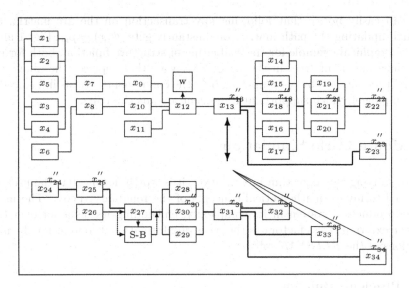

Fig. 1. Pattern chart: new system's state after a transaction in X_{24}

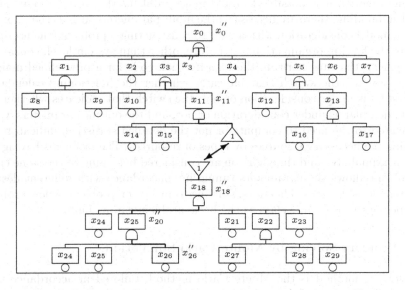

Fig. 2. Fault tree: new system's state after a transaction in X_{26}

Therefore $P = A^{(k+1)}$ for which $A^{(k+1)} = A^{(k)}$. The matrix P, (i.e. *path matrix*), is the simplest way for defining a structure function:

$$\Phi(\underline{x}) = p_{ij} = \begin{cases} 1, & \text{if there is a path from the node } i \text{ to the node } j \\ 0, & \text{otherwise.} \end{cases}$$

One can easily notice that following any transaction on the arc matrix, and through updating the path matrix, one instantly gets $\Phi(\underline{x}) = p_{ij}$. Fig. 1 and 2 show a graphical example for the evaluating of structure function, both for pattern chart and fault tree, in the RELSYAN way, that replaces the above consuming time method by re-computing only the components' status influenced by the component which supports at a moment a staus changing.

4   Monte Carlo Simulations

Before starting the algorithm presentation, it is worth noting that the user will be saved to invoke mathematical concepts of the reliability theory. The model of the production system in its complexity, embedding its behavior over time and the external stress factors, is the basis of informational content for the main functions of the *RELSYAN* software.

4.1   Problems Building

Having a model, it is possible to build more problems that are different. For a problem building, there are necessary: the input parameters such as the maximal area of model consideration with a view to constructing a problem (the user may specify the level in organizational structure, other than system-level, considered as implicit), the time horizon, the identification code for a pre-defined maintenance strategy, the switch that indicates whether the functioning calendar of the system is under consideration or not, the switch that indicates whether the production plan is under consideration or not, and the output parameters which are related to the need to compute or not the detailed statistical indicators expressing the necessary resources by types of resources. The problems having the same maximal area and time horizon are considered belonging to the same class. A problem allows the parameters tuning in concordance with different desired maintenance strategies. Therefore, starting from a given problem, a new problem that belongs to the same class as the basic problem can be built.

4.2   Problem Solving by Monte Carlo Simulation

The solving method is the Monte Carlo method, tailored in accordance with the specific demands of the maintenance activity. All types of maintenance jobs: corrective (direct, postponed, curative), preventive (regular, conditional, systematic, cyclical, indicative, critical, limitative), mixed (functional, cellular, permissive, cyclical, indicative), spare parts acquisition policies and modalities to use the available manpower can be under consideration.

The main idea of the simulation model will be first underlined. We assume that the system consists of independent components, in stochastic sense. Each component C_i has its own *clock time* $T[i]$ (or operational time) specified separately for different types of maintenance (corrective, preventive, and mixed).

In order to describe the correct dynamic behavior of the system, we will use the technique of the *variable increment clock time*, which is calculated during the simulation run as [2,10]:

$$clock = \min_{1 \leq i \leq n} T[i]. \tag{1}$$

In short, the simulation algorithm looks as follows: initially the clock is zero and the system components are set in some given states which are either read into the computer or (better) generated from a specified distribution together with their *life times*, i.e. the times components remain in these initial states; the logical succession of states is also specified, e.g. is: 1, 0, 1, 0, ... etc. Then the clock time is advanced according to the next event rule (i.e. given by (1)). The components are then analyzed and processed by updating statistics of interest as: counting some events (number of values with $\Phi(\underline{x}) = 1$, number of times when financial resources are available for maintenance which can be used for estimating *system's sufficiency* etc.) updating sums of *repair times* or *waiting times* or quantities of resources (e.g. spare parts or manpower) necessary to estimate the *maintainability*, updating costs of inventories of spare parts involved in reliability management and so on. After simulating a trajectory, the required reliability characteristics are determined. Then, using a large number of trajectories, the estimates as *arithmetic means*, empirical variances or other statistics are calculated giving more accurate information.

A structured description of the simulation algorithm, named *RELSIM*, is given in the following:

```
begin
    call INITP;
    { This module initializes the simulation variables
      and inputs the parameters of the model}
    call INITSTATALL;
    { This module initializes all statistics of the model}
    call PRELIMSIM;
    { This module simulates a small number NS of histories
      in order to define the basic limits of the histograms and
      contingency tables}
    call DEFINETABLES;
    { This module determines the limits of all
      tables and initializes the frequencies}
    for nrhist:=1 to N do {N is the number of histories to be
simulated}
        begin
            call INITHIST; { This module initializes data necessary
                             to simulate a history}
```

call INITCLOCK; {*Determine the initial clock of the history*}

while $clock \leq TMAX$ {*TMAX is time-duration of a simulated history*} do
  begin
    call MAINREL; {*This is the main module*}
    call UPDATECLOCK; {*This module updates the clock*}
  end;
  call UPDATEHIST; {*This updates statistics for the current history*}
end;
call OUTPUTSIM; {*This module calculates the output of simulation*}
end.

The procedure INITP is almost self explanatory; it initializes all parameters of the simulation model. But there is a necessity of some comments to be made on INITSTATALL, PRELIMSIM, and DEFINETABLES. While each history is simulated, it is produced *one sampling value* of some reliability characteristics (e.g. availability, maintainability, etc.). By simulating many histories we produce *samples* on these characteristics. The simulation estimates mean values, variances, and *correlation coefficients* of these characteristics as well as their *empirical* probability distributions (e.g. *histograms or contingency tables* for pairs of such characteristics). All statistics used to calculate these estimates are initialized in INITSTATALL. Here is a counter recording the number of system failures. Its initial value is zero. The PRELIMSIM simulates initially a small number of histories necessary to define limits of intervals of histograms. For instance if for a variable Y the histogram has the limits of intervals $a_0, a_1, a_2, \ldots, a_k$, the PRELIMSIM simulates NS histories (the NS, $NS < N$, is a small integer) producing the sample $Y_1, Y_2, \ldots, Y_{NS}$ and the DEFINETABLES calculates the elements of the histogram:

$$a_1 = \min_{1 \leq i \leq NS} Y_i, \qquad a_{k-1} = \max_{1 \leq i \leq NS} Y_i, \qquad \delta = (a_{k-1} - a_1)/(k-2),$$

$$a_i = a_1 + (i-1) * \delta, \qquad 1 \leq i \leq k-1.$$

The limits a_0 and a_k will be updated during the simulation of the following histories. Procedures INITHIST, INITCLOCK, and UPDATECLOCK are also self explanatory; the INITHIST initializes the history and INITCLOCK initializes the clocks of components and determines the first value of the system clock. The last one for instance *updates* the clock according to formula (1).

The procedure MAINREL is one very important. Here are some comments on it. This procedure selects first the *event* to be processed. If the component whose clock is equal to the simulation clock in (1) is failed, then first it is *repaired* (put in state 0), the structure function Φ is calculated and if $\Phi = 0$ then the counter of failures is updated; otherwise the component is put to run (i.e. is in state 1). In each case the corresponding times (repair or run) are simulated as random variates and all sums are updated. For each current value of the clock, appropriate statistical indicators (see bellow) are recorded and / or updated. The procedure is complex and it can be expanded by computing various indicators, both of technical or economical nature, and by considering diverse maintenance or inspection specifications. It could be itself a huge computer package. In case of repair, all situations deriving from maintenance policy are performed and corresponding characteristics are calculated or updated. When necessary, various probability distributions, involving even correlated variables, and other discrete random events are simulated. (See for instance [10].)

The procedure UPDATEHIST *updates* all statistics with the output values produced by the previous history and *tabulates* the corresponding characteristics (in histograms or contingency tables) [4,5,8,9].

The procedure OUTPUTSIM calculates the estimates of mean values, variances and correlation coefficients, calculates the economic importance of the components, estimates the reliability of the system, and displays the calculated values. The output will offer estimations of the Reliability, Availability, Maintainability (RAM) indicators. The offered statistics are of three kinds: *basics, complementary* and *details*. Basic Statistical Indicators are: Reliability, Availability, Maintainability, Spare parts costs, Manpower costs. Complementary Statistical Indicators are: Unavailability, Number of system failures, Down time, Down time due to failures, Down time due to corrective maintenance, Down time due to preventive maintenance, Down time due to lack of spare parts, Down time due to lack of manpower, Total cost, Production loss cost due to unavailability, Maintenance cost, Cost due to corrective maintenance, Cost due to preventive maintenance, Cost due to lack of spare parts, Cost due to lack of manpower.

5   Conclusions

In order to obtain a clearer image of *RELSYAN* performances, an implementation result will be presented for a model of medium size and complexity (500 nodes, 6 imbrications levels, all kinds of micro-structures, 30% k-out-of-n micro-structures, two fault trees supporting events of exogenous nature with 125 nodes). The simulation was done over a year. The input data were offered by the software company SIVECO-ROMANIA, which has done aplications for some factories in Romania. This one year history is not relevant for testing; relevant is the fact that were simulated 10 different maintenance strategies, 5000 histories (ensuring good estimates), and all the computing capabilities of the product were taken into consideration. It was noticed that the span of evaluating the system function, after the changes of variable's states, took 15% out of the total running time. The performance was considered very well because, on a Pentium 4 computer, this task takes only one hour. To the authors' knowledge, no famous product is available on the market, which can concurrently consider, during a Monte Carlo simulation process, a hybrid model consisting of a pattern chart and more fault trees. The situation is the same for the couple system function, Monte Carlo simulation. As one can see, to put simulation at work is obtained a good maintenance strategy and, in the same time, a good resources (spare parts and manpower) policy, which is very usefull in maintenance management.

References

1. Barlow, R.E., Prochan, F.: Mathematical Theory of Reliability. John Wiley and Sons, New York (1965)
2. Cassandras Ch.: Discrete Events Systems Modelling and Performance Analysis, Richard D. Irvin and Aksen Associates Inc. (1993)
3. Fishman, S.G.: Monte Carlo Concepts, Algorithms and Applications. Springer, Berlin (1996)
4. Gerstbakh, I.B.: Statistical Reliability Theory. Marcel Dekker, Inc., New York (1989)
5. Hoyland, A., Raussand, M.: System Reliability Theory. Models and Statistical Methods. John Wiley and Sons, Inc., New York (1994)
6. Resteanu, C., et al.: Building Pattern Charts and Fault Trees for Reliability, Availability and Maintainability Analysis. In: The Proceedings of the 12th European Simulation Multi-Conference, Society for Computer Simulation International, Manchester, United Kingdom, pp. 211–215 (1998)
7. Ripley, B.D.: Stochastic Simulation. John Wiley and Sons, New York (1987)
8. Shaked, M., Shantikunar, G.J.: Reliability and Maintainability. In: Heyman, D.P., Sobel, M.J. (eds.) Handbook in OR & MS, vol. 2, pp. 653–713. Elsevier Science Publishers B.V (North Holland), Amsterdam (1990)
9. Sherwin, D.J., Bosche, A.: The Reliability, Availability and Productivness of Systems. Chapman and Hall, London (1993)
10. Vaduva, I.: Fast Algorithms for Computer Generation of Random Vectors Used in Reliability and Applications. Preprint No. 1603, Januar 1994, Technische Hochschule Darmstadt, Germany (1994)

Monte Carlo Algorithm for Mobility Calculations in Thin Body Field Effect Transistors: Role of Degeneracy and Intersubband Scattering

V. Sverdlov, E. Ungersboeck, and H. Kosina

Institute for Microelectronics, Technical University of Vienna,
Gusshausstrasse 27–29, A 1040 Vienna, Austria
sverdlov@iue.tuwien.ac.at

Abstract. We generalize the Monte Carlo algorithm originally designed for small signal analysis of the three-dimensional electron gas to quasi-two-dimensional electron systems. The method allows inclusion of arbitrary scattering mechanisms and general band structure. Contrary to standard Monte Carlo methods to simulate transport, this algorithm takes naturally into account the fermionic nature of electrons via the Pauli exclusion principle. The method is based on the solution of the linearized Boltzmann equation and is exact in the limit of negligible driving fields. The theoretically derived Monte Carlo algorithm has a clear physical interpretation. The diffusion tensor is calculated as an integral of the velocity autocorrelation function. The mobility tensor is related to the diffusion tensor via the Einstein relation for degenerate statistics. We demonstrate the importance of degeneracy effects by evaluating the low-field mobility in contemporary field-effect transistors with a thin silicon body. We show that degeneracy effects are essential for the correct interpretation of experimental mobility data for field effect transistors in single- and double-gate operation mode. In double-gate structures with (100) crystal orientation of the silicon film degeneracy effects lead to an increased occupation of the higher subbands. This opens an additional channel for elastic scattering. Increased intersubband scattering compensates the volume inversion induced effect on the mobility enhancement and leads to an overall decrease in the mobility per channel in double-gate structures.

1 Introduction

Monte Carlo is a well-established numerical method to solve the Boltzmann transport equation. Traditionally, the so called forward Monte Carlo technique [5] is used to find the distribution function. Within this approach, particles are moving on classical trajectories determined by Newton's law. The motion along the trajectory is interrupted by scattering processes with phonons and impurities. Scattering is modeled as a random process. The duration of a free flight, the scattering mechanism and the state after scattering are selected randomly from

I. Lirkov, S. Margenov, and J. Waśniewski (Eds.): LSSC 2007, LNCS 4818, pp. 157–164, 2008.

a given probability distribution which is characteristic of the scattering process. This technique of generation sequences of free flights and scattering events appears to be so intuitively transparent that it is frequently interpreted as a direct emulation of transport process. Due to the Pauli exclusion principle, scattering into an occupied state is prohibited. Therefore, scattering rates depend on the probability that the final state is occupied, given by the distribution function, which is the solution of the Boltzmann equation. Dependence of scattering rates on the solution makes the Boltzmann transport equation nonlinear. In many cases the occupation numbers are small and can be safely neglected in transport simulations for practically used devices. With downscaling of semiconductor devices continuing, the introduction of double-gate (DG) silicon-on-insulator field-effect transistors (FETs) with ultra-thin (UTB) silicon body seems increasingly likely [6]. Excellent electrostatic channel control makes them perfect candidates for the far-end of ITRS scaling [1]. However, in UTB FETs degeneracy effects are more pronounced, and their proper incorporation becomes an important issue for accurate transport calculations.

Different approaches are known to include degeneracy effects into Monte Carlo algorithms. One method is to compute the occupation numbers self-consis- tently [2,8]. This approach is applicable not only to mobility simulations at equilibrium but also for higher driving fields [7]. When the distribution function is close to the equilibrium solution, the blocking factor can be approximated with the Fermi-Dirac distribution function [3]. A similar technique to account for degeneracy effects was recently reported in [14].

In this work we use a Monte Carlo algorithm originally developed for three-dimensional simulations [10] which was recently generalized to a quasi-two-dimen- sional electron gas [11,4]. This method incorporates degeneracy effects *exactly* in the limit of vanishing driving fields and is valid for arbitrary scattering mechanisms and for general band structure. We demonstrate that in UTB DG FETs degeneracy effects lead to a qualitatively different mobility behavior than in the classical simulations. Degeneracy results in higher occupation of upper subbands which substantially increases intersubband scattering in (100) UTB DG FETs, resulting in a mobility decrease.

2   Simulation Method

In order to obtain the low-field mobility in UTB FETs, we compute the response of a quasi-two dimensional electron system to the small electric field $\mathbf{E}(t)$. The system is described by a set of subband functions $\psi_n(z)$ in the confinement direction and dispersions $E_n(\mathbf{k})$ relating subband energies to the two-dimensional quasi-momentum $\mathbf{k} = (k_x, k_y)$ along $\mathrm{Si/SiO_2}$ interfaces. Representing the distribution function as $f(E_n(\mathbf{k}, t)) = f_0(E_n(\mathbf{k})) + \delta f_n(\mathbf{k}, t)$, where $f_0(E_n(\mathbf{k}))$ is the equilibrium Fermi-Dirac distribution and $\delta f_n(\mathbf{k}, t)$ is a small perturbation, we arrive to the system of coupled linearized subband equations:

$$\frac{\partial \delta f_n(\mathbf{k}, t)}{\partial t} = -e\mathbf{E}(t)\nabla_{\mathbf{k}} f_0(E_n(\mathbf{k})) + Q_n[\delta f], \qquad (1)$$

where Q_n is the scattering operator of the linearized Boltzmann equation

$$Q_n[\delta f] = \sum_m \int \frac{d^2\mathbf{k}'}{(2\pi)^2} (\Lambda_{nm}(\mathbf{k}, \mathbf{k}')\delta f_m(\mathbf{k}', t) - \Lambda_{mn}(\mathbf{k}', \mathbf{k})\delta f_n(\mathbf{k}, t)). \quad (2)$$

The scattering rates $\Lambda_{mn}(\mathbf{k}, \mathbf{k}')$ in (2) are related to the rates $S_{mn}(\mathbf{k}, \mathbf{k}')$ of the original Boltzmann equation via

$$\Lambda_{mn}(\mathbf{k}', \mathbf{k}) = (1 - f_0(E_m(\mathbf{k}')))S_{mn}(\mathbf{k}', \mathbf{k}) + f_0(E_m(\mathbf{k}'))S_{mn}(\mathbf{k}, \mathbf{k}'), \quad (3)$$

where $f_0(E)$ is the Fermi-Dirac distribution function, and $E_n(\mathbf{k})$ is the total energy in the n-th subband. The equation for the perturbation has a form similar to the Boltzmann equation, with two important differences: (i) a source term which depends on the small driving field and is proportional to the derivative of the equilibrium function is present, and (ii) it has renormalized scattering rates which enforce the equilibrium solution of the homogeneous equation (1) to be $f_0(E_n(\mathbf{k}))(1 - f_0(E_n(\mathbf{k})))$, and not $f_0(E_n(\mathbf{k}))$.

In order to calculate the mobility, a subband Monte Carlo method is used to solve the system (1). Following the procedure outlined in [10], we assume the time dependence of the driving field to be a set of instantaneous delta-like pulses:

$$\mathbf{E}(t) = \mathbf{E}_0\tau \sum_i \delta(t - t_i), \quad (4)$$

In (4) $\mathbf{E}_0$ is the value of the field averaged over a long simulation time T

$$\mathbf{E}_0 = \frac{1}{T} \int_0^T dt\, \mathbf{E}(t).$$

Then τ is the average period between the delta-pulses.

We compute the current response $\mathbf{I}_i(t)$ produced by an electric field pulse at the moment t_i as

$$\mathbf{I}_i = eH(t - t_i) \sum_m \int \frac{d^2\mathbf{k}}{(2\pi)^2} \mathbf{v}_m(\mathbf{k})\delta f_m(\mathbf{k}, t - t_i), \quad (5)$$

where $\mathbf{v}$ is the velocity, and $H(t)$ is the Heaviside function. The instanteneous current density $\mathbf{J}(t) = \sum_i \mathbf{I}_i$ is calculated as the sum over current densities $\mathbf{I}_i$ produced by all pulses i. The current density value averaged over the long time T is expressed as $\mathbf{J} = \left(\sum_i \int_0^T dt\mathbf{I}_i(t)\right)/T$. The low field mobility is defined as $\mu_{\alpha\beta} = J_\alpha/(enE_\beta)$, where the direction of β-axis coincides with the direction of $\mathbf{E}_0$, and n is the carrier concentration. Now the mobility can be easily computed using a single-particle Monte Carlo technique.

The method can be illustrated as follows. The diffusion tensor $D_{\alpha\beta}$ is calculated as an integral of the velocity autocorrelation function [9]

$$D_{\alpha\beta} = \int_0^\infty d\tau \langle v_\alpha(t)v_\beta(t + \tau)\rangle, \quad (6)$$

where angular brackets denote the time averaging over the stochastic dynamics determined by the rates $\Lambda_{mn}(\mathbf{k}, \mathbf{k}')$ of the *linearized* multi-subband Boltzmann scattering integral in case of degenerate statistics. The mobility tensor $\tilde{\mu}_{\alpha\beta}$ is related to the diffusion tensor via the Einstein relation for degenerate statistics

$$\tilde{\mu}_{\alpha\beta} = eD_{\alpha\beta}\frac{1}{n}\frac{\delta n}{\delta E_F}, \tag{7}$$

where E_F is the Fermi level.

In order to compute the mobility, we accumulate three temporary estimators t, w_β, and $\nu_{\alpha\beta}$ during the Monte Carlo simulations:

(i) initialize $t = 0$, $w_\beta = 0$, $\nu_{\alpha\beta} = 0$, and start the particle trajectory with the stochastic dynamics determined by the scattering rates $\Lambda_{mn}(\mathbf{k}, \mathbf{k}')$ from (3) of the *linearized* multi-subband Boltzmann equations;

(ii) before each scattering event update $\nu_{\alpha\beta}$, w_β, and t:

$$t = t + \frac{\tau(j)}{1 - f(E(j))},$$
$$w_\beta = w_\beta + v_\beta(j)\tau(j),$$
$$\nu_{\alpha\beta} = \nu_{\alpha\beta} + \tau(j)v_\alpha(j)w_\beta(j);$$

(iii) When t is sufficiently large, compute the mobility tensor as

$$\tilde{\mu}_{\alpha\beta} = \frac{e}{k_B T}\frac{\nu_{\alpha\beta}}{t},$$

where $v_\alpha(j)$ denotes the α-component of the velocity, $E(j)$ is the particle energy, $f(E)$ is the Fermi-Dirac function, and $\tau(j)$ is the time of j-th free flight. The convergence of the method is improved by resetting $w_\beta = 0$ each time a velocity randomizing scattering event occurs.

3   Degeneracy Effects and Intersubband Scattering

We demonstrate the importance of degeneracy effects by evaluating the low-field mobility in inversion layers and in UTB FETs. The phonon-limited mobility in inversion layers shows a different behavior if the Pauli exclusion principle is taken into account. However, if surface roughness scattering is included, the relative difference decreases, and the universal mobility curve can be reproduced equally well using both degenerate and nondegenerate statistics [14], as shown in Fig. 1.

In UTB FETs degeneracy effects are expected to be more pronounced. We consider as example a 3 nm thick (100) UTB FET. The nondegenerate statistics is assured by using the rates $S_{mn}(\mathbf{k}', \mathbf{k})$ of the original Boltzmann equation in the Monte Carlo algorithm described above. Results of mobility calculations for single-gate (SG) and DG structures, with and without degeneracy effects taken into account in the Monte Carlo simulations are summarized in Fig. 2. Mobility in a DG FET is plotted as function of the carrier concentration per channel, or $N_{DG}/2$. When degeneracy effects are neglected, the DG mobility is superior as

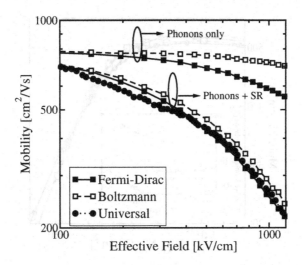

Fig. 1. Effective mobility of a Si inversion layer at (100) interface computed with Boltzmann (open symbols) and Fermi-Dirac (filled symbols) statistics reproduces well the universal mobility curve [12] (circles). Phonon-limited mobility for degenerate and nondegenerate statistics is also shown.

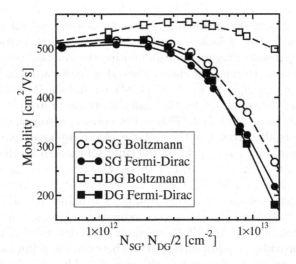

Fig. 2. Mobility in 3 nm thick (100) SG (circles) and DG (squares) structures computed with Boltzmann (open symbols) and Fermi-Dirac (filled symbols) statistics

compared to the SG mobility. When the degeneracy effects are included, behavior of the DG mobility is qualitatively different. Namely, the DG mobility becomes lower than the SG mobility at high carrier concentrations, in agreement with experimental data [13].

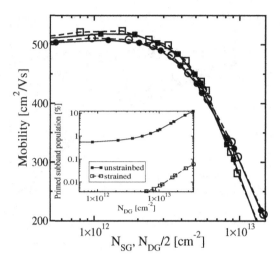

Fig. 3. Mobility in (100) 3 nm thick DG (squares) and SG (circles) structures computed with (open symbols) and without (filled symbols) in-plane biaxial stress of 1.6 GPa. Inset: occupation of primed subbands in relaxed (filled symbols) and biaxially stressed (open symbols) DG structure.

The difference between the mobility values for degenerate and nondegenerate statistics shown in Fig. 2 looks surprising. Indeed, at high carrier concentrations the principal scattering mechanism limiting the low-field mobility is elastic surface roughness scattering. For elastic scattering the forward and inverse scattering rates are equal: $S_{mn}^{el}(\mathbf{k}', \mathbf{k}) = S_{nm}^{el}(\mathbf{k}, \mathbf{k}')$, so that the Pauli blocking factor cancels out from the equations for the elastic scattering rates (3), and degeneracy effects seem to be irrelevant. This is not correct, however, since the Pauli blocking factor is also present in the inelastic electron-phonon part of the total scattering integral and ensures the equilibrium solution to be the Fermi-Dirac distribution function. In case of Fermi-Dirac statistics the Fermi level in a DG FET is higher than in a SG FET, due to twice as high carrier concentration for the same gate voltage [11]. This results in a higher occupation of upper subbands. To study the influence of the occupation of *primed* subbands on the mobility lowering in (100) DG FETs we apply a biaxial stress of 1.6 GPa. This level of stress provides an additional splitting between the primed and unprimed subbands high enough to depopulate the primed ladder completely. Results of the mobility simulation in 3 nm DG and SG structures, with biaxial stress applied, are shown in Fig. 3 together with the results for the unstrained structure. Both mobilities in strained and unstrained structures are similar in the whole range of concentrations. The inset displays the population of primed subbands in a 3 nm DG structure, showing that the primed ladder in a strained FET is practically depopulated. Since the mobilities of strained and unstrained UTB FETs are almost equivalent for both SG and DG structures, it then follows that the higher occupation of primed subbands is not the reason for the DG mobility

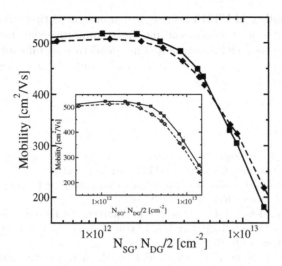

Fig. 4. Mobility in 3 nm thick (100) structures, computed with and without (Inset) inter-subband scattering. Higher carrier concentration in a UTB DG structure (squares) at the same gate voltage pushes the Fermi-level up and opens additional inter-subband scattering channels between unprimed subbands. It decreases the mobility in (100) UTB DG FETs below its SG values (diamonds) at high carrier concentrations.

lowering in (100) DG UTB structures. Another consequence of twice as high carrier concentration in a DG UTB FET is the higher occupation of upper *unprimed* subbands. When the carrier energy is above the bottom E_1 of the next unprimed subband, and intensive elastic intersubband scattering occurs. This additional scattering channel leads to a step-like increase in the density of after-scattering states, which results in higher scattering rates.

To demonstrate the importance of intersubband scattering for mobility calculations, we artificially switch off the scattering between unprimed subbands. We consider degenerate statistics and restore screening. Results of the mobility calculations for a 3 nm thick UTB structure, with and without intersubband scattering are shown in Fig. 4. Without intersubband scattering, the DG FET mobility is higher than the corresponding SG mobility, in analogy to nondegenerate results. It confirms our finding that as soon as the additional intersubband scattering channel becomes activated, the DG mobility value sinks below the SG mobility.

4   Conclusion

The Monte Carlo algorithm originally designed for small signal analysis of the three-dimensional electron gas response is generalized to quasi-two-dimensional electron systems. In the limit of vanishing driving fields the method includes degeneracy effects exactly. The method is valid for arbitrary scattering rates and includes realistic band structure. The Monte Carlo method is applied to compute the low-field mobility in UTB FETs. It is demonstrated that degeneracy effects play a significant role in compensating the volume inversion induced mobility enhancement in (100) DG structures. They lead to a significant occupation of higher subbands in the unprimed ladder, which results in increased intersubband scattering and mobility lowering.

References

1. International Technology Roadmap for Semiconductors — 2005 edn.(2005), http://www.itrs.net/Common/2005ITRS/Home2005.htm
2. Bosi, S., Jacoboni, C.: Monte Carlo High Field Transport in Degenerate GaAs. J. Phys. C: Solid State Phys. 9, 315–319 (1976)
3. Fischetti, M.V., Laux, S.E.: Monte Carlo Analysis of Electron Transport in Small Semiconductor Devices Including Band-Structure and Space-Charge Effects. Physical Review B 38, 9721–9745 (1988)
4. Jungemann, C., Pham, A.T., Meinerzhagen, B.: A Linear Response Monte Carlo Algorithm for Inversion Layers and Magnetotransport. In: Proc. Intl. Workshop Comput. Electronics, pp. 13–14 (May 2006)
5. Kosina, H., Nedjalkov, M., Selberherr, S.: Theory of the Monte Carlo Method for Semiconductor Device Simulation. IEEE Trans. Electron Devices 47, 1899–1908 (2000)
6. Likharev, K.K.: Sub-20-nm Electron Devices. In: Morkoc, H. (ed.) Advanced Semiconductor and Organic Nano-Techniques, pp. 239–302. Academic Press, New York (2003)
7. Lucci, L., et al.: Multi-Subband Monte-Carlo Modeling of Nano-MOSFETs with Strong Vertical Quantization and Electron Gas Degeneration. In: IEDM Techn. Dig., pp. 531–534 (2005)
8. Lugli, P., Ferry, D.K.: Degeneracy in the Ensemble Monte Carlo Method for High Field Transport in Semiconductors. IEEE Trans. Electron Devices 32, 2431–2437 (1985)
9. Reggiani, L., et al.: Diffusion and fluctuations in a nonequilibrium electron gas with electron-electron collisions. Phys. Rev. B 40, 12209–12214 (1989)
10. Smirnov, S., et al.: Monte Carlo Method for Modeling of Small Signal Response Including the Pauli Exclusion Principle. J. Appl. Phys. 94, 5791–5799 (2003)
11. Sverdlov, V., et al.: Mobility for High Effective Field in Double-Gate and Single-Gate SOI for Different Substrate Orientations. In: Proc. EUROSOI 2006, pp. 133–134 (March 2006)
12. Takagi, S.I., et al.: On the Universality of Inversion Layer Mobility in Si MOSFET's: Part I — Effects of Substrate Impurity Concentration. IEEE Trans. Electron Devices 41, 2357–2362 (1994)
13. Uchida, K., Koga, J., Takagi, S.: Experimental Study on Carrier Transport Mechanisms in Double- and Single-Gate Ultrathin-Body MOSFETs — Coulomb Scattering, Volume Inversion, and δT_{SOI}-induced Scattering. In: IEDM Techn. Dig., pp. 805–808 (2003)
14. Ungersboeck, E., Kosina, H.: The Effect of Degeneracy on Electron Transport in Strained Silicon Inversion Layer. In: Proc. Intl. Conf. on Simulation of Semiconductor Processes and Devices, pp. 311–314 (2005)

Part IV

Operator Splittings, Their Application and Realization

A Parallel Combustion Solver within an Operator Splitting Context for Engine Simulations on Grids

Laura Antonelli[1], Pasqua D'Ambra[1], Francesco Gregoretti[1], Gennaro Oliva[1], and Paola Belardini[2]

[1] Institute for High-Performance Computing and Networking (ICAR)-CNR
Via Pietro Castellino 111, I-80131 Naples, Italy
{laura.antonelli,pasqua.dambra,francesco.gregoretti,
gennaro.oliva}@na.icar.cnr.it
[2] Istituto Motori (IM)-CNR
Via Marconi, 8 I-80125 Naples, Italy
p.belardini@im.cnr.it

Abstract. Multidimensional engine simulation is a very challenging field, since many thermofluid processes in complex geometrical configurations have to be considered. Typical mathematical models involve the complete system of unsteady Navier-Stokes equations for turbulent multi-component mixtures of ideal gases, coupled to equations for modeling vaporizing liquid fuel spray and combustion. Numerical solutions of the full system of equations are usually obtained by applying an operator splitting technique that decouples fluid flow phenomena from spray and combustion, leading to a solution strategy for which a sequence of three different sub-models have to be solved. In this context, the solution of the combustion model is often the most time consuming part of engine simulations. This work is devoted to obtain high-performance solution of combustion models in the overall procedure for simulation of engines in a distributed heterogeneous environment. First experiments of multi-computer simulations on realistic test cases are discussed.

1 Introduction

The design of modern engines relies on sophisticated technologies devoted to reduce pollutant emissions and combustion noise. The impact on engine modeling is the need of accurately simulating highly complex, different physical-chemical phenomena occurring in each engine cycle. Mathematical models for the description of the overall problem typically involve unsteady Navier-Stokes equations for turbulent multi-component mixtures of ideal gases, coupled with suitable equations for fuel spray and combustion modeling. The solution of the overall model usually relies on an operator splitting technique, where different physical phenomena are decoupled, and different sub-models are separately solved on a 3d computational grid.

I. Lirkov, S. Margenov, and J. Waśniewski (Eds.): LSSC 2007, LNCS 4818, pp. 167–174, 2008.

In recent years much attention has been addressed to the combustion models, by introducing detailed chemical reaction models, where the number of the chemical species and the reactions to be considered reach also several hundreds. Therefore, the numerical solution of chemistry has become one of the most computationally demanding parts in simulations, thus leading to the need of efficient combustion solvers. The typical main computational kernel in this framework is the solution of systems of non-linear Ordinary Differential Equations (ODEs), characterized by a very high stiffness degree. The chemical reactions do not introduce any coupling among grid cells, therefore combustion models show an instrinsic parallelism to be exploited for making possible even more detailed chemical models and advanced solution methods, as reported in [2,3]. In this work we focus on the efficient implementation of a distributed combustion solver in heterogeneous environments, by using MJMS, a Multi-site Job Management System for execution of MPI applications in a Grid environment [5]. The combustion solver is based on the CHEMKIN-II package for managing detailed chemistry and on a multi-method ODE solver for the solution of the ODE systems arising from the chemical reaction model. The software is interfaced with the sequential KIVA3V-II code [1] for the simulation of the entire engine cycle. The paper is organized as follows. In Section 2 we report the complete system of unsteady compressible Navier-Stokes equations used in the KIVA3V-II code and we describe the detailed combustion model used for our simulations. In Section 3 we describe the main features of the combustion solver, both in terms of employed numerical algorithms and software and in terms of the distributed implementation. In Section 4 we analyze results of simulations on a realistic test case. Some conclusions and future work are included in Section 5.

2   Mathematical Models for Engine Simulations

Mathematical models for engine simulations, such as that employed into KIVA3V-II, usually solve the following system of equations:

- Species continuity:

$$\frac{\partial \rho_m}{\partial t} + \nabla \cdot (\rho_m \mathbf{u}) = \nabla \cdot [\rho D \nabla (\frac{\rho_m}{\rho})] + \dot{\rho}_m^c + \dot{\rho}_m^s \delta_{ml}$$

where ρ_m is the mass density of species m, ρ is the total mass density, $\mathbf{u}$ is the fluid velocity, $\dot{\rho}_m^c$ is a source term due to combustion, $\dot{\rho}_m^s$ is a source term due to spray and δ is the Dirac delta function.

- Total mass conservation:

$$\frac{\partial \rho}{\partial t} + \nabla \cdot (\rho \mathbf{u}) = \dot{\rho}^s$$

- Momentum conservation:

$$\frac{\partial (\rho \mathbf{u})}{\partial t} + \nabla \cdot (\rho \mathbf{u} \, \mathbf{u}) =$$
$$-\frac{1}{\alpha^2} \nabla p - A_0 \nabla (\frac{2}{3} \rho k) + \nabla \cdot \overline{\sigma} + \mathbf{F}^s + \rho \mathbf{g}$$

where $\bar{\sigma}$ is the viscous stress tensor, $\mathbf{F}^s$ is the rate of momentum gain per unit volume due to spray and $\mathbf{g}$ is the constant specific body force. The quantity A_0 is zero in laminar calculations and unity when turbulence is considered.

– Internal energy conservation:

$$\frac{\partial(\rho I)}{\partial t} + \nabla \cdot (\rho I \mathbf{u}) =$$
$$-p\nabla \cdot \mathbf{u} + (1 - A_0)\bar{\sigma} : \nabla \mathbf{u} - \nabla \cdot \mathbf{J} + A_0 \rho \epsilon + \dot{Q}^c + \dot{Q}^s$$

where I is the specific internal energy, the symbol : indicates the matrix product, $\mathbf{J}$ is the heat flux vector, $\dot{Q}^c$ and $\dot{Q}^s$ are the source terms due to combustion heat release and spray interactions.

Two additional transport equations are considered. These are the standard $K - \epsilon$ equations for the turbulence with terms due to interaction with spray. Suitable initial and boundary conditions are added to the equations.

In the above mathematical model, the source term $\dot{\rho}_m^c$ defines the contribution of the chemical reactions to the variation in time of mass density for species m. Its computation is decoupled from the solution of the Navier-Stokes equations, following a linear operator splitting that separates contribution of chemistry and contribution of spray from equations driving the turbulent fluid flow. Unlike the original KIVA3V-II code, where a reduced chemical reaction mechanism is solved, we consider a complex and detailed chemical reaction model for Diesel engine simulations. The model is based on a recent detailed kinetic scheme, which involves 62 chemical species and 285 reactions. The kinetic scheme considers the H abstraction and the oxidation of a N-dodecane, with production of alchil-peroxy-radicals, followed by the ketoydroperoxide branching. In the model the fuel pirolysis determines the chetons and olefins formation. Moreover, a scheme of soot formation and oxidation is provided, together with a classical scheme of NOx formation. The reaction system is expressed by the following system of non-linear ODEs:

$$\dot{\rho}_m = W_m \sum_{r=1}^{R}(b_{mr} - a_{mr})\dot{\omega}_r(\rho_1, \ldots, \rho_m, T), \qquad m = 1, \ldots, M, \qquad (1)$$

where R is the number of chemical reactions involved in the system, $\dot{\rho}_m$ is the production rate of species m, W_m is its molecular weight, a_{mr} and b_{mr} are integral stoichiometric coefficients for reaction r and $\dot{\omega}_r$ is the kinetic reaction rate.

Production rate terms can be separated into creation rates and destruction rates [7]:

$$\dot{\rho}_m = \dot{C}_m - \dot{D}_m, \qquad m = 1, \ldots M, \qquad (2)$$

where $\dot{C}_m$, $\dot{D}_m$ are the creation and the destruction rate of species m respectively. It holds

$$\dot{D}_m = \frac{X_m}{\tau_m^c}, \qquad m = 1, \ldots M, \qquad (3)$$

where X_m is the molar concentration of species m and τ_m^c is the *characteristic time* of species m, that is, the time needed by species m to reach equilibrium state. Expression (3) shows that the eigenvalues of the Jacobian matrix of the right-hand side of system (1) are related to the characteristic times of species involved in the combustion model. Detailed reaction models involve a great number of intermediate species and no equilibrium assumption is made. Thus, the overall reaction systems include species varying on very different timescales; this motivates the high stiffness degree that typically characterizes ODE systems arising in this framework. Note that, in order to take into account the effects of the turbulence during the combustion, we consider an interaction model between complex kinetics and turbulence which preserves the locality of the reaction process with respect to the grid cells [3].

3   Distributed Solution of Combustion Model

In this Section we describe the main features of the software component we developed for distributed solution of complex chemical reaction schemes in simulation of Diesel engines. The computational kernel is the solution of a system of non-linear ODEs per each grid cell of a 3d computational domain, at each time step of an operator splitting procedure. Since reaction schemes do not introduce any coupling among the grid cells, the solution of the ODE systems is a so-called inherently distributed problem, and we can exploit modern features of Grid environments in order to obtain high-performance solution of reaction schemes in large-scale simulations.

In our software, the systems of stiff ODEs are solved by means of a multi-method solver, based on an adaptive combination of a 5-stages Singly Diagonally Implicit Runge-Kutta (SDIRK) method [6] and variable coefficient Backward Differentiation Formulas (BDF) [4]. Results on the use of the multi-method solver in the solution of detailed chemical kinetics arising from multidimensional Diesel engine simulations are reported in [3].

Note that physical stiffness is strongly related to local conditions, therefore, when adaptive solvers are considered in a distributed environment, also including heterogeneous resources, data partitioning and process allocation become critical issues for computational load balancing and reduction of idle times.

In order to reduce the impact of local stiffness and adaptive solution strategies on a possible computational load imbalance, our software component supports a data distribution where systems of ODEs related to contiguous cells are assigned to different processors. To this aim, grid cells are reordered according to a permutation of indices, deduced by a pseudo-random sequence, and the ODE systems per each grid cell are distributed among the available processes, following the new order of the grid cells. Furthermore, in order to take into account possible load imbalance due to the use of heterogeneous resources, the ODE systems are distributed on the basis of CPU performances, as we explain in the following, so that faster processors get more workload. Our software is written in Fortran and it is based on CHEMKIN-II [7], a software package for managing large models of chemical reactions in the context of simulation software. It provides a database and a software library for computing model parameters involved in system (1). The parallel software component for combustion modeling is also interfaced with the sequential KIVA3V-II code, in order to properly test it within real simulations.

The distributed implementation of our combustion solver relies on a job management system, named MJMS (Multi-site Jobs Management System) [5]. This system interacts with the pre-webservices portions of the Globus Toolkit 4.0 for job submission and monitoring, with the MPICH-G2 implementation of MPI for interprocess communication on Grids and with the Condor-G system for job allocation and execution. MJMS allows the users to submit execution requests for multi-site parallel applications which consist of multiple distributed processes running on one or more potentially heterogeneous computing resources in different locations. The processes are mapped on the available resources according to requirements and preferences specified by the user in order to meet the application needs. In our context we require multiple parallel computing resources for the execution of the combustion solver. Once the resources have been selected, MJMS makes available the information about the number of processors for each computing resource and the corresponding CPU performances, then our application, taking into account the above information, configures itself at run time in order to perform a balanced load distribution. This data distribution is achieved through the algorithm described as follow. Let $p_1, ..., p_n$ be the number of processors respectively for the n computing resources $c_1, ..., c_n$ selected by the system and let $cpow_1, ..., cpow_n$ be the corresponding CPU performances. The algorithm computes

$$pow_{min} = \min\{cpow_1, \ldots, cpow_n\}$$

and $k_1, ..., k_n$ such that

$$k_i = \frac{cpow_i}{pow_{min}}, \qquad i = 1, \ldots, n$$

Then it computes the factor

$$f = \frac{1}{\sum_{i=1}^{n} p_i k_i}$$

so that each processor of the c_i computing resource get the fraction $f \cdot k_i$ of the total workload, i.e. each processor, at each time step, owns the data for solving the ODE systems related to $f \cdot k_i \cdot ncells$ computational grid cells, where $ncells$ is the total number of active grid cells. Note that in the previous algorithm we neglect possible performance degradations due to network heterogeneity and interprocess communication and synchronization, which we are going to consider for future experiments.

4   Numerical Experiments

In this Section we show preliminary results concerning distributed simulations performed on a prototype, single cylinder Diesel engine, having characteristics

Table 1. Multijet 16V Engine Characteristics

Bore[mm]	82.0
Stroke[mm]	90
Compression ratio	16.5:1
Engine speed	1500 rpm
Displacement[cm^3]	475
Valves per cylinder	4
Injector	microsac 7 holes Φ 0.140 mm
Injection apparatus	Bosch Common Rail III generation

Fig. 1. Computational grid

similar to the 16 valves EURO IV Fiat Multijet. Main engine parameters are reported in Table 1. Our typical computational grid is a 3d cylindrical sector representing a sector of the engine cylinder and piston-bowl. It is formed by about 3000 cells, numbered in counter-clockwise fashion on each horyzontal layer, from bottom-up (Figure 1).

The structure of the active computational grid changes within each simulation of the entire engine cycle in order to follow the piston movement into the cylinder. The limit positions of the piston, that is the lowest point from which it can leave and the highest point it can reach, are expressed with respect to the so called crank angle values and they correspond to -180° and 0°.

ODE systems have been solved by means of the multi-method solver we developed. In the stopping criteria, both relative and absolute error control tolerances were considered; at this purpose, we defined two vectors, *rtol* and *atol*, respectively. In all the experiments here analyzed *atol* values were fixed in dependence of the particular chemical species. The reason motivating this choice relies on the very different concentrations characterizing chemical species involved in detailed reaction models. All the components of *rtol* were set to 10^{-3}, in order to satisfy the application accuracy request.

We carried out our preliminary experiments using a small Grid testbed composed of two Linux clusters:

- *Vega*: a Beowulf-class cluster of 16 nodes connected via Fast Ethernet, operated by the Naples Branch of the Institute for High-Performance Computing and Networking (ICAR-CNR). Each processor is equipped with a 1.5 GHz Pentium IV processor and a RAM of 512 MB.
- *Imbeo*: a Beowulf-class cluster of 16 nodes connected via Fast Ethernet, operated by the Engine Institute (IM-CNR). Eight nodes are equipped with a 2.8GHz Pentium IV processor, while the others have a 3.8GHz Pentium IV processor. All the processors have a RAM of 1 GB.

The Globus Toolkit 4.0 and MPICH-G2 rel. 1.2.5.2 are installed on both clusters as Grid middleware; MPICH 1.2.5 based on the ch_p4 device is the local version of MPI library.

MJMS gathers the information about the CPU performances provided by the Globus Information System as *ClusterCPUsSpecFloat* value and the application configures itself at run time by using those information as $cpow_i$ values. Note that this feature becomes more important in a general Grid environment, where available resources are not known in advance. The *ClusterCPUsSpecFloat* value obtained for the single processor of Vega is *534 Mflops*, while a value of about *1474 Mflops* is obtained for the processors of Imbeo. Therefore, the workload factor K_i is 1 for the first ones (processors of cluster c_1) and nearly 3 for the latter ones (processors of cluster c_2). This means that adding the 16 processors of Vega to the 16 processors of Imbeo can be seen as an increase of about 25% of the computational power of Imbeo.

In our experiments, we ran engine simulations for the Crank angle interval $[-180, 40]$ both on each cluster, using the local version of MPI, and also on the Grid testbed via MPICH-G2. For the distributed simulations we analyzed the performances, in terms of the total execution times, running the code for increasing values of the workload factor K_2, starting from the value 1, corresponding to a 50% of workload for each resource, in terms of the total number of grid cells assigned to the resources. As expected, we obtained performance improvements with respect to the best parallel performance for $K_2 \geq 3$. The improvement is about 15.5% of the performance obtained on 16 processors of Imbeo, when $K_2 = 3$ is used. The total execution times obtained on each cluster and also for the distributed simulation, when we choose $K_2 = 3$, are reported in Table 2.

Table 2. Total execution times in seconds

Vega	4307
Imbeo	1480
Grid (Vega+Imbeo)	1250

5   Conclusions

In this work we present preliminary results of some experiments in using heterogeneous distributed resources for high-performance solution of detailed chemical reaction models in engine simulations. The inherently distributed chemical reaction model is solved in a Grid environment composed of different Linux clusters operated in different sites. Distribution of workload is based on a strategy that take into account both model features, such as local stiffness and use of adaptive solvers, and platform features, such as different computational power of processors, in order to get performance in distributed engine simulations. Preliminary results on a small size test case are encouraging, therefore, future work will be devoted to improve the model to estimate process workload, also taking into account data communication and synchronization, and to analyze the performance of our approach on larger problems and larger Grid platforms.

References

1. Amsden, A.A.: KIVA-3V: A Block-Structured KIVA Program for Engines with Vertical or Canted Valves, Los Alamos National Laboratory Report No. LA-13313-MS (1997)
2. Belardini, P., et al.: The Impact of Different Stiff ODE Solvers in Parallel Simulation of Diesel Combustion. In: Yang, L.T., et al. (eds.) HPCC 2005. LNCS, vol. 3726, pp. 958–968. Springer, Heidelberg (2005)
3. Belardini, P., et al.: Introducing Combustion-Turbulence Interaction in Parallel Simulation of Diesel Engines. In: Gerndt, M., Kranzlmüller, D. (eds.) HPCC 2006. LNCS, vol. 4208, pp. 1–10. Springer, Heidelberg (2006)
4. Brown, P.N., Byrne, G.D., Hindmarsh, A.C.: VODE: A Variable Coefficient ODE Solver. SIAM J. Sci. Stat. Comput 10(5), 1038–1051 (1989)
5. Frey, J., et al.: Multi-site Jobs Management System (MJMS): A Tool to manage Multi-site MPI Applications Execution in Grid Environment. In: Proceedings of the HPDC'15 Workshop on HPC Grid Programming Environments and Components (HPC-GECO/CompFrame), IEEE Computer Society, Los Alamitos (to appear)
6. Hairer, E., Wanner, G.: Solving Ordinary Differential Equations II. Stiff and Differential-Algebraic Problems, 2nd edn. Springer Series in Comput. Mathematics, vol. 14. Springer, Heidelberg (1996)
7. Kee, R.J., Rupley, F.M., Miller, J.A.: Chemkin-II: A Fortran Chemical Kinetics Package for the Analysis of Gas-phase Chemical Kinetics, SAND89–8009, Sandia National Laboratories (1989)

Identifying the Stationary Viscous Flows Around a Circular Cylinder at High Reynolds Numbers

Christo I. Christov[1], Rossitza S. Marinova[2], and Tchavdar T. Marinov[1]

[1] Dept. of Math., University of Louisiana at Lafayette, LA 70504-1010, USA
[2] Dept. of Math. & Computing Sci., Concordia Univ. College of Alberta
7128 Ada Boul., Edmonton, AB, Canada T5B 4E4
christov@louisiana.edu, rossitza.marinova@concordia.ab.ca,
marinov@louisiana.edu

Abstract. We propose an approach to identifying the solutions of the steady incompressible Navier-Stokes equations for high Reynolds numbers. These cannot be obtained as initial-value problems for the unsteady system because of the loss of stability of the latter. Our approach consists in replacing the original steady-state problem for the Navier-Stokes equations by a boundary value problem for the Euler-Lagrange equations for minimization of the quadratic functional of the original equations. This technique is called Method of Variational Imbedding (MVI) and in this case it leads to a system of higher-order partial differential equations, which is solved by means of an operator-splitting method. As a featuring example we consider the classical flow around a circular cylinder which is known to lose stability as early as for Re = 40. We find a stationary solution with recirculation zone for Reynolds numbers as large as Re = 200. Thus, new information about the possible hybrid flow regimes is obtained.

1 Introduction

Navier-Stokes (N-S) equations, describing the flows of viscous incompressible liquid, exhibit reach phenomenology. Especially challenging are the flows for high Reynolds numbers when the underlying stationary flow loses stability and a complex system of transients takes place leading eventually to turbulence. In high Reynolds numbers regimes, the steady solution still exists alongside with the transients, but cannot be reached via numerical approximations of the standard initial-boundary value problem. It is important for the theory of the N-S model, to find the shape of the stationary solution even when it loses stability.

To illustrate the point of the present work, we consider the classical flow around a circular cylinder which has attracted the attention because of the early instability and intriguing transitions to turbulence. The flow becomes unstable as early as Re = 40 and the stationary regime is then replaced by an unsteady laminar flow called "Kármán vortex street, for 40 < Re < 100. With further increase of the Reynolds number the experiments show that the flow ends up in the turbulent regime around Re = 180. The appearance of 3D instabilities

I. Lirkov, S. Margenov, and J. Waśniewski (Eds.): LSSC 2007, LNCS 4818, pp. 175–183, 2008.
© Springer-Verlag Berlin Heidelberg 2008

around Re $= 200$ was also confirmed by direct numerical simulations [9]. In spite of the many numerical calculations of the flow past a circular cylinder, accurate steady-state solutions for very large Reynolds numbers up to about 700 have been obtained only by Fornberg [7,8]. In his works, Fornberg reached high values of Reynolds number by means of a smoothing technique, which means that the problem is still not rigorously solved and is open to different approaches.

Although some agreement between theoretical, numerical and experimental results exists, there is a need for further work in this classical problem. To answer some of the above questions, we present here a new approach to identify the two-dimensional steady-state solution of N-S for the flow around a circular cylinder.

2   Problem Formulation

Consider the two-dimensional steady flow past a circular cylinder. The governing equations and the boundary conditions are presented in dimensionless form and polar coordinates (r, φ). The N-S equations read

$$\Omega = u_r \frac{\partial u_r}{\partial r} + \frac{u_\varphi}{r} \frac{\partial u_r}{\partial \varphi} - \frac{u_\varphi^2}{r} + \frac{\partial p}{\partial r} - \frac{1}{Re} \left(D u_r - \frac{2}{r^2} \frac{\partial u_\varphi}{\partial \varphi} \right) \tag{1}$$

$$\Phi = u_r \frac{\partial u_\varphi}{\partial r} + \frac{u_\varphi}{r} \frac{\partial u_\varphi}{\partial \varphi} + \frac{u_\varphi u_r}{r} + \frac{1}{r} \frac{\partial p}{\partial \varphi} - \frac{1}{Re} \left(D u_\varphi + \frac{2}{r^2} \frac{\partial u_r}{\partial \varphi} \right) \tag{2}$$

$$\Xi = \frac{\partial u_r}{\partial r} + \frac{u_r}{r} + \frac{1}{r} \frac{\partial u_\varphi}{\partial \varphi} = 0, \tag{3}$$

where $u_r = u(r, \varphi)$ and $u_\varphi = v(r, \varphi)$ are the velocity components parallel respectively to the polar axes r and φ; $p = p(r, \varphi)$ is the pressure. Furthermore, $D \equiv \frac{\partial^2}{\partial r^2} + \frac{1}{r} \frac{\partial}{\partial r} - \frac{1}{r^2} + \frac{1}{r^2} \frac{\partial^2}{\partial \varphi^2}$ is the so-called *Stokesian*. As usually, the Reynolds number (Re $= U_\infty d/\nu$) is based on the cylinder diameter $d = 2a$, velocity at infinity U_∞, with ν standing for the kinematic coefficient of viscosity. In terms of dimensionless variables, the cylinder surface is represented by $r = 1$ and the velocity at infinity is taken equal to the unity, i.e., $U_\infty = 1$.

The boundary conditions reflect the non-slipping at the cylinder surface and the asymptotic matching with the uniform outer flow at infinity, i.e. at certain large enough value of the radial coordinate, say, r_∞. Due to the flow symmetry the computational domain may be reduced. Thus

$$u_r(1, \varphi) = u_\varphi(1, \varphi) = 0,$$
$$u_r(r_\infty, \varphi) = \cos \varphi, \quad u_\varphi(r_\infty, \varphi) = - \sin \varphi,$$
$$u_\varphi = \frac{\partial u_r}{\partial \varphi} = \frac{\partial p}{\partial \varphi} = 0 \quad \text{at} \quad \varphi = 0 \text{ and } \varphi = \pi. \tag{4}$$

3   Method of Variational Imbedding (MVI)

For tackling inverse and incorrect problems Christov [1] developed the already mentioned MVI. Consider the imbedding functional of the governing equations (1)–(3)

$$\mathcal{J}(u_r, u_\varphi, p) = \int_0^\pi \int_1^\infty \left(\Phi^2 + \Omega^2 + X^2 \right) r \, dr d\varphi. \tag{5}$$

The idea of MVI is to solve the equations of Euler-Lagrange, which are the necessary conditions of the minimization of the functional with respect to u_r, u_ϕ, and p, respectively

$$\left[\frac{1}{Re} \left(\frac{\partial^2}{\partial r^2} - \frac{\partial}{\partial r} \frac{1}{r} - \frac{1}{r^2} + \frac{1}{r^2} \frac{\partial^2}{\partial \varphi^2} \right) + \frac{\partial}{\partial r} u_r + \frac{\partial}{\partial \varphi} \frac{u_\varphi}{r} - \frac{\partial u_r}{\partial r} \right] (r\Omega)$$
$$- \left(\frac{1}{Re} \frac{2}{r^2} \frac{\partial}{\partial \varphi} + \frac{u_\varphi}{r} + \frac{\partial u_\varphi}{\partial r} \right) (r\Phi) + \left(\frac{\partial}{\partial r} - \frac{1}{r} \right) (r\Xi) = 0, \tag{6}$$

$$\left[\frac{1}{Re} \left(\frac{\partial^2}{\partial r^2} - \frac{\partial}{\partial r} \frac{1}{r} - \frac{1}{r^2} + \frac{1}{r^2} \frac{\partial^2}{\partial \varphi^2} \right) + \frac{\partial}{\partial r} u_r + \frac{\partial}{\partial \varphi} \frac{u_\varphi}{r} - \frac{1}{r} \frac{\partial u_\varphi}{\partial \varphi} - \frac{u_r}{r} \right] (r\Phi)$$
$$+ \left(\frac{1}{Re} \frac{2}{r^2} \frac{\partial}{\partial \varphi} - \frac{1}{r} \frac{\partial u_r}{\partial \varphi} + \frac{2u_\varphi}{r} \right) (r\Omega) + \frac{1}{r} \frac{\partial}{\partial \varphi} (r\Xi) = 0. \tag{7}$$

$$\frac{\partial}{\partial r} (r\Omega) + \frac{1}{r} \frac{\partial}{\partial \varphi} (r\Phi) = 0. \tag{8}$$

The last equation after acknowledging the continuity equation becomes the well-known Poisson equation for pressure

$$\frac{1}{r} \frac{\partial}{\partial r} r \frac{\partial p}{\partial r} + \frac{1}{r^2} \frac{\partial^2 p}{\partial \varphi^2} - \frac{2}{r} \left(\frac{\partial u_\varphi}{\partial \varphi} \frac{\partial u_r}{\partial r} - \frac{\partial u_\varphi}{\partial r} \frac{\partial u_r}{\partial \varphi} + u_\varphi \frac{\partial u_\varphi}{\partial r} + u_r \frac{\partial u_r}{\partial r} \right) = 0. \tag{9}$$

To Eq. (4) we add also the so-called natural boundary conditions for minimization, which in this case reduce to $\Phi = \Omega = \Xi = 0$ for $r = a$ and $r = r_\infty$.

4   Interpretation of the MVI System and Implementation

Preserving the implicit nature of the system is of crucial importance because of the implicit nature of the boundary conditions, which involves the continuity equation but does not have explicit condition on pressure. To this end we introduce the vector unknown and right-hand sides

$$\theta = \text{Column}[\Phi, u_\phi, \Omega, u_r, p], \quad F = \text{Column}[F^\Phi, F^{u_\varphi}, F^\Omega, F^{u_r}, F^p], \tag{10}$$

where

$$F^\Phi = \frac{2}{Re \cdot r^2}\frac{\partial \Omega}{\partial \varphi} + u_r \frac{\partial \Phi}{\partial r} + \frac{u_\varphi}{r}\frac{\partial \Phi}{\partial \varphi} + \frac{2u_\varphi \Omega}{r} + \frac{1}{r}\frac{\partial X}{\partial \varphi} + \Phi\frac{\partial u_r}{\partial r} - \frac{\Omega}{r}\frac{\partial u_r}{\partial \varphi},$$

$$F^{u_\varphi} = \frac{2}{Re.r^2}\frac{\partial u_r}{\partial \varphi} - \left(u_r\frac{\partial u_\varphi}{\partial r} + \frac{u_\varphi}{r}\frac{\partial u_\varphi}{\partial \varphi} + \frac{u_\varphi u_r}{r}\right),$$

$$F^\Omega = -\frac{2}{Re.r^2}\frac{\partial \Phi}{\partial \varphi} + u_r\frac{\partial \Omega}{\partial r} + \frac{u_\varphi}{r}\frac{\partial \Omega}{\partial \varphi} - \frac{u_\varphi \Phi}{r} + \frac{\partial X}{\partial r} - \Phi\frac{\partial u_\varphi}{\partial r} - \Omega\frac{\partial u_r}{\partial r},$$

$$F^{u_r} = -\frac{2}{Re.r^2}\frac{\partial u_\varphi}{\partial \varphi} - \left(u_r\frac{\partial u_r}{\partial r} + \frac{u_\varphi}{r}\frac{\partial u_r}{\partial \varphi} - \frac{u_\varphi^2}{r}\right),$$

$$F^p = -\frac{2}{r}\left(\frac{\partial u_\varphi}{\partial \varphi}\frac{\partial u_r}{\partial r} - \frac{\partial u_\varphi}{\partial r}\frac{\partial u_r}{\partial \varphi} + u_\varphi\frac{\partial u_\varphi}{\partial r} + u_r\frac{\partial u_r}{\partial r}\right).$$

We denote the linear differential operators as $\Lambda_r = \frac{\partial}{\partial r}$, $\Lambda_\varphi = \frac{1}{r}\frac{\partial}{\partial \varphi}$ and

$$\Lambda_{rr} = \frac{1}{r}\frac{\partial}{\partial r}r\frac{\partial}{\partial r}, \quad \hat{\Lambda}_{rr}\frac{1}{Re}(\lambda_{rr} - \frac{1}{r^2}), \quad \Lambda_{\varphi\varphi} = \frac{1}{r^2}\frac{\partial^2}{\partial \varphi^2}, \quad \hat{\Lambda}_{\phi\phi}\frac{1}{Re}\Lambda_{\phi\phi}.$$

Then the matrix operators have the form

$$\mathbb{L}_1 = \begin{pmatrix} \hat{\Lambda}_{rr} & 0 & 0 & 0 & 0 \\ \frac{1}{2} & \hat{\Lambda}_{rr} & 0 & 0 & 0 \\ 0 & 0 & \hat{\Lambda}_{rr} & 0 & 0 \\ 0 & 0 & \frac{1}{2} & \hat{\Lambda}_{rr} & -\Lambda_r \\ 0 & 0 & 0 & 0 & \Lambda_{rr} \end{pmatrix} \quad \mathbb{L}_2 = \begin{pmatrix} \hat{\Lambda}_{\varphi\varphi} & 0 & 0 & 0 & 0 \\ \frac{1}{2} & \hat{\Lambda}_{\varphi\varphi} & 0 & 0 & -\Lambda_\varphi \\ 0 & 0 & \hat{\Lambda}_{\varphi\varphi} & 0 & 0 \\ 0 & 0 & \frac{1}{2} & \hat{\Lambda}_{\varphi\varphi} & 0 \\ 0 & 0 & 0 & 0 & \Lambda_{\varphi\varphi} \end{pmatrix}. \quad (11)$$

The original system under consideration is non-linear but the way we introduced the non-homogeneous vector $\boldsymbol{F}$ hints at the obvious linearization: we invert just the linear operators $\mathbb{L}_i$. The iterative procedure is based on the operator splitting and the novel element here is that we perform the splitting in the vector form of the system. We generalize the second Douglas scheme [6], sometimes called the scheme of stabilizing correction (see also [15]) in the form

$$\frac{\boldsymbol{\theta}^{n+\frac{1}{2}} - \boldsymbol{\theta}^n}{\sigma} = \mathbb{L}_1\boldsymbol{\theta}^{n+\frac{1}{2}} + \mathbb{L}_2\boldsymbol{\theta}^n + \boldsymbol{F}^n, \quad \frac{\boldsymbol{\theta}^{n+1} - \boldsymbol{\theta}^{n+\frac{1}{2}}}{\sigma} = \mathbb{L}_2\boldsymbol{\theta}^{n+1} - \mathbb{L}_2\boldsymbol{\theta}^n, \quad (12)$$

or, which is the same

$$(I - \sigma\mathbb{L}_1)\boldsymbol{\theta}^{n+\frac{1}{2}} = (I + \sigma\mathbb{L}_2)\boldsymbol{\theta}^n + \sigma\boldsymbol{F}^n, \quad (I - \sigma\mathbb{L}_2)\boldsymbol{\theta}^{n+1} = \boldsymbol{\theta}^{n+\frac{1}{2}} - \tau\mathbb{L}_1\boldsymbol{\theta}^n, \quad (13)$$

where σ is the increment of the artificial time, and L_1 and L_2 are one-dimensional operators. The superscript n stands for the "old" time stage, $n + \frac{1}{2}$ for the intermediate step and $(n + 1)$ – for the "new" step.

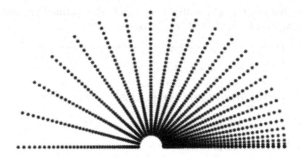

Fig. 1. Grid-points distribution

The scheme of stabilizing correction approximates in the full time step the backward Euler scheme and, therefore, for linear systems it is unconditionally stable. For the nonlinear problem under consideration it retains its strong stability allowing us to choose rather large time increments σ and to obtain the steady solution with fewer steps with respect to the artificial time.

We chose finite difference method with symmetric differences for the spatial discretization. The number of points in r direction is M, and in ϕ direction – N. The pointwise values of the set functions that comprise the vector of unknowns θ are arranged in the same order and as a result we get a vector of dimension $5M$ or $5N$. Note that each of the five equations of the system on each half-time step is represented by a tri-diagonal algebraic system. However, in order to preserve the coupling, we render it to an eleven-diagonal system for the properly arranged vector of unknowns. Thus we are able to impose the boundary conditions as they stand: two conditions on a velocity component and no conditions on pressure. As a result, the scheme is fully implicit with respect to the boundary conditions which is extremely beneficial for the stability. We have used this idea in different algorithms for solving the N-S equation and it proved crucial for the effectiveness of the scheme ([11,13]).

We use both uniform and non-uniform grids in order to evaluate better the approximation properties of the algorithm. The special non-uniform grid

$$r_i = \exp\left[(i-1)\frac{R-1}{N_r-1}\right], \quad \varphi_j = \frac{1}{\pi}\left[(j-1)\frac{\pi}{N_\varphi-1}\right]^2,$$

takes into account the *a-priori* information of the regions of large gradients of the flow and it is presented in Fig. 1. The grid is staggered for p in direction φ, which ensures the second order of approximation of the pressure boundary condition on the lines of symmetry $\varphi = 0$ and $\varphi = \pi$. For the same reason the grid for u_r and Ω is staggered in both directions. The grid-lines in direction r are denser near the cylinder and sparser far from the body. The grid in direction φ is chosen to be denser behind the body. All boundary conditions are imposed implicitly with the second order of approximation.

Table 1. Results for C_p, C_f, C_D and $p(1,\pi) - p(1,0)$, obtained on uniform grid with different h_φ and h_r and fixed Re = 4, $r_\infty = 16$

Grid	C_p	C_f	C_D	$p(1,\pi) - p(1,0)$
51×26	2.9732	2.3695	5.3427	1.9533
101×51	2.9652	2.5203	5.4855	1.9420
201×101	2.9623	2.5812	5.5435	1.9384

5   Scheme Validation

Since we use artificial time and splitting scheme, the first thing to verify is that the result for the steady problem does not depend on the time increment. We computed the solution for Re = 40 on the non-uniform grid using three different artificial-time steps: $\sigma = 0.1$, 0.01, 0.001 and have found that after the stationary regime is reached, the results for the three different time steps do not differ more than 10^{-6}, which is of the order of the criterion of convergence of the iterations.

The next important test is the verification of the spatial approximation of the scheme. We have conducted a number of calculations with different values of mesh parameters in order to confirm the practical convergence and the approximation of the difference scheme. In these tests, the mesh is uniform in both directions with spacings h_r and h_φ. The uniform grid is adequate only for Re ≤ 40, but it is enough for the sake of this particular test. Results for some of the important characteristics as obtained with different spacings h_r and h_φ are presented in Table 1 for $r_\infty = 16$ and Re = 4.

The third test is to find the magnitude of r_∞ for which the solution is adequate. Clearly, the boundary conditions should be posed as close to the body as possible to save computational resources. On the other hand, the boundary condition has to be at a sufficiently large distance from the end of the separation bubble. We examined the dependence of solution on r_∞ for Re = 20 when a separation is known to exist. Table 2 presents some of the flow characteristics as computed with different r_∞. The impact of r_∞ is significant for small values, but the results converge with the increase of r_∞. It is clear that $r_\infty \geq 20$ presents a large enough computational domain, so the further change of the parameters with the increase of r_∞ is insignificant.

Table 2. Separation angle φ_{sep}, bubble length L and width W, pressure drag coefficient C_p, friction drag coefficient C_f, difference $p_{\text{diff}} = p(1,\pi) - p(1,0)$, max $|\omega(1,\varphi_j)|$ as functions of r_∞ grid for Re = 20 and uniform grid with $h_r = 0.07$ and $h_\varphi = \pi/100$

r_∞	φ_{sep}	L	W	C_p	C_f	p_{diff}	max $|\omega(1,\varphi_j)|$
4.5	0.636(36.45°)	1.81	0.581	2.5105	1.4345	1.8219	6.61
8	0.717(41.06°)	2.35	0.689	1.7368	1.0701	1.3945	5.15
15	0.744(42.63°)	2.68	0.728	1.4208	0.9098	1.0706	4.45
22	0.749(42.94°)	2.78	0.749	1.3308	0.8621	1.0056	4.24

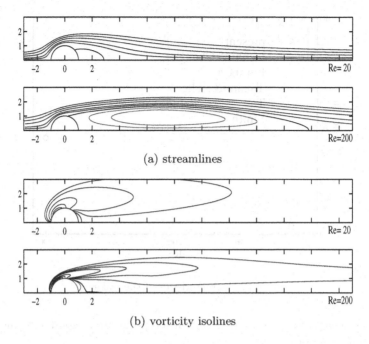

(a) streamlines

(b) vorticity isolines

Fig. 2. Flow patterns for $Re = 20$ and $Re = 200$

6   Results

The above described algorithm yielded stable computations for Reynolds numbers as high as Re = 200, much higher than the threshold of instability, which is believed to be around Re = 40. In Fig. 2 we present the streamlines and the vorticity distribution for two Reynolds numbers.

Our results indicate that the flow separation and formation of a recirculation zone behind the body appears first for Re = 10. The length of the wake (separation bubble) becomes longer and wider with increasing the Reynolds number. Our numerical results with Re = 20 on a non-uniform grid ($r_\infty \approx 88$) differ less than 10% from those with the uniform grid 301×101, where $r_\infty = 22$. The reason for the relatively high discrepancies are the large spacings of the uniform grid near the surface of the body, in particular in the r-direction, where $h_r = 0.07$. In this case the value of $r_\infty = 22$ is not sufficiently large as well.

Fornberg [7] has found that the wake bubble (region of recirculating flow) has eddy length $L \propto Re$, width $W \propto \sqrt{Re}$ up to Re = 300, and $W \propto Re$ beyond that. Smith [14] and Peregrine [12] have performed theoretical work which gives a fresh interpretation of Fornberg's results. There are several discrepancies between the theories of Smith and Peregrine, some of which are a matter of interpretation. These are unlikely to be resolved without further analysis and computational work. Our works is a contribution to this direction. The characteristics of the wake are presented in Fig. 3. Similarly, the values of the separation

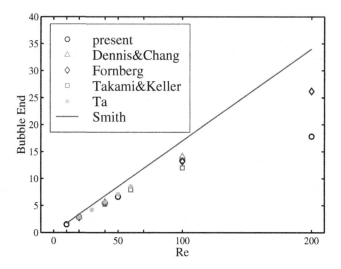

Fig. 3. Characteristics of the wake

angle φ_{sep} measured from the rear stagnation point are in good agreement with the computations of [5].

7 Conclusion

The results of the present work validate the proposed approach in the sense that the possibility of finding the steady solution even when it is unstable physically is clearly demonstrated. It should be noted that within the truncation error, our approach is exact, and no additional procedures, like filtering or smoothing, are applied. Steady solutions have been computed in primitive variables formulation for Re ≤ 200. The Reynolds numbers range of our work is wider than the others with the exception of Fornberg's work in stream function/vorticity function formulation, where an *ad hoc* smoothing is used.

Acknowledgement

The work of R. Marinova was partially supported by a grant from MITACS.

References

1. Christov, C.I.: A method for identification of homoclinic trajectories. In: Proc. 14th Spring Conf. Union of Bulg, Mathematicians, Sunny Beach, pp. 571–577 (1985)
2. Christov, C.I., Marinova, R.S.: Numerical investigation of high-Re stationary viscous flow around circular cylinder as inverse problem. Bulg. J. Meteorology Hydrology 5(3–4), 105–118 (1994)

3. Christov, C.I., Marinova, R.S.: Implicit scheme for Navier-Stokes equations in primitive variables via vectorial operator splitting. In: Griebel, M., et al. (eds.) Notes on Num. Fluid Mech., Vieweg, Germany, vol. 62, pp. 251–259 (1997)
4. Christov, C.I., Marinova, R.S.: Numerical solutions for steady flow past a circular cylinder via method of variational imbedding. Annuaire de l'Universite de Sofia 'St. Kl. Ohridski' 90, 177–189 (1998)
5. Dennis, S.C.R., Gau–Zu, C.: Numerical solutions for steady flow past a circular cylinder at Reynolds numbers up to 100. J. Fluid Mech. 42(3), 471–489 (1970)
6. Douglas, J., Rachford, H.H.: On the numerical solution of heat conduction problems in two and three space variables. Trans, Amer. Math. Soc. 82, 421–439 (1956)
7. Fornberg, B.: Steady flow past a circular cylinder up to Reynolds number 600. J. Comput. Phys. 61, 297–320 (1985)
8. Fornberg, B.: Steady incompressible flow past a row of circular cylinders. J. Fluid Mech. 225, 655–671 (1991)
9. Karniadakis, G.E., Triantafyllou, G.S.: Three-dimensional dynamics and transition to turbulence in the wake of bluff objects. J. of Fluid Mech. 238, 1–30 (1992)
10. Mallison, G.D., de Vahl Davis, G.: The method of false transients for the solution of coupled elliptic equations. J. Comp. Phys. 12, 435–461 (1973)
11. Marinova, R.S., Christov, C.I., Marinov, T.T.: A fully coupled solver for incompressible Navier-Stokes equations using coordinate operator splitting. Int. J. Comp. Fluid Dynamics 17(5), 371–385 (2003)
12. Peregrine, D.H.: A note on steady high-Reynolds-number flow about a circular cylinder. J. Fluid Mech. 157, 493–500 (1985)
13. Smagulov, S., Christov, C.I.: An iterationless implementation of boundary condition for vorticity function (preprint). Inst. Theor. Appl. Mech., Russian Acad. Sci., Novosibirsk 20 (1980) (in Russian)
14. Smith, F.T.: A structure for laminar flow past a bluff body at high Reynolds number. J. Fluid Mech. 155, 175–191 (1985)
15. Yanenko, N.N.: Method of Fractional Steps. Gordon and Breach, London (1971)

On the Richardson Extrapolation as Applied to the Sequential Splitting Method

István Faragó[1] and Ágnes Havasi[2]

[1] Eötvös Loránd University, Pázmány P. s. 1/c
[2] Eötvös Loránd University, Pázmány P. s. 1/a,
H-1117 Budapest, Hungary

Abstract. It is known from the literature that applying the same ODE solver by using two different step sizes and combining appropriately the obtained numerical solutions at each time step we can increase the convergence order of the method. Moreover, this technique allows us to estimate the absolute error of the underlying method. In this paper we apply this procedure, widely known as Richardson extrapolation, to the sequential splitting, and investigate the performance of the obtained scheme on several test examples.

1 Introduction

It has been shown [15] that the order of a time discretization method can be increased by the technique of Richardson extrapolation [2]. Namely, if one solves a problem by a numerical method of order p by using some time step τ_1, and then applies the same method with a different time step τ_2, the results can be combined to give a method of order $p + 1$. Moreover, the combination of the two methods in this way allows us to estimate the global error of the underlying method [13].

The above idea can be applied to any time discretization method, so it can also be used to increase the order of a splitting method. The application of operator splitting is a necessary step during the solution of many real-life problems. If the required accuracy is high, then even if we use a higher-order numerical method for solving the sub-problems, we need to use a higher-order splitting scheme as well [4].

The Richardson extrapolation has been successfully applied to the second-order Marchuk-Strang splitting [5,14]. In this paper we apply this method to the first-order sequential splitting. In Section 2 we present the basic idea of the Richardson extrapolation for increasing the order of a numerical method. In Section 3 the Richardson-extrapolated sequential splitting is introduced. In Section 4 numerical experiments are given for checking the theoretically derived convergence order. We present numerical results for the matrix case as well as for a more realistic stiff reaction-diffusion problem.

I. Lirkov, S. Margenov, and J. Waśniewski (Eds.): LSSC 2007, LNCS 4818, pp. 184–191, 2008.

2   Increasing the Order

Consider the Cauchy problem

$$\left.\begin{array}{l} \dfrac{du(t)}{dt} = Au(t), \quad t \in (0, T] \\[2ex] u(0) = u_0 \end{array}\right\} \tag{1}$$

where X is a Banach space, $u : [0, T] \to X$ is the unknown function, A is a linear operator $X \to X$, and $u_0 \in X$ a given initial function.

Assume that we apply some convergent numerical method of order p to solving the problem (1). Let $y_\tau(t^*)$ denote the numerical solution at a fixed time level t^* on a mesh with step size τ. Then we have

$$u(t^*) = y_\tau(t^*) + \alpha(t^*)\tau^p + \mathcal{O}(\tau^{p+1}). \tag{2}$$

On the meshes with step sizes $\tau_1 < \tau$ and $\tau_2 < \tau$ the equalities

$$u(t^*) = y_{\tau_1}(t^*) + \alpha(t^*)\tau_1^p + \mathcal{O}(\tau^{p+1}) \tag{3}$$

and

$$u(t^*) = y_{\tau_2}(t^*) + \alpha(t^*)\tau_2^p + \mathcal{O}(\tau^{p+1}) \tag{4}$$

hold, respectively. Our aim is to get a mesh function with accuracy $\mathcal{O}(\tau^{p+1})$. For the intersection of the above two meshes we define a mesh function $y_{comb}(t^*)$ as follows:

$$y_{comb}(t^*) = c_1 y_{\tau_1}(t^*) + c_2 y_{\tau_2}(t^*). \tag{5}$$

Let us substitute (3) and (4) into (5). Then we get

$$y_{comb}(t^*) = (c_1 + c_2)u(t^*) - (c_1\tau_1^p + c_2\tau_2^p)\alpha(t^*) + \mathcal{O}(\tau^{p+1}). \tag{6}$$

From (6) one can see that a necessary condition for the combined method to be convergent is that the relation

$$c_1 + c_2 = 1 \tag{7}$$

holds. Moreover, we will only have a convergence order higher than p if

$$c_1\tau_1^p + c_2\tau_2^p = 0. \tag{8}$$

The solution of system (7)–(8) is $c_1 = -\tau_2^p/(\tau_1^p - \tau_2^p)$, $c_2 = 1 - c_1$. For example, if $\tau_2 = \tau_1/2$, then for $p = 1$ we have $c_1 = -1$ and $c_2 = 2$, and for $p = 2$ we have $c_1 = -1/3$ and $c_2 = 4/3$ (cf. [10], p. 331).

If the original method is convergent (as we assumed) then y_{comb} is also convergent, which implies stability according to Lax's equivalence theorem.

The application of the same method by using two different time steps allows us to estimate the global error of the underlying method, see e.g. [13], p. 513.

Formulas (3) and (4) allow us to determine the coefficient $\alpha(t^*)$ approximately. Let us subtract (3) from (4). Then we get

$$0 = y_{\tau_2}(t^*) - y_{\tau_1}(t^*) + \alpha(t^*)(\tau_2^p - \tau_1^p) + \mathcal{O}(\tau^{p+1}).$$

Expressing $\alpha(t^*)$ gives

$$\alpha(t^*) = \frac{y_{\tau_2}(t^*) - y_{\tau_1}(t^*)}{\tau_1^p - \tau_2^p} + \frac{\mathcal{O}(\tau^{p+1})}{\tau_1^p - \tau_2^p}.$$

The second term on the right-hand side is $\mathcal{O}(\tau)$, so the ratio $\hat{\alpha}(t^*) := (y_{\tau_2}(t^*) - y_{\tau_1}(t^*))/(\tau_1^p - \tau_2^p)$ approximates $\alpha(t^*)$ to the first order in τ. Then the absolute errors of the methods (3) and (4) can be approximated by the expressions $\hat{\alpha}(t^*)\tau_1^p$ and $\hat{\alpha}(t^*)\tau_2^p$, respectively, to the order $\mathcal{O}(\tau^{p+1})$ (a posteriori error estimates).

3   Richardson-Extrapolated Splittings

The Richardson extrapolation can be used for any time discretization method, i.e., also for splitting methods. Operator splitting is widely used for solving complex time-dependent models, where the stationary (elliptic) part consists of a sum of several structurally simpler sub-operators. The classical splitting methods include the sequential splitting, the Marchuk–Strang (MS) splitting [6,7,18] and the recently re-developed symmetrically weighted sequential (SWS) splitting [3,17]. Further, newly constructed schemes include the first-order additive splitting and the iterative splitting [8].

In this paper we will focus on the sequential splitting. Assume that the operator of problem (1) is decomposed into a sum $A_1 + A_2$. Denote by $S_1(t_n, \tau)$ the solution operator belonging to the sub-problem defined by A_1 on $[t_n, t_n + \tau]$, and by $S_2(t_n, \tau)$ that defined by A_2 on $[t_n, t_n + \tau]$. We remark that S_1 and S_2 may be the exact solution operators of the sub-problems in the splitting [1,7] as well as their numerical solution operators if the sub-problems are solved numerically [4]. If $y_{\text{seq}}(t_n)$ denotes the solution obtained by the sequential splitting at time level t_n, then the solution at time t_{n+1} reads

$$y_{\text{seq}}(t_{n+1}) = S_2(t_n, \tau)S_1(t_n, \tau)y_{\text{seq}}(t_n).$$

Since the sequential splitting is a first-order time discretization method, therefore by the choice $c_1 = -1$ and $c_2 = 2$ the Richardson-extrapolated sequential splitting

$$y_{\text{Ri}}(t_{n+1}) = \{-S_2(t_n, \tau)S_1(t_n, \tau) + 2(S_2(t_n, \tau/2)S_1(t_n, \tau/2))^2\}y_{\text{Ri}}(t_n).$$

has second order.

Here we also introduce briefly the second-order MS and SWS splittings, since we will use them in our numerical comparisons. The MS splitting is defined as

$$y_{\text{MS}}(t_{n+1}) = S_1(t_n + \tau/2, \tau/2)S_2(t_n, \tau)S_1(t_n, \tau/2)y_{\text{MS}}(t_n),$$

and the SWS splitting as

$$y_{\text{SWS}}(t_{n+1}) = \frac{1}{2}(S_2(t_n, \tau)S_1(t_n, \tau) + S_1(t_n, \tau)S_2(t_n, \tau))y_{\text{SWS}}(t_n).$$

4   Numerical Experiments

In this section some numerical experiments done in Matlab are presented in order to confirm our theoretical results. We will check the convergence order of the Richardson-extrapolated sequential splitting in matrix examples (by exact/numerical solution of the sub-problems) as well as in a stiff reaction-diffusion problem.

4.1   Order Analysis in the Matrix Case

We considered the Cauchy problem

$$
\begin{cases}
c'(t) = Ac(t), \quad t \in [0, 1] \\
\\
c(0) = c_0
\end{cases}
\tag{9}
$$

with

$$
A = \begin{bmatrix} -7 & 4 \\ -6 & -4 \end{bmatrix} \quad \text{and} \quad c_0 = (1, 1)^T.
\tag{10}
$$

We decomposed matrix A as

$$
A = A_1 + A_2 = \begin{bmatrix} -6 & 3 \\ -4 & 1 \end{bmatrix} + \begin{bmatrix} -1 & 1 \\ -2 & -5 \end{bmatrix}.
\tag{11}
$$

In the first group of experiments the sub-problems were solved exactly. We applied the Richardson-extrapolated sequential splitting, the SWS splitting and the MS splitting with decreasing time steps τ to problem (9). The obtained error norms at the end of the time interval are shown in Table 1. One can conclude that while all the methods have second order, the extrapolated splitting performs better for each time step than the SWS splitting, and almost as well as the MS splitting. If we assume that the sequential splitting has already been applied for some time step τ, then to complete the extrapolated splitting (one sequential splitting with halved step size) takes practically equally as much time as the SWS splitting. However, both methods require more CPU time for the same time step than the MS splitting. (Here we assumed that all computations are

Table 1. Comparing the errors of the solutions obtained by the Richardson-extrapolated sequential splitting, the SWS splitting and the MS splitting in example (9)

τ	Ri seq.	SWS	MS
1	9.6506e-2	1.0301e-1	7.8753e-2
0.1	5.8213e-4	1.2699e-3	3.3685e-4
0.01	5.9761e-6	1.2037e-5	3.3052e-6
0.001	5.9823e-8	1.1974e-7	3.3046e-8
0.0001	5.9829e-10	1.1967e-9	3.3047e-10

Table 2. Comparing the errors of the solutions obtained by the Richardson-extrapolated sequential splitting and the SWS splitting in example (9), when the sub-problems are solved by the explicit Euler method

τ	Ri sequential	SWS splitting
1	2.7494e+1	1.2657e+1
0.1	2.6149e-3 *(9.511e-5)*	5.4703e-3 *(4.322e-4)*
0.01	1.2927e-5 *(4.944e-3)*	7.2132e-4 *(1.319e-1)*
0.001	1.2322e-7 *(9.531e-3)*	7.3991e-5 *(1.026e-1)*
0.0001	1.2264e-9 *(9.954e-3)*	7.4173e-6 *(1.002e-1)*

performed sequentially.) In order that we compare equally expensive methods, some of the further comparisons will be restricted to the SWS splitting.

In the second group of experiments we combined the Richardson-extrapolated sequential splitting and the SWS splitting with numerical methods. So in this case S_1 and S_2 are numerical solution operators.

The results obtained by the explicit Euler method in the case of problem (9) are shown in Table 2. Here and in the following tables the numbers in parentheses are the ratios by which the errors decreased in comparison with the error corresponding to the previous step size. In this case the extrapolated splitting shows second-order convergence, while the SWS splitting has only first order. This is understandable, since the sequential splitting applied together with a first-order numerical method has first order, and the application of the Richardson-extrapolation to this method must give second order. However, the SWS splitting applied together with a first-order numerical method has only first order. The results for the implicit Euler method, not shown here, were similar as for the explicit Euler method.

If the sub-problems are solved by a second-order numerical method, then the order achieved by the SWS splitting will be two. Since the sequential splitting combined with any numerical method will have first order, the extrapolated version is expected to have second order when combined with a second-order method. All this is confirmed by the results presented in Table 3.

Table 3. Comparing the errors of the solutions obtained by the Richardson-extrapolated sequential splitting and the SWS splitting in example (9), when the sub-problems are solved by the midpoint method

τ	Ri sequential	SWS splitting
1	2.9205e+1	4.0007e+1
0.1	4.4780e-4 *(1.533e-5)*	9.0087e-4 *(2.252e-5)*
0.01	2.4508e-6 *(5.473e-3)*	4.8727e-6 *(5.409e-3)*
0.001	2.3227e-8 *(9.478e-3)*	4.6428e-8 *(9.528e-3)*
0.0001	2.3101e-10 *(9.946e-3)*	4.6206e-10 *(9.952e-3)*

4.2   Order Analysis in a Stiff Diffusion-Reaction Problem

We move on to a more complex model problem, studied in [8,11], and investigate the effect of stiffness on the accuracy of the Richardson-extrapolated sequential splitting. Stiffness has been shown to reduce the order of the MS splitting from two to one [9,16,19], which gives rise to the question of how to obtain stiff convergence of order two with a splitting method [16].

Consider the diffusion-reaction equations

$$\left. \begin{aligned} \frac{\partial u}{\partial t} &= D_1 \frac{\partial^2 u}{\partial x^2} - k_1 u + k_2 v + s_1(x) \\[2mm] \frac{\partial v}{\partial t} &= D_2 \frac{\partial^2 v}{\partial x^2} + k_1 u - k_2 v + s_2(x), \end{aligned} \right\} \tag{12}$$

where $0 < x < 1$ and $0 < t \leq T = \frac{1}{2}$, and the initial and boundary conditions are defined as follows:

$$\begin{cases} u(x,0) = 1 + \sin(\frac{1}{2}\pi x), \\ v(x,0) = \frac{k_1}{k_2} u(x,0), \end{cases} \quad \begin{cases} u(0,t) = 1, \\ v(0,t) = \frac{k_1}{k_2}, \\ \frac{\partial u}{\partial x}(1,t) = \frac{\partial v}{\partial x}(1,t) = 0. \end{cases} \tag{13}$$

We used the following parameter values: $D_1 = 0,1$, $D_2 = 0$, $k_1 = 1$, $k_2 = 10^4$, $s_1(x) \equiv 1$, $s_2(x) \equiv 0$. The reference solution for the discretized problem was computed by the Matlab's ODE45 solver.

We solved problem (12)–(13) by the Richardson-extrapolated sequential splitting, the SWS splitting and the MS splitting. The differential operator defined by the right-hand side was split into the sum $D + R$, where D contained the discretized diffusion and the inhomogeneous boundary conditions, and R the reaction and source terms. The spatial discretization of the diffusion terms and the big difference in the magnitude of the reaction rates give rise to stiffness, also indicated by the big operator norms $\|D\| = \mathcal{O}(10^3)$ and $\|R\| = \mathcal{O}(10^4)$. We emphasize that here both sub-operators are stiff, while most studies are restricted to the case where one of the operators is stiff, the other is non-stiff, see e.g. [19].

The sub-problems were solved by two different time integration methods: 1) the implicit Euler method and 2) the two-step DIRK method. Tables 4–5 show the maximum norms of the errors at the end of the time interval for methods 1) and 2). For the implicit Euler method the extrapolated sequential splitting shows the expected second-order convergence only in the sequence R-D. The errors obtained in the sequence D-R are at least one magnitude higher, moreover, here we only obtained first-order convergence (order reduction). The worst results were produced by the SWS splitting, which, as expected, behaves as a first-order method, just like the MS splitting. For the MS splitting we only give the errors for the sequence D-R-D, which generally produced better results.

Table 5 illustrates that it is not worth combining the extrapolated sequential splitting with a second-order numerical method. Note that the order of the SWS and MS splittings did not increase to two (order reduction). We only obtained factors close to 0.25 in the case of the MS splitting for the smallest time steps.

Table 4. Comparing the errors of the solutions obtained by the Richardson-extrapolated sequential, SWS and MS(D-R-D) splittings in the reaction-diffusion problem (12)–(13) for the implicit Euler method

τ	Ri D-R	Ri R-D	SWS	MS
1/10	7.07e-3	5.86e-4	4.80e-2	1.26e-2
1/20	3.67e-3 *(0.52)*	1.78e-4 *(0.30)*	2.40e-2 *(0.50)*	7.52e-3 *(0.60)*
1/40	1.86e-3 *(0.51)*	4.96e-5 *(0.28)*	1.20e-2 *(0.50)*	4.28e-3 *(0.57)*
1/80	9.21e-4 *(0.50)*	1.31e-5 *(0.26)*	6.02e-3 *(0.50)*	2.35e-3 *(0.55)*
1/160	4.59e-4 *(0.50)*	3.22e-6 *(0.25)*	3.01e-3 *(0.50)*	1.26e-3 *(0.54)*
1/320	2.22e-4 *(0.48)*	6.52e-7 *(0.20)*	1.50e-3 *(0.50)*	6.63e-4 *(0.53)*

Table 5. Comparing the errors of the solutions obtained by the Richardson-extrapolated sequential, SWS and MS(D-R-D) splittings in the reaction-diffusion problem (12)–(13) for the two-step DIRK method

τ	Ri D-R	Ri R-D	SWS	MS
1/10	2.13e-2	3.86e-3	8.43e-2	1.91e-2
1/20	1.06e-2 *(0.50)*	1.86e-3 *(0.48)*	3.99e-2 *(0.47)*	9.48e-3 *(0.50)*
1/40	4.86e-3 *(0.46)*	8.95e-4 *(0.48)*	1.86e-2 *(0.47)*	4.42e-3 *(0.47)*
1/80	2.51e-3 *(0.52)*	3.47e-4 *(0.39)*	8.55e-3 *(0.46)*	2.25e-3 *(0.51)*
1/160	1.10e-3 *(0.44)*	9.50e-5 *(0.27)*	3.92e-3 *(0.46)*	1.01e-3 *(0.45)*
1/320	3.71e-4 *(0.34)*	2.04e-5 *(0.21)*	1.80e-3 *(0.46)*	3.76e-4 *(0.37)*
1/640	9.42e-5 *(0.25)*	8.54e-6 *(0.42)*	8.45e-4 *(0.47)*	1.15e-4 *(0.31)*

5   Conclusions

We applied Richardson extrapolation for increasing the convergence order of the first-order sequential splitting. The computer experiments with matrix examples confirmed the theoretically derived second-order convergence of the method. The extrapolated sequential splitting proved to be competitive both with the MS and SWS splittings. Moreover, when combined with a first-order numerical method, it still has second order, which is not true for the other two splitting schemes.

The method was also tested on a fully stiff diffusion-reaction system. While the traditional second-order splittings suffered from order reduction, the extrapolated method in the sequence R-D (reaction-diffusion) was able to produce second-order convergence.

Acknowledgements

Á. Havasi is a grantee of the Bolyai János Scholarship. This work was supported by Hungarian National Research Founds (OTKA) N. F61016.

References

1. Bjørhus, M.: Operator splitting for abstract Cauchy problems. IMA Journal of Numerical Analysis 18, 419–443 (1998)
2. Brezinski, C., Redivo Zaglia, M.: Extrapolation Methods. Theory and Practice. North-Holland, Amsterdam (1991)
3. Csomós, P., Faragó, I., Havasi, Á.: Weighted sequential splittings and their analysis. Comp. Math. Appl. 50, 1017–1031 (2005)
4. Csomós, P., Faragó, I.: Error analysis of the numerical solution obtained by applying operator splitting. Mathematical and Computer Modelling (to appear)
5. Descombes, S., Schatzman, M.: On Richardson Extrapolation of Strang's Formula for Reaction-Diffusion Equations, Equations aux Drives Partielles et Applications, articles ddis a Jacques-Louis Lions, Gauthier-Villars Elsevier, Paris, 429–452 (1998)
6. Dimov, I., et al.: L-commutativity of the operators in splitting methods for air pollution models. Annales Univ. Sci. Sec. Math. 44, 127–148 (2001)
7. Faragó, I., Havasi, Á.: Consistency analysis of operator splitting methods for C_0-semigroups. Semigroup Forum 74, 125–139 (2007)
8. Faragó, I., Gnandt, B., Havasi, Á.: Additive and iterative operator splitting methods and their numerical investigation. Comput. Math. Appl (to appear 2006)
9. Hundsdorfer, W., Verwer, J.G.: A note on splitting errors for advection-reaction equations. Appl. Numer. Math. 18, 191–199 (1994)
10. Hundsdorfer, W., Verwer, J.G.: Numerical solution of time-dependent advection-diffusion-reaction equations. Springer, Berlin (2003)
11. Hundsdorfer, W., Portero, L.: A note on iterated splitting schemes. J. Comput. Appl. Math (2006)
12. Price, P.J., Dormand, J.R.: High order embedded Runge-Kutta formulae. J. Comput. Appl. Math. 7, 67–85 (1981)
13. Quarteroni, A., Sacco, R., Saleri, F.: Numerical Mathematics. Springer, New York (2000)
14. Salcedo-Ruíz, J., Sánchez-Bernabe, F.J.: A numerical study of stiffness effects on some high order splitting methods. Revista Mexicana de Física 52(2), 129–134 (2006)
15. Shampine, L., Watts, H.: Global error estimation for ordinary differential equations. ACM Transactions on Mathematical Software 2, 172–186 (1976)
16. Jin, S.: Runge-Kutta methods for hyperbolic conservation laws with stiff relaxation terms. J. Comput. Phys. 122, 51–67 (1995)
17. Strang, G.: Accurate partial difference methods I: Linear Cauchy problems. Archive for Rational Mechanics and Analysis 12, 392–402 (1963)
18. Strang, G.: On the construction and comparison of difference schemes. SIAM J. Numer. Anal 5(3) (1968)
19. Verwer, J.G., Sportisse, B.: A note on operator splitting in a stiff linear case, MAS-R9830, CWI (1998)

A Penalty-Projection Method Using Staggered Grids for Incompressible Flows

C. Févrière[1], Ph. Angot[2], and P. Poullet[1]

[1] Groupe de Recherche en Info. et Math. Appli. des Antilles et de la Guyane,
Université des Antilles et de la Guyane,
97159 Pointe-à-Pitre, Guadeloupe F.W.I.
Carine.Fevriere@univ-ag.fr, Pascal.Poullet@univ-ag.fr
[2] Laboratoire d'Analyse Topologie et Probabilités, Université de Provence,
39 rue F. Joliot-Curie, 13453 Marseille Cédex 13
angot@latp.cmi.univ-mrs.fr

Abstract. We deal with the time-dependent Navier-Stokes equations with Dirichlet boundary conditions on all the domain or, on a part of the domain and open boundary conditions on the other part. It is shown numerically that a staggered mesh with penalty-projection method yields reasonable good results for solving the above mentioned problem. Similarly to the results obtained recently by other scientists using finite element method (FEM) [1] and [2] (with the rotational pressure-correction method for the latter), we confirm that the penalty-projection scheme with spatial discretization of the Marker And Cell method (MAC) [3] is compatible with our problem.

1 Introduction

The numerical simulation of the time-dependent Navier-Stokes equations for incompressible flows is CPU-time consuming. In fact, at each time step the velocity and the pressure are coupled by the incompressibility constraint. There are various ways to discretize the time-dependent Navier-Stokes equations. However the most popular is using projection methods, like pressure-correction methods. This family of methods has been introduced by Chorin and Temam ([4,5]) in the late sixties. They are time-marching techniques based on a fractional step technique that may be viewed as a predictor-corrector strategy aiming at uncoupling viscous diffusion and incompressibility effects. The interest of pressure-correction projection methods is that the velocity and the pressure are computed separately. In fact, in a first step, we solve the momentum balance equation to obtain an intermediate velocity and then, this intermediate velocity is projected on a space of solenoidal vector fields. Using this method, a numerical error named the splitting error appears and several papers have been written to estimate this error.

In [6], Shen introduced another approach which consists of constraining the divergence of the intermediate velocity field by adding in the first step of the scheme an augmentation term built from the divergence constraint (of the same

I. Lirkov, S. Margenov, and J. Waśniewski (Eds.): LSSC 2007, LNCS 4818, pp. 192–200, 2008.
© Springer-Verlag Berlin Heidelberg 2008

form as in augmented lagrangian methods [7]). And recently, some authors applied this penalty-projection method in a different way than it has been designed, and obtained with finite element approximation, reasonably good results [1]. We will show that the results obtained with different pressure-correction schemes using finite difference approximation with staggered mesh are as accurate as these obtained by the same approach but using FEM.

This paper is organized as follows:
Firstly, after we recall some preliminaries in section 2, we present the penalty pressure-correction schemes with time discretization. In section 3, we show the main numerical results: we compare the penalty projection scheme to reference algorithms with prescribed velocity on Dirichlet boundary and open boundary conditions. And finally, in the last section we report concluding remarks.

2   Formulation of the Problem

2.1   The Continuous Unsteady Navier-Stokes Problem

In this paper, we study numerical approximations with respect to time and space of the time-dependent Navier-Stokes equations which read as follows:

$$\begin{cases} \frac{\partial u}{\partial t} - \nu \Delta u + (u \cdot \nabla)u + \nabla p = f & \text{in} \quad \Omega \times [0, T] \\ \nabla \cdot u = 0 & \text{in} \quad \Omega \times [0, T], \end{cases} \tag{1}$$

for which, we add the following boundary and initial conditions:

$$\begin{cases} u = u_{\Gamma_D} & \text{on } \Gamma_D \times [0, T], \\ \nabla u \cdot n - pn = f_N & \text{on } \Gamma_N \times [0, T] \end{cases} \quad \text{and } u \mid_{t=0} = u_0 \text{ in } \Omega. \tag{2}$$

f is a smooth term source and u_0 stands for an initial velocity field, n is the normal vector and ν stands for the dynamic viscousity of fluid. Ω is an open set connected and bounded in $\mathbb{R}^2$ representing the domain of the fluid, Γ, a sufficiently smooth set, representing its boundary. We assume that the boundary Γ can be splitted into two sets Γ_N and Γ_D where the subscripts $_N$ and $_D$ stand for Neumann and Dirichlet boundary conditions. Then, the following non trivial partition holds : $\Gamma = \Gamma_N \cup \Gamma_D$, $\Gamma_D \cap \Gamma_N = \emptyset$, with $meas(\Gamma_N) \neq \emptyset$, and $meas(\Gamma_D) \neq \emptyset$. On Γ_D the velocity set to the value u_{Γ_D} whereas the force per unit area exerted at each point of the boundary Γ_N is given, equals to f_N.

2.2   The Non Incremental and Incremental Projection Schemes

Projection methods are time-marching techniques composed of two substeps for each time step in order to solve efficiently some problems coming from CFD. Whereas the first substep takes the viscous effects into account, the incompressibility constraint is secondly treated. To describe projection methods, one needs to introduce the following orthogonal decomposition:

$$L^2(\Omega) = \mathrm{H} \oplus \mathrm{H}^\perp, \quad \text{where} \quad \mathrm{H} = \{v \in [L^2(\Omega)]^2, \ \nabla \cdot v = 0, \ v \cdot n \mid_{\Gamma_D} = u_{\Gamma_D} \cdot n\}, \tag{3}$$

and $H^\perp$ is the orthogonal complement of H in $L^2(\Omega)$. Classical results shown that functions of $H^\perp$ can be characterized by functions of $L^2(\Omega)$ which can be written like the gradient of a function of $H^1(\Omega)$ [8].

To introduce some projection methods often mentioned in litterature as pressure-correction methods, let us set notations to develop semi-discrete formulations with respect to the time variable. Let $\Delta t > 0$ be a time step and for $0 \le k \le K = [T/\Delta t]$, set $t_k = k\Delta t$ such that $0 = t_0 < t_1 < \cdots < t_K$ is a uniform partition of the time interval of computation. Let also $\boldsymbol{f}^{k+1}$ be $\boldsymbol{f}(t^{k+1})$.

The first algorithm introduced by Chorin and Temam consists of decoupling each time step into two substeps. Using implicit Euler time stepping, the algorithm is as follows:

Set $u^0 = u_0$, for $k \ge 0$, compute first $\tilde{\boldsymbol{u}}^{k+1}$ by solving the equation accounting for the viscous effects:

$$\frac{1}{\Delta t}(\tilde{\boldsymbol{u}}^{k+1} - \boldsymbol{u}^k) - \nu\Delta\tilde{\boldsymbol{u}}^{k+1} + (\boldsymbol{u}^k \cdot \nabla)\tilde{\boldsymbol{u}}^{k+1} = \boldsymbol{f}^{k+1}, \qquad \tilde{\boldsymbol{u}}^{k+1}\mid_{\Gamma_{\mathrm{D}}} = \boldsymbol{u}_{\Gamma_{\mathrm{D}}}^{k+1}. \quad (4)$$

Following that, compute (u^{k+1}, p^{k+1}) by projecting the intermediate velocity $\tilde{\boldsymbol{u}}^{k+1}$ onto H, the space of vanishing divergence. Due to the decomposition (3) and its characterization, this substep can be written as follows:

$$\frac{1}{\Delta t}(\boldsymbol{u}^{k+1} - \tilde{\boldsymbol{u}}^{k+1}) + \nabla p^{k+1} = 0, \quad \nabla \cdot \boldsymbol{u}^{k+1} = 0, \quad \boldsymbol{u}^{k+1} \cdot \boldsymbol{n}\mid_{\Gamma_{\mathrm{D}}} = \boldsymbol{u}_{\Gamma_{\mathrm{D}}}^{k+1} \cdot \boldsymbol{n}.$$

The last algorithm can be improved by adding in its first substep a value of pressure gradient already computed. Thus, this algorithm known as incremental form of pressure-correction reads as follows:

Set $u^0 = u_0$ and $p^0 = p_0$ for $k \ge 0$, compute first $\tilde{\boldsymbol{u}}^{k+1}$ by solving the problem accounting for the viscous effects:

$$\frac{1}{\Delta t}(\tilde{\boldsymbol{u}}^{k+1} - \boldsymbol{u}^k) - \nu\Delta\tilde{\boldsymbol{u}}^{k+1} + (\boldsymbol{u}^k \cdot \nabla)\tilde{\boldsymbol{u}}^{k+1} + \nabla p^k = \boldsymbol{f}(t^{k+1}), \qquad (5)$$

$$\tilde{\boldsymbol{u}}^{k+1}\mid_{\Gamma_{\mathrm{D}}} = \boldsymbol{u}_{\Gamma_{\mathrm{D}}}^{k+1}, \qquad (\nabla\boldsymbol{u}^{k+1} \cdot \boldsymbol{n} - p\boldsymbol{n})\mid_{\Gamma_{\mathrm{N}}} = \boldsymbol{f}_{\mathrm{N}}^{k+1}. \qquad (6)$$

The second substep consists also of projecting $\tilde{\boldsymbol{u}}^{k+1}$ onto H orthogonally with respect to L^2. But for a sake of efficiency, we deal with the following elliptic problem, obtained after applying the divergence operator:

$$\Delta\phi = \frac{1}{\Delta t}\nabla \cdot \tilde{\boldsymbol{u}}^{k+1}, \quad \text{with } \nabla\phi \cdot \boldsymbol{n}\mid_{\Gamma_{\mathrm{D}}} = 0, \quad \phi\mid_{\Gamma_{\mathrm{N}}} = 0. \qquad (7)$$

The computed pressure can be recovered with the solution of (7) by the next formula, where χ is a coefficient equal to 0 or 1:

$$p^{k+1} = p^k + \phi - \chi\nu\nabla \cdot \tilde{\boldsymbol{u}}^{k+1}. \qquad (8)$$

The choice $\chi = 0$ yields the standard version of the algorithm whereas $\chi = 1$ yields the rotational version.

2.3   The Penalty Projection Scheme

Firstly introduced by Shen in [9], this form yields to find the intermediate velocity $\tilde{\boldsymbol{u}}^{k+1}$, solution of the following elliptic problem at time t^{k+1}. For example, using the Backward Difference Formula of second order (BDF2) for the approximation of the time derivative, the time semi-discrete scheme can be described as follows:

$$\frac{3\tilde{\boldsymbol{u}}^{k+1} - 4\boldsymbol{u}^k + \boldsymbol{u}^{k-1}}{2\Delta t} - \nu\Delta\tilde{\boldsymbol{u}}^{k+1} + ((2\boldsymbol{u}^k - \boldsymbol{u}^{k-1})\cdot\nabla)\tilde{\boldsymbol{u}}^{k+1}$$

$$+\nabla p^k - r\nabla(\nabla\cdot\tilde{\boldsymbol{u}}^{k+1}) = \boldsymbol{f}^{k+1}, \qquad (9)$$

$$\tilde{\boldsymbol{u}}^{k+1}\mid_{\Gamma_D} = \boldsymbol{u}_{\Gamma_D}^{k+1}, \qquad (\nabla\tilde{\boldsymbol{u}}^{k+1}\cdot\boldsymbol{n} - p\boldsymbol{n}),\mid_{\Gamma_N} = \boldsymbol{f}_N^{k+1}.$$

Then, the projection step consists of projecting onto H orthogonally with respect to L^2 the intermediate velocity, which takes the following form:

$$\frac{3\boldsymbol{u}^{k+1} - 3\tilde{\boldsymbol{u}}^{k+1}}{2\Delta t} + \nabla\phi = 0, \qquad (10)$$

$$\nabla\cdot\boldsymbol{u}^{k+1} = 0, \qquad (11)$$

$$\boldsymbol{u}^{k+1}\cdot\boldsymbol{n}\mid_{\Gamma_D} = \boldsymbol{u}_{\Gamma_D}^{k+1}\cdot\boldsymbol{n}, \qquad (12)$$

$$\phi\mid_{\Gamma_N} = 0. \qquad (13)$$

Practically, we also take the divergence of (10) and using (12) into (10), we obtain the following Poisson problem for ϕ :

$$\Delta\phi = \frac{3}{2\Delta t}\nabla\cdot\tilde{\boldsymbol{u}}^{k+1}, \qquad (14)$$

$$(\nabla\phi\cdot\boldsymbol{n})\mid_{\Gamma_D} = 0, \qquad \phi\mid_{\Gamma_N} = 0. \qquad (15)$$

Due to the relation (10) we obtain the end-of-step velocity which reads as follows:

$$\boldsymbol{u}^{k+1} = \tilde{\boldsymbol{u}}^{k+1} - \frac{2\Delta t}{3}\nabla\phi. \qquad (16)$$

To recover an expression for an approximation of the pressure at time t^{k+1}, first adding (9) to (10), one obtains an expression of the discrete momentum balance equation at time t^{k+1}:

$$\frac{3\tilde{\boldsymbol{u}}^{k+1} - 4\boldsymbol{u}^k + \boldsymbol{u}^{k-1}}{2\Delta t} - \nu\Delta\tilde{\boldsymbol{u}}^{k+1} + ((2\boldsymbol{u}^k - \boldsymbol{u}^{k-1})\cdot\nabla)\tilde{\boldsymbol{u}}^{k+1}$$

$$+\nabla(p^k - r\nabla\cdot\tilde{\boldsymbol{u}}^{k+1} + \phi) = \boldsymbol{f}^{k+1}. \qquad (17)$$

Then, the equation (17) suggests the following expression of p^{k+1}, where r is a parameter to be specified:

$$p^{k+1} = p^k + \phi - r\nabla\cdot\tilde{\boldsymbol{u}}^{k+1}. \qquad (18)$$

There are two forms (the standard and the rotational one) of the penalty pressure-correction scheme which can be summarized by the following algorithm.

Assuming that $\boldsymbol{u}^{k-1}, \boldsymbol{u}^k, p^k$ are known, the viscous step consists of computing $\tilde{\boldsymbol{u}}^{k+1}$ by:

$$\frac{3\tilde{\boldsymbol{u}}^{k+1} - 4\boldsymbol{u}^k + \boldsymbol{u}^{k-1}}{2\varDelta t} - \nu\varDelta\tilde{\boldsymbol{u}}^{k+1} + ((2\boldsymbol{u}^k - \boldsymbol{u}^{k-1}) \cdot \nabla)\tilde{\boldsymbol{u}}^{k+1}$$

$$+\nabla p^k - r_1\nabla(\nabla \cdot \tilde{\boldsymbol{u}}^{k+1}) = \boldsymbol{f}^{k+1}, \quad (19)$$

$$\tilde{\boldsymbol{u}}^{k+1}\,|_{\varGamma_{\mathrm{D}}} = \boldsymbol{u}_{\varGamma_{\mathrm{D}}}^{k+1}, \qquad (\nabla\tilde{\boldsymbol{u}}^{k+1} \cdot \boldsymbol{n} - p\boldsymbol{n})\,|_{\varGamma_{\mathrm{N}}} = \boldsymbol{f}_{\mathrm{N}}^{k+1}. \qquad (20)$$

Then, the projection step consists of computing ϕ, the solution of the next elliptic problem:

$$\varDelta\phi = \frac{3}{2\varDelta t}\nabla \cdot \tilde{\boldsymbol{u}}^{k+1}, \qquad (21)$$

$$(\nabla\phi \cdot n)\,|_{\varGamma_{\mathrm{D}}} = 0, \qquad \phi\,|_{\varGamma_{\mathrm{N}}} = 0. \qquad (22)$$

Therefore, the approximation of the velocity and pressure at time t^{k+1} is straightforward:

$$\boldsymbol{u}^{k+1} = \tilde{\boldsymbol{u}}^{k+1} - \frac{2\varDelta t}{3}\nabla\phi, \qquad p^{k+1} = p^k + \alpha\phi - r_2\nabla \cdot \tilde{\boldsymbol{u}}^{k+1},$$

where r_1, r_2 and α are parameters to be specified.

The choice of the parameters r_1, r_2 and α leads to different methods. For example, setting $r_1 = r_2 = 0$ and $\alpha = 1$ leads to the so-called standard incremental method. Whereas $r_1 = 0$, $r_2 = \nu$ and $\alpha = 1$ leads to the so-called rotational incremental method introduced by Van Kan in [10]. The choice $r_1 = r_2 = r$ and $\alpha = 1$ leads to the standard penalty projection method and the rotational penalty projection method is obtained by the choice $r_1 = r$, $r_2 = r + \nu$ and $\alpha = 1$.

3   Main Results

We illustrate in this section the convergence properties of the pressure-correction algorithm using BDF2 to march in time and MAC method for the spatial discretization of the problem. We also do a comparative study between the penalty projection method and some pressure-correction schemes often used in the litterature for the solution of unstationnary imcompressible flow problems. The results trend to prove that a Navier-Stokes flow with either Dirichlet boundary conditions or open boundary conditions can be well computed with this penalty-projection method combined with staggered mesh.

3.1   A Stokes Flow with Dirichlet Boundary Conditions

To compute a Stokes flow with Dirichlet boundary conditions, we consider a square domain $\varOmega =]0, 1[^2$ with Dirichlet boundary conditions on the velocity. We assume that the exact solution $(\boldsymbol{u}, p)$ of the Stokes problem (1)-(2) is:

$$\boldsymbol{u}(x, y, t) = (\sin(x+t)\sin(y+t), \cos(x+t)\cos(y+t)), \qquad p(x, y, t) = \cos(x-y+t),$$

which define the right hand side of the balance momentum equation.

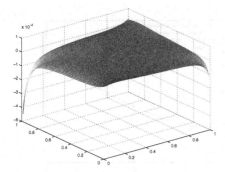

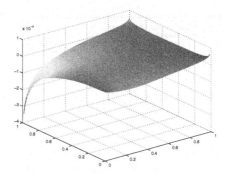

Fig. 1. Pressure error at T=1 in $]0,1[^2$ for the standard form of the penalty pressure-correction method, with $r = 1$

Fig. 2. Pressure error at T=1 in $]0,1[^2$ for the standard form of the penalty pressure-correction method, $r = 100$

First, to check the pressure error at $T = 1$, we set $h = 1/80$ whereas $\Delta t = 1/160$. With the standard form of the incremental pressure-correction method, we obtain the pressure error of $1.26 \ 10^{-2}$, whereas with the rotational form, it equals to $3.73 \ 10^{-3}$.

With the penalty pressure-correction, we made the same test at (Figure 1) and we obtain that the error of pressure fields for the penalty form with r ranging between 1 and 10 are equivalent to the error of pressure field for the rotational form of the incremental pressure-correction method. But, for higher values of the r parameter, like $r = 100$ (*cf.* Fig. 2), the error is well reduced (the vertical range is approximatively divided by 10).

At Figure 3, we plot the l^∞−norm of the error of the pressure as a function of the time step Δt. The error is measured at $T = 2$, after we made series of computation on the unit square domain with the mesh size h equal to $1/160$. The results of the error of the computed pressure prove that $3/2$ is the convergence rate for the approximate pressure in rotational form. This result conforms with those which has been reported in [11]. Also, as it has been shown elsewhere [12] that the error of pressure for the penalty form with r ranging between 1 and 10 seems to be almost the same comparing to those computed from the rotational form. Moreover, we notice that the convergence rate for the penalty-projection scheme is smaller than those computed by the rotational form. In fact, the higher the parameter of the penalty-projection method is, the smaller the error of the pressure is.

3.2   A Stokes Flow with Open Boundary Conditions

We consider the unit square as our computation domain $\Omega =]0,1[^2$ with open boundary conditions on the velocity. We take the exact solution $(\boldsymbol{u}, p)$ of (1)-(2) to be:

$$\boldsymbol{u}(x,y,t) = (\sin(x)\sin(y+t), \cos(x)\cos(y+t)), \qquad p(x,y,t) = \cos(x)\sin(y+t),$$

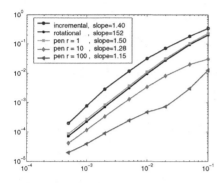
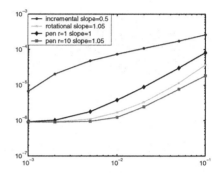

Fig. 3. Convergence rates on pressure in l^∞-norm at T=2, for the Stokes flow with Dirichlet boundary conditions

Fig. 4. Convergence rates on pressure in l^∞-norm at T=1, for the Stokes flow with open boundary conditions

satisfying the boundary conditions: $(p\mathbf{n} - \nabla\mathbf{u} \cdot \mathbf{n}) \, |_{\Gamma_N} = 0$, $\Gamma_N = \{(x, y) \in \Gamma,$ $x = 0\}$.

At Figure 4, we compare the l^∞-norm of the pressure error as a function of Δt for different pressure-correction methods. The error is measured at $T = 1$ after we performed computations on a square domain with the mesh size h equal to 1/250.

For each time step, we obtain that the penalty-projection method (with $r \geq 10$) is more accurate than the incremental method. Moreover, a space convergence order of 1/2 and 1 is observed respectively for the incremental and the rotational projection method, whereas for the penalty-projection methods the convergence order is of 1. Our results match with those obtained by [1] with FEM.

3.3   Taylor-Green Vortices

Now, we treat a well-known benchmark which consists of a periodic flow governed by Navier-Stokes equations. This particular case of non-forcing flow has the advantage of owning analytic solutions. Then, let us consider a square domain $\Omega =]1/8, 5/8[^2$ with Dirichlet boundary conditions on the velocity. For the tests we use the following exact solution: $(\mathbf{u}, p)$ of (1)-(2) to be:

$$\begin{cases} \mathbf{u}(x, y, t) = \left(-\cos(2\pi x)\sin(2\pi y)\exp(-8\pi^2\mu t), \sin(2\pi x)\cos(2\pi y)\exp(-8\pi^2\mu t)\right), \\ p(x, y, t) = -\dfrac{(\cos(4\pi x) + \cos(4\pi y))}{4}\exp(-16\pi^2\mu t), \end{cases}$$

At Figure 5, we compare the L^2-norm of the pressure error as a function of Δt for different pressure-correction methods. The error is measured at $T = 1$ after we performed computations on a square domain with the mesh size h equal to 1/80.

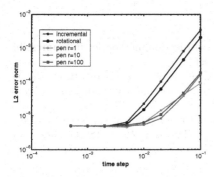
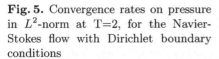

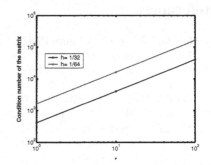

Fig. 5. Convergence rates on pressure in L^2-norm at T=2, for the Navier-Stokes flow with Dirichlet boundary conditions

Fig. 6. Condition number of the matrix of the viscous step for a flow governed by NSE endowed with Dirichlet boundary conditions, $\delta t = 0.05$

All the curves show the decrease of the error with the time step. For the small values of this time step, a plateau is observed, which corresponds to the space discretization error. We can observe that the rate of convergence agrees with the theoretical studies.

4   Concluding Remarks

The numerical results show that with the penalty-projection method, the computations are more accurate than these obtained with incremental projection schemes. Our results of either the Stokes flow with Dirichlet boundary conditions, and the Stokes flow with open boundary conditions, or the Taylor-Green vortices match with the ones which have been reported in [13] and in [1]. In one hand, these results prove the compatibility of the spatial discretization by MAC mesh and the projection methods for our problems with various boundary conditions. In other hand, the penalty-projection scheme is more time-consuming than an incremental projection scheme for a given time-step. This fact can be confirmed by computing the condition number of the matrix of the viscous step for various values of the r parameter. The computations from several experiments of three values of r ($r = 1$, $r = 10$, $r = 100$) show that the condition number increases with the r parameter. From Figure 6, one could think that the condition number varies as a power of r. This might be confirmed by further next works. Nevertheless, our experiment with the penalty-projection suffers from a lack of well-designed preconditioner to speed-up the solver of the viscous step.

Acknowledgements

The authors are pleased to acknowledge the Centre Commun de Calcul Intensif of UAG where computational tests have been performed.

References

1. Jobelin, M., et al.: A Finite Element penalty-projection method for incompressible flows. J. Comput. Phys. 217(2), 502–518 (2006)
2. Guermond, J.L., Minev, P., Shen, J.: Error analysis of pressure-correction schemes for the time-dependent Stokes equations with open boundary conditions. SIAM J. Numer. Anal. 43(1), 239–258 (2005)
3. Harlow, F., Welch, J.: Numerical calculation of time-dependent viscous incompressible flow of fluid with free surfaces. J. E. Phys. Fluids 8, 2181–2189 (1965)
4. Chorin, A.J.: Numerical solution of the Navier-Stokes equations. Math. Comput. 22, 745–762 (1968)
5. Temam, R.: Sur l'approximation de la solution des équations de Navier-Stokes par la méthode à pas fractionnaires (II). Arch. for Rat. Mech. Anal. 33, 377–385 (1969)
6. Shen, J.: On error estimates of projection methods for Navier-Stokes equations: Second-order schemes. Math. Comput. 65(215), 1039–1065 (1996)
7. Fortin, M., Glowinski, R.: Augmented Lagrangian Methods: Applications to the numerical solution of boundary value problems. North-Holland, Amsterdam (1983)
8. Temam, R.: Navier-Stokes Equations, AMS Chelsea Publishing (2001)
9. Shen, J.: On error estimates of some higher order projection and penalty-projection methods for Navier-Stokes equations. Numer. Math. 62, 49–73 (1992)
10. Van Kan, J.: A second-order accurate pressure-correction scheme for viscous incompressible flow. SIAM J. on Scient. Stat. Comput. 7(3), 870–891 (1986)
11. Guermond, J.L., Shen, J.: On the error estimates for the rotational pressure-correction projection methods. Math. Comput. 73(248), 1719–1737 (2004)
12. Février, C. et al.: An accurate projection method for incompressible flows, An Int. Conf. of the Carribbean Academy of Science, Le Gosier (2006)
13. Guermond, J.L., Minev, P., Shen, J.: An overview of projection methods for incompressible flows. Comput. Meth. Appl. Mech. Engrg. 195(44–47), 6011–6045 (2006)

Qualitatively Correct Discretizations in an Air Pollution Model

K. Georgiev[1] and M. Mincsovics[2]

[1] Institute for Parallel Processing, Bulgarian Academy of Sciences, Sofia, Bulgaria
georgiev@parallel.bas.bg
[2] Institute of Mathematics, Eötvös Loránd University, Budapest, Hungary
mikka76@freemail.hu

Abstract. We deal with one subproblem of an air pollution model, the horizontal diffusion, which can be mathematically described by a linear partial differential equation of parabolic type. With different space discretization schemes (like a FDM, FEM), and using the θ-method for time discretization we get a one-step algebraic iteration as a numerical model. The preservation of characteristic qualitative properties of different phenomena is an increasingly important requirement in the construction of reliable numerical models. For that reason we analyze the connection between the shape and time-monotonicity in the continuous and the numerical model, and we give the necessary and sufficient condition to fulfil this property.

1   Introduction

When constructing numerical models, engineers and scientists involved in scientific computing must take into consideration the qualitative properties of the original phenomenon. Numerical models should not give results that contradict to the physical reality. For example in air-pollution models, which are an indispensable tool of environmental protection, the solution methods should not result in negative concentration values. These values would have no physical meaning. Certain choices of the time steps in the numerical schemes result in qualitatively correct numerical models, while others do not. Even unconditionally stable schemes, like the Crank-Nicolson or the implicit Euler scheme, can produce qualitative deficiencies, see [4].

The main goal of this paper is to define a property which is essential for the diffusion sub-model, and we give the necessary and sufficient condition for this property.

1.1   Air Pollution Models and Application of Splitting Techniques

Operator splitting methods are frequently applied to the solution of air pollution problems ([11]). The governing equation is split into several subproblems according to the physical processes involved, and these subproblems are solved cyclically ([10]). In this way, a qualitatively incorrect solution of the diffusion

I. Lirkov, S. Margenov, and J. Waśniewski (Eds.): LSSC 2007, LNCS 4818, pp. 201–208, 2008.

subproblem can cause a series of problems in the solving process of the other subproblems, because the solution of the first problem will be input to the second one.

Large air pollution models are normally described by systems of partial differential equations (PDE's):

$$\frac{\partial c_s}{\partial t} = -\frac{\partial(uc_s)}{\partial x} - \frac{\partial(vc_s)}{\partial y} - \frac{\partial(wc_s)}{\partial z} \tag{1}$$

$$+\frac{\partial}{\partial x}\left(\kappa_x \frac{\partial c_s}{\partial x}\right) + \frac{\partial}{\partial y}\left(\kappa_y \frac{\partial c_s}{\partial y}\right) + \frac{\partial}{\partial z}\left(\kappa_z \frac{\partial c_s}{\partial z}\right)$$

$$+E_s - (\lambda_{1s} + \lambda_{2s})c_s + Q_s(c_1, c_2, \ldots, c_q), \quad s = 1, 2, \ldots, q \ ,$$

where (i) the concentrations of the chemical species are denoted by c_s, (ii) u, v and w are wind velocities, (iii) κ_x, κ_y and κ_z are diffusion coefficients, (iv) the emission sources are described by E_s, (v) λ_{1s} and λ_{2s} are deposition coefficients and (vi) the chemical reactions are denoted by $Q_s(c_1, c_2, \ldots, c_q)$ (e.g. [11]).

It is difficult to treat the system of PDE's (1) directly. This is the reason for using different kinds of splitting. A splitting procedure, based on ideas proposed in [6] and [8], leads, for $s = 1, 2, \ldots, q$, to five sub-models, representing the horizontal advection, the horizontal diffusion (together with the emission terms), the chemistry, the deposition and the vertical exchange. We will investigate the diffusion submodel.

2   Connection Between the Shape and the Time-Monotonicity

In the following we will denote by I the identity operator. The matrices will be typeset in boldface, for example $\mathbf{A}$, in the fully discretized case we will use the following style: A, and the operators in the continuous case will be typeset in normal font: A.

In this section we define two qualitative properties of the solution, namely the shape and time-monotonicity, and we investigate their relation. We present these properties and the effect of time discretization on them through examples. Our examples will range from simple to complex. We consider the diffusion equation in the following form.

$$\left.\begin{aligned} D_t u(x,t) &= D_x^2 u(x,t) + f(x,t), \quad x \in \Omega, \ t > 0 \\[2mm] u(x,t) &= g(x,t), \quad x \in \partial\Omega, \ t > 0 \ . \end{aligned}\right\} \tag{2}$$

Here $u \in C^{2,1}(Q_T) \cap C(\overline{Q}_T)$ is the unknown function, $Q_T = \Omega \times (0,T)$ with some $T > 0$. Let $\Omega = (0,1)$ in the examples, but we remark that all works in higher dimension, too.

Definition 1. *Let U denote the set of solutions of (2) for the different (sufficiently smooth) initial functions u_0. We define with $S(Y)_{U,\rho}$ a subset of U projected on the time level ρ for a given operator Y in the following way:*

$$S(Y)_{U,\rho} := \{u(x,t)|_{t=\rho} : u(x,t) \in U,\, Yu(x,t)|_{t=\rho} \geq 0\}\ ,$$

and we denote for a given $\tau \in \mathbb{R}$ the shift (in time) operator with L_τ:

$$L_\tau(u(x,t)|_{t=\rho}) := u(x,t)|_{t=\rho+\tau} \quad for \quad u(x,t) \in U\ .$$

2.1 Example 1

Let $u(0,t) = u(1,t) = 0$, $t > 0$ and $f(x,t) = 0$ in (2).

Note that $S(-D_t)_{U,\rho}$ is the set of the time-decreasing solutions on the ρ-th time level, similarly $S(-D_x^2)_{U,\rho}$ is the set of the concave solutions on the ρ-th time level. In this case the following result is important:

$$L_\tau S(-D_x^2)_{U,\rho} \subseteq S(-D_x^2)_{U,\rho+\tau} \quad for \quad \tau > 0\ .$$

Namely the solution preserves the concavity (shape), see [2]. We remark that $L_\tau S(-D_x^2)_{U,\rho} \subseteq S(-D_x^2)_{U,\rho}$ for $\tau > 0$, since $S(-D_x^2)_{U,\rho+\tau} \subseteq S(-D_x^2)_{U,\rho}$ for $\tau > 0$. Here we cannot write equality due to the problem of the backward solvability. From the relation

$$S(-D_x^2)_{U,\rho} = S(-D_t)_{U,\rho} \quad for \quad \rho \leq T \tag{3}$$

we have

$$L_\tau S(-D_t)_{U,\rho} \subseteq S(-D_t)_{U,\rho+\tau} \quad for \quad \tau > 0\ .$$

We remark that if $u(x,t)|_{t=\rho} = \sum_{k=1}^{\infty} \xi_k \sin k\pi x$ and $\xi_1 > 0$, then there exists $\omega \geq 0$: $L_\omega(u(x,t)|_{t=\rho}) \in S(-D_x^2)_{\rho+\omega}$ (and $\in S(-D_t)_{\rho+\omega}$), see [3].

The relation (3) presents the connection between the concavity (shape) and the time-monotonicity. We are interested in how this relation changes due to time distretization. Let us imagine the time distretization as if we could see the solution not continuously, but on certain time levels, namely we choose a time step τ and

$$\Sigma := \{k\tau : k \in \mathbb{N},\, k\tau \leq T\}$$

is the set of the visible time levels. The problem arises from this, because the solution may decrease from one time level to the next even if it was not concave on the first one. At the same time the following two statements are true:

- If a solution on some time level is concave, then it will decrease in the next step.
- With decrease we can only get a concave solution on some time level.

To describe this we approximate (in time) the operator $-D_t$ with $\frac{1}{\tau}(I - L_\tau)$ (and D_x^2 with itself), thus the corresponding approximated subsets are $S(I - L_\tau)_{U,\rho}$ and $S(-D_x^2)_{U,\rho}$ for $t \in [\rho, \rho+\tau)$. Therefore the property corresponding to (3) in the time-discretized case is:

$$S(I - L_\tau)_{U,\rho} \supseteq S(-D_x^2)_{U,\rho} \supseteq S((I - L_\tau)L_\tau^{-1})_{U,\rho} \quad \forall \rho \in \Sigma\ . \tag{4}$$

2.2   Example 2

Let $u(x,t) = g(x)$ for $x \in \partial\Omega$ and $f(x,t) = f(x)$ in (2).

We proceed similarly as in Subsection 2.1. The only difference is that the operator which is responsible for the shape will be $\hat{Q}$, defined as follows: $\hat{Q}u(x,t) := -D_x^2 u(x,t) - f(x)$. Therefore the property corresponding to (3) in the continuous and the time-discretized case can be formulated simply.

$$\mathcal{S}(\hat{Q})_{U,\rho} = \mathcal{S}(-D_t)_{U,\rho} \quad \text{for} \quad \rho \leq T \ , \tag{5}$$

$$\mathcal{S}(I - L_\tau)_{U,\rho} \supseteq \mathcal{S}(\hat{Q})_{U,\rho} \supseteq \mathcal{S}((I - L_\tau)L_\tau^{-1})_{U,\rho} \quad \forall \rho \in \Sigma \ . \tag{6}$$

We give the following trivial statement without proof.

Lemma 1. *The following statements are equivalent.*
(i) $\mathcal{S}(\hat{Q})_{U,\rho} = \mathcal{S}(-D_t)_{U,\rho} \quad \forall \rho \leq T$.
(ii) $\mathcal{S}(I - L_\tau)_{U,\rho} \supseteq \mathcal{S}(\hat{Q})_{U,\rho} \supseteq \mathcal{S}((I - L_\tau)L_\tau^{-1})_{U,\rho} \quad \forall \tau > 0, \forall \rho \in \Sigma$.

Clearly, this statement is valid for Example 1, too.

We remark that if $f(x,t)$ or $g(x,t)$ depends on t, then we can approximate $\hat{Q}$ with $\hat{Q}_{\text{approx}}$, $\hat{Q}_{\text{approx}}u(x,t) = -D_x^2 u(x,t) - C(g(x,t), g(x,t+\tau), f(x,t+\frac{1}{2}\tau))$ at every $[t, t+\tau]$, $t \in \Sigma$ with some operator C. But substituting $\hat{Q}$ with $\hat{Q}_{\text{approx}}$ in (6) we get only an approximation of the property (6).

In the next section we show what all this means for the diffusion submodel discretized by the finite difference or finite element method in space and by the θ-method in time.

3   The Diffusion Submodel

We will investigate the diffusion submodel in the form of the following initial boundary value problem:

$$C\frac{\partial c(\mathbf{x},t)}{\partial t} - \nabla(\kappa\nabla c(\mathbf{x},t)) = f(\mathbf{x},t), \quad (\mathbf{x},t) \in Q_{t_{\max}} := \Omega \times (0, t_{\max}), \tag{7}$$

$$c(\mathbf{x},t) = g(\mathbf{x},t), \quad \mathbf{x} \in \partial\Omega, \ 0 \leq t \leq t_{\max}, \tag{8}$$

$$c(\mathbf{x},0) = c_0(\mathbf{x}), \quad \mathbf{x} \in \Omega, \tag{9}$$

where (8) is the boundary condition and (9) is the initial condition. The sufficiently smooth unknown function $c = c(\mathbf{x},t)$ is defined in $\overline{\Omega} \times [0, t_{\max}]$, where Ω is a d-dimensional domain and $t_{\max} > 0$ is a fixed real number. The symbol $\partial\Omega$ denotes the boundary of Ω. As usual, ∇ stands for the nabla operator. The function $C : \Omega \to \mathbb{R}$ has the property $0 < C_{\min} \leq C \leq C_{\max}$. The bounded function $\kappa : \Omega \to \mathbb{R}$ fulfils the property $0 < \kappa_{\min} \leq \kappa \leq \kappa_{\max}$ and has continuous first derivatives. The function $g : \Gamma_{t_{\max}} \to \mathbb{R}$ is continuous on the lateral surface of $Q_{t_{\max}}$ denoted by $\Gamma_{t_{\max}}$. Furthermore, the function $f : Q_{t_{\max}} \to \mathbb{R}$ is bounded in $Q_{t_{\max}}$.

3.1   Discretization, One-Step Iteration

Using the finite difference method (on rectangular domain) or the Galerkin finite element method to the semidiscretization, and the θ-method to get a fully discretized model, we arrive at an algebraic one-step iteration in a partitioned form:

$$[\mathbf{A}_0|\mathbf{A}_\partial] \begin{bmatrix} \mathbf{u}^{n+1} \\ \mathbf{g}^{n+1} \end{bmatrix} = [\mathbf{B}_0|\mathbf{B}_\partial] \begin{bmatrix} \mathbf{u}^n \\ \mathbf{g}^n \end{bmatrix} + \tau \, \mathbf{f}^{n+1/2} \; . \tag{10}$$

In Equation (10), the vectors $\mathbf{u}^n$, $\mathbf{g}^n$, $\mathbf{g}^{n+1}$, and $\mathbf{f}^{n+1/2}$ are known: $\mathbf{u}^n$ is known from the previous time level (originally from the initial condition), $\mathbf{g}^n$ and $\mathbf{g}^{n+1}$ are given from the boundary condition, and $\mathbf{f}^{n+1/2}$ can be computed from the source function f. τ is the time step. The matrices come from the spatial discretization, see [4,5].

Remark 1. It can be shown, that $\mathbf{A}_0 \in \mathbb{R}^{N \times N}$ is regular for both the finite difference and the Galerkin finite element method (with corresponding basis functions), see [5], which yields that the numerical solution does exist and it is unique for each setting of the source and the initial and boundary conditions. In the case of the finite difference method $\mathbf{A}_0$ is an M-matrix (see [1]) and in the case of the Galerkin finite element method it is a Gram matrix. We can tell the same about the matrix $\mathbf{Q} = \mathbf{A}_0 - \mathbf{B}_0 \in \mathbb{R}^{N \times N}$. We suppose in the following that $\mathbf{B}_0 \in \mathbb{R}^{N \times N}$ is also regular.

In an unpartitioned form (10) looks as follows

$$\mathbf{u}^{n+1} = \mathbf{A}_0^{-1}\mathbf{B}_0\mathbf{u}^n - \mathbf{A}_0^{-1}\mathbf{A}_\partial\mathbf{g}^{n+1} + \mathbf{A}_0^{-1}\mathbf{B}_\partial\mathbf{g}^n + \tau\mathbf{A}_0^{-1}\mathbf{f}^{n+1/2} \; . \tag{11}$$

With the notations $\mathbf{H} = \mathbf{A}_0^{-1}\mathbf{B}_0$, $\mathbf{h}^n = \mathbf{A}_0^{-1}(-\mathbf{A}_\partial\mathbf{g}^{n+1} + \mathbf{B}_\partial\mathbf{g}^n - \tau\mathbf{f}^{n+1/2})$, $\mathbf{L}^n\mathbf{x} = \mathbf{H}\mathbf{x} + \mathbf{h}^n$ (11) has the form

$$\mathbf{u}^{n+1} = \mathbf{L}^n\mathbf{u}^n \; . \tag{12}$$

$\mathbf{L}^n$ is an approximation of L_τ on the n-th time level. Note that $\mathbf{L}^n$ is invertible for every n due to Remark 1.

3.2   Connection Between the Shape and the Time-Monotonicity

We denote the sequences $(\mathbf{u}^n)$ for the different initial vectors with U. Note that for a given operator $Y : \mathbb{R}^n \to \mathbb{R}^n$ the equality $\mathcal{S}(Y)_{\mathsf{U},\rho} = \mathcal{S}(Y)_{\mathsf{U},\rho+\tau} = \{\mathbf{x} \in \mathbb{R}^n : Y\mathbf{x} \geq 0\}$ holds, since $\mathbf{L}^n$ is invertible. This explains the following definition.

Definition 2. *Let* $Y : \mathbb{R}^n \to \mathbb{R}^n$ *be an operator, then we define the subset* $\mathcal{S}(Y) \subseteq \mathbb{R}^n$ *as follows:*

$$\mathcal{S}(Y) := \{\mathbf{x} \in \mathbb{R}^n : Y\mathbf{x} \geq 0\} \; .$$

When Y is linear, i.e., $\mathbf{Y} \in \mathbb{R}^{n \times n}$ is a matrix, then $\mathcal{S}(\mathbf{Y})$ is a cone. If $\mathbf{Y}$ is a nonsingular matrix, then $\mathcal{S}(\mathbf{Y})$ is a proper cone, and $\mathcal{S}(\mathbf{Y}) = \mathbf{Y}^{-1}\mathbb{R}_+^n$, where

$\mathbb{R}_+^n$ denotes the nonnegative orthant, shows that it is a simplicial cone [1]. For example $\mathcal{S}(\mathbf{I}) = \mathbb{R}_+^n$.

First we investigate the Example of Subsection 2.1, discretized both in space and time. Here we have $\mathbf{L}^n = \mathbf{H}$, due to the homogeneous boundary condition and to the lack of the source. Let us notice that here $\mathbf{Q} = -N^2 tridiag[-1, 2-1] \in \mathbb{R}^{N \times N}$ is the discrete analogue of the second-order differential operator in a space variable.

The concavity of the vector $\mathbf{x} \in \mathbb{R}^n$ is defined as $\mathbf{Qx} \geq 0$, which is in accordance with the continuous case. Therefore, the set of the concave vectors is $\mathcal{S}(\mathbf{Q})$. Similarly, the time-decreasing of a vector $\mathbf{x} \in \mathbb{R}^n$ with respect to the iteration can be defined as $\mathbf{x} \geq \mathbf{Hx}$ or equivalently $(\mathbf{I} - \mathbf{H})\mathbf{x} \geq 0$. Therefore, the set of the time-decreasing vectors is $\mathcal{S}(\mathbf{I} - \mathbf{H})$ (it is not depending on n).

Consequently, we call the discrete model correct if it possesses the discrete analogue properties formulated in Subsection 2.1. The property of concavity preserving is discussed in [2,3,9], and the time-monotonicity in [3,9]. We are interested now in the connection between the concavity (shape) and time-monotonicity. The expectations can be formulated as follows.

Definition 3. *The discrete model corresponding to the Example of 2.1 is said to be correct from the point of view of the connection between concavity and time-monotonicity if the following statement holds:*

$$\mathcal{S}(\mathbf{I} - \mathbf{H}) \supseteq \mathcal{S}(\mathbf{Q}) \supseteq \mathcal{S}((\mathbf{I} - \mathbf{H})\mathbf{H}^{-1}) \ .$$

With this we change over to the general case: we will see what is the corresponding definition, and then we will already give the conditions for the discrete model to fulfil this property.

The operator $\hat{\mathbf{Q}}^n \mathbf{x} = (\mathbf{A}_0 - \mathbf{B}_0)\mathbf{x} + (\mathbf{A}_\partial \mathbf{g}^{n+1} - \mathbf{B}_\partial \mathbf{g}^n + \tau \mathbf{f}^{n+1/2}) = \mathbf{Qx} - \mathbf{A}_0 \mathbf{h}^n$ is responsible for the shape. The property corresponding to Definition 3 is exactly as follows: $\mathcal{S}(\mathbf{I} - \mathbf{L}^n) \supseteq \mathcal{S}(\hat{\mathbf{Q}}^n) \supseteq \mathcal{S}((\mathbf{I} - \mathbf{L}^n)(\mathbf{L}^n)^{-1})$ for all $n \in \mathbb{N}$. However we will see that the realization of this property does not depend on n. Therefore it is enough to check for the two operators defined as

$$\hat{\mathbf{Q}}\mathbf{x} := (\mathbf{A}_0 - \mathbf{B}_0)\mathbf{x} + (\mathbf{A}_\partial \mathbf{g}' - \mathbf{B}_\partial \mathbf{g} + \tau \hat{\mathbf{f}}) = \mathbf{Qx} - \mathbf{A}_0 \mathbf{h} \ ,$$

$$\mathbf{L}\mathbf{x} := \mathbf{A}_0^{-1}\mathbf{B}_0 \mathbf{x} + \mathbf{A}_0^{-1}(-\mathbf{A}_\partial \mathbf{g}' + \mathbf{B}_\partial \mathbf{g} - \tau \hat{\mathbf{f}}) = \mathbf{Hx} + \mathbf{h} \ . \tag{13}$$

This means that

$$(\mathbf{I} - \mathbf{L})\mathbf{x} = \mathbf{A}_0^{-1}\hat{\mathbf{Q}}\mathbf{x} \ . \tag{14}$$

Definition 4. *The discrete model corresponding to (7), (8) is said to be correct from the point of view of the connection between shape and time-monotonicity, if the following statement holds:*

$$\mathcal{S}(\mathbf{I} - \mathbf{L}) \supseteq \mathcal{S}(\hat{\mathbf{Q}}) \supseteq \mathcal{S}((\mathbf{I} - \mathbf{L})\mathbf{L}^{-1}) \ .$$

Now we have arrived at the point to investigate under what assumption the discrete model is correct. The next lemma will be helpful in this.

Lemma 2. *Let* $X, Y : V \to V$ *be operators, where* V *is a vector space over an ordered field. If* X *is bijective, then the following statements are equivalent.*
(i) $\mathcal{S}(X) \subseteq \mathcal{S}(Y)$.
(ii) $YX^{-1} \geq 0$.

Proof. It follows from the following equivalent statements.
$a \in \mathcal{S}(X) \Rightarrow a \in \mathcal{S}(Y)$. $Xa \geq 0 \Rightarrow Ya \geq 0$. While X is bijective, we can introduce b as $a = X^{-1}b$. $b \geq 0 \Rightarrow YX^{-1}b \geq 0$. □

Theorem 1. *(a) The following statements are equivalent.*
(a1) $\mathcal{S}(I - L) \supseteq \mathcal{S}(\hat{Q})$.
(a2) $\mathbf{A}_0^{-1} \geq 0$.

(b) The following statements are equivalent.
(b1) $\mathcal{S}(\hat{Q}) \supseteq \mathcal{S}((I - L)L^{-1})$.
(b2) $\mathbf{B}_0 \geq 0$.

Proof. (a) It follows from Remark 1, Lemma 2 and Equation (14).
(b) Using Remark 1, (13) and (14) we can write that $(I - L)\mathbf{x} = (I - H)\mathbf{x} - \mathbf{h}$ and $\hat{Q}\mathbf{x} = \mathbf{A}_0[(I - H)\mathbf{x} - \mathbf{h}]$. Using Lemma 2 we get that (b1) is equivalent to $\mathbf{x} \geq 0 \Rightarrow \mathbf{A}_0\{(I - H)[H(I - H)^{-1}(\mathbf{x} + \mathbf{h}) + \mathbf{h}] - \mathbf{h}\} \geq 0$. While H and $(I - H)^{-1}$ commute we can write it in a simplified form: $\mathbf{x} \geq 0 \Rightarrow \mathbf{B}_0\mathbf{x} \geq 0$, which proves the statement. □

4   Remarks and Summary

Remark 2. In the discrete case the connection between shape and time monotonicity is in a close relation with the nonnegativity preservation property (DNP) and the maximum-minimum principle (DMP), since the necessary and sufficient conditions to fulfil the DMP (and DNP, too) in the general case are (P1) $-\mathbf{A}_0^{-1}\mathbf{A}_\partial \geq 0$ and (P2) $\mathbf{A}_0^{-1}\mathbf{B} \geq 0$, see [4,5].

Let us consider Example 2 with homogeneous Dirichlet boundary condition $(g(x,t) = 0)$. Since $f(x,t)$ does not depend on time, therefore to it corresponds a discretized elliptic problem $\mathbf{Q}\mathbf{u} = \tau\mathbf{f}$. The iteration $\mathbf{u}^{n+1} = \mathbf{A}_0^{-1}\mathbf{B}_0\mathbf{u}^n + \tau\mathbf{A}_0^{-1}\mathbf{f}$, which solves it, is based on a so-called regular splitting of $\mathbf{Q}$ if $\mathbf{A}_0^{-1} \geq 0$, $\mathbf{B}_0 \geq 0$, a weak regular splitting if $\mathbf{A}_0^{-1} \geq 0$, $\mathbf{A}_0^{-1}\mathbf{B}_0 \geq 0$, ([1]) and a weak splitting if $\mathbf{A}_0^{-1}\mathbf{B}_0 \geq 0$, ([7]).

Remark 3. In this case (P1) and (P2) get a simplified form, namely $\mathbf{A}_0^{-1} \geq 0$ and $\mathbf{A}_0^{-1}\mathbf{B}_0 \geq 0$. This means that the DNP and DMP corresponds to the weak regular splitting.
The correct discrete model from the point of view of the connection between shape and time-monotonicity corresponds to the regular splitting.
It is easy to check that $\mathcal{S}(I - L) \supseteq \mathcal{S}((I - L)L^{-1})$ is equivalent to $\mathbf{A}_0^{-1}\mathbf{B}_0 \geq 0$, which means that this qualitative propery corresponds to the weak splitting.

Summary

We defined a property, the connection between the shape and time monotonicity of the diffusion submodel. We formulated this property in the time-discretized case, too. After that we gave the necessary and sufficient conditions to fulfil this property in a practical case, when we used the finite difference or finite element method for space discretization and the θ-method for time discretization. Finally we pointed out the connection between the qualitative properies and matrix splittings.

Acknowledgments

This research was supported by the Bulgarian IST Center of Competence in 21st century — BIS-21++, funded by the European Commission in FP6 INCO via Grant 016639/2005.

References

1. Berman, A., Plemmons, R.J.: Nonnegative Matrices in the Mathematical Sciences. Academic Press, New York (1979)
2. Faragó, I., Pfeil, T.: Preserving concavity in initial-boundary value problems of parabolic type and its numerical solution. Periodica Mathematica Hungarica 30, 135–139 (1995)
3. Faragó, I., et al.: A hővezetési egyenlet és numerikus megoldásának kvalitatív tulajdonságai I. Az elsőfokú közelítések nemnegativitása. Alkalmazott Matematikai Lapok 17, 101–121 (1993)
4. Faragó, I., Horváth, R.: A Review of Reliable Numerical Models for Three-Dimensional Linear Parabolic Problems. Int. J. Numer. Meth. Engng. 70, 25–45 (2007)
5. Faragó, I., Horváth, R., Korotov, S.: Discrete Maximum Principle for Linear Parabolic Problems Solved on Hybrid Meshes. Appl. Num. Math. 53, 249–264 (2005)
6. Marchuk, G.I.: Mathematical modeling for the problem of the environment. Studies in Mathematics and Applications, vol. 16. North-Holland, Amsterdam (1985)
7. Marek, I., Szyld, D.B.: Comparison theorems for weak splittings of bounded operators. Numer. Math. 58, 389–397 (1990)
8. McRae, G.J., Goodin, W.R., Seinfeld, J.H.: Numerical solution of the atmospheric diffusion equations for chemically reacting flows. Journal of Computational Physics 45, 1–42 (1984)
9. Mincsovics, M.: Qualitative analysis of the one-step iterative methods and consistent matrix splittings. Special Issue of Computers and Mathematics with Applications (accepted)
10. Verwer, J.G., Hundsdorfer, W., Blom, J.G.: Numerical Time Integration for Air Pollution Models, Report of CWI, MAS-R9825 (1982)
11. Zlatev, Z.: Computer Treatment of Large Air Pollution Models. Kluwer Academic Publishers, Dordrecht-Boston-London (1995)

Limit Cycles and Bifurcations in a Biological Clock Model

Bálint Nagy

Department of Mathematical Analysis,
College of Dunaújváros, Hungary

Abstract. A three-variable dynamical system describing the circadian oscillation of two proteins (PER and TIM) in cells is investigated. We studied the saddle-node and Hopf bifurcation curves and distinguished four cases according to their mutual position in a former article. Other bifurcation curves were determined in a simplified, two-variable model by Simon and Volford [6]. Here we show a set of bifurcation curves that divide the parameter plane into regions according to topological equivalence of global phase portraits, namely the global bifurcation diagram, for the three-variable system. We determine the Bautin-bifurcation point, and fold bifurcation of cycles numerically. We also investigate unstable limit cycles and the case when two stable limit cycles exist.

Keywords: limit cycle, bifurcation, circadian rhythm model.

1 Introduction

Chronobiological rhythms can be observed in the physiology and behavior of animals entrained to the 24h cycle of light and darkness. Discoveries show that there is an internal biological clock at molecular level. Two proteins (PER and TIM) are thought to be responsible for this mechanism. Leloup and Goldbeter [3] introduced a model for this phenomenon. Tyson *et al.* [7] developed the original model applying a new positive feedback-loop. The original six variable model can be reduced to three differential equations [7] in which the state variables are the concentration of mRNA (M), the protein (P_1) and the dimer (P_2) [7]:

$$\dot{M} = \frac{\nu_m}{1 + \left(\frac{P_2}{P_c}\right)^2} - k_m M \tag{1}$$

$$\dot{P_1} = \nu_p M - \frac{k_1 P_1}{J_p + P_1 + r P_2} - k_3 P_1 - 2k_a P_1^2 + 2k_d P_2 \tag{2}$$

$$\dot{P_2} = k_a P_1^2 - (k_d + k_3) P_2 - \frac{k_2 P_2}{J_p + P_1 + r P_2} \tag{3}$$

Equation (1) includes the inhibition of the mRNA transcription by the dimer. The reaction constants k_1, k_2, k_3 are related to the phosphorylation and the reaction constants k_a, k_d belong to the dimerization process. The ratio of enzyme substrate dissociation constant is r. ν_m is the maximum rate of synthesis of

I. Lirkov, S. Margenov, and J. Waśniewski (Eds.): LSSC 2007, LNCS 4818, pp. 209–216, 2008.

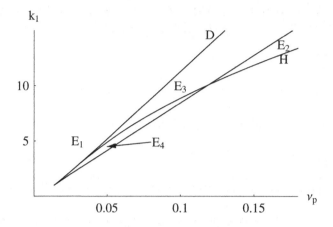

Fig. 1. D–curve, H–curve and E_1, E_2, E_3, E_4 domains for $P_c = 0.5$

mRNA, k_m is the first-order rate constant for mRNA degradation, ν_p is the rate constant for translation of mRNA, P_c is the dimer concentration at the half-maximum transcription rate and J_p is the Michaelis constant for protein kinase (DBT) [7]. With additional simplifying assumptions the system can be reduced to two equations [7]. The two variable model was studied by Simon and Volford in [6].

System (1)–(3) consists of three variables and eleven parameters. k_1 and ν_p are linearly involved, hence these can be used as control parameters, the other nine are fixed at the values taken from [7]. In a former article [4] we gave the saddle-node, i.e. the D- and the Hopf-, i.e. the H-curves paramterized by the state variable P_1 by the parametric representation method (PRM) [5]. We also investigated the qualitative shape of these curves and determined how the curves change if the parameters are varied. Crossing the D-curve the number of stationary points changes, while crossing the H-curve the stability changes. Since the curves are parametrized by P_1 one can determine (ν_p, k_1) point of curves by the value of P_1. We also discussed the connection between these curves, especially the common points of the two curves, the Takens-Bogdanov bifurcation points are determined. The system has the most complex behavior if $P_c > P_c^{\circledast} \approx 0.26$. We studied the stationary points and found that for example for $P_c = 0.5$ there are four regions in the (ν_p, k_1) plane according to the number and type of stationary points, see Figure 1. If $(\nu_p, k_1) \in E_1$ the system (1)–(3) has one stationary point and it is stable, if $(\nu_p, k_1) \in E_2$ the system has one unstable stationary point [4]. Stability and the number of periodic orbits change crossing the H-curve.

Deciding whether the Hopf bifurcation is subcritical or supercritical we calculate Lyapunov coefficient σ in the (1)–(3) three-variable system. The coefficient σ determines the stability. If $\sigma > 0$, there is a subcritical Hopf-bifurcation at the appropriate point of the H-curve and the periodic orbit is unstable. If $\sigma < 0$, there is a supercritical Hopf-bifurcation, the periodic orbit is stable. Bautin-bifurcation (generalized Hopf-bifurcation) [2] occurs, if $\sigma = 0$.

The number of periodic orbits can change as well when a stable and an unstable periodic orbit coalesce and disappear, i.e. the fold bifurcation of cycles. The F_c-curve consists of points, where a fold bifurcation of periodic orbits occurs. From the Bautin-bifurcation point an F_c curves starts as it is stated by the general theory [2].

In Section 2 we show how a center manifold reduction can be carried out to reduce the three-variable case to the two-variable one. In Section 3 we apply this method for system (1)–(3), hence we determine the Bautin bifurcation point, and fold bifurcation of cycles numerically. We determine the region where two stable limit cycles exist, hence bistability arises.

2   Andronov-Hopf Bifurcation

In this section we show how to calculate Lyapunov coefficient σ in a three variable model.

In [1] it is shown that in the system

$$\begin{pmatrix} \dot{x} \\ \dot{y} \end{pmatrix} = \begin{pmatrix} 0 & -\omega \\ \omega & 0 \end{pmatrix} \begin{pmatrix} x \\ y \end{pmatrix} + \begin{pmatrix} F(x,y) \\ G(x,y) \end{pmatrix} \tag{4}$$

with $F(0) = G(0) = 0$ and $DF(0) = DG(0) = 0$, the normal form calculation yields

$$\sigma = \frac{1}{16}[F_{xxx} + F_{xxy} + G_{xxy} + G_{yyy}]+$$
$$+ \frac{1}{16\omega}[F_{xy}(F_{xx} + F_{yy}) - G_{xy}(G_{xx} + G_{yy}) - F_{xx}G_{xx} + F_{yy}G_{yy}], \tag{5}$$

where F_{xy} denotes $(\partial^2 F/\partial x \partial y)(0,0)$, etc.

Let us consider the system

$$\dot{X}(t) = f(X(t)), \tag{6}$$

where $X(t) \in \mathbb{R}^3$ and J is the Jacobian of f.

Let (x_s, y_s, z_s) be the stationary point of system (6). To determine σ, first we translate this stationary point into the origin with the transformation

$$\bar{x} = x - x_s, \quad \bar{y} = y - y_s, \quad \bar{z} = z - z_s. \tag{7}$$

The stability of the stationary points can be determined by

$$g = TrJ(A_{11} + A_{22} + A_{33}) - DetJ.$$

It can be proved, that if the Jacobian of the system (6) has two pure imaginary eigenvalues, then $g = 0$. Here TrJ and $DetJ$ denote the trace and the determinant of J, and A_{11}, A_{22}, A_{33} are the corresponding minors of J. It can be shown that $g = 0$ and sgn $DetJ = $ sgn TrJ imply that the Jacobian has two pure imaginary eigenvalues hence a Hopf–bifurcation exists. However, if $g = 0$ and sgn $DetJ \neq $ sgn TrJ, then the Jacobian has only real eigenvalues, thus there is no Hopf-bifurcation.

Let $g = 0$ and sgn $DetJ = $ sgn TrJ, hence the Jacobian at the origin has one real eigenvalue (λ_1), and two pure imaginary eigenvalues (λ_2, λ_3), $\lambda_2 = \bar{\lambda}_3$ hence $\lambda_2 + \lambda_3 = 0$.

Let the eigenvectors of the Jacobian be $\mathbf{v}_1$, $\mathbf{v}_2$, $\mathbf{v}_3$, and $T = (\mathbf{v}_1, \mathbf{v}_2, \mathbf{v}_3)^T$, where $(\cdot)^T$ denotes the transposed of a matrix.

Now we introduce the new variable $U = T^{-1}X$.

Hence the system (6) is transformed to a diagonal one:

$$\dot{U} = T^{-1}JTU + a(U), \tag{8}$$

where $T^{-1}JT = \begin{pmatrix} \lambda_1 & 0 & 0 \\ 0 & \lambda_2 & 0 \\ 0 & 0 & \lambda_3 \end{pmatrix}$, $U = \begin{pmatrix} u \\ v \\ w \end{pmatrix}$, $a(U) = T^{-1}f(TU) - T^{-1}JTU$.

We will approximate the center manifold W^c of system (8) with

$$u = q(v, w) = \alpha v^2 + \beta vw + \gamma w^2. \tag{9}$$

Remark 1. Since our aim is to determine whether the Hopf-bifurcation is supercritical or subcritical, its enough to approximate u with second order terms [1].

Differentiating (9), we have

$$\dot{u} = (2\alpha v + \beta w)\dot{v} + (\beta v + 2\gamma w)\dot{w}. \tag{10}$$

Let us write (8) into the form

$$\dot{u} = \lambda_1 u + a_1(u, v, w) \tag{11}$$
$$\dot{v} = \lambda_2 v + a_2(u, v, w) \tag{12}$$
$$\dot{w} = \lambda_3 w + a_3(u, v, w) \tag{13}$$

where

$$a_1 = c_1 w^2 + c_2 vw + c_3 v^2 + c_4 u^2 + c_5 uv + c_6 uw + \mathcal{O}(u^3, v^3, w^3). \tag{14}$$

Substituting (11)–(13) into (10), we get

$$\lambda_1 u + a_1(u, v, w) = (2\alpha v + \beta w)(\lambda_2 v + a_2(u, v, w)) \\ + (2\gamma w + \beta v)(\lambda_3 w + a_3(u, v, w)). \tag{15}$$

Let us observe, that if we substitute (9) into (14) the terms $c_4 u^2$, $c_5 wv$, $c_6 w^2$ will not contain v^2, wv, w^2. Hence at the second order $a_1 = c_1 w^2 + c_2 vw + c_3 w^2$. Moreover, neither $(2\alpha v + \beta w)a_2(u, v, w)$, nor $(2\gamma w + \beta v)a_3(u, v, w)$ contain v^2, wv, w^2, because a is the nonlinear part of f.

Now we use (9) and write (15) into the form

$$\lambda_1(\alpha v^2 + \beta vw + \gamma w^2) + c_1 w^2 + c_2 vw + c_3 v^2 = 2\alpha\lambda_2 v^2 + 2\gamma\lambda_3 w^2. \qquad (16)$$

Equating the coefficients of v^2, we have $\alpha\lambda_1 + c_3 = 2\alpha\lambda_2$, hence

$$\alpha = \frac{c_3}{2\lambda_2 - \lambda_1}.$$

The value of β and γ can be calculated similarly. Hence we have proved the following statement:

Lemma 1. *In (9)*

$$\alpha = \frac{c_3}{2\lambda_2 - \lambda_1}, \qquad \beta = \frac{c_2}{-\lambda_1}, \qquad \gamma = \frac{c_1}{2\lambda_3 - \lambda_1}.$$

Introducing $\bar{a}(v,w) = a(q(v,w), v, w)$ we can approximate the system (8) in the center manifold:

$$\begin{pmatrix} \dot{v} \\ \dot{w} \end{pmatrix} = \begin{pmatrix} \lambda_2 & 0 \\ 0 & \lambda_3 \end{pmatrix} \cdot \begin{pmatrix} v \\ w \end{pmatrix} + \begin{pmatrix} \bar{a}_2(v,w) \\ \bar{a}_3(v,w) \end{pmatrix}. \qquad (17)$$

Let

$$v = \xi - i \cdot \eta, \qquad w = \xi + i \cdot \eta. \qquad (18)$$

The system for ξ, η takes the form

$$\begin{pmatrix} \dot{\xi} \\ \dot{\eta} \end{pmatrix} = \begin{pmatrix} 0 & -\omega \\ \omega & 0 \end{pmatrix} \cdot \begin{pmatrix} \xi \\ \eta \end{pmatrix} + \begin{pmatrix} F(\xi, \eta) \\ G(\xi, \eta) \end{pmatrix}. \qquad (19)$$

Proposition 1. *For F, G in (19) we have*

$$F(\xi, \eta) = \frac{1}{2}(\bar{a}_2(\xi - i \cdot \eta, \xi + i \cdot \eta) + \bar{a}_3(\xi - i \cdot \eta, \xi + i \cdot \eta)), \qquad (20)$$

$$G(\xi, \eta) = \frac{1}{2i}(\bar{a}_3(\xi - i \cdot \eta, \xi + i \cdot \eta) - \bar{a}_2(\xi - i \cdot \eta, \xi + i \cdot \eta)). \qquad (21)$$

Proof. For brevity we show (20) and use the notation $\bar{a}_i = \bar{a}_i(\xi - i \cdot \eta, \xi + i \cdot \eta)$, $i = 2, 3$. Since (18)

$$\xi = \frac{1}{2}(v + w), \qquad \eta = \frac{1}{2i}(w - v).$$

Using $\lambda_2 + \lambda_3 = 0$ and introducing $\omega = (\lambda_2 - \lambda_3)i/2$

$$\begin{aligned} \dot{\xi} &= \frac{1}{2}(\dot{v} + \dot{w}) = \\ &= \frac{1}{2}(\lambda_2(\xi - i \cdot \eta) + \bar{a}_2 + \lambda_3(\xi + i \cdot \eta) + \bar{a}_3) = \\ &= \frac{1}{2}((\lambda_2 + \lambda_3)\xi + i \cdot \eta(\lambda_3 - \lambda_2) + \bar{a}_2 + \bar{a}_3 = \\ &= \frac{1}{2}(0 - 2\omega\eta + \bar{a}_2 + \bar{a}_3) = -\omega\eta + \frac{1}{2}(\bar{a}_2 + \bar{a}_3). \qquad (22) \end{aligned}$$

(21) can be seen similarly.

3   Determining the Bautin Bifurcation Point

In this section we apply the method showed in Section 2 for the (1)–(3) model. Let us consider the system (1)-(3). With $X = (M, P_1, P_2)^T$ (1)–(3) takes the form of (6). Let us suppose that $(\nu_p, k_1) \in E_1 \cup E_2$ (see Figure 1), hence the system has one stationary point, with coordinates (x_s, y_s, z_s). First we translate this stationary point into the origin with transformation (7), that is we introduce the new variables $\bar{M}, \bar{P}_1, \bar{P}_2 : \bar{M} = M - x_s$, $\bar{P}_1 = P_1 - y_s$, $\bar{P}_2 = P_2 - z_s$. Dropping the bars we get:

$$\dot{M} = \frac{\nu_m}{1 + \left(\frac{P_2 + z_s}{P_c}\right)^2} - k_m(M + x_s) \tag{23}$$

$$\dot{P}_1 = \nu_p(M + x_s) - \frac{k_1(P_1 + y_s)}{J_p + P_1 + y_s + r(P_2 + z_s)} - \\ -k_3(P_1 + y_s) - 2k_a(P_1 + y_s)^2 + 2k_d(P_2 + z_s) \tag{24}$$

$$\dot{P}_2 = k_a(P_1 + y_s)^2 - (k_d + k_3)(P_2 + z_s) - \\ -\frac{k_2(P_2 + z_s)}{J_p + P_1 + y_s + r(P_2 + z_s)} \tag{25}$$

The Jacobian of the system (23)–(25) takes the form

$$J = \begin{pmatrix} a_{11} & 0 & a_{13} \\ \nu_p & k_1\bar{c}_1 + \bar{c}_2 & k_1\bar{c}_3 + \bar{c}_4 \\ 0 & a_{32} & a_{33} \end{pmatrix},$$

where

$$a_{11} = -k_m, \quad a_{13} = -\frac{2\nu_m(P_2 + z_s)P_c^2}{((P_2 + z_s)^2 + P_c^2)^2},$$

$$\bar{c}_1 = -\frac{J_p + r(P_2 + z_s)}{(J_p + P_1 + y_s + r(P_2 + z_s))^2}, \quad \bar{c}_2 = -k_3 - 4k_a(P_1 + y_s),$$

$$\bar{c}_3 = \frac{(P_1 + y_s)r}{(J_p + P_1 + y_s + r(P_2 + z_s))^2}, \quad \bar{c}_4 = 2k_d,$$

$$a_{32} = 2k_a(P_1 + y_s) + \frac{k_2(P_2 + z_s)}{(J_p + P_1 + y_s + r(P_2 + z_s))^2},$$

$$a_{33} = -k_3 - k_d - k_2\frac{J_p + P_1 + y_s}{(J_p + P_1 + y_s + r(P_2 + z_s))^2}.$$

Hence we can determine the eigenvalues and eigenvectors of $J(0, 0, 0)$, and the matrix T. To calculate T, we used *Mathematica*. Introducing new variables (u, v, w) where $(M, P_1, P_2)^T = T \cdot (u, v, w)^T$ we can transform the system into a diagonal form, and approximate the center manifold with (9). Now we have the system transformed into the form (4).

We determined the H–curve with the PRM in [4], hence (ν_p, k_1) are parameterized by P_1, thus using the method above σ can be determined for any P_1, i.e.

for any (ν_p, k_1) from (5) using (20) and (21). For example $P_1 = 0.07$ yields $\nu_p \approx$ 0.43, $k_1 \approx 22.92$ and $\sigma \approx -0.003$, hence in the point $(0.43, 22.92)$ supercritical Hopf-bifurcation occurs. $P_1 = 0.055$ yields $(\nu_p, k_1) \approx (0.15, 11.81)$ and $\sigma \approx 0.009$, hence there is a subcritical Hopf-bifurcation in the point $(0.15, 11.81)$. Our numerical investigation shows, that for $\check{P}_1 \approx 0.065$ there is a Bautin-bifurcation point, i.e. $\sigma = 0$.

In system (1)–(3) the number of periodic orbits can change in different ways. The first is Hopf bifurcation, along the points of the H-curve. The value of σ determines the type of Hopf bifurcation [2]. If $\sigma > 0$, then the bifurcation is subcritical and an unstable limit cycle arises or disappears. If $\sigma < 0$, then the bifurcation is supercritical, and a stable limit cycle emerges or disappears.

The number of periodic orbits can change as well crossing an F_c-curve, when a fold bifurcation of periodic orbits occurs.

4   Bifurcation Diagram

In this section we summarize what we know about the bifurcation curves, the stationary points and the limit cycles of system (1)–(3).

Our numerical investigation shows, that system (1)–(3) has two F_c-curves. F_{c1} is under the H-curve and F_{c2} starts from the Bautin point B with coordinates $(\nu_p^B, k1^B)$ [2].

Now we give the bifurcation diagram for $P_c = 0.5$ in the region $E_1 \cup E_2$, see Figure 2. The D-, H-, F_{c1}-, F_{c2}-curves divide the (ν_p, k_1) parameter plane into E_1^1, E_1^2, E_2^1, E_2^2 regions. We discuss how the number of limit cycles changes varying ν_p and k_1, see Figure 2.

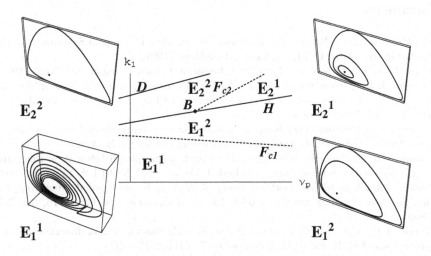

Fig. 2. Schematic figure for region $E_1 \cup E_2$ with bifurcation curves and typical phase portraits

Let $\nu_p < \nu_p^B$, and $(\nu_p, k_1) \in E_1^1$. System (1)–(3) has a stable stationary point (as we showed in [4]), and there is no limit cycle. Increasing k_1 we cross the F_{c1} curve where two cycles emerge, we get to region E_1^2. There is a stable stationary point, since we did not get through the D-curve and we found, that there are two limit cycles. A stable and an unstable one, since we crossed a fold bifurcation curve (F_{c1}). Crossing the H-curve (ν_p, k_1) is in region E_2^2. Since $\nu_p < \nu_p^B$ on the H-curve an unstable limit cycle disappear, $\sigma > 0$, thus subcritical Hopf-bifurcation occurs. In E_2^2 there is an unstable stationary point and a stable limit cycle.

Let us change (ν_p, k_1) in E_2^2 to gain $\nu_p > \nu_p^B$. If we cross the F_{c2} fold bifurcation curve, two limit cycles — a stable and an unstable one — arise: we get to region E_2^1. In E_2^1 there are three limit cycles — two stable and an unstable one — and there is an unstable stationary point. Crossing the H-curve with $\nu_p > \nu_p^B$ supercritical Hopf bifurcation occurs, hence a stable limit cycle disappears and the unstable stationary point becomes stable. We get one stable stationary point and two limit cycles — a stable and an unstable one — as it is stated before in region E_1^2.

We found that two stable limit cycles occur in region E_2^1 if (ν_p, k_1) lies between the H- and F_{c1}-curves. This bistability can be important in biological systems. Depending on the initial condition the system may approach two alternative stable limit cycles. Bistable systems can approach one of two stable periodic orbits, hence may undergo sudden transitions, corresponding to switches from one stable limit cycle to the other because of small changes in the initial conditions.

Acknowledgment. The author is grateful to S. L. P. for the valuable discussions.

References

1. Guckenheimer, J., Holmes, P.: Nonlinear Oscillations, Dynamical Systems, and Bifurcations of Vector Fields. Springer, Heidelberg (1997)
2. Kuznetsov, Y.A.: Elements of applied bifurcation theory. Springer, New York (1995)
3. Leloup, J.-C., Goldbeter, A.: A model for circadian rhythms in Drosophilia incorporating the formation of a complex between PER and TIM proteins. J. Biol. Rhythms 13, 70–87 (1998)
4. Nagy, B.: Comparison of the bifurcation curves of a two-variable and a three-variable circadian rhythm model. Appl. Math. Modelling (under publication)
5. Simon, P.L., Farkas, H., Wittmann, M.: Constructing global bifurcation diagrams by the parametric representation method. J. Comp. Appl. Math 108, 157–176 (1999)
6. Simon, P.L., Volford, A.: Detailed study of limit cycles and global bifurcation diagrams in a circadian rhythm model. Int. J. Bifurcation and Chaos 16, 349–367 (2006)
7. Tyson, J.J., et al.: A simple model of circadian rhythms based on dimerization and proteolysis of PER and TIM. Biophys. J. 77, 2411–2417 (1999)
8. Wilhelm, T., Heinrich, R.: Mathematical analysis of the smallest reaction system with Hopf bifurcation. J. Math. Chem. 19, 111–130 (1996)

Large Matrices Arising in Traveling Wave Bifurcations

Peter L. Simon[*]

Department of Applied Analysis and Computational Mathematics,
Eötvös Loránd University Budapest, Hungary
simonp@cs.elte.hu

Abstract. Traveling wave solutions of reaction-diffusion (semilinear parabolic) systems are studied. The number and stability of these solutions can change via different bifurcations: saddle-node, Hopf and transverse instability bifurcations. Conditions for these bifurcations can be determined from the linearization of the reaction-diffusion system. If an eigenvalue (or a pair) of the linearized system has zero real part, then a bifurcation occurs. Discretizing the linear system we obtain a matrix eigenvalue problem. It is known that its eigenvalues tend to the eigenvalues of the original system as the discretization step size goes to zero. Thus to obtain bifurcation curves we have to study the spectra of large matrices. The general bifurcation conditions for the matrices will be derived. These results will be applied to a reaction-diffusion system describing flame propagation.

1 Introduction

Let us consider the equation

$$\partial_t u = D\Delta u + f(u), \tag{1}$$

where $u : \mathbb{R}_+ \times \mathbb{R}^2 \to \mathbb{R}^m$, D is an $m \times m$ diagonal matrix with positive elements and $f : \mathbb{R}^m \to \mathbb{R}^m$ is a differentiable function. The traveling wave solution of this equation (moving in the x direction) has the form $u(t, x, y) = U(x - ct)$, where

$$U : \mathbb{R} \to \mathbb{R}^m, \qquad U(-\infty) = U_- \in \mathbb{R}^m, \quad U(\infty) = U_+ \in \mathbb{R}^m.$$

For this function we have

$$DU''(z) + cU'(z) + f(U(z)) = 0, \qquad (z = x - ct). \tag{2}$$

Several bifurcations of the traveling wave U can be determined from the spectrum of the linear differential operator obtained by linearizing (1) around U, the details will be given in Section 2. We will refer to this operator as linearized operator. The spectrum of this operator has been widely investigated. The position of the essential spectrum has been estimated, see e.g. [3], and Weyl's lemma

---

[*] This work was supported by the Hungarian National Science Foundation OTKA No. T049819.

I. Lirkov, S. Margenov, and J. Waśniewski (Eds.): LSSC 2007, LNCS 4818, pp. 217–224, 2008.

in [8]. In [7] an algorithm is given for the determination of the exact position of the essential spectrum. In this paper we concentrate on the eigenvalues of the linearized operator that can be determined only numerically. Discretizing the linearised operator its spectrum can be approximated by the eigenvalues of a large matrix. It is known that for a traveling wave solution of a reaction-diffusion system the eigenvalues of the matrix approximate the spectrum of the original differential operator [1,4].

In Section 2 the linearization around a traveling wave is derived. Then the three most important bifurcations of traveling waves are introduced and sufficient conditions for these bifurcations in terms of the linearized operator are presented. In Section 3 the matrix that is the discretization of the operator is derived. Then the sufficient conditions of the bifurcations are expressed for the matrix. In Section 4 a flame propagation model is studied. This model consists of two reaction diffusion equations, the traveling wave solutions of which were determined in [6]. Here the bifurcations of these traveling waves are investigated by using the bifurcation conditions for the discretization of the system.

2   Bifurcations of Traveling Waves

The stability of U can be determined by linearization. Put $u(t, x, y) = U(x - ct) + v(t, z, y)$. Then the linearized equation for v takes the form

$$\partial_t v = D\Delta v + c\partial_z v + f'(U(z))v. \tag{3}$$

Here we used the linear approximation, i.e. instead of $f(U + v)$ we put $f(U) + f'(U)v$. If the solution of this equation starting from a sufficiently small initial value tends to zero as t tends to infinity, then the traveling wave solution U is said to be linearly stable. It is important to note, that shifting the function U we get another traveling wave solution, i.e. $\overline{U}(z) = U(z + z_0)$ is also a solution of (2) for any $z_0 \in \mathbb{R}$. Hence the traveling wave solution cannot be asymptotically stable in the usual sense. This is reflected in the fact that the function $v(t, z, y) = U'(z)$ is always a solution of (3), which obviously does not tend to zero.

Thus for the linear stability of the traveling wave solution one has to study the spectrum of the linear differential operator in the right hand side of (3). If the spectrum is in the left half of the complex plane then the traveling wave solution is said to be linearly stable. If it has points in the right half of the complex plane then the traveling wave solution is said to be linearly unstable. The spectrum of this differential operator consists of two parts: the isolated eigenvalues and the essential spectrum. The position of the later can be determined theoretically, in some cases it can be determined explicitly [6,7]. The eigenvalues can be obtained only numerically. In this paper we study the eigenvalues and determine them numerically. In the case of the most important bifurcations as a parameter in the system is varied an eigenvalue crosses the imaginary axis and the corresponding eigenfunction does not depend on y or depends on y periodically. If an eigenvalue crosses the imaginary axis at the origin and the corresponding eigenfunction does not depend on y, then we have a so-called saddle-node bifurcation. In this case the number of traveling wave solutions changes by two. If a pair of complex eigenvalues crosses the imaginary axis and the corresponding eigenfunction does not depend on y, then we have a so-called Hopf bifurcation. In this case the number of traveling wave solutions does not change, however, it loses its stability and a wave propagating with periodically changing velocity appears. If a pair of complex eigenvalues crosses the imaginary axis and the corresponding eigenfunction depends on y periodically, then we have a so-called transverse instability bifurcation. In this case the number of traveling wave solutions does not change, however, it loses its stability and a non planar traveling wave appears.

Thus in order to study these bifurcations it is enough to determine the solutions of (3) in the form $v(t, z, y) = V(z) \exp(iky + \sigma t)$. Substituting this form into (3) we get for V

$$DV''(z) + cV'(z) + (f'(U(z)) - Dk^2)V(z) = \sigma V(z) \tag{4}$$
$$V(\pm\infty) = 0.$$

For a given real value of k this is an eigenvalue problem. The eigenvalue is σ, the eigenfunction is V. Let us denote by $\sigma(k)$ the eigenvalue with largest real part. The plot of $\sigma(k)$ against k is called a **dispersion curve**. For instability we require a range of k over which $\text{Re } \sigma(k) > 0$. If (4) has a pair of purely imaginary eigenvalues for $k = 0$, then Hopf bifurcation occurs.

3   Discretization

In order to study transverse instability we plot the dispersion curves for different values of the parameters. We solve the eigenvalue problem with finite difference discretization. First the problem is truncated, i.e. we solve (4) subject to $V(-l) = V(l) = 0 \in \mathbb{R}^m$ with some l large enough. In [1] it is shown that the eigenvalues of the truncated problem tend to those of the original problem as $l \to \infty$. Then choosing a sufficiently large number l we discretize the truncated problem with finite differences on a grid of N points:

$$-l < z_1 < z_2 < \ldots < z_N < l.$$

The discretization of the i-th coordinate function V_i is

$$v_i = (V_i(z_1), V_i(z_2), \ldots, V_i(z_N))^T \in \mathbb{R}^N.$$

(In this Section v will denote the discretization and not the solution of the linearization of the differential equation.) The derivatives of V will be approximated by finite differences, that is we introduce the $N \times N$ matrices A and B for which

$$(V_i'(z_1), V_i'(z_2), \ldots, V_i'(z_N))^T = Av_i$$
$$(V_i''(z_1), V_i''(z_2), \ldots, V_i''(z_N))^T = Bv_i.$$

System (4) consists of m equations and holds in N grid points, yielding $m \cdot N$ equations. The first equation of the system written in the N grid points and using the finite difference approximation of the derivatives takes the form

$$D_1 B v_1 + c A v_1 + Q_{11} v_1 + Q_{21} v_2 + \ldots + Q_{m1} v_m - D_1 k^2 v_1 = \sigma v_1 \qquad (5)$$

where Q_{ij} is the $N \times N$ diagonal matrix with elements

$$\partial_i f_j(U(z_1)), \partial_i f_j(U(z_2)), \ldots \partial_i f_j(U(z_N)).$$

Similarly, we can write the second equation of the system at the grid points:

$$D_2 B v_2 + c A v_2 + Q_{12} v_1 + Q_{22} v_2 + \ldots + Q_{m2} v_m - D_2 k^2 v_2 = \sigma v_2. \qquad (6)$$

Finally, the discretization of the m-th equation takes the form

$$D_m B v_m + c A v_m + Q_{1m} v_1 + Q_{2m} v_2 + \ldots + Q_{mm} v_m - D_m k^2 v_m = \sigma v_m. \qquad (7)$$

Let us introduce the column vector v with $m \cdot N$ elements as

$$v = (v_1^T, v_2^T, \ldots, v_m^T)^T$$

and the $m \cdot N \times m \cdot N$ matrices M and P as

$$M = \begin{pmatrix} D_1 B + c A + Q_{11} & Q_{21} & \cdots & Q_{m1} \\ Q_{12} & D_2 B + c A + Q_{22} & \cdots & Q_{m2} \\ \vdots & \vdots & \ddots & \vdots \\ Q_{1m} & Q_{2m} & \cdots & D_m B + c A + Q_{mm} \end{pmatrix}$$

$$P = \begin{pmatrix} D_1 I & 0 & \cdots & 0 \\ 0 & D_2 I & \cdots & 0 \\ \vdots & \vdots & \ddots & \vdots \\ 0 & 0 & \cdots & D_m I \end{pmatrix}$$

where I stands for the $N \times N$ unit matrix. Then the discretization of (4) is the $m \cdot N$ dimensional matrix eigenvalue problem

$$(M - \kappa P) v = \sigma v \qquad (8)$$

where κ stands for k^2. It can be shown that the eigenvalues of the matrix $M - \kappa P$ tend to those of the truncated problem as $N \to \infty$ [4]. We can increase the values of l and N until the required accuracy is achieved.

The matrices M and P involve the parameters of the reaction-diffusion system. If for some values of the parameters M has a pair of pure imaginary eigenvalues, then at these parameter values Hopf-bifurcation occurs. If for some κ the eigenvalue problem (8) has an eigenvalue σ with positive real part, then the traveling wave is transversally unstable.

Thus we have to solve the matrix eigenvalue problem (8) as the value of κ is varied. Let us denote by ν an eigenvalue of M. In the next Lemma we show that under suitable conditions there is a differentiable function s_ν that associates to κ an eigenvalue $\sigma = s_\nu(\kappa)$ of (8) and $s_\nu(0) = \nu$.

Lemma 1. *Let ν be an eigenvalue of the matrix $M \in \mathbb{R}^{n \times n}$, for which $\dim \ker$ $(M - \nu I) = \dim \ker(M^T - \nu I) = 1$. Let us denote by u_0 and v_0 the eigenvectors (with unit length) that span these one-dimensional eigenspaces. If $\langle u_0, v_0 \rangle \neq 0$, then there exists $\delta > 0$ and there exist differentiable functions $s : (-\delta, \delta) \to \mathbb{C}$ and $\mathcal{U} : (-\delta, \delta) \to \mathbb{C}^n$, such that $\mathcal{U}(0) = u_0$, $s(0) = \nu$ and for all $\kappa \in (-\delta, \delta)$ we have*

$$(M - \kappa P)\mathcal{U}(\kappa) = s(\kappa)\mathcal{U}(\kappa). \tag{9}$$

Moreover,

$$s'(0) = -\frac{\langle Pu_0, v_0 \rangle}{\langle u_0, v_0 \rangle}. \tag{10}$$

Proof. The implicit function theorem will be applied to the function $F : \mathbb{R}^n \times \mathbb{C} \times \mathbb{R} \to \mathbb{R}^n \times \mathbb{C}$

$$F(u, \sigma, \kappa) = \left((M - \kappa P)u - \sigma u, \|u\|^2 - 1\right)^T.$$

According to the assumptions $F(u_0, \nu, 0) = 0 \in \mathbb{R}^n \times \mathbb{C}$ and in the case $F(u, \sigma, \kappa) = 0 \in \mathbb{R}^n \times \mathbb{C}$ the vector u is an eigenvector with unit length and σ is an eigenvalue in (8). In order to apply the implicit function theorem we have to show that

$$\det \partial_{(u,\sigma)} F(u_0, \nu, 0) \neq 0,$$

that is $\ker \mathcal{A} = 0 \in \mathbb{R}^n \times \mathbb{C}$, where

$$\mathcal{A} = \partial_{(u,\sigma)} F(u_0, \nu, 0) = \begin{pmatrix} M - \nu I & -u_0 \\ 2u_0^T & 0 \end{pmatrix}.$$

Assume that $\mathcal{A} \cdot (v, \alpha)^T = 0$, then $(M - \nu I)v = \alpha u_0$ and $\langle u_0, v \rangle = 0$. Multiplying the first equation by v_0 and using that $(M - \nu I)^T v_0 = 0$ we get

$$0 = \langle v, (M - \nu I)^T v_0 \rangle = \alpha \langle u_0, v_0 \rangle.$$

Because of the assumption $\langle u_0, v_0 \rangle \neq 0$ we get $\alpha = 0$. Now we have to show that $v = 0$. Since $(M - \nu I)v = \alpha u_0 = 0$ and the eigenspace is one-dimensional there exists $\beta \in \mathbb{R}$ such that $v = \beta u_0$. Hence from $\langle u_0, v \rangle = 0$ we get $\beta = 0$, since $\|u_0\| \neq 0$. Then $\beta = 0$ implies $v = 0$ yielding $\ker \mathcal{A} = 0$. Hence the implicit function theorem ensures the existence of the functions s and $\mathcal{U}$.

It remains to prove (10). Differentiating (9) and substituting $\kappa = 0$ we get

$$M\mathcal{U}'(0) - P\mathcal{U}(0) = s(0)\mathcal{U}'(0) + s'(0)\mathcal{U}(0).$$

Multiplying this equation by v_0 and using $s(0) = \nu$ we get

$$\langle \mathcal{U}'(0), (M^T - \nu I)v_0 \rangle - \langle P\mathcal{U}(0), v_0 \rangle = s'(0)\langle \mathcal{U}(0), v_0 \rangle.$$

Using $(M^T - \nu I)v_0 = 0$ and $\mathcal{U}(0) = u_0$ we get the desired expression for $s'(0)$. ◇

The significance of $s'(0)$ is the following. If the traveling wave is stable in the 1D geometry, i.e. M has a zero eigenvalue (because of the translational invariance) and the other eigenvalues have negative real part, and $s'(0) > 0$, then the wave is transversally unstable, since (8) has positive eigenvalues for some positive values of κ. If the traveling wave is stable in the 1D geometry and $s'(0) < 0$, then the wave is transversally stable at least for small positive values of κ, since (8) has positive eigenvalues. In fact, in many problems $s(\kappa) < 0$ for all κ. That is the wave is transversally stable if $s'(0) < 0$, and transversally unstable if $s'(0) > 0$. Therefore the transverse instability appears when $s'(0) = 0$, i.e. this is the condition for the transverse instability bifurcation.

Thus the bifurcation curves can be obtained numerically as follows. For the Hopf-bifurcation curve we have to determine those parameter values for which M has a pair of pure imaginary eigenvalues. For the transverse instability bifurcation curve we have to determine those parameter values, for which

$$\langle Pu_0, v_0 \rangle = 0, \tag{11}$$

where $Mu_0 = 0$ and $M^T v_0 = 0$. In the next Section we will show some examples where these bifurcation values are determined.

4   Application

Now we apply the procedure presented above in a flame propagation model. We consider the simple case of a first-order reaction, when the reactant (fuel) is burnt to an inert product during an exothermic reaction in the presence of heat loss, see [2,5,6,9]. Here the longitudinal and transverse instability of a planar laminar premixed flame moving in the direction of the x coordinate axis is studied in this model case with linear heat loss.

The dimensionless equations governing our model are

$$\partial_t a = L_A^{-1} \left(\partial_x^2 a + \partial_y^2 a \right) - a g(\theta) \tag{12}$$

$$\partial_t \theta = \partial_x^2 \theta + \partial_y^2 \theta + a g(\theta) - \gamma \theta \tag{13}$$

where a is the concentration of the fuel, θ is temperature (both are dimensionless), L_A is the Lewis number (ratio of thermal to mass diffusivity), γ is the heat loss parameter,

$$g(\theta) = \varepsilon^{-2} \exp(\frac{\theta - 1}{\varepsilon \theta}) \tag{14}$$

is the scaled Arrhenius function (determining the reaction rate to a given value of the temperature) and $1/\varepsilon$ is the activation energy. We will consider plane wave solutions moving along the positive direction of the x co-ordinate axis, hence we assume that ahead of the combustion front $a \to 1$, and $\theta \to 0$ since the ambient temperature is assumed to be zero in order to avoid the so-called cold boundary problem, see e.g. [2,6,9].

In order to get the traveling wave equations (2) in this case let us substitute $a(x, y, t) = a_0(z)$ and $\theta(x, y, t) = \theta_0(z)$ into (12)–(13), where $z = x - ct$. Then we get

$$L_A^{-1} a_0'' + c a_0' - a_0 g(\theta_0) = 0 \tag{15}$$
$$\theta_0'' + c\theta_0' + a_0 g(\theta_0) - \gamma \theta_0 = 0 \tag{16}$$

(where primes denote differentiation with respect to z).

The main question concerning problem (15)–(16) is the number and stability of traveling wave solutions. Three different bifurcations of traveling waves may occur in this model. Increasing the value of the heat loss parameter the number of traveling wave solutions changes from two to zero through saddle-node bifurcation at a critical value of γ. That is there is no propagating flame when the heat loss is strong enough. The traveling wave can loose its stability via Hopf bifurcation as the Lewis number or γ is varied. Considering the traveling wave in 2D geometry it can be stable under 1D perturbations but unstable for perturbations which are transversal to the direction of its propagation. This phenomenon is called transverse instability and yields cellular flames.

After linearization around the traveling wave (4) takes the form

$$L_A^{-1} A_0'' + c A_0' - (L_A^{-1} k^2 + \sigma + f(\theta_0)) A_0 - a_0 f'(\theta_0) T_0 = 0 \tag{17}$$
$$T_0'' + c T_0' - (k^2 + \sigma + \gamma - a_0 f'(\theta_0)) T_0 + f(\theta_0) A_0 = 0 \tag{18}$$

on $-\infty < \zeta < \infty$, subject to

$$A_0 \to 0, \quad T_0 \to 0 \quad \text{as} \quad \zeta \to \pm\infty. \tag{19}$$

For a given real value of k this is an eigenvalue problem. The eigenvalue is σ, the eigenfunctions are (A_0, T_0). We solve this eigenvalue problem numerically with finite difference discretization.

The Hopf-bifurcation points can be obtained for $\kappa = 0$, i.e. the eigenvalues of the matrix M have to be determined. For a fixed value of L_A we can determine that value of γ for which M has a pair of pure imaginary eigenvalues. Changing the value of L_A we get the Hopf-bifurcation curve in the (γ, L_A) parameter plane. It is shown in Figure 1. In this figure the saddle-node bifurcation curve and its common point with the Hopf-curve, the Takens-Bogdanov bifurcation point (TB) are also shown, these were computed in [6].

The transverse instability bifurcation point can be determined from condition (11) as follows. First we fix a value of γ and define a function which associates to a given value of L_A the left hand side of (11). For the computations we used MATLAB. With the given values of γ and L_A we solve first the discretization of (17)–(19) and get the values of a_0 and θ_0 at the grid points. These determine the matrix M in (8). Then we determine the eigenvectors of M and M^T belonging to the eigenvalue zero, these are u_0 and v_0 that are used to compute the left hand side of (11). Then we apply a root finding algorithm to get the bifurcation value of L_A for which (11) holds. This way we get a point of the bifurcation curve. Increasing the value of γ starting from zero we determine the bifurcation value of

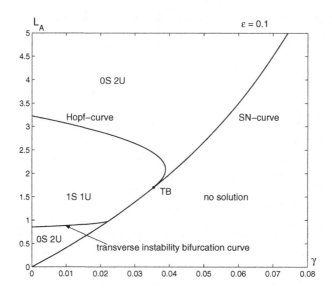

Fig. 1. The saddle-node, Hopf and transverse instability bifurcation curves. The number of stable and unstable planar flame solutions is indicated in the different regions determined by the bifurcation curves.

L_A to each γ, hence we obtain the transverse instability bifurcation curve, shown in Figure 1. The bifurcation curves divide the (γ, L_A) parameter plane into four regions where the number and type of planar flame solutions are constant. The number and stability of them are also given in the figure.

References

1. Beyn, W.J., Lorenz, J.: Stability of traveling waves: Dichotomies and eigenvalue conditions on finite intervals. Numer. Func. Anal. Optim. 20, 201–244 (1999)
2. Giovangigli, V.: Nonadiabatic plane laminar flames and their singular limits. SIAM J. Math. Anal 21, 1305–1325 (1990)
3. Henry, D.: Geometric theory of semilinear parabolic equations. Springer, Heidelberg (1981)
4. Keller, H.B.: Numerical methods for two-point boundary-value problems. Blaisdell Publishing Company (1968)
5. Lasseigne, D.G., Jackson, T.L., Jameson, L.: Stability of freely propagating flames revisited. Combust. Theory Modelling 3, 591–611 (1999)
6. Simon, P.L., et al.: Evans function analysis of the stability of non-adiabatic flames. Combust. Theory Modelling 7, 545–561 (2003)
7. Simon, P.L.: On the structure of spectra of travelling waves. Elect. J. Qual. Th. Diff. Equ. 15, 1–19 (2003)
8. Smoller, J.A.: Shock waves and reaction diffusion equations. Springer, Heidelberg (1983)
9. Weber, R.O., et al.: Combustion waves for gases (Le = 1) and solids (Le → ∞). Proc. Roy. Soc. Lond. A 453, 1105–1118 (1997)

Part V

Recent Advances in Methods and Applications for Large Scale Computations and Optimization of Coupled Engineering Problems

Parallel Implementation of LQG Balanced Truncation for Large-Scale Systems[*]

Jose M. Badía[1], Peter Benner[2], Rafael Mayo[1], Enrique S. Quintana-Ortí[1], Gregorio Quintana-Ortí[1], and Alfredo Remón[1]

[1] Depto. de Ingeniería y Ciencia de Computadores, Universidad Jaume I, 12.071–Castellón, Spain
{badia,mayo,quintana,gquintan,remon}@icc.uji.es
[2] Fakultät für Mathematik, Technische Universität Chemnitz, 09107 Chemnitz, Germany
benner@mathematik.tu-chemnitz.de

Abstract. Model reduction of large-scale linear time-invariant systems is an ubiquitous task in control and simulation of complex dynamical processes. We discuss how LQG balanced truncation can be applied to reduce the order of large-scale control problems arising from the spatial discretization of time-dependent partial differential equations. Numerical examples on a parallel computer demonstrate the effectiveness of our approach.

Keywords: model reduction, LQG balancing, algebraic Riccati equation, Newton's method, parallel algorithms.

1 Introduction

We consider linear time-invariant (LTI) dynamical systems in state-space form:

$$
\begin{aligned}
\dot{x}(t) &= Ax(t) + Bu(t), \quad t > 0, \quad x(0) = x^0, \\
y(t) &= Cx(t) + Du(t), \quad t \geq 0,
\end{aligned}
\tag{1}
$$

where $A \in \mathbb{R}^{n \times n}$, $B \in \mathbb{R}^{n \times m}$, $C \in \mathbb{R}^{p \times n}$, $D \in \mathbb{R}^{p \times m}$, and $x^0 \in \mathbb{R}^n$ is the initial state of the system. In particular, we are interested in LTI systems arising from the semi- or spatial discretization of a control problem for a time-dependent linear partial differential equation (PDE) like the instationary heat or convection-diffusion equations. In such a situation, the state matrix A is typically large and sparse and, often, $n \gg m, p$. The number of states, n, is known as the state-space dimension (or the order) of the system. Applying the Laplace transform to (1) and assuming $x^0 = 0$, we obtain the input-output mapping in frequency domain,

$$
y(s) = G(s)u(s), \quad s \in \mathbb{C},
\tag{2}
$$

---

[*] This research has been partially supported by the DAAD programme Acciones Integradas HA2005-0081. J. M. Badía, R. Mayo, E. S. Quintana-Ortí, G. Quintana-Ortí, and A. Remón were also supported by the CICYT project TIN2005-09037-C02-02 and FEDER.

I. Lirkov, S. Margenov, and J. Waśniewski (Eds.): LSSC 2007, LNCS 4818, pp. 227–234, 2008.

where G is the associated transfer function matrix (TFM) defined by $G(s) = C(sI - A)^{-1}B + D$.

For purposes of control and simulation, it is often desirable to reduce the order of the system; see, e.g., [1,6,11]. Specifically, the task of model reduction is to find a reduced-order LTI system,

$$\dot{\hat{x}}(t) = \hat{A}\hat{x}(t) + \hat{B}\hat{u}(t), \quad t > 0, \quad \hat{x}(0) = \hat{x}^0,$$
$$\hat{y}(t) = \hat{C}\hat{x}(t) + \hat{D}\hat{u}(t), \quad t \geq 0, \tag{3}$$

of order r, $r \ll n$, with associated TFM $\hat{G}(s) = \hat{C}(sI - \hat{A})^{-1}\hat{B} + \hat{D}$ which approximates $G(s)$. Using (2), this becomes an approximation problem for TFMs:

$$y(s) - \hat{y}(s) = G(s)u(s) - \hat{G}(s)u(s) = (G(s) - \hat{G}(s))u(s),$$

so that in an appropriate norm,

$$\|y - \hat{y}\| \leq \|G - \hat{G}\|\|u\|.$$

There are plenty of approaches for computing a reduced-order model. Here we focus on balancing-related methods due to their favorable system theoretic properties. Among these, linear-quadratic Gaussian balanced truncation (LQG BT) [7] can be used as a model reduction method for unstable plants, but also provides a closed-loop balancing technique as it directly provides a reduced-order LQG compensator. Compared with the standard model reduction method of balanced truncation, the only difference is that the controllability and observability Gramians, W_c and W_o respectively, are replaced by the stabilizing solutions of the dual algebraic Riccati equations (AREs)

$$AW_c + W_cA^T - W_cC^TCW_c + BB^T = 0, \tag{4}$$
$$A^TW_o + W_oA - W_oBB^TW_o + C^TC = 0, \tag{5}$$

associated with the regulator and filter AREs used in LQG control design.

In this paper we describe a coarse-grain parallel algorithm for model reduction, extending the applicability of LQG BT methods to sparse systems with up to $\mathcal{O}(10^5)$ states. Specifically, we will investigate the parallel solution of sparse large-scale AREs as this is the main computational step in LQG BT model reduction. Our derivations will be based on [3], where we developed a parallel algorithm based on the combination of the Newton-Kleinman iteration for the (generalized) AREs and the low-rank ADI method for the (generalized) Lyapunov equation. We will also discuss the quality of the obtained LQG reduced-order model.

Numerical experiments on a large parallel architecture consisting of Intel Itanium 2 processors will illustrate the numerical performance of this approach and the potential of the parallel algorithms for model reduction of large-scale sparse systems via LQG BT.

2   LQG Balanced Truncation

In this section we briefly review the main computational steps necessary to compute a LQG reduced-order model. We will assume that the LTI system (1) is stabilizable and detectable, i.e., there exist matrices $K \in \mathbb{R}^{m \times n}$ and $L \in \mathbb{R}^{n \times p}$ such that both $A - BK$ and $A - LC$ are stable. Under these assumptions, the stabilizing solutions of the AREs (4) and (5) are known to be positive semidefinite. (By stabilizing solutions we mean the unique solutions for which $A - BB^T W_o$ and $A - W_c C^T C$ are stable. For further details on AREs and their solutions, consult [9].) Therefore, W_c and W_o can be factored as $W_c = S^T S$ and $W_o = R^T R$. A possibility here are the Cholesky factors, but in our algorithm we will rather employ low-rank factors; see below.

Consider now the singular value decomposition (SVD) of the product

$$SR^T = U \Sigma V^T = [\, U_1 \; U_2 \,] \begin{bmatrix} \Sigma_1 & \\ & \Sigma_2 \end{bmatrix} [\, V_1 \; V_2 \,]^T, \tag{6}$$

where U and V are orthogonal matrices, and $\Sigma = \mathrm{diag}(\gamma_1, \gamma_2, \ldots, \gamma_n)$ is a diagonal matrix containing the singular values $\gamma_1, \gamma_2, \ldots, \gamma_n$ of SR^T, which are called the *LQG characteristic values* of the system.

In complete analogy to classical balanced truncation, LQG BT now determines the reduced-order model of order r as $(\hat{A}, \hat{B}, \hat{C}, \hat{D}) = (T_L A T_R, T_L B, C T_R, D)$ with the truncation operators T_L and T_R given by

$$T_L = \Sigma_1^{-1/2} V_1^T R \quad \text{and} \quad T_R = S^T U_1 \Sigma_1^{-1/2}. \tag{7}$$

This method provides a realization $\hat{G}$ which satisfies

$$\|\Delta_{\mathrm{lqg}}\|_\infty = \|G - \hat{G}\|_\infty \le 2 \sum_{j=r+1}^{n} \frac{\gamma_j}{\sqrt{1 + \gamma_j^2}}, \tag{8}$$

where $\|\,.\,\|$ denotes the L_∞-norm of transfer functions without poles on the imaginary axis.

3 Newton's Method for the ARE

In this section we review a variant of Newton's method, described in [13], which delivers a full-rank approximation of the solution of large-scale sparse AREs. Starting from an initial solution X_0, Newton's method for the ARE (5) proceeds as follows:

Newton's method
repeat with $j := 0, 1, 2, \ldots$
   1) $K_j := B^T X_j$
   2) $\bar{C}_j := \begin{bmatrix} C \\ K_j \end{bmatrix}$
   3) Solve for X_{j+1}:
         $(A - BK_j)^T X_{j+1} + X_{j+1}(A - BK_j) + \bar{C}_j^T \bar{C}_j = 0$
until $\|X_{j+1} - X_j\| < \tau \|X_j\|$

with the tolerance threshold τ defined by the user.

Provided $A - BB^T X_0$ is stable (i.e., all its eigenvalues lie on the open left half plane), this iteration converges quadratically to the desired symmetric positive semidefinite solution of the ARE [8], $W_o = \lim_{j \to \infty} X_j$. A line search procedure in [5] can be used to accelerate initial convergence.

In real large-scale applications, $m, p \ll n$, and A is sparse, but the solution matrix W_o is in general dense and, therefore, impossible to construct explicitly. However, W_o is often of low-numerical rank and thus can be approximated by a full-rank factor $\bar{R} \in \mathbb{R}^{n \times \bar{l}}$, with $\bar{l} \ll n$, such that $\bar{R}\bar{R}^T \approx W_o$. The method described next aims at computing this "narrow" factor $\bar{R}$.

Let us review how to modify Newton's method in order to avoid explicit references to X_j. Assume for the moment that, at the beginning of iteration j, we maintain $\bar{R}_j \in \mathbb{R}^{n \times \bar{l}_j}$ such that $\bar{R}_j \bar{R}_j^T = X_j$. Then, in the first step of the iteration, we can compute $K_j := B^T X_j = (B^T \bar{R}_j)\bar{R}_j^T$, which initially requires a (dense) matrix product, $M := B^T \bar{R}_j$, at a cost of $2\bar{l}_j mn$ flops (floating-point arithmetic operations); and then a (dense) matrix product, $M\bar{R}_j^T$, with the same cost. Even for large-scale problems, as m is usually a small order constant, this represents at most a quadratic cost. In practice, $\bar{l}_j$ usually remains a small value during the iteration so that this cost becomes as low as linear.

The key of this approach lies in solving the Lyapunov equation in the third step for a full-rank factor $\bar{R}_{j+1}$, such that $\bar{R}_{j+1}\bar{R}_{j+1}^T = X_{j+1}$. We describe how to do so in the next section.

4   Low Rank Solution of Lyapunov Equations

In this section we introduce a generalization of the Lyapunov solver proposed in [10,12], based on the cyclic *low-rank alternating direction implicit* (LR-ADI) iteration. In particular, consider the Lyapunov equation to be solved at each iteration of Newton's method

$$(A - BK)^T X + X(A - BK) + \bar{C}^T \bar{C} = 0, \tag{9}$$

where, for simplicity, we drop all subindices in the expression. Here, $A \in \mathbb{R}^{n \times n}$, $B \in \mathbb{R}^{n \times m}$, $K \in \mathbb{R}^{m \times n}$, and $\bar{C} \in \mathbb{R}^{(p+m) \times n}$. Recall that we are interested in finding a full-rank factor $R \in \mathbb{R}^{n \times \bar{l}}$, with $\bar{l} \ll n$, such that $RR^T \approx X$. Then, in iteration j of Newton's method, $\bar{R}_{j+1} := R$ and $\bar{l}_{j+1} := \bar{l}$.

The LR-ADI iteration, tailored for equation (9), can be formulated as follows:

> **LR-ADI iteration**
> 1) $V_0 := ((A - BK)^T + \sigma_1 I_n)^{-1}\bar{C}^T$
> 2) $R_0 := \sqrt{-2\,\alpha_1}\, V_0$
> `repeat with` $j := 0, 1, 2, \ldots$
>   3) $V_{j+1} := V_j - \delta_j((A - BK)^T + \sigma_{j+1} I_n)^{-1}V_j$
>   4) $R_{j+1} := [R_j \, , \, \eta_j V_{j+1}]$
> `until` $\|\eta_j V_{j+1}\| < \tau \|R_{j+1}\|$

In the iteration, $\{\sigma_1, \sigma_2, \ldots\}$, $\sigma_j = \alpha_j + \beta_j\,\jmath$, is a cyclic set of (possibly complex) shift parameters (that is, $\sigma_j = \sigma_{j+t_s}$ for a given period t_s), $\eta_j = \sqrt{\alpha_{j+1}/\alpha_j}$, and $\delta_j = \sigma_{j+1} + \overline{\sigma_j}$, with $\overline{\sigma_j}$ the conjugate of σ_j. The convergence rate of the LR-ADI iteration strongly depends on the selection of the shift parameters and is super-linear at best [10,12,14].

At each iteration the column dimension of R_{j+1} is increased by $(p + m)$ columns with respect to that of R_j so that, after $\bar{j}$ iterations, $R_{\bar{j}} \in \mathbb{R}^{n \times \bar{j}(p+m)}$. For details on a practical criterion to stop the iteration, see [2,12]. Note that the LR-ADI iteration does not guarantee a full colum rank for R_j.

From the computational view point, the iteration only requires the solution of linear systems of the form

$$((A - BK)^T + \sigma I_n)V = W \quad \Leftrightarrow \quad ((A + \sigma I_n) - BK)^T V = W, \qquad (10)$$

for V. Now, even if A is sparse (and therefore, so is $\bar{A} := A + \sigma I_n$), the coefficient matrix of this linear system is not necessarily sparse. Nevertheless, we can still exploit the sparsity of A by relying on the *Sherman-Morrison-Woodbury (SMW) formula*

$$(\bar{A} - BK)^{-1} = \bar{A}^{-1} + \bar{A}^{-1}B(I_m - K\bar{A}^{-1}B)^{-1}K\bar{A}^{-1}.$$

Specifically, the solution V of (10) can be obtained following the next five steps:

SMW formula
1) $V := \bar{A}^{-T}W$ 4) $T := TF^{-1}$
2) $T := \bar{A}^{-T}K^T$ 5) $V := V + T(B^TV)$
3) $F := I_m - B^TT$

Steps 1 and 2 require the solution of two linear systems with sparse coefficient matrix $\bar{A}$. The use of direct solvers is recommended here as iterations j and $j + t$ of the LR-ADI method share the same coefficient matrices for the linear system. The remaining three steps operate with dense matrices of small-order; specifically, $F \in \mathbb{R}^{m \times m}$, $T \in \mathbb{R}^{n \times m}$ so that Steps 3, 4, and 5 only require $2m^2n$, $2m^3/3 + m^2n$, and $4mn(m + p)$ flops, respectively.

5   Parallel Implementation

The major part of the cost of the LQG BT model reduction method lies with the Lyapunov equations that need to be solved during Newton's iteration for the ARE. We next describe the parallelization of this stage.

The two main operations involved in the LR-ADI iteration for the Lyapunov equation are the factorization of $\bar{A} - BK$ and the subsequent solution of linear systems with the corresponding triangular factors. Let F_j (factorization) and S_j (triangular solves) denote these two operations. Figure 1 (left) illustrates the data dependencies existing in the LR-ADI iteration. Note the dependency between the j-th factorization and the solutions of the triangular linear systems in iterations j, $j + t_s$, $j + 2t_s$, ...

Now, as all factorizations are independent, we can compute them in parallel using n_p processors (P1, P2,...); on the other hand, the triangular solves need to be computed sequentially, as the result of iteration j is needed during iteration

Fig. 1. Left: Data dependencies graph among factorizations (F1,F2,...) and triangular solves $(S1, S2, ...)$; $t_s=5$ shifts and $l_c=7$ iterations. Right: Mapping of the operations (F1,F2,... for the factorizations, and $S1, S2, ...$ for the triangular solves) to $n_p=4$ processes in the coarse-grain parallel algorithm; $t_s=5$ shifts and $l_c=7$ iterations.

$j + 1$; see Figure 1 (right). Therefore, we can overlap factorizations (F1–F4 in the figure), factorizations and triangular solves (F5 with both S2 and S3 in the figure), but not triangular solves.

The organization of the coarse-grain parallel scheme only requires efficient serial routines for the factorization of (sparse) linear systems and the solution of the corresponding triangular linear systems. The only communication is between "neighbour" processes, which need to transfer the solution of the current iteration step. A detailed description of the parallelization of the LR-ADI iteration can be found in [4].

6   Numerical Examples

All the experiments presented in this section were performed on a ccNUMA SGI Altix 350 platform with $n_p=16$ processors using IEEE double-precision floating-point arithmetic ($\varepsilon \approx 2.2204e-16$). In order to solve the sparse linear systems, we employed the multifrontal massively parallel sparse direct solver in package MUMPS 4.6.2 (http://graal.ens-lyon.fr/MUMPS/). For dense linear algebra operations, we employed LAPACK and the MKL 8.1 implementation of BLAS.

In the evaluation of the parallel LQG BT method, we employ two examples:

heat1D: This problem models the variation of the temperature in a 1D thin rod with a single source of heat.

heat2D: This model corresponds to a linear 2D heat equation with homogeneous Dirichlet boundary and point control/observation, resulting from a FD discretization on a uniform $g \times g$ grid.

We first evaluate the numerical quality of the reduced order models obtained via the LQG BT method. For that purpose, in Figure 2 we compare the frequency response of the original systems with that of the reduced-order realizations for the heat1D example (with $n=400$, $m=p=1$) and the heat2D example (with a grid dimension $g=20$, resulting in a system of order $n=400$, and $m=p=1$). For both examples, the order of the reduced model was set to $r=30$, and $t_s=10$ shifts were selected. A good approximation is achieved at most frequencies with a quite

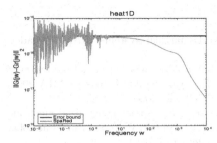 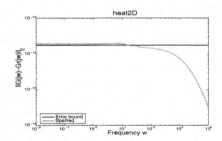

Fig. 2. Frequency responses obtained with the original system and reduced-order realization (of order r=30) computed with the parallel LQG BT model for the heat1D (left) and heat2D (right) examples

Table 1. Execution times of the parallel LR-ADI solver for the Lyapunov equation associated with the the heat1D (left) and heat2D (right) examples

Example	n_p=1	n_p=2	n_p=3	n_p=4	n_p=5	n_p=10
heat1D	113.21	75.27	–	61.34	58.61	–
heat2D	109.12	72.72	59.87	51.37	47.94	40.42

reduced-order model in both cases. Only for the heat1D, there is a deviation from frequencies starting at 10^3; however, by then the deviation is below the machine precision ($\varepsilon \approx 2.2e\text{-}16$).

Table 1 reports the execution time (in seconds) of the parallel LR-ADI solver for the (first) Lyapunov equations arising in the heat1D and heat2D examples of order n=200,000 and 160,000, respectively. The number of shifts t_s was set to 20, l_c=50 iterations were allowed, and the number of processes/processors employed in the evaluation was varied from n_p=1 to 10. The parallel results report a mild reduction the execution time of the LR-ADI solver. We note that this reduction is highly dependent on several factors. First, the sparsity pattern of the state matrix A and the efficiency of the sparse solver, as that determines the costs of the factorization and triangular solve stages. A high ratio of factorization time to triangular solve time benefits the parallelism of the parallel LQG BT algorithm. Second, the number of iterations required for convergence compared with the number of shifts employed during the iterations. A high value of t_s and a small number of iterations benefit the parallelism of our approach.

7   Concluding Remarks

We have provided evidence in support of the benefits of the LQG BT method for model reduction of large-scale systems. However, further experiments are due before general conclusions can be extracted for the numerical accuracy and parallelism of this approach.

References

1. Antoulas, A.: Approximation of Large-Scale Dynamical Systems. SIAM, Philadelphia (2005)
2. Badía, J.M., et al.: Parallel algorithms for balanced truncation model reduction of sparse systems. In: Dongarra, J., Madsen, K., Waśniewski, J. (eds.) PARA 2004. LNCS, vol. 3732, pp. 267–275. Springer, Heidelberg (2006)
3. Badía, J.M., et al.: Parallel solution of large-scale and sparse generalized algebraic Riccati equations. In: Nagel, W.E., Walter, W.V., Lehner, W. (eds.) Euro-Par 2006. LNCS, vol. 4128, pp. 710–719. Springer, Heidelberg (2006)
4. Badía, J.M., et al.: Balanced truncation model reduction of large and sparse generalized linear systems. Technical Report CSC/06-04, Technical University of Chemnitz, Chemnitz, 09107 Chemnitz, Germany (2006)
5. Benner, P., Byers, R.: An exact line search method for solving generalized continuous-time algebraic Riccati equations. IEEE Trans. Automat. Control 43(1), 101–107 (1998)
6. Benner, P., Mehrmann, V., Sorensen, D. (eds.): MFCS 1976. LNCS, vol. 45. Springer, Heidelberg (2005)
7. Jonckheere, E., Silverman, L.: A new set of invariants for linear systems—application to reduced order compensator. IEEE Trans. Automat. Control AC-28, 953–964 (1983)
8. Kleinman, D.: On an iterative technique for Riccati equation computations. IEEE Trans. Automat. Control AC-13, 114–115 (1968)
9. Lancaster, P., Rodman, L.: The Algebraic Riccati Equation. Oxford University Press, Oxford (1995)
10. Li, J.-R., White, J.: Low rank solution of Lyapunov equations. SIAM J. Matrix Anal. Appl. 24(1), 260–280 (2002)
11. Obinata, G., Anderson, B.: Model Reduction for Control System Design. Communications and Control Engineering Series. Springer, London (2001)
12. Penzl, T.: A cyclic low rank Smith method for large sparse Lyapunov equations. SIAM J. Sci. Comput. 21(4), 1401–1418 (2000)
13. Penzl, T.: Lyapack Users Guide. Technical Report SFB393/00-33, Sonderforschungsbereich 393 Numerische Simulation auf massiv parallelen Rechnern, TU Chemnitz, 09107 Chemnitz, FRG (2000), available from http://www.tu-chemnitz.de/sfb393/sfb00pr.html
14. Wachspress, E.: The ADI model problem, Available from the author (1995)

Finite Element Solution of Optimal Control Problems Arising in Semiconductor Modeling

Pavel Bochev and Denis Ridzal

Computational Mathematics and Algorithms Department,
Sandia National Laboratories*, Albuquerque, NM 87185-1320, USA
{pbboche,dridzal}@sandia.gov

Abstract. Optimal design, parameter estimation, and inverse problems arising in the modeling of semiconductor devices lead to optimization problems constrained by systems of PDEs. We study the impact of different state equation discretizations on optimization problems whose objective functionals involve flux terms. Galerkin methods, in which the flux is a derived quantity, are compared with mixed Galerkin discretizations where the flux is approximated directly. Our results show that the latter approach leads to more robust and accurate solutions of the optimization problem, especially for highly heterogeneous materials with large jumps in material properties.

1 Introduction

Common objectives in the modeling of semiconductor devices are, e.g., to control the current flow over a contact by changing the so called doping profile of the device (optimal design problem), or to characterize an unknown doping profile based on measurements of the current flow (inverse problem).

In either case, the resulting PDE constrained optimization problems call for objective functionals that involve flux terms in their definition. Depending on whether the state equation is discretized by a Galerkin or a mixed Galerkin method, flux terms can have fundamentally different representations. For instance, the Galerkin method approximates the scalar concentration variables by finite element subspaces of $H^1(\Omega)$, and the flux is a derived quantity, while in the mixed method the flux is approximated directly by subspaces of $H(\text{div}, \Omega)$.

While numerical solution of optimization problems arising in semiconductor device modeling has been previously addressed in the literature [6,8], there are virtually no studies on how discretization choices impact the accuracy and the robustness of the numerical approximation. To a degree, our work is motivated by earlier studies [7,1] of optimization problems governed by advection-dominated PDEs, which showed that stabilization of the state equations and stabilization of the optimality system yield different solutions of the optimization problem.

* Sandia is a multiprogram laboratory operated by Sandia Corporation, a Lockheed Martin Company, for the United States Department of Energy under contract DE-AC04-94-AL85000.

I. Lirkov, S. Margenov, and J. Waśniewski (Eds.): LSSC 2007, LNCS 4818, pp. 235–242, 2008.

However, in contrast to these papers, our main focus is on how different discrete formulations of the state equation may affect the optimization problem.

Our report is organized as follows. The model optimization problem is described in Section 2. Section 3 states the Galerkin and mixed Galerkin discretizations of the optimization problem (1). Numerical results contrasting these methods, and a discussion of the results, are presented in Section 4.

2   Model Optimization Problem

A common objective in the design of semiconductor devices is to match the current J measured at a portion Γ_o of the Dirichlet boundary (see Fig. 1) to a prescribed value $\widehat{J}$, while allowing for "small" (controlled) deviations of the doping profile u from a reference doping profile $\widehat{u}$. For a complete formulation of such optimization problems, constrained by the drift-diffusion semiconductor equations, we refer to [6,8,9].

The primary goal of this work is to study how different discretizations of the state equations impact the solution of the optimization problem. This question can be investigated on a much simpler model, and so, we restrict our attention to the following linear–quadratic elliptic optimization[1] problem,

$$\text{minimize}   \frac{1}{2}\|\nabla y \cdot \nu - \nabla \widehat{y} \cdot \nu\|^2_{-1/2,\Gamma_o} + \frac{\alpha}{2}\|u - \widehat{u}\|^2_{0,\Omega} \tag{1a}$$

subject to

$$-\nabla \cdot (k(x)\nabla y(x)) = f(x) + u(x) \qquad \text{in } \Omega \tag{1b}$$
$$y(x) = y_D(x) \qquad \text{on } \Gamma_D \tag{1c}$$
$$(k(x)\nabla y(x)) \cdot \nu = g(x) \qquad \text{on } \Gamma_N, \tag{1d}$$

where $\alpha, k(x) > 0$, $\Omega \subset \mathbb{R}^d$, $d = 1, 2, 3$ is a bounded domain, and Γ_D, Γ_N are the Dirichlet and Neumann parts of $\partial\Omega$. We assume that $\Gamma_D \neq \emptyset$ and $\Gamma_N = \partial\Omega/\Gamma_D$. Model (1) follows from the full problem considered in [9] by assuming that the electron and hole densities are given functions.

3   Discretization of the Optimization Problem

We consider two discretizations of (1) that differ by their choice of finite element methods for the state equation. In each case we discuss computation of the flux terms in the objective functional, necessary to complete the discretization of the optimization problem.

For more details regarding the solution of the discrete optimization problem, or the existence and uniqueness of optimal solutions of (1) we refer to [9,6,8].

---

[1] As it is customary in the optimal control context, we refer to the variables $y(x)$ in (1) as the *state variables*, whereas the doping profile $u(x)$ will play the role of the *control variables*. Equation (1b) is known as the *state equation*. Additionally, we will often abbreviate $y(x)$, $u(x)$, etc., by y, u, etc.

3.1   Galerkin Discretization

We define the state and control spaces $Y = \{y \in H^1(\Omega) : y = y_D \text{ on } \Gamma_D\}$, $U = L^2(\Omega)$, and the space of test functions $V_0 = \{v \in H^1(\Omega) : v = 0 \text{ on } \Gamma_D\}$. The weak form of (1) is to find $y \in Y, u \in U$, which solve the problem

$$\text{minimize}   \frac{1}{2}\|\nabla y \cdot \nu - \nabla \widehat{y} \cdot \nu\|^2_{-1/2,\Gamma_o} + \frac{\alpha}{2}\|u - \widehat{u}\|^2_{0,\Omega} \tag{2a}$$

subject to

$$a(y, v) + b(u, v) = (f, v) + \langle g, v\rangle_{\Gamma_N}, \quad \forall v \in V_0 \tag{2b}$$

where $\langle \cdot, \cdot\rangle_\star$ denotes the duality pairing between $H^{-1/2}(\star)$ and $H^{1/2}(\star)$, and

$$a(y, v) = \int_\Omega k\nabla y \cdot \nabla v \, dx, \qquad b(u, v) = -\int_\Omega uv \, dx, \qquad (f, v) = \int_\Omega fv \, dx.$$

The finite element discretization of the state equation is obtained in the usual manner by restricting (2b) to finite element subspaces $Y_h \subset Y$, $V_{0,h} \subset V_0$ and $U_h \subset U$ of the state, test, and control spaces, respectively.

When using a Galerkin method for the state equation, discretization of the objective functional (2a) requires additional attention, because the flux $\nabla y \cdot \nu$ appearing in (2a) is not approximated directly by the method. A standard approach to discretizing the flux term $\|\nabla y \cdot \nu - \nabla \widehat{y} \cdot \nu\|^2_{-1/2,\Gamma_o}$ would be to restrict it to the finite element space Y_h for the states, and then use a weighted L^2 norm to approximate the norm on $H^{-1/2}(\Gamma_o)$:

$$\|\nabla y \cdot \nu - \nabla \widehat{y} \cdot \nu\|^2_{-1/2,\Gamma_o} \approx h\|\nabla y_h \cdot \nu - \nabla \widehat{y}_h \cdot \nu\|^2_{0,\Gamma_o}. \tag{3}$$

While this discretization of the flux term is consistent, in the sense that every instance of the state y in the optimization problem is approximated by the same finite element basis, it may not be the best possible choice for this term. This is certainly true if the state equation is solved separately and ∇y_h is used to approximate the flux.

It is well-known that a more accurate flux approximation can be obtained by postprocessing the finite element solution, instead of simply taking its derivative. One such technique is the variational flux approximation (VFA) [2,5,10,11]. It is based on the Green's formula and has been applied to optimization problems by Berggren et. al. in [3]. In the VFA approach, the standard flux $\nabla y_h \cdot \nu$ is replaced by a more accurate, C^0 approximation λ_h, obtained by solving the equation

$$\int_{\Gamma_o} \lambda_h v_h \, dl = k^{-1}\left(a(y_h, v_h) + b(u_h, v_h) - (f, v_h) - \int_{\Gamma\backslash\Gamma_o} k\nabla y_h \cdot \nu v_h \, dl\right). \tag{4}$$

Using VFA we approximate the flux term as follows:

$$\|\nabla y \cdot \nu - \nabla \widehat{y} \cdot \nu\|^2_{-1/2,\Gamma_o} \approx h\|\lambda_h - \widehat{\lambda}_h\|^2_{0,\Gamma_o}. \tag{5}$$

For implementation details of VFA we refer to the above papers.

Remark 1. The use of VFA in our optimization problem differs substantially from its use as a postprocessing technique and in the optimization problem of [3]. When (4) is used to postprocess a given finite element solution (y_h, u_h), the right hand side in (4) involves only known quantities. In [3], VFA is used in an already defined optimality system to improve the accuracy of the solution. In contrast, in our case, the VFA approximation *changes* the optimization problem, because the discretization of $\|\nabla y \cdot \nu - \nabla \widehat{y} \cdot \nu\|^2_{-1/2,\Gamma_o}$ by (5) makes this term a function of *both* the unknown state y_h and the unknown control u_h.

3.2   Mixed Galerkin Discretization

A mixed Galerkin method for the state equation is defined by using its equivalent first-order system form

$$\begin{cases} \nabla \cdot p + u = -f \text{ in } \Omega \\ k^{-1} p - \nabla y = 0 \text{ in } \Omega \end{cases} \quad \text{and} \quad \begin{aligned} y &= y_D \text{ on } \Gamma_D \\ (k\nabla y) \cdot \nu &= g \text{ on } \Gamma_N. \end{aligned} \tag{6}$$

For the variational formulation of the optimization problem, we introduce[2] the state spaces $Y = L^2(\Omega)$ and $P = H_{g,N}(\text{div}, \Omega)$, the control space $U = L^2(\Omega)$, and the trial spaces $V = L^2(\Omega)$ and $Q_0 = H_{0,N}(\text{div}, \Omega)$. To simplify the notation we write $\widehat{p} = k\,\nabla\widehat{y}$. The weak form of (1), using the mixed Galerkin discretization of (6), is to find $y \in Y, p \in P, u \in U$, which solve the problem

$$\text{minimize} \quad \frac{1}{2}\|k^{-2}(p \cdot \nu - \widehat{p} \cdot \nu)\|^2_{-1/2,\Gamma_o} + \frac{\alpha}{2}\|u - \widehat{u}\|^2_{0,\Omega}, \tag{7a}$$

subject to

$$\begin{aligned} a(p, q) + b(q, y) &= \langle y_d, q \cdot \nu \rangle_{\Gamma_D} & \forall q \in Q_0 \\ b(p, v) + c(u, v) &= -(f, v) & \forall v \in V, \end{aligned} \tag{7b}$$

where $(\cdot, \cdot)$, and $\langle \cdot, \cdot \rangle_*$ were defined in Sec. 3.1, and

$$a(p, q) = \int_\Omega k^{-1} p \cdot q\, dx, \qquad b(q, y) = \int_\Omega (\nabla \cdot q) y\, dx, \qquad c(u, v) = \int_\Omega uv\, dx.$$

The mixed finite element discretization of the state equation follows by restricting (7b) to finite element subspaces $Y_h \subset Y$, $P_h \subset P$, $U_h \subset U$, $V_h \subset V$, and $Q_{0,h} \subset Q_0$, for the states, controls, and the respective test functions in (7b). We recall that the pairs (Y_h, P_h) and $(Q_{0,h}, V_h)$ are subject to an inf-sup stability condition [4].

In contrast to the Galerkin approach in Sec. 3.1, in the mixed method the flux is approximated directly by p_h. As a result, the flux term in the objective functional can be discretized as follows:

$$\|k^{-2}(p \cdot \nu - \widehat{p} \cdot \nu)\|^2_{-1/2,\Gamma_o} \approx h\|k^{-2}(p_h \cdot \nu - \widehat{p}_h \cdot \nu)\|^2_{0,\Gamma_o}. \tag{8}$$

---

[2] We recall that $H(\text{div}, \Omega) = \left\{ q \in \left[L^2(\Omega)\right]^2 : \nabla \cdot q \in L^2(\Omega) \right\}$; $H_{0,N}(\text{div}, \Omega)$ is the subspace of all fields in $H(\text{div}, \Omega)$ whose normal component vanishes on Γ_N, and $H_{g,N}(\text{div}, \Omega)$ are the fields in $H(\text{div}, \Omega)$ whose normal component on Γ_N equals g.

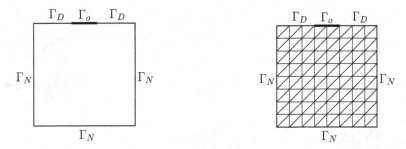

Fig. 1. Computational domain (left) and its partition into finite elements (right plot)

Table 1. $\mathcal{J}_F$, $\mathcal{J}_u$ and $\mathcal{J}$ denote the values of the flux term, the control term and their sum (the total value of the objective functional)

	Example 1			Example 2		
	GM	Mixed	GM-VFA	GM	Mixed	GM-VFA
$\mathcal{J}_F$	1.99e-06	1.88e-06	1.95e-04	6.42e-09	2.51e-09	2.70e-09
$\mathcal{J}_u$	1.10e-08	1.10e-08	3.81e-07	1.12e-03	1.08e-03	1.11e-03
$\mathcal{J}$	2.00e-06	1.89e-06	1.95e-04	1.12e-03	1.08e-03	1.11e-03
	Example 3			Example 4		
	GM	Mixed	GM-VFA	GM	Mixed	GM-VFA
$\mathcal{J}_F$	6.07e-05	8.10e-10	2.83e-07	1.06e+01	2.56e-07	6.29e-07
$\mathcal{J}_u$	7.17e-05	4.62e-05	4.50e-03	3.63e+00	4.57e-03	7.19e-03
$\mathcal{J}$	1.32e-04	4.62e-05	4.50e-03	1.42e+01	4.57e-03	7.19e-03

4   Numerical Results

The computational domain $\Omega = [-1, 1]^2$, its finite element partition, and the boundary part Γ_o are shown in Fig. 1. All numerical results were obtained on a 32×32 mesh with 2048 triangles, 1082 vertices, and 3136 edges. For the Galerkin method in Sec. 3.1, we use standard C^0, piecewise linear approximation spaces. The mixed Galerkin method for the state equation was implemented using the lowest order Raviart-Thomas element for p_h and piecewise constant finite elements for y_h.

In the examples below we compare finite element solutions of the optimization problem using the standard Galerkin method, the Galerkin method with VFA, and the mixed method. The most distinguishing characteristic of the examples used in our study are the differences in the corresponding diffusivity (i.e. permittivity, in the case of semiconductors) profiles $k(x)$. For all examples we use $f(x) = 0$, $y_D = 0$, $\widehat{u} = 1$, and $\alpha = 6.25 \cdot 10^{-4}$. The Neumann data g is set to 0 on the left and right sides of Ω and to $-k(x)$ on the bottom.

Example 1. The desired flux is $\nabla \widehat{y} \cdot \nu = 1$ and $k(x) = 10^2$ in Ω.
Example 2. The desired flux is $\nabla \widehat{y} \cdot \nu = 1$ and $k(x) = 10^{-2}$ in Ω.

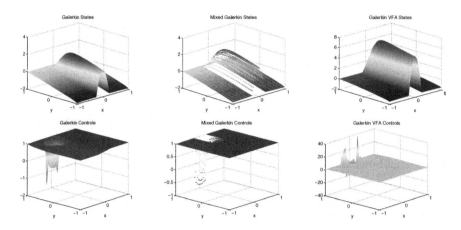

Fig. 2. Galerkin, mixed Galerkin, and Galerkin VFA optimal states (top row) and optimal controls (bottom row) for Ex. 3

Example 3. The desired flux is $\nabla \widehat{y} \cdot \nu = 1$ and

$$k(x) = \begin{cases} 10 & \text{in } [-1, -0.25] \times [-1, 1] \\ 10^{-2} & \text{in } [-0.25, 0.25] \times [-1, 1] \\ 10 & \text{in } [0.25, 1] \times [-1, 1], \end{cases}$$

Example 4. The desired flux is $\nabla \widehat{y} \cdot \nu = 100$ and $k(x)$ is as in Ex. 3.

Objective functional values for the four examples are summarized in Table 1. The data for Ex. 1-2 shows that for constant $k(x)$ all three discretizations perform at a comparable level. Nevertheless, the mixed Galerkin method does consistently outperform the other two discretizations, albeit by a small margin. We also observe that the VFA approach gives better results (i.e., closer to the mixed Galerkin results) for $k(x) \ll 1$, while the standard flux approximation does better for $k(x) \gg 1$.

The data for Ex. 3-4 shows a completely different situation, as far as the standard flux approximation is concerned. For these two examples the mixed Galerkin method clearly outperforms the standard Galerkin discretization of the optimization problem, especially when the desired flux value is large, as in Ex. 4.

Using the VFA approach, the Galerkin discretization fares better, however, the objective functional values remain less accurate than those computed with the mixed method. These observations are also confirmed by the plots in Fig. 2-3. We see that for Example 4 the states and controls computed by the standard Galerkin method are grossly inaccurate. We also note that among all three methods the controls computed by the mixed method exhibit the most robust behavior. An interesting feature of the controls for the VFA approach is their oscillatory nature. This could be a problem in some specific applications where the controls have to be implemented in real materials.

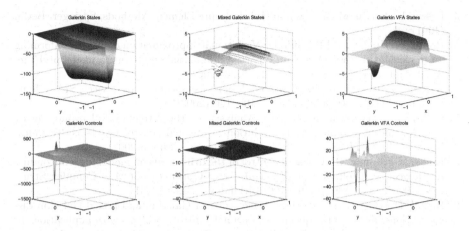

Fig. 3. Galerkin, mixed Galerkin, and Galerkin VFA optimal states (top row) and optimal controls (bottom row) for Ex. 4

Based on the numerical data, we can conclude that for problems with heterogeneous material properties the mixed Galerkin method offers the most robust performance and the most accurate results. The worst performer is the standard Galerkin method, which may yield state and control approximations that are many orders of magnitude less accurate than those computed by the mixed method. Thus, we cannot recommend the standard Galerkin discretization as a reliable approach to solving optimization problems whose objective functionals involve flux terms. Instead, for such problems, one should use the mixed Galerkin discretization whenever possible. If, for whatever reason, the use of the mixed method is not feasible, then the Galerkin discretization of the state equations should be combined with the VFA approach in order to improve robustness and accuracy.

Acknowledgement

The authors wish to thank Martin Berggren for valuable discussions about the VFA approach and for pointing out the reference [3].

References

1. Abraham, F., Behr, M., Heinkenschloss, M.: The effect of stabilization in finite element methods for the optimal boundary control of the oseen equations. Finite Elements in Analysis and Design 41, 229–251 (2004)
2. Babuška, I., Miller, A.: The post processing approach in the finite element method, part 1: calculation of displacements, stresses and other higher derivatives of the displacements. Internat. J. Numer. Methods. Engrg. 34, 1085–1109 (1984)
3. Berggren, M., Glowinski, R., Lions, J.L.: computational approach to controllability issues for flow-related models (II): Control of two-dimensional, linear advection-diffusion and Stokes models. INJCF 6(4), 253–274 (1996)

4. Brezzi, F., Fortin, M.: Mixed and Hybrid Finite Element Methods. Springer, Berlin (1991)
5. Brezzi, F., Hughes, T.J.R., Süli, E.: Variational approximation of flux in conforming finite element methods for elliptic partial differential equations: a model problem. Rend. Mat. Acc. Lincei ser. 9 12, 159–166 (2001)
6. Burger, M., Pinnau, R.: Fast optimal design of semiconductor devices. SIAM Journal on Applied Mathematics 64(1), 108–126 (2003)
7. Collis, S.S., Heinkenschloss, M.: Analysis of the Streamline Upwind/Petrov Galerkin method applied to the solution of optimal control problems. Technical Report CAAM TR02-01, CAAM, Rice University (March 2002)
8. Hinze, M., Pinnau, R.: Optimal control of the drift-diffusion model for semiconductor devices. In: Hoffmann, K.H., et al. (eds.) Optimal Control of Complex Structures. ISNM, vol. 139, pp. 95–106. Birkhäuser, Basel (2001)
9. Ridzal, D., Bochev, P.: A comparative study of Galerkin and mixed Galerkin discretizations of the state equation in optimal control problems with applications to semiconductor modeling. Technical Report SAND2007, Sandia National Laboratories (2007)
10. Wheeler, J.A.: Simulation of heat transfer from a warm pipeline buried in permafrost. In: Proceedings of the 74th National Meeting of the American Institute of Chemical Engineers (1973)
11. Wheeler, M.F.: A Galerkin procedure for estimating the flux for two-point boundary value problems. SIAM Journal on Numerical Analysis 11(4), 764–768 (1974)

Orthogonality Measures and Applications in Systems Theory in One and More Variables

Adhemar Bultheel[1], Annie Cuyt[2], and Brigitte Verdonk[2]

[1] Dept of Computer Science, Katholieke Universiteit Leuven,
Celestijnenlaan 200A, B3001 Heverlee, Belgium
adhemar.bultheel@cs.kuleuven.be
[2] Dept of Mathematics and Computer Science, University of Antwerp
Middelheimlaan 1, B2020 Antwerp, Belgium
{annie.cuyt,brigitte.verdonk}@ua.ac.be

Abstract. The representation or order reduction of a rational transfer function by a linear combination of orthogonal rational functions offers several advantages, among which the possibility to work with prescribed poles and hence the guarantee of system stability. Also for multidimensional linear shift-invariant systems with infinite-extent impulse response, stability can be guaranteed a priori by the use of a multivariate Padé-type approximation technique, which is again a rational approximation technique. In both the one- and multidimensional case the choice of the moment functional with respect to which the orthogonality of the functions in use is imposed, plays a crucial role.

1 Introduction

Let $\{c_i\}_{i\in\mathbb{N}}$ be a sequence of complex numbers and let c be a linear functional defined on the space of polynomials $\mathbb{C}[t]$ with complex coefficients $c(t^i) = c_i$, $i = 0, 1, \ldots$ Then c is called the moment functional determined by the moment sequence $\{c_i\}_{i\in\mathbb{N}}$. By means of c a formal series development of $h(z)$ with coefficients c_i (for instance a transfer function $h(z)$ with impulse response c_i for $i = 0, 1, 2, \ldots$) can be viewed as

$$h(z) = \sum_{i=0}^{\infty} c_i z^i = c(1) + c(t)z + c(t^2)z^2 + \ldots = c\left(\frac{1}{1 - tz}\right). \tag{1}$$

Let $\mathcal{L}_n = \mathrm{span}\{1, \ldots, t^n\}$ denote the space of polynomials of degree n and let ∂V denote the exact degree of a polynomial $V(t) \in \mathbb{C}[t]$. A sequence of polynomials $\{V_m(z)\}_{m\in\mathbb{N}}$ is called orthogonal with respect to the moment functional c provided that $V_m \in \mathcal{L}_m \setminus \mathcal{L}_{m-1}$ and

$$c\left(t^i V_m(t)\right) = 0, \quad i = 0, \ldots, m - 1, \quad c\left(V_m^2(t)\right) \neq 0. \tag{2}$$

For an arbitrary polynomial $V_m(z) \in \mathcal{L}_m$ with coefficients b_i, we can also construct the associated polynomial $W_{m-1}(z)$ by

$$W_{m-1}(z) = c\left(\frac{V_m(t) - V_m(z)}{t - z}\right). \tag{3}$$

I. Lirkov, S. Margenov, and J. Waśniewski (Eds.): LSSC 2007, LNCS 4818, pp. 243–250, 2008.
© Springer-Verlag Berlin Heidelberg 2008

Then $W_{m-1} \in \mathcal{L}_{m-1}$ with coefficients $a_i = \sum_{j=0}^{m-1-i} c_j b_{i+j+1}$ [3, p. 10]. It can be proven [3, p. 11] that the polynomials $\tilde{W}_{m-1}(z) = z^{m-1}W_{m-1}(1/z)$ and $\tilde{V}_m(z) = z^m V_m(1/z)$ satisfy [3, p. 11]

$$\left(h - \tilde{W}_{m-1}/\tilde{V}_m\right)(z) = \sum_{i=q}^{\infty} d_i z^i \tag{4}$$

with $q = m$. In this way it is easy to obtain rational approximants $\tilde{W}_{m-1}/\tilde{V}_m$ for a given transfer function $h(z)$. Choosing $V_m(z)$ in (4) allows to control the poles of the rational approximant. If however $V_m(z)$ is fixed by the orthogonality conditions (2), then $q = 2m$ in (4) and many more moments are matched, but the control over the poles is lost.

We recall that a system is called bounded-input bounded-output (BIBO) stable if the output signal is bounded whenever the input signal is bounded. Since stability is guaranteed when the rational approximant has all its poles inside the unit disk or polydisk, respectively, the aim is to obtain a rational function either in one or in more variables, that has this property.

Rational approximants of higher numerator degree can be obtained in the following way. If we define a linear functional $c^{(k)}(t^i) = c_{k+i}$ and set

$$h(z) = \sum_{i=0}^{k} c_i z^i + z^{k+1} h_k(z), \tag{5}$$

$$W_{m-1}^{(k+1)}(z) = c^{(k+1)} \left(\frac{V_m(t) - V_m(z)}{t - z} \right) \tag{6}$$

$$\tilde{W}_{m+k}(z)/\tilde{V}_m(z) = \sum_{i=0}^{k} c_i z^i + z^{k+1} \tilde{W}_{m-1}^{(k+1)}(z)/\tilde{V}_{m-1}(z) \tag{7}$$

then (4) generalizes to

$$\left(h - \tilde{W}_{m+k}/\tilde{V}_m\right)(z) = \sum_{i=q}^{\infty} d_i z^i \tag{8}$$

with $q = m + k + 1$ for arbitrary polynomials $V_m(z)$ and $q = 2m + k + 1$ when $V_m(z)$ satisfies the orthogonality conditions (2).

In (1)–(4), a linear functional is defined in terms of the impulse response, and the rational function that approximates the transfer function matches as many of the initial impulse response coefficients, also called Markov parameters, as possible. This corresponds to a good approximation of the transient behaviour of the system for small time. In this paper we discuss the generalization of the steps (1)–(4), namely

- defining a linear functional c using information collected from the transfer function h as in (1),
- computing a numerator polynomial $\tilde{W}_{m-1}(z)$ as in (3), possibly in combination with (2) for the denominator polynomial $\tilde{V}_m(z)$,
- setting up a sequence of rational approximants to the transfer function $h(z)$ as in (4) or (8),

in two ways.

The approximation of $h(z)$ in (4) may be improved when also the steady state of the system is approximated. This means that the transfer function is not only approximated in the neighborhood of $z = 0$, but also near $z = \infty$. The coefficients of the series expansion at infinity are called time moments. Matching some of the Markov parameters and some of the time moments corresponds to rational approximation in two points. This idea can be generalized as follows. Instead of approximating in just two points, one can find a rational approximant interpolating in several points, some of which may coincide [5].

Instead of considering one variable z, one can also study multidimensional systems and transfer functions, which arise in problems like computer-aided tomography, image processing, image deblurring, seismology, sonar and radar applications, and many other problems. As an illustration we consider the filtering of signals, which is concerned with the extraction and/or enhancement of information contained in either a one-dimensional or multidimensional sequence of measurements. Noises can be filtered from spoken messages as well as from picture images.

The former generalization is dealt with in Section 2 while the latter is introduced in Section 3. In both sections the aim is to provide an a priori stable rational approximant since for model reduction techniques the issue of stability of the reduced rational system is an important one. In Section 2 the rational approximants are obtained by combining generalizations of (3) and (2), while in Section 3 the approximants are constructed for appropriately chosen denominator polynomials in combination with (3) for the numerator.

2   Orthogonal Rational Functions Analytic Outside the Unit Disk

In frequency domain methods, it is assumed that the information about the system transfer function is not given by moments defined in 0 and infinity or at arbitrary points in the complex plane, but they are given in the frequency domain, which for a discrete time system is the complex unit circle $\mathbb{T}$. So what can be measured are not the samples of the transfer function h, but samples of its power spectrum $|h(z)|^2$ for many values of $z \in \mathbb{T}$. Using autocorrelation techniques, one actually knows the coefficients of the Fourier series $|h(z)|^2 = \sum_{k \in \mathbb{Z}} c_k z^k$ where $z = e^{i\omega}$. We can now define a moment functional for the Laurent polynomials by setting $c(t^i) = c_{-i}$ for all $i \in \mathbb{Z}$. Since we are working on $\mathbb{T}$ we reformulate the orthogonality conditions (2) as follows. A sequence of polynomials $\{V_m(z)\}_{m \in \mathbb{N}}$ is orthogonal with respect to the moment functional c provided that $V_m(z) \in \mathcal{L}_m \setminus \mathcal{L}_{m-1}$ and

$$c\left(t^i V_{m*}(t)\right) = 0, \qquad i = 0, \ldots, m-1, \quad c\left(V_m(t)V_{m*}(t)\right) \neq 0, \qquad (9)$$

where for any function f we define $f_*(z) = \overline{f(1/\overline{z})}$.[1]

---

[1] Observe that for $t \in \mathbb{T}$, $f_*(t) = \overline{f(t)}$.

Knowing c, the problem is still to approximate h. The original approach [9,8] is to construct an autoregressive (AR) approximant, i.e., one of the form $R_m(z) = Kz^m/\tilde{V}_m(z)$ with K a constant and V_m the orthogonal polynomial with respect to c. A relation similar to (4) can be derived, not for the transfer function h but for "half the Fourier series" $\Omega(z) = c_0 + 2\sum_{k>0} c_k z^k$.

We now give a generalization that results in an approximant R_m with a non-constant numerator (ARMA model). Consider a sequence α_k with $|\alpha_k| < 1$ for all k. The classical AR case will pop up as the special situation where all $\alpha_k = 0$. The α_k will turn out to play a multiple role (i) as the zeros of the approximant R_m, (ii) as the reciprocals of the poles of the orthogonal rational functions (which generalize the orthogonal polynomials) and (iii) as the interpolation points for the multipoint version of the approximation (4) to Ω.

Consider the kernel $D(t, z) = \frac{(t+z)}{(t-z)}$ with formal expansion [4, p. 240]

$$D(t, z) = 1 + 2\sum_{k=1}^{\infty} a_k(t)z\pi_{k-1}(z), \quad a_k(t) = \frac{1}{\pi_k(t)}, \quad \pi_k(z) = \prod_{i=1}^{k}(z - \alpha_i). \quad (10)$$

Then, assuming for simplicity of notation but without loss of generality, that $\int_{-\pi}^{\pi}|h(e^{i\theta})|^2 d\theta = 1$, we get, at least formally,

$$\Omega(z) = \int_{-\pi}^{\pi} D(e^{i\theta}, z)|h(e^{i\theta})|^2 d\theta = c_0 + 2\sum_{k=1}^{\infty} c_k z\pi_{k-1}(z),$$

$$c_0 = 1, \quad c_k = \int_{-\pi}^{\pi} a_k(e^{i\theta})|h(e^{i\theta})|^2 d\theta, \quad k = 1, 2, \dots. \quad (11)$$

Observe that if all $\alpha_k = 0$, then $\pi_k(z) = z^k$, and the c_k are the trigonometric moments, in other words the Fourier coefficients of $|h(z)|^2$ for $z \in \mathbb{T}$. Since the definition of Ω does not depend on the choice of the α_k, we can see that $\Omega(z)$ is the same as introduced above. It is an analytic function in $|z| < 1$.

For general prefixed α_k, let $a_k(t)$ be given by (10) and c_k be given by (11). One can define a linear functional c on the space $\mathcal{L} = \text{span}\{1, a_1(t), a_2(t), \dots\}$ via $c(a_k) = c_k$, $k = 0, 1, 2, \dots$. For negative k, we set $a_{-k} = a_{k*}$, so that $c_{-k} = \bar{c}_k$, and using partial fraction expansion, we may even assume that the linear functional c is defined on $\mathcal{L} \cdot \mathcal{L}_* = \mathcal{L} + \mathcal{L}_*$ where $\mathcal{L}_* = \{f : f_* \in \mathcal{L}\}$ by the relation $c(a_k) = c_k$, $k \in \mathbb{Z}$.

The sequence of orthogonal polynomials becomes a sequence of orthogonal rational functions with polynomials as a special case. A sequence of *rational functions* $V_m(z) \in \mathcal{L}_m = \text{span}\{1, a_1(z), \dots, a_m(z)\}$ is called orthogonal with respect to the moment functional c defined on $\mathcal{L} \cdot \mathcal{L}_*$ as outlined above, provided that $V_m \in \mathcal{L}_m \setminus \mathcal{L}_{m-1}$ and that the rational functions V_m satisfy the relations (9) with t^i replaced by the rational basis $a_i(t) = 1/\pi_i(t)$ from (10). It turns out that for $m \geq 1$, the associated functions W_m defined by

$$W_m(z) = c\Big(D(t, z)(V_m(t) - V_m(z))\Big)$$

also belong to $\mathcal{L}_m$ [4, Eq. (4.21)].

Moreover one has interpolation properties of the following type [4, Theorem 6.1.4]

$$\Omega(z) - \frac{W_m(z)}{V_m(z)} = z\pi_m(z)g_+(z), \quad g_+ \text{ analytic in } |z| < 1$$

$$\Omega(z) - \frac{W_{m*}(z)}{V_{m*}(z)} = (z\pi_m(z))_* g_-(z), \quad g_- \text{ analytic in } |z| > 1, \text{ including } \infty.$$

The ratios are well defined because the linear functional is positive definite, which implies that the zeros of V_m are all in $|z| > 1$ and hence, the zeros of V_{m*} are all in $|z| < 1$. In case all $\alpha_k = 0$ (the polynomial or AR case), we match the first m coefficients of the series expansion of Ω in $z = 0$ and in $z = \infty$ respectively. In general, as the above interpolation properties show, the AR interpolation conditions in $z = 0$ are replaced by ARMA interpolation conditions in $z = 0, \alpha_1, \ldots, \alpha_m$ and the AR interpolation conditions in $z = \infty$ are distributed over the points $z = \infty, 1/\overline{\alpha}_1, \ldots, 1/\overline{\alpha}_m$ taking multiplicity into account. This is multipoint moment matching.

To come to the original problem of approximating h itself, one makes use of the determinant formula [4, Theorem 4.2.6]

$$\frac{1}{2}\left(W_m(z)V_{m*}(z) + W_{m*}(z)V_m(z)\right) = \frac{1 - |\alpha_m|^2}{(1/z - \overline{\alpha}_m)(z - \alpha_m)}.$$

Recall that for $z \in \mathbb{T}$, $\Omega(z)$ is the real part of $|h(z)|^2$, in other words $|h(e^{i\omega})|^2 = \frac{1}{2}(\Omega(e^{i\omega}) + \Omega_*(e^{i\omega}))$. It then follows, after dividing the previous relation for $z = e^{i\omega}$ by $V_m(z)V_{m*}(z) = |V_m(z)|^2$, that

$$|h(e^{i\omega})|^2 \approx \left|\frac{K}{(e^{i\omega} - \alpha_m)V_m(e^{i\omega})}\right|^2 = \left|\frac{K\pi_{m-1}(e^{i\omega})}{P_m(e^{i\omega})}\right|^2, \quad V_m(e^{i\omega}) = \frac{P_m(e^{i\omega})}{\pi_m(e^{i\omega})},$$

with $K = \sqrt{1 - |\alpha_m|^2}$. Knowing that h is analytic in $|z| > 1$ if the system is stable, we can approximate it by $K\pi_{m-1}/P_m$ as described above. Note that the α_k which are chosen as the poles of space $\mathcal{L}_m$ of rational functions are interpolation points when approximating Ω and that they now show up as the zeros of the approximating transfer function.

3   Homogeneous Padé-Type Approximants Analytic Outside the Unit Polydisk

To deal with multivariate polynomials and functions we switch between the cartesian and the spherical coordinate system. The cartesian coordinates $X = (x_1, \ldots, x_n) \in \mathbb{C}^n$ are then replaced by $X = (\xi_1 z, \ldots, \xi_n z)$ with $\xi_k \in \mathbb{C}, z \in \mathbb{R}$ where the directional vector $\xi = (\xi_1, \ldots, \xi_n)$ belongs to the unit sphere $S_n = \{\xi : \|\xi\|_p = 1\}$. Here $\|\cdot\|_p$ denotes one of the usual Minkowski norms. While ξ contains the directional information of X, the variable z contains the (possibly signed) distance information. With the multi-index $\kappa = (\kappa_1, \ldots, \kappa_n) \in \mathbb{N}^n$ the notations $X^\kappa, \kappa!$ and $|\kappa|$ respectively denote

$$X^\kappa = x_1^{\kappa_1} \ldots x_n^{\kappa_n},$$
$$\kappa! = \kappa_1! \ldots \kappa_n!,$$
$$|\kappa| = \kappa_1 + \ldots + \kappa_n,$$

and with X, we associate its signed distance

$$\mathrm{sd}(X) = \mathrm{sgn}(x_1) \|X\|_p.$$

Note that it is always possible to choose ξ such that $z = \mathrm{sd}(X)$. For the sequel of the discussion we need some more notation. We denote by $\mathbb{C}[\xi] := \mathbb{C}[\xi_1, \ldots, \xi_n]$ the linear space of n-variate polynomials in ξ_k with complex coefficients, by $\mathbb{C}(\xi) := \mathbb{C}(\xi_1, \ldots, \xi_n)$ the commutative field of rational functions in ξ_k with complex coefficients, by $\mathbb{C}(\xi)[z]$ the linear space of polynomials in the variable z with coefficients from $\mathbb{C}(\xi)$ and by $\mathbb{C}[\xi][z]$ the linear space of polynomials in the variable z with coefficients from $\mathbb{C}[\xi]$.

Let us introduce the linear functional C acting on the signed distance variable z as $C(z^i) = c_i(\xi)$, where $c_i(\xi)$ is a homogeneous expression of degree i in the $\xi_k : c_i(\xi) = \sum_{|\kappa|=i} c_\kappa \xi^\kappa$. Then C is a multivariate moment functional with multidimensional moments c_κ. Multivariate orthogonality with respect to the linear functional C can be defined [6]. The n-variate polynomials under investigation here, are of the form

$$V_m(X) = V_m(z) = \sum_{i=0}^{m} B_{m-i}(\xi) z^i, \qquad B_{m-i}(\xi) = \sum_{|\kappa|=m-i} b_\kappa \xi^\kappa.$$

The function $V_m(X)$ is a polynomial of degree m in z with polynomial coefficients from $\mathbb{C}[\xi]$. The coefficients $B_0(\xi), \ldots, B_m(\xi)$ are homogeneous polynomials in the parameters ξ_k. The function $V_m(X)$ does itself not belong to $\mathbb{C}[X]$, but as $V_m(X)$ can be viewed as $V_m(z)$, it belongs to $\mathbb{C}[\xi][z]$. Therefore the function $V_m(X)$ can be coined a spherical polynomial: for every $\xi \in S_n$ we can identify the function $V_m(X)$ with the polynomial $V_m(z)$ of degree m in the variable $z = \mathrm{sd}(X)$.

With an arbitrarily chosen $V_m(X)$ we can associate the function $W_{m-1}(X)$ defined by

$$W_{m-1}(X) = W_{m-1}(z) = C\left(\frac{V_m(t) - V_m(z)}{t - z}\right).$$

One can show [2] that $W_{m-1}(X)$ is a polynomial of degree $m-1$ in z, but not that it is a polynomial in X. Instead it belongs to $\mathbb{C}[\xi][z]$ and has the form

$$W_{m-1}(X) = W_{m-1}(z) = \sum_{i=0}^{m-1} A_{m-1-i}(\xi) z^i.$$

For the polynomials $\tilde{V}_m(X)$ and $\tilde{W}_{m-1}(X)$ defined by

$$\tilde{V}_m(X) = \tilde{V}_m(z) = z^m V_m(z^{-1}) = \sum_{k=0}^{m} \tilde{B}_k(\xi) z^k = \sum_{k=0}^{m} \left(\sum_{|\kappa|=k} \tilde{b}_\kappa X^\kappa \right),$$

$$\tilde{W}_{m-1}(X) = \tilde{W}_{m-1}(z) = z^{m-1} W_{m-1}(z^{-1}) = \sum_{k=0}^{m-1} \tilde{A}_k(\xi) z^k = \sum_{k=0}^{m-1} \left(\sum_{|\kappa|=k} \tilde{a}_\kappa X^\kappa \right),$$

and belonging to $\mathbb{C}[X]$, it can be proved that [2]

$$\left(f \tilde{V}_m - \tilde{W}_{m-1} \right)(X) = \left(f \tilde{V}_m - \tilde{W}_{m-1} \right)(z) = \sum_{i=m}^{\infty} d_i(\xi) z^i = \sum_{i=m}^{\infty} \left(\sum_{|\kappa|=i} d_\kappa X^\kappa \right).$$

As in (5)–(8), the linear functional C and the rational function $\tilde{W}_{m-1}/\tilde{V}_m(X)$ can be generalized to $C^{(k)}$ and $\tilde{W}_{m+k}/\tilde{V}_m(X)$.

Now let us consider a multidimensional LSI system with IIR [7] and transfer function $H(X) = F(Y)/G(Y)$ where $F(Y)$ and $G(Y)$ are polynomials in the variables $y_i = x_i^{-1}$ and $Y = (y_1, \ldots, y_n)$. In terms of the impulse response c_κ (without loss of generality we restrict ourselves to support on the first quadrant), we have:

$$H(X) = \sum_{|\kappa|=0}^{\infty} c_\kappa Y^\kappa.$$

The system is stable if $G(Y)$ has all its zeroes strictly inside the unit polydisc. A stable identification or model order reduction of $H(X)$ can be given by $\tilde{W}_{m+k}(Y)/\tilde{V}_m(Y)$ provided $\tilde{V}_m(Y)$ has all its zeroes inside the unit polydisc [1].

Let us illustrate the above in the context of IIR filter design ($n = 2$). An ideal lowpass filter can be specified by the frequency response ($x_1 = \exp(it_1), x_2 = \exp(it_2)$)

$$H\left(e^{it_1}, e^{it_2}\right) = \begin{cases} 1 & (t_1, t_2) \in T \subset [-\pi, \pi] \times [-\pi, \pi] \\ 0 & (t_1, t_2) \notin T \end{cases}$$

where T is usually a symmetric domain. For $T = [-\pi/8, \pi/8] \times [-\pi/8, \pi/8]$ we have for instance

$$c_{\kappa_1, \kappa_2} = \frac{\sin\left(\frac{\pi}{8}\kappa_1\right)}{\pi\kappa_1} \frac{\sin\left(\frac{\pi}{8}\kappa_2\right)}{\pi\kappa_2}.$$

Let us in addition impose the quadrant symmetry conditions

$$\frac{\tilde{W}_{m+k}}{\tilde{V}_m}\left(e^{it_1}, e^{it_2}\right) = \frac{\tilde{W}_{m+k}}{\tilde{V}_m}\left(e^{-it_1}, e^{it_2}\right) = \frac{\tilde{W}_{m+k}}{\tilde{V}_m}\left(e^{it_1}, e^{-it_2}\right)$$

$$= \frac{\tilde{W}_{m+k}}{\tilde{V}_m}\left(e^{-it_1}, e^{-it_2}\right).$$

Then with the choice

$$V_2(Y) = 1.94145 z^2 - 1.30911(\xi_1 + \xi_2)z + 0.340194(\xi_1^2 + \xi_2^2) + 0.000033\xi_1\xi_2$$

250   A. Bultheel, A. Cuyt, and B. Verdonk

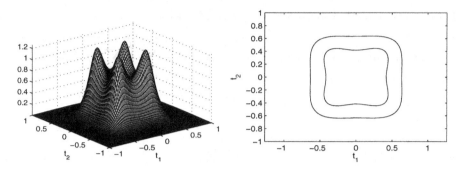

Fig. 1. Frequency response $\tilde{W}_2(e^{it_1}, e^{it_2})/$ **Fig. 2.** Contours of frequency response
$\tilde{V}_2(e^{it_1}, e^{it_2})$

which guarantees a stable filter, we find

$$\frac{\tilde{W}_2(Y)}{\tilde{V}_2(Y)} =$$

$$\frac{303.351 + 91.0655(\xi_1 + \xi_2)z^{-1} + 126.936(\xi_1^2 + \xi_2^2)z^{-2} - 110.584\xi_1\xi_2 z^{-2}}{358.005\,(1.94145 - 1.30911(\xi_1 + \xi_2)z^{-1} + 0.340194(\xi_1^2 + \xi_2^2)z^{-2} + 0.000033\xi_1\xi_2 z^{-2})}$$

The frequency response $\tilde{W}_2(e^{it_1}, e^{it_2})/\tilde{V}_2(e^{it_1}, e^{it_2})$ is shown in Figure 1. The contour lines $|(\tilde{W}_2/\tilde{V}_2)(e^{it_1}, e^{it_2})| = 0.1$ and $|(\tilde{W}_2/\tilde{V}_2)(e^{it_1}, e^{it_2})| = 0.5$ are shown in Figure 2.

References

1. Abouir, J., Cuyt, A.: Stable multi-dimensional model reduction and IIR filter design. Int. J. Computing Science and Mathematics 1(1), 16–27 (2007)
2. Benouahmane, B., Cuyt, A.: Properties of multivariate homogeneous orthogonal polynomials. J. Approx. Theory 113, 1–20 (2001)
3. Brezinski, C.: Padé type approximation and general orthogonal polynomials. In: ISNM 50, Birkhauser Verlag, Basel (1980)
4. Bultheel, A., et al.: Orthogonal rational functions. In: Cambridge Monographs on Applied and Computational Mathematics, vol. 5, Cambridge University Press, Cambridge (1999)
5. Bultheel, A., Van Barel, M.: Padé techniques for model reduction in linear system theory. In: J. Comput. Appl. Math., vol. 14, pp. 401–438 (1986)
6. Cuyt, A., Benouahmane, B., Verdonk, B.: Spherical orthogonal polynomials and symbolic-numeric Gaussian cubature formulas. In: Bubak, M., et al. (eds.) ICCS 2004. LNCS, vol. 3037, pp. 557–560. Springer, Heidelberg (2004)
7. Dudgeon, D., Mersereau, R.: Multidimensional digital signal processing. Prentice Hall, Englewood Cliffs (1984)
8. Grenander, U., Szegő, G.: Toeplitz forms and their applications. University of California Press, Berkley (1958)
9. Wiener, N.: Extrapolation, interpolation and smoothing of time series. Wiley, New York (1949)

DNS and LES of Scalar Transport in a Turbulent Plane Channel Flow at Low Reynolds Number

Jordan A. Denev[1], Jochen Fröhlich[1], Henning Bockhorn[1],
Florian Schwertfirm[2], and Michael Manhart[2]

[1] Institute for Technical Chemistry and Polymer Chemistry, University of Karlsruhe,
Kaiserstraße 12, D-76128 Karlsruhe, Germany
denev@ict.uni-karlsruhe.de
http://www.ict.uni-karlsruhe.de
[2] Department of Hydromechanics, Technical University of Munich,
Arcisstraße 21, 80333 Munich, Germany

Abstract. The paper reports on DNS and LES of plane channel flow at
$Re_\tau = 180$ and compares these to a DNS with a higher order convection
scheme. For LES different subgrid-scale models like the Smagorinsky, the
Dynamic Smagorinsky and the Dynamic Mixed Model were used with
the grid being locally refined in the near-wall region. The mixing of a
passive scalar has been simulated with two convection schemes, central
differencing and $HLPA$. The latter exhibits numerical diffusion and the
results with the central scheme are clearly superior. LES with this scheme
reproduced the budget of the scalar variance equation reasonably well.

1 Introduction

Turbulent mixing of scalar quantities is a phenomenon observed in environmental
flows as well as in abundant engineering applications of chemical, nuclear power,
pharmaceutical or food industries. Their simulation requires reliable models for
turbulent mixing processes. After discretization, however, the physical and the
numerical model interact in a complex way, which is not fully understood so far.
To address these issues, this paper presents results from both, Direct Numerical
Simulations (DNS) and Large Eddy Simulations (LES) of fully developed plane
channel flow at a friction Reynolds number of $Re_\tau = 180$. This is a prototypical
flow frequently used to study physical and numerical modelling of wall-bounded
flows. The first DNS of this configuration was performed by Kim *et al.* [4].

In an earlier paper by the present authors [2] the impact of local grid refinement near the walls on the LES modelling of the flow field was investigated.
In the present paper we extend this approach and focus on the modelling of a
transported scalar.

2 Numerical Methods and Simulation Details

The turbulent channel flow between two parallel plates is simulated for a nominal
friction Reynolds number $Re_\tau = 180$, defined by the friction velocity U_τ and the

I. Lirkov, S. Margenov, and J. Waśniewski (Eds.): LSSC 2007, LNCS 4818, pp. 251–258, 2008.
© Springer-Verlag Berlin Heidelberg 2008

channel half-width h. In the computation, the corresponding bulk Reynolds number $Re_b = 2817$ was imposed by instantaneously adjusting a spatially constant volume force in each time step, so that in fact U_τ and hence Re_τ is a result of the simulation. The computational domain in streamwise, wall-normal and spanwise direction extends over $L_x = 6.4h$, $L_y = 2.0h$ and $L_z = 3.2h$, respectively. Periodic boundary conditions were imposed for the streamwise and the spanwise direction. At the walls, a no-slip condition was applied for both LES and DNS, together with Van Driest damping for the Smagorinsky model. Dirichlet boundary conditions at the walls were imposed for the scalar, i.e. $C(x, 0, z) = 1.0$ and $C(x, 2h, z) = -1.0$.

Two different numerical codes have been applied for the present work. Both utilize a finite volume method for incompressible fluid on block-structured grids together with a Runge-Kutta (RK) scheme in time. A Poisson equation is solved for the pressure-correction. The code $MGLET$ has been developed at the TU Munich and uses staggered Cartesian grids with a 6th order central discretization scheme (CDS) in space and a 3rd order RK scheme [5,8]. The code $LESOCC2$, developed at the University of Karlsruhe, can handle curvilinear collocated grids and employs discretizations of second order in space and time [3]. In the present study, DNS results from this code were first compared to those obtained with the higher-order discretization of $MGLET$. Subsequently, LES were carried out with different subgrid-scale (SGS) models as described, e.g., in [6]. These comprise the Smagorinsky Model (SM) with constant $C_S = 0.1$, the Dynamic Smagorinsky Model (DSM), and the Dynamic Mixed Model (DMM), see Table 1. The first case in Table 1, denoted $DNS - 6O$ (sixth order discretization in space for both convection and diffusion), has been calculated with $MGLET$, all others with $LESOCC2$. The LES filter is not accounted for in the present notation.

An equation governing the transport of a scalar quantity with Schmidt number $Sc = 1$ is also considered in the present study. The concentration C, is regarded as passive, i.e. it does not influence the fluid flow. The unresolved turbulent transport of C is modelled by an eddy diffusivity $D_t = \nu_t/Sc_t$, where ν_t is the SGS eddy viscosity and Sc_t the turbulent Schmidt number, here set equal to 0.6. This also holds for the mixed model.

With the present equations and Dirichlet boundary conditions the concentration fulfils a maximum condition. The extrema are attained on the boundaries, so that the scalar is restricted to the interval $C \in [-1; 1]$ for physical reasons (the lower bound -1 was chosen here instead of 0 for technical reasons). This boundedness does not neccessarily carry over to the discretized solution. In numerous studies bounded convection schemes are therefore applied to guarantee the boundedness of the numerical solution which is not guaranteed with a central scheme. In [3] and related work the $HLPA$ scheme developed in [9] was used for this purpose. $HLPA$ determines the convective flux by a blending between second-order upwinding and first-order upwinding as

$$F_{i+\frac{1}{2}} = \begin{cases} UC_i + U(C_{i+1} - C_i)\Theta_{i+\frac{1}{2}} & , 0 < \Theta_{i+\frac{1}{2}} < 1 \\ UC_i & , \text{else} \end{cases} , \quad \Theta_{i+\frac{1}{2}} = \frac{C_i - C_{i-1}}{C_{i+1} - C_{i-1}}, \quad (1)$$

Table 1. Overview over the runs discussed. The nomenclaure is defined in the text.

Case	CV_{tot}	Δ_x^+	y_1^+	Δ_z^+	SGS	SCS	t_{av}	U_τ	C_τ
DNS									
DNS-6O	1,407,120	9.1	0.68	7.2	-	CDS-6O	544	0.064018	-0.041478
DNS	1,407,120	9.1	0.68	7.2	-	HLPA	638	0.062237	-0.039232
DNS-F	10,866,960	4.5	0.34	3.6	-	HLPA	537	0.061821	-0.041106
DNS-CDS	1,407,120	9.1	0.68	7.2	-	CDS	745	0.062487	-0.042083
LES									
HLPA-SM	258,688	29.8	0.37	14.9	SM	HLPA	615	0.067434	-0.038431
CDS-SM	258,688	29.8	0.37	14.9	SM	CDS	643	0.066032	-0.047632
CDS-DSM	258,688	29.8	0.37	14.9	DSM	CDS	650	0.060801	-0.042553
CDS-DMM	258,688	29.8	0.37	14.9	DMM	CDS	646	0.070095	-0.043067

where U is the velocity at the cell boundary $i + 1/2$. This scheme was employed for the present study for the convection term of the concentration equation in cases DNS and $DNS - F$, while still using second order CDS in the momentum equation (see the column for the scalar convection scheme (SCS) in Table 1).

A special feature of the code $LESOOC2$ is the possibility of block-wise local grid refinement (LGR). This allows to use a fine grid close to the walls without excessively refining in the center of the channel, so that CPU time and storage are not overly increased. LGR is utilized near the walls up to a distance $y_{ref} = h/8$ equal to $y_{ref}^+ = 22.5$ from the wall with a refinement ratio of 2 in both, $x-$ and $z-$direction. In $y-$direction the grid is stretched uniformly by a factor 1.03 throughout the channel. As observed in a previous study [2] the turbulent characteristics of the flow exhibit some visible changes at the block-interface, and this issue will also be addressed later in this paper. Table 1 presents information on the numerical grids, i.e. the total number of control volumes of the entire grid, CV_{tot}, and the dimensionless size of the control volumes in $x-$ and $z-$direction, respectively. In case of LGR, which is used with all LES cases, this is the unrefined spacing used in the core of the flow. Furthermore, y_1^+ indicates the distance of the wall-adjacent point from the wall.

3   Results from DNS

Statistical data for all computations in the present work have been collected over averaging times t_{av} larger than 540 dimensionless time units $t_b = h/U_b$, where U_b is the bulk velocity of the flow. Table 1 shows the results obtained for the friction velocity and the reference concentration defined as

$$U_\tau = \sqrt{\frac{\tau_w}{\rho}} \qquad , \qquad C_\tau = \frac{D}{U_\tau} \left(\frac{\partial \langle C \rangle}{\partial y} \right)_{y=0} , \tag{2}$$

respectively, with D being the laminar diffusion coefficient (the turbulent diffusion coefficient vanishes at the wall). In the present section four DNS cases

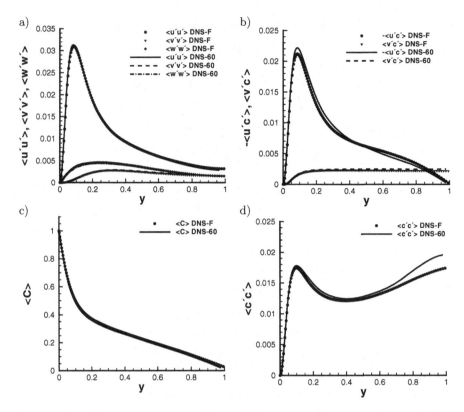

Fig. 1. Comparison of the results from cases $DNS - 6O$ and $DNS - F$: a) normal turbulent stresses $\langle u'u' \rangle$, $\langle v'v' \rangle$ and $\langle w'w' \rangle$; b) turbulent scalar fluxes $\langle u'c' \rangle$ and $\langle v'c' \rangle$; c) mean scalar $\langle C \rangle$; d) scalar variance $\langle c'c' \rangle$

are compared, which allows to identify the role of the numerical discretization scheme and the grid resolution. Case $DNS - 6O$ is chosen as a reference case. In [7] these data were compared with the classical ones of [4] showing excellent agreement. The run DNS was performed with $LESOCC2$ on the collocated equivalent of this grid with the second order method. Due to the lower order these results (not reproduced here) were unsatisfactory showing deviations of up to 18% from the reference data. Therefore, the grid was refined by a factor of 2 in each direction (case $DNS - F$) to compensate for the lower-order discretization. The comparison with $DNS - 6O$ is presented in Figure 1. The turbulent stresses and the time-averaged scalar match very well. The turbulent scalar flux $\langle u'c' \rangle$ and the scalar variance $\langle c'c' \rangle$ exhibit differences, which for the latter mainly appear in the middle of the channel.

In order to further elucidate the role of the numerical scheme, case DNS has been repeated employing the CDS scheme of second order instead of the $HLPA$ scheme. These results (not depicted here for lack of space) show a clear improvement for the scalar variance $\langle c'c' \rangle$ compared to case DNS, and also when compared to case $DNS - F$, as the difference with respect to the reference data in the middle of the channel decreases. The scalar flux $\langle v'c' \rangle$ for $DNS - CDS$ matches perfectly well with $DNS - 6O$ and the agreement of $\langle u'c' \rangle$ is practically as good as for $DNS - F$ in Figure 1. It should also be noticed that the value for the reference scalar C_τ for this case is closer to the value of $DNS - 6O$ than that of DNS. This considerable improvement from case DNS to $DNS - CDS$ shows that for the flow considered the central differencing scheme appears clearly superior compared to the $HLPA$ scheme.

4   Results from LES

The results of the previous section were obtained with DNS, i.e. on fine grids and without any turbulence model. Now we turn to LES for which numerical and modelling errors interact in a complex way. The grid used for these LES is much coarser in the core region of the flow (see Δx^+ and Δz^+ in Table 1), while in the vicinity of the wall ($y < 1/8h$) it is of similar cell size as in the DNS cases (195,000 control volumes in the region of refinement).

The results obtained with the CDS for the convective terms of the scalar transport equation confirm the findings of a previous paper by the authors [2], in which a higher Reynolds number was considered, and where the Smagorinsky model performed better than the other two models. In the present investigation, $CDS - DMM$ shows slightly better results than $CDS - DSM$. This assertion is mainly based on the behaviour of the averaged scalar and the scalar variance near the wall. $CDS - DSM$ on the other hand shows the most accurate LES value for C_τ.

To address the impact of the convection scheme, results for the cases $CDS - SM$ and $HLPA - SM$ are presented together with the reference case $DNS - 6O$ in Figure 2. These results again show the superiority of the CDS, which is more pronounced in the proximity of the wall. The results also demonstrate the diffusive characteristics of the $HLPA$ scheme. The presence of additional diffusion is noticed in the averaged scalar distribution by an increased value and an almost linear distribution for the region away from the wall. Furthermore, the turbulence quantities such as the scalar fluxes and the scalar variance are underestimated near the wall, i.e. damped by the numerical diffusion.

5   Transport Equation of the Scalar Variance

Finally, an evaluation of the terms in the budget of the scalar variance was carried out. In the case of the Smagorinsky model the equation for the resolved scalar variance reads

$$0 = \underbrace{-2\langle c'v' \rangle \frac{\partial \langle C \rangle}{\partial y}}_{P_C} \underbrace{-2\left\langle (D+D_t)\left(\frac{\partial c'}{\partial x_k}\right)^2 \right\rangle}_{E_C} + \underbrace{\frac{\partial}{\partial y}\left((D+D_t)\frac{\partial \langle c' \rangle^2}{\partial y}\right)}_{D_C} \underbrace{-\frac{\partial}{\partial y}(\langle c'^2 v' \rangle)}_{T_C}$$

Here, Pc denotes the production by the mean concentration gradients, Ec the scalar dissipation, Dc the diffusion transport term (comprises molecular and,

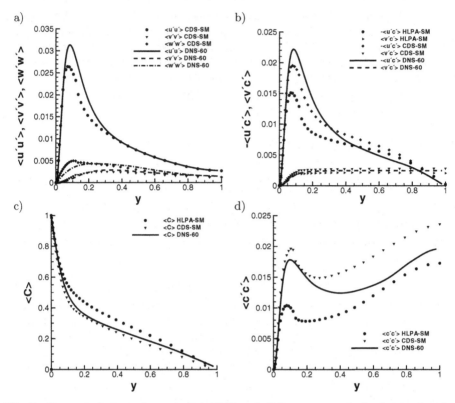

Fig. 2. Computations performed with LES and different numerical schemes for the scalar: cases $CDS-SM$ and $HLPA-SM$ compared with the reference case $DNS-6O$. a),b),c) and d) as in the previous figure.

in the case of LES, eddy-diffusion) and Tc turbulent transport by the normal velocity fluctuation. In the case of dynamic LES-models, additional terms appear in the balance due to the fluctuation of the model parameter. These are however negligible compared to the other terms of the equation [1]. In the case of DNS the tubulent diffusivity is omitted.

Figure 3 shows the comparison of the different terms, constituting the budget of the scalar variance for $DNS - 6O$, $DNS - CDS$ and $CDS - SM$. While for the first two cases the match is very good, the case $CDS - SM$ shows some differences. As expected, the magnitude of the terms in the middle of the channel is underestimated, which is due to the fact that only part of the turbulent spectrum is resolved with LES. The values at the wall on the other hand are quite accurately reproduced for all terms of the above equation due to the fine grid near the walls. For $CDS - SM$ the Figure 3b shows some artefacts at the block boundary separating the refined and the coarse grid. They are present only in those terms containing a derivative of a correlation term normal to the wall, i.e. in Ec, Dc and Tc. The reason is that the abrupt changes in the subgrid-filter size and the numerical resolution cause inevitable irregularities in the turbulent

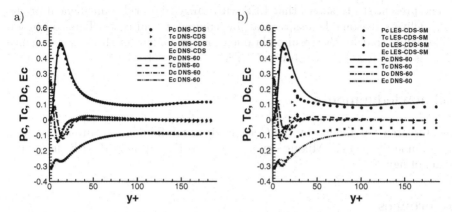

Fig. 3. Terms in the budget of the scalar variance $\langle c'c' \rangle$. All terms are normalized by $D/(C_\tau^2 U_\tau^2)$ and explained in the text. a) Case $DNS - CDS$ (symbols) compared to reference case $DNS - 6O$ (lines). b) LES with $CDS - SM$ (symbols) compared to $DNS - 6O$ (lines).

quantities at the two sides of the block boundary which modify the derivative operator applied normal to the wall. Apart from this, the agreement between the $CDS - SM$ and the DNS cases is reasonably good. It can hence be concluded that in the present case LES (case $CDS - SM$) is capable of qualitatively, and to some extent also quantitatively, reproducing the terms in the budget for the scalar variance.

6 Conclusions

Different numerical and modeling issues have been studied when calculating fluid flow and passive scalar distribution in a plane turbulent channel flow. DNS with CDS of sixth and second order accuracy have been compared. It has been shown, that the second-odrer scheme achieved the desired accuracy (shown by the sixth-order CDS) only after the numerical grid has been refined twice in each spatial direction.

Comparison of two schemes for the scalar, unbounded CDS and non-linear, monotonous upstream-weighted $HLPA$ showed superiority of the CDS scheme, while the results with $HLPA$ were found to suffer from numerical diffusion. This is in line with the general attitude when modelling the SGS terms in the LES-momentum equation. Usually, a non-dissipative scheme is preferred and dissipation entirely introduced by the laminar viscous terms and the SGS model. Additional numerical dissipation without modifying the SGS model is avoided. The same is observed here for the scalar transport. It should however not be concluded that the CDS is best for any LES involving a passive scalar. In other simulations of the present authors concerned with a jet in crossflow this scheme led to numerical instability. More appropriate schemes to maintain boundedness of the scalar are needed.

The present study shows that LES with tangential grid refinement near the walls delivers reasonable accuracy at low computational costs. This conclusion is also supported by the results obtained for the budget of the scalar variance which is reproduced reasonably well with the present LES.

Acknowledgements

This research was funded by the German Research foundation through priority programme SPP-1141 "Mixing devices". Carlos Falconi helped with the preparation of figures.

References

1. Calmet, I., Magnaudet, J.: Large-eddy simulation of high-schmidt number mass transfer in a turbulent channel flow. Phys. Fluids 9(2), 438–455 (1997)
2. Fröhlich, J., et al.: On the impact of tangential grid refinement on subgrid-scale modeling in large eddy simulation. In: Boyanov, T., et al. (eds.) NMA 2006. LNCS, vol. 4310, pp. 550–557. Springer, Heidelberg (2007)
3. Hinterberger, C.: Dreidimensionale und tiefengemittelte Large–Eddy–Simulation von Flachwasserströmungen. PhD thesis, Institute for Hydromechanics, University of Karlsruhe (2004)
4. Kim, J., Moin, P., Moser, R.: Turbulence statistics in fully developed channel flow at low Reynolds number. J. Fluid Mech. 177, 133–166 (1987)
5. Manhart, M.: A zonal grid algorithm for DNS of turbulent boundary layers. Computers and Fluids 33(3), 435–461 (2004)
6. Sagaut, P.: Large Eddy Simulation for Incompressible Flows: An Introduction, 2nd edn. Springer, Berlin (2002)
7. Schwertfirm, F., Manhart, M.: ADM Modelling for Semi-Direct Numerical Simulation of Turbulent Mixing and Mass Transport. In: Humphrey, J.A.C., et al. (eds.) Fourth International Symposium. On Turbulence and Shear Flow Phenomena, vol. 2, pp. 823–828. Williamsburg, Virginia (2005)
8. Schwertfirm, F., Manhart, M.: DNS of passive scalar transport in turbulent channel flow at high Schmidt numbers. In: Hanjalic, K., Nagano, Y., Jakrilic, S. (eds.) Turbulence, Heat and Mass Transfer 5, Dubrovnik, Coratia, pp. 289–292 (2006)
9. Zhu, J.: A low-diffusive and oscillation-free convection scheme. Communications in applied numerical methods 7, 225–232 (1991)

Adaptive Path Following Primal Dual Interior Point Methods for Shape Optimization of Linear and Nonlinear Stokes Flow Problems

Ronald H.W. Hoppe[1,2], Christopher Linsenmann[1,2], and Harbir Antil[1]

[1] University of Houston, Department of Mathematics
http://www.math.uh.edu/~rohop/
[2] University of Augsburg, Institute for Mathematics
http://scicomp.math.uni-augsburg.de

Abstract. We are concerned with structural optimization problems in CFD where the state variables are supposed to satisfy a linear or nonlinear Stokes system and the design variables are subject to bilateral pointwise constraints. Within a primal-dual setting, we suggest an all-at-once approach based on interior-point methods. The discretization is taken care of by Taylor-Hood elements with respect to a simplicial triangulation of the computational domain. The efficient numerical solution of the discretized problem relies on adaptive path-following techniques featuring a predictor-corrector scheme with inexact Newton solves of the KKT system by means of an iterative null-space approach. The performance of the suggested method is documented by several illustrative numerical examples.

1 Introduction

Simplified problems in shape optimization have already been addressed by Bernoulli, Euler, Lagrange and Saint-Venant. However, it became its own discipline during the second half of the last century when the rapidly growing performance of computing platforms and the simultaneously achieved significant improvement of algorithmic tools enabled the appropriate treatment of complex problems (cf. [1,3,6,9,13,14,15] and the references therein). The design criteria in shape optimization are determined by a goal oriented operational behavior of the devices and systems under consideration and typically occur as nonlinear, often non convex, objective functionals which depend on the state variables describing the operational mode and the design variables determining the shape. The state variables often satisfy partial differential equations or systems thereof representing the underlying physical laws. Technological aspects are taken into account by constraints on the state and/or design variables which may occur both as equality and inequality constraints in the model.

Shape optimization problems associated with fluid flow problems play an important role in a wide variety of engineering applications [13]. A typical setting is the design of the geometry of the container of the fluid, e.g., a channel, a

I. Lirkov, S. Margenov, and J. Waśniewski (Eds.): LSSC 2007, LNCS 4818, pp. 259–266, 2008.
© Springer-Verlag Berlin Heidelberg 2008

reservoir, or a network of channels and reservoirs such that a desired flow velocity and/or pressure profile is achieved. The solution of the problem amounts to
the minimization of an objective functional that depends on the state variables
(velocity, pressure) and on the design variables which determine the geometry
of the fluid filled domain. The state variables are supposed to satisfy the underlying fluid mechanical equations, and there are typically constraints on the
design variables which restrict the shape of the fluid filled domain to that what
is technologically feasible.

The typical approach to shape optimization problems relies on a separate
treatment of the design issue and the underlying state equation what is called
alternate approximation in [1]: For a given initial design the state equation is
solved, followed by a sensitivity analysis that leads to an update of the design
variables. This process is iteratively repeated until convergence. Moreover, many
methods, e.g., those based on the concept of shape derivatives [6,15], only use
first order information by employing gradient type techniques. In this paper, we
focus on a so-called *all-at-once approach* where the numerical solution of the
discretized state equation is an integral part of the optimization routine (cf., e.g.,
[4,5,10,12]). Moreover, we use second order information by means of primal-dual
interior-point methods. In particular, we consider an adaptive path-following
technique for the shape optimization of stationary flow problems as described
by a linear or nonlinear Stokes system in channels where the objective is to
design the lateral walls such that a desired velocity and/or pressure profile is
obtained. The design variables are chosen as the control points of a Bézier curve
representation of the lateral walls.

The paper is organized as follows: Section 2 is devoted to the setup of the shape
optimization problem including its finite element discretization by Taylor-Hood
elements. In section 3, we focus on the primal-dual interior-point approach and
a path-following predictor-corrector type continuation method with an adaptive
choice of the continuation parameter. Finally, in section 4 we illustrate the application of the algorithm for the design of a channel with a backward facing step
assuming a linear Stokes regime and for the shape optimization of the inlet and
outlet boundaries of the ducts of an electrorheological shock absorber, where the
states satisfy a nonlinear Stokes equation.

2   Shape Optimization of Stationary Stokes Flow

We consider Stokes flow in a bounded domain $\Omega(\alpha) \subset \mathbb{R}^2$ with boundary
$\Gamma(\alpha) = \Gamma_{in}(\alpha) + \Gamma_{out}(\alpha) + \Gamma_{lat}(\alpha)$ consisting of the inflow, the outflow and
the lateral boundaries with $\mathbf{n}$ and $\mathbf{t}$ denoting the outward unit normal and unit
tangential vector, respectively. Here, $\alpha = (\alpha_1, \cdots, \alpha_m)^T \in \mathbb{R}^m$ is the vector
of design variables which are chosen as the Bézier control points of a Bézier
curve representation of $\Gamma(\alpha)$ and which are subject to upper and lower bounds
$\alpha_i^{min}, \alpha_i^{max}, 1 \leq i \leq m$. The state variables are the velocity $\mathbf{u}$ and the pressure
p. Given desired velocity and pressure profiles $\mathbf{u^d}$ and p^d, an inflow u_{in} at the

inflow boundary $\Gamma_{in}(\alpha)$ and weighting factors $\kappa_i \geq 0, 1 \leq i \leq 2, \kappa_1 + \kappa_2 > 0$, the shape optimization problem can be stated as follows:

$$\text{minimize} \quad J(\mathbf{u}, p, \alpha) = \frac{\kappa_1}{2} \int_{\Omega(\alpha)} |\mathbf{u} - \mathbf{u^d}|^2 dx + \frac{\kappa_2}{2} \int_{\Omega(\alpha)} |p - p^d|^2 dx, \quad (1a)$$

$$\text{subject to} \quad -\nabla \cdot \boldsymbol{\sigma}(\mathbf{u}) = 0 \quad \text{in } \Omega(\alpha), \quad (1b)$$

$$\nabla \cdot \mathbf{u} = 0 \quad \text{in } \Omega(\alpha),$$

$$\boldsymbol{\sigma}(\mathbf{u}) = -p\mathbf{I} + g(\mathbf{u}, \mathbf{D}(\mathbf{u}))\mathbf{D}(\mathbf{u}), \quad (1c)$$

$$\mathbf{n} \cdot \mathbf{u} = u_{in} \quad \text{on } \Gamma_{in}(\alpha),$$

$$\mathbf{n} \cdot \mathbf{u} = 0 \quad \text{on } \Gamma_{lat}(\alpha),$$

$$\mathbf{t} \cdot \mathbf{u} = 0 \quad \text{on } \Gamma(\alpha),$$

$$\alpha_i^{min} \leq \alpha_i \leq \alpha_i^{max}, \ 1 \leq i \leq m. \quad (1d)$$

We note that in the constitutive equation (1c) the tensor $\mathbf{D}(\mathbf{u})$ stands for the rate of deformation tensor $\mathbf{D}(\mathbf{u}) := (\nabla\mathbf{u} + (\nabla\mathbf{u})^T)/2$ and $g(\mathbf{u}, \mathbf{D}(\mathbf{u}))$ denotes the viscosity function which is given by $g(\mathbf{u}, \mathbf{D}(\mathbf{u})) = \nu$ for linear Stokes flow and depends nonlinearly on $\mathbf{u}, \mathbf{D}(\mathbf{u})$ in the nonlinear regime.

We choose $\hat{\alpha} \in K$ as a reference design and refer to $\hat{\Omega} := \Omega(\hat{\alpha})$ as the associated reference domain. Then, the actual domain $\Omega(\alpha)$ can be obtained from the reference domain $\hat{\Omega}$ by means of an isomorphism

$$\Omega(\alpha) = \Phi(\hat{\Omega}; \alpha), \quad (2)$$

$$\Phi(\hat{x}; \alpha) = (\Phi_1(\hat{x}; \alpha), \Phi_2(\hat{x}; \alpha))^T, \ \hat{x} = (\hat{x}_1, \hat{x}_2)^T$$

with continuous components $\Phi_i, 1 \leq i \leq 2$. Due to the reference domain, finite element approximations of (1) can be performed with respect to $\hat{\Omega}$ without being forced to remesh any time the design parameters are changed.

We introduce $(\mathcal{T}_h(\hat{\Omega}))_\mathbb{N}$ as a shape regular family of simplicial triangulations of $\hat{\Omega}$. In view of (2), these triangulations induce an associated family $(\mathcal{T}_h(\Omega(\alpha)))_\mathbb{N}$ of simplicial triangulations of the actual physical domains $\Omega(\alpha)$. For the discretization of the velocity $\mathbf{u}$ and the pressure p we use Taylor-Hood P2/P1 elements. We refer to $\mathbf{u}_h^d \in \mathbb{R}^{n_1}$ and $p_h^d \in \mathbb{R}^{n_2}$ as the vectors representing the L^2-projections of $\mathbf{u^d}, p$ onto the respective finite element spaces giving rise to the discrete objective functional

$$J_h(\mathbf{y}_h, \alpha) := \frac{\kappa_1}{2} (\mathbf{u}_h - \mathbf{u}_h^d)^T I_{1,h}(\alpha)(\mathbf{u}_h - \mathbf{u}_h^d) + \frac{\kappa_2}{2} p_h^T I_{2,h}(\alpha)p_h,$$

where $\mathbf{y}_h := (\mathbf{u}_h, p_h)^T$ and $I_{i,h}(\alpha), 1 \leq i \leq 2$, are the associated mass matrices. Further, denoting by

$$S_h(\mathbf{y}_h, \alpha) := \begin{pmatrix} A_h(\mathbf{u}_h, \alpha) & B_h^T(\alpha) \\ B_h(\alpha) & 0 \end{pmatrix} \begin{pmatrix} \mathbf{u}_h \\ p_h \end{pmatrix} = \begin{pmatrix} \mathbf{g}_{1,h} \\ \mathbf{g}_{2,h} \end{pmatrix} =: \mathbf{g}_h, \quad (3)$$

the Taylor-Hood approximation of the Stokes system (1a), the discretized shape optimization problem can be stated as

$$\text{minimize} \quad J_h(\mathbf{y}_h, \alpha), \tag{4a}$$

$$\text{subject to} \quad S_h(\mathbf{y}_h, \alpha) = \mathbf{g}_h, \tag{4b}$$

$$\alpha_i^{min} \le \alpha_i \le \alpha_i^{max} , \ 1 \le i \le m. \tag{4c}$$

For notational convenience, in the sequel we will drop the discretization subindex h.

3   Path-Following Primal-Dual Interior-Point Method

We use a primal-dual interior-point method where the inequality constraints (4c) are coupled by logarithmic barrier functions with a barrier parameter $\beta = 1/\mu > 0$, $\mu \to \infty$, resulting in the following parameterized family of minimization subproblems

$$\inf_{\mathbf{y}, \alpha} B(\mathbf{y}, \alpha, \mu) := J(\mathbf{y}, \alpha) - \frac{1}{\mu} \sum_{i=1}^{m} [\ln(\alpha_i - \alpha_i^{\min}) + \ln(\alpha_i^{\max} - \alpha_i)] \tag{5}$$

subject to (4b). Coupling (4b) by a Lagrange multiplier $\boldsymbol{\lambda} = (\boldsymbol{\lambda_u}, \lambda_p)^T$, we are led to the saddle point problem

$$\inf_{\mathbf{y}, \alpha} \sup_{\boldsymbol{\lambda}} L^{(\mu)}(\mathbf{y}, \boldsymbol{\lambda}, \alpha) = B^{(\mu)}(\mathbf{y}, \alpha) + \langle S(\mathbf{y}, \alpha) - \mathbf{g}, \boldsymbol{\lambda} \rangle. \tag{6}$$

The central path $\mu \longmapsto x(\mu) := (\mathbf{y}(\mu), \boldsymbol{\lambda}(\mu), \alpha(\mu))^T$ is given as the solution of the nonlinear system

$$F(x(\mu), \mu) = \begin{pmatrix} L_{\mathbf{y}}^{(\mu)}(\mathbf{y}, \boldsymbol{\lambda}, \alpha) \\ L_{\boldsymbol{\lambda}}^{(\mu)}(\mathbf{y}, \boldsymbol{\lambda}, \alpha) \\ L_{\alpha}^{(\mu)}(\mathbf{y}, \boldsymbol{\lambda}, \alpha) \end{pmatrix} = 0 , \tag{7}$$

which represents the first order necessary optimality conditions for (5).
For the solution of (7) we use an adaptive path-following predictor-corrector strategy following strategies developed in [7].

Predictor Step: The predictor step relies on tangent continuation along the trajectory of the Davidenko equation

$$F_{\mathbf{x}}(\mathbf{x}(\mu), \mu) \, \mathbf{x}'(\mu) = -F_\mu(\mathbf{x}(\mu), \mu) . \tag{8}$$

Given some approximation $\tilde{\mathbf{x}}(\mu_k)$ at $\mu_k > 0$, compute $\tilde{\mathbf{x}}^{(0)}(\mu_{k+1})$, where $\mu_{k+1} = \mu_k + \Delta\mu_k^{(0)}$, according to

$$F_x(\tilde{\mathbf{x}}(\mu_k), \mu_k) \, \delta\mathbf{x}(\mu_k) = - F_\mu(\tilde{\mathbf{x}}(\mu_k), \mu_k) , \tag{9a}$$

$$\tilde{\mathbf{x}}^{(0)}(\mu_{k+1}) = \tilde{\mathbf{x}}(\mu_k) + \Delta\mu_k^{(0)} \, \delta\mathbf{x}(\mu_k) . \tag{9b}$$

We use $\Delta\mu_0^{(0)} = \Delta\mu_0$ for some given initial step size $\Delta\mu_0$, whereas for $k \geq 1$ the predicted step size $\Delta\mu_k^{(0)}$ is chosen by

$$\Delta\mu_k^{(0)} := \left(\frac{\|\Delta\mathbf{x}^{(0)}(\mu_k)\|}{\|\tilde{\mathbf{x}}(\mu_k) - \tilde{\mathbf{x}}^{(0)}(\mu_k)\|} \frac{\sqrt{2}-1}{2\Theta(\mu_k)} \right)^{1/2} \Delta\mu_{k-1} , \tag{10}$$

where $\Delta\mu_{k-1}$ is the computed continuation step size, $\Delta\mathbf{x}^{(0)}(\mu_k)$ is the first Newton correction (see below), and $\Theta(\mu_k) < 1$ is the contraction factor associated with a successful previous continuation step.

Corrector step. As a corrector, we use Newton's method applied to

$$F(\mathbf{x}(\mu_{k+1}), \mu_{k+1}) = 0$$

with $\tilde{\mathbf{x}}^{(0)}(\mu_{k+1})$ from (9) as a start vector. In particular, for $\ell \geq 0$ and $j_\ell \geq 0$ we compute $\Delta x^{(j_\ell)}(\mu_{k+1})$ according to

$$F'(\tilde{x}^{(j_\ell)}(\mu_{k+1}), \mu_{k+1}) \, \Delta x^{(j_\ell)}(\mu_{k+1}) = - F(\tilde{x}^{(j_\ell)}(\mu_{k+1}), \mu_{k+1})$$

and $\overline{\Delta x}^{(j_\ell)}(\mu_{k+1})$ as the associated simplified Newton correction

$$F'(\tilde{x}^{(j_\ell)}(\mu_{k+1}), \mu_{k+1}) \, \overline{\Delta x}^{(j_\ell)}(\mu_{k+1}) = - F(\tilde{x}^{(j_\ell)}(\mu_{k+1}) + \Delta x^{(j_\ell)}(\mu_{k+1}), \mu_{k+1}) .$$

We monitor convergence of Newton's method by means of

$$\Theta^{(j_\ell)}(\mu_{k+1}) := \|\overline{\Delta x}^{(j_\ell)}(\mu_{k+1})\| / \|\Delta x^{(j_\ell)}(\mu_{k+1})\| .$$

In case of successful convergence, we accept the current step size and proceed with the next continuation step. However, if the monotonicity test

$$\Theta^{(j_\ell)}(\mu_{k+1}) < 1 \tag{11}$$

fails for some $j_\ell \geq 0$, the continuation step has to be repeated with the reduced step size

$$\Delta\mu_k^{(\ell+1)} := \left(\frac{\sqrt{2}-1}{g(\Theta^{(j_\ell)})} \right)^{1/2} \Delta\mu_k^{(\ell)} , \quad g(\Theta) := \sqrt{\Theta+1} - 1 \tag{12}$$

until we either achieve convergence or for some prespecified lower bound $\Delta\mu_{min}$ observe

$$\Delta\mu_k^{(\ell+1)} < \Delta\mu_{min} .$$

In the latter case, we stop the algorithm and report convergence failure.

The Newton steps are realized by an inexact Newton method featuring right-transforming iterations (cf., e.g., [10,12]). The derivatives occurring in the KKT conditions and the Hessians are computed by automatic differentiation [8].

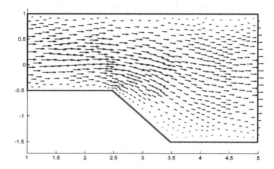

Fig. 1. Backward facing step with final shape and computed velocity field

Table 1. Backward facing step: Convergence history of the continuation method

k	μ	$\Delta\mu$	corr.	ℓ_k	$\|\Delta\alpha\|_2$	$\|\alpha - \alpha^*\|_\infty$	J	Θ
0	100.0	(300.0)	–	–	–	9.0e-01	2.6e+00	–
1	100.0	(300.0)	yes	1	1.2e+00	1.7e-01	9.6e-01	0.58
				2	8.8e-01	1.7e-01	1.3e-01	618.42
1	100.0	425.5	no	1	2.1e-01	3.3e-02	4.3e-04	0.11
2	525.5	417.1	no	1	1.2e-01	3.3e-02	2.3e-03	0.41
				2	3.3e-02	2.5e-02	2.3e-03	0.58
				3	1.6e-02	2.4e-02	2.0e-03	0.92
				4	2.0e-02	2.9e-03	1.6e-05	0.43
				5	5.7e-04	3.2e-03	2.5e-05	–
3	942.6	323.5	no	1	2.9e-03	3.6e-03	5.1e-05	0.34
4	1266.1	283.7	no	1	1.4e-03	3.5e-03	4.9e-05	0.27
5	1549.8	593.1	no	1	1.7e-04	2.9e-03	3.3e-05	0.05
6	2142.9	2265.3	no	1	1.3e-04	2.1e-03	1.7e-05	0.01
7	4408.2	–	–	–	–	2.0e-04	1.9e-07	–

4   Numerical Simulation Results

As a first example, we consider linear Stokes flow with viscosity $\nu = 1$ and given inflow u_{in} in a channel with a backward facing step. The initial shape corresponds to a 90° step, whereas the desired velocity profile $\mathbf{u^d}$ has been chosen according to the final shape as shown in Fig. 1. We have used a total of six Bézier control points with given lower and upper bounds.

Table 1 contains the convergence history. Here, k counts the continuation steps, μ and $\Delta\mu$ stand for the values of the continuation parameter and continuation steplength, respectively. The following column 'corr.' indicates whether a correction was necessary, ℓ_k counts the inner iterations, $\|\Delta\alpha\|_2$ refers to the ℓ_2-norm of the increments in the design variables, $\|\alpha - \alpha^*\|_\infty$ stands for the maximal distance to the optimal design, and J denotes the value of the objective

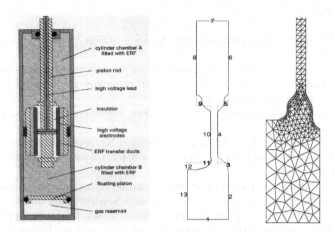

Fig. 2. Electrorheological shock absorber (left), Bézier curve representation of the inlet and outlet boundary of the right part of the fluid chamber (middle), and optimal design of the outlet boundary in rebound mode (right)

functional. Finally, Θ is the quantity used in the monotonicity test to check contractivity. For further details we refer to [2].

As an example for a shape optimization problem associated with nonlinear Stokes flow, we consider the optimization of the inlet and outlet boundaries of the ducts in an electrorheological shock absorber (cf. Fig. 2 (left). Such shock absorbers are based on an electrorheological fluids (ERF). An ERF is a suspension of small electrically polarizable particles dissolved in nonconducting liquids which under the influence of an outer electric field changes its viscosity within a few milliseconds in a reversible way. The viscosity function in (1c) is of the form $g(I(\mathbf{u}), |\mathbf{E}|)$, where $I(\mathbf{u})$ is the second invariant of the rate of strain tensor and $|\mathbf{E}|$ stands for the electric field strength (for details see [11]). The issue is to avoid too large pressure fluctuations at the boundaries of the duct both in the compression and the rebound mode. We have chosen a desired pressure profile p^d and used Bézier curve representations of the inlet and outlet boundaries as illustrated in Fig. 2 (middle). A computed optimal shape of the outlet boundary in the rebound mode is shown in Fig. 2 (right). For details we refer to [10].

Acknowledgements

The first two authors acknowledge support by the NSF under Grant-No. DMS-0511611 and by the German National Science Foundation within the Priority program SPP1253 'Optimization with Partial Differential Equations'. The work of the third author is supported by the Texas Computational and Learning Center TLC2.

References

1. Allaire, G.: Shape Optimization by the Homogenization Method. Springer, Heidelberg (2002)
2. Antil, H., Hoppe, R.H.W., Linsenmann, C.: Path-following primal-dual interior-point methods for shape optimization of stationary flow problems. Journal of Numerical Mathematics (to appear 2007)
3. Bendsøe, M.P.: Optimization of Structural Topology, Shape, and Material. Springer, Berlin (1995)
4. Biros, G., Ghattas, O.: Parallel Lagrange-Newton-Krylov-Schur methods for PDE-constrained optimization. Part i: The Krylov-Schur solver. SIAM J. Sci. Comp (to appear 2004)
5. Biros, G., Ghattas, O.: Parallel Lagrange-Newton-Krylov-Schur methods for PDE-constrained optimization. Part ii: The Lagrange-Newton solver and its application to optimal control of staedy viscous flows. SIAM J. Sci. Comp (to appear 2004)
6. Delfour, M.C., Zolesio, J.P.: Shapes and Geometries: Analysis, Differential Calculus and Optimization. SIAM, Philadelphia (2001)
7. Deuflhard, P.: Newton Methods for Nonlinear Problems. Affine Invariance and Adaptive Algorithms. Springer, Berlin (2004)
8. Griewank, A.: Evaluating Derivatives, Principles and Techniques of Automatic Differentiation. SIAM, Phildelphia (2000)
9. Haslinger, J., Mäkinen, R.A.E.: Introduction to Shape Optimization: Theory, Approximation, and Computation. SIAM, Philadelphia (2004)
10. Hoppe, R.H.W., Linsenmann, C., Petrova, S.I.: Primal-dual Newton methods in structural optimization. Comp. Visual. Sci. 9, 71–87 (2006)
11. Hoppe, R.H.W., Litvinov, W.G.: Problems on electrorheological fluid flows. Communications in Pure and Applied Analysis 3, 809–848 (2004)
12. Hoppe, R.H.W., Petrova, S.I.: Primal-dual Newton interior point methods in shape and topology optimization. Numerical Linear Algebra with Applications 11, 413–429 (2004)
13. Mohammadi, B., Pironneau, O.: Applied Shape Optimization for Fluids. Oxford University Press, Oxford (2001)
14. Rozvany, G.: Structural Design via Optimality Criteria. Kluwer, Dordrecht (1989)
15. Sokolowski, J., Zolesio, J.P.: Introduction to Shape Optimization. Springer, Berlin (1992)

Analytical Effective Coefficient and First-Order Approximation to Linear Darcy's Law through Block Inclusions

Rosangela F. Sviercoski and Bryan J. Travis

Los Alamos National Laboratory
Los Alamos, NM 87545, USA
rsvier@lanl.gov

Abstract. We present an analytical form for the effective coefficient to linear diffusion equations such that the heterogeneous coefficients are periodic and rapidly oscillating and can be defined as step functions describing inclusions in a main matrix. The new contribution comes from using an analytical approximation for the solution of the well known periodic cell-problem. By defining a correction to the given approximation, the analytical effective coefficient, the zeroth-order approximation in $H_0^1(Ømega)$ and the first-order in $L^2(\Omega)$ are readily obtained. The known results for effective coefficient are obtained as particular cases, including the geometric average for the checkerboard structure of the medium. We demonstrate numerically that our proposed approximation agrees with the classical theoretical results in homogenization theory. This is done by applying it to problems of interest in flow in porous media, for cases where the contrast ratio between the inclusion and the main matrix are 10:1, 100:1, 1000:1, and 1:10, respectively.

1 Introduction

Many problems of fundamental and practical importance needs a multiple-scale approach. For example, a major problem in modeling natural porous media is to obtain an accurate description of flow and transport behavior, in spite of the intrinsic heterogeneity of geological formations. For composite materials, the heterogeneous distribution of particles or fibers gives rise to fluctuations in the thermal or electrical conductivity. Accurate numerical solution of these models requires a very finely divided computational mesh, something that is frequently infeasible to consider. However, if one is interested in analyzing the system from a macrostructure point of view, it is possible to simplify the models in such a way that the phenomena of interest remain adequately described. The simplified equations are called homogenized equations, and the procedure of replacing the original system is called homogenization. The effective coefficient plays a crucial role in the process as it represents the heterogeneous medium in a simplified way. Many homogenization procedures have been proposed — see for example, Renard and De Marsily [9] and Milton [8].

I. Lirkov, S. Margenov, and J. Waśniewski (Eds.): LSSC 2007, LNCS 4818, pp. 267–274, 2008.

The homogenization approach, that is the subject of this paper, considers the separation of scales and performs a two-scale asymptotic expansion into the original equation, following works by Tartar [14], Bensoussan, et. al [2], Sanchez-Palencia [10] and Keller [7] (among others). The robustness of this approach may be limited, as one needs to solve numerically the cell-problem of the auxiliary variable [9]. However, our proposed procedure simplifies a typical numerical computation by avoiding computing the solution of the cell-problem and by adding an analytical version of the basis functions to get the first-order approximation. It also provides an accurate upper bound estimate for the error (UPE) implied by using the zeroth-order solution. It can be considered as the analytical version of MsFEM, from Hou and Wu [4], when this is applied to the particular case of scale separation, since we introduce explicit analytical basis functions to obtain the fine scale approximation. The basis functions are defined on the whole domain, and it can be used to simplify the MsFEM method by avoiding the need for the oversampling, as it is a local-global upscaling procedure. This method has the advantage of portability as it can be used with any existing elliptic solvers.

We study, without loss of generality, the limit as $\varepsilon \to 0$, of the two-scale asymptotic approximation:

$$u^\varepsilon(x) = u^0(x,y) + \varepsilon u^1(x,y) + \varepsilon^2 u^2(x,y) + \dots \quad (1)$$

of the solution of the linear boundary value problem (BVP):

$$\begin{cases} \nabla \cdot (K^\varepsilon(x) \nabla u^\varepsilon(x)) = 1 & x \in \Omega \\ u^\varepsilon(x) = 0 & x \in \partial\Omega \end{cases} \quad (2)$$

where the flux density $q^\varepsilon(x)$ is from Darcy's law:

$$q^\varepsilon(x) = -K^\varepsilon(x)\nabla u^\varepsilon(x) \quad (3)$$

where $K^\varepsilon(x)$, defined over $\Omega = \cup_\varepsilon \Omega^\varepsilon$, at each Ω^ε, is the step function:

$$K^\varepsilon(x) = \begin{cases} \xi_1 & \text{if } x \in \Omega_c^\varepsilon \\ \xi_2 & \text{if } x \in \Omega^\varepsilon \setminus \Omega_c^\varepsilon \end{cases} \quad (4)$$

with ξ_1 being the value on the inclusion Ω_c^ε, centered symmetrically on each Ω^ε and $\xi_1 : \xi_2$ is the inclusion ratio.

This paper achieves four goals:

- It presents an analytical way of obtaining the homogenized coefficient K^0.
- It presents an upscaled version of (3), if $K_s^\varepsilon(x)$ is given as (4), by incorporating heterogeneity features into the gradient and flux sequences.
- It presents an analytical form for the first-order approximation, in $L^2(\Omega)$, for (2) and (3).
- It demonstrates numerically the convergence results for the proposed approximations, as it had been theoretically proved in the homogenization literature, ([14,2,10,7]).

The approach uses an analytical approximation belonging to $L^2(\Omega)$, proposed in Sviercoski et. al [11], for the well known periodic cell-problem [2,10,7]:

$$\nabla \cdot (K^\varepsilon (x) \nabla w_i^\varepsilon (x)) = -\nabla \cdot (K^\varepsilon (x) e_i) \tag{5}$$

with e_i the coordinate direction. In Sviercoski et. al [13], we presented a corrector and obtained the effective coefficient and the first-order approximation for the linear case.

The paper is organized as follows: In section 2, we briefly review homogenization theory and the contributions regarding this analytical approach. In section 3, we present the comparison with other known results for the effective coefficient and the algorithm's convergence properties by applying the results to the BVP (2) with oscillating coefficient given by square inclusions (in 2-D) in a primary matrix.

2   Diffusion in Periodic Media

The procedure described in homogenization literature to obtain the upscaled limit of equation (2), is based on the substitution of the expansion (1) into (2) and equating $\varepsilon-$ like powers. By doing so, the first subproblem that needs to be solved, in $H^1(Y)$, for the local variable $y = \varepsilon^{-1}x$, is the periodic cell-problem (5) . By averaging procedures, one obtains the effective coefficient:

$$K^0 = \int_Y K(y) \left(\delta_{ij} + \sum_{i=1}^{n} \nabla_y w_i(y) \right) dY \tag{6}$$

where $\delta_{i,j}$ is the *Kronecker delta*. The homogenized equation that approximates (2) is then found to be:

$$\sum_{i,j=1}^{n} K_{ij}^0 \partial_{x_i x_j} u^0(x) - 1 = 0 \tag{7}$$

This limiting equation is possible by choosing a weakly converging sequence (oscillating test functions) and using the compensated compactness theory presented in Tartar [14]. Then one has the following theorem:

Theorem 1. *The sequence $u^\varepsilon(x)$ of solutions of (2) converges weakly in $H^1(\Omega)$ to a limit $u^0(x)$ which is the unique solution of the homogenized problem (7).*

Proof. See, for example, Bensoussan, et. al [2] and Sanchez-Palencia, [10].   □

2.1   Analytical Approach

There are known cases, in the literature, when the solution to (5) can be obtained (see [13]). The one of special interest we review here is when the geometry describes checkerboard structures, then K^0 is the diagonal tensor with entries given by the geometric average of the eigenvalues (see [6], p.37).

When $K^\varepsilon(x)$ assumes the form of (4) then it has been shown in [11] that:

$$\tilde{w}_i(y) = \left[\int_0^{y_i} \frac{dy_i}{K(y)} \left(\int_0^1 \frac{dy_i}{K(y)} \right)^{-1} - y_i \right] \tag{8}$$

is an approximation in $L^2(\Omega)$ to the solution $w_i(y) \in H^1(Y)$. In [13], we proposed a corrector to this approximation, leading to define K^0, as the diagonal matrix:

$$K^0 = diag\left(c_1 \int_Y R_1 dY, c_2 \int_Y R_2 dY, ..., c_n \int_Y R_n dY \right) \tag{9}$$

where $\int_Y R_i(y) dY = \tilde{K}^0$ is the arithmetic average of the harmonic average at each $i-$component, and:

$$c_i = ||C_i(y)||_2 = \left\| 1 + \frac{\partial \tilde{w}_i}{\partial y_i} \right\|_2 = \left\| K(y) \left(\int_0^1 \frac{dy_i}{K(y)} \right)^{-1} \right\|_2 \tag{10}$$

where c_i measures the lack of smoothness, incurred by computing K^0 using the approximation (8). The diagonal form of (9) is a consequence of the fact that $K^\varepsilon(x)$ has its center of mass at half of the period, as presented in Sviercoski et. al ([11,12,13]). However, in [12], it has been shown that simply using the effective coefficient in the gradient and flux sequences does give the averaging behavior but does not get a bounded sequence in $L^2(\Omega)$. In order to ensure a bounded sequence, we need to incorporate the heterogeneity features into the gradient and flux sequences. To obtain that, we define in [13], the following correction:

$$\nabla u^\varepsilon(x) = [\delta_{ij} - C_{ii} + C_{ii} C_{ii}^\varepsilon(x)] \nabla u^0(x) + \sum_{i=1}^n \tilde{w}_i(y) \nabla \left(\frac{\partial u^0}{\partial x_i} \right) + ...$$
$$= P^\varepsilon(x) \nabla u^0(x) + \sum_{i=1}^n \tilde{w}_i(y) \nabla \left(\frac{\partial u^0}{\partial x_i} \right) + ... \tag{11}$$

where $C = diag(c_1, .., c_n)$ and the flux. With the flux (3), being approximated as:

$$K^\varepsilon(x) \nabla u^\varepsilon(x) \approx K^0(x) P^\varepsilon(x) \nabla u^0(x) \tag{12}$$

Furthermore, the first-order approximation to $u^\varepsilon(x)$, is then:

$$u^\varepsilon(x) \approx u^0(x, y) + \varepsilon u^1(x, y) = u^0(x, y) + \sum_{i=1}^n C_{ii} \tilde{w}_i(y) \frac{\partial^2}{\partial x_i^2} u^0 = u^{fo}(x) \tag{13}$$

which allows us to get the upper bound estimate for the error implied in using $u^0(x)$:

$$Error = ||u^\varepsilon(x) - u^0(x)||_2 \leq \left\| \sum_{i=1}^n C_{ii} \tilde{w}_i \frac{\partial u^0(x)}{\partial x_i} \right\|_2 = UPE \tag{14}$$

and the first-order approximation for the flux and gradient:

$$\nabla u^\varepsilon \approx P^\varepsilon(x) \nabla u^0 + \sum_{i=1}^n C_{ii} \tilde{w}_i \frac{\partial^2}{\partial x_i^2} u^0 = \nabla u^{fo}(x) \tag{15}$$

$$K^\varepsilon \nabla u^\varepsilon \approx K^0 \left(P^\varepsilon(x) \nabla u^0 + \sum_{i=1}^n C_{ii} \tilde{w}_i \frac{\partial^2}{\partial x_i^2} u^0 \right) \tag{16}$$

Table 1. K^0 from (9) and the geometric average, K^g, for the checkerboard structure

$\xi_1 : \xi_2$	K^g	$C \times \tilde{K}^0 = K^0$
5:20	10	**1.0725**×9.31=9.98
1:10	3.16	**1.1732**×2.60=3.05
2:8	4	**1.0725**×3.722=3.99
4:16	8	**1.0725**×7.4451=7.985

Table 2. K^0 from (9), K^{num} from [3], and K^{bb} from [5] with various shapes of inclusion

Shape	K^{num}	K^{bb}	$C \times \tilde{K}^0 = K^0$
square (ratio 10:1)	1.548	1.598	**1.0937**×1.4091=1.5411
circle (ratio 10:1)	1.516	1.563	**1.0785**×1.403=1.5131
lozenge (ratio 10:1)	1.573	1.608	**1.069**× 1.417=1.5148

Table 3. K^0 from (9) and $K^{\#}$, obtained from numerical solution of eq. (5) [1]. K^h and K^a are the harmonic and arithmetic averages.

Test	K^h	K^a	$K^{\#}$	$C \times \tilde{K}^0 = K^0$
Test 1 (ratio 1:10)	3.09	8.5	6.52	**1.093**×5.91=6.459
Test 3 (ratio 1:100)	3.89	76.0	59.2	**1.1378**×51=58.03

3 Numerical Results

3.1 Comparing K^0 with Published Numerical Results

Tables 1–3 show how C plays an important role in obtaining a more accurate value for K^0.

3.2 Zeroth-Order Approximation for the Linear Case

We now demonstrate numerically the convergence properties of the approximations for coefficient functions defined in (4) with ratios as 10:1, 100:1, 1000:1 and 1:10. Table 4 shows the comparison between the heterogeneous solution, $u^\varepsilon(x)$ of the BVP (2) and its zeroth-order approximation, given by solution of the BVP (7), with $u^0(x) = 0$ on $\partial\Omega$. The procedure for obtaining the table is outlined: **(a)** Compute numerically $u^\varepsilon(x)$, $\nabla u^\varepsilon(x)$ and $K^\varepsilon \nabla u^\varepsilon(x)$ on a given mesh. **(b)** Compute K^0 using (9), then $u^0(x)$ from (7) and $\nabla u^0(x)$ on the same mesh as in the heterogeneous. **(c)** Compute analytically $\tilde{w}_i^\varepsilon(x)$ and $C^\varepsilon(x)$ at the same mesh as $u^\varepsilon(x)$, in order to obtain UPB (14). **(d)** Compute the gradient error from (11) and the flux error from (12).

3.3 First-Order Approximation

We use zeroth−order approximation of BVP (2) from Table 4 to obtain the first-order approximation to BVP (2) from (13), (15), and (16). The summary of the approximations is on Table 5 and an illustration is in Fig. 1.

Table 4. Zeroth order approximation from (7) using (9) and and ratios (a) 10:1, (b) 100:1, (c) 1000:1 and (d) 1:10 on $[0,1]^2$ with $K^0 = \mathbf{1.0937} \times 1.4091 = 1.5411$, $K^0 = \mathbf{1.139} \times 1.4901 = 1.6972$, $K^0 = \mathbf{1.1441} \times 1.5 = 1.7161$, and $K^0 = \mathbf{1.093} \times 5.91 = 6.459$, respectively. Note how the error is decaying linearly on the first column and also that UPE is a reliable upper bound. The errors do not increase significantly as the ratio increases.

ε	$\left\|\left\|u^\varepsilon - u^0\right\|\right\|_2$	UPE	$\left\|\left\|\nabla u^\varepsilon - P^\varepsilon \nabla u^0\right\|\right\|_2$	$\left\|\left\|K^\varepsilon \nabla u^\varepsilon - P^\varepsilon K^0 \nabla u^0\right\|\right\|_2$	grid
$(0.5)^1$	1.10e-2	1.61e-2	4.67e-2	1.87e-1	130X130
$(0.5)^2$	4.92e-3	7.48e-3	4.31e-2	1.61e-1	130X130
$(0.5)^3$	2.13e-3	3.74e-3	3.62e-2	1.58e-1	130X130
$(0.5)^4$	9.48e-4	1.80e-3	3.11e-2	1.57e-1	130X130
$(0.5)^5$	5.13e-4	9.33e-4	3.31e-2	1.57e-1	402X402
$(0.5)^1$	1.293e-2	1.82e-2	5.86e-2	2.53e-1	130X130
$(0.5)^2$	5.736e-3	8.47e-3	5.45e-2	2.18e-1	130X130
$(0.5)^3$	2.441e-3	4.23e-3	4.41e-2	2.16e-1	130X130
$(0.5)^4$	1.223e-3	2.05e-3	3.68e-2	2.14e-1	130X130
$(0.5)^5$	6.995e-4	1.06e-3	3.90e-2	2.14e-1	402X402
$(0.5)^1$	1.32e-2	1.85e-2	6.00e-2	2.62e-1	182X182
$(0.5)^2$	5.84e-3	8.65e-3	5.61e-2	2.26e-1	182X182
$(0.5)^3$	2.51e-3	4.36e-3	4.73e-2	2.25e-1	230X230
$(0.5)^4$	1.14-3	2.16e-3	4.16e-2	2.23e-1	230X230
$(0.5)^5$	6.90e-4	1.09e-3	4.07e-2	2.26e-1	230X230
$(0.5)^1$	3.35e-3	3.83e-3	6.69e-2	3.00e-1	130X130
$(0.5)^2$	8.81e-4	1.78e-3	4.37e-2	2.80e-1	130X130
$(0.5)^3$	3.16e-4	8.9e-4	3.85e-2	2.78e-1	130X130
$(0.5)^4$	1.69e-4	4.31e-4	3.75e-3	2.79e-1	130X130
$(0.5)^5$	9.06e-5	2.22e-4	3.75e-3	2.78e-1	402X402

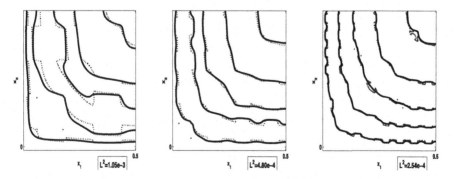

Fig. 1. Quarter zone of the fine scale (solid) and first-order (dashed), $u^{fo}(x)$, obtained by (13) for $\varepsilon = (0.5)^3, (0.5)^4, (0.5)^5$, respectively, from Table 5 (a). Observe how the dashed lines capture the oscillations of the solid line.

Table 5. First-Order Approximation from (13), on $[0, 1]^2$, for the and ratios (a) 10:1, (b) 100:1, (c) 1000:1 and (d) 1:10, respectively. Note how the first column is going linearly to zero and it is about half of the UPE from Table 4. Also, the gradient and flux sequences are bounded implying the weak convergence.

ε	$\left\|\left\|u^\varepsilon - u^{fo}\right\|\right\|_2$	$\left\|\left\|\nabla u^\varepsilon - \nabla u^{fo}\right\|\right\|_2$	$\left\|\left\|K^\varepsilon \nabla u^\varepsilon - K^0 \nabla u^{fo}\right\|\right\|_2$	grid
$(0.5)^1$	4.43e-3	3.62e-2	1.80e-1	130X130
$(0.5)^2$	2.31e-3	4.12e-2	1.60e-1	130X130
$(0.5)^3$	1.05e-3	3.59e-2	1.58e-1	130X130
$(0.5)^4$	4.80e-4	3.11e-2	1.57e-1	180X180
$(0.5)^5$	2.54e-4	3.20e-2	2.57e-1	180X180
$(0.5)^1$	5.13e-3	5.30e-2	2.42e-1	130X130
$(0.5)^2$	2.75e-3	5.33e-2	2.17e-1	130X130
$(0.5)^3$	1.24e-3	4.38e-2	2.16e-1	130X130
$(0.5)^4$	6.61e-3	3.68e-2	2.14e-1	182X182
$(0.5)^5$	3.98e-3	3.90e-2	2.14e-1	182X182
$(0.5)^1$	5.18e-3	5.98e-2	2.50e-1	182X182
$(0.5)^2$	2.80e-3	5.40e-2	2.25e-1	182X182
$(0.5)^3$	1.28e-3	4.68e-2	2.25e-1	230X230
$(0.5)^4$	5.86e-4	4.15e-2	2.23e-1	230X230
$(0.5)^5$	3.77e-4	4.07e-2	2.26e-1	230X230
$(0.5)^1$	3.48e-3	6.41e-2	2.94e-1	130X130
$(0.5)^2$	7.80e-4	4.32e-2	2.78e-1	130X130
$(0.5)^3$	1.81e-4	3.84e-2	2.78e-1	130X130
$(0.5)^4$	1.01e-4	3.75e-3	2.79e-1	130X130
$(0.5)^5$	5.47e-5	3.75e-3	2.78e-1	402X402

4   Conclusion and Future Work

This work represents one step towards obtaining the effective coefficient and the first-order approximation, by analytical means, for more general geometries and random media. A continuation of these results applied for particular cases of nonlinear equations is in progress as well as a comparison of this formulation with experimental results. Future work includes applying these results to random media. Another future application is to use the zeroth order approximation as an initial guess for iterative methods for solving linear and nonlinear systems, obtained by discretizing heterogeneous coefficient equations, therefore improving accuracy and convergence rates for these and more general geometries. The ultimate goal is to apply the method to multi-phase systems where diffusion is the driving process.

Acknowledgments. The authors are thankful to Peter A. Popov for providing the numerical code *FEM-O-MATIC*. This work is part of the Chevron/LANL Cooperative Research and Development Agreement (CRADA).

References

1. Amaziane, B., Bourgeat, A., Koebbe, J.: Numerical Simulation and Homogenization of Two-Phase Flow in Heterogeneous Porous Media. Transport in Porous Media 6, 519–547 (1991)
2. Bensoussan, A., Lions, J.L., Papanicolau, G.: Asymptotic Analysis for Periodic Structures. North Holland, Amsterdam (1978)
3. Bourgat, J.F.: Numerical Experiments of the Homogenization Method for Operators with Periodic Coefficient. In: Glowinski, R., Lions, J.L. (eds.) Comp. Met. in Appl. Sc. and Eng., Versailles, December 5–9, vol. I, p. 330. Springer, Heidelberg (1978)
4. Hou, T.Y., Wu, X.H.: A Multiscale Finite Element Method for Elliptic Problems in Composite Materials and Porous Media. J.C.P. 134, 169–189 (1997)
5. Moulton, J.D., Dendy, J.E., Hyman, J.M.: The Black box Multigrid Numerical Homogenization Algorithm. J.C.P. 142, 80–108 (1998)
6. Jikov, V.V., Kozlov, S.M., Oleinik, O.A.: Homogenization of Differential Operators and Integral Functionals. Springer, Heidelberg (1994)
7. Keller, J.B.: Darcy's Law for Flow in Porous Media and the Two-Space Method. In: Sternberg, R.L. (ed.) Nonlinear Partial Differential Equations in Engineering and Applied Sciences, pp. 429–443. Marcel Dekker, New York (1980)
8. Milton, G.W.: The Theory of Composites, Cambridge (2002)
9. Renard, P., De Marsily, G.: Calculating Equivalent Permeability: A Review. Adv. Water Resources 20(5–6), 253–278 (1997)
10. Sanchez-Palencia, E.: Non-Homogeneous Media and Vibration Theory. Lecture Notes in Physics, vol. 129. Springer, Berlin (1980)
11. Sviercoski, R.F., Winter, C.L., Warrick, A.W.: Analytical Approximation for the Generalized Laplace's Equation with Step Function Coefficient. SIAM Journal of Applied Mathematics (in press, 2007)
12. Sviercoski, R.F., Travis, B.J.: Analytical Effective Coefficient for Block Inclusions: Numerical Results for Linear Flow. Transport in Porous Media (Submitted 2007)
13. Sviercoski, R.F., Travis, B.J., Hyman, J.M.: Analytical Effective Coefficient and First-Order Approximation for Linear Flow through Block Permeability Inclusions. Comp. Math. Appl (in press, 2007)
14. Tartar, L.: Quelques Remarques sur L'homogeneisation. In: Fujita, H. (ed.) Functional Analysis and Numerical Analysis, Proc. Japan-France Seminar, pp. 468–482. Japanese Society for the Promotion of Science (1976)

Part VI
Control Systems

Optimal Control for Lotka-Volterra Systems with a Hunter Population

Narcisa Apreutesei[1] and Gabriel Dimitriu[2]

[1] Technical University "Gh. Asachi", Department of Mathematics,
700506 Iaşi, Romania
napreut@net89mail.dntis.ro
[2] University of Medicine and Pharmacy "Gr. T. Popa",
Department of Mathematics and Informatics,
700115 Iaşi, Romania
dimitriu.gabriel@gmail.com

Abstract. Of concern is an ecosystem consisting of a herbivorous species and a carnivorous one. A hunter population is introduced in the ecosystem. Suppose that it acts only on the carnivorous species and that the number of the hunted individuals is proportional to the number of the existing individuals in the carnivorous population. We find the optimal control in order to maximize the total number of individuals (prey and predators) at the end of a given time interval. Some numerical experiments are also presented.

1 Introduction

Consider the prey-predator system of differential equations

$$\begin{cases} y_1' = y_1 (a_1 - b_1 y_2) \\ y_2' = y_2 (-a_2 + b_2 y_1), \end{cases}$$

where a_1, a_2, b_1, b_2 are positive constants. It describes the dynamics of an ecosystem composed by two populations which coexist: a prey population and a predator one [5,7,10]. We denoted by $y_1(t)$ and $y_2(t)$ the number of the individuals of the two populations, the prey and the predators, respectively.

Suppose now that a hunter population is introduced in the ecosystem and it acts only on the predators. At each moment t, the number of the hunted individuals is assumed to be proportional to the total number of the existing predators. Let $u(t)$ be the proportionality factor, $0 \leq u(t) \leq 1$. Thus the dynamics of the new ecosystem is described by the system of ordinary differential equations

$$\begin{cases} y_1' = y_1 (a_1 - b_1 y_2) \\ y_2' = y_2 (-a_2 - u + b_2 y_1) . \end{cases} \tag{1}$$

We study this system on a finite time interval $[0, T]$ and interpret $u : [0, T] \to [0, 1]$ as a control function. One associates some initial conditions:

$$y_1(0) = y_1^0 > 0, \qquad y_2(0) = y_2^0 > 0 . \tag{2}$$

I. Lirkov, S. Margenov, and J. Waśniewski (Eds.): LSSC 2007, LNCS 4818, pp. 277–284, 2008.
© Springer-Verlag Berlin Heidelberg 2008

Problem $(1) - (2)$ has a unique solution (y_1, y_2), which is positive [7].

Our goal is to find the optimal control u and the corresponding state (y_1, y_2) such that the total number of individuals in the end of the time interval $[0, T]$ is maximal. In other words, we have to maximize $y_1(T) + y_2(T)$, i.e. to solve the following optimal control problem:

$$\inf \{-y_1(T) - y_2(T)\}, \tag{3}$$

where $0 \leq u(t) \leq 1$, $t \in [0, T]$ and (y_1, y_2) verifies $(1) - (2)$.

Other similar control problems for systems of ODE were treated in [2,5,6,10]. In [1], an optimal control problem is analyzed in connection with a partial differential system. The densities of the populations depend not only on the time moment, but also on the position in the habitat. Some basic concepts and methods of the optimal control theory can be read in [3]. Mathematical models in population dynamics are presented and studied in [4,7]. About the number of switching points of the optimal control, the reader may refer to [8,9].

In Section 2, we apply Pontrjagin's maximum principle to write the optimality system and to find thus the form of the optimal control u. One proves that it is pointwise continuous and takes on only the values 0 and 1. A discussion of the number of the commutation points of u is presented.

Section 3 is devoted to some numerical experiments. There are analyzed several cases corresponding to different values of the parameters (the initial conditions of the ecosystem (y_1^0, y_2^0), and the positive constants a_1, a_2, b_1 and b_2).

2 Necessary Optimality Conditions

Denote by y the pair (y_1, y_2) and by p the adjoint variable, with its components p_1 and p_2. Therefore we have

$$y = \begin{pmatrix} y_1 \\ y_2 \end{pmatrix}, \quad p = \begin{pmatrix} p_1 \\ p_2 \end{pmatrix}, \quad f(y, u) = \begin{pmatrix} y_1(a_1 - b_1 y_2) \\ y_2(-a_2 - u + b_2 y_1) \end{pmatrix}$$

and the cost functional

$$l(y(T)) = -y_1(T) - y_2(T).$$

The adjoint system in our case is $p' = -f_y^* \cdot p$, the transversality condition is $p(T) = -\nabla l(y(T))$, and the last optimality condition is $f_u^* \cdot p \in N_{[0,1]}(u)$. Here $N_{[0,1]}(u)$ denotes the normal cone to the closed convex set $[0, 1]$ at point u, while f_y^*, f_u^* are the adjoint matrices of the Jacobian matrices f_y, f_u. In detail, this can be written as

$$\begin{cases} p_1' = -a_1 p_1 + y_2(b_1 p_1 - b_2 p_2) \\ p_2' = (a_2 + u)p_2 + y_1(b_1 p_1 - b_2 p_2) \\ p_1(T) = p_2(T) = 1 \end{cases} \tag{4}$$

and since

$$N_{[0,1]}(u) = \begin{cases} 0, & 0 < u < 1 \\ \mathbb{R}_-, & u = 0 \\ \mathbb{R}_+, & u = 1, \end{cases}$$

by the last optimality condition, it follows that the optimal control u has the form

$$u(t) = \begin{cases} 0, & \text{if } y_2 p_2 > 0 \\ 1, & \text{if } y_2 p_2 < 0. \end{cases} \tag{5}$$

Put $z_1 = y_1 p_1$, $z_2 = y_2 p_2$. Taking into account the adjoint system and the transversality conditions, we obtain

$$\begin{cases} z_1' = -b_2 y_1 z_2, \ t \in [0, T] \\ z_2' = b_1 y_2 z_1, \ t \in [0, T] \end{cases}, \qquad \begin{cases} z_1(T) = y_1(T) > 0 \\ z_2(T) = y_2(T) > 0. \end{cases} \tag{6}$$

Therefore, we can write

$$u(t) = \begin{cases} 0, & \text{if } z_2 > 0 \\ 1, & \text{if } z_2 < 0. \end{cases} \tag{7}$$

We now analyze the number of the commutation points for u, that is the number of the points in $(0, T)$, where z_2 changes its sign.

By [8,9], we conclude that the number of switching points is finite; more exactly the system satisfies the so-called bang-bang property with bounds on the number of switchings.

Since $z_2(T) = y_2(T) > 0$ and $z_2'(T) = b_1 y_1(T) y_2(T) > 0$, observe that in a left neighborhood of the final point T, the function z_2 is positive and strictly increasing. Analogously, we find that $z_1 > 0$ and z_1 is strictly decreasing in a left neighborhood of T. Moreover, $z_1 > 0$ if and only if $z_2' > 0$, while $z_2 > 0$ if and only if $z_1' < 0$.

The system in (z_1, z_2) has the coefficients continuous on subintervals of $[0, T]$. This implies that z_2 has a finite number of zeros in $(0, T)$, say $\tau_n < \tau_{n-1} < ... < \tau_2 < \tau_1$.

We have the following cases.

If z_2 does not change its sign on $[0, T]$, then it is positive and thus $u(t) = 0$, for all $t \in [0, T]$.

If z_2 changes its sign on $[0, T]$, then let τ_1 be the closest to T from all its zeros. This means that $z_2(t) > 0$ on $(\tau_1, T]$, $z_2(\tau_1) = 0$, and $z_2(t) < 0$ at least on a left neighborhood $(\tau_1 - \varepsilon_1, \tau_1)$ of τ_1. Then z_1 is decreasing on $(\tau_1, T]$ and increasing on $(\tau_1 - \varepsilon_1, \tau_1)$.

If z_1 does not change its sign at the left side of τ_1, then $z_1(t) > 0$ and $z_2'(t) > 0$, $t \in [0, \tau_1)$, and consequently $z_2 < 0$ on $[0, \tau_1)$ and $z_2 > 0$ on $(\tau_1, T]$. We have $u(t) = 0$ on $(\tau_1, T]$ and $u(t) = 1$ on $[0, \tau_1)$.

If z_1 changes its sign, then let θ_1 be the closest to T from all its switching points: $z_1(t) > 0$ on $(\theta_1, \tau_1]$, $z_1(\theta_1) = 0$, and $z_1(t) < 0$ at least in a left neighborhood $(\theta_1 - \xi_1, \theta_1)$ of θ_1. Then, z_2 is monotonically increasing on $(\theta_1, \tau_1]$ (so $z_2 < 0$) and monotonically decreasing on $(\theta_1 - \xi_1, \theta_1)$. Hence z_1 cannot change its sign on (θ_1, τ_1).

If $z_2 < 0$ on $[0, \theta_1)$, then $u(t) = 1$ on $[0, \tau_1)$ and $u(t) = 0$ on $(\tau_1, T]$. If this is not the case, then let τ_2 be the next switching point of z_2 ($\tau_2 < \theta_1 < \tau_1$), i.e. $z_2 < 0$ on (τ_2, θ_1), $z_2(\tau_2) = 0$, and $z_2 > 0$ on $(\tau_2 - \varepsilon_2, \tau_2)$.

If $z_1 < 0$ on $[0, \tau_2)$, then $u(t) = 0$ on $[0, \tau_2) \cup (\tau_1, T]$. If z_1 still has some switching points at the left side of τ_2, then let θ_2 be the closest to τ_2. So, $\theta_2 < \tau_2 < \theta_1 < \tau_1$.

Continuing this reasoning, let τ_{2k+1} be the switching points of z_2 such that $z_2 < 0$ at the left side and $z_2 > 0$ at the right side, and let τ_{2k} be the switching points of z_2 such that $z_2 > 0$ at the left side and $z_2 < 0$ at the right side of τ_{2k}.

If $n = 2m$, then the optimal control u has the form

$$u(t) = \begin{cases} 0, & t \in [0, \tau_{2m}) \cup (\tau_{2m-1}, \tau_{2m-2}) \cup ... \cup (\tau_1, T] \\ 1, & t \in (\tau_{2m}, \tau_{2m-1}) \cup ... \cup (\tau_2, \tau_1). \end{cases} \quad (8)$$

If $n = 2m + 1$, then

$$u(t) = \begin{cases} 0, & t \in (\tau_{2m+1}, \tau_{2m}) \cup ... \cup (\tau_1, T] \\ 1, & t \in [0, \tau_{2m+1}) \cup (\tau_{2m}, \tau_{2m-1}) \cup ... \cup (\tau_2, \tau_1). \end{cases} \quad (9)$$

The optimal state (y_1, y_2) is the solution of system $(1) - (2)$ corresponding to $u = 0$ and $u = 1$, respectively.

Our conclusion can be expressed in the following result.

Theorem 1. *If a_1, a_2, b_1, $b_2 > 0$ are given constants, then the optimal control u is piecewise continuous and takes on only the values 0 and 1.*

If p_1 and p_2 are the adjoint variables and $\tau_n < \tau_{n-1} < ... < \tau_2 < \tau_1$ are the zeros in $(0, T)$ of function p_2, then u has the form (8) (if $n = 2m$) or (9) (if $n = 2m + 1$).

3   Computational Issue

This section presents numerical results of the optimal control problem (3) associated to the prey-predator system (1)-(2) and gives some interpretations. We set in our numerical approach $T = 1$, and used 100 discretization points in time interval $[0, T]$.

Both systems (the ecosystem (1)-(2) and its adjoint (4) corresponding to the optimal control problem (3)) were solved by finite differences using Matlab routines. Five cases have been analyzed corresponding to different sets of the parameters a_1, a_2, b_1, b_2 and of the initial conditions (y_1^0, y_2^0). These values are indicated in each figure included in this section.

In the first two cases (Case I and II) there is no switching point of the control $u(t)$ between 0 and 1. According to the theoretical result stated in the previous section, this situation appears when the variable z_2 does not change its sign on $[0, T]$ (see Figures 1 and 2).

The following two cases (Case III and IV) present situations in which the control variable u has one and two switching points, respectively. Again, we remark in the plots (see Figures 3 and 4) that the variable z_2 changes its sign once (see Figure 3) and twice (see Figure 4).

The last case (Case V) illustrates the situation when the variable z_2 modifies its sign five times (see the lower subplot in Figure 5). Therefore, the control u indicates the same number of switching points, which again is in agreement with the theoretical reasoning.

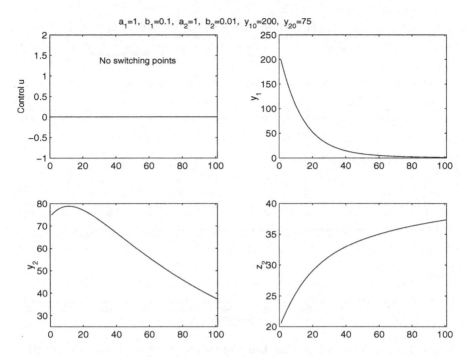

Fig. 1. The control u and the time behavior for (y_1, y_2) and the variable z_2 (Case I)

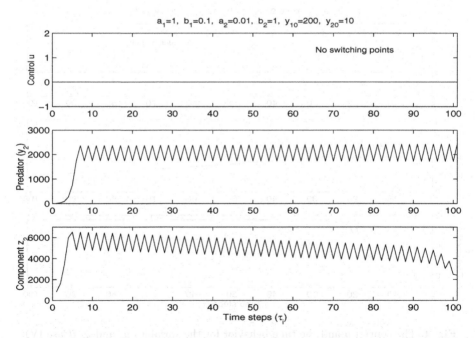

Fig. 2. The control u and the time behavior for the variables y_2 and z_2 (Case II)

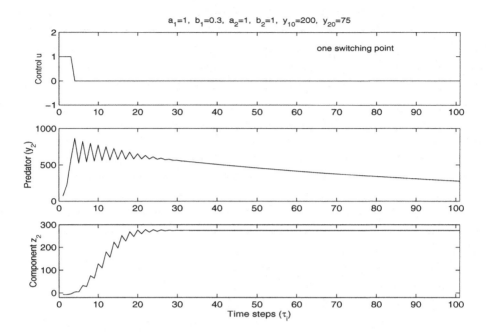

Fig. 3. The control u and the time behavior for the variables y_2 and z_2 (Case III)

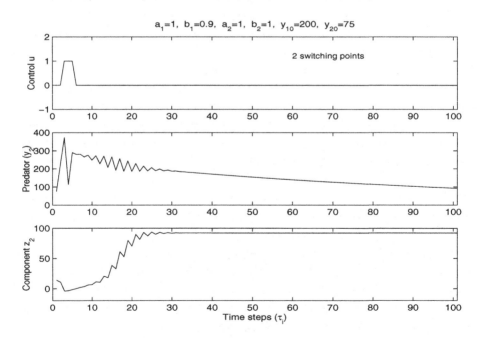

Fig. 4. The control u and the time behavior for the variables y_2 and z_2 (Case IV)

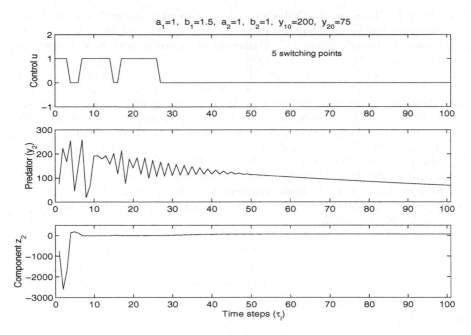

Fig. 5. The control u and the time behavior for the variables y_2 and z_2 (Case V)

4 Conclusions

In this paper we analyzed an ecosystem by means of an optimal control problem. In order to maximize the total number of individuals (prey and predators) from the ecosystem at the final moment T, the hunter should act on the predators, alternatively. Periods of time when the hunter population acts continuously ($u = 1$) should alternate with periods when the hunter does not hunt predators at all ($u = 0$). We say that the optimal control u is bang-bang. At least in a neighbourhood of the final time T, the hunter should be inactive ($u = 0$). Numerical experiments show that the optimal control u can be zero on the whole interval $[0, T]$, or it can admit several points of commutation. Unfortunately, we could not establish yet any connection between the number of switching points for the control variable u and certain values of the parameters a_1, a_2, b_1, and b_2. The numerical estimates show that we can have different number of switching points for the optimal control.

References

1. Apreutesei, N.: An optimal control problem for Lotka-Volterra system with diffusion. Bul. Inst. Polytechnic Iaşi 41(48), fasc. 1–2, 31–41 (1998)
2. Apreutesei, N.: Necessary optimality conditions for a Lotka-Volterra three species system. Math. Modelling Natural Phen. 1(1), 123–135 (2006)

3. Barbu, V.: Mathematical Methods in Optimization of Differential Systems. Kluwer Academic Publishers, Dordrecht (1994)
4. Brauer, F., Castillo-Chavez, C.: Mathematical Models in Population Biology and Epidemiology. Springer, Berlin (2000)
5. Haimovici, A.: A control problem related to a Volterra prey- predator system. In: Anal. Ştiinţ. Univ. Al. I. Cuza Iaşi, Supliment la tomul 25, s. I, pp. 33–41 (1979)
6. Haimovici, A.: A control problem for a Volterra three species system. Mathematica — Revue d'Analyse Numérique et de Théorie de l'Approx 23(46),(1) 35–41 (1980)
7. Murray, J.D.: Mathematical Biology, 3rd edn. Springer, Berlin (2002)
8. Sussmann, H.: A bang-bang theorem with bounds on the number of switchings. SIAM J. Control Optimization 17, 629–651 (1979)
9. Sussmann, H.: Bounds on the number of switchings for trajectories of piecewise analytic vector fields. J. Differ. Equations 43, 399–418 (1982)
10. Yosida, S.: An optimal control problem of the prey-predator system. Funck. Ekvacioj 25, 283–293 (1982)

Modeling Supply Shocks in Optimal Control Models of Illicit Drug Consumption

Roswitha Bultmann[1], Jonathan P. Caulkins[2], Gustav Feichtinger[3], and Gernot Tragler[4]

[1] Vienna University of Technology (TU Wien), Institute for Mathematical Methods in Economics (IWM), Research Unit for Operations Research and Nonlinear Dynamical Systems (ORDYS), A-1040 Wien, Austria
e0025955@student.tuwien.ac.at
[2] Carnegie Mellon University, Qatar Campus & Heinz School of Public Policy and Management, Pittsburgh PA 15237, USA
caulkins@andrew.cmu.edu
[3] TU Wien, IWM, ORDYS, A-1040 Wien, Austria
or@eos.tuwien.ac.at
[4] TU Wien, IWM, ORDYS, A-1040 Wien, Austria
tragler@eos.tuwien.ac.at

Abstract. There is empirical evidence that drug prices have significant impact on demand. For instance, emergency department mentions of various drugs vary in proportion to price raised to a (negative) exponent, which in economists' terms is a constant price elasticity model. This relationship holds even for abrupt spikes in price induced by sudden shortages such as the recent Australian heroin drought. It seems natural to ask how, if at all, drug policy should be varied to take advantage of the opportunity offered by such supply disruptions. We address this question by analyzing a two-stage optimal control model parameterized with data on the current U.S. cocaine epidemic. The number of users and drug control spending are the state and control variables, respectively. The aim is to minimize the discounted stream of the social costs arising from drug consumption plus the control costs. We focus on scenarios with multiple steady states and DNSS-thresholds separating different basins of attraction.

1 Introduction

Optimal control models have been applied to problems of illicit drug consumption for more than a decade (see, e.g., [1], [14], [2]). This paper emerged from the observations that from time to time supply shocks create sharp spikes in the purity-adjusted prices of illicit drugs, and economists have demonstrated empirically that drug use varies inversely with price ([8]). Use responds to price even in the short-run. For instance, the recent Australian heroin drought caused an almost eightfold increase in price within a few months, with prices dropping part-way back again afterwards. Over this period there is a strong correlation between the number of times ambulances responded to a heroin overdose and

I. Lirkov, S. Margenov, and J. Waśniewski (Eds.): LSSC 2007, LNCS 4818, pp. 285–292, 2008.

purity-adjusted heroin prices raised to a power (known as a constant-elasticity price relation) [10]. The corresponding price elasticity in this case is estimated to be -1.12, implying that a 1% increase in price per pure gram was associated with a 1.12% decrease in ambulance call-outs. In a parallel study with U.S. data, price changes explained 97.5% of the variation in emergency department mentions for cocaine and 95% of the variation for heroin, with the price elasticities of demand being estimated at -1.30 and -0.84, respectively [4]. Crane et al. ([5]) documented similar correlations with other indicators, including proportions of people testing positive for cocaine among both arrestees and workers subject to workplace drug-testing.

Assuming now that a country with a drug problem will at some point have its supply significantly disrupted, the natural question to ask is, what should the dynamic response of its drug policy be to that disruption?

2   The Model

We address this question by analyzing a one-state one-control model parameterized with data from the current U.S. cocaine epidemic, where the number of users $(A(t))$ and treatment spending $(u(t))$ provide the state and control variables at time t, respectively. The aim is to minimize the discounted sum of the social costs arising from drug consumption plus the control costs, over an infinite planning horizon, where the control (called treatment) increases outflow from the drug use state. Retail price is modeled as a parameter, influencing current demand as well as initiation into and desistance from drug use.

Supply disruptions are modeled as temporary shifts in price. That is, we assume that the retail price takes an abnormally high or low value p_s from the beginning of the time horizon $(t = 0)$ up to some given time $t = T < \infty$. (Supply shortages with corresponding increases in price are the more common and interesting case from a policy perspective, but the analysis is similar for so-called gluts that temporarily depress prices.) At time $t = T$ the retail price switches back to its base level p_b $(p_b \neq p_s)$. In other words, we decompose the infinite horizon problem into a (first) finite stage of duration T with the retail price equal to p_s and a (second) stage of infinite horizon with the retail price equal to p_b.

Expressed in mathematical terms, we consider the following two-stage optimal control model

$$\min_{u(t) \geq 0} \int_0^T e^{-rt} \left(\kappa p_s^{-\omega} A(t) + u(t) \right) dt \; + \; \int_T^\infty e^{-rt} \left(\kappa p_b^{-\omega} A(t) + u(t) \right) dt$$

subject to

$$\dot{A}(t) = \bar{k} p_s^{-a} A(t) \left(\bar{A} - A(t) \right) - c u(t)^z A(t)^{1-z} - \mu p_s^b A(t) \qquad 0 \leq t < T,$$

$$\dot{A}(t) = \bar{k} p_b^{-a} A(t) \left(\bar{A} - A(t) \right) - c u(t)^z A(t)^{1-z} - \mu p_b^b A(t) \qquad T < t < \infty,$$

where r is the time discount rate; κ is the social cost per unit of consumption; ω is the absolute value of the short run price elasticity of demand; $\bar{k}$ is a constant governing the rate of initiation; $\bar{A}$ is the maximum number of users; a is the absolute value of the elasticity of initiation with respect to price; c is the treatment efficiency proportionality constant; z is a parameter reflecting diminishing returns of treatment; μ is the baseline rate at which users quit without treatment; and b is the elasticity of desistance with respect to price.

3   Analysis

To analyze our model we consider both stages separately. The shock period (i.e., the first stage of our model) is described as a finite horizon optimal control model with objective functional

$$J_T = \int_0^T e^{-rt} \left(\kappa p_s^{-\omega} A(t) + u(t) \right) dt + e^{-rT} S(A(T)) \quad \longmapsto \quad \min,$$

where the salvage value function $S(A(T))$ describes the costs of having $A(T)$ users at time T. According to the maximum principle the following transversality condition must hold at time T:

$$\lambda(T) = S_A(A^*(T)), \tag{1}$$

where $A^*(T)$ denotes the optimal value of A at time T and $S_A(.)$ denotes the derivative of $S(.)$ with respect to A.

The second stage, which describes what happens after the shock has ended, is modeled as an optimal control problem with infinite time horizon. We connect the two stages using the transversality condition (1). As a reasonable choice for the salvage function $S(A(T))$ we take the optimal objective functional value of the second stage (value function). The derivative of the value function with respect to the state variable A exists everywhere except for at most one point (i.e., the DNSS-threshold, which will be defined a few lines later) and is given by the costate variable λ, so condition (1) reduces to

$$\lambda^{(p_s)}(T^-) = \lambda^{(p_b)}(T^+), \tag{2}$$

where the superscripts (p_s) and (p_b) refer to the first and second stage, respectively, and

$$T^- = \lim_{t \uparrow T} t \quad \text{and} \quad T^+ = \lim_{t \downarrow T} t.$$

A note is appropriate here. One may argue that the problem considered here is not really a two-stage problem because the discontinuity that occurs at time T is only with respect to the time. For such systems the classical maximum principle applies and the co-state variable is absolutely continuous, so that (2) holds automatically. However, for the numerical analysis described in what follows we prefer to consider the problem as a two-stage problem, which also ensures time-invariance in the second stage.

The problem is now solved using Pontryagin's maximum principle (see, e.g., [7]), where the usual necessary optimality conditions must hold. For the special case $z = 0.5$, it is possible to derive all possible steady states of the canonical system analytically, which are given by

$$\hat{A}_1 = 0,$$

$$\hat{A}_{2/3} = \frac{1}{3\bar{k}p_b^{-a}} \left[2\varDelta - r \pm \sqrt{(\varDelta + r)^2 - 3c^2\kappa p_b^{-\omega}} \right]$$

with the corresponding values of the costates

$$\hat{\lambda}_1 = \frac{2}{c^2} \left(r - \varDelta - \sqrt{(r - \varDelta)^2 + c^2\kappa p_b^{-\omega}} \right),$$

$$\hat{\lambda}_{2/3} = \frac{2}{3c^2} \left[-(\varDelta + r) \pm \sqrt{(\varDelta + r)^2 - 3c^2\kappa p_b^{-\omega}} \right],$$

where $\varDelta := \bar{k}p_b^{-a}\bar{A} - \mu p_b^b$. A complete stability analysis of these steady states is also possible, but omitted here for brevity.

Varying a parameter such as κ (the social cost per unit of consumption), we find that either there is a unique long-run optimal steady state or there are two optimal steady states, with their basins of attraction being separated by a so-called DNSS-threshold. (For their contributions in pointing to, promoting, and proving the existence of these thresholds, they are named after Dechert and Nishimura [6], Skiba [13], and Sethi [11,12].)

For the remainder of this paper, we fix the parameter values as summarized in Table 1. These values are reasonable for the current U.S. cocaine epidemic (for further details see [3]), and they are in a range such that a DNSS-threshold exists, separating the basins of attraction of a low-level and a high-level equilibrium. We will focus on droughts ($p_s > p_b$) and note that gluts ($p_s < p_b$) yield fairly symmetric results.

As in the long run it is optimal to approach one of the two saddle points of the canonical system of the second phase (cf. [7]), we compute the stable manifolds associated with them. To obtain the solution to the first phase problem we have to solve a boundary value problem (BVP) consisting of the system dynamics of the first phase with the boundary conditions $A(0) = A_0$ and (2).

In other words, we look for trajectories that satisfy $A(0) = A_0$ and cross one of the stable manifolds of the second stage problem at time T. Compound solutions consisting of a solution of the BVP continuously connected with a stable manifold of the second phase problem are candidates for the optimal solution

Table 1. Parameter values derived from data for the current U.S. cocaine epidemic

$r = 0.04$	$\kappa = 3.396819914$	$p_b = 0.12454$	$\omega = 0.5$
$\bar{k} = 0.00000001581272$	$\bar{A} = 16{,}250{,}000$	$a = 0.25$	$c = 0.043229$
$z = 0.5$	$\mu = 0.181282758$	$b = 0.25$	

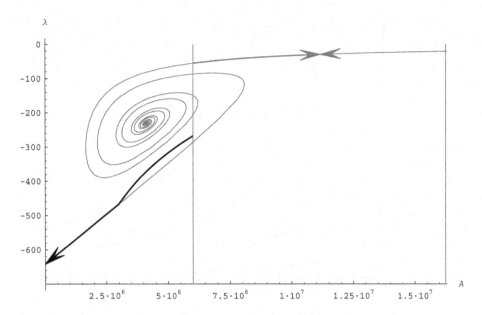

Fig. 1. Optimal paths starting at $A_0 = 6,000,000$ with shock ($p_s = 10 \cdot p_b$, $T = 2$; black curve) and without shock ($T = 0$; dark gray curve)

of the two-stage problem. Comparison of the utility functional values yields the optimal solution.

Figure 1 shows the optimal trajectories in state-costate space for both the "drought" and "no drought" cases. In the absence of a drought, the best strategy is simply to accommodate an increase in drug use up to the high-level equilibrium. Driving down drug use, while possible, is too expensive to be worthwhile. However, with a drought helping to suppress use, it is optimal to push drug use down to its low-level equilibrium. Note that in the drought case use does not only decline because of the drought. Rather, the optimal policy response to the drought starting at 2.5 million users is to greatly increase drug control spending, in effect following a strategy of attacking the market when it is weak. If the drought were created by a supply control intervention and the control is interpreted as treatment, this shows that in certain circumstances supply- and demand-control efforts can be complements, not substitutes, as is often implicitly assumed in debates about the relative merits of supply- vs. demand-side interventions.

Results of this sort pertain as long as the drought is large enough. If the price (p_s) or duration (T) of the drought are small enough, the optimal policy is to moderate the drug problem. If they are large enough the optimal policy is essentially to eradicate the drug epidemic. In either case the drought helps improve the optimal utility functional value.

What sort of shock is of the greatest value to policy makers, a short but intense one or a smaller one that is sustained longer? Figure 2 shows combinations

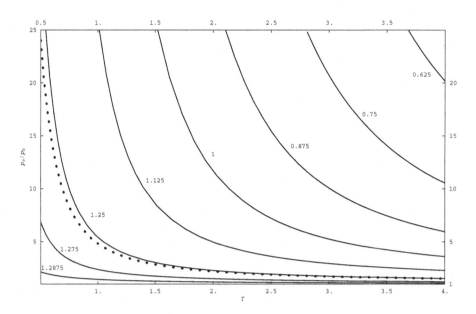

Fig. 2. Level curves of the value function for varying values of the shock length T (abscissa) and the price p_s (ordinate). The values of these level curves are divided by $2 \cdot 10^9$ for ease of exposition. The initial state $A_0 = 6,000,000$ is such that below the dotted curve, convergence is to the high-level steady state, while above the dotted curve, convergence is to the low-level steady state. Minimal costs decrease from lower-left to upper-right.

of drought durations (abscissa) and intensities (ordinate) that yield the same objective functional value. Long and intense droughts (upper-right) are best. Short and mild droughts (lower-left) are the worst, and those to the lower left of the dotted-line are not large enough to make it worth pushing drug use down to the low-level equilibrium.

The figure suggests that duration and intensity are complements. If the drought is very sharp and short, the policy maker would prefer a drought half as intense but twice as long. Conversely, if the drought is very mild but long, the policy maker would prefer one that is twice as intense and half as long. More generally, the area under the price spike curve (height times duration) is not a sufficient statistic for the value of the drought.

4   Extensions

There are multiple avenues for further work of this sort. Some would preserve the basic two-stage structure but elaborate on the underlying model by adding additional drug use states (e.g., to differentiate between light and heavy users) and/or controls (e.g., prevention programs, domestic enforcement that can further drive up prices, etc.). Others would preserve the drug model but elaborate on the modeling of the supply shocks.

The price-spikes induced by supply shocks look more like skewed triangles or decaying exponentials than rectangles. Considering more than two stages would allow more realistic approximations to the actual price variations.

Sometimes it may be sensible to presume policy makers know the duration of a shortage (e.g., when yields for an annual crop such as opium poppies is reduced by a literal drought), but other times the duration of the drought will not be known until it is over. In those cases, there would be a probability distribution governing the length of the shock T.

Sometimes the shock is created not by poor growing conditions or by happenstance but rather is the result of deliberate policy action (arguably that pertains for the 1989/1990 shock to cocaine markets). So in some cases T could be a control variable. However, even when market disruptions are caused by source or transit zone interventions, policy makers do not have complete control over their duration, so T could also be modeled as being stochastic but with the probability distribution affected by drug control spending.

Another intriguing possibility would consider the case of multiple drugs, only one of which is subject to a price shock, in a model that recognizes that many users consume multiple drugs and have some willingness to substitute one drug for another. Such a model would allow one to address questions such as how heroin treatment spending should respond to a cocaine supply shock?

Hence, this model only scratches the surface of multi-stage modeling of illegal markets. It seems to us that the necessary optimality conditions presented in [9] will be helpful for many of these extensions.

Acknowledgements

This research was partly financed by the Austrian Science Fund (FWF) under contract No. P18527-G14. We thank two anonymous referees for their helpful comments and suggestions to improve an earlier version of this paper.

References

1. Behrens, D.A., et al.: Optimal control of drug epidemics: prevent and treat – but not at the same time? Management Science 46(3), 333–347 (2000)
2. Behrens, D.A., et al.: Why present-oriented societies undergo cycles of drug epidemics. Journal of Economic Dynamics and Control 26, 919–936 (2002)
3. Bultmann, R., et al.: How should drug policy respond to market disruptions. In: Contemporary drug problems (forthcoming)
4. Caulkins, J.P.: Drug Prices and Emergency Department Mentions for Cocaine and Heroin. American Journal of Public Health 91(9), 1446–1448 (2001)
5. Crane, B.D., Rivolo, A.R., Comfort, G.C.: An Empirical Examination of Counterdrug Interdiction Program Effectiveness. Institute for Defense Analysis, Alexandria, Virginia (1997)

6. Dechert, W.D., Nishimura, K.: A complete characterization of optimal growth paths in an aggregated model with a non-concave production function. Journal of Economic Theory 31, 332–354 (1983)

7. Feichtinger, G., Hartl, R.F.: Optimale Kontrolle ökonomischer Prozesse: Anwendungen des Maximumprinzips in den Wirtschaftswissenschaften. Walter de Gruyter, Berlin, New York (1986)

8. Grossman, M.: Individual Behaviors and Substance Abuse: The Role of Price. NBER Working Paper No. 10948, National Bureau of Economic Research, Cambridge, MA (2004)

9. Makris, M.: Necessary conditions for infinite-horizon discounted two-stage optimal control problems. Journal of Economic Dynamics and Control 25, 1935–1950 (2001)

10. Moore, T.J., et al.: Heroin Markets in Australia: Current Understanding and Future Possibilities. In: DPMP Monograph Series, Turning Point Alcohol and Drug Centre (2005)

11. Sethi, S.P.: Nearest Feasible Paths in Optimal Control Problems: Theory, Examples, and Counter examples. Journal of Optimization Theory and Applications 23(4), 563–579 (1977)

12. Sethi, S.P.: Optimal Advertising Policy with the Contagion Model. Journal of Optimization Theory and Applications 29(4), 615–627 (1979)

13. Skiba, A.K.: Optimal growth with a convex-concave production function. Econometrica 46, 527–539 (1978)

14. Tragler, G., Caulkins, J.P., Feichtinger, G.: Optimal dynamic allocation of treatment and enforcement in illicit drug control. Operations Research 49(3), 352–362 (2001)

Multicriteria Optimal Control and Vectorial Hamilton-Jacobi Equation

Nathalie Caroff

Laboratoire MANO, Université de Perpignan,
52 avenue Paul Alduy, 66860 Perpignan cedex, France
`caroff@univ-perp.fr`

Abstract. In this paper we investigate a multicriteria optimal control problem associated to a preference relation based on the lexicographic order. We extend different notions of non-smooth analysis and control and show that the vector Value function is the unique vector lower semi-continuous solution to a suitable system of Hamilton-Jacobi equations in the sense of contingent solution or equivalently in the sense of extended viscosity solution.

1 Introduction

We consider a decision maker who has to optimize several objectives. This problem can be described in terms of an optimization problem for a vector valued function associated to a preference relation of the decision maker based on the lexicographic order of $\mathbb{R}^n$. We write $(x_1, \cdots, x_n) \preccurlyeq (y_1, \cdots, y_n)$ if $x = y$ or for some $j \leq n$ and all $i \leq j$, $x_i = y_i$ and $x_j < y_j$. In this context the graph of the preference relation is not necessarily closed, contrary to what is usually assumed in this kind of problem (see for instance [8,9,10]) and scalarization plays no rules.

We underline that our problem is different from those studied in [10] where it was noticed that the preference determinated by the lexicographical order of the vector is not continuous. Furthermore, the results proved there assume that the graph of the preference relation is closed, while those pertaining to the lexicographic order are proved under an hypothesis which does not hold for the lexicographic order. As a matter of fact little is known on this type of problem.

More precisely we consider the following control system where $x \in \mathbb{R}^d$ is a solution of

$$\begin{cases} x'(t) = f(t, x(t), u(t)), \ u(t) \in U \\ x(t_0) = x_0 \end{cases} \tag{1}$$

and the following optimal control problem

$$\text{minimize} \quad \left(\varPhi_i(x(T)) \right)_{i=\{1,\cdots,n\}} \text{ over the solution set to (1)},$$

where $T > 0$, $I = \{1, \cdots, n\}$, f maps $[0, T] \times \mathbb{R}^d \times U$ to $\mathbb{R}^d$ and each $\varPhi_i$ maps $\mathbb{R}^d$ to $\mathbb{R}$.

I. Lirkov, S. Margenov, and J. Waśniewski (Eds.): LSSC 2007, LNCS 4818, pp. 293–299, 2008.

The lexicographic order is total but the infimum of an unbounded set is not well-defined. The good notion in this case will be an extension of the Painlevé-Kuratowski upper limit of a set. The notions of vector upper and lower limit of a set will be defined in the next section. We extend different notions of nonsmooth analysis and control to vector valued functions. The value function associated with this problem, which we call the vector Value function, is defined as follows: for all $(t_0, x_0) \in [0, T] \times \mathbb{R}^d$

$$V(t_0, x_0) = \left(V_i(t_0, x_0) \right)_{i \in I} = \overrightarrow{\liminf}_{x \in S_{[t_0, T]}(x_0)} \left(\Phi_i(x(T)) \right)_{i \in I}. \qquad (2)$$

where $S_{[t_0, T]}(x_0) = \{x \text{ solution to } (1)\}$.

We extend the notions of viscosity subsolutions and supersolutions and show that the vector Value function is the only extended vector valued lower semicontinuous function satisfying a system of Hamilton-Jacobi equations.

$$-\frac{\partial V}{\partial t}(t, x) + H\left(t, x, -\frac{\partial V}{\partial x}(t, x)\right) = 0, \quad V(T, \cdot) = \Phi(\cdot)$$

where the Hamiltonian H is defined by

$$H(t, x, p) = \overrightarrow{\limsup_{u \in U}} \left(\langle p_i, f(t, x, u) \rangle \right)_{i \in I}.$$

The outline of paper is as follows. Section 2 is devoted to definitions and extensions of notions of non-smooth analysis to the vectorial case and in section 3 we prove that the vector Value function is the unique solution to a system of Hamilton-Jacobi equations in the sense of contingent solution and viscosity solution.

2   Vector Lower Semicontinuity

Definition 1. *Let C be a subset of $\mathbb{R}^n$. Consider for $i \in I$ the functions Π_i and p_i defined on $\mathbb{R}^n$ with values, respectively, in $\mathbb{R}^i$ and $\mathbb{R}$, given by $\Pi_i(x_1, \cdots, x_n) = (x_1, \cdots, x_i)$ and $p_i(x_1, \cdots, x_n) = x_i$.*

We define $\overrightarrow{\inf} C = (c_i)_{i \in I}$ as follows :

$$\text{let } C_0 = C \text{ and } c_1 = \inf_{x \in C_0} x_1$$

for $1 \leq i \leq n - 1$ let

$$C_i = \begin{cases} C_{i-1} \cap p_i^{-1}(c_i) & \text{if} \quad \exists \, \overline{x} \in C_{i-1} \text{ such that } p_i(\overline{x}) = c_i \\ C_{i-1} & \text{otherwise} \end{cases}$$

and

$$c_{i+1} = \liminf_{\substack{x \in C_i \\ \Pi_i(x) \longrightarrow (c_1, \cdots, c_i)}} p_{i+1}(x).$$

Similarly, $\overrightarrow{\sup} C = (c_i)_{i \in I}$ *is defined by*

$$c_{i+1} = \limsup_{\substack{x \in C_i \\ \Pi_i(x) \longrightarrow (c_1, \cdots, c_i)}} p_{i+1}(x)$$

where C_i *is defined as above.*

By convention we say that
$\overrightarrow{\inf} C = -\infty$ *if there exists* $i \in I$ *such that* $c_i = -\infty$, $\overrightarrow{\inf} \emptyset = +\infty$, $\overrightarrow{\sup} C = +\infty$
if there exists $i \in I$ *such that* $c_i = +\infty$ *and* $\overrightarrow{\sup} \emptyset = -\infty$.

Remark 1. This definition is a generalization of the notion of infimum for sets, when $n = 1$ it reduces to the usual infimum.

Let X be a metric space.

Definition 2. *Let* $\Phi = (\Phi_i)_{i \in I} : X \mapsto \mathbb{R}^n \cup \{+\infty\}$ *be a vector valued extended function. We define*

$$\overrightarrow{\liminf_{x \longrightarrow x_0}} \Phi(x) = (c_i)_{i \in I} \ \text{as follows :}$$

$$\begin{cases} c_1 = \liminf_{x \longrightarrow x_0} \Phi_1(x) \\ \text{and for all } i > 1 \quad c_i = \liminf_{\substack{x \longrightarrow x_0, \, x \in C_{i-1} \\ \Pi_{i-1}(\Phi(x)) \longrightarrow (c_1, \cdots, c_{i-1})}} \Phi_i(x) \end{cases}$$

where $C_0 = \{\Phi(x) \mid x \in X\}$ *and for* $i \geq 1$, C_i *is defined as in the previous definition and with the same convention.*

Proposition 1

1. *If* $A \subset B \subset \mathbb{R}^n$ *then* $\overrightarrow{\inf} B \preccurlyeq \overrightarrow{\inf} A$.
2. *Let* $F, \Phi : X \mapsto \mathbb{R}^n \cup \{+\infty\}$ *such that* $\forall x \in X, \ F(x) \preccurlyeq \Phi(x)$ *then*

$$\overrightarrow{\liminf_{y \longrightarrow x}} F(x) \preccurlyeq \overrightarrow{\liminf_{y \longrightarrow x}} \Phi(x).$$

3. *If* $F, \Phi : X \mapsto \mathbb{R}^n \cup \{+\infty\}$ *then*

$$\overrightarrow{\liminf_{y \longrightarrow x}} F(x) + \overrightarrow{\liminf_{y \longrightarrow x}} \Phi(x) \preccurlyeq \overrightarrow{\liminf_{y \longrightarrow x}} (F + \Phi)(x).$$

Remark 2. Let $\Phi : \mathbb{R} \mapsto \mathbb{R}^2 \cup \{+\infty\}$ be defined by $\Phi(x) = (x, -1)$ and let $(x_n) = (\frac{1}{n}) \in \mathbb{R}^{\mathbb{N}}$ then $0 \preccurlyeq \Phi(x_n)$, but $\overrightarrow{\liminf}_{n \longrightarrow +\infty} \Phi(x_n) = (0, -1) \preccurlyeq 0$.

Definition 3. *Let* $\Phi : X \mapsto \mathbb{R}^n \cup \{+\infty\}$ *be a vector valued extended function. We define the domain of* Φ *by*

$$Dom(\Phi) = \{x \in \mathsf{X} \mid \Phi(x) \neq +\infty\}$$

Definition 4. *Let* $\Phi : X \mapsto \mathbb{R}^n \cup \{+\infty\}$ *be a vector valued extended function. Then* Φ *is vector lower semicontinuous if for all* $x_0 \in Dom(\Phi)$

$$\Phi(x_0) \preccurlyeq \overrightarrow{\liminf_{x \longrightarrow x_0}} \Phi(x).$$

Remark 3.   1. If for all $i \in I$, $\Phi_i : X \mapsto \mathbb{R}$ is lower semicontinuous, then $(\Phi_i)_{i \in I}$
   is vector lower semicontinuous.

2. Conversely if $\Phi : X \mapsto \mathbb{R}^n$ is vector lower semicontinuous, then V_1 is lower
   semicontinuous but, for $i > 1$, V_i is not necessary lower semicontinuous. For
   example, consider $\Phi : \mathbb{R} \mapsto \mathbb{R}^2$ defined by

$$\Phi(x) = \begin{cases} (1,0) & \text{if } x \neq 0 \\ (0,1) & \text{if } x = 0 \end{cases}.$$

Definition 5. *The lexicographical epigraph of* $\Phi : X \mapsto \mathbb{R}^n \cup \{+\infty\}$ *is defined
by*

$$Epi\,(\Phi) = \{(x,y) \in X \times \mathbb{R}^n : \Phi(x) \preccurlyeq y\}.$$

Remark 4. Let $\Phi : X \mapsto \mathbb{R}^n \cup \{+\infty\}$ be a vector valued extended lower semicontinuous. Then the lexicographical epigraph and the set $\{x \in \mathbb{R}^d \mid \Phi(x) \preccurlyeq \lambda\}$
are not necessarily closed, as one can see from the previous example.

Proposition 2. *Let* $\Phi : X \mapsto \mathbb{R}^n \cup \{+\infty\}$ *be a vector valued lower semicontinuous function defined on a compact set* X. *Then there exists* $\bar{x}$ *such that*
$\Phi(\bar{x}) = \overline{\inf}\{\Phi(x) \mid x \in X\}.$

Definition 6. *Let* $\Phi = (\Phi_i) : X \mapsto \mathbb{R}^n \cup \{+\infty\}$ *be a vector valued extended
function,* $v \in X$ *and* $x_0 \in X$ *be such that* $\Phi(x_0) \neq +\infty$. *We define the contingent
vector epiderivative of* Φ *at* x_0 *in the direction* v *by*

$$D_\uparrow^n \Phi(x_0)\,(v) = \varliminf_{\substack{h \longrightarrow 0+ \\ v' \longrightarrow v}} \left(\frac{\Phi(x_0 + hv') - \Phi(x_0)}{h} \right)$$

and the contingent vector hypoderivative of Φ *at* x_0 *in the direction* v *by*

$$D_\downarrow^n \Phi(x_0)\,(v) = \varlimsup_{\substack{h \longrightarrow 0+ \\ v' \longrightarrow v}} \left(\frac{\Phi(x_0 + hv') - \Phi(x_0)}{h} \right).$$

Definition 7. *Let* $\Phi = (\Phi_i)_{i \in I} : X \mapsto \mathbb{R}^n \cup \{+\infty\}$ *be a vector valued extended
function,* $v \in X$ *and* $x_0 \in X$ *be such that* $\Phi(x_0) \neq +\infty$. *The vector subdifferential
of* Φ *at* x_0 *is the set defined by*

$$\partial_-^n \Phi(x_0) = \left\{ (p_i)_{i \in I} \in X^n \Big| \varliminf_{x \longrightarrow x_0} \left(\frac{\Phi_i(x) - \Phi_i(x_0) - \langle p_i, x - x_0 \rangle}{\|x - x_0\|} \right)_{i \in I} \succcurlyeq 0 \right\}$$

and the vector superdifferential of Φ *at* x_0 *is the set defined by*

$$\partial_+^n \Phi(x_0) = \left\{ (p_i)_{i \in I} \in X^n \Big| \varlimsup_{x \longrightarrow x_0} \left(\frac{\Phi_i(x) - \Phi_i(x_0) - \langle p_i, x - x_0 \rangle}{\|x - x_0\|} \right)_{i \in I} \preccurlyeq 0 \right\}.$$

Theorem 1. *Let* $\Phi : X \mapsto \mathbb{R}^n \cup \{+\infty\}$ *be a vector valued extended function.
Then*

$$\partial_-^n \Phi(x_0) = \left\{ (p_i)_{i \in I} \in X^n \mid \forall v \in X \; D_\uparrow^n \Phi(x_0)\,(v) \succcurlyeq \left(\langle p_i, v \rangle \right)_{i \in I} \right\}.$$

Proposition 3. *Let* $\Phi : X \mapsto \mathbb{R}^n \cup \{+\infty\}$ *be a vector valued extended function. Then*

$$\forall\, x \in Dom(\Phi),\ Epi(D_{\uparrow}^n \Phi(x)) = T_{Epi(\Phi)}(x, \Phi(x))\ and$$

$$Hypo(D_{\uparrow}^n \Phi(x)) \subset T_{Hypo(\Phi)}(x, \Phi(x)).$$

3   Contingent Solutions and Viscosity Solutions of Vectorial Hamilton-Jacobi Equation

We impose the following assumptions

$$\begin{cases} i) & f \text{ is continuous and for all } R > 0,\ \exists\, c_R \in L^1(0, T) \text{ such that} \\ & \text{for almost all } t,\ \forall\, u \in U,\ f(t, \cdot, u) \text{ is } c_R(t) \text{ - Lipschitz on } B_R(0). \\[2mm] ii) & \exists\, k \in L^1(0, T) \text{ such that for almost all } t \in [0, T], \\ & \forall\, x \in \mathbb{R}^d,\ \sup_{u \in U} \| f(t, x, u) \| \le k(t)(1 + \|x\|). \\[2mm] iii) & \forall\, (t, x) \in [0, T] \times \mathbb{R},\ f(t, x, U) \text{ is convex, compact.} \end{cases} \qquad (3)$$

$$\Phi : \mathbb{R}^d \mapsto \mathbb{R}^n \cup \{+\infty\} \text{ is vector lower semicontinuous.} \qquad (4)$$

where $B_R(0)$ denotes the ball of radius R.

With these assumptions the control system (1) may be replaced by the differential inclusion

$$x'(t) \in F(t, x(t)) \text{ almost everywhere}$$

where $F(t, x) = \{f(t, x, u) \mid u \in U\}$ satisfies the following conditions:

$$\begin{cases} i) & F \text{ is continuous and has nonempty convex compact images.} \\[2mm] ii) & \exists\, k \in L^1(0, T) \text{ such that for almost all } t \in [0, T] \\ & \forall\, x \in \mathbb{R}^d,\ \sup_{v \in F(t,x)} \|v\| \le k(t)(1 + \|x\|). \\[2mm] iii) & \forall\, R > 0,\ \exists\, c_R \in L^1(0, T) \text{ such that for a.e } t \in [0, T] \\ & F(t, \cdot) \text{ is } c_R(t)\text{-Lipschitz on } B_R(0). \end{cases} \qquad (5)$$

Proposition 4. *Assume (3) and (4), let V be defined by (2). Then V is vector valued lower semicontinuous with values in $\mathbb{R}^n \cup \{+\infty\}$. Moreover for any $(t_0, x_0) \in [0, T] \times \mathbb{R}^d$ there exists a control-pair $(\overline{x}, \overline{u})$ solution of (1) such that $V(t_0, x_0) = \Phi(\overline{x}(T))$.*

The two following lemmas will be needed in the proof of theorem (2)

Lemma 1. *Let $V : [0, T] \times \mathbb{R}^d \mapsto \mathbb{R}^n \cup \{+\infty)$ be a vector valued extended lower semicontinuous function. Assume that : (1) F is upper semicontinuous, (2) $F(t, x)$ is nonempty convex and compact for all $(t, x) \in Dom(V)$, and (3) for some $k \in L^1(0, T)$ and for all $(t, x) \in Dom(V)$*

$$\sup_{v \in F(t,x)} \|v\| \le k(t)(1 + \|x\|).$$

Then the two following statements are equivalent

i) $\forall\,(t,x)\in Dom(V)$ with $t<T$, $\overrightarrow{\underset{v\in F(t,x)}{\inf}}\,D_{\uparrow}^{n}\,V(t,x)(1,v)\preccurlyeq 0.$

ii) For all $(t_0,x_0)\in[0,T]\times\mathbb{R}^d$, there exists $x\in S_{[t_0,T]}(x_0)$ such that,

for all $t\in[t_0,T]$ $V(t,x(t))\preccurlyeq V(t_0,x_0).$

Lemma 2. *Let* $V:[0,T]\times\mathbb{R}^d\mapsto\mathbb{R}^n\cup\{+\infty\}$ *be a vector valued extended lower semicontinuous function. If* F *satisfies (5), then the two statements below are equivalent*

i) $\forall\,(t,x)\in Dom(V)$ with $t>0$, $\overrightarrow{\underset{v\in F(t,x)}{\sup}}\,D_{\uparrow}^{n}V(t,x)(-1,-v)\preccurlyeq 0.$

ii) For all $x\in S_{[t_0,T]}(x_0)$ and all $t\in[t_0,T]$, $V(t_0,x_0)\preccurlyeq V(t,x(t)).$

Theorem 2. *Assume (3). Then the vector valued function* V *defined by (2) is the unique vector valued lower semicontinuous function from* $[0,T]\times\mathbb{R}^d$ *into* $\mathbb{R}^n\cup\{+\infty\}$ *such that* $V(T,\cdot)=\Phi(\cdot)$ *and for all* $(t,x)\in Dom(V)$ *the following holds*

$$\begin{cases} \text{for } 0\le t<T\ \ \overrightarrow{\underset{v\in F(t,x)}{\inf}}\,D_{\uparrow}^{n}V(t,x)(1,v)\preccurlyeq 0.\\ \text{for } 0<t\le T\ \ \overrightarrow{\underset{v\in F(t,x)}{\sup}}\,D_{\uparrow}^{n}V(t,x)(-1,-v)\preccurlyeq 0. \end{cases}$$

Definition 8. *The Hamiltonian associated to problem (1), (2) is the vector valued function* $H:[0,T]\times\mathbb{R}^d\times(\mathbb{R}^d)^n\longrightarrow\mathbb{R}^n$ *given by*

$$H(t,x,p)=\overrightarrow{\underset{u\in U}{\sup}}\,\big(\langle p_i,f(t,x,u)\rangle\big)_{i\in I}\,.$$

Next consider the vectorial Hamilton-Jacobi equation

$$-\frac{\partial V}{\partial t}(t,x)+H\left(t,x,\frac{\partial V}{\partial x}(t,x)\right)=0,\quad V(T,\cdot)=\Phi(\cdot). \tag{6}$$

We now extend the notions of viscosity solutions to this vectorial problem but the notion of subsolution replaced by undersolution. In the vectorial case the notion of viscosity solution is too strong since a vector extended lower and upper continuous solution $V=(V_i)_{i\in I}$ would have to be continuous, but when there exist several optimal trajectories, it is obvious that generally even V_2 is not continuous but it is only lower semicontinuous.

Definition 9. *A vector valued extended lower semicontinuous function* $V:$ $[0,T]\times\mathbb{R}^d\mapsto\mathbb{R}^n\cup\{+\infty\}$ *is called a vector viscosity supersolution to (6) if for all* $t\in]0,T[$ *and* $x\in\mathbb{R}^d$ *such that* $(t,x)\in Dom(V)$ *we have*

$$\forall(p_t,p_x)\in\partial_{-}^{n}V(t,x),\quad -p_t+H(t,x,-p_x)\succcurlyeq 0.$$

A vector valued extended lower semicontinuous function $V:[0,T]\times\mathbb{R}^d\mapsto\mathbb{R}^n\cup$ $\{+\infty\}$ *is called a vector viscosity undersolution to (6) if for all* $t\in]0,T[$ *and* $x\in\mathbb{R}^d$ *such that* $(t,x)\in Dom(V)$ *we have*

$$\forall(p_t,p_x)\in\partial_{-}^{n}V(t,x),\quad -p_t+H(t,x,-p_x)\preccurlyeq 0.$$

A vector valued extended lower semicontinuous function $V : [0,T] \times \mathbb{R}^d \mapsto \mathbb{R}^n \cup \{+\infty\}$ is called a vector viscosity solution to (6) if it is a vector viscosity super and under solution.

We now caracterize the solution to a system of Hamilton-Jacobi equations in the sense of viscosity and contingent solution

Theorem 3. *Assume (5) and consider a vector extended lower semicontinuous function $V : [0,T] \times \mathbb{R}^d \mapsto \mathbb{R}^n \cup \{+\infty\}$. Then the following statements are equivalent*

i) V is the vector Value function given by (2).

ii)
$$
\begin{cases}
\forall (t,x) \in [0,T] \times \mathbb{R}^d, \ \forall (p_t, p_x) \in \partial_-^n V(t,x) \quad -p_t + H(t,x,-p_x) = 0. \\[2mm]
\forall x \in \mathbb{R}^d, \quad V(0,x) = \varliminf_{\substack{t \longrightarrow 0_+ \\ x' \longrightarrow x}} V(t,x'). \\[2mm]
\forall x \in \mathbb{R}^d, \quad V(T,x) = \varliminf_{\substack{t \longrightarrow T_- \\ x' \longrightarrow x}} V(t,x').
\end{cases}
$$

iii)
$$
\begin{cases}
\text{for } 0 \le t < T \quad \overrightarrow{\inf_{v \in F(t,x)}} D_\uparrow^n \Phi(t,x)(1,v) \preccurlyeq 0. \\[2mm]
\text{for } 0 < t \le T \quad \overrightarrow{\sup_{v \in F(t,x)}} D_\uparrow^n(\Phi)(t,x)(-1,-v) \preccurlyeq 0.
\end{cases}
$$

References

1. Aubin, J.P., Cellina, A.: Differential Inclusions. Springer, Berlin (1984)
2. Aubin, J.P., Frankowska, H.: Set-valued analysis. Birkhaüser, Berlin (1991)
3. Barron, E.N., Jensen, R.: Semicontinuous viscosity solutions for Hamilton-Jacobi equations with convex Hamiltonians. Partial Differential Equations 15, 1713–1742 (1990)
4. Borwein, J.M., Nieuwenhuis, J.W.: Two kind of Normality in vector optimization. Mathematical Programming 28, 185–191 (1984)
5. Clarke, F.H.: Optimization and non smooth analysis. Wiley-Interscience, New York (1983)
6. Crandall, M.G., Lyons, P.L.: Viscosity solutions of Hamilton-Jacobi equations. Transactions of American Mathematical Society 277, 1–42 (1983)
7. Frankowska, H.: Lower semicontinuous solutions of Hamilton-Jacobi-Bellman equations. SIAM J. Control Optim. 31, 257–272 (1993)
8. Imbert, C., Volle, M.: On vectorial Hamilton-Jacobi equations. Well-posedness in optimization and related topics. Control Cybernet 31, 493–506 (2002)
9. Volle, M.: Duality Principle for optimization problems Dealing with the Difference of vector-valued Convex Mappings. J. Optim. Theory Appl. 114, 223–241 (2002)
10. Zhu, Q.J.: Hamiltonian necessary conditions for a multiobjective optimal control problem with endpoint constraints. SIAM J. Control Optim. 39, 97–112 (2000)

Descent-Penalty Methods for Relaxed Nonlinear Elliptic Optimal Control Problems

Ion Chryssoverghi[1] and Juergen Geiser[2]

[1] Department of Mathematics, National Technical University of Athens
Zografou Campus, 15780 Athens, Greece
`ichris@central.ntua.gr`
[2] Department of Mathematics, Humboldt University
Unter den Linden 6, D-10099 Berlin, Germany
`geiser@mathematik.hu-berlin.de`

Abstract. An optimal control problem is considered, described by a second order elliptic partial differential equation, jointly nonlinear in the state and control variables, with high monotone nonlinearity in the state, and with control and state constraints. Since no convexity assumptions are made, the problem may have no classical solutions, and so it is reformulated in the relaxed form. The relaxed problem is discretized by a Galerkin finite element method for state approximation, while the controls are approximated by elementwise constant relaxed ones. The first result is that relaxed accumulation points of sequences of admissible and extremal discrete controls are admissible and extremal for the continuous relaxed problem. We then propose a mixed conditional descent-penalty method, applied to a fixed discrete relaxed problem, and also a progressively refining version of this method that reduces computing time and memory. We show that accumulation points of sequences generated by the fixed discretization (resp. progressively refining) method are admissible and extremal for the discrete (resp. continuous) relaxed problem. Numerical examples are given. This paper proposes relaxed discretization and optimization methods instead of the corresponding classical methods presented in [5]. Considered here problems are with not necessarily convex control constraint sets, and with state constraints and cost functionals depending also on the state gradient. Also, the results of Sections 1 and 2 generalize those of [8] w.r.t. the assumptions made.

1 The Continuous Optimal Control Problems

Let Ω be a bounded domain in $\mathbb{R}^d$, with Lipschitz boundary Γ. Consider the nonlinear elliptic state equation
$$Ay + f(x, y(x), w(x)) = 0 \text{ in } \Omega, \quad y(x) = 0 \text{ on } \Gamma,$$
where A is the second order elliptic differential operator
$$Ay := -\sum_{j=1}^{d} \sum_{i=1}^{d} (\partial/\partial x_i)[a_{ij}(x)\partial y/\partial x_j].$$
The state equation will be interpreted in the following weak form
$$y \in V := H_0^1(\Omega) \text{ and } a(y, v) + \int_\Omega f(x, y(x), w(x))v(x)dx = 0, \ \forall v \in V,$$

I. Lirkov, S. Margenov, and J. Waśniewski (Eds.): LSSC 2007, LNCS 4818, pp. 300–308, 2008.
© Springer-Verlag Berlin Heidelberg 2008

where $a(\cdot, \cdot)$ is the usual bilinear form on $V \times V$ associated with A

$$a(y, v) := \sum_{i,j=1}^{d} \int_{\Omega} a_{ij}(x)(\partial y/\partial x_i)(\partial v/\partial x_j)dx.$$

Define the set of classical controls $W := \{w : \Omega \to U \mid w \text{ measurable}\}$, where U is a compact (not necessarily convex) subset of $\mathbb{R}^{\nu}$, and the functionals
$G_m(w) := \int_{\Omega} g_m(x, y(x), \nabla y(x), w(x))dx, \quad m = 0, ..., q.$
The continuous classical optimal control problem P is to minimize $G_0(w)$ subject to the constraints $w \in W$, $G_m(w) = 0$, $m = 1, ..., p$, $G_m(w) \leqslant 0$, $m = p+1, ..., q$.

Next, define the set of relaxed controls (or Young measures, see [13,12])
$R := \{r : \Omega \to M_1(U) \mid r \text{ weakly measurable}\} \subset L_w^{\infty}(\Omega, M(U)) \equiv L^1(\Omega, C(U))^*,$
where $M(U)$ (resp. $M_1(U)$) is the set of Radon (resp. probability) measures on U. The set R is endowed with the relative weak star topology, and R is convex, metrizable and compact. If each classical control $w(\cdot)$ is identified with its associated Dirac relaxed control $r(\cdot) := \delta_{w(\cdot)}$, then W may also be regarded as a subset of R, and W is thus dense in R. For a given $\phi \in L^1(\Omega; C(U)) \equiv B(\bar{\Omega}, U; \mathbb{R})$, where $B(\bar{\Omega}, U; \mathbb{R})$ denotes the set of Caratheodory functions in the sense of Warga [13], and $r \in R$, we shall write, for simplicity

$$\phi(x, r(x)) := \int_U \phi(x, u)r(x)(du).$$

The continuous relaxed optimal control Problem $\bar{P}$ is then defined by replacing w by r, with the above notation, and W by R in the continuous classical problem.

We suppose that the coefficients a_{ij} satisfy the ellipticity condition

$$\sum_{i,j=1}^{d} a_{ij}(x)z_i z_j \geqslant \alpha_0 \sum_{i=1}^{d} z_i^2, \quad \forall z_i, z_j \in \mathbb{R}, \quad x \in \Omega,$$

with $\alpha_0 > 0$, $a_{ij} \in L^{\infty}(\Omega)$, and that the functions f, f_y (resp $g_m, g_{my}, g_{my'}$) are defined on $\Omega \times \mathbb{R} \times U$ (resp. on $\Omega \times \mathbb{R}^{d+1} \times U$), measurable for fixed y, u (resp. y, y', u), continuous for fixed x, and satisfy
$|f(x, y, u)| \leqslant c_1(1 + |y|^{\rho-1}), \ 0 \leqslant f_y(x, y, u) \leqslant c_2(1 + |y|^{\rho-2})$, in $\Omega \times \mathbb{R} \times U$,
$|g_m(x, y, y', u)| \leqslant c_3(1 + |y|^{\rho} + |y'|^2)$,
$|g_{my}(x, y, y', u)| \leqslant c_4(1 + |y|^{\rho-1} + |y'|^{\frac{2(\rho-1)}{\rho}})$,
$|g_{my'}(x, y, y', u)| \leqslant c_5(1 + |y|^{\frac{\rho}{2}} + |y'|)$, in $\Omega \times \mathbb{R}^{d+1} \times U$,
with $c_i \geqslant 0$, $2 \leqslant \rho < +\infty$ if $d = 1$ or 2, $2 \leqslant \rho < \frac{2d}{d-2}$ if $d \geqslant 3$.
For every $r \in R$, the state equation has a unique solution $y := y_r \in V$ (see [2]). The results of this section can be proved by using the techniques of [8,13].

Theorem 1. *If the continuous relaxed problem is feasible, then it has a solution.*

Lemma 1. *Dropping m in G_m, g_m, the directional derivative of G is given by*
$DG(r, \bar{r} - r) = \lim_{\alpha \to 0^+} \{[G(r + \alpha(\bar{r} - r)) - G(r)]/\alpha\}$
$= \int_{\Omega} H(x, y(x), z(x), r'(x) - r(x))dx$, *for $r, \bar{r} \in R$,*
where the Hamiltonian H is defined by
$H(x, y, y', z, u) := -z f(x, y, u) + g(x, y, y', u),$

and the adjoint state $z := z_r \in V$ (resp. $z := z_w$) satisfies the linear adjoint equation
$$a(v, z) + (f_y(y, r)z, v) = (g_y(y, \nabla y, r), v) + (g_{y'}(y, \nabla y, r), \nabla v), \ \forall v \in V, \ \text{with } y := y_r.$$

Theorem 2. (Necessary Conditions for Optimality) If $r \in R$ is optimal for Problem $\bar{P}$, then r is extremal, i.e. there exist multipliers $\lambda_m \in \mathbb{R}$, $m = 0, ..., q$, with $\lambda_0 \geqslant 0$, $\lambda_m \geqslant 0$, $m = p+1, ..., q$, $\sum_{m=0}^{q} |\lambda_m| = 1$, such that
$$\sum_{m=0}^{q} \lambda_m DG_m(r, \bar{r} - r) \geqslant 0, \ \forall \bar{r} \in R,$$
$$\lambda_m G_m(r) = 0, \quad m = p+1, ..., q \ (transversality \ conditions).$$

The above inequalities are equivalent to the relaxed pointwise minimum principle
$$H(x, y(x), \nabla y(x), z(x), r(x)) = \min_{u \in U} H(x, y(x), \nabla y(x), z(x), u), \ \text{a.e. in } \Omega,$$

where the complete Hamiltonian H and adjoint z are defined with $g := \sum_{m=0}^{q} \lambda_m g_m$.

2   Discretization

We suppose in what follows that Ω is a polyhedron, for simplicity. For each integer $n \geqslant 0$, let $\{E_i^n\}_{i=1}^{N^n}$ be an admissible regular partition of $\bar{\Omega}$ into elements (e.g. d-simplices), with $h^n = \max_i[\text{diam}(E_i^n)] \to 0$ as $n \to \infty$. Let $V^n \subset V$ be the subspace of functions that are continuous on $\bar{\Omega}$ and multilinear (or linear for d-simplices) on each element E_i^n. The set of discrete relaxed (resp. classical) controls $R^n \subset R$ (resp. $W^n \subset W$) is defined as the subset of relaxed (resp. classical) controls that are equal, on each element $\overset{o}{E_i^n}$, to a constant probability measure on U (resp. constant value in U). Clearly, $W^n \subset R^n$. We endow R^n with the weak star topology of $M_1(U)^N$. For a given discrete control $r^n \in R^n$, the discrete state $y^n := y_{r^n}^n \in V^n$ is the solution of the discrete state equation
$$a(y^n, v^n) + (f(y^n, r^n), v^n) = 0, \ \forall v^n \in V^n.$$
For every $r^n \in R^n$, the discrete state equation (a nonlinear system) has a unique solution $y^n \in V^n$ (see [9]), and can be solved by iterative methods. The discrete functionals are defined by
$$G_m^n(r^n) = \int_\Omega g_m(x, y^n, \nabla y^n, r^n) dx, \ m = 0, ..., q.$$
The discrete control constraint is $r^n \in R^n$ and the discrete state constraints are
$$G_m^n(r^n) = \varepsilon_m^n, \ m = 1, ..., p, \ \text{and } G_m^n(r^n) \leqslant \varepsilon_m^n, \ \varepsilon_m^n \geqslant 0, \ m = p+1, ..., q,$$
where the feasibility perturbations ε_m^n are chosen numbers converging to zero, to be defined later. The discrete relaxed optimal control Problem $\bar{P}^n$ is to minimize $G_0^n(r^n)$ subject to $r^n \in R^n$ and to the above state constraints. The results of this section can be proved by using the techniques of [8].

Theorem 3. The operator $r^n \mapsto y^n$, from R^n to V^n, and the functionals $r^n \mapsto G_m^n(r^n)$ on R^n, are continuous. For every n, if Problem $\bar{P}^n$ is feasible, then it has a solution.

Lemma 2. *Dropping the index m, the directional derivative of G^n is given by $DG^n(r^n, \bar{r}^n - r^n) = \int_\Omega H(x, y^n, \nabla y^n, z^n, \bar{r}^n - r^n)dx,$*
where the discrete adjoint state $z^n := z_{r^n}^n \in V^n$ satisfies the discrete adjoint equation
$$a(z^n, v^n) + (z^n f_y(y^n, r^n), v^n) = (g_y(y^n, \nabla y^n, r^n), v^n) + (g_{y'}(y^n, \nabla y^n, r^n), \nabla v^n),$$
$\forall v^n \in V^n$, with $y^n := y_{r^n}^n$.
Moreover, the operator $r^n \mapsto z_{r^n}^n$, from R^n to V^n, and the functional $(r^n, \bar{r}^n) \mapsto DG^n(r^n, \bar{r}^n - r^n)$ on $R^n \times R^n$, are continuous.

Theorem 4. *(Discrete Necessary Conditions for Optimality) If $r^n \in R^n$ is optimal for Problem $\bar{P}^n$, then r^n is discrete extremal, i.e. there exist multipliers $\lambda_m^n \in \mathbb{R}$, $m = 0, ..., q$, with $\lambda_0^n \geqslant 0$, $\lambda_m^n \geqslant 0$, $m = p+1, ..., q$, $\sum_{m=0}^{q} |\lambda_m^n| = 1$, such that*
$$\sum_{m=0}^{q} \lambda_m^n DG_m^n(r^n, \bar{r}^n - r^n) = \int_\Omega H^n(x, y^n, \nabla y^n, z^n, \bar{r}^n - r^n)dx \geqslant 0, \ \forall \bar{r}^n \in R^n,$$
$$\lambda_m^n[G_m(r^n) - \varepsilon_m^n] = 0, \ m = p+1, ..., q,$$
where H^n and z^n are defined with $g := \sum_{m=0}^{q} \lambda_m^n g_m$. The above inequalities are equivalent to the discrete relaxed elementwise minimum principle
$$\int_{E_i^n} H^n(x, y^n, \nabla y^n, z^n, r^n)dx = \min_{u \in U} \int_{E_i^n} H^n(x, y^n, \nabla y^n, z^n, u)dx, \ i = 1, ..., N^n.$$

Proposition 1. *For every $r \in R$, there exists a sequence $(w^n \in W^n \subset R^n)$ of discrete classical controls, regarded as relaxed ones, that converges to r in R.*

Lemma 3. *(Consistency) (i) If the sequence $(r^n \in R^n)$ converges to $r \in R$ in R, then $y^n \to y_r$ in V strongly, $G^n(r^n) \to G(r)$, and $z^n \to z_r$ in $L^p(\Omega)$ strongly (and in V strongly, if the functionals do not depend on ∇y).*
(ii) If the sequences $(r^n \in R^n)$, $(\bar{r}^n \in R^n)$ converge to $r, \bar{r}$ in R, then $DG^n(r^n, \bar{r}^n - r^n) \to DG(r, \bar{r} - r)$.

We suppose in what follows that Problem $\bar{P}$ is feasible. We now examine the behavior in the limit of extremal discrete controls. We shall construct sequences of perturbations (ε_m^n) that converge to zero and such that the discrete problem $\bar{P}^n$ is feasible for every n. Let $r'^n \in R^n$ be any solution of the following auxiliary minimization problem without state constraints
$$c^n := \min_{r^n \in R^n} \{ \sum_{m=1}^{p} [G_m^n(r^n)]^2 + \sum_{m=p+1}^{q} [\max(0, G_m^n(r^n))]^2 \},$$
and set $\varepsilon_m^n := G_m^n(r'^n)$, $m = 1, ..., p$, $\varepsilon_m^n := \max(0, G_m^n(r'^n))$, $m = p+1, ..., q$. It can be easily shown that $c^n \to 0$, hence $\varepsilon_m^n \to 0$. Then clearly $\bar{P}^n$ is feasible for every n, for these ε_m^n. We suppose in what follows that the ε_m^n are chosen as in the above minimum feasibility procedure.

Theorem 5. *For each n, let r^n be admissible and extremal for Problem $\bar{P}^n$. Then every accumulation point of the sequence (r^n) is admissible and extremal for Problem $\bar{P}$.*

3   Discrete Relaxed Descent-Penalty Methods

Let (M_m^l), $m = 1, ..., q$, be positive increasing sequences such that $M_m^l \to \infty$ as $l \to \infty$, and define the penalized discrete functionals

$$G^{nl}(r^n) := G_0^n(r^n) + \{ \sum_{m=1}^{p} M_m^l [G_m^n(r^n)]^2 + \sum_{m=p+1}^{q} M_m^l [\max(0, G_m^n(r^n))]^2 \}/2.$$

Let $b, c \in (0, 1)$, and let (β^l), (ζ_k) be positive sequences, with (β^l) decreasing and converging to zero, and $\zeta_k \leqslant 1$. The algorithm described below contains two options. In the case of the progressively refining version, we suppose in addition that each element $E_{i'}^{n+1}$ is a subset of some E_i^n, in which case $W^n \subset W^{n+1}$.

Algorithm

Step 1. Set $k := 0$, $l := 1$, choose a value of n and an initial control $r_0^{n1} \in R^n$.

Step 2. Find $\bar{r}_k^{nl} \in R^n$ such that
$d_k := DG^{nl}(r_k^{nl}, \bar{r}_k^{nl} - r_k^{nl}) = \min_{r'^n \in R^n} DG^{nl}(r_k^{nl}, r'^n - r_k^{nl})$.

Step 3. If $|d_k| > \beta^l$, go to Step 4. Else, set $r^{nl} := r_k^{nl}$, $\bar{r}^{nl} := \bar{r}_k^{nl}$, $d^l := d_k$, then:
In Version A: Set $r_k^{n,l+1} := r_k^{nl}$. In Version B: Set $r_k^{n+1,l+1} := r_k^{nl}$, $n := n + 1$.
In both versions, set $l := l + 1$ and go to Step 2.

Step 4. (Modified Armijo Step Search) Find the lowest integer value $s \in \mathbb{Z}$, say $\bar{s}$, such that $\alpha(s) = c^s \zeta_k \in (0, 1]$ and $\alpha(s)$ satisfies the inequality
$G^{nl}(r_k^{nl} + \alpha(s)(\bar{r}_k^{nl} - r_k^{nl})) - G^{nl}(r_k^{nl}) \leqslant \alpha(s)bd_k$,
and then set $\alpha_k := \alpha(\bar{s})$.

Step 5. Choose any $r_{k+1}^{nl} \in R^n$ such that
$G^{nl}(r_{k+1}^{nl}) \leqslant G^{nl}(r_k^{nl} + \alpha_k(\bar{r}_k^{nl} - r_k^{nl}))$,
set $k := k + 1$, and go to Step 2.

This Algorithm contains two versions:
Version A: n is a constant integer chosen in Step 1, i.e. a fixed discretization is chosen, and the G_m^n, $m = 1, ..., q$, are replaced by the perturbed ones $\tilde{G}_m^n = G_m^n - \varepsilon_m^n$.
Version B: This is a progressively refining method, i.e. $n \to \infty$, in which case we can take $n = 1$ in Step 1, hence $n = l$ in the Algorithm. This version has the advantage of reducing computing time and memory, and also of avoiding the computation of minimum feasibility perturbations ε_m^n (see Section 2). It is justified by the fact that finer discretizations become progressively more essential as the iterate gets closer to an extremal control.

One can easily see that a classical control $\bar{r}_k^{nl}$ in Step 2 can be found for every k by minimizing w.r.t. $u \in U$ the integral $\int_{E_i^n} H(x, y^n, \nabla y^n, z^n, u)dx$ on the element E_i^n (practically by using some numerical integration rule), independently for each $i = 1, ..., M$. On the other hand, by the definition of the directional derivative and since $b, c \in (0, 1)$ and $d_k \neq 0$, the Armijo step α_k in Step 4 can be found for every k.

With r^{nl} defined in Step 3, define the sequences of multipliers
$$\lambda_m^{nl} = M_m^l G_m^n(r^{nl}), \ m = 1, ..., p, \ \lambda_m^{nl} = M_m^l \max(0, G_m^n(r^{nl})), \ m = p+1, ..., q.$$

Theorem 6. *i) In Version B, let (r^{nl}) be a subsequence of the sequence generated by the Algorithm in Step 3 that converges to some $r \in R$, as $l \to \infty$ (hence $n \to \infty$). If the sequences of multipliers (λ_m^{nl}) are bounded, then r is admissible and extremal for Problem $\bar{P}$.*
(ii) In Version A, let (r^{nl}), n fixed, be a subsequence of the sequence generated by the Algorithm in Step 3 that converges to some $r^n \in R^n$ as $l \to \infty$. If the sequences (λ_m^{nl}) are bounded, then r^n is admissible and discrete extremal for Problem $\bar{P}^n$.

Proof. It can first be shown by contradiction, similarly to Theorem 5.1 in [7], that $l \to \infty$ in the Algorithm, hence $d^l \to 0$, $e^l \to 0$, in Step 3, and $n \to \infty$ in Version B.
(i) Let (r^{nl}) be a subsequence (same notation) of the sequence generated by the Algorithm in Step 3 that converges to some $r \in R$ as $l, n \to \infty$. Suppose that the sequences (λ_m^{nl}) are bounded and (up to subsequences) that $\lambda_m^{nl} \to \lambda_m$. By Lemma 3, we have
$$0 = \lim_{l \to \infty} \frac{\lambda_m^{nl}}{M_m^l} = \lim_{l \to \infty} G_m^n(r^{nl}) = G_m(r), \quad m = 1, ..., p,$$
$$0 = \lim_{l \to \infty} \frac{\lambda_m^{nl}}{M_m^l} = \lim_{l \to \infty} [\max(0, G_m^n(r^{nl}))] = \max(0, G_m(r)), \ m = p+1, ..., q,$$
which show that r is admissible. Now let any $\tilde{r} \in R$ and, by Proposition 1, let $(\tilde{r}^n \in R^n)$ be a sequence of discrete controls that converges to $\tilde{r}$. By Step 2, we have
$$\sum_{m=0}^{q} \lambda_m^{nl} DG_m^n(r^{nl}, \tilde{r}^n - r^{nl}) \geqslant d^l,$$
where $\lambda_0^{nl} := 1$. Since $|d^l| \leqslant \beta^l$ by Step 3, we have $d^l \to 0$. By Lemma 3, we can pass to the limit as $l, n \to \infty$ in the above inequality and obtain the first optimality condition
$$\sum_{m=0}^{q} \lambda_m DG_m(r, \tilde{r} - r) \geqslant 0, \ \forall \tilde{r} \in R.$$
By construction of the λ_m^{nl}, we clearly have $\lambda_0 = 1$, $\lambda_m \geqslant 0$, $m = p+1, ..., q$, $\sum_{m=0}^{q} |\lambda_m| := c \geqslant 1$, and we can suppose that $\sum_{m=0}^{q} |\lambda_m| = 1$, by dividing the above inequality by c. On the other hand, if $G_m(r) < 0$, for some index $m \in [p+1, q]$, then for sufficiently large l we have $G_m^{nl}(r^{nl}) < 0$ and $\lambda_m^l = 0$, hence $\lambda_m = 0$, i.e. the transversality conditions hold. Therefore, r is also extremal.
(ii) The admissibility of the limit control r^n is proved as in (i). Passing here to the limit in the inequality resulting from Step 2, as $l \to \infty$, for n fixed, and using Theorem 3 and Lemma 2, we obtain, similarly to (i)
$$\sum_{m=0}^{q} \lambda_m D\tilde{G}_m^n(r^n, \tilde{r}^n - r^n) = \sum_{m=0}^{q} \lambda_m DG_m^n(r^n, \tilde{r}^n - r^n) \geqslant 0, \ \forall r'^n \in R^n,$$
and the discrete transversality conditions

$\lambda_m^n \tilde{G}_m^n(r^n) = \lambda_m^n[G_m^n(r^n) - \varepsilon_m^n] = 0, \; m = p+1, ..., q,$

with multipliers λ_m^n as in the discrete optimality conditions.

The Algorithm can be practically implemented as follows. Suppose that the integrals involving f and g_m, $m = 0, ..., q$, are calculated with sufficient accuracy by some numerical integration rule, involving (usually a small number) μ nodes x_{ji}^n, $j = 1, ..., \mu$, on each element E_i^n, of the form

$$\int_{E_i^n} \phi(x)dx \approx \text{meas}(E_i^n) \sum_{j=1}^{\mu} C_j \phi(x_{ji}^n).$$

We first choose the initial discrete control in Step 1 to be of Gamkrelidze type, i.e. equal on each E_i^n to a convex combination of $(\mu + q + 1) + 1$ Dirac measures on U concentrated at $(\mu + q + 1) + 1$ points of U. Suppose, by induction, that the control r_k^{nl} computed in the Algorithm is of Gamkrelidze type. Since the control $\bar{r}_k^{nl}$ in Step 2 is chosen to be classical (see above), i.e. elementwise Dirac, the resulting control $\tilde{r}_k^{nl} := (1 - \alpha_k)r_k^{nl} + \alpha_k\bar{r}_k^{nl}$ in Step 5 is elementwise equal to a convex combination of $(\mu + q + 1) + 2$ Dirac measures. Using now a known property of convex hulls of finite vector sets, we can construct a Gamkrelidze control r_{k+1}^{nl} equivalent to $\tilde{r}_k^{nl}$, i.e. such that the following $\mu + q + 1$ equalities (equality in $\mathbb{R}^{\mu+q+1}$) hold for each $i = 1, ..., M$

$f(x_{ji}^n, \tilde{y}^{nl}(x_{ji}^n), r_{k+1,i}^{nl}) = f(x_{ji}^n, \tilde{y}^{nl}(x_{ji}^n), \tilde{r}_{ki}^{nl}), \; j = 1, ..., \mu,$

$\text{meas}(E_i^n) \sum_{j=1}^{\mu} C_j g_m(x_{ji}^n, \tilde{y}^{nl}(x_{ji}^n), \nabla\tilde{y}^{nl}(x_{ji}^n), r_{k+1,i}^{nl})$

$= \text{meas}(E_i^n) \sum_{j=1}^{\mu} C_j g_m(x_{ji}^n, \tilde{y}^{nl}(x_{ji}^n), \nabla\tilde{y}^{nl}(x_{ji}^n), \tilde{r}_{ki}^{nl}), \; m = 0, ..., q,$

where $\tilde{y}^{nl}$ corresponds to $\tilde{r}_k^{nl}$, by selecting only $(\mu + q + 1) + 1$ appropriate points in U among the $(\mu + q + 1) + 2$ ones defining $\tilde{r}_k^{nl}$, for each i. Then the control r_{k+1}^{nl} clearly yields the same discrete state and functionals as $\tilde{r}_k^{nl}$. Therefore, the constructed control r_k^{nl} is of Gamkrelidze type for every k. Finally, discrete Gamkrelidze controls computed as above can then be approximated by piecewise constant classical controls using a standard procedure (see [6]), by subdividing here the elements in appropriate subsets (e.g. subelements) whose measures are proportional to the Gamkrelidze coefficients. For various approximation and optimization methods applied to relaxed optimal control problems, see [1,3,4,6,7,8,10,11,14] and the references therein.

4   Numerical Examples

Example 1. Let $\Omega = (0,1)^2$. Define the reference classical control and state
$\bar{w}(x) := \min(1, -1 + 1.5(x_1 + x_2))$, $\bar{y}(x) := 8x_1x_2(1 - x_1)(1 - x_2)$,
and consider the optimal control problem with state equation
$-\Delta y + y^3/3 + (2 + w - \bar{w})y - \bar{y}^3/3 - 2\bar{y} - 16[x_1(1 - x_1) + x_2(1 - x_2)] = 0$ in Ω,
$y(x) = 0$ on Γ,
nonconvex control constraint set $U := \{-1\} \cup [0,1]$, and nonconvex cost
$G_0(w) := \int_\Omega \{0.5[(y - \bar{y})^2 + |\nabla y - \nabla\bar{y}|^2] - w^2 + 1\}dx$.
One can easily verify that the unique optimal relaxed control r is given by

$$r(x)\{1\} = [\bar{w}(x) - (-1)]/2 = \begin{cases} 1, & \text{if } -1 + 1.5(x_1 + x_2) \geqslant 1 \\ < 1, & \text{if } -1 + 1.5(x_1 + x_2) < 1 \end{cases}$$

$r(x)\{-1\} = 1 - r(x)\{1\}$,

for $x \in \Omega$, with optimal state $\bar{y}$ and cost 0. Note also that the optimal cost value 0 can be approximated as closely as desired by using a classical control, as W is dense in R, but cannot be attained for such a control because the control values $u \in (-1, 0)$ of $\bar{w}$ do not belong to U. The Algorithm, without penalties, was applied to this problem using triangular elements (half-squares of fixed edge size $h = 1/80$), the second order three edge-midpoints rule for numerical integration, and with Armijo parameters $b = c = 0.5$. After 90 iterations in k, we obtained the following results:

$G_0^n(r_k^n) = 2.966 \cdot 10^{-4}$, $d_k = -6.733 \cdot 10^{-7}$, $\varepsilon_k = 3.331 \cdot 10^{-4}$,

where ε_k is the discrete max state error at the vertices of the triangles.

Example 2. Introducing here the state constraint
$G_1(u) := \int_\Omega (y - 0.22) dx = 0$,
in Example 1 and applying the penalized Algorithm, we obtained, after 210 iterations in k, the results:
$G_0^n(r_k^{nl}) = 3.736895449878 \cdot 10^{-4}$, $G_1^n(r_k^{nl}) = 1.102 \cdot 10^{-6}$, $d_k = -4.361 \cdot 10^{-6}$.

The progressively refining version of the algorithm was also applied to the above problems, with successive step sizes $h = 1/20$, $1/40$, $1/80$, in three equal iteration periods, and yielded results of similar accuracy, but required here less than half the computing time.

Finally, we remark that the relaxed methods proposed here usually exhibit slower convergence than the classical methods presented in [5], where a gradient projection method was used, but have the advantage of being applicable to problems with non-convex control constraint sets.

References

1. Bartels, S.: Adaptive approximation of Young measure solution in scalar non-convex variational problems. SIAM J. Numer. Anal. 42, 505–629 (2004)
2. Bonnans, J., Casas, E.: Un principe de Pontryagine pour le contrôle des systèmes semilineaires elliptiques. J. Diff. Equat. 90, 288–303 (1991)
3. Cartensen, C., Roubíček, T.: Numerical approximation of Young measure in non-convex variational problems. Numer. Math. 84, 395–415 (2000)
4. Casas, E.: The relaxation theory applied to optimal control problems of semilinear elliptic equations. In: Proc. of the 17th Conf. on Systems, Modeling and Optimization, Prague, 1995, Chapman and Hall, Boca Raton (1996)
5. Chryssoverghi, I.: Mixed discretization-optimization methods for nonlinear elliptic optimal control problems. In: Boyanov, T., et al. (eds.) NMA 2006. LNCS, vol. 4310, pp. 287–295. Springer, Heidelberg (2007)
6. Chryssoverghi, I., et al.: Mixed Frank-Wolfe penalty method with applications to nonconvex optimal control problems. J. Optim. Theory Appl. 94, 311–334 (1997)
7. Chryssoverghi, I., Coletsos, J., Kokkinis, B.: Discretization methods for nonconvex optimal control problems with state constraints. Numer. Funct. Anal. Optim. 26, 321–348 (2005)

8. Chryssoverghi, I., Kokkinis, B.: Discretization of nonlinear elliptic optimal control problems. Syst. Control Lett. 22, 227–234 (1994)
9. Lions, J.-L.: Certain methods of solution of nonlinear boundary value problems, English Translation, Holt, Rinehart and Winston New York (1972)
10. Mach, J.: Numerical solution of a class of nonconvex variational problems by SQP. Numer. Funct. Anal. Optim. 23, 573–587 (2002)
11. Matache, A.-M., Roubíček, T., Schwab, C.: Higher-order convex approximations of Young measures in optimal control. Adv. Comp. Math. 19, 73–79 (2003)
12. Roubíček, T.: Relaxation in Optimization Theory and Variational Calculus, Walter de Gruyter Berlin (1997)
13. Warga, J.: Optimal control of Differential and Functional Equations. Academic Press, New York (1972)
14. Warga, J.: Steepest descent with relaxed controls. SIAM J. Control Optim., 674–682 (1977)

Approximation of the Solution Set of Impulsive Systems

Tzanko Donchev

Department of Mathematics
University of Architecture and Civil Engineering
1 Hr. Smirnenski, 1046 Sofia, Bulgaria
tzankodd@gmail.com

Abstract. We investigate discrete approximation of the solution set of an impulsive differential inclusions with not fixed time of impulses (jumps) in finite dimensional Euclidean space. The right-hand side is assumed to be almost upper semi continuous and one sided Lipschitz. The fact that the impulsive times are not fixed posses problems and in the paper we study several variants of the Euler method in case of autonomous and not autonomous systems. The accuracy (in appropriate metric) of all considered variants is $O(\sqrt{h})$. The results can be applied to impulsive optimal control problems.

1 Preliminaries

We study discrete approximations of the following impulsive differential inclusion:

$$\dot{x}(t) \in F(t, x(t)), \quad x(0) = x_0 \text{ a.e. } t \in I = [0, 1], \quad t \neq \tau_i(x(t)),$$
$$\Delta x|_{t=\tau_i(x)} = S_i(x), \quad i = 1, \dots, p, \tag{1}$$

where $F : I \times \mathbb{R}^n \rightrightarrows \mathbb{R}^n$ is a multifunction with nonempty compact and convex values. Here $S_i : \mathbb{R}^n \to \mathbb{R}^n$ are impulse functions and $\tau_i : \mathbb{R}^n \to \mathbb{R}$ are functions which determine the times of impulses. The piecewise continuous function $x(\cdot)$ is said to be solution of (1) if:

1) it is absolutely continuous on $(\tau_i(x), \tau_{i+1}(x))$ $(i = 1, \dots, p)$ and satisfies (1) for almost all $t \in I$,

2) it has (possible) jumps on $t = \tau_i(x(t))$ $(x(t)$ is the value of the solution before the jump), defined by: $\Delta x|_{t=\tau_i(x(t))} = x(t+0) - x(t) = S_i(x(t))$.

We refer to [1,2,9,13,14] for the theory of impulsive systems.

There are also a lot of papers devoted to approximation of the solution and of the reachable set of systems without impulses. We mention only [6,8], survey papers [7,10] and the references therein. This paper is one of the first (to our knowledge), where numerical approximations of the solution set of differential inclusions with non-fixed time of impulses is studied, although approximation of the solution set has been studied (see [11], where measure-driven systems are considered and the references therein).

I. Lirkov, S. Margenov, and J. Waśniewski (Eds.): LSSC 2007, LNCS 4818, pp. 309–316, 2008.

This paper is a natural continuation of [4] (in case of finite dimensional state space).

The multifunction $G : \mathbb{R}^n \rightrightarrows \mathbb{R}^n$ with nonempty compact values is said to be upper semi continuous (USC) at x_0 when for every $\varepsilon > 0$ there exists $\delta > 0$ such that $F(x_0 + \delta \mathbb{B}) \subset F(x_0) + \varepsilon \mathbb{B}$, where $\mathbb{B}$ is the closed unit ball. The multifunction $F : I \times \mathbb{R}^n \rightrightarrows \mathbb{R}^n$ is said to be almost USC when for every $\varepsilon > 0$ there exists a compact $I_\varepsilon \subset I$ with meas $(I \setminus I_\varepsilon) < \varepsilon$ such that $F(\cdot, \cdot)$ is USC on $I_\varepsilon \times \mathbb{R}^n$. For the compact set A we let $\sigma(l, A) = \max\limits_{a \in A} \langle l, a \rangle$ – the support function.

Standing hypotheses
We suppose that $F(\cdot, \cdot)$ is almost USC with nonempty convex and compact values and moreover:

A1. $\tau_i(\cdot)$ are Lipschitz with a constant N and moreover, $\tau_i(x) \geq \tau_i(x + S_i(x))$.

A2. $\tau_i(x) < \tau_{i+1}(x)$ for every $x \in \mathbf{R}^n$.

A3. There exists a constant C such that $\|F(t, x)\| \leq C$ for every $x \in \mathbf{R}^n$ a.e. $t \in I$ and $NC < 1$.

A4. The functions $S_i : \mathbb{R}^n \to \mathbb{R}^n$ are Lipschitz with constant μ such that $C\mu < 1$.

Notice that without loss of generality one can assume that $\mu = N$.
We refer to [3] for every concepts used here but not explicitly defined.
Given $\varepsilon > 0$ we study:

$$\dot{x}(t) \in F(t, x(t) + \varepsilon \mathbb{B}), \ x(0) = x_0 \text{ a.e. } t \in I = [0, 1], \ t \neq \tau_i(x(t)),$$
$$\Delta x|_{t = \tau_i(x(t))} = S_i(x), \ i = 1, \ldots, p. \tag{2}$$

Lemma 1. *Given $\varepsilon > 0$. Under **A1** – **A3** every solution of (2) intersects every surface $t = \tau_i(x(t))$ at most once.*

The proof is contained in [4] (where follows essentially [12]).

Corollary 1. *Under the conditions of Lemma 1 there exists a constant $\lambda > 0$ such that for every solution $y(\cdot)$ of (1) $\tau_{i+1}(y(t)) - \tau_i(y(t)) \geq \lambda$ for $i = 1, 2, \ldots, p - 1$.*

Proof. Suppose the contrary. Let there exist a sequence $\{y^k(\cdot)\}_{k=1}^\infty$ such that

$$\min_i \left(\tau_{i+1}(y^k(t)) - \tau_i(y^k(t)) \right) \to 0 \text{ as } k \to \infty. \tag{3}$$

Denote by τ_i^k the i–th impulse of $y^k(\cdot)$. Passing to subsequences if necessary we may assume that $\lim\limits_{k \to \infty} \tau_i^k = \tau_i$. Since $y^k(\cdot)$ are C Lipschitz on every $[\tau_i^k, \tau_{i+1}^k]$ there exists a subsequence converging to a solution $y(\cdot)$ of (1) with impulsive points τ_i for $i = 1, 2, \ldots, p$. Due to (3) either $\tau_i(x(\tau_i) + S_i(x(\tau_i)) = \tau_{i+1}$ for some i – contradiction to **A1**, or $\tau_i = \tau_{i+1}$ – contradiction to **A2**.

The following theorem is proved in [4] in more general form:

Theorem 1. *Under the standing hypotheses the problem (1) has a nonempty solution set and every solution exists on the whole interval I.*

Further we assume that

$$\sigma(x - y, F(t, x)) - \sigma(x - y, F(t, y)) \leq L|x - y|^2$$

for every $x, y \in \mathbb{R}^n$, which is called one sided Lipschitz (OSL) condition.

When the right-hand side is OSL the approximation of the solution set (in $C(I, \mathbb{R}^n)$) in case without impulses is studied in [5].

Let $\Delta_k = \left\{ t_j = \dfrac{j}{k} \right\}$, $j = 0, 1, \ldots, k$ be a uniform grid of $J_\upsilon := [0, 1 + \upsilon]$. To (1) we juxtapose the following discrete system:

For $j = 0, 1, \ldots, k - 1$ we let $x_j^k = \lim\limits_{t \uparrow t_j} x^k(t)$ and for $t \in [t_j, t_{j+1})$:

$$\dot{x}^k(t) \in F(t, x_j^k), \quad x^k(t_j) = x_j^k. \tag{4}$$

If $\exists\, t \in [t_j, t_{j+1})$, with $t = \tau_i(x(t))$, then we consider on $[t_{i+1}, 1]$ the problem (1) with initial condition $x_{j+1}^k = x^k(t_{j+1}) + S_i(x)$. And for it we juxtapose the discrete system (4). Of course $x^k(0) = x_0$.

2   Approximation of the Solution Set

First we study $C(I, \mathbb{R}^n)$ approximation of the solution set of (1) with discrete trajectories (the solution set of (4)).

We will use the following known lemma (Lemma 2 of [12]).

Lemma 2. *Let $a_1, a_2, b, d \geq 0$ and let $\delta_0^+ = \delta_0^- = \delta_0$. If for $i = 1, 2, \ldots, p$*

$$\delta_i^+ \leq a_1 \delta_i^- + d, \quad \delta_i^- \leq a_2 \delta_{i-1}^+ + b$$

then $\delta_i^- \leq (ad + b) \sum\limits_{j=0}^{i-1} (a_1 a_2)^j + \delta_0 a_1 a_2$, where $\delta_0^+ \geq 0$.

Theorem 2. *(Lemma of Filippov–Plis) Under the **standing hypotheses** and OSL condition there exists a constant C such that if $\varepsilon > 0$ is sufficiently small, then for any solution $x(\cdot)$ of (2) there exists a solution $y(\cdot)$ of (1) with $|x(t) - y(t)| \leq C\sqrt{\varepsilon}$ on $I \setminus \bigcup\limits_{i=1}^{p} [\tau_i^-, \tau_i^+]$, $\tau_i^+ < \tau_{i+1}^-$ and $\sum\limits_{i=1}^{p} (\tau_i^+ - \tau_i^-) \leq C\sqrt{\varepsilon}$. Here $\tau_i^- = \min\{\tau_i(x(\cdot)), \tau_i(y(\cdot))\}$, $\tau_i^+ = \max\{\tau_i(x(\cdot)), \tau_i(y(\cdot))\}$.*

Proof. Let $y(\cdot)$ be a solution of (2). It is easy to see that $\dot{y}(t) \in F(t, \bar{y}(t))$, where $|y(t) - \bar{y}(t)| < \varepsilon$. We are looking for a solution $x(\cdot)$ of (1), satisfying the condition of the theorem. First we assume that $x(\cdot)$ and $y(\cdot)$ intersect the impulse surfaces successively, i.e. $x(\cdot)$ do not intersect the $i+1$-th surface before $y(\cdot)$ to intersect i-th and vice versa. Let $L > 0$. On the interval $[0, \tau_1^-)$ define

$$\Gamma(t, z) := \left\{ u \in F(t, x) : \langle \bar{y}(t) - x, \dot{y}(t) - u \rangle \le L \left(|y(t) - x| + \varepsilon \right)^2 \right\}.$$

It is easy to see that $\Gamma(\cdot, \cdot)$ is almost USC with nonempty convex and compact values. Let $x(\cdot)$ be a solution of:

$$\dot{x}(t) \in \Gamma(t, x(t)), \quad x(0) = y(0). \tag{5}$$

One has that:

$$\frac{1}{2}\frac{d}{dt}|x(t) - y(t)| \le L \left(|x(t) - y(t)| + \varepsilon \right)^2.$$

Thus $\dfrac{d}{dt}|x(t) - y(t)| \le 4L \left(|x(t) - y(t)| + \varepsilon^2 \right)^2$. If $a_2 = \sqrt{\max\limits_{t \in I} e^{4Lt} \displaystyle\int_0^t e^{-4L\tau}\, d\tau}$

then $|x(t) - y(t)| \le a_2 \sqrt{\varepsilon^2 + 2C\varepsilon}$

It is easy to see that $|x(t) - y(t)| \le \sqrt{2C\varepsilon}$ when $L \le 0$.

We let $b := \max\{a_1\sqrt{\varepsilon^2 + 2C\varepsilon}, \sqrt{2C\varepsilon}\}$ and derive

$$\delta_1^- = |y(\tau_1^-) - x(\tau_1^-)| \le a_2 \delta_0^+ + b.$$

When $\tau_1(x(t)) > \tau_1(y(t))$ the proof is similar.

One can prove also that for every interval $[\tau_i^+, \tau_{i+1}^-]$

$$\delta_{i+1}^- = |y(\tau_{i+1}^-) - x(\tau_{i+1}^-)| \le a_2 \delta_i^+ + b.$$

We need an estimate for δ_i^+. If $\tau_1(x(t)) < \tau_1(y(t))$ we have:

$$|x(\tau_1^-) - y(\tau_1^+)| \le |x(\tau_1^-) - y(\tau_1^-)| + |y(\tau_1^-) - y(\tau_1^+)| \le \delta_1^- + C|\tau_1^- - \tau_1^+|.$$

Furthermore $|\tau_1^- - \tau_1^+| = |\tau(x(\tau_1^-)) - \tau(y(\tau_1^+))| \le N|x(\tau_1^-) - y(\tau_1^+)| \le N[\delta_1^- + C(\tau_1^+ - \tau_1^-)]$. Therefore $|\tau_1^- - \tau_1^+| \le \dfrac{N\delta_1^-}{1 - CN}$. Obviously

$$\delta_1^+ = |x(\tau_1^+) - y(\tau_1^+ + 0)| \le |x(\tau_1^-) - y(\tau_1^-)| + \int_{\tau_i^-}^{\tau_i^+} |\dot{x}(t) - \dot{y}(t)|\, dt +$$

$$|I_1(x(\tau_i^-)) - I_1(y(\tau_i^+))| \le \delta_1^- + 2C(\tau_1^+ - \tau_1^-)$$
$$+N|x(\tau_1^-) - y(\tau_1^+)| \le \delta_1^- + 2C(\tau_1^+ - \tau_1^-) +$$
$$N[\delta_1^- + C(\tau_1^+ - \tau_1^-)] \le (1+N)\delta_1^- + \frac{N(2+N)C}{1 - CN}\delta_1^- = \frac{1 + N + CN}{1 - CN}\delta_1^-.$$

Denote $a_1 := \dfrac{1 + N + NC}{1 - NC}$. One can prove by induction that

$$\delta_i^+ \le a_1 \delta_i^-, \quad \delta_{i+1}^- \le a_2 \delta_i^+ + b, \quad |\tau_i^- - \tau_i^+| \le \frac{N\delta_i^-}{1 - CN}. \tag{6}$$

Therefore $\delta_l^- \leq \dfrac{N}{1-CN} a_2 \sqrt{C\varepsilon} \sum_{i=1}^{l} (1+N+CN)^{i-1}(p+1-l)$. Thus there exists

a constant $K > 0$ such that $|x(t) - y(t)| \leq K\sqrt{\varepsilon}$ on $I \setminus \bigcup_{i=1}^{p}[\tau_i^-, \tau_i^+]$. Furthermore

$\displaystyle\sum_{i=1}^{p}(\tau_i^+ - \tau_i^-) \leq \sum_{i=1}^{p} \dfrac{N\delta_i^-}{1-CN}$ and hence $\displaystyle\sum_{i=1}^{p}(\tau_i^+ - \tau_i^-) \leq K\sqrt{\varepsilon}$.

Due to **A1**, **A2** there exists a constant $\lambda > 0$ such that $\tau_{i+1}^- - \tau_i^- \geq \lambda$, thanks to Corollary 1. It follows from (6) that $\tau_i^+ - \tau_i^- < \delta$ for sufficiently small $\varepsilon > 0$.

Now we study the discrete approximations. First we prove a lemma for approximation of the solutions of (1) with the solutions of:

$$\dot{x}^k(t) \in F(t, x_j^k), \quad x_{j+1}^k = \lim_{t \uparrow t_{j+1}} x(t).$$

If $\tau_i(x(t)) \in [t_j, t_{j+1})$, then $x^k(\tau_i(x) + 0) = x^k(\tau_i(x) - 0) + S_i(x)$. $\quad$ (7)

Recall that we use the same partition Δ_k as in (4).

Lemma 3. *Let $\varepsilon > 0$ be fixed. Then for k big enough for every solution $y(\cdot)$ of*

(1) there exists a solution $x(\cdot)$ of (4) such that $|x(t) - y(t)| \leq \varepsilon$ on $I \setminus \bigcup_{i=1}^{p}[\tau_i^-, \tau_i^+]$,

$\tau_i^+ < \tau_{i+1}^-$ and $\displaystyle\sum_{i=1}^{p}(\tau_i^+ - \tau_i^-) \leq \varepsilon$. *Here $\tau_i^- < \tau_i^+$ are the i-th impulses of $x(\cdot)$*

and $y(\cdot)$. Furthermore one can choose k such that $\varepsilon = O\left(\sqrt{\dfrac{1}{k}}\right)$

Proof. We define $x^k(\cdot)$ successive on the intervals $[t_i^k, t_{i+1}^k]$. Let $x_i^k = x^k(t_i)$. Define the multifunction:

$$G_i(t, z) := \{u \in F(t, x_i^k) : \langle y(t) - x_i^k, \dot{y}(t) - u \rangle \leq L|y(t) - x_i^k|^2\}. \quad (8)$$

It is easy to see that $G_i(\cdot, \cdot)$ is almost USC with nonempty convex and compact values. Thus there exists a solution $z(\cdot)$ of

$$\dot{z}(t) \in G_i(t, z(t)), \quad z(t_i) = x_i^k. \quad (9)$$

If $\tau_i(z(t)) \neq t$ for every $t \in [t_i^k, t_{i+1}^k]$ and every i we let $x^k(t) := z(t)$ on $[t_i^k, t_{i+1}^k]$. If $\tau_l(z(t^*)) = t^* \in [t_i^k, t_{i+1}^k)$ then we let $x^k(t) = z(t)$ on $[t_i^k, t^*)$ and $x^k(t^* + 0) = x^k(t^*) + S_l(x^k(t^*))$. Now we define

$$G_i(t, z) := \{u \in F(t, x^k(\tau_l)) : \langle y(t) - x^k(\tau_l), \dot{y}(t) - u \rangle \leq L|y(t) - x^k(\tau_l)|^2\}. \quad (10)$$

We let $x(t) = z(t)$ on $[\tau_l, t_{i+1}^k]$, where $z(\cdot)$ is a solution of (9) with x_i^k replaced by $x^k(\tau_l)$. One can define $x^k(\cdot)$ on the whole interval I.

As in the proof of Theorem 2 we assume first that $x^k(\cdot)$ and $y(\cdot)$ intersect the impulse surfaces successively. For $t \in [\tau_i^+, \tau_{i+1}^-]$ we have

$$\langle y(t) - x^k(t), \dot{y}(t) - \dot{x}^k(t) \rangle \leq L|y(t) - x^k(t)|^2 + \frac{4C^2|L|}{k}. \tag{11}$$

Denote $a_2 = \sqrt{\max 4C^2|L|e^{2Lt} \int_0^t e^{-\tau} d\tau}$. If $b = \max\left\{ a_2\sqrt{\frac{1}{k}}, \sqrt{\frac{2C}{k}} \right\}$ then $\delta_{i+1}^- = |y(t_{i+1}^-) - x(t_{i+1}^-)| \leq a_2\delta_i^+ + b$. Using similar fashion as in the proof of Theorem 2 we can show that $\delta_1^+ \leq \dfrac{1 + N + NC}{1 - NC}\delta_1^-$. If $a_1 := \dfrac{1 + N + NC}{1 - NC}$, then it is easy to show by induction that:

$$\delta_i^+ \leq a_1\delta_i^-, \; \delta_{i+1}^- \leq a_2\delta_i^+ + b, \; |\tau_i^- - \tau_i^+| \leq \frac{N\delta_i^-}{1 - NC}.$$

Due to Lemma 2 for sufficiently large k there exists a constant $M > 0$ (not depending on k) such that for every solution $y(\cdot)$ of (1) (obtained after (8)–(10)) there exists an approximate solution $x^k(\cdot)$ such that:

$$|y(t) - x^k(t)| \leq \frac{M}{\sqrt{k}} \text{ on } I \setminus \left(\bigcup_{i=1}^p [\tau_i^-, \tau_i^+] \right)$$

and $\displaystyle\sum_{i=1}^p (\tau_i^+ - \tau_i^-) \leq \frac{M}{\sqrt{k}}$.

It remains to show that for every $x^k(\cdot)$ obtained after (8)–(10) there exists a solution $z^k(\cdot)$ of (7) such that $|x^k(t) - z^k(t)| \leq O\left(\sqrt{\frac{1}{k}}\right)$ on $I \setminus \bigcup_{i=1}^p [\tau_i^-, \tau_i^+]$ and

$$\sum_{i=1}^p (\tau_i^+ - \tau_i^-) \leq O\left(\sqrt{\frac{1}{k}}\right).$$

Obviously for sufficiently large k every approximate solution $x^k(\cdot)$ has no more than 1 impulse on $[t_i, t_{i+1})$ for $i = 0, 1, \ldots, k-1$.

On every $[t_i, t_{i+1})$ we define the multifunction:

$$G(t, z) := \{ u \in F(t, z_i^k) : \langle x^k(t) - z^k(t_i), \dot{x}^k(t) - u \rangle \leq L|x^k(t) - z^k(t_i)|^2 \},$$

where $z_0^k = x_0$ and $z_i^k := \lim_{t \uparrow t_i} z^k(t)$. Furthermore $z^k(\cdot)$ is a solution on $[t_i, t_{i+1}]$ of

$$\dot{z}(t) \in G(t, z(t)), \; z(t_i) = z_i^k,$$

for $i = 0, 1, \ldots, k-1$. We have to show that $|x^k(t) - z^k(t)| \leq O\left(\sqrt{\frac{1}{k}}\right)$ on

$$I \setminus \bigcup_{i=1}^p [\tau_i^-, \tau_i^+] \text{ and } \sum_{i=1}^p (\tau_i^+ - \tau_i^-) \leq O\left(\sqrt{\frac{1}{k}}\right).$$

If $x^k(\cdot)$ is without impulses on $[t_i, t_{i+1}]$, then

$$\langle x^k(t) - z^k(t), \dot{x}^k(t) - \dot{z}^k(t) \rangle \leq L|x^k(t) - z^k(t)|^2 + \frac{4C}{k}. \tag{12}$$

Let τ_i^-, τ_i^+, δ_i^- and δ_i^+ be the above. It is easy to show by induction that $\delta_{i+1}^- \leq a_2\delta_i^+ + b$. Also $|\tau_i^- - \tau_i^+| \leq \dfrac{N\delta_i^-}{1 - NC} + \dfrac{1}{k}$ and $\delta_i^+ \leq a_1\delta_i^- + \dfrac{2C}{k}$. The proof is completed thanks to Lemma 2.

Using the same fashion as in the proof of Lemma 3 as a corollary of Theorem 2 one can proof:

Theorem 3. *Under the conditions of Theorem 2 we have (for k big enough):*
1) For every solution $x(\cdot)$ of (1) there exists a solution $y(\cdot)$ of (4) such that

$$|x(t) - y(t)| \leq O\left(\sqrt{\frac{1}{k}}\right), \ \forall t \in I \setminus \bigcup_{i=1}^{p}[\tau_i^-, \tau_i^+], \ \sum_{i=1}^{p}(\tau_i^+ - \tau_i^-) \leq O\left(\sqrt{\frac{1}{k}}\right) \tag{13}$$

2) For every solution $y(\cdot)$ of (4) there exists a solution $x(\cdot)$ of (1) such that (13) holds.
In the both cases $\tau_i^+ < \tau_{i+1}^-$.

Consider now the autonomous variant of (1),i.e.

$$\dot{x}(t) \in F(x(t)), \ x(0) = x_0 \ \text{a.e.} \ t \in I = [0,1], \ t \neq \tau_i(x(t)),$$
$$\Delta x|_{t=\tau_i(x)} = S_i(x), \ \tau_i(x) \leq \tau_{i+1}(x), \ i = 1, \ldots, p, \tag{14}$$

where $F : \mathbf{R}^n \rightrightarrows \mathbf{R}^n$.
    The corresponding to (4) system has the form:

$$\dot{x}^k(t) = v_j \in F(x_j^k), \ x^k(t) = x_j^k + v_j(t - t_j). \tag{15}$$

If $\exists \, t \in [t_j, t_{j+1})$, with $t = \tau_i(x(t))$, then we consider on $[\tau_{i+1}, 1]$ the problem (1) with initial condition $x_{j+1}^k = x^k(t_{j+1}) + S_i(x)$. And for it we juxtapose the discrete system (15). Of course $x^k(0) = x_0$ (x_j^k is defined before (4)).
    The following theorem is a corollary of Theorem 3:

Theorem 4. *If all the conditions of Theorem 2, then for k big enough we have:*
1) For every solution $x(\cdot)$ of (14) there exists a solution $y(\cdot)$ of (15) such that

$$|x(t) - y(t)| \leq O\left(\sqrt{\frac{1}{k}}\right), \ t \in I \setminus \bigcup_{i=1}^{p}[\tau_i^-, \tau_i^+], \ \sum_{i=1}^{p}(\tau_i^+ - \tau_i^-) \leq O\left(\sqrt{\frac{1}{k}}\right) \tag{16}$$

2) For every solution $y(\cdot)$ of (15) there exists a solution $x(\cdot)$ of (14) such that (16) holds.
In the both cases $\tau_i^+ < \tau_{i+1}^-$.

Acknowledgement

This work was supported by the Australian Research Council Discovery-Project Grant DP0346099.

References

1. Aubin, J.-P.: Impulsive Differential Inclusions and Hybrid Systems: A Viability Approach, Lecture Notes, Univ. Paris (2002)
2. Benchohra, M., Henderson, J., Ntouyas, S.: Impulsive Differential Equations and Inclusions. Hindawi Publishing Company, New York (in press)
3. Deimling, K.: Multivalued differential equations. de Gruyter Series in Nonlinear Analysis and Applications. Walter de Gruyter & Co., Berlin (1992)
4. Donchev, T.: Impulsive differential inclusions with constrains. EJDE 66, 1–12 (2006)
5. Donchev, T., Farkhi, E.: Euler approximation of discontinuous one-sided Lipschitz convex differential inclusions. In: Ioffe, A., Reich, S., Shafrir, I. (eds.) Calculus of Variations and Differential Equations, pp. 101–118. Chapman & Hall/CRC, Boca Raton, New York (1999)
6. Dontchev, A., Farkhi, E.: Error estimates for discretized differential inclusions. Computing 41, 349–358 (1989)
7. Dontchev, A., Lempio, F.: Difference methods for differential inclusions: a survey. SIAM Rev. 34, 263–294 (1992)
8. Grammel, G.: Towards fully discretized differential inclusions. Set Valued Analysis 11, 1–8 (2003)
9. Lakshmikantham, V., Bainov, D., Simeonov, P.: Theory of Impulsive Differential Equations. World Scientific, Singapore (1989)
10. Lempio, F., Veliov, V.: Discrete approximations of differential inclusions. Bayreuther Mathematische Schriften, Heft 54, 149–232 (1998)
11. Pereira, F., Silva, G.: Necessary Conditions of Optimality for Vector-Valued Impulsive Control Problems. System Control Letters 40, 205–215 (2000)
12. Plotnikov, V., Kitanov, N.: On Continuous dependence of solutions of impulsive differential inclusions and impulse control problems. Kybern. Syst. Analysis 5, 143–154 (2002) (in Russian) English translation: Cybernetics System Analysis 38, 749–758 (2002)
13. Plotnikov, V., Plotnikov, A., Vityuk, A.: Differential Equations with Multivalued Right-Hand Side. Asymptotic Methods, Astro Print Odessa (Russian) (1999)
14. Samoilenko, A., Peresyuk, N.: Differential Equations with Impulsive Effects. World Scientific, Singapore (1995)

Lipschitz Stability of Broken Extremals in Bang-Bang Control Problems

Ursula Felgenhauer

Brandenburgische Technische Universität Cottbus, Institut für Mathematik,
PF 101344, 03013 Cottbus, Germany
felgenh@tu-cottbus.de
http://www.math.tu-cottbus.de/~felgenh/

Abstract. Optimal *bang-bang* controls appear in problems where the system dynamics linearly depends on the control input. The principal control structure as well as switching points localization are essential solution characteristics. Under rather strong optimality and regularity conditions, for so-called *simple* switches of (only) one control component, the switching points had been shown being differentiable w.r.t. problem parameters. In case that *multiple* (or: simultaneous) switches occur, the differentiability is lost but Lipschitz continuous behavior can be observed e.g. for double switches. The proof of local structural stability is based on parametrizations of broken extremals via certain backward shooting approach. In a second step, the Lipschitz property is derived by means of nonsmooth Implicit Function Theorems.

1 Introduction

In the paper, we consider ODE driven optimal control problems with system dynamics depending linearly on the control input. In case when the control is vector-valued and underlying bound constraints, extremals are often of bang-bang type. The structure of the control functions and, particularly, switching points localization are essential solution characteristics.

In recent years, substantial results on second-order optimality conditions have been obtained [1,9,10,11,12]. Important early results are given in [14,13]. The solution stability under parameter perturbation was investigated e.g. in [7,3,5]. Optimality and stability conditions therein include: (i) bang-bang regularity assumptions (finite number of switches, excluding e.g. endpoints), (ii) strict bang-bang properties (nonvanishing time derivatives of switching functions at switching points e.g.), (iii) assumption of simple switches (switch of no more than one control component at each time), (iv) appropriate second-order conditions (positive definiteness of related quadratic forms e.g.).

Stability properties for the switching points localization had been obtained from the so-called deduced finite-dimensional problem ([7,5]) using standard sensitivity results from nonlinear programming, or from a shooting type approach applied to first-order conditions from Pontryagin's maximum principle (e.g. [3,6]). In particular, the results cover differentiability of switching points w.r.t. parameters under conditions (i), (ii) for linear state systems $\dot{x} = Ax + Bu$,

I. Lirkov, S. Margenov, and J. Waśniewski (Eds.): LSSC 2007, LNCS 4818, pp. 317–325, 2008.
© Springer-Verlag Berlin Heidelberg 2008

cf. [3], and differentiable behavior and local uniqueness of structure for semi-linear systems $\dot{x} = f(x) + Bu$ under (i), (ii), (iv), cf. [6]. In the given paper, Lipschitz behavior (and possible lack of differentiability) will be proved for systems $\dot{x} = f(x) + g(x)u$ with simultaneous switch of two control components.

Notations. Let R^n be the Euclidean vector space with norm $|\cdot|$, and scalar product written as $(a, b) = a^T b$. Superscript T is generally used for transposition of matrices resp. vectors. The Lebesgue space of order p of vector-valued functions on $[0, 1]$ is denoted by $L_p(0, 1; R^k)$. $W_p^l(0, 1; R^k)$ is the related Sobolev space, and norms are given as $\|\cdot\|_p$ and $\|\cdot\|_{l,p}$, $(1 \le p \le \infty, l \ge 1)$, resp. In several places, Lie brackets $[g, f] = \nabla_x g\, f - \nabla_x f\, g$ occur. The symbol ∇_x denotes (partial) gradients whereas ∂_x is used for (partial) generalized derivative in sense of Clarke. By $conv M$ resp. $cl\, M$ the convex hull and closure of a set M are determined. For characterizing discontinuities, jump terms are given as $[v]^s = v(t_s + 0) - v(t_s - 0)$ e.g. The index s will become clear from the context.

2 Bang-Bang Extremals and Generalized Shooting System

Consider the following optimal control problem where the control function enters the state equation linearly, and pointwise control bounds are assumed:

$$(\mathbf{P}_h) \qquad \min J_h(x, u) = k(x(1), h)$$

$$\text{s.t.} \qquad \dot{x}(t) = f(x(t), h) + g(x(t), h)\, u(t) \qquad \text{a.e. in } [0, 1], \qquad (1)$$

$$x(0) = a(h), \qquad (2)$$

$$|u_i(t)| \le 1, \quad i = 1, \ldots, m, \qquad \text{a.e. in } [0, 1]. \qquad (3)$$

The parameter h is assumed to be real-valued and chosen close to $h_0 = 0$. Further, all data functions are supposed to be sufficiently smooth w.r.t. their respective variables. In particular, we assume that f, g and k are twice continuously differentiable w.r.t. x and, together with their derivatives, are Lipschitz continuous functions of (x, h).

The state-control pair $(x, u) \in W_\infty^1(0, 1; R^n) \times L_\infty(0, 1; R^m)$ is called *admissible* for (P_h) if the state equation (1) together with the initial condition (2) and the control constraints (3) hold with parameter h.

The admissible pair (x, u) is an *extremal* for (P_h), if first-order necessary optimality conditions from Pontryagin's Maximum Principle (PMP) hold. It includes the adjoint system

$$\dot{p}(t) = -A(x(t), u(t), h)^T p(t), \quad p(1) = \nabla_x k(x(1), h) \qquad (4)$$

with $A(x, u, h) = \nabla_x (f(x, h) + g(x, h)\, u)$, and the maximum condition

$$u(t) \in arg \max_{|v_i| \le 1} \{ -p(t)^T g(x(t), h)\, v \}.$$

Using the subdifferential notation and defining

$$Sign\, y = \partial |y| = \begin{cases} sign\, y & \text{if } y \ne 0 \\ [-1, +1] & \text{for } y = 0 \end{cases}, \qquad y \in R,$$

one may equivalently write

$$u_j(t) \in -Sign\left(p(t)^T g_j(x(t),h)\right), \qquad j = 1, \ldots, m. \tag{5}$$

With these notations, extremals for problem (P_h) can be characterized e.g. as solutions of the following *backward* shooting system:

$$\begin{aligned}
\dot{x}(t) &= f(x(t),h) + g(x(t),h)u, & x(1) &= z, \\
\dot{p}(t) &= -A(x(t),u(t),h)^T p(t), & p(1) &= \nabla_x k(z), \\
\sigma(t) &= g(x(t),h)^T p(t), & u(t) &\in -Sign\,\sigma(t).
\end{aligned} \tag{6}$$

For given parameter value h and terminal state guess $z \in R^n$, a solution of the *generalized* differential-algebraic system (6) will consist of the functions $x = x(t,z,h)$, $u = u(t,z,h)$ and $p = p(t,z,h)$. The pair (x,u) herein is a stationary solution for (P_h) if the boundary condition (2) holds:

$$T(z,h) = x(0,z,h) - a(h) = 0. \tag{7}$$

The vector function $\sigma = g(x,h)^T p$ in (6) is called the *switching* function related to (x,u). If, for some $j \le m$, $\sigma_j \equiv 0$ on a certain interval then this part of the control trajectory is called a *singular arc*.

In order to analyze the stability properties of optimal control structure in (P_h) for h near $h_0 = 0$, we make assumptions on existence and structure of (a possibly local) reference solution (x^0, u^0) of (P_0):

Assumption 1.   (bang-bang regularity)
The pair (x^0, u^0) is a solution such that u^0 is piecewise constant and has no singular arcs. For every j and $\sigma_j^0 = g_j(x^0,0)^T p$, the set $\Sigma_j = \{t \in [0,1] : \sigma_j^0(t) = 0\}$ is finite, and $0, 1 \notin \Sigma_j$.

Notice that, under the given smoothness assumptions, $\sigma^0 \in W_\infty^1(0,1;R^m) \subset C([0,1])$. However, in general the function is not continuously differentiable on $[0,1]$. For $t \notin \bigcup \Sigma_k$, from $\dot{\sigma} = \dot{p}^T g + p^T \nabla_x g\, \dot{x}$ we see that

$$\dot{\sigma}_j^0(t) = p(t)^T [g_j, f](t) + \sum_{k \ne j} u_k^0(t)\, p(t)^T [g_j, g_k](t) \tag{8}$$

(where $[\cdot, \cdot]$ stand for Lie brackets, and the functions f, g etc. are evaluated along $x^0(t)$ with $h = 0$ fixed). The formula shows that, for all t, the function σ^0 has one-sided time-derivatives but $\dot{\sigma}_j^0$ may be discontinuous for $t \in \Sigma_k$, $k \ne j$.

Remark. The above piecewise representation for $\dot{\sigma}^0$ yields, in particular, that $\sigma^0 \in C^1([0,1])$ in case of so-called semilinear state systems where $g = g(h)$ is independent of x (cf. [5,6]). Further, $\dot{\sigma}_j^0$ will be continuous in all simple switching points $t \in \Sigma_j$ where $\Sigma_j \cap \Sigma_k = \emptyset$ $\forall k \ne j$ (see e.g. [7]).

In case of differentiable switching function components, a common stability condition consists in requiring all zeros being regular, i.e. $\dot{\sigma}_j^0(t_s) \ne 0$ for each $t_s \in \Sigma_j$. The natural generalization to piecewise differentiable case is

Assumption 2a. (strict bang-bang property)
For every $j = 1, \ldots, m$, for all $t_s \in \Sigma_j$: $\dot{\sigma}_j^0(t_s + 0) \cdot \dot{\sigma}_j^0(t_s - 0) > 0.$

Optimality and stability investigations in e.g. [1,12,9] considered reference control functions having only *simple* switches, i.e. the case where $\Sigma_i \cap \Sigma_j = \emptyset \; \forall \, i \neq j$. We will call $t_s \in \Sigma_j \cap \Sigma_i$, $i \neq j$, a *double* switching point if $t_s \notin \Sigma_k$ for arbitrary $k \notin \{i, j\}$, and a *multiple* switching point in general. Alternatively, one could speak of *simultaneous* switches of two or more control components at certain time. The notation has to be distinguished from situations where a switching point is a *multiple zero* of the related switching function component: the latter is excluded by Assumption 2a. For the purpose of the given paper we add

Assumption 2b. (double switch restriction)
All $t_s \in \bigcup \Sigma_k$ related to $h = 0$ are switching points of at most two control components.

As a consequence from (8), we obtain the jump condition for $\dot{\sigma}_j^0$ at $t_s \in \Sigma_j \cap \Sigma_i$:

$$\left[\dot{\sigma}_j^0 \right]^s = \dot{\sigma}_j^0(t_s + 0) - \dot{\sigma}_j^0(t_s - 0) = \left[u_i^0 \right]^s p(t_s)^T [g_j, g_i] (t_s). \qquad (9)$$

3   Local Stability of Bang-Bang Structure

The shooting system (6), (7) introduced in the preceding section, at $h = 0$ admits the solution $x = x^0$, $u = u^0$ with the related adjoint function p. For general h, solution existence and uniqueness cannot be guaranteed without additional assumptions which are considered in forthcoming sections. However, for an assumed solution (x^h, u^h) near (x^0, u^0) one can provide conditions ensuring that (i) the control u^h is of bang-bang type, and (ii) the switching structure of both u^h and u^0 coincide.

Theorem 1. *Suppose Assumptions 1, 2a, and 2b hold true. Further, let $(x^h, u^h, p^h) \in W_\infty^1 \times L_\infty \times W_\infty^1$ with $z = z^h = x^h(1)$ be a solution of (6) satisfying the initial condition (7) and the estimate*

$$\|x^h - x^0\|_\infty + \|u^h - u^0\|_1 < \epsilon. \qquad (10)$$

If $\delta = |h| + \epsilon$ is sufficiently small then the following relations hold for (x^h, u^h, p^h) together with $\sigma^h = g(x^h, h)^T p^h$:
(i) $u^h(t) = -\text{sign } \sigma^h(t)$ *almost everywhere on $[0, 1]$, and u^h has the same switching structure and number of switching points as u^0,*
(ii)   *if $\sigma_j^h(t_s^h) = 0$, then $\dot{\sigma}_j^h(t_s^h + 0) \cdot \dot{\sigma}_j^h(t_s^h - 0) > 0$ $(j = 1, \ldots, m)$.*   □

For the proof, the following standard estimates are useful:

$$\| x^h - x^0 \|_{1,1} + \| p^h - p \|_{1,1} + \| \sigma^h - \sigma \|_{1,1} = O(\delta) \qquad (11)$$

with $\delta = |h| + \epsilon$. In L_∞ sense, they follow from Gronwall's Lemma applied to $(x^h - x^0)$ resp. $(p^h - p^0)$. Additional L_1 estimates for time derivative terms are then deduced from state and adjoint equations in (6).

Proof. Let $t_s, t_{s'}$ be switching points from $\bigcup \Sigma_k$. By Assumption 1, a radius $\rho > 0$ exists such that

$$B_\rho(t_s) \cap B_\rho(t_{s'}) = \emptyset \quad \text{for all } t_s \neq t_{s'}, \qquad 0, 1 \notin B_\rho(t_s) \text{ for all } t_s.$$

For $t_s \in \Sigma_j$, define σ_j^+, σ_j^- as smooth continuations of σ_j^0 from $(t_s, t_s + \rho)$ (resp. $(t_s - \rho, t_s)$) to $B_\rho(t_s)$ by setting $\sigma_j^\pm(t_s) = \sigma_j^0(t_s) = 0$ and integrating

$$\dot\sigma_j^\pm(t) = p(t)^T \left[g_j^0, f^0 \right](t) + \sum_{k \neq j} u_k^0(t_s \pm 0) \, p(t)^T \left[g_j^0, g_k^0 \right](t)$$

(where f^0, g_k^0 etc. denote f or g_k evaluated along $x = x^0$), cf. (8).

From the continuity properties of σ together with Assumption 2a and (9), we see that there exist $\mu > 0$, $r \in (0, \rho)$ such that

$$|\sigma_k^0(t)| \geq \mu \qquad \forall t \text{ with } dist\{t, \Sigma_k\} \geq r, \ \forall k,$$
$$|\dot\sigma_k^\pm(t)| \geq \mu \qquad \forall t \text{ with } dist\{t, \Sigma_k\} \leq 2r, \ \forall k. \tag{12}$$

Let $\delta = |h| + \epsilon$ be sufficiently small: Due to (11), outside of $B_r(t_s)$, $t_s \in \Sigma_k$, the functions σ_k^0 and σ_k^h then are of the same sign, and

$$|\sigma_k^h(t)| \geq \mu/2 \qquad \forall t \text{ with } dist\{t, \Sigma_k\} \geq r, \ \forall k.$$

Thus, the functions may vanish only inside intervals B_r, i.e. near switching points of the reference control u^0. By (6), each of the functions σ_j has a time-derivative in L_∞ given almost everywhere on $[0, 1]$ by

$$\dot\sigma_j^h = (p^h)^T \left[g_j^h, f^h \right] + \sum_{k \neq j} u_k^h \cdot (p^h)^T \left[g_j^h, g_k^h \right]. \tag{13}$$

Consider first the case that $t_s \in \Sigma_j$ is a simple switch of u^0, and let $t \in B_r(t_s)$: In this situation, all terms in the representation (13) are continuous on $B_r(t_s)$ and, for sufficiently small δ,

$$|\dot\sigma_j^h(t)| \geq \mu/2 \qquad \forall t \text{ with } |t - t_s| \leq r$$

follows from (11). Moreover, in $B_r(t_s)$, $\dot\sigma_j^h$ and $\dot\sigma_j^0$ will be of the same sign so that σ_j^h is strictly monotone there. Consequently, it will have an unique zero on this interval (as does σ_j^0).

Now, let $t_s \in \Sigma_j \cap \Sigma_i$ be a double switch of u^0: by (13), we may write

$$\dot\sigma_j^h = u_i^h (p^h)^T \left[g_j^h, g_i^h \right] + \omega_j^h$$

where ω_j^h and $(p^h)^T [g_j^h, g_i^h]$ are continuous on $B_r(t_s)$. Since the control values $u_i^h(t)$ belong to $[-1, 1] = conv\{u_i^0(t_s - 0), u_i^0(t_s + 0)\}$ we have

$$dist\left\{ \dot\sigma_j^h(t), conv\{\dot\sigma_j^+(t), \dot\sigma_j^-(t)\} \right\} = O(\delta).$$

If δ is chosen sufficiently small then it follows from (12) that, on $B_r(t_s)$, $\dot\sigma_j^h$ has the same sign as both of $\dot\sigma_j^\pm$, and $|\dot\sigma_j^h(t)| \geq \mu/2$ there. Again, σ_j^h turns out to be strictly monotone in $B_r(t_s)$ and will have an unique and regular zero there.

Summing up, the conclusions (i) and (ii) of the Theorem follow.   □

4   Generalized Second-Order Condition

In the last section it was shown that locally, under condition (10), any extremal described by the shooting system (6), (7) corresponds to a bang-bang control with the same prinicipal switching behavior as u^0. Notice further that, for h, δ sufficiently small, the restriction $\|x - x^0\|_\infty + \|p - p^0\|_\infty = O(\delta + |h|)$ together with Assumption 1 fixes the value $u(1)$ for any stationary solution related to (P_h). Therefore, locally one may uniquely determine extremals by solving the so-called *deduced* finite-dimensional problem formulated in terms of the switching vector $\Sigma = (\Sigma_1, \ldots, \Sigma_m) \in R^L$, cf. [1,7].

In order to allow for double or general multiple switches, in [4] the enumeration of switching points by double indices $\alpha = (j, s)$ was introduced, where j points to the switching control component, and s enumerates the elements in Σ_j in increasing order: $0 = t_{j0} < \ldots < t_{js} < t_{j,s+1} < \ldots < t_{j,l(j)} < t_{j,l(j)+1} = 1$ for all $j = 1, \ldots, m$. Vectors Σ satisfying the given monotonicity requirement form an open set $D_\Sigma \subset R^L$ for $L = \sum_{j=1}^m l(j)$.

Each $\Sigma = (\tau_{js}) \in D_\Sigma$ sufficiently close to Σ^0 can be used to construct an admissible pair $x = x(t, \Sigma, h)$, $u = u(t, \Sigma)$ for (P_h) by the following rules:

$$u_j(t, \Sigma) \equiv (-1)^{l(j)-s} u_j^0(1) \qquad \text{for } t \in (\tau_{js}, \tau_{j,s+1}), \qquad (14)$$

$$\dot{x}(t) = f(x(t), h) + g(x(t), h)\, u(t, \Sigma), \qquad x(0) = a(h). \qquad (15)$$

Inserting $x(1, \Sigma, h)$ into the objective functional we obtain the *deduced* finite-dimensional problem related to (P_h):

$$\min \phi_h(\Sigma) = k(x(1, \Sigma, h), h) \qquad \text{w.r.t. } \Sigma \in D_\Sigma. \qquad (16)$$

For the reference parameter $h_0 = 0$, the vector $\Sigma = \Sigma^0$ is a local solution.

Under the given smoothness assumptions and for small parameter values $h \approx 0$, the functionals ϕ_h are continuously differentiable w.r.t. Σ but may not belong to $C^2(D_\Sigma)$: in general, $\nabla_\Sigma \phi_h$ is only piecewise continuously differentiable in those parts of D_Σ where all t_{js} are simple. Instead of classical Hessian, Clarke's generalized derivatives will thus be used below (see [2], also [8]).

For sake of simplicity, the calculation of $\nabla_\Sigma \phi_h$ and $\partial_\Sigma(\nabla_\Sigma \phi_h)$ will be carried out for the case that at most one double switch occurs, e.g. $t_\alpha = t_\beta$ for some $\alpha = (j, s)$, $\beta = (i, r)$. On the following sets, $\nabla_\Sigma^2 \phi_h$ is continuous w.r.t. (Σ, h):

$$D^1 = \{\Sigma \in D_\Sigma : \tau_\alpha < \tau_\beta\}, \quad D^2 = \{\Sigma \in D_\Sigma : \tau_\alpha > \tau_\beta\}.$$

At points $\Sigma' \in cl D^1 \cap cl D^2$ (i.e. $\tau'_\alpha = \tau'_\beta$), one can find $\partial_\Sigma(\nabla_\Sigma \phi_h(\Sigma'))$ by a selection representation for generalized derivatives, [8]:

$$\partial_\Sigma(\nabla_\Sigma \phi_h(\Sigma')) = conv\{ \lim_{\Sigma \in D^1, \Sigma \to \Sigma'} \nabla_\Sigma^2 \phi_h(\Sigma), \lim_{\Sigma \in D^2, \Sigma \to \Sigma'} \nabla_\Sigma^2 \phi_h(\Sigma) \}. \quad (17)$$

For convenience, define an adjoint function $p = p(\cdot, \Sigma, h)$ related to (15),(14) by

$$\dot{p} = -A(x(\Sigma, h), u(\Sigma), h)^T p, \quad p(T) = \nabla_x k(x(T, \Sigma, h), h) \qquad (18)$$

and, in addition, set $\sigma = \sigma(\cdot, \Sigma, h) = g(x(\cdot, \Sigma, h))^T p(\cdot, \Sigma, h)$.

Let us further define auxiliary functions related to $\nabla_\Sigma(x,p)$ and $\nabla_\Sigma\phi$: The matrix function $A[t] = A(x(t,\Sigma,h), u(t,\Sigma),h)$ given by $\nabla_x(f+gu)$ from (4) is bounded and piecewise continuous. For given (x,u) and h, it is used to define fundamental matrix solutions for linearized state and adjoint equations:

$$\dot{\Psi}(t) = A[t]\Psi(t), \quad \dot{\Phi}(t) = -A[t]^T\Phi(t), \quad \Psi(0) = \Phi(0) = I. \tag{19}$$

By direct calculation, $\Phi^{-1} = \Psi^T$ is checked for arbitrary $t \in [0,1]$. Further, it is easy to see that p from (18) satisfies $p(t) = \Phi(t)\Psi(1)^T\nabla_x k$.

The functions $x = x(t,\Sigma,h)$ and $p = p(t,\Sigma,h)$ almost everywhere have first order derivatives $\eta_\alpha = \partial x/\partial\tau_\alpha$ and $\rho_\alpha = \partial p/\partial\tau_\alpha$ w.r.t. switching points τ_α, $\alpha = (j,s)$. They solve the following multi-point boundary value problems (cf. [4]):

$$\dot\eta_\alpha = A\,\eta_\alpha, \qquad\qquad \eta_\alpha(0) = 0, \quad [\eta_\alpha]^s = -[\nabla_p H]^\alpha, \tag{20}$$
$$\dot\rho_\alpha = -A^T\rho_\alpha - C\,\eta_\alpha, \quad \rho_\alpha(1) = \nabla_x^2 k \cdot \eta_\alpha(1), \quad [\rho_\alpha]^s = [\nabla_x H]^\alpha. \tag{21}$$

The terms $g, \nabla g, k$ herein are evaluated along $x = x(\cdot, \Sigma, h)$, and $C = C[t]$ stands for $\nabla_x^2 H[t]$. The switching terms are

$$[\nabla_p H]^\alpha = g_j[t_\alpha]\,[u_j^0]^s, \qquad [\nabla_x H]^\alpha = \nabla_x g_j[t_\alpha]^T p(t_\alpha)\,[u_j^0]^s. \tag{22}$$

If $t_\alpha = t_{js}$ is a simple switch for u, then we have $[\nabla H]^\alpha = [\nabla H]^s$ – the full jump of ∇H at t_α. For multiple switches, in general this is not true.

After these preliminaries, the derivatives of ϕ_h may be calculated. In the first step we obtain

$$\frac{\partial\phi_h}{\partial\tau_\alpha} = \nabla_x k^T \eta_\alpha(1) = -[u_j^0]^s \nabla_x k^T \Psi(1)\Phi(t_\alpha)^T g_j[t_\alpha] = -[u_j^0]^s (p^T g_j)[t_\alpha]. \tag{23}$$

Repeating differentiation, consider first mixed second-order terms with $\alpha \neq \beta$: in $D^{1,2}$ we have $t_\alpha \neq t_\beta$ so that we get continuous representation

$$\frac{\partial}{\partial\tau_\beta}\left(\frac{\partial}{\partial\tau_\alpha}\phi_h\right) = -[u_j^0]^s\left\{\rho_\beta(t_\alpha)^T g_j[t_\alpha] + p(t_\alpha)^T\nabla_x g_j[t_\alpha]\eta_\beta(t_\alpha)\right\}.$$

In symmetric form, it may be written as (cf. [4], or [9]):

$$\frac{\partial}{\partial\tau_\beta}\left(\frac{\partial}{\partial\tau_\alpha}\phi_h\right) = \eta_\beta(1)^T\nabla_x^2 k\,\eta_\alpha(1) + \int_{t_{\alpha\beta}}^1 \eta_\beta(s)^T\nabla_x^2 H[s]\,\eta_\alpha(s)\,ds$$
$$- [\nabla_x H]^\beta \eta_\alpha(t_\beta) - [\nabla_x H]^\alpha \eta_\beta(t_\alpha). \tag{24}$$

Notice that, due to (22), the last terms will have jump discontinuities at $t_\alpha = t_\beta$.

Next consider $\partial^2\phi_h/\partial t_\alpha^2$ in $D^1\cup D^2$: from (23) we obtain one-sided derivatives,

$$\frac{\partial^\pm}{\partial\tau_\alpha}\left(\frac{\partial}{\partial\tau_\alpha}\phi_h\right) = -[u_j^0]^s\left\{\dot\sigma_j(t_\alpha) + \rho_\alpha^\pm(t_\alpha)^T g_j[t_\alpha] + p(t_\alpha)^T\nabla_x g_j[t_\alpha]\eta_\alpha^\pm(t_\alpha)\right\}.$$

As long as t_α does not coincide with any other control switching point from Σ the derivative $\dot\sigma$ is continuous at t_α. The remaining switches cancel out by

(20)-(21), so that $\partial^2 \phi_h / \partial t_\alpha^2$ is continuous in D^1 resp. D^2. Thus, for $t_\alpha \neq t_\beta$, one may give the derivative a final form similar to (24)

$$\frac{\partial^2}{\partial \tau_\alpha^2} \phi_h = \eta_\alpha(1)^T \nabla_x^2 k\, \eta_\alpha(1) + \int_{t_\alpha}^1 \eta_\alpha(s)^T \nabla_x^2 H[s]\, \eta_\alpha(s)\, ds$$
$$- \left[u_j^0 \right]^s \dot{\sigma}_j(t_\alpha) - \left[\nabla_x H \right]^\alpha \eta_\alpha(t_\alpha + 0). \qquad (25)$$

In the limit for $t_\alpha - t_\beta \to \pm 0$, one has to replace $\dot{\sigma}_j(t_\alpha)$ by $\dot{\sigma}_j(t_\alpha \pm 0)$ resp.

After partial second-order derivatives and their limits have been found, one can compose $\partial_\Sigma (\nabla_\Sigma \phi_h(\Sigma'))$ for $\Sigma' \in cl D^1 \cap cl D^2$ in accordance to (17).

As it was mentioned in the beginning, for $h = 0$ the vector Σ^0 solves (16). We will assume for stability investigation that, at Σ^0, the following strong second-order condition holds with Clarke's generalized Hessian [2]:

Assumption 3. $\exists\, c > 0$: $\quad v^T(Q\,v) \geq c\,|v|^2 \quad$ for all $v \in R^L$ and each matrix $Q \in \partial_\Sigma (\nabla_\Sigma \phi_0)(\Sigma^0)$.

5   Lipschitz Continuity of Bang-Bang Structure

For further analysis of stability properties of switching points consider the stationary point map

$$Z(\Sigma, h) = \nabla_\Sigma \phi_h(\Sigma) = 0 \qquad (26)$$

for h near $h_0 = 0$, cf. (23). Then the following stability result holds:

Theorem 2. Let (P_0) have a bang-bang solution (x^0, u^0) with switching points Σ^0 such that Assumptions 1, 2a and 2b are fulfilled. Then, a neighborhood $\mathcal{H}$ of $h_0 = 0$ exists such that

(i)   In R^L there exists a neighborhood S of Σ^0 such that $\forall\, h \in \mathcal{H}$ equation (26) has an unique solution $\Sigma = \Sigma(h)$. As a function of h, $\Sigma = \Sigma(h)$ is Lipschitz continuous on $\mathcal{H}$.

(ii) $\forall\, h \in \mathcal{H}$: $\Sigma = \Sigma(h) \in D_\Sigma$.
In particular, t_s^h may be (an at most) double switch belonging to $\Sigma_j(h) \cap \Sigma_i(h)$ only if there is a neighboring $t_s^0 \in \Sigma_j^0 \cap \Sigma_i^0$, too.

(iii)   All matrices Q in $\partial_\Sigma (\nabla_\Sigma \phi_h)(\Sigma(h))$ are positive definite with lower eigenvalue bound $c' > 0$ independent of $h \in \mathcal{H}$. The vector $\Sigma(h)$ thus is a strict local minimizer of ϕ_h from (16).   □

Proof. The first statement of the Theorem is proved by Clarke's version of Implicit Function Theorem ([2], §7.1): by Assumption 2b, every matrix Q in $\partial_\Sigma Z(\Sigma^0, 0) = \partial_\Sigma (\nabla_\Sigma \phi_0)(\Sigma^0)$ is of maximal rank so that (i) is valid.

In order to prove part (ii) it is sufficient to remember the monotonicity properties defining switching vectors in D_Σ. Since by (i), $|\Sigma(h) - \Sigma^0| = O(h)$, for h sufficiently close to zero the relations follow from continuity arguments.

Finally, consider $\partial_\Sigma Z(\Sigma(h), h) = \partial_\Sigma (\nabla_\Sigma \phi_h)(\Sigma(h))$:
With given $\Sigma = \Sigma(h)$, construct $u^h = u(\cdot, \Sigma(h))$, $x^h = x(\cdot, \Sigma(h), h)$ by (14), (15) and accomplish this admissible pair by adjoint and switching functions p^h, σ^h,

cf. (18). Defining k^h, A^h, C^h and H^h analogously, one can solve related systems of type (19) and (20) in order to find $\eta_\alpha^h = \partial x^h / \partial t_\alpha$.

Matrices from $\partial_\Sigma Z(\Sigma(h), h)$ again have the form of convex combinations of limits of the Hessians for $\phi_h(\Sigma(h))$ from sets of type D^j, whose elements are described by formulas (24), (25) with the appropriate new data functions. As far as, by (i) and the general smoothness assumptions on (P_h), all elements of the related Hessians continuously depend on $h \in \mathcal{H}$ near $h_0 = 0$, the positive definiteness remains to be true for the perturbed matrices with certain $c' = c - \mathrm{O}(h)$. □

Conclusion. The main results on local structural stability of bang-bang extremals and Lipschitz continuity of switching points have been proved in the paper for case of at most double switches. A generalization to multiple switches of higher order requires thorough strengthening of assumptions 2 and 3.

In contrast to situations where all switching points are simple, it could not been proved yet whether the given assumptions are sufficient optimality conditions for extremals from (6) and (7). Work on this problem is in progress.

References

1. Agrachev, A., Stefani, G., Zezza, P.L.: Strong optimality for a bang-bang trajectory. SIAM J. Control Optim. 41, 991–1014 (2002)
2. Clarke, F.: Optimization and nonsmooth analysis. Wiley, New York (1983)
3. Felgenhauer, U.: On stability of bang-bang type controls. SIAM J. Control Optim. 41(6), 1843–1867 (2003)
4. Felgenhauer, U.: Optimality and sensitivity for semilinear bang-bang type optimal control problems. Internat. J. Appl. Math. Computer Sc. 14(4), 447–454 (2004)
5. Felgenhauer, U.: Optimality properties of controls with bang-bang components in problems with semilinear state equation. Control & Cybernetics 34(3), 763–785 (2005)
6. Felgenhauer, U.: Primal-dual stability approach for bang-bang optimal controls in semilinear systems. Internat. J. Appl. Math. Computer Sc. (to appear)
7. Kim, J.R., Maurer, H.: Sensitivity analysis of optimal control problems with bang-bang controls. In: 42nd IEEE Conference on Decision and Control, Hawaii, vol. 4, pp. 3281–3286 (2003)
8. Klatte, D., Kummer, B.: Nonsmooth equations in optimization. Kluwer Acad. Publ., Dordrecht (2002)
9. Maurer, H., Osmolovskii, N.P.: Equivalence of second-order optimality conditions for bang-bang control problems. Control & Cybernetics 34(3), 927–950 (2005)
10. Milyutin, A.A., Osmolovskii, N.P.: Calculus of variations and optimal control. Amer. Mathem. Soc. Providence, Rhode Island (1998)
11. Noble, J., Schättler, H.: Sufficient conditions for relative minima of broken extremals in optimal control theory. J. Math. Anal. Appl. 269, 98–128 (2002)
12. Osmolovskii, N.P.: Second-order conditions for broken extremals. In: Ioffe, A., et al. (eds.) Calculus of variations and optimal control, Res. Notes Math., vol. 411, pp. 198–216. Chapman & Hall/CRC, Boca Raton, FL (2000)
13. Sarychev, A.V.: First- and second-order sufficient optimality conditions for bang-bang controls. SIAM J. Control Optim. 35(1), 315–340 (1997)
14. Veliov, V.: On the bang-bang principle for linear control systems. Dokl. Bolg. Akad. Nauk 40, 31–33 (1987)

On State Estimation Approaches for Uncertain Dynamical Systems with Quadratic Nonlinearity: Theory and Computer Simulations

Tatiana F. Filippova and Elena V. Berezina

Institute of Mathematics and Mechanics,
Russian Academy of Sciences,
16 S. Kovalevskaya str., GSP-384, 620219 Ekaterinburg, Russia
ftf@imm.uran.ru, lnx@imm.uran.ru
http://www.imm.uran.ru

Abstract. The paper deals with the problems of state estimation for nonlinear dynamical control system described by differential equations with unknown but bounded initial condition. The nonlinear function in the right-hand part of a differential system is assumed to be of quadratic type with respect to the state variable. Basing on the well-known results of ellipsoidal calculus developed for linear uncertain systems we present the modified state estimation approaches which use the special structure of the dynamical system.

1 Introduction

In many applied problems the evolution of a dynamical system depends not only on the current system states but also on uncertain disturbances or errors in modelling. There are many publications devoted to different aspects of treatment of uncertain dynamical systems (e.g., [5,9,10,11,12,13,15]). The model of uncertainty considered here is deterministic, with set-membership description of uncertain items which are taken to be unknown with prescribed given bounds.

The paper deals with the problems of control and state estimation for a dynamical control system

$$\dot{x}(t) = A(t)x + f(x(t)) + G(t)u(t), \quad x \in R^n, \quad t_0 \leq t \leq T, \tag{1}$$

with unknown but bounded initial condition

$$x(t_0) = x_0, \quad x_0 \in X_0, \quad X_0 \subset R^n, \tag{2}$$

$$u(t) \in U, \quad U \subset R^m \text{ for a.e. } t \in [t_0, T]. \tag{3}$$

Here matrices $A(t)$ and $G(t)$ (of dimensions $n \times n$ and $n \times m$, respectively) are assumed to be continuous on $t \in [t_0, T]$, X_0 and U are compact and convex. The nonlinear n-vector function $f(x)$ in (1) is assumed to be of quadratic type

$$f(x) = (f_1(x), \ldots, f_n(x)), \quad f_i(x) = x'B_i x, \quad i = 1, \ldots, n, \tag{4}$$

where B_i is a constant $n \times n$ - matrix $(i = 1, \ldots, n)$.

I. Lirkov, S. Margenov, and J. Waśniewski (Eds.): LSSC 2007, LNCS 4818, pp. 326–333, 2008.
© Springer-Verlag Berlin Heidelberg 2008

Consider the following differential inclusion [6] related to (1)–(3)

$$\dot{x}(t) \in A(t)x(t) + f(x(t)) + P(t) \quad \text{for a.e. } t \in [t_0, T], \tag{5}$$

where $P(t) = G(t)U$.

Let absolutely continuous function $x(t) = x(t, t_0, x_0)$ be a solution to (5) with initial state x_0 satisfying (2). The differential system (1)–(3) (or equivalently, (5)–(2)) is studied here in the framework of the theory of uncertain dynamical systems (differential inclusions) through the techniques of trajectory tubes $X(\cdot, t_0, X_0) = \{x(\cdot) = x(\cdot, t_0, x_0) \mid x_0 \in X_0\}$ of solutions to (1)–(3) with their t-cross-sections $X(t) = X(t, t_0, X_0)$ being the reachable sets (the informational sets) at instant t for control system (1)–(3).

Basing on the well-known results of ellipsoidal calculus [12,2,3] developed for linear uncertain systems we present the modified state estimation approaches which use the special structure of the control system (1)–(4) and combine advantages of estimating tools mentioned above. Examples and numerical results related to procedures of set-valued approximations of trajectory tubes and reachable sets are also presented.

2   Preliminaries

2.1   Basic Notations

In this section we introduce the following basic notations. Let R^n be the n–dimensional Euclidean space and $x'y$ be the usual inner product of $x, y \in R^n$ with prime as a transpose, $\| x \| = (x'x)^{1/2}$. We denote as $B(a, r)$ the ball in R^n, $B(a, r) = \{x \in R^n : \| x - a \| \leq r\}$, I is the identity $n \times n$-matrix. Denote by $E(a, Q)$ the ellipsoid in R^n, $E(a, Q) = \{x \in R^n : (Q^{-1}(x - a), (x - a)) \leq 1\}$ with center $a \in R^n$ and symmetric positively definite $n \times n$–matrix Q.

Let $h(A, B) = \max\{h^+(A, B), h^-(A, B)\}$, be the Hausdorf distance for A, $B \subset R^n$, with $h^+(A, B)$ and $h^-(A, B)$ being the Hausdorf semidistances between A and B, $h^+(A, B) = \sup\{d(x, B) \mid x \in A\}$, $h^-(A, B) = h^+(B, A)$, $d(x, A) = \inf\{\|x - y \| \mid y \in A\}$.

One of the approaches that we will discuss here is related to evolution equations of the funnel type [7,11,14,17]. Note first that we will consider the Caratheodory–type solutions $x(\cdot)$ for (5)–(2), i.e. absolutely continuous functions $x(t)$ which satisfy the inclusion (5) for a. e. $t \in [t_0, T]$. Assume that all solutions $\{x(t) = x(t, t_0, x_0) \mid x_0 \in X_0\}$ are extendable up to the instant T that is possible under some additional conditions ([6], §7, Theorem 2). The precise value of T depending on studied system data will be given later.

Let us consider the particular case of the funnel equation related to (5)–(2)

$$\lim_{\sigma \to +0} \sigma^{-1} h \left(X(t + \sigma), \bigcup_{x \in X(t)} \{x + \sigma(A(t)x + f(x)) + \sigma P(t)\} \right) = 0, \ t \in [t_0, T], \tag{6}$$

$$X(t_0) = X_0. \tag{7}$$

Under above mentioned assumptions the following theorem is true (details may be found in [7,11,14,17]).

Theorem 1. *The nonempty compact-valued function* $X(t) = X(t, t_0, X_0)$ *is the unique solution to the evolution equation (6)-(7).*

Other versions of funnel equation (6) are known written in terms of semidistance h^+ instead of h [12]. The solution to the h^+-equations may be not unique and the "maximal" one (with respect to inclusion) is studied in this case. Mention here also the second order analogies of funnel equations for differential inclusions and for control systems based on ideas of Runge-Kutta scheme [5,15,16]. Discrete approximations for differential inclusions through a set-valued Euler's method were developed in [1,4,8]. Funnel equations for differential inclusions with state constraints were studied in [11], the analogies of funnel equations for impulsive control systems were given in [7].

2.2   Trajectory Tubes in Uncertain Systems with Quadratic Nonlinearity

Computer simulations related to modelling in nonlinear problems of control under uncertainty and based on the funnel equations require in common difficult and long calculations with additional quantization in the state space. Nevertheless in some simpler cases it is possible to find reachable sets $X(t; t_0, X_0)$ (more precisely, to find their ϵ–approximations in Hausdorf metrics) basing on the Theorem 1.

Example 1. Consider the following differential system:

$$\begin{cases} \dot{x}_1 = -x_1, \\ \dot{x}_2 = 0.5x_2 + 3(0.25x_1^2 + x_2^2), \end{cases} \quad 0 \le t \le T. \tag{8}$$

Here we take the case when the uncertainty in the system is defined by uncertain initial states x_0 that are unknown but belong to the following ellipsoid

$$X_0 = E(0, Q_0), \quad Q_0 = \begin{pmatrix} 4 & 0 \\ 0 & 1 \end{pmatrix}. \tag{9}$$

The trajectory tube $X(\cdot)$ found on the base of Theorem 1 is given at Fig. 1 (here $T = 0.2$).

Computational analysis of systems even under quadratic nonlinearity may be more difficult if cross-sections $X(t)$ (the reachable sets) of the trajectory tube $X(\cdot)$ are not convex.

Example 2. Consider the following differential system:

$$\begin{cases} \dot{x}_1 = 2x_1, \\ \dot{x}_2 = 2x_2 + x_2^2, \end{cases} \quad 0 \le t \le T. \tag{10}$$

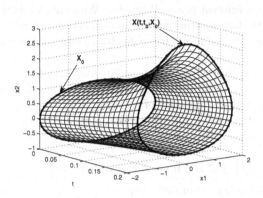

Fig. 1. Trajectory tube $X(\cdot)$ for the uncertain system with quadratic nonlinearity

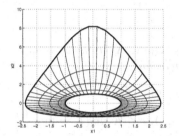 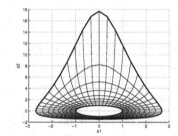

Fig. 2. The reachable set $X(t)$ for $0 \le t \le t^* = 0.44$ (the left picture) and for $0 \le t \le T = 0.5$ (the right picture)

Here an unknown state x_0 belongs to the ball $B(0,1) \subset R^2$. The left picture at Fig. 2 shows that $X(t)$ keeps the convexity property till the instant $t^* = 0.5\ln(1 + \sqrt{2}) \simeq 0.44$. For later time $t > t^*$ the reachable set $X(t)$ loses the convexity as shown at the right picture of Fig. 2. Mention that in this example the value t^* is calculated by analyzing the properties of the curve

$$\left(\frac{x_1}{e^{2t}}\right)^2 + \left(\frac{2x_2}{e^{2t}(2 + x_2) - x_2}\right)^2 = 1$$

that describes the boundary of $X(t)$.

3   Results

The methods of constructing the external (or "upper" with respect to inclusion) estimates of trajectory tubes $X(\cdot, t_0, X_0)$ of a differential system with uncertainty may be based on the combination of ellipsoidal calculus [2,3,12] and the techniques of evolution funnel equations. It should be noted that external ellipsoids approximated the trajectory tube may be chosen in various ways and several minimization criteria are well-known. We consider here the ellipsoidal techniques related to the construction of external estimates with minimal volume (details of this approach and motivations for linear control systems may be found in [2,12]).

Here we assume for simplicity that $U = \{0\}$ and therefore $P(t) = \{0\}$ in (5)-(2), matrices B_i $(i = 1, ..., n)$ are symmetric and positively definite, $A(t) \equiv A$. We may assume that all trajectories of the system (5)-(2) belong to a bounded domain $D = \{x \in R^n : \| x \| \leq K\}$ where the existence of such constant $K > 0$ follows from classical theorems of the theory of differential equations and differential inclusions [6].

From the structure (4) of the function f we have two auxiliary results.

Lemma 1. *The following estimate is true*

$$\| f(x) \| \leq N, \qquad N = K^2 (\sum_{i=1}^{n} \lambda_i^2)^{1/2},$$

where λ_i is the maximal eigenvalue for matrix B_i $(i = 1, ..., n)$.

Lemma 2. *For all $t \in [t_0, T]$ the inclusion $X(t) \subset X^*(t)$ holds where $X^*(\cdot)$ is a trajectory tube of the linear differential inclusion*

$$\dot{x} \in Ax + B(0, N), \quad x_0 \in X_0. \tag{11}$$

Theorem 2. *Let $t_0 = 0$, $a_0 = 0$, $X_0 = B(a_0, r)$, $r \leq K$ and $t_* = \min \left\{ \frac{K-r}{\sqrt{2M}}; \frac{1}{L}; T \right\}$. Then for all $t \in [t_0, t_*]$ the following inclusion is true*

$$X(t, t_0, X_0) \subset E(a(t), Q^+(t)) \tag{12}$$

where $M = K\sqrt{\lambda} + N$, $L = \sqrt{\lambda} + 2K \left(\sum_{i=1}^{n} \lambda_i^2 \right)^{1/2}$ with λ and λ_i being the maximal eigenvalues of matrices AA' and B_i $(i = 1, ..., n)$, respectively, and where vector function $a(t)$ and matrix function $Q^+(t)$ satisfy the equations

$$\dot{a} = Aa, \quad a(t_0) = a_0 \tag{13}$$

$$\dot{Q}^+ = AQ^+ + Q^+ A^T + qQ^+ + q^{-1}G,$$
$$q = \{n^{-1} Tr((Q^+)^{-1}G)\}^{1/2}, G = N^2 I, \tag{14}$$
$$Q^+(t_0) = Q_0 = r^2 I.$$

Proof. Applying the ellipsoidal techniques [2,12], Lemmas 1-2 and comparing systems (5)-(2) and (11) we come to the inclusion (12).

Example 3. Results of computer simulations based on the above theorem for the differential system

$$\begin{cases} \dot{x_1} = x_1, \\ \dot{x_2} = 3x_2 + x_2^2, \end{cases} \tag{15}$$

with $X_0 = B(0, 1)$ are given at Fig. 3. The pictures at Fig. 3 show that the approach presented in Theorem 2 is appropriate only for enough small values of $t \in [t_0, T]$.

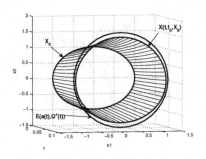
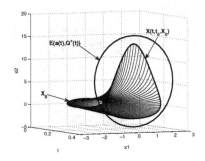

Fig. 3. Reachable sets $X(t, t_0, X_0)$ and their estimates $E(a(t), Q^+(t))$ at instants $t = 0.1$ and $t = 0.4$.

Let us discuss another estimation approach based on techniques of evolution funnel equations. Consider the following system

$$\dot{x} = Ax + \tilde{f}(x)d, \quad x_0 \in X_0, \ t_0 \leq t \leq T, \tag{16}$$

where $x \in R^n$, $\|x\| \leq K$, d is a given n-vector and a scalar function $\tilde{f}(x)$ has a form $\tilde{f}(x) = x'Bx$ with $B = \text{diag}(b_1^2, ..., b_n^2)$. The following theorem presents an easy computational tool to find estimates of $X(t)$ by step-by-step procedures.

Theorem 3. Let $X_0 = E(0, B^{-1})$. Then for all $\sigma > 0$

$$X(t_0 + \sigma, t_0, X_0) \subseteq E(a(\sigma), Q(\sigma)) + o(\sigma)B(0, 1), \tag{17}$$

where $a(\sigma) = \sigma d$ and $Q(\sigma) = (I + \sigma A)B^{-1}(I + \sigma A)'$.

Proof. The funnel equation for (16) is

$$\lim_{\sigma \to +0} \sigma^{-1} h\left(X(t + \sigma, t_0, X_0), \bigcup_{x \in X(t, t_0, X_0)} \left\{ x + \sigma(Ax + \tilde{f}(x)d) \right\} \right) = 0, \ t \in [t_0, T],$$

$$X(t_0, t_0, X_0) = X_0. \tag{18}$$

If $x_0 \in \partial X_0$ where ∂X_0 means the boundary of X_0, we have $\tilde{f}(x_0) = 1$ and ([2])

$$\bigcup_{x_0 \in \partial X_0} \{(I + \sigma A)x_0 + \sigma \tilde{f}(x_0)d\} = \bigcup_{x_0 \in \partial X_0} \{(I + \sigma A)x_0 + \sigma d\} \subseteq E(a(\sigma), Q(\sigma)). \tag{19}$$

It is not difficult to check that the inclusion (19) is true also for any interior point $x_0 \in \text{int} X_0$, so we have

$$\bigcup_{x_0 \in X_0} \{(I + \sigma A)x_0 + \sigma \tilde{f}(x_0)d\} \subseteq E(a(\sigma), Q(\sigma)). \tag{20}$$

Applying Theorem 1 and taking into account the inclusion (20) we come to the estimate (17).

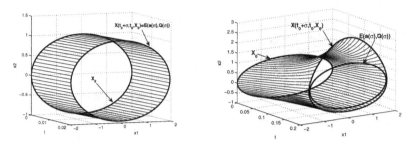

Fig. 4. Trajectory tube $X(t_0 + \sigma, t_0, X_0)$ and $E(a(\sigma), Q(\sigma))$ for $\sigma = 0.02$ and $\sigma = 0.2$

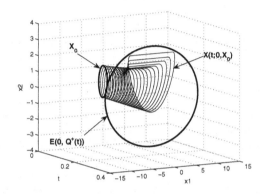

Fig. 5. Trajectory tube $x(t) = X(t, t_0, X_0)$ and its ellipsoidal estimate $E(0, Q^*(t))$ for the uncertain control system

Example 4. Consider the system (8)-(9). Fig. 4 presents the results of the computer simulations based on Theorem 3.

Note that the last approach requires less computations then the first one given by Theorem 2 and provides better upper estimates of the trajectory tube.

Results similar to above Theorems 2–3 may be derived also for the system (5). In this case the final ellipsoidal estimate is more complicated because the technique includes an additional operation of approximating externally the sum of two ellipsoids, the first one related to nonlinearity (as in Theorems 2-3) and the second one due to presence of $P = E(q_p, Q_p)$. We illustrate the estimation by the following example.

Example 5. Consider the following control system:

$$\begin{cases} \dot{x}_1 &=& 6x_1 + u_1, \\ \dot{x}_2 &=& x_1^2 + x_2^2 + u_2, \end{cases} \quad 0 \le t \le T, \tag{21}$$

$$X_0 = B(0,1), \ P(t) \equiv U = B(0,1). \tag{22}$$

The trajectory tube $X(t)$ and its estimate $E(0, Q^*(t)$ are given at Fig. 5 (here $T = 0.27$).

Acknowledgments. The research was supported by the Russian Foundation for Basic Research (RFBR) under Project 06-01-00483.

References

1. Chahma, I.A.: Set-valued discrete approximation of state-constrained differential inclusions. Bayreuth. Math. Schr. 67, 3–162 (2003)
2. Chernousko, F.L.: State Estimation for Dynamic Systems. Nauka, Moscow (1988)
3. Chernousko, F.L., Ovseevich, A.I.: Properties of the optimal ellipsoids approximating the reachable sets of uncertain systems. J. Optimization Theory Appl. 120(2), 223–246 (2004)
4. Dontchev, A.L., Farkhi, E.M.: Error estimates for discretized differential inclusions. Computing 41(4), 349–358 (1989)
5. Dontchev, A.L., Lempio, F.: Difference methods for differential inclusions: a survey. SIAM Review 34, 263–294 (1992)
6. Filippov, A.F.: Differential Equations with Discontinuous Right-hand Side. Nauka, Moscow (1985)
7. Filippova, T.F.: Sensitivity Problems for Impulsive Differential Inclusions. In: Proc. of the 6th WSEAS Conference on Applied Mathematics, Corfu, Greece (2004)
8. Häckl, G.: Reachable sets, control sets and their computation. Augsburger Mathematisch-Naturwissenschaftliche Schriften. PhD Thesis, University of Augsburg, Augsburg, 7 (1996)
9. Krasovskii, N.N., Subbotin, A.I.: Positional Differential Games. Nauka, Moscow (1974)
10. Kurzhanski, A.B.: Control and Observation under Conditions of Uncertainty. Nauka, Moscow (1977)
11. Kurzhanski, A.B., Filippova, T.F.: On the theory of trajectory tubes — a mathematical formalism for uncertain dynamics, viability and control. In: Kurzhanski, A.B. (ed.) Advances in Nonlinear Dynamics and Control: a Report from Russia. Progress in Systems and Control Theory, vol. 17, pp. 122–188. Birkhauser, Boston (1993)
12. Kurzhanski, A.B., Valyi, I.: Ellipsoidal Calculus for Estimation and Control. Birkhauser, Boston (1997)
13. Kurzhanski, A.B., Veliov, V.M. (eds.): Set-valued Analysis and Differential Inclusions. Progress in Systems and Control Theory, vol. 16. Birkhauser, Boston (1990)
14. Panasyuk, A.I.: Equations of attainable set dynamics. Part 1: Integral funnel equations. J. Optimiz. Theory Appl. 64(2), 349–366 (1990)
15. Veliov, V.M.: Second order discrete approximations to strongly convex differential inclusions. Systems and Control Letters 13, 263–269 (1989)
16. Veliov, V.: Second-order discrete approximation to linear differential inclusions. SIAM J. Numer. Anal. 29(2), 439–451 (1992)
17. Wolenski, P.R.: The exponential formula for the reachable set of a Lipschitz differential inclusion. SIAM J. Control Optimization 28(5), 1148–1161 (1990)

Using the Escalator Boxcar Train to Determine the Optimal Management of a Size-Distributed Forest When Carbon Sequestration Is Taken into Account

Renan Goetz[1], Natali Hritonenko[2], Angels Xabadia[1], and Yuri Yatsenko[3]

[1] Department of Economics, University of Girona,
Campus de Montilivi, Girona 17071, Spain
`renan.goetz@udg.edu`
[2] Department of Mathematics, Prairie View A&M University,
Box 519, Prairie View, TX 77446, USA
[3] College of Business and Economics, Houston Baptist University,
7502 Fondren Road, Houston, TX 77074, USA

Abstract. We present a theoretical model to determine the optimal management of a size-distributed forest. The decision model is given in form of a distributed optimal control problem that cannot be solved analytically. Thus, the paper presents a numerical technique that allows transforming the original distributed control problem into an ordinary control problem. The method has the advantage that it does not require programming numerical algorithms but rather can be implemented with standard commercial optimization packages such as GAMS. The empirical application of our model for the case of forest management allows determining the selective cutting regime when carbon sequestration is taken into account.

1 Introduction

The analysis of the optimal management of renewable resources has a very long history within economics. Traditionally, the renewable resource is considered as a non-structured population, i.e. all individuals are identical. However, the economic literature has now recognized that the assumption of non-structured populations does not allow to address correctly the two key issues of the management of renewable natural resources - the determination of the optimal replacement periods, and the optimal long-run allocation of the total population among the different values of the structuring variable (Kennedy, 1987). In the case of forestry, economic analysis based on non structured populations cannot correctly determine the optimal replacement period since it does not take account of the fact that the birth, growth and death processes of the population depend on the structuring characteristic of each individual.

Moreover, the distribution of the individual characteristics over the entire population is important to model correctly the multiple services that forests may

I. Lirkov, S. Margenov, and J. Waśniewski (Eds.): LSSC 2007, LNCS 4818, pp. 334–341, 2008.

offer: amenity and recreational values, natural habitat of wildlife, mushrooms, protection of watersheds, carbon sequestration etc. A simple example is the production and timber and sequestration of carbon. The maximization of the net benefits with respect to timber calls for growing a reduced number of high value trees while maximization of the net benefits with respect to carbon requires maximizing the standing biomass of the forest. Thus, a specific distribution of the individual characteristics over the entire forest is necessary to favor a particular forest service.

In this paper we present a theoretical model to determine the optimal selective-logging regime of a size-distributed forest. The law of motion of the economic model is governed by a partial integrodifferential equation that describes the evolution of the forest stock over time. Given the complexity of the problem it is not possible to obtain an analytical solution. To solve the problem numerically we employ a technique known as the "Escalator Boxcar Train". This technique has been utilized so far to solve partial differential or partial integrodifferential equations describing the evolution of biological populations. In this paper we provide the necessary changes to utilize this method within the context of a distributed optimal control problem. The usefulness of the "Escalator Boxcar Train" is illustrated by determining the optimal selective-logging regime of a privately owned forest when carbon sequestration is taken into account.

The paper is organized as follows. The following section presents the economic decision problem. Section 3 presents the numerical technique of the "Escalator Boxcar Train" and derives the necessary changes that allow to employ it for the solution of the distributed optimal control problem. Section 4 briefly illustrates the proposed numerical technique. The paper ends with a summary.

2   The Economic Model

Denote the diameter of a tree by $l \in \Omega$, $\Omega \equiv [l_0, l_m)$, where l_0 and l_m indicate the biological minimum and maximum size of a tree. The decision problem of the forest owner, given a planning horizon of t_1, can be stated as:

$$\max_{u(t,l),p(t,l_0)} \int_0^{t_1} \int_{l_0}^{l_m} B(u(t,l), x(t,l)) exp^{-rt} \, dl \, dt - \int_0^{t_1} C(p(t,l_0)) exp^{-rt} \, dt \ (1)$$

$$+ \int_{l_0}^{l_m} S^{t_1}(x(t_1,l)) exp^{-rt_1} \, dl$$

subject to the constraints

$$\frac{\partial x(t,l)}{\partial t} \equiv -\frac{\partial (g(e(t),l) \, x(t,l))}{\partial l} - \delta(e(t),l) \, x(t,l) - u(t,l), \tag{2}$$

$$x(t_0,l) = x_0(l), \quad g(e(t),l_0) x(t,l_0) = p(t,l_0), \quad u(t,l) \geq 0, \quad p(t,l_0) \geq 0, \tag{3}$$

where $x(t,l)$ is the state variable that characterizes the distribution density of trees, and $u(t,l)$ $p(t)$ are the control variables that denote the flux of logged trees, and the flux of trees planted with diameter l_0 respectively, $g(e(t),l)$ is the growth rate of trees, $\delta(e(t), l)$ is the instantaneous mortality rate, $e(t)$ denotes the competition between individuals in the forest, that is measured by the basal area, that is, $e(t) = \int_{l_0}^{l_m} \pi \left(\frac{l}{2}\right)^2 x(t, l) \, dl$, and r denotes the discount rate. The function $B(u(t, l), x(t, l)) exp^{-rt}$ presents the discounted net benefits of forest management, including the benefits from timber and sequestered carbon. The strictly convex function $C(p) exp^{-rt} \in C^2$ captures the discounted cost of planting trees, the function $S^{t_1}(x) exp^{-rt_1} \in C^1$ the value of the standing trees at the end of the planning horizon. The term $x_0(l)$ denotes the initial distribution density of trees, and the restriction $g(e(t), l_0) x(t, l_0) = p(t, l_0)$ requires that the flux of the change in diameter at l_0 multiplied the tree density coincides with the total amount of trees planted with diameter l_0. Finally, it is required that the control variables are nonnegative. Equation (2) provides the equation of motion for size-structured populations in the case of a managed forest with no biological reproduction (i.e., all young trees are planted).

3   The Numerical Approach

To numerically solve the distributed optimal control problem we propose a method named the Escalator Boxcar Train (de Roos, 1988). In contrast to other available methods (Feichtinger, Prskawetz and Veliov, 2004), the Escalator Boxcar Train, EBT, can be implemented with standard computer software utilized for solving mathematical programming problems.

The partial integrodifferential equation (2) describes the time evolution of the population density-function over the domain Ω of the individual structural variable l. For the derivation of the EBT method we will assume that $e(t)$ is constant such that the partial integrodifferential equation is a partial differential equation. However, as de Roos (1988) specifies, the method is also applicable in the case where the environment is not constant over time. Let Ω be partitioned into $n + 1$ subdomains $\Omega_i(t = 0) = [l_i(0), l_{i+1}(0)), i = 0, 1, 2, \cdots, n$ at the initial point of time of the planning horizon and define $\Omega_i(t)$ as $\Omega_i(t) = \{l(t, t = 0, l_0) | l_0 \in \Omega_i(0)\}$, i.e., $\Omega_i(t)$ defines the evolution of the tree diameter over time, such that all trees that form a cohort in a subdomain at t=0 stay together throughout their life.

Let the total number of trees, $X_i(t)$, and the average diameter of the trees, $L_i(t)$ within the subdomain $\Omega_i, i = 1, 2, \cdots, n$, that we call internal cohorts, be defined as

$$X_i(t) = \int_{\Omega_i} x(t, l) dl, \quad \text{and} \quad L_i(t) = \frac{1}{X_i(t)} \int_{\Omega_i} l \, x(t, l) dl. \tag{4}$$

To take account of the control variable, let U_i denote the amount of cut trees in the subdomain Ω_i. That is,

$$U_i = \int_{\Omega_i} u(t, l) \, dl. \tag{5}$$

the total number of trees and their average diameter totally characterizes the population within the subdomain or cohort i. The total population is thus a collection of cohorts. Mathematically, the distribution density $x(t, l)$ is approximated by a set of delta functions of size $X_i(t)$ at diameter $L_i(t)$. Using Leibniz rule, the change in time of $X_i(t)$ is given by:

$$\frac{dX_i(t)}{dt} = \frac{d}{dt}\int_{l_i}^{l_{i+1}} x(t,l)dl = \int_{l_i}^{l_{i+1}} \frac{\partial x(t,l)}{\partial t}dl + x(t, l_{i+1})\frac{dl_{i+1}}{dt} - x(t, l_i)\frac{dl_i}{dt} \qquad (6)$$

Since $\dfrac{dl_i}{dt} = g(e, l_i)$ and $\dfrac{dl_{i+1}}{dt} = g(e, l_{i+1})$ by definition, and using the equality presented in equation (2), equation (6) leads to:

$$\frac{dX_i(t)}{dt} = -\int_{l_i}^{l_{i+1}} \Big(\delta(e, l)x(t, l) + u(t, l)\Big)dl \qquad (7)$$

In a similar fashion we obtain the ordinary differential equation describing the dynamics of $L_i(t)$. From now on, the arguments of the functions will be suppressed unless it is required for an unambiguous notation.

$$\begin{aligned}
\frac{dL_i}{dt} &= \frac{d}{dt}\frac{1}{X_i}\int_{l_i}^{l_{i+1}} l\,x\,dl \\
&= \frac{1}{X_i}\left(\int_{l_i}^{l_{i+1}} l\frac{\partial x}{\partial t}dl + l_{i+1}x\frac{dl_{i+1}}{dt} - l_i\,x\frac{dl_i}{dt}\right) - \frac{1}{X_i^2}\int_{l_i}^{l_{i+1}} l\,x\,dl\frac{dX_i}{dt} \\
&= \frac{1}{X_i}\int_{l_i}^{l_{i+1}} l\frac{\partial x}{\partial t}dl + \frac{1}{X_i}\int_{l_i}^{l_{i+1}} \frac{\partial(l\,g\,x)}{\partial l}dl - \frac{1}{X_i}L_i\frac{dX_i}{dt} \\
&= \frac{1}{X_i}\int_{l_i}^{l_{i+1}} l\Big(-\delta x - u\Big)dl + \frac{1}{X_i}\int_{l_i}^{l_{i+1}} g\,x\,dl - \frac{1}{X_i}L_i\frac{dX_i}{dt} \\
&= \frac{1}{X_i}\int_{l_i}^{l_{i+1}} g\,x\,dl - \frac{1}{X_i}\int_{l_i}^{l_{i+1}} (l - L_i)\Big(\delta x + u\Big)dl \qquad (8)
\end{aligned}$$

The ordinary differential equations for $X_i(t)$ and $L_i(t)$ described in equations (7) and (8) do not form a solvable system because they involve weighted integrals over the density function $x(t, l)$. To obtain a closed solvable system the functions $\delta(e, l)$ and $g(e, l)$ are approximated by their first order Taylor expansion around $l = L_i(t)$. Higher order terms involving squares and higher powers of $(l - L_i)$ are neglected. In this way, equation (7) can be approximated to:

$$\frac{dX_i(t)}{dt} \simeq -\delta(e, L_i)X_i(t) - U_i(t). \qquad (9)$$

And equation (8) is approximated to:

$$\frac{dL_i(t)}{dt} \simeq g(e, L_i) - \frac{1}{X_i}\left(\int_{l_i}^{l_{i+1}} l\,u\,dl - L_iU_i\right). \qquad (10)$$

Define $\bar{U}_i$ as the average diameter of logged trees in cohort i, i.e., $\bar{U}_i = \frac{1}{U_i}$ $\int_{l_i}^{l_{i+1}} l\, u(t, l)\, dl$, and substitute it into equation (10). Assuming that the cut trees have in average the same diameter than the standing trees in the cohort, that is, $\bar{U}_i = L_i$, equation (10) leads to

$$\frac{dL_i(t)}{dt} = g(e, L_i). \tag{11}$$

Equations (9) and (11) describe the dynamics of internal cohorts but they do not account for the plantation of new trees. The boundary cohort is characterized by

$$X_0(t) = \int_{l_0}^{l_1} x(t, s)\, ds \quad \text{and} \quad \hat{L}_0(t) = \int_{l_0}^{l_1} (l - l_0)x(t, l)\, dl. \tag{12}$$

Differentiation of $X_0(t)$ and $\hat{L}_0(t)$ with respect to time, and employing first order Taylor approximations of the functions $g(e, l)$ and $\delta(e, l)$ leads to a set of ordinary differential equations that describe the dynamics of the boundary cohort. They are given by

$$\frac{dX_0}{dt} \simeq -\delta(e, l_0)X_0(t) - \frac{d}{dl}\delta(e, l_0)\hat{L}_0(t) + p(t)$$

$$\frac{d\hat{L}_0}{dt} \simeq g(e, l_0)X_0(t) + \frac{d}{dl}g(e, l_0)\hat{L}_0(t) - \delta(e, l_0)\hat{L}_0(t). \tag{13}$$

In this way, the partial differential equation and its boundary condition (2) governing the dynamics of the distribution density can be decoupled to a system of ordinary differential equations in $X_i(t)$ and $L_i(t)$ described by equations (9), (11), and (13).

Once the partial integrodifferential equation describing the evolution of the system has been transformed into a system of ordinary differential equations, the objective function needs to be transformed also. Thus, the approximate decision problem is given by:

$$\max_{\bar{U}(t), p(t)} \int_0^{t_1} \hat{B}(\bar{X}(t), \bar{L}(t), \bar{U}(t))exp^{-rt}\, dt - \int_0^{t_1} C(p(t))exp^{-rt}\, dt \tag{D'}$$
$$+ \quad S^{t_1}(\bar{X}(t_1), \bar{L}(t_1))exp^{-rt_1}$$

subject to the constraints

$$\frac{dX_i(t)}{dt} = -\delta(E(t), L_i)X_i(t) - U_i(t), \qquad \frac{dL_i(t)}{dt} = g(E(t), L_i)$$

$$\frac{dX_0}{dt} = -\delta(E(t), l_0)X_0(t) - \frac{d}{dl}\delta(E(t), l_0)\hat{L}_0 + p(t)$$

$$\frac{d\hat{L}_0}{dt} = g(E(t), l_0)X_0(t) + \frac{d}{dl}g(E(t), l_0)\hat{L}_0(t) - \delta(E(t), l_0)\hat{L}_0(t)$$

$$X_i(0) = \bar{X}^0, \quad g(E(t), l_0)\, X_0(t) = p(t), \quad U_i(t), p(t) \geq 0 \quad U_i(t) \leq X_i(t),$$

where $\bar{X}$, $\bar{L}(t)$ and $\bar{U}$ denote the vectors $\bar{X} = (X_0, \cdots, X_n)$, $\bar{L} = (L_0, \cdots, L_n)$ and $\bar{U} = (U_1, \cdots, U_n)$ respectively, $\bar{X}^0$ the initial density of each cohort, and $E(t)$ is the approximation of $e(t)$, that is, $E(t) = \sum_i^n \pi \left(\frac{L_i}{2}\right)^2 X_i$.

It is only left one additional transformation. Since the value of the upper interval bound of the boundary cohort, l_1, increases over time, all newly planted trees have a length widening interval $[l_0, l_1]$. Thus, the boundary cohort cannot be continued indefinitely, because the range would become larger and larger and the approximation would break down. Therefore, the cohorts have to be renumbered at regular time intervals $\triangle t$. This renumbering operation transforms the current boundary cohort into an internal cohort and initializes a new, empty boundary cohort realizing the following operations

$$X_i(k\triangle t^+) = X_{i-1}(k\triangle t^-), \; L_i(k\triangle t^+) = L_{i-1}(k\triangle t^-), \; X_1(k\triangle t^+) = X_0(k\triangle t^-)$$

$$L_1(k\triangle t^+) = l_0 + \frac{\hat{L}_0(k\triangle t^-)}{X_0(k\triangle t^-)}, \; X_0(k\triangle t^+) = 0, \; \hat{L}_0(k\triangle t^+) = 0, \; k = 1, 2, \cdots \; (14)$$

Once this transformation has been applied, the system of differential equations may be converted into a system of difference equations that can be solved using standard software such as GAMS (General Algebraic Modelling System, (Brooke, Kendrick and Meeraus, 1992)).

4 Empirical Study

The purpose of the empirical analysis is to illustrate the applicability of the EBT method by determining the optimal selective-logging regime of a diameter-distributed forest, that is, the selective logging regime that maximizes the discounted private net benefits from timber production and carbon sequestration of a stand of *pinus sylvestris* (Scots pine), over time horizon of 300 years.

The initial diameter distribution of the forest is presented in Table 1 and parameters and functions used are defined in Table 2. The thinning and planting period, $\triangle t$, is set equal to 10 years, which is a common practice for a *pinus sylvestris* forest.

The optimization problem was programmed in GAMS, and the Conopt2 solver was employed to find the numerical solution. Optimizations with different random initializations of the control variables were carried out to assure that solutions are independent from the initially chosen values for the numerical optimization technique. The empirical analysis calculates the optimal selective-logging regime, given the initial diameter distribution of the trees described in Table 1 and a discount rate of 2%. Table 3 summarizes the results for the case where carbon sequestration is not considered and for the case where it is considered.

Table 1. Initial diameter distribution of the trees used in the empirical analysis

Number of trees	287	218	167	115	68	35	16	7	3	1	1
Average diameter (cm)	7.5	12.5	17.5	22.5	27.5	32.5	37.5	42.5	47.5	52.5	57.5

Table 2. Parameters and functions used in the empirical analysis

growth function:	$g(E,l) = (80 - l)(0.0068983 - 0.00003107\,E)$
mortality rate:	$\delta(E,l) = 0.01$
marketable volume (m^3):	$(0.699 + 0.000411L)(0.0002949L^{2.167563})$
timber price (euros/m^3):	$min[-23.24 + 13.63L_i^{1/2}, 86.65]$
carbon price (euros/Ton of sequestered CO$_2$):	15
logging costs:	23.4 euros/m^3 + 3.6 euros/ha
maintenance costs (euros/ha)^a:	$0.07X + 1.18\,10^{-5}X^2$
planting costs (euros/tree):	0.60

^{a} Maintenance costs depend on the total number of trees per hectare, X.

Table 3. Optimal selective-logging regime (where a discount rate of 2% is assumed)

	Optimal management for timber			Optimal management for timber and carbon			
Year	Number of trees$^{(a)}$	Net timber revenue$^{(b)}$ (Euros/ha)	Discounted net benefit (Euros/ha)	Number of trees	Net timber revenue	Carbon revenue	Discounted net benefit (Euros/ha)
					(Euros/ha)	(Euros/ha)	(Euros/ha)
0	890	796.89	14.24	906	476.09	-44.37	-394.79
10	948	587.71	-175.24	1056	402.72	153.62	-262.71
20	977	615.18	-135.54	1128	760.82	205.97	37.21
30	993	892.85	33.96	1155	1296.09	193.43	305.54
40	997	1428.42	259.82	1155	1938.56	137.95	511.83
50	965	2543.29	628.85	1128	2494.13	45.09	596.15
60	936	2629.98	547.05	1138	1856.15	-69.68	255.01
70	845	3318.32	638.49	1040	3744.92	42.23	728.32
...	...	...	...	...	...	...	...
300	926	2018.79	3.18	1120	2241.93	-17.19	3.41

$^{(a)}$ The number of trees in the forest is calculated just after the trees are planted, and before the thinning takes place.
$^{(b)}$ All monetary values apart from the discounted net benefit are expressed as current values.

It shows that the incorporation of carbon sequestration benefits in the decision problem leads to an increase in the optimal number of trees in the forest. As a consequence, the average diameter of these trees is lower[1], leading to a decrease in revenues from timber sale. However, the revenues from the sequestering of carbon compensates the smaller timber revenues and the total sum of discounted net benefits (NPV) of the forest over 300 years is 2972.02 euros per hectare, which is a 10% higher than the NPV obtained when the carbon sequestration is not accounted for.

5 Summary

This paper presents a theoretical model that allows determining the optimal management of a diameter-distributed forest where the growth process of the

---

[1] Due to space limitations, the average diameter is not shown in Table 2.

trees depends not only on individual characteristics but also on the distribution of the individual characteristics over the entire population. The corresponding economic decision problem can be formulated as a distributed optimal control problem where the resulting necessary conditions include a system of partial integrodifferential equation that cannot be solved analytically. For this reason, the utilization of a numerical method (Escalator Boxcar Train) is proposed. The Escalator Boxcar Train method allows to transform the partial integrodifferential equation into a set of ordinary differential equations and thereby to approximate the distributed optimal control problem by a standard optimal control problem. In contrast to the existing literature, the resulting optimization problem can be solved numerically utilizing standard mathematical programming techniques and does not require the programming of complex numerical algorithms.

An empirical analysis is conducted to show how the EBT method can be implemented, determining the optimal selective-logging regime of a diameter-distributed forest when carbon sequestration is taken into account.

References

1. Brooke, A., Kendrick, D., Meeraus, A.: GAMS: A User's Guide, release 2.25. The Scientific Press, San Francisco (1992)
2. de Roos, A.: Numerical methods for structured population models: The Escalator Boxcar Train. Numer Methods Partial Diff. Equations 4, 173–195 (1988)
3. Feichtinger, G., Prskawetz, A., Veliov, V.: Age-structured optimal control in population economics. Theor. Pop. Biol. 65, 373–387 (2004)
4. Kennedy, J.: Uncertainty, irreversibility and the loss of agricultural land: a reconsideration. J. Agric. Econ. 38, 75–81 (1987)

On Optimal Redistributive Capital Income Taxation

Mikhail I. Krastanov[1] and Rossen Rozenov[2]

[1] Institute of Mathematics and Informatics, Bulgarian Academy of Sciences,
Acad. G. Bonchev, Bl. 8, 1113 Sofia, Bulgaria
krast@math.bas.bg
[2] Department of Mathematics and Informatics, Sofia University,
James Bourchier Boul. 5, 1126 Sofia, Bulgaria
rossen_rozenov@yahoo.com

Abstract. The problem of optimal redistributive capital income taxation in a differential game setup is studied. Following the influential works by Judd [3] and Chamley [1], it has been quite common in the economic literature to assume that the optimal limiting tax on capital income is zero. Using a simple model of capital income taxation, proposed originally by Judd, we show that the optimal tax can be different from zero under quite general assumptions. The main result is a sufficient condition for obtaining an appropriate solution to a differential game.

1 Motivation

Some twenty years ago Judd [3] and Chamley [1] independently arrived at the conclusion that the optimal tax on capital income converges to zero in the long run. In the simple version of his optimal redistributive taxation model Judd considered three types of agents – workers, who supply labor inelastically and consume all their income; capitalists who only receive rents and do not work; and government which finances its consumption either through capital income taxation, or through lump-sum taxes imposed on workers. Chamley used a slightly different setup. He derived the zero optimal tax result based on the representative agent model with an additively separable utility function, isoelastic in consumption.

The Judd-Chamley findings have become widely accepted in public economics. Yet, some economists have questioned the validity of the zero limiting capital income tax in different contexts. For example, Kemp et al. [4] have demonstrated that in the original open-loop formulation of Judd's model the optimal tax rate on capital income may not converge to a stable equilibrium and the optimal trajectories may contain closed orbits. When the feedback Stackelberg formulation is considered, the equilibrium tax rate could be a positive or a negative number.

Lansing [5] provided a counter-example to the simple redistributive tax model examined by Judd and argued that with logarithmic utility function of capitalists the optimal tax is non-zero. However, he claimed that the obtained result is specific for this functional form and will not be true in the general case.

I. Lirkov, S. Margenov, and J. Waśniewski (Eds.): LSSC 2007, LNCS 4818, pp. 342–349, 2008.
© Springer-Verlag Berlin Heidelberg 2008

In this paper we show that under very general assumptions about the utility and production functions the optimal tax can be different from zero. We prove a proposition which provides a method for the explicit computation of the optimal capital income tax and illustrate how the proposition works by examples with CRRA utility functions. What distinguishes our approach from the rest of the literature is that we derive the solution to the capital income tax problem using sufficient conditions for optimality, whereas previous results have been based on necessary conditions only. In terms of formulation, we take the Lansing version of Judd's optimal redistributive tax model and show that when one takes into account the dependence of the tax rate on capital, the optimal tax may not be zero and this result is not restricted to the logarithmic utility function. While Kemp et al. [4] also show that the closed-loop formulation can lead to positive or negative taxes, their model setup is different from the one we study. Some of the more important differences are that in their model there is no depreciation and the workers' income consists of wages only (it does not include transfers from the government).

2   Model Setup

We consider an economy populated by three types of infinitely-lived agents – capitalists, workers and a government. Capitalists do not work and receive income only from their capital ownership. Capital is used for the production of a single good which can be either consumed or saved and invested. Workers supply labor inelastically for which they receive wages and in addition, the government grants them a lump-sum transfer or subtracts a lump-sum tax from their labor income. The production technology is described by the function $F(k)$ with the standard properties that $F(0) = 0$, $F'(k)$ is positive and decreasing in k and $F''(k)$ is negative. Capital depreciates at a constant rate $\delta > 0$ and $F(k) - \delta k$ is output net of depreciation. We assume that capital is bounded in this model. Such an assumption is also made in [3] and it seems to hold true for some reasonable specifications of production and utility functions as will be shown later in the example. For simplicity, we consider the case where there is only one representative capitalist.

We begin by stating the static profit maximization conditions for the competitive firm:

$$r = F'(k) \tag{1}$$

$$w = F(k) - kF'(k), \tag{2}$$

where r denotes the interest rate and w is the wage paid to the worker. Since in this setup workers do not save, their consumption is equal to their income. The workers' income x is composed of their wage earnings plus the government transfer T (or minus the lump-sum tax depending on the sign of T) which is calculated as the tax rate τ times the capital income:

$$x(k, \tau) = w + T = F(k) - kF'(k) + \tau k(F'(k) - \delta).$$

In this setting the problem that the representative capitalist faces is to maximize his lifetime utility $U(c)$ subject to the capital accumulation constraint. The utility function $U(c)$ is assumed to be increasing in consumption c and concave. Admissible controls $c(t)$ are those non-negative integrable functions for which the corresponding trajectory $k(t)$ takes non-negative values. Formally the problem is written as [1]

$$\int_0^\infty e^{-\rho t} U(c(t)) dt \to \max \tag{3}$$

with respect to $c(t)$ and subject to

$$\dot{k}(t) = (F'(k(t)) - \delta)(1 - \tau^*(t))k(t) - c(t) \tag{4}$$

$$k(0) = k_0 > 0.$$

Here $F'(k(t))$ is the rental rate at time t, $\tau^*(t)$ is the capital income tax rate, which is determined by the government and which the capitalist takes as given, and $\rho > 0$ is the time preference parameter. To simplify the notations, let $g(k) = (F'(k) - \delta)k$ and $h(k) = F'(k) - \delta - \rho$. In this notation equation (4) can be written as $\dot{k}(t) = (1 - \tau^*(t))g(k(t)) - c(t)$.

The problem that the government faces is to find a tax rule $\tau^*(t)$ that maximizes welfare which in this model is the weighted sum of the instantaneous utilities of the capitalist $U(c)$ and the worker $V(x)$. The utility function $V(x)$ is also assumed to be increasing and concave. Formally, the government solves:

$$\int_0^\infty e^{-\rho t} [\gamma V(x(k(t), \tau(t))) + U(c^*(t))] dt \to \max \tag{5}$$

with respect to $\tau(t)$ and subject to

$$\dot{k}(t) = (F'(k(t)) - \delta)(1 - \tau(t))k(t) - c^*(t) \tag{6}$$

$$k(0) = k_0 > 0,$$

where c^* is the solution of the optimal control problem (3)–(4) and γ is a parameter which determines how much weight the government puts on the wellbeing of the workers. While no specific restrictions on γ are needed, it seems reasonable to assign positive values to this parameter as typically governments care about the workers. Using the definition of the function $g(k)$ equation (6) becomes $\dot{k}(t) = (1 - \tau(t))g(k(t)) - c^*(t)$ and the workers' income is $x = F(k) - kF'(k) + \tau g(k)$.

Note that the optimization problems of the government and the representative capitalist are interrelated. When the capitalist decides about his consumption, the decision is influenced by the tax policy of the government. Similarly, when the government determines the optimal capital income tax, it must take into account the response of the capitalist in terms of consumption/saving behavior. Thus, the optimal capital income taxation problem naturally fits into the differential games

---

[1] The integrals below are understood in the Riemann sense.

framework. We shall say that a pair of functions (c^*, τ^*) is a Nash equilibrium for the considered differential game (cf. [2]) iff c^* and τ^* are solutions of the optimal control problems (3)-(4) and (5)-(6), respectively. If c^* and τ^* depend on k the corresponding Nash equilibrium is called a *feedback Nash equilibrium*.

Below we formulate and prove a proposition which provides a method for explicit computation of the solution to the optimal capital income tax problem and to other problems with similar structure. The proof utilizes the sufficiency conditions obtained by Seierstad and Sydsaeter in [6]. More specifically, we use a combination of Theorem 8 and Theorem 10 and replace their transversality condition with $\lim_{t \to \infty} e^{-\rho t} \pi(t) k(t) = 0$ since both the state variable k and the co-state variable π take non-negative values. The proof is based on the approach proposed in [7].

Proposition 1. *Let $\bar{\tau}(k)$, $k \in [0, +\infty)$, be solution of the following differential equation:*

$$U'(\bar{c})g(k) = \alpha \quad \text{with the initial condition } \bar{\tau}(k_0) = \tau_0, \tag{7}$$

where α is a constant and

$$\bar{c}(k) = \frac{g(k)[\rho + g(k)\bar{\tau}'(k)]}{g'(k)} \geq 0. \tag{8}$$

Let the solution $\bar{k}(t)$, $t \in [0, +\infty)$, of the following differential equation

$$\dot{k}(t) = (1 - \bar{\tau}(k(t)))g(k(t)) - \bar{c}(t) \quad \text{with the initial condition } k(0) = k_0 > 0 \tag{9}$$

exist and be bounded. We set

$$H^{c*}(k, \pi) := \max_{c \geq 0} H^c(k, c, \pi) = \max_{c \geq 0}[U(c) + \pi((1 - \bar{\tau}(k))g(k) - c)],$$

$$H^{g*}(k, \lambda) := \max_{\tau} H^g(k, \tau, \lambda) = \max_{\tau}[U(\bar{c}(k)) + \gamma V(x(\tau, k)) + \lambda((1 - \tau)g(k) - \bar{c}(k))]$$

and assume that $H^{c*}(\cdot, \pi)$ and $H^{g*}(\cdot, \lambda)$ are concave functions for each $\pi > 0$ and $\lambda > 0$. We set

$$\bar{\pi}(t) = U'(\bar{c}(\bar{k}(t))) \quad \text{and} \quad \bar{\lambda}(t) := \gamma V'(x(\bar{\tau}(\bar{k}(t)), \bar{k}(t))), \quad t \in [0, \infty),$$

and assume that $H^{c*}(\bar{k}(t), \bar{\pi}(t)) = H^c(\bar{k}(t), \bar{c}(\bar{k}(t)), \bar{\pi}(t))$ and $H^{g*}(\bar{k}(t), \bar{\lambda}(t)) = H^g(\bar{k}(t), \bar{\tau}(\bar{k}(t)), \bar{\lambda}(t))$ for almost all $t \in [0, \infty)$. Assume in addition that the following relationship is satisfied for every k:

$$\gamma V''(x(\bar{\tau}(k), k)) \left[h(k) - g'(k)(1 - \bar{\tau}(k)) + \frac{\bar{c}(k)g'(k)}{g(k)} \right]((1 - \bar{\tau}(k))g(k) - \bar{c}(k))$$

$$= \gamma V'(x(\bar{\tau}(k), k))[\bar{c}'(k) - h(k)] - U'(\bar{c}(k))\bar{c}'(k). \tag{10}$$

Then the feedback Nash equilibrium of the game is given by the functions

$$c^*(k) = \bar{c}(k), \quad \tau^*(k) = \bar{\tau}(k).$$

Remark 1. For example, to ensure the assumption that the solution of (9) is bounded, we can calculate the value $\theta(k_0) := (1 - \bar{\tau}(k_0))g(k_0) - \bar{c}(k_0)$ of the right-hand side of (9) at the initial point k_0. Clearly, for $\theta(k_0) \leq 0$, the solution of (9) will belong to the interval $[0, k_0]$ and will be bounded. Let us consider the case when $\theta(k_0) > 0$. If there exists a point $\hat{k} > k_0$ with $\theta(\hat{k}) = 0$, then the solution of (9) will belong to the interval $[k_0, \hat{k}]$, and will be also bounded. The existence of such $\hat{k}$ can be checked easily once we have $\bar{c}$ and $\bar{\tau}$. It is important to note, however, that the proposition remains valid even in some cases when capital grows unboundedly. An important example is the one with logarithmic utility functions $U(c)$ and $V(x)$ and a linear production function $F(k) = Ak$. Direct application of Proposition (1) leads to the solution $c^* = \rho k, \tau^* = \dfrac{\gamma\rho}{(1+\gamma)(A - \delta)}$.

If the parameters of the model are such that $A - \delta - \rho - \dfrac{\gamma\rho}{\gamma+1} > 0$, capital will increase indefinitely. The transversality conditions though, will still hold as $k(t)\pi(t) = \dfrac{1}{\rho}$ and $k(t)\lambda(t) = \dfrac{\gamma+1}{\rho}$.

Remark 2. Equations (7) and (8) determine a family of tax functions which depend on the parameter α. The value of this parameter is found by solving (10) with respect to α provided that a positive solution exists.

Proof. Let $c^*(t), t \in [0, +\infty)$, be an admissible control and let the corresponding trajectory $k^*(t)$ be well defined on the interval $[0, +\infty)$. Then c^* will be a solution of the optimal control problems (3)-(4) whenever the following sufficient optimality conditions[2] hold true:

$$U'(c^*(t)) = \pi(t) \tag{11}$$

$$\dot{\pi}(t) = \pi(t)\left(\rho - g'(k^*(t)) + \bar{\tau}'(k^*(t))g(k^*(t)) + g'(k^*(t))\bar{\tau}(k^*(t))\right) \tag{12}$$

plus the concavity of $H^{c*}(\cdot, \pi)$ and the transversality condition

$$\lim_{t \to \infty} e^{-\rho t} k^*(t)\pi(t) = 0.$$

Taking the time derivative of (7) with $k := \bar{k}(t)$ gives:

$$\dot{\bar{\pi}}(t)g(\bar{k}(t)) + \bar{\pi}(t)g'(\bar{k}(t))\dot{\bar{k}}(t) = 0$$

$$\dot{\bar{\pi}}(t) = -\frac{\bar{\pi}(t)g'(\bar{k}(t))\dot{\bar{k}}(t)}{g(\bar{k}(t))}.$$

By substituting $\dot{\bar{k}}(t)$ with the right-hand side of the differential equation (4) with $c = \bar{c}(\bar{k}(t))$, we get

$$\dot{\bar{\pi}} = -\frac{\bar{\pi}g'(\bar{k})}{g(\bar{k})}\left((1 - \bar{\tau}(\bar{k}))g(\bar{k}) - \frac{g(\bar{k})(\rho + \bar{\tau}'(\bar{k})g(\bar{k}))}{g'(\bar{k})}\right),$$

---

[2] In fact, for this problem the constraint on the state variable $k \geq 0$ is not binding since $\bar{k}(t) = 0$ is a solution of the capital equation. Because of the uniqueness of solution, given that $k_0 > 0$ the value $\bar{k} = 0$ is never reached and the Lagrangian reduces to the Hamiltonian.

which is exactly (12). Thus, $c^*(t) = \bar{c}(\bar{k}(t))$ satisfies the sufficient conditions for optimality.

The transversality condition is ensured by the fact that capital in this model is bounded and so is the co-state variable which depends on k. Next we turn to the optimization problem of the government.

Let $\tau^*(t)$, $t \in [0, +\infty)$, be an admissible control and let the corresponding trajectory $k^*(t)$ be well defined on the interval $[0, +\infty)$. Then τ^* will be a solution of the optimal control problems (5)-(6) whenever the following sufficient optimality conditions hold true:

$$\lambda(t) = \gamma V'(x(\tau^*(t), k^*(t)) \tag{13}$$

$$\dot{\lambda} = \lambda\rho - \gamma V'(x)x'_k - U'(\bar{c}(k^*))\bar{c}'_k(k^*) - \lambda(1 - \tau^*(t))g'(k^*) + \lambda\bar{c}'_k(k^*),$$

which can be rewritten using the definition for $h(k)$ as

$$\dot{\lambda}(t) = \lambda(t)[\bar{c}'_k(k^*(t)) - h(k^*(t))] - U'(\bar{c}(k^*(t)))\bar{c}'_k(k^*(t)), \tag{14}$$

plus the concavity of $H^{g*}(\cdot, \lambda)$ and the transversality condition

$$\lim_{t\to\infty} e^{-\rho t}k^*(t)\lambda(t) = 0.$$

Taking the time derivative of $\bar{\lambda}(t)$ defined earlier, we obtain that

$$\dot{\bar{\lambda}}(t) = \gamma V''(x(\bar{\tau}(\bar{k}(t)), \bar{k}(t)))[x'_k(\bar{\tau}(\bar{k}(t)), \bar{k}(t)) + x'_\tau(\bar{\tau}(\bar{k}(t)), \bar{k}(t))\bar{\tau}'(\bar{k}(t))]\dot{\bar{k}}(t).$$

After computing the respective derivatives and using (8), this can be written as

$$\dot{\bar{\lambda}} = \gamma V''(x(\bar{\tau}(\bar{k}), \bar{k}))\left[h(\bar{k}) - (1 - \bar{\tau}(\bar{k}))g'(\bar{k}) + \frac{\bar{c}(\bar{k})g'(\bar{k})}{g(\bar{k})}\right]((1-\bar{\tau}(\bar{k}))g(\bar{k}) - \bar{c}(\bar{k})).$$

Finally, substituting the right-hand side of the above expression with the right-hand side of (10) we obtain equation (14). Thus, $\tau^*(t) = \bar{\tau}(\bar{k}(t))$ also satisfies the sufficient conditions for optimality. Again, like in the problem of the representative capitalist, the transversality condition is satisfied due to the boundedness of capital which is non-negative. $\qquad\square$

3   An Example

To demonstrate how the proposition works we provide a simple example for which the optimal capital income tax at infinity is obtained explicitly and it is different from zero. The example uses the following production and utility functions: $F(k) = \dfrac{k^\sigma}{\sigma} + \delta k$, $U(c) = \dfrac{c^{1-\sigma}}{1-\sigma}$, $V(x) = \dfrac{x^{1-\sigma}}{1-\sigma}$, where σ is some number between zero and one. In this formulation we have that $g(k) = k^\sigma$. Condition (7) with $\alpha > 0$ becomes $\bar{c}^{-\sigma}k^\sigma = \alpha$, from where if we denote $\beta = \alpha^{-\frac{1}{\sigma}}$, it follows that $\bar{c} = \beta k$. So based on (7) and (8) we can find the optimal tax rule:

$$\beta k = \frac{k^\sigma}{\sigma k^{\sigma-1}}(\rho + k^\sigma \bar{\tau}'(k))$$

$$\bar{\tau}'(k) = \frac{\sigma\beta - \rho}{k^\sigma}, \quad \bar{\tau}(k) = \frac{\sigma\beta - \rho}{1 - \sigma}k^{1-\sigma} + \tau_0.$$

Assume in addition that $\bar{\tau}(k_0) = \dfrac{\sigma\beta - \rho}{1 - \sigma}k_0^{1-\sigma} + 1 - \dfrac{1}{\sigma}$. Then the optimal tax is

$$\bar{\tau}(k) = \frac{\sigma\beta - \rho}{1 - \sigma}k^{1-\sigma} + 1 - \frac{1}{\sigma}.$$

The income of the workers is $x = \dfrac{\sigma\beta - \rho}{1 - \sigma}k$ and the differential equation for k becomes $\dot{k} = \dfrac{k^\sigma}{\sigma} - \dfrac{\beta - \rho}{1 - \sigma}k$, from where

$$k(t)^{1-\sigma} = \frac{1 - \sigma + e^{(\rho-\beta)t}(\sigma(\beta - \rho)k_0^{1-\sigma} - 1 + \sigma)}{\sigma(\beta - \rho)}.$$

Having found $k(t)$ and $\bar{\tau}(k(t))$, we are in a position to calculate the limiting capital income tax $\bar{\tau}_\infty$ which is defined as the limit of $\bar{\tau}(\bar{k}(t))$ as $t \to \infty$. We have that

$$\bar{k}_\infty^{1-\sigma} := \lim_{t\to\infty} \bar{k}(t)^{1-\sigma} = \frac{1 - \sigma}{\sigma(\beta - \rho)}$$

and accordingly,

$$\bar{\tau}_\infty = \frac{\sigma(\rho - 2\beta) + \beta}{\sigma(\rho - \beta)}.$$

To ensure that $\bar{c}(k)$ and $\bar{\tau}(k)$ as obtained above are indeed the solution to the differential game we need to verify two more conditions – condition (10) and the concavity of the maximized Hamiltonian functions. Using that $\dfrac{V''(x)x}{V'(x)} = -\sigma$, condition (10) simplifies to

$$\gamma(1 - \sigma)^{\sigma-1}\left(1 - 2\sigma + \frac{\rho}{\beta}\right) = \left(\frac{\sigma\beta - \rho}{\beta}\right)^\sigma. \tag{15}$$

The latter represents an equality from which the parameter β (and therefore α) can be computed.

The maximized Hamiltonians for the problems of the capitalist and the government are $H^{c*}(k, \pi) = \dfrac{\sigma\pi^{\frac{\sigma-1}{\sigma}}}{1 - \sigma} + \dfrac{\pi k^\sigma}{\sigma} - \dfrac{\sigma\beta - \rho}{1 - \sigma}\pi k$ and $H^{g*}(k, \lambda) = \dfrac{\lambda^{\frac{\sigma-1}{\sigma}}\gamma^{\frac{1}{\sigma}}}{1 - \sigma} + \dfrac{(\beta k)^{1-\sigma}}{1 - \sigma} + \dfrac{\lambda k^\sigma}{\sigma} - \lambda^{\frac{\sigma-1}{\sigma}}\gamma^{\frac{1}{\sigma}} - \lambda\beta k$, respectively, and they are concave.

From Proposition 1 it follows that $\bar{c} = c^*$ and $\bar{\tau} = \tau^*$. Below we show that for different values of the parameters of the model in this example we can have either positive or negative limiting capital income taxes.

Case 1. Positive limiting tax. Let $\sigma = 0.8$, $\rho = 0.25$ and $\gamma = 7.1895$. Then direct calculations yield $\tau_\infty^* = 0.3333$ and $k_\infty^* = 12.8601$. If, for example, $k_0 = 1$, the corresponding tax at time zero will be $\tau^*(1) = 0.10$. Thus, the government begins with a relatively small tax on capital income and as time elapses the tax rate gradually increases to approach $1/3$. Of course, the above results will be valid only if condition (15) is satisfied which requires that $\beta = 0.4$.

Case 2. Negative limiting tax. Now let $\sigma = 0.5$, $\rho = 0.5$, and $\gamma = \sqrt{2}$. In this case $\beta = 2$. Assume also that $k_0 = 1.21$. With these values of the parameters the government again begins at time zero with a tax rate of 10 percent of capital income as can be directly computed. However, with time, unlike in the previous case when the tax rate increases, here it decreases to the equilibrium value $\tau_\infty^* = -0.3333$ corresponding to $k_\infty^* = 0.4444$. The intuition is that as capital becomes smaller and smaller at some point in time it becomes optimal from the welfare perspective to subsidize capital in order to sustain production. Note that the government budget is balanced in every period since the moment that the capital income tax becomes negative, or in other words a subsidy is granted to the owners of capital, the government takes exactly the same amount from the workers in the form of a lump-sum tax.

References

1. Chamley, C.: Optimal Taxation of Capital Income in General Equilibrium with Infinite Lives. Econometrica 54(3), 607–622 (1986)
2. Dockner, E., et al.: Differential Games in Economics and Management Science. Cambridge University Press, Cambridge (2000)
3. Judd, D.: Redistributive Taxation in a Simple Perfect Foresight Model. Journal of Public Economics 28, 59–83 (1985)
4. Kemp, L., van Long, N., Shimomura, K.: Cyclical and Noncyclical Redistributive Taxation. International Economic Review 34(2), 415–429 (1993)
5. Lansing, K.: Optimal Redistributive Capital Taxation in a Neoclassical Growth Model. Journal of Public Economics 73, 423–453 (1999)
6. Seierstad, A., Sydsaeter, K.: Sufficient Conditions in Optimal Control Theory. International Economic Review 18(2), 367–391 (1977)
7. Xie, D.: On Time Inconsistency: A Technical Issue in Stackelberg Differential Games. Journal of Economic Theory 76(2), 412–430 (1997)

Numerical Methods for Robust Control

P.Hr. Petkov[1], A.S. Yonchev[1], N.D. Christov[2], and M.M. Konstantinov[3]

[1] Technical University of Sofia, 1000 Sofia, Bulgaria
[2] Université des Sciences et Technologies de Lille,
59655 Villeneuve d'Ascq, France
[3] University of Architecture, Civil Engineering and Geodesy, 1046 Sofia, Bulgaria

Abstract. A brief survey on the numerical properties of the methods for $\mathcal{H}_\infty$ design and μ-analysis and synthesis of linear control systems is given. A new approach to the sensitivity analysis of LMI – based $\mathcal{H}_\infty$ design is presented that allows to obtain linear perturbation bounds on the computed controller matrices. Some results from a detailed numerical comparison of the properties of the different available $\mathcal{H}_\infty$ optimization methods are then presented. We also discuss the sensitivity of the structured singular value (μ) that plays an important role in the robust stability analysis and design of linear control systems.

1 Introduction

In this paper we present a brief survey on the numerical properties of the methods for $\mathcal{H}_\infty$ design and μ-analysis and synthesis of linear control systems. While there are several results in the sensitivity and error analysis of the Riccati – based $\mathcal{H}_\infty$ optimization methods, the numerical properties of the Linear Matrix Inequalities (LMI) – based $\mathcal{H}_\infty$ design methods are not studied in depth up to the moment. This is due to the fact that the solution of the LMI involved is done by complicated optimization methods which makes difficult the derivation of error bounds. In this paper we present a new approach to the sensitivity analysis of LMI – based $\mathcal{H}_\infty$ design that allows to obtain linear perturbation bounds on the computed controller matrices. A comparison of the sensitivity bounds for the Riccati – based and LMI – based $\mathcal{H}_\infty$ design methods shows that at the price of larger volume of computations the LMI – based methods are free from the numerical difficulties typical for the Riccati – based methods near the optimum. Some results from a detailed numerical comparison of the properties of the different available $\mathcal{H}_\infty$ optimization methods are then presented. We also discuss the sensitivity of the structured singular value (μ) that plays an important role in the robust stability analysis and design of linear control systems.

For brevity, we consider only the discrete-time case $\mathcal{H}_\infty$ optimization problem, the conclusions concerning the continuous-time case being essentially the same.

We use the following notations: $\mathbb{R}^{m \times n}$ – the space of real $m \times n$ matrices; $\mathbb{R}^n = \mathbb{R}^{n \times 1}$; I_n – the identity $n \times n$ matrix; $M^\top$ – the transpose of M; $\|M\|_2$ – the spectral norm of M; $\|M\|_F = \sqrt{\operatorname{tr}(M^\top M)}$ – the Frobenius norm of M; $\|M\|_\infty := \sup_{\operatorname{Re} s \geq 0} \|M(s)\|_2$; $\operatorname{vec}(M) \in \mathbb{R}^{mn}$ – the column-wise vector representation of $M \in \mathbb{R}^{m \times n}$. The notation ":=" stands for "equal by definition".

I. Lirkov, S. Margenov, and J. Waśniewski (Eds.): LSSC 2007, LNCS 4818, pp. 350–357, 2008.

2   Numerical Methods for $\mathcal{H}_\infty$ Optimization

Consider a generalized linear discrete-time plant, described by the equations

$$\begin{aligned}
x_{k+1} &= Ax_k + B_1w_k + B_2u_k \\
z_k &= C_1x_k + D_{11}w_k + D_{12}u_k \\
y_k &= C_2x_k + D_{21}w_k + D_{22}u_k
\end{aligned} \qquad (1)$$

where $x_k \in \mathcal{R}^n$ is the state vector, $w_k \in \mathcal{R}^{m_1}$ is the exogenous input vector (the disturbance), $u_k \in \mathcal{R}^{m_2}$ is the control input vector, $z_k \in \mathcal{R}^{p_1}$ is the controlled vector, and $y_k \in \mathcal{R}^{p_2}$ is the measurement vector. The transfer function matrix of the system will be denoted by

$$P(z) = \left[\begin{array}{c|cc} A & B_1 & B_2 \\ \hline C_1 & D_{11} & D_{12} \\ C_2 & D_{21} & D_{22} \end{array} \right].$$

The '$\mathcal{H}_\infty$ suboptimal discrete-time control problem' is to find an internally stabilizing controller $K(z)$ such that, for a pre-specified positive value of γ, it is fulfilled that

$$\|F_\ell(P, K)\|_\infty < \gamma \qquad (2)$$

where $F_\ell(P, K)$ is the lower linear fractional transformation (LFT) on $K(z)$.

In '$\mathcal{H}_\infty$ optimization control problem' one tries to find the infimum of γ (further denoted by γ_{opt}) which satisfies (2). This infimum is difficult to find analytically in the general case. It is usually computed numerically by a bisection procedure involving some method for suboptimal design. The solution of the $\mathcal{H}_\infty$ optimization control problem corresponds to the best disturbance attenuation at the controlled output of the closed-loop system.

There are three groups of methods for solving the discrete-time $\mathcal{H}_\infty$ optimization problems.

1. *Methods based on a bilinear transformation to a continuous-time problem* [1]
These methods allow to solve the discrete-time $\mathcal{H}_\infty$ problem with the algorithms and software for solving continuous-time problems. They implement a bilinear transformation and its inverse so that the accuracy of the final solution depends on the condition number of this transformation. Usually such methods perform badly for high order problems and are considered unreliable from numerical point of view. Note that it is possible to derive formulas for the discrete-time controller implementing analytically the bilinear transformation [8]. Unfortunately, these formulas are very complicated and we were unable to find a stabilizing controller applying them to our examples.

2. *Riccati equations based methods* [1,6]
These methods include the solution of two matrix algebraic Riccati equations and represent an efficient way for solving the $\mathcal{H}_\infty$ suboptimal problem requiring a volume of computational operations proportional to n^3. Their implementation,

however, is restricted only to 'regular' plants, i.e. plants for which the matrices D_{12} and D_{21} are of full rank and the transfer functions from controls to controlled outputs and from disturbance to measured outputs have no invariant zeros on the unit circle. Although various extensions are proposed for the 'singular' cases, the numerical difficulties in such cases are not overcame. Also, for γ approaching the optimum value γ_{opt} the Riccati equations become ill-conditioned which may affect the accuracy of the controller matrices obtained.

The numerical properties of the Riccati based $\mathcal{H}_\infty$ design are relatively well studied (see,for instance, [3]).

3. LMI-based methods [4,5,11]
These methods are based on the solution of three or more LMIs derived from Lyapunov based criteria. The main advantage of these methods is that they are implemented without difficulties to 'singular' plants the only assumption being the stabilizability and detectability of the triple (A, B_2, C_2). The LMI approach yields a finite-dimensional parametrization of all $\mathcal{H}_\infty$ controllers which allows to exploit the remaining freedom for controller order reduction and for handling additional constraints on the closed-loop performance. The LMIs are solved by convex optimization algorithms which require a volume of computational operations proportional to n^6 [2,9]. This fact restricts the implementation of LMI-based methods to relatively low order plants in contrast to the Riccati equations based methods.

3   Sensitivity Analysis of the LMI-Based $\mathcal{H}_\infty$ Optimization

Consider a proper discrete-time plant $P(z)$ with state equations in the form (1) with $D_{22} = 0$ and let $\mathcal{N}_{12}$ and $\mathcal{N}_{21}$ denote the null spaces of (B_2^T, D_{12}^T) and (C_2, D_{21}), respectively. The discrete-time suboptimal $\mathcal{H}_\infty$ problem (2) is solvable if and only if there exist two symmetric matrices $R, S \in \mathcal{R}^{n \times n}$ satisfying the following system of three LMIs [5]

$$\begin{bmatrix} \mathcal{N}_{12} & 0 \\ \hline 0 & I \end{bmatrix}^T \left[\begin{array}{cc|c} ARA^T - R & ARC_1^T & B_1 \\ C_1 RA^T & -\gamma I + C_1 RC_1^T & D_{11} \\ \hline B_1^T & D_{11}^T & -\gamma I \end{array} \right] \begin{bmatrix} \mathcal{N}_{12} & 0 \\ \hline 0 & I \end{bmatrix} < 0, \qquad (3)$$

$$\begin{bmatrix} \mathcal{N}_{21} & 0 \\ \hline 0 & I \end{bmatrix}^T \left[\begin{array}{cc|c} A^T S A - S & A^T S B_1 & C_1^T \\ B_1^T S A & -\gamma I + B_1^T S B_1 & D_{11}^T \\ \hline C_1 & D_{11} & -\gamma I \end{array} \right] \begin{bmatrix} \mathcal{N}_{21} & 0 \\ \hline 0 & I \end{bmatrix} < 0, \qquad (4)$$

$$\begin{bmatrix} R & I \\ I & S \end{bmatrix} > 0. \qquad (5)$$

Computing solutions (R, S) of the LMI system (3)-(5) is a convex optimization problem. The sensitivity of the LMIs under consideration, subject to variations in the plant data, may affect the accuracy of the matrices R and S and hence the accuracy of controller matrices. It is not clear up to the moment how LMIs sensitivity is connected to the sensitivity of the given $\mathcal{H}_\infty$ suboptimal problem.

In what follows, we present briefly a sensitivity analysis of the LMI approach to $\mathcal{H}_\infty$ optimization. First we perform sensitivity analysis of the LMI (4). Its structure allows us to consider only the following part

$$
\begin{aligned}
&(\mathcal{N}_{21} + \Delta\mathcal{N}_{21})^\top \\
&* \left\{ \left[\begin{matrix} (A + \Delta A)^\top (S^* + \Delta S)(A + \Delta A) - (S^* + \Delta S) & 0 \\ (B_1 + \Delta B_1)^\top (S^* + \Delta S)(A + \Delta A) & 0 \end{matrix} \right] \right. \\
&+ \left. \left[\begin{matrix} 0 & (A + \Delta A)^\top (S^* + \Delta S)(B_1 + \Delta B_1) \\ 0 & -\gamma_{opt} I - \Delta\gamma_{opt} I + (B_1 + \Delta B_1)^\top (S + \Delta S)(B_1 + \Delta B_1) \end{matrix} \right] \right\} \\
&*(\mathcal{N}_{21} + \Delta\mathcal{N}_{21}) := \bar{\mathcal{H}}^* + \Delta\bar{\mathcal{H}}_1 < 0,
\end{aligned}
\tag{6}
$$

and to study the effect of the perturbations ΔA, ΔB_1, ΔB_2, ΔC_1, ΔC_2, ΔD_{11}, ΔD_{12}, ΔD_{21}, ΔD_{22} and $\Delta\gamma_{opt}$ on the perturbed LMI solution $S^* + \Delta S$, where S^* and ΔS are the nominal solution of LMI (4) and the perturbation, respectively. The essence of our approach is to perform sensitivity analysis of the LMI (4) in a similar manner as for a proper matrix equation after introducing a suitable right hand part, which is slightly perturbed.

The matrix $\bar{\mathcal{H}}^*$ is obtained using the nominal LMI

$$
\mathcal{N}_{21}^\top \left[\begin{matrix} A^\top S^* A - S^* & A^\top S^* B_1 \\ B_1^\top S^* A & -\gamma_{opt} I + B_1^\top S^* B_1 \end{matrix} \right] \mathcal{N}_{21} := \bar{\mathcal{H}}^* < 0.
\tag{7}
$$

The matrix $\Delta\bar{\mathcal{H}}_1$ is due to the data and closed-loop performance perturbations, the rounding errors and the sensitivity of the interior point method that is used to solve the LMIs. It is important to mention that the $(1,2)$, $(2,1)$, $(2,2)$ blocks of the LMIs (3) and (4) pose constraints on the size of the perturbations $\Delta B_1, \Delta C_1, \Delta D_{11}$ and $\Delta\gamma_{opt}$ since the introduced right hand part matrix must be negative definite.

The perturbed relation (6) may be written as

$$
\bar{\mathcal{H}}^* + \Delta\bar{\mathcal{H}}_1 = \mathcal{N}_{21}^\top \bar{\mathcal{V}} \mathcal{N}_{21} + \mathcal{N}_{21}^\top \bar{\mathcal{V}} \Delta\mathcal{N}_{21} + \Delta\mathcal{N}_{21}^\top \bar{\mathcal{V}} \mathcal{N}_{21} + \Delta\mathcal{N}_{21}^\top \bar{\mathcal{V}} \Delta\mathcal{N}_{21},
\tag{8}
$$

where

$$
\begin{aligned}
\bar{\mathcal{V}} = &\left[\begin{matrix} A^\top S^* A - S^* + A^\top \Delta S A - \Delta S + A^\top S^* \Delta A + \Delta A^\top S^* A & 0 \\ B_1^\top S^* A + B_1 \Delta S A + B_1^\top S^* \Delta A + \Delta B_1^\top S^* A & 0 \end{matrix} \right] + \\
&\left[\begin{matrix} 0 & A^\top S^* B_1 + A^\top \Delta S B_1 + A^\top S^* \Delta B_1 + \Delta A^\top S^* B_1 \\ 0 & -\gamma_{opt} I - \Delta\gamma_{opt} I + B_1^\top S^* B_1 + B_1^\top \Delta S^* B_1 + B_1^\top S^* \Delta B_1 + \Delta B_1^\top S^* B_1 \end{matrix} \right].
\end{aligned}
$$

Here the terms of second and higher order are neglected and we use the relation (7) to obtain the following expression

$$
\begin{aligned}
\Delta\bar{\mathcal{H}}_1 = &\mathcal{N}_{21}^\top \Psi_S \mathcal{N}_{21} + \mathcal{N}_{21}^\top (\bar{\mathcal{H}}^* + \Psi_S) \Delta\mathcal{N}_{21} + \Delta\mathcal{N}_{21}^\top (\bar{\mathcal{H}}^* + \Psi_S) \mathcal{N}_{21} + \\
&\Delta\mathcal{N}_{21}^\top (\bar{\mathcal{H}}^* + \Psi_S) \Delta\mathcal{N}_{21},
\end{aligned}
\tag{9}
$$

where

$$\bar{\mathcal{H}}^* = \mathcal{N}_{21}^\top \mathcal{H}^* \mathcal{N}_{21}, \ \Psi_S = \Theta_S + \Lambda_S, \ \Theta_S = \begin{bmatrix} A^\top \Delta S A - \Delta S & A^\top \Delta S B_1 \\ B_1^\top \Delta S A & B_1^\top \Delta S B_1 \end{bmatrix},$$

$$\Lambda_S = \begin{bmatrix} A^\top S^* \Delta A + \Delta A^\top S^* A & A^\top S^* \Delta B_1 + \Delta A^\top S^* B_1 \\ B_1^\top S^* \Delta A + \Delta B_1^\top S^* A & B_1^\top S^* \Delta B_1 + \Delta B_1^\top S^* B_1 - \Delta \gamma_{\mathrm{opt}} I \end{bmatrix}.$$

Neglecting the second and higher order terms in (9) one obtains

$$\Delta \bar{\mathcal{H}}_1 = \mathcal{N}_{21}^\top \Theta_S \mathcal{N}_{21} + \mathcal{N}_{21}^\top \Lambda_S \mathcal{N}_{21} + \Delta \mathcal{N}_{21}^\top \mathcal{H}^* \mathcal{N}_{21} + \mathcal{N}_{21}^\top \mathcal{H}^* \Delta \mathcal{N}_{21}. \quad (10)$$

Using the fact that $\|\mathrm{vec}(M)\|_2 = \|M\|_{\mathcal{F}}$ we can finally obtain relative perturbation bound for the solution S^* of the LMI (10) in the form

$$\frac{\|\Delta S\|_2}{\|S^*\|_{\mathcal{F}}} \leq k_A \frac{\|\Delta A\|_{\mathcal{F}}}{\|A\|_{\mathcal{F}}} + k_{B_1} \frac{\|\Delta B_1\|_{\mathcal{F}}}{\|B_1\|_{\mathcal{F}}} + k_\gamma \frac{|\Delta \gamma_{\mathrm{opt}}|}{|\gamma_{\mathrm{opt}}|}$$
$$+ k_{CD} \frac{\|[\Delta C_2, \Delta D_{21}]\|_{\mathcal{F}}}{\|[C_2, D_{21}]\|_{\mathcal{F}}} + k_{\mathcal{H}} \frac{\|\Delta \bar{\mathcal{H}}_1\|_{\mathcal{F}}}{\|\bar{\mathcal{H}}^*\|_{\mathcal{F}}}, \quad (11)$$

where $k_A, k_{B_1}, k_\gamma, k_{CD}, k_{\mathcal{H}}$ may be considered as individual relative condition numbers of the LMI (4) with respect to the perturbations in the data.

In a similar way it is possible to find relative perturbation bound for the solution R of the LMI (3). The bounds on ΔR and ΔS are then used to find perturbation bounds on the controller matrices.

4   A Comparison of the Available Methods for $\mathcal{H}_\infty$ Design

In this section we present in brief the results from a numerical comparison of the methods for discrete-time $\mathcal{H}_\infty$ design as implemented in Robust Control Toolbox of MATLAB [1] and SLICOT library [10].

As a test example we consider a family of discrete-time $\mathcal{H}_\infty$ optimization problems for a system with

$$A = \mathrm{diag}\,\{0.2, 0.3, 0.4, 0.5, 0.6, 0.7, 1 - \alpha, -1 + \alpha\}\,,$$

$$B_1 = \begin{bmatrix} 0.1 & 0 & 0 \\ 0 & -0.1 & 0 \\ 0 & 0 & 0.1 \\ 0 & 0 & 0 \\ 0 & 0 & 0 \\ 0 & 0 & 0 \\ 0 & 0 & 0 \\ 0 & 0 & 0 \end{bmatrix}, \ B_2 = \begin{bmatrix} 0 & 0 \\ 0 & 0 \\ 0 & 0 \\ 0 & 0 \\ 0 & 0 \\ 0 & 0 \\ 1 & 0 \\ 0 & 1 \end{bmatrix},$$

$$C_1 = \begin{bmatrix} 1 & 0 & 0 & 0 & 0 & 0 & 0 & 0 \\ 0 & -1 & 0 & 0 & 0 & 0 & 0 & 0 \\ 0 & 0 & 1 & 0 & 0 & 0 & 0 & 0 \end{bmatrix}, \ C_2 = \begin{bmatrix} 0 & 0 & 0 & 0 & 0 & 0 & 1 & 0 \\ 0 & 0 & 0 & 0 & 0 & 0 & 0 & 1 \end{bmatrix},$$

$$D_{11} = \begin{bmatrix} 0.1 & 0 & 0 \\ 0 & 0.1 & 0 \\ 0 & 0 & 0.1 \end{bmatrix}, \quad D_{12} = \begin{bmatrix} 0 & 0 \\ \beta & 0 \\ 0 & \beta \end{bmatrix}, \quad D_{21} = \begin{bmatrix} 0 & \eta & 0 \\ 0 & 0 & \eta \end{bmatrix}, \quad D_{22} = \begin{bmatrix} 0 & 0 \\ 0 & 0 \end{bmatrix}.$$

With the decreasing of α two open-loop poles will approach the unit circle which creates difficulties for the methods based on Riccati equations solution. If the parameter β (η) is zero then the regularity conditions are violated. Changing parameters α, β, η allows to reveal the numerical properties of the different methods for $\mathcal{H}_\infty$ optimization.

The numerical results obtained lead to the following conclusions.

- The experiments show a serious disadvantage of the routines based on the Riccati equations solutions. The condition numbers of the Riccati equations in some cases approach $1/eps$ (*eps* is the relative machine precison) while in the same cases the solution obtained by the LMI based method do not show singularities and allows to obtain smaller values of γ_{opt}.
- The fulfillment of regularity conditions is important only for the methods based on the Riccati equations solution. The violation of these conditions doesn't affect the method based on LMI solution.
- Although the LMI based method produces the best solution it is necessary to take into account that the numerical properties of the second step of this method - the computation of controller matrices - are not studied well. A numerical analysis showing some type of stability of the method is still needed.

5   Numerical Aspects of the μ-Analysis and Synthesis

The structure of an uncertain control system is shown in Figure 1 where $M \in \mathcal{C}^{n \times n}$ and Δ is the block-diagonal matrix representing the uncertainty.

The robust stability analysis of the system, shown in Figure 1, is done by using the so called *structured singular value* (μ)[13]. The structured singular value is a powerful tool in the robust analysis and design of control systems. In particular, the size of the smallest destabilizing perturbation for a linear time invariant system is inversely proportional to μ. The value of μ may be also connected to the robust performance of a control system. Unfortunately, in the general case the structured singular value can not be computed analytically and its determination is done by a complicated numerical method [12]. The function

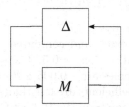

Fig. 1. M-Δ feedback connection

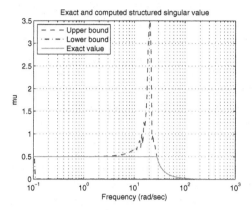

Fig. 2. Exact and computed structured singular value

μ can be bounded above by the solution of a convex optimization problem, and in general there exists a gap between μ and its convex bound. This gap may introduce an unnecessary pessimism in the robust analysis and design. In what follows, we show that by an appropriate choice of the system parameters, the gap between μ and its upper bound may be made arbitrarily large.

Consider a second order system described by the equation

$$\frac{d^2y}{dt^2} + a_1\frac{dy}{dt} + a_2 y = 0,$$

where the coefficients a_1, a_2 are subject to uncertainty. Specifically, we shall suppose that

$$a_1 = \bar{a}_1(1 + p_1\delta_1), \quad |\delta_1| \leq 1,$$

$$a_2 = \bar{a}_2(1 + p_2\delta_1), \quad |\delta_2| \leq 1$$

where $\bar{a}_1$, $\bar{a}_2$ are the nominal values of the coefficients, p_1, p_2 are the relative uncertainties in a_1, a_2, respectively and δ_1, δ_2 are real uncertainties.

For this system it is possible to compute the exact value of the structured singular value. After some lengthly derivation, it is possible to show that in the given case the frequency response of μ is given by

$$\mu = \frac{1}{\max\left\{|-1/p_1|, |(-\bar{a}_2 + \omega^2)/(\bar{a}_2 p_2)|\right\}} \tag{12}$$

where ω is the frequency in rad/s.

In Figure 2 we show the exact and computed structured singular value for $\bar{a}_1 = 0.01$, $\bar{a}_2 = 400.0025$ and $p_1 = p_2 = 0.5$. The system under consideration is lightly damped (the system poles are $-0.005 \pm 20.0001j$) and it is assumed that the relative changes in the coefficients are less than 50 %. The computed value of μ is found by the function **mussv** from the Robust Control Toolbox of MATLAB. It is seen from Figure 2 that the upper bound computed exceeds the exact value of μ several times. This may be an indication of a high sensitivity of the structured singular value or may be associated with some numerical difficulties of its computation. From the computed upper bound one may conclude that the system does not achieve robust stability (the supremum of μ is greater than 1). In fact, the system remains stable for relative changes in the coefficients less than 100 %.

References

1. Balas, G., et al.: Robust Control Toolbox User's Guide. The MathWorks, Inc., Natick, MA (2006)
2. Ben-Tal, A., Nemirovski, A.: Lectures on Modern Convex Optimization. In: MPS-SIAM Series on Optimization, SIAM, Philadelphia (2001)
3. Christov, N.D., Konstantinov, M.M., Petkov, P.Hr.: New perturbation bounds for the continuous-time H_∞ - optimization problem. In: Li, Z., Vulkov, L.G., Waśniewski, J. (eds.) NAA 2004. LNCS, vol. 3401, pp. 232–239. Springer, Heidelberg (2005)
4. Gahinet, P.: Explicit controller formulas for LMI-based H_∞ synthesis. Automatica 32, 1007–1014 (1996)
5. Gahinet, P., Apkarian, P.: A linear matrix inequality approach to H_∞ control. Int. J. Robust Non. Contr. 4, 421–448 (1994)
6. Green, M., Limebeer, D.J.N.: Linear Robust Control. Prentice Hall, Englewood Cliffs, NJ (1995)
7. Higham, N., et al.: Sensitivity of computational control problems. IEEE Control Syst. Magazine 24, 28–43 (2004)
8. Iglesias, P., Glover, K.: State-space approach to discrete-time H_∞-control. International J. of Control 54, 1031–1074 (1991)
9. Nesterov, Y., Nemirovski, A.: Interior–Point Polynomial Algorithms in Convex Programming. SIAM, Phildadelphia (1994)
10. Petkov, P.Hr., Gu, D.-W., Konstantinov, M.M.: Fortran 77 Routines for H_∞ and H_2 Design of Discrete-Time Linear Control Systems. NICONET Report 1999-5, available electronically at http://www.win.tue.nl/NICONET
11. Yonchev, A., Konstantinov, M., Petkov, P.: Linear Matrix Inequalities in Control Theory. Demetra, Sofia (in Bulgarian) (2005) ISBN 954-9526-32-1
12. Young, P.M., Newlin, M.P., Doyle, J.C.: Computing bounds for the mixed μ problem. Int. J. of Robust and Nonlinear Control 5, 573–590 (1995)
13. Zhou, K., Doyle, J.C., Glover, K.: Robust and Optimal Control, p. 07458. Prentice-Hall, Upper Saddle River, New Jersey (1995)

Runge-Kutta Schemes in Control Constrained Optimal Control

Nedka V. Pulova

Department of Mathematics, Technical University-Varna
Studentska str. 1, Varna 9010, Bulgaria
vpulov@yahoo.com

Abstract. We consider an optimal control problem with control constraints satisfying conditions of smoothness and coercivity. By applying two Runge-Kutta schemes: one to the differential equation and a different one to the adjoint differential equation that is related to the maximum principle for optimal control, we show that the error of approximation taken with respect to the mesh spacing is of second order. This extends previous results where the same scheme was applied to the prime and to the adjoint equation. This extends previously known results where the same discretization scheme is applied to the primal and to the adjoint equation.

Keywords: optimal control, Runge-Kutta scheme, numerical solution.

1 Introduction

In this paper we consider the terminal optimal control problem with control constraints assuming conditions of coercivity and smoothness. Our goal is to find a discrete approximation, using Runge-Kutta formulae whose error is of second order with respect to the step size h. Estimates of the same order for the problem considered here were obtained earlier in [1]. In contrast to [1], in this paper we consider a larger class of discretizations employing two Runge-Kutta schemes to the state and costate equations that may be different from each other. When the optimal control has time derivative of bounded variation, we prove the existence of an approximate solution, which error estimate is $O(h^2)$, as in [1], but for larger class of discretizations. In our proof we apply the fixed point argument from [1].

Section 2 contains a preliminary material, while section 3 provides the proof of our main result.

2 Preliminaries

We consider the following optimal control problem

$$\text{minimize } C(x(1)) \tag{1}$$
$$\text{subject to } \dot{x}(t) = f(x(t), u(t)), \ u(t) \in U, \text{ a.e. } t \in [0, 1],$$
$$x(0) = x_0, \ x \in W^{1,\infty}, \ u \in L^\infty,$$

I. Lirkov, S. Margenov, and J. Waśniewski (Eds.): LSSC 2007, LNCS 4818, pp. 358–365, 2008.
© Springer-Verlag Berlin Heidelberg 2008

where the state $x(t) \in \mathbf{R}^n$, the control $u(t) \in U$, $f : \mathbf{R}^n \times \mathbf{R}^m \rightarrow \mathbf{R}^n$, $C : \mathbf{R}^n \rightarrow \mathbf{R}$ and the set $U \subset \mathbf{R}^m$ is closed and convex. The two main assumptions are formulated below.

Smoothness. The problem (1) has a local solution $(x^*, u^*) \in W^{2,\infty} \times W^{1,\infty}$. There exist an open set $\Omega \subset \mathbf{R}^n \times \mathbf{R}^m$ and a ball $B_\rho(x^*(t), u^*(t)) \subset \Omega$ for $t \in [0, 1]$, such that the second derivative of f is Lipschitz continuous in Ω, while the second derivative of C is Lipschitz continuous in $B_\rho(x^*(1))$.

Here $B_a(z)$ is the closed ball centered at z with radius a and $W^{m,\infty}(\mathbf{R}^n)$ denotes the Sobolev space of vector-valued measurable functions $x : [0, 1] \rightarrow \mathbf{R}^n$ whose j-th derivative belongs to $L^\infty(\mathbf{R}^n)$ for all $0 \leq j \leq m$ with norm $\|x\|_{W^{m,\infty}}$

$$= \sum_{j=1}^{m} \|x^{(j)}\|_{L^\infty}.$$

Under Smoothness assumption, there exists an associated adjoint function $\psi^* \in W^{2,\infty}$, for which the minimum principle is satisfied at (x^*, ψ^*, u^*):

$$\dot{x}(t) = f(x(t), u(t)) \text{ for all } t \in [0, 1], \ x(0) = x_0,$$
$$\dot{\psi}(t) = -\nabla_x H(x(t), \psi(t), u(t)) \text{ for all } t \in [0, 1], \ \psi(1) = \nabla C(x(1)),$$
$$-\nabla_u H(x(t), \psi(t), u(t)) \in N_U(u(t)), \ u(t) \in U \text{ for all } t \in [0, 1],$$

where H is the Hamiltonian defined by $H(x, \psi, u) = \psi f(x, u)$, and ψ is a row vector in $\mathbf{R}^n$. The set $N_U(u)$ is a set of row vectors

$$N_U(u) = \{w \in \mathbf{R}^m : w(v - u) \leq 0 \text{ for all } v \in U\}.$$

Coercivity. There exists a constant $\alpha > 0$ such that

$$\mathcal{B}(x, u) \geq \alpha \|u\|_{L^2}^2 \text{ for all } (x, u) \in \mathcal{M},$$

where the set $\mathcal{M}$ the quadratic form $\mathcal{B}$, and the matrices $A(t)$, $B(t)$, $C(t)$, $Q(t)$, $R(t)$, $S(t)$ are defined by

$$\mathcal{M} = \{(x, u) : x \in W^{1,2}, u \in L^2, \dot{x}(t) = Ax + Bu,$$
$$x(0) = 0, u(t) \in U - U \text{ a.e. } t \in [0, 1]\},$$

$$\mathcal{B}(x, u) = \frac{1}{2}(x(1)^T V x(1) + \langle x, Qx \rangle + \langle u, Ru \rangle + 2 \langle x, Su \rangle),$$

$$A(t) = \nabla_x f(x^*(t), u^*(t)), \quad B(t) = \nabla_u f(x^*(t), u^*(t)), \quad V = \nabla^2 C(x^*(1)),$$

$$Q(t) = \nabla_{xx} H(w^*(t)), \quad R(t) = \nabla_{uu} H(w^*(t)), \quad S(t) = \nabla_{xu} H(w^*(t));$$

Here, $\langle \cdot, \cdot \rangle$ is the usual L^2 inner product, $w^* = (x^*, \psi^*, u^*)$. Coercivity represents a second-order sufficient condition for strict local optimality. We need the following abstract result [1,2].

Proposition 1. *Let $\mathcal{X}$ be a Banach space and let $\mathcal{Y}$ be a normed space with norms in both spaces denoted by $\| \cdot \|$. Let $\mathcal{F} : \mathcal{X} \to 2^{\mathcal{Y}}$ let $\mathcal{L} : \mathcal{X} \to \mathcal{Y}$ be a bounded linear operator, and let $\mathcal{T} : \mathcal{X} \to \mathcal{Y}$ with $\mathcal{T}$ continuously Frechet differentiable in $B_r(\omega^*)$ for some $\omega^* \in \mathcal{X}$ and $r > 0$. Suppose that the following conditions hold for some $\delta \in \mathcal{Y}$ and scalars $\epsilon, \lambda,$ and $\sigma > 0$:*

(P1) $\mathcal{T}(\omega^*) + \delta \in \mathcal{F}(\omega^*)$;
(P2) $\|\nabla \mathcal{T}(\omega) - \mathcal{L}\| \leq \epsilon$ for all $\omega \in \mathcal{B}_r(w^*)$;
(P3) The map $(\mathcal{F} - \mathcal{L})^{-1}$ is single valued and Lipschitz continuous in $\mathcal{B}_r(\pi), \pi = (\mathcal{T} - \mathcal{L})(\omega^*)$, with Lipschitz constant λ;

If $\epsilon\lambda < 1, \epsilon r \leq \sigma, \|\delta\| \leq \sigma$ and $\|\delta\| \leq (1 - \lambda\epsilon)r/\lambda$, then there exists a unique $\omega \in B_r(\omega^*)$ such that $\mathcal{T}(\omega) \in \mathcal{F}(\omega)$. Moreover we have the estimate

$$\|\omega - \omega^*\| \leq \frac{\lambda}{1 - \lambda\epsilon}\|\delta\|.$$

We approximate the optimality system for (1) by the following discrete time system

$$\mathbf{x}'_k = \sum_{i=1}^{s} b_i f(y_{ki}, \mathbf{u}_{ki}), \ \mathbf{x}_0 = a, \tag{2}$$

$$y_{ki} = \mathbf{x}_k + h \sum_{j=1}^{s} a_{ij} f(y_{kj}, \mathbf{u}_{kj}), \tag{3}$$

$$\psi'_k = -\sum_{i=1}^{s} \overline{b}_i z_{ki} \nabla_x f(y_{ki}, \mathbf{u}_{ki}), \ \psi_N = \nabla C(\mathbf{x}_N), \tag{4}$$

$$z_{ki} = \psi_{k+1} + h \sum_{j=1}^{s} \overline{a}_{ij} z_{kj} \nabla_x f(y_{kj}, \mathbf{u}_{kj}), \tag{5}$$

$$-\sum_{j \in N_i} \overline{b}_j z_{kj} \nabla_u f(y_{kj}, \mathbf{u}_{kj}) \in N_U(\mathbf{u}_{ki}), \tag{6}$$

where $\mathbf{x}'_k = (\mathbf{x}_{k+1} - \mathbf{x}_k)/h$, $\psi'_k = (\psi_{k+1} - \psi_k)/h$; ψ_k and z_{ki} are row vectors in $\mathbf{R}^n$, and $y_{ki}, z_{ki}, \mathbf{u}_{ki}$ are approximations, respectively, to the state, costate, and control variables at the moment $(t_k + \sigma_i h) \in [0, 1]$, $0 \leq \sigma_i \leq 1$, $N_i = \{j \in [1, s] : \sigma_j = \sigma_i\}$, $1 \leq i \leq s$, $0 \leq k \leq N - 1$. Theorem 7.2 in [3] [p. 214] implies that there exist positive constants γ and β, $\beta \leq \rho$, such that, if $h < \gamma$ whenever $(\mathbf{x}_k, \mathbf{u}_{ki}, \psi_{k+1}) \in B_\beta(x^*(t), u^*(t), \psi^*(t))$ for some t, then each one of the equations (3) and (5) has a unique solution $y_{ki} = y_i(\mathbf{x}_k, \mathbf{u}_k)$, $z_{ki} = z_i(\mathbf{x}_k, \mathbf{u}_k, \psi_{k+1})$ and $y_{ki}, z_{ki} \in B_\rho(x^*(t), u^*(t), \psi^*(t))$.
We assume that the following conditions hold

$$\text{(a)} \ \sum_{i=1}^{s} b_i = 1, \ \text{(b)} \ \sum_{i=1}^{s} b_i c_i = \frac{1}{2}, \ c_i = \sum_{j=1}^{s} a_{ij}, \ \text{(c)} \ \sum_{i=1}^{s} b_i \sigma_i = \frac{1}{2}, \tag{7}$$

$$\text{(a) } \sum_{i=1}^{s} \bar{b}_i = 1, \text{ (b) } \sum_{i \in N_l} \bar{b}_i \bar{c}_i = \sum_{i \in N_l} \bar{b}_i (1 - \sigma_i), \ \bar{c}_i = \sum_{j=1}^{s} \bar{a}_{ij}, \text{ (c) } \sum_{i=1}^{s} \bar{b}_i \sigma_i = \frac{1}{2}, \quad (8)$$

$$\text{(a) } \sum_{i \in N_l} \bar{b}_i c_i = \sum_{i \in N_l} \bar{b}_i \sigma_i, \text{ (b) } \sum_{i \in N_l} b_i = \sum_{i \in N_l} \bar{b}_i > 0, \quad (9)$$

where $l \in [1, s]$, $0 \leq \sigma_i \leq 1$. Conditions (7) are related to the state equations while the conditions (8) are related to the costate equations. The parts (a) and (c) of (8) and conditions (a) and (c) of (7) are identical with the conditions (19) in [1]. They allow us to apply the numerical integration result formulated in [1] (Proposition 4.1). The coefficients of the state and costate equations are related with each other through the conditions (9). The condition (8c) follows from (7c) and (9b) as well as (8a) follows from (7a) and (9b). We will use the notations $b = (b_1, b_2, \cdots, b_s)^T$, $\bar{b} = (\bar{b}_1, \bar{b}_2, \cdots, \bar{b}_s)^T$, and

$$\tau(v; h) = \int_0^1 \omega(v, [0, 1]; t, h) \, dt, \quad \tau_k(v; h) = \int_{t_k}^{t_{k+1}} \omega(v, [t_k, t_{k+1}]; t, h) \, dt$$

for the averaged modulus of smoothness of v over the intervals $[0, 1]$ and $[t_k, t_{k+1}]$ with modulus of continuity

$$\omega(v, J; t, h) = \sup\{|v(s_1) - v(s_2)| : s_1, s_2 \in [t - h/2, t + h/2] \cap J\};$$

$|\cdot|$ denotes the Euclidean norm for vectors.

3   Main Result

Theorem 1. *If the coefficients of the Runge-Kutta integration scheme satisfy conditions (7)-(9), and if smoothness and coercivity conditions hold, then for all sufficiently small h, there exists a solution $(\mathbf{x}^h, \psi^h, \mathbf{u}^h)$ of (2)-(6) such that*

$$\max_{\substack{0 \leq k \leq N \\ 1 \leq i \leq s}} |\mathbf{x}_k^h - x^*(t_k)| + |\psi_k^h - \psi^*(t_k)| + |\mathbf{u}_{ki}^h - u^*(t_k + \sigma_i h)| \leq ch(h + \tau(\dot{u}^*; h)).$$

Proof. The proof is based on the Proposition 1. In the L^∞ discrete space $\mathcal{X}$ with elements $\omega = (\mathbf{x}, \psi, \mathbf{u})$, where

$$\mathbf{x} = (\mathbf{x}_0, \ldots, \mathbf{x}_N), \ \mathbf{x}_k \in R^n,$$
$$\psi = (\psi_0, \ldots, \psi_N), \ \psi_k \in R^n,$$
$$\mathbf{u} = (\mathbf{u}_0, \ldots, \mathbf{u}_{N-1}), \ \mathbf{u}_k \in \mathbf{U},$$

and $\mathbf{U} = \{(\mathbf{u}_1, \mathbf{u}_2, \ldots, \mathbf{u}_s) \in \mathbf{R}^{ms} : \mathbf{u}_i \in U \text{ for each } 1 \leq i \leq s \text{ and } \mathbf{u}_i = \mathbf{u}_j \text{ for every } j \in N_i\}$, we consider the maps

$$
\mathcal{T}(\omega) = \begin{pmatrix} \mathbf{x}'_k - \sum_{i=1}^{s} b_i f(y_{ki}, \mathbf{u}_{ki}), \ 0 \le k \le N-1 \\[2ex] \psi'_k + \sum_{i=1}^{s} \overline{b}_i z_{ki} \nabla_x f(y_{ki}, \mathbf{u}_{ki}), \ 0 \le k \le N-1 \\[2ex] - \sum_{j \in N_i} \overline{b}_j z_{kj} \nabla_u f(y_{kj}, \mathbf{u}_{kj}), \ 1 \le i \le s, \ 0 \le k \le N-1 \\[2ex] \psi_N - \nabla C(\mathbf{x}_N) \end{pmatrix},
$$

$$
\mathcal{F}(\omega) = \begin{pmatrix} 0 \\ 0 \\ N_U(\mathbf{u}_{k1}) \times N_U(\mathbf{u}_{k2}) \times \cdots \times N_U(\mathbf{u}_{ks}), \ 0 \le k \le N-1 \\ 0 \end{pmatrix}
$$

and the linear operator

$$
\mathcal{L}(\omega) = \begin{pmatrix} \mathbf{x}'_k - A_k \mathbf{x}_k - B_k \mathbf{u}_k \mathbf{b}, \ 0 \le k \le N-1 \\[2ex] \psi'_k + \psi_{k+1} A_k + (Q_k \mathbf{x}_k + S_k \mathbf{u}_k \overline{\mathbf{b}})^T, \ 0 \le k \le N-1 \\[2ex] - \sum_{j \in N_i} \overline{b}_j (\mathbf{u}_{kj}^T R_k + \mathbf{x}_k^T S_k + \psi_{k+1} B_k), \ 1 \le i \le s, \ 0 \le k \le N-1 \\[2ex] \psi_N - \mathbf{x}_N^T V \end{pmatrix}.
$$

Here and hereafter we use the notations: $A_k = A(t_k)$, $B_k = B(t_k)$, etc. The above three operators map the space $\mathcal{X}$ into the space $\mathcal{Y}$ consisting of 4-tuples of finite sequences in $L^1 \times L^1 \times L^\infty \times \mathbf{R}^n$. Here $L^1(R^{Nn})$ and $L^\infty(R^m)$ are discrete spaces with norms defined by $\|z\|_{L^1} = \sum_{i=0}^{N} h|z_i|$, $\|z\|_{L^\infty} = \sup_{0 \le i \le N} |z_i|$, $z_i \in R^n$. The point $\omega^* \in \mathcal{X}$ is the sequence with elements $\omega_k^* = (x_k^*, \psi_k^*, u_k^*)$, where $x_k^* = x^*(t_k)$, $\psi_k^* = \psi^*(t_k)$, $u_{ki}^* = u^*(t_k + \sigma_i h)$. Now we are going to show that conditions (P1)-(P3) are fulfilled.

1. The proof of condition (P1) consists of four parts, that refer to the corresponding components of the operators $\mathcal{T}$ and $\mathcal{F}$. In the first part, i. e. for the proof of the estimate of the state residual conditions (7) are used, which can be seen in [1]. The fourth part is straightforward from $\psi_N^* = \nabla C(x_N^*)$. For the estimate of the costate residual (the second part) we use the Lipschitz continuity of the functions $\nabla_x f$, x^*, u^*, ψ^* to obtain

$$
\psi_{ki}^* := \psi^*(t_k + \sigma_i h) =
$$
$$
\psi_{k+1}^* + (\sigma_i - 1)h\dot{\psi}_{k+1}^* + O(h^2) = \psi_{k+1}^* + (1 - \sigma_i)h\psi_{k+1}^* \nabla_x f_k + O(h^2),
$$
$$
z_{ki}^* := \psi_{k+1}^* + h\sum_{j=1}^{s} \overline{a}_{ij} z_{kj}^* \nabla_x f(y_{kj}^*, u_{kj}^*) = \psi_{k+1}^* + h\sum_{j=1}^{s} \overline{a}_{ij} \psi_{k+1}^* \nabla_x f_k + O(h^2),
$$

which together with (8) (b) yield $\sum\limits_{i=1}^{s} \bar{b}_i z_{ki}^* = \sum\limits_{i=1}^{s} \bar{b}_i \psi_{ki}^* + O(h^2)$, where $x_{ki}^* = x^*(t_k + \sigma_i h)$, $y_{ki}^* = y_k(x_k^*, u_k^*)$, $\nabla_x f_k = \nabla_x f(x^*(t_k), u^*(t_k))$. It is clear that $|y_{ki}^* - x_{ki}^*| = O(h)$, $|y_{ki}^* - x_k^*| = O(h)$, $|z_{ki}^* - \psi_{ki}^*| = O(h)$, $|u_{ki}^* - u_k^*| = O(h)$ and $|\nabla_x f(y_{ki}^*, u_{ki}^*) - \nabla_x f(x_{ki}^*, u_{ki}^*)| = O(h)$. By the help of $\sum\limits_{i=1}^{s} \bar{b}_i \nabla_x f(y_{ki}^*, u_{ki}^*) = \sum\limits_{i=1}^{s} \bar{b}_i \nabla_x f(x_{ki}^*, u_{ki}^*) + O(h^2)$ (the proof of this estimate makes use of (9) (a)), and $\dot{\psi}^* = -\nabla_x H(x^*, \psi^*, u^*)$, we can write

$$\sum_{i=1}^{s} \bar{b}_i z_{ki}^* \nabla_x f(y_{ki}^*, u_{ki}^*) =$$

$$\sum_{i=1}^{s} \bar{b}_i \psi_{ki}^* \nabla_x f(x_{ki}^*, u_{ki}^*) + \sum_{i=1}^{s} \bar{b}_i (z_{ki}^* - \psi_{ki}^*) \nabla_x f(y_{ki}^*, u_{ki}^*) +$$

$$\sum_{i=1}^{s} \bar{b}_i \psi_{ki}^* (\nabla_x f(y_{ki}^*, u_{ki}^*) - \nabla_x f(x_{ki}^*, u_{ki}^*)) = \sum_{i=1}^{s} \bar{b}_i \psi_{ki}^* \nabla_x f(x_{ki}^*, u_{ki}^*) +$$

$$\sum_{i=1}^{s} \bar{b}_i (z_{ki}^* - \psi_{ki}^*) \nabla_x f_k + \sum_{i=1}^{s} \bar{b}_i (z_{ki}^* - \psi_{ki}^*) M_{ki} +$$

$$\sum_{i=1}^{s} \bar{b}_i \psi_k^* (\nabla_x f(y_{ki}^*, u_{ki}^*) - \nabla_x f(x_{ki}^*, u_{ki}^*)) -$$

$$\sum_{i=1}^{s} \bar{b}_i \sigma_i h \nabla_x H(t_k + \theta h)(\nabla_x f(y_{ki}^*, u_{ki}^*) - \nabla_x f(x_{ki}^*, u_{ki}^*)) =$$

$$\sum_{i=1}^{s} \bar{b}_i \psi_{ki}^* \nabla_x f(x_{ki}^*, u_{ki}^*) + O(h^2),$$

where M_{ki} is a matrix whose (p, q)-th element is

$$(M_{ki})_{pq} = \nabla_x \frac{\partial f^p}{\partial x^q}(t_k + \theta_{pq} h)(y_{ki}^* - x_k^*) + \nabla_u \frac{\partial f^p}{\partial x^q}(t_k + \theta_{pq} h)(u_{ki}^* - u_k^*),$$

where $0 < \theta_{pq} < 1$, $\frac{\partial f^p}{\partial x^q}(t) = \frac{\partial f^p}{\partial x^q}(x^*(t), u^*(t))$ (the upper index p indicates the p-th component of the vector $x \in R^n$) and $\nabla_x H(t) = \nabla_x H(x(t), \psi(t), u(t))$.

Using Proposition 4.1 from [1] under conditions (8a) and (8c) we obtain that

$$\psi_k^{*\prime} + \sum_{i=1}^{s} \bar{b}_i z_{ki}^* \nabla_x f(y_{ki}^*, u_{ki}^*) =$$

$$-\frac{1}{h} \int_{t_k}^{t_{k+1}} \psi^*(t) \nabla_x f(x^*(t), u^*(t)) \, dt + \sum_{i=1}^{s} \bar{b}_i \psi_{ki}^* \nabla_x f(x_{ki}^*, u_{ki}^*) + O(h^2) \leq$$

$$c \tau_k(\dot{u}^*; h) + O(h^2).$$

Hence, the L^1 norm of the second component of $\mathcal{T}(\omega^*) - \mathcal{F}(\omega^*)$ satisfies the following inequality:

$$\sum_{i=0}^{N-1} h \left| \psi_k^{*\prime} + \sum_{i=1}^{s} \bar{b}_i z_{ki}^* \nabla_x f(y_{ki}^*, u_{ki}^*) \right| \leq ch(h + \tau(\dot{u}^*; h)).$$

In order to estimate the control residual we first mention that the condition (9) (b) gives $-\sum_{j \in N_i} \bar{b}_j \psi_{kj}^* \nabla u f(x_{kj}^*, u_{kj}^*) \in N_U(u_{ki}^*)$. Using the estimate $\sum_{j \in N_i} \bar{b}_j z_{kj}^* = \sum_{j \in N_i} \bar{b}_j \psi_{kj}^* + O(h^2)$ that is straightforward from (8b), and also the estimate $\sum_{j \in N_i} \bar{b}_j \nabla u f(y_{kj}^*, u_{kj}^*) = \sum_{j \in N_i} \bar{b}_j \nabla u f(x_{kj}^*, u_{kj}^*) + O(h^2)$, which can be proven with the help of (9a), by analogy with the costate case we obtain that

$$\sum_{j \in N_i} \bar{b}_j z_{kj}^* \nabla u f(y_{kj}^*, u_{kj}^*) = \sum_{j \in N_i} \bar{b}_j \psi_{kj}^* \nabla u f(x_{kj}^*, u_{kj}^*) + O(h^2).$$

Finally, we achieve the estimate

$$\min \left\{ \left| y + \sum_{j \in N_i} \bar{b}_j z_{kj}^* \nabla_u f(y_{kj}^*, u_{kj}^*) \right| : y \in N_U(u_{ki}^*) \right\} = O(h^2)$$

and conclude that

$$\|\mathcal{T}(w^*) - \mathcal{F}(w^*)\| \leq ch(h + \tau(\dot{u}^*; h)).$$

2. The next estimate,

$$\|\nabla \mathcal{T}(\omega) - \mathcal{L}\| \leq c(\|w - w^*\| + h),$$

is derived by estimating the differences between the matrices of these two operators, which is achieved by making use of the fact that the elements of the matrices are Lipschitz continuous functions on the closed ball $B_\beta(x^*, u^*, \psi^*)$.

3. In order to prove (P3) we have to make sure (see [4], Theorem 3, p. 79) that for a given parameter $\pi = (\mathbf{p}, \mathbf{q}, \mathbf{r}, \mathbf{s})$, under conditions (9b), the inclusion

$$\mathcal{L}(w) + \pi \in \mathcal{F}(w),$$

is exactly the necessary condition for the linear quadratic problem:

$$\text{minimize } \overline{\mathcal{B}}^h(\mathbf{x}, \mathbf{u}) + \mathbf{s}^T \mathbf{x}_N + h \sum_{k=0}^{N-1} \left(\mathbf{q}_k^T \mathbf{x}_k + \sum_{i=1}^{s} \mathbf{r}_{ki}^T \mathbf{u}_{ki} \right)$$

$$\text{subject to } \mathbf{x}_k' = A_k \mathbf{x}_k + B_k \mathbf{u}_k \mathbf{b} - \mathbf{p}_k, \ \mathbf{x}_0 = 0, \ \mathbf{u}_k \in \mathbf{U},$$

where

$$\overline{\mathcal{B}}^h(\mathbf{x}, \mathbf{u}) = \frac{1}{2} \left(\mathbf{x}_N^T V \mathbf{x}_N + h \sum_{k=0}^{N-1} \left(\mathbf{x}_k^T Q_k \mathbf{x}_k + 2\mathbf{x}_k^T S_k \mathbf{u}_k \overline{\mathbf{b}} + \sum_{i=1}^{s} \bar{b}_i \mathbf{u}_{ki}^T R_k \mathbf{u}_{ki} \right) \right).$$

If the quadratic form $\overline{\mathcal{B}}^h$ satisfies the discrete coercivity condition

$$\overline{\mathcal{B}}^h(\mathbf{x}, \mathbf{u}) \geq \alpha \|u\|_{L^2}^2 \text{ for all } (\mathbf{x}, \mathbf{u}) \in \mathcal{M}^h, \tag{10}$$

where $\alpha > 0$ is independent of h and

$$\mathcal{M}^h = \{(\mathbf{x}, \mathbf{u}) : \mathbf{x}_k' = A_k \mathbf{x}_k + B_k \mathbf{u}_k \mathbf{b}, \ \mathbf{x}_0 = 0, \ \mathbf{u}_k \in \mathbf{U} - \mathbf{U}\},$$

then the problem has unique solution that depends Lipschitz continuously on π [5] (Lemma 1).

If (7a), (8a), (9b) hold, then the quadratic form $\overline{\mathcal{B}}^h$ satisfies (10). In order to prove this let us consider the quadratic form $\mathcal{B}^h(\mathbf{x}, \mathbf{u})$ obtained from $\overline{\mathcal{B}}^h(\mathbf{x}, \mathbf{u})$ by replacing $\overline{b}$ and $\overline{b}_i$ with b and b_i, respectively. Using condition (9b), and

$$\mathbf{x}_k^T S_k \mathbf{u}_k \overline{\mathbf{b}} - \mathbf{x}_k^T S_k \mathbf{u}_k \mathbf{b}$$
$$= \mathbf{x}_k^T S_k \left(\mathbf{u}_{k1}\overline{b}_1 + \cdots + \mathbf{u}_{ks}\overline{b}_s - \mathbf{u}_{k1}b_1 - \cdots - \mathbf{u}_{ks}b_s\right) = 0,$$

$$\sum_{i=1}^{s} \left(\overline{b}_i \mathbf{u}_{ki}^T R_k \mathbf{u}_{ki} - b_i \mathbf{u}_{ki}^T R_k \mathbf{u}_{ki}\right) = 0,$$

we can make the conclusion that the equalities

$$\overline{\mathcal{B}}^h(\mathbf{x}, \mathbf{u}) = \mathcal{B}^h(\mathbf{x}, \mathbf{u}) +$$

$$\frac{h}{2}\sum_{k=0}^{N-1}\left(\mathbf{x}_k^T S_k \mathbf{u}_k \overline{\mathbf{b}} - \mathbf{x}_k^T S_k \mathbf{u}_k \mathbf{b} + \sum_{i=1}^{s}\left(\overline{b}_i \mathbf{u}_{ki}^T R_k \mathbf{u}_{ki} - b_i \mathbf{u}_{ki}^T R_k \mathbf{u}_{ki}\right)\right) = \mathcal{B}^h(\mathbf{x}, \mathbf{u})$$

hold. Taking into account that for sufficiently small h the quadratic form $\mathcal{B}^h(\mathbf{x}, \mathbf{u})$ satisfies the discrete coercivity condition (see formula (47) in [1]) we complete the proof.

To complete the proof we choose the constant quantities in Proposition 1 in such a way that the inequalities therein are satisfied. $\square$

References

1. Dontchev, A.L., Hager, W.W., Veliov, V.M.: Second-order Runge-Kutta Approximations in Control Constrained Optimal Control. SIAM J. Numer. Anal. 36, 202–226 (2000)
2. Dontchev, A.L., Hager, W.W.: Lipschitzian Stability in Nonlinear Control and Optimization. SIAM J. Control Optim. 31, 569–603 (1993)
3. Hairer, E., Norsett, S.P., Wanner, G.: Solving Ordinary Differential Equations: I Nonstiff problems. Mir, Moscow (1990) (in Russian)
4. Joffe, A., Tihomirov, V.: Theory of Extremal Problems. Nauka, Moscow (1974) (in Russian)
5. Hager, W.W.: Multiplier Methods for Nonlinear Optimal Control. SIAM J. Numer. Anal. 27, 1061–1080 (1990)

Optimal Control of a Class of Size-Structured Systems*

Oana Carmen Tarniceriu[1] and Vladimir M. Veliov[2,3]

[1] Faculty of Mathematics, "Al. I. Cuza" University
tarniceriuoana@yahoo.co.uk
[2] Institute Mathematical Methods in Economics, Vienna University of Technology,
Wiedner Hauptstr. 8–10/115, A-1040 Vienna, Austria
http://www.eos.tuwien.ac.at/OR/Veliov
[3] Institute of Mathematics and Informatics, Bulgarian Academy of Sciences

Abstract. Optimality conditions of Pontryagin's type are obtained for an optimal control problem for a size-structured system described by a first order PDE where the differential operator depends on a control and on an aggregated state variable. Typically in this sort of problems the state function is not differentiable, even discontinuous, which creates difficulties for the variational analysis. Using the method of characteristics (which are control and state dependent for the considered system) the problem is reformulated as an optimization problem for a heterogeneous control system, investigated earlier by the second author. Based on this transformation, the optimality conditions are obtained and a stylized meaningful example is given where the optimality conditions allow to obtain an explicit differential equation for the optimal control.

1 Introduction

In this paper we consider the following optimal control problem:

$$\max_{v(t) \in V} \left\{ \int_0^\omega L(s, x(T, s)) \, ds + \int_0^T K(t, y(t), v(t)) \, dt \right\}, \qquad (1)$$

subject to the constraints

$$x_t(t, s) + g(t, s, y(t), v(t)) x_s(t, s) = f(t, s, x(t, s), y(t), v(t)), \quad (t, s) \in D, \qquad (2)$$

$$x(0, s) = x^0(s), \quad x(t, 0) = 0, \quad s \in \Omega, \ t \in [0, T], \qquad (3)$$

$$y(t) = \int_\Omega h(t, s, x(t, s)) \, ds, \qquad t \in [0, T], \qquad (4)$$

Here $t \in [0, T]$, $s \in \Omega := [0, \omega]$ ($T > 0$ and $\omega > 0$ are fixed), $D = [0, T] \times \Omega$; $x : D \mapsto R$ and $y : [0, T] \mapsto \mathbf{R}^m$ are state variables, $v : [0, T] \mapsto V \subset \mathbf{R}^r$ is a

---

* The first author was supported for this research by the grant CNCSIS 1416/2005, and the second — by the Austrian Science Foundation (FWF) under grant P18161.

I. Lirkov, S. Margenov, and J. Waśniewski (Eds.): LSSC 2007, LNCS 4818, pp. 366–373, 2008.

control variable, L, K, f, h, x_0 are given functions of the respective arguments. Informally x_t and x_s mean the partial derivatives of x, although these will not necessarily exist, therefore the strict meaning of the system will be explained in the next section. Here t is interpreted as time, s – as the "size" of the agents in a population, then $x(t, \cdot)$ is the size-density of the population, $y(t)$ is a vector of some aggregated quantities related to the population (such as the total number of agents, if $h = x$, the total "biomass", if $h = sx$, etc.). The meaning of the control $v(t)$ in the population context is the intensity of supply of food, water, heating, etc., in absolute or relative (to the population size) units. In the same context $x^0(s)$ is the density of the initially existing agents. The function g is interpreted as the growth "velocity" of the agents of size s at time t, given the values of the aggregated quantities $y(t)$ and of the control $v(t)$. The function f represents the in-/out-flow due to mortality, migration, etc.; f includes also the term $-g_s x$ that appears since we have taken $g x_s$ in the l.-h. side of (2) instead of $[gx]_s$ (which results from the standard model micro-foundation). Of course, several other interpretations are possible and well known in the contexts of physics or economics.

In the present paper we deliberately restrict the generality of the model by excluding distributed (harvesting) controls, inflow of agents of zero size (which is often present in this sort of models), more general objective function, etc. There are several reasons for this restriction: size limitation of the paper, numerous technical difficulties that arise in some more general considerations, more transparency that will help to exhibit our main point: derivation of optimality conditions that hold also in case of a discontinuous state function x. The last point is important, especially in the control context, since x is *typically* discontinuous. In our restricted model this happens due to possible discontinuity of x^0 and/or inconsistency of the initial and boundary conditions in (3). In case of controllable boundary conditions or of distributed control of the in-/out-flow such discontinuity (or at least non-differentiability) may be created endogenously.

Despite of the large amount of literature on size-structured systems ([6,4,1,2,5], to mention a few) the control theory for such systems is, to our knowledge, not developed (in contrast to the age-structured systems, which could be viewed as a special case). On the other hand, recent research in management of renewable resources and in air quality protection faces challenging control problems for size-structured systems. The present paper is aimed to make a step in the development of a respective control theory and the corresponding numerical methods.

The paper is organized as follows. The strict formulation of the problem is given in Section 2. In Section 3 the problem is transformed to an optimization problem for a heterogeneous control system [9], which is used in Section 4 to obtain Pontryagin's type optimality conditions. Section 5 presents the solution of a stylized meaningful example, which can be used for numerical tests.

2   Main Assumptions and Strict Formulation of the Problem

The following will be assumed further on in the paper.

Standing assumptions. The set $V \subset \mathbf{R}^r$ is convex and compact. The functions f, g, h, L, K are defined on the projections of the set $D \times \mathbf{R} \times \mathbf{R}^m \times V$, corresponding to the indicated variables, with values in $\mathbf{R}$ (respectively, in $\mathbf{R}^m$ for h), are measurable in t and continuous in v (locally uniformly in the rest of the variables). The derivatives f_s, f_x, f_y, g_s, g_y, g_{ss}, g_{sy}, h_s, h_x, L_s, L_x, K_y exist, are locally bounded, continuous in v and locally Lipschitz in (s, x, y) (uniformly in the rest of the variables). In addition, $|f(t, s, x, y, v)| + |h(t, s, x)| \leq C(1 + |x|)$, where C is a constant, $f(t, s, 0, y, v) = 0$, $h(t, s, 0) = 0$, $L(s, 0) = 0$, $g(t, 0, y, v) > 0$, $g(t, \omega, y, v) \leq 0$. Moreover, $x^0 : \Omega \mapsto \mathbf{R}$ is measurable and bounded, $x^0(s) = 0$ for $s > s^0$, where $s^0 < \omega$.

Due to the assumptions for g we have that for given measurable functions y and v the solution of the equation

$$\dot{c}(t) = g(t, c(t), y(t), v(t)), \quad c(0) = \sigma \in \Omega$$

exists on D and is unique. We denote this solution by $c_{y,v}[\sigma](t)$, $t \in [0, T]$. Then for every fixed t the mapping $\sigma \longmapsto c_{y,v}[\sigma](t)$ is a diffeomorphism between $S_0 := [0, s^0]$ and $S_{y,v}(t) := [c_{y,v}[0](t), c_{y,v}[s^0](t)]$.

For fixed measurable y and v and $\sigma \in S_0$ we consider the equation

$$\dot{\xi}(t) = f(t, c_{y,v}[\sigma](t), \xi(t), y(t), v(t)), \quad \xi(0) = x^0(\sigma).$$

According to the assumptions for f this solution exists on $[0, T]$ and is unique. Denote it by $\xi_{y,v}[\sigma](t)$. Define the function $x(t, s)$ for $t \in [0, T]$ and $s \in S_{y,v}(t)$ as $x(t, s) = (\xi_{y,v}[s])^{-1}(t)$, with the inversion of the above diffeomorphism in the r.-h. side. Then the following equation holds:

$$\frac{\mathrm{d}}{\mathrm{d}t} x(t, c_{y,v}[\sigma](t)) = f(t, c_{y,v}[\sigma](t), x(t, c_{y,v}[\sigma](t)), y(t), v(t)), \quad x(0, \sigma) = x^0(\sigma). \quad (5)$$

We extend the definition of x by setting $x(t, s) = 0$ if $s \notin S_{y,v}(t)$. Due to the assumptions for x^0, f, and the zero boundary condition in (3), this definition of x as a solution of (2) (for fixed y and v) coincides with the usual definition of a "mild" solution (that is, a solution along the characteristics) systematically developed i.e. in [6]. We mention that this definition of a solution of (2) can be well micro-founded in the population context without assuming differentiability and even continuity of x.

Then the meaning of a solution of (2)–(4) for a fixed measurable v is clear: x has to satisfy (2) in the sense just explained (involving (5)), and in addition equation (4) has to be satisfied. In order to make use of functional spaces (which is convenient for proving existence theorems, for example), both (5) and (4) are required to be satisfied almost everywhere, which does not affect the objective function (1).

3 Reformulation of the Problem in the Framework of Heterogeneous Control Systems

We shall reformulate problem (1)–(4) in the framework of *heterogeneous control systems* as presented in [9], which allows for an easy utilization of the classical method of variations for obtaining necessary optimality conditions.

Applying Corollary 2 of Theorem 102, Chapter 4 in [8] we change the variable $s = c_{y,v}[\sigma](t)$ to get (due to $h(s,0) = 0$)

$$y(t) = \int_{S_{y,v}(t)} h(t, s, x(t, s))\, \mathrm{d}s = \int_{S_0} h(t, c_{y,v}[\sigma](t), x(t, c_{y,v}[\sigma](t))) \frac{\partial c_{y,v}[\sigma](t)}{\partial \sigma}\, \mathrm{d}\sigma.$$

Similarly (due to $L(s,0) = 0$),

$$\int_\Omega L(s, x(T, s))\, \mathrm{d}s = \int_{S_0} L(c_{y,v}[\sigma](T), x(T, c_{y,v}[\sigma](T))) \frac{\partial c_{y,v}[\sigma](T)}{\partial \sigma}\, \mathrm{d}\sigma.$$

The term $z_{y,v}[\sigma](t) := \frac{\partial c_{y,v}[\sigma](t)}{\partial \sigma}$ satisfies the equation

$$\frac{\mathrm{d}}{\mathrm{d}t} z_{y,v}[\sigma](t) = g_s(t, c_{y,v}[\sigma](t), y(t), v(t)) z_{y,v}[\sigma](t), \quad z_{y,v}[\sigma](0) = 1.$$

Thus we can reformulate problem (1)–(4) in the following way, where X, C, Z, are considered as distributed state variables, and y and v are as before:

$$\max_{v(t) \in V} \left\{ \int_{S_0} L(C(T, \sigma), X(T, \sigma)) Z(T, \sigma)\, \mathrm{d}\sigma + \int_0^T K(t, y(t), v(t))\, \mathrm{d}t \right\} \qquad (6)$$

subject to

$$\begin{aligned}
\dot{X}(t, \sigma) &= f(t, C(t, \sigma), X(t, \sigma), y(t), v(t)), & X(0, \sigma) &= x^0(\sigma), \\
\dot{C}(t, \sigma) &= g(t, C(t, \sigma), y(t), v(t)), & C(0, \sigma) &= \sigma, \\
\dot{Z}(t, \sigma) &= g_s(t, C(t, \sigma), y(t), v(t)) Z(t, \sigma), & Z(0, \sigma) &= 1, \\
y(t) &= \int_{S_0} h(t, C(t, \sigma), X(t, \sigma)) Z(t, \sigma)\, \mathrm{d}\sigma,
\end{aligned} \qquad (7)$$

where "dot" means differentiation with respect to t. Clearly the connection between x and the new state variable X is that $x(t, c_{y,v}[\sigma](t)) = X(t, \sigma)$. Problem (6)–(7) is a special case of the heterogeneous control problem investigated in [9]. Namely, let us consider the following problem

$$\max_{v(t) \in V} \left\{ \int_S l(\sigma, R(T, \sigma))\, \mathrm{d}\sigma + \int_0^T \int_S M(t, \sigma, R(t, \sigma), Y(t), v(t))\, \mathrm{d}\sigma\, \mathrm{d}t \right\} \qquad (8)$$

subject to

$$\dot{R}(t, \sigma) = F(t, \sigma, R(t, \sigma), Y(t), v(t)), \quad R(0, \sigma) = R^0(\sigma), \qquad (9)$$

$$Y(t) = \int_S H(t, \sigma, R(t, \sigma))\, \mathrm{d}\sigma, \qquad (10)$$

where now S is a measurable set of parameters in a finite dimensional space, $R(t, \sigma) \in \mathbf{R}^n$ is a distributed state variable, $Y(t) \in \mathbf{R}^m$ is an aggregated state variable (not necessarily having the same meaning as in the previous considerations). Clearly, this model extends to (6), (7) by setting

$$R = (X, C, Z), \quad S = S^0, \quad F = (f, g, g_s Z), \quad H = hZ, \quad l = LZ, \quad M = K. \quad (11)$$

A solution of (9)–(10) (and the adjoint system that will appear below), given a measurable $v(t)$, is a pair (R, Y) of measurable and bounded functions on $[0, T] \times S$, such that $X(\cdot, \sigma)$ is absolutely continuous for a.e. $s \in S$, and all the equations in (9)–(10) are satisfied for a.e. $t \in [0, T]$ and $\sigma \in S$. The optimality in problem (8)–(10) has the usual meaning with all measurable selections v for which a corresponding bounded solution (R, Y) exists on $[0, T] \times S$ (it will be unique on the assumptions below) considered as admissible controls. This definition is consistent with the one for the particular case of a size-structured system.

Theorem 1. *Assume that S is a compact measurable set with positive measure in a finite-dimensional Euclidean space; $V \subset \mathbf{R}^r$ is nonempty and compact, the functions F, H, l, M, R^0 are defined on the respective projections of the set $[0, T] \times S \times \mathbf{R}^n \times \mathbf{R}^m \times V$ (corresponding to the variables on which they depend), measurable in (t, s), differentiable in R, Y, Lipschitz in v (locally uniformly in the rest of the variables), the derivatives with respect to R and Y are locally bounded, and locally Lipschitz in R, Y (uniformly in the rest of the variables).*

Let $(v(t), R(t, \sigma), Y(t))$ be an optimal solution of problem (8)–(10), and let v be bounded. Then the following adjoint system

$$- \dot{\lambda}(t, \sigma) = \lambda(t, \sigma) F_R(t, \sigma) + M_R(t, \sigma) + \eta(t) H_R(t, \sigma), \quad \lambda(T, \sigma) = l_R(\sigma), (12)$$

$$\eta(t) = \int_S [\lambda(t, \sigma) F_Y(t, \sigma) + M_Y(t, \sigma)] \, d\sigma \qquad (13)$$

has a unique solution[1] (λ, η) on $[0, T] \times S$. In the above equations the argument (t, σ) stays for $(t, \sigma, R(t, \sigma), Y(t), v(t))$, while $l_R(\sigma) := l_R(\sigma, R(T, \sigma))$. Moreover, for a.e. $t \in [0, T]$ the value $v(t)$ maximizes on V the function

$$v \longrightarrow \int_S [M(t, \sigma, R(t, \sigma), Y(t), v) + \lambda(t, \sigma) F(t, \sigma, R(t, \sigma), Y(t), v)] \, d\sigma. \qquad (14)$$

This theorem follows from a slight improvement of [9, Theorem 1] (by making use of [9, Proposition 1]). Since no detailed proofs are presented in [9], such will be given in a full-size paper in preparation by the authors, including a more general size-structured problem.

4 Maximum Principle

We apply Theorem 1 with the specifications (11) arising from the original problem (1)–(4). In this case $\lambda = (\lambda^X, \lambda^C, \lambda^Z)$ and having in mind (11) the adjoint system takes the form

---

[1] The co-state variables λ and η are row-vectors, while X and Y are column-vectors.

$$-\dot{\lambda}^X = \lambda^X f_x + \eta h_x Z, \qquad\qquad \lambda^X(T,\sigma) = L_x(\sigma) Z(T,\sigma),$$
$$-\dot{\lambda}^C = \lambda^X f_s + \lambda^C g_s + \lambda^Z g_{ss} Z + \eta h_s Z, \qquad \lambda^C(T,\sigma) = L_s(\sigma) Z(T,\sigma),$$
$$-\dot{\lambda}^Z = \lambda^Z g_s + \eta h, \qquad\qquad \lambda^Z(T,\sigma) = L(\sigma),$$
$$\eta(t) = \int_{S_0} [\lambda^X f_y + \lambda^C g_y + \lambda^Z g_{sy} Z]\, \mathrm{d}\sigma + K_y,$$

where we skip the arguments which are evaluated along the optimal solution:
$\lambda^X := \lambda^X(t,\sigma)$, ..., $f_x := f_x(t, C(t,\sigma), X(t,\sigma), Y(t), v(t))$, $L_x(\sigma) := L_x(C(T,\sigma),$
$X(T,\sigma))$ etc. The function (14) to be maximized takes the form

$$v \longrightarrow K + \int_{S_0} [\lambda^X f + \lambda^C g + \lambda^Z g_s Z]\, \mathrm{d}\sigma.$$

Changing the variables $\lambda^x Z = \lambda^X$, $\lambda^c Z = \lambda^C$, $\lambda^z = \lambda^Z$, and dividing by Z (which is nonzero due to its differential equation), we obtain the following optimality condition.

Theorem 2. *Let (x, y, v) be an optimal solution of problem (1)–(4). Then for almost every t the value $v(t)$ maximizes the function*

$$v \longrightarrow \left\{ K(t, y(t), v) + \int_{S_0} [\lambda^x(t,\sigma) f + \lambda^c(t,\sigma) g + \lambda^z(t,\sigma) g_s] Z(t,\sigma)\, \mathrm{d}\sigma \right\}, \quad (15)$$

where $(\lambda^x, \lambda^c, \lambda^z, \eta)$ are uniquely determined by the equations

$$-\dot{\lambda}^x(t,\sigma) = \lambda^x(t,\sigma)(f_x + g_s) + \eta(t) h_x, \qquad\qquad \lambda^x(T,\sigma) = L_x(\sigma),$$
$$-\dot{\lambda}^c(t,\sigma) = \lambda^x(t,\sigma) f_s + 2\lambda^c(t,\sigma) g_s + \lambda^z(t,\sigma) g_{ss} + \eta(t) h_s, \quad \lambda^c(T,\sigma) = L_s(\sigma),$$
$$-\dot{\lambda}^z(t,\sigma) = \lambda^z(t,\sigma) g_s + \eta(t) h, \qquad\qquad \lambda^z(T,\sigma) = L(\sigma),$$

$$\eta(t) = \int_{S_0} [\lambda^x(t,\sigma) f_y + \lambda^c(t,\sigma) g_y + \lambda^z(t,\sigma) g_{sy}] Z(t,\sigma)\, \mathrm{d}\sigma + K_y(t, y(t), v(t)).$$

The functions with missing arguments are evaluated at $(t, C(t,\sigma), X(t,\sigma), y(t), v)$ in (15), and at $(t, C(t,\sigma), X(t,\sigma), y(t), v(t))$ in the rest of the equations (wherever applicable). The relation between the original state function $x(t,s)$ and the function $X(t,\sigma)$ is that $x(t, C(t,\sigma)) = X(t,\sigma)$, on the set $\{C(t,\sigma) : t \in [0,T], \sigma \in S_0\}$, and $x(t,s) = 0$ otherwise.

Thus the original system and the optimality conditions are both represented in terms of ODEs. In addition, the proof of Theorem 1 (not presented here) applied with the specifications (11) implies the following. Denote by $J(v)$ the objective value (1) corresponding to a control function $v \in \mathcal{V} := \{v \in L_2(0,T) : v(t) \in V\}$. Then, assuming that the derivatives with respect to v in the r.-h. side below exist, the functional J is differentiable, and

$$J'(v)(t) = K_v(t, y(t), v(t)) + \int_{S_0} [\lambda^x(t,\sigma) f_v + \lambda^c(t,\sigma) g_v + \lambda^z(t,\sigma) g_{sv}] Z(t,\sigma)\, \mathrm{d}\sigma,$$

where all the variables are evaluated as in Theorem 2, but $v \in \mathcal{V}$ is arbitrary (not necessarily the optimal one). This makes it possible, in principle, to apply a gradient projection method for numerical approximation of the optimal solution. One can use the discretization scheme proposed in [9] for approximating the objective function and its gradient. However, the issue of convergence is not simple, especially in the non-coercive case, as the recent paper [7] shows even for purely ODE optimal control systems (see also [3] for convergence analysis).

5   An Example

In the example below we obtain explicit differential equations for the optimal control, which can be used for numerical tests. The explicit solution is possible due to independence of the growth function g on y.

Consider the problem

$$\max_{v(t) \geq 0} \left\{ \int_0^\omega sx(T, s)\, \mathrm{d}s - \int_0^T \left[by(t) + \frac{d}{2}v(t)^2 \right] \mathrm{d}t \right\},$$

$$x_t + \alpha(\omega - s)v(t)x_s = \alpha v(t)x(t, s), \quad x(0, s) = x^0(s),$$

$$y(t) = \int_0^\omega s^\kappa x(t, s)\, \mathrm{d}s,$$

where b, d, α are positive constants, κ is either 0 or 1, and x^0 is a non-negative measurable function, which equals zero outside $S_0 := [0, s_0] \subset [0, \omega)$. Obviously the standing assumptions are satisfied. The differential equation represents a conservation law, and the overall model has an obvious interpretation in the population context.

Using equations (7) we obtain

$$X(t, \sigma) = \frac{x^0(\sigma)}{\varphi(t)}, \quad C(t, \sigma) = \omega - (\omega - \sigma)\varphi(t), \quad Z(t, \sigma) = \varphi(t),$$

where $\varphi(t) = e^{-\alpha \int_0^t v(\theta)\, \mathrm{d}\theta}$. Since the adjoint equations in Theorem 2 are linear, we can express λ^x, λ^c and λ^z in terms of X, C and Z. After somewhat tedious and long calculations one finds an expression for the function to be maximized in (15), involving only the "free" value v and the optimal control $v(t)$, together with the data of the problem. Maximizing in v, and assuming that the resulting values are non-negative (which naturally happens under mild and natural — in view of the interpretation of the problem — conditions) we obtain in the case $\kappa = 0$ that $v(t) = v^*$, where v^* is the unique solution of the equation

$$v = \frac{\alpha}{d} \int_{S_0} (\omega - \sigma)x^0(\sigma)\, \mathrm{d}\sigma e^{-\alpha T v}.$$

The fact that v is constant results from the fact that the maintenance costs $by(t)$ depend only on the size of the population, which is constant under the conservation low assumed in the model.

In the case $\kappa = 1$ (which is more realistic, since the term $by(t)$ represents maintenance costs proportional to the total biomass, rather to the population size) we obtain that the optimal control $v(t)$ satisfies the nonlinear (and nonevolutionary) integral equation

$$v(t) = \frac{\alpha}{d} \int_{S_0} (\omega - \sigma) x^0(\sigma) \, d\sigma \left[e^{-\alpha \int_0^T v(\theta) \, d\theta} - b \int_t^T e^{-\alpha \int_0^\tau v(\theta) \, d\theta} \, d\tau \right].$$

Introducing

$$\mu = e^{-\alpha \int_0^T v(\theta) \, d\theta} \tag{16}$$

as an unknown parameter we obtain an equation of the form

$$v(t) = p\mu - pb \int_t^T e^{-\alpha \int_0^\tau v(\theta) \, d\theta} \, d\tau,$$

where

$$p = \frac{\alpha}{d} \int_{S_0} (\omega - \sigma) x^0(\sigma) \, d\sigma,$$

Obviously the solution $v(t)$ is twice differentiable, and differentiating two times we obtain the following boundary value problem for a second order differential equation for v:

$$\ddot{v}(t) + \alpha \dot{v}(t) v(t) = 0, \quad v(T) = \mu p, \quad \dot{v}(0) = bp.$$

The solution depends on μ, which has to be determined from equation (16).

References

1. Ackleh, A., Deng, K., Wang, X.: Competitive exclusion and coexistence for a quasi-linear size-structured population model. Mathematical Biosciences 192, 177–192 (2004)
2. Abia, L.M., Angulo, O., Lopez-Marcos, J.C.: Size-structured population dynamics models and their numerical solutions. Discrete Contin. Dyn. Syst. Ser. B4(4), 1203–1222 (2004)
3. Chryssoverghi, I.: Approximate Gradient Projection Method with General Runge-Kutta Schemes and Piecewise Polynomials Controls for Optimal Control Problems. Control and Cybernetics 34(2), 425–451 (2005)
4. Farkas, J.Z.: Stability conditions for a nonlinear size-structured model. Nonlinear Analysis 6(5), 962–969 (2005)
5. Kato, N., Sato, K.: Continuous dependence results for a general model of size dependent population dynamics. J. Math. Anal. Appl. 272, 200–222 (2002)
6. Kato, N.: A general model of size-dependent population dynamics with nonlinear growth rate. J. Math. Anal. Appl. 297, 234–256 (2004)
7. Nikol'skii, M.S.: Convergence of the gradient projection method in optimal control problems. Computational Mathematics and Modeling 18(2), 148–156 (2007)
8. Schwartz, L.: Analyse Mathématique. Hermann (1967)
9. Veliov, V.M.: Newton's method for problems of optimal control of heterogeneous systems. Optimization Methods and Software 18(6), 689–703 (2003)

Part VII
Environmental Modelling

Modelling Evaluation of Emission Scenario Impact in Northern Italy

Claudio Carnevale, Giovanna Finzi, Enrico Pisoni, and Marialuisa Volta

Department of Electronics for Automation
University of Brescia
Via Branze 38, 25123 Brescia, Italy
enrico.pisoni@ing.unibs.it

Abstract. In this work, the multiphase model TCAM has been applied to evaluate the impact of three different emission scenarios on PM10 concentrations in northern Italy. This domain, due to high industrial and residential site, to a close road net and to frequently stagnating meteorological conditions, is often affected by severe PM10 levels, far from the European standard laws. The impact evaluation has been performed in terms of both yearly mean value and 50 $\mu g/m^3$ exceedance days in 9 points of the domain, chosen to be representative for the chemical and meteorological regimes of the domain. The results show that all the three emission reduction scenarios improve air quality all over the domain, and in particular in the area with higher emission density.

1 Introduction

Multiphase models can simulate the physical-chemical processes involving secondary pollutants in the troposphere, allowing to assess the effectiveness of sustainable emission control strategies. In this paper, the chemical and transport model TCAM [1] is introduced and applied, as a module of GAMES (Gas Aerosol Modelling Evaluation System) integrated modelling system [2], including the emission model POEM-PM [3], the CALMET meteorological model [4], a preprocessor providing the initial and boundary conditions required by the model and the System Evaluation Tool (SET). The modelling system has been validated over northern Italy in the frame of CityDelta project. The model has been used to asses the effectiveness of three different emission control strategies in the frame of CityDelta project [5]. The first emission scenario is related to the emission reduction expected up to 2020 assuming the European Current Legislation (CLE), while the second one is based on the Most Feasible emission Reduction (MFR) that could be obtained using the best available technology. Finally, the third scenario is equal to the CLE one, but the PM2.5 emission are drop to zero inside the Milan metropolitan area.

2 Model Description

TCAM (Transport and Chemical Aerosol Model) is a multiphase three-dimensional Eulerian model, in a terrain-following co-ordinate system [3]. The model

I. Lirkov, S. Margenov, and J. Waśniewski (Eds.): LSSC 2007, LNCS 4818, pp. 377–384, 2008.

formalizes the physical and chemical phenomena involved in the formation of secondary air pollution by means of the following mass-balance equation:

$$\frac{\partial C_i}{\partial t} = T_i + R_i + D_i + S_i, \tag{1}$$

where C_i $[\mu g \ m^{-3}]$ is the concentration of the i-th species, T_i is the transport/diffusion term $[\mu g \ m^{-3} \ s^{-1}]$, R_i $[\mu g \ m^{-3} \ s^{-1}]$ is the multiphase term, D_i $[\mu g \ m^{-3} \ s^{-1}]$ includes the wet and dry deposition and S_i is the emission term.

To solve equation (1), TCAM implements a split operator technique [6] allowing to separately treat the horizontal transport, the vertical phenomena (including transport-diffusion, emissions and deposition) and the chemistry.

The advection scheme implemented in TCAM is derived by CALGRID model [7]. The module is based on a finite differences scheme and it solves horizontal transport of both gas and aerosol species. The module describes the convective and the turbulent transport, neglecting the molecular diffusion processes [8] solving the PDE transport equation using chapeau functions [9] and the non linear Forester filter [10].

The dry deposition phenomenon is described by the equation [8]:

$$F_i = C_i \cdot v_{di}, \tag{2}$$

where F_i $[\mu g \ m^{-2} \ s^{-1}]$ is the removed pollutant flux, C_i is the concentration of the i-th species at ground level and v_{di} $[m \ s^{-1}]$ is i-th species deposition velocity.

Wet deposition (for both gas and aerosol species) is described by the equation [8]:

$$\frac{\partial C_i}{\partial t} = -\Lambda_p \cdot C_i, \tag{3}$$

where Λ_p $[s^{-1}]$ is the scavenging coefficient defined distinctly for gas (Λ_{gas}) and aerosol (Λ_{pm}) species. For gases, two components are calculated: (1) the uptake of ambient gas concentration into falling precipitation, which can occur within and below clouds, and (2) the collection by precipitation of cloud droplets containing dissolved gas species. For particles, separate in-cloud and below-cloud scavenging coefficients are determined. Within clouds, all aerosol mass is assumed to exist in cloud droplets (all particles are activated as condensation nuclei), so scavenging is due to the efficient collection of cloud droplets by precipitation. Below clouds, dry particles are scavenged by falling precipitation with efficiency depending on particle size.

TCAM allows the simulation of gas chemistry using both the *lumped structure* (Carbon Bond 90 [11]) and the *lumped molecule* (SAPRC90 [12], SAPRC97 [13], SAPRC99 [14]) approach. In order to describe the mass transfer between gas and aerosol phase, the COCOH97 [15], an extended version of SAPRC97 mechanism is implemented in the model. The ODE chemical kinetic system is solved by means of the Implicit-Explicit Hybrid (IEH) solver [16], that splits the species in fast and slow ones, according to their reaction velocity. The system of fast species is solved by means of the implicit Livermore Solver for Ordinary Differential Equations (LSODE) [17,16] implementing an Adams predictor-corrector method in the non-stiff case [18], and the Backward Differentiation Formula method in the stiff case [18]. The slow species system is solved by the Adams-Bashfort method [18]. The aerosol module implemented in TCAM is coupled with the COCOH97 gas phase chemical mechanism. TCAM describes the most relevant aerosol processes: the condensation, the evaporation [8], the nucleation of H_2SO_4 [19] and the aqueous oxidation of SO_2 [8]. The aerosol module describes the particles by means of a fixed-moving approach; a generic particle is represented with an internal core containing the non volatile material, like elemental carbon, crustal and dust. The core dimension of each size class is established at the beginning of the simulation and is held constant during the simulation. The volatile material is supposed to reside in the outer shell of the particle whose dimension is evaluated by the module at each time step on the basis of the total mass and of the total number of suspended particles. The aerosol module describes the dynamics of 21 chemical compounds: twelve inorganic species (H_2O, $SO_4=$, NH_4+, $Cl-$, NO_3-, $Na+$, $H+$, $SO_2(aq)$, $H_2O_2(aq)$, $O_3(aq)$, elemental carbon and other), and 9 organics, namely a generic primary organic species and 8 classes of secondary organic species. Each chemical species is split in n (namely $n = 10$) size bins, so that the prognostic variables of the aerosol module are $21n$. The estimation of equilibrium pressures of the condensing inorganic species is computed by means of the SCAPE2 thermodynamic module [20], while the *Condensible Organic Compounds* included in COCOH97 mechanism are considered as fully condensed due to their very low volatility.

Water is assumed to be always in equilibrium between the gas and the aerosol phases. This assumption is also made by other model developers [21]. Moreover, in the case of the TCAM model, the gas phase water content is known from the relative humidity, provided by the meteorological model. As a consequence the gas phase water concentration is known for the whole simulation at each grid point and does not participate to the balance equation [8].

3   Simulations Setup

The model has been applied to a 300×300 km^2 domain placed in northern Italy (Figure 1), including the Lombardia region as well as portions of Piemonte, Liguria, Veneto and Emilia-Romagna. The site, centred on the Milan metropolitan area, is characterized by complex terrain, by high industrial and urban emissions and by a close road net.

The domain has been horizontally divided into 5×5 km^2 grid cells and vertically in 11 varying levels ranging from 20 to 3900 meters above ground level. The 2004 simulation has been performed. The input data are provided to the model by meteorological, emission and boundary condition GAMES preprocessors, starting from data shared by JRC-IES in the frame of CityDelta-CAFE exercise [5].

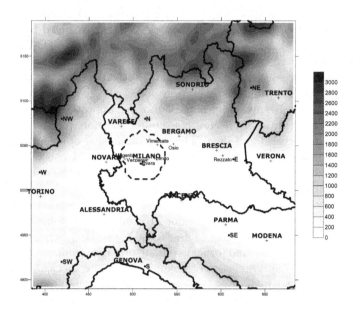

Fig. 1. Simulation domain

3.1   Emission Data

The emission fields have been estimated by means of POEM-PM model [3] processing two inventories: the Lombardia Region inventory, with a 5×5 km^2 resolution, and the EMEP (European Monitoring and Evaluation Programme) one [22], following a resolution of 50×50 km^2. The inventories include yearly emission data of NOx, VOC, CO, NH_3, SOx, $PM10$ and $PM2.5$ for each CORINAIR sector. Temporal modulation is performed using monthly, weekly and hourly profiles provided by EMEP [22]. Speciation profiles for organic compounds are defined mapping UK classes (227 species) into SAROAD ones [3]. Chemical characterization of emitted PM has been performed using EMEP profiles [22], provided by JRC. Size distribution of emitted particles has been obtained using profiles defined in [23]. The simulations concern the base case (2004) and three different emission scenario at 2020 (Table 1): (1) the CLE scenario, computed applying to the emission the current legislation up to 2020, (2) the MFR (Most

Table 1. Total emission for the base case (kTons/year/domain) and emission scenario reduction

Scenario	NOx	VOC	PM2.5	PM10	SO2	NH3
Base Case	208	258	30	38	63	80
2020 CLE	54%	58%	57%	53%	54%	6%
2020 MFR	74%	58%	57%	53%	54%	6%
2020 CLE+City	54%	58%	77%	68%	54%	6%

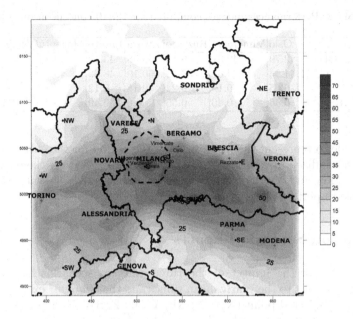

Fig. 2. PM10 mean concentration ($\mu g/m^3$ computed by TCAM model for the base case

Feasible Reduction) scenario, in which the emissions are computed supposing that for each pollutant the best emission reduction technology is applied and (3) the CLE+City scenario, which considers the CLE scenario with PM2.5 set to 0 in the Milan metropolitan area (see dot line in Figure 1). Table 1 highlights the heavy emission reduction estimated using the current legislation scenario (close to 50% with the exception of ammonia), while the most feasible reduction scenario implies only an extra reduction of NOx with respect the CLE one. The CLE+City scenario is equal to CLE scenario with the exception of PM2.5 (and consequently of PM10), which shows a reduction of about the 80% (compared to the 50% of the CLE emission) with respect to the base case, due to the switch off of the Milan metropolitan area emissions.

4   Base Case Results

Figure 2 presents the mean PM10 concentration computed for the base case. The spatial distribution of yearly mean concentration shows higher concentrations in the Po Valley and in the south-east area of the domain where high $NH3$ emissions favour the formation of secondary inorganic aerosol.

The validation of the aerosol phase simulation results has been performed comparing computed (TCAM) and observed (OBS) 2004 daily mean concentration in a set of stations selected to be representative of the chemical and meteorological regimes over the domain (Figure 2). Table 2 highlightes that the

Table 2. Performances indexes computed for PM10 concentration series

Index	Osio	Vimercate	Rezzato	Juvara	Limito	Magenta	Verziere
Mean OBS	42.79	42.31	49.11	50.98	43.19	49.29	51.83
Mean TCAM	44.94	34.08	37.32	60.19	48.62	40.16	69.43
NME	-0.05	0.19	0.24	-0.18	-0.12	0.18	-0.33
CORR	0.46	0.36	0.55	0.6	0.49	0.47	0.64

model is able to represent the mean value of the period for the entire year, with values of normalized mean error (NME) lower than 0.25, with the exception of Rezzato station and Verziere. The values of correlation coefficient (CORR) is comparable to performances of the CityDelta models [5].

5   Emission Scenario Impacts

The evaluation of the impact of the three different scenarios has been performed with respect to the yearly mean values and the $50\mu g/m^3$ exceedance days (ECC50) in 9 selected point (NW, N, NE, W, C, E, SW, S, SE) representative of the different meteorological and chemical regimes in the domain (Figure 1). Both the indicators present in the base case values out of the current air quality standard limits[1] in the center east of the domain (C, E, SE). In terms of

Table 3. Scenario impact assessment for mean concentration and number of $50\mu g/m^3$ exceedances

Index	Scenario	NW	N	NE	W	C	E	SW	S	SE
Mean										
	Base Case	8.9	17.1	5.5	24.3	67.5	36.3	20.2	12.2	35.3
	CLE-Base Case	-4.8	-7.5	-2.1	-14.1	-18.2	-17.1	-9.1	-4.5	-14.1
	MFR-Base Case	-4.6	-9.5	-2.6	-13.1	-37.1	-18.4	-11.3	-6.1	-19.2
	CLE+City-Base Case	-4.8	-7.3	-2.0	-14.1	-24.1	-17.0	-9.0	-4.5	-14.3
ECC50										
	Base Case	1	15	0	31	210	87	18	2	85
	CLE-Base Case	-1	-14	0	-30	-97	-80	-16	-1	-72
	MFR-Base Case	-1	-15	0	-30	-192	-76	-18	-2	-78
	CLE+City-Base Case	-1	-14	0	-30	-138	-80	-16	-1	-72

mean values (Table 3), the three emission scenarios have very similar impacts in the areas where base case concentrations are lower than $30\mu g/m^3$. The differences are markable in the Milan metropolitan area (C point) where MFR and CLE+City scenarios show reductions consistently higher than CLE. The impact of the 3 scenarios on the number of exceedance days (Table 3) is noticeable in all the points of the domain. The impacts are quite similar, with the exception of

---

[1] Yearly mean lower than $50\mu g/m^3$ and exceedance days lower than 35.

the C point, where the MFR shows a reduction of 192 days with respect to the base case, while the CLE shows a reduction of 97 days. In this point, the local impact of the CLE+City scenario could be highlighted, with 40 exceedance days less than the CLE one. For each scenarios, the number of exceedance days in the higher concentration areas exceed (C point) or are very close (E, SE points) to the 2020 air quality standard of 7 days per year. It is important to note that for both the indicators the impact of the CLE and CLE+City scenarios outside the Milan metropolitan area is the same, suggesting that local emission reduction has effect only in close to the area of the intervention.

6   Conclusion

The work presents the formulation and the application of the Transport and Chemical Aerosol Model (TCAM) over a northern Italy domain. The first part of the work presents the results of the validation phase, showing that the model is able to correctly reproduce measured daily PM10 concentration series, in terms of both mean values and correlation coefficient. The second part highlights the evaluation of the impact of three different emission scenarios over the domain at 2020. The results show that all the scenarios have a high impact on the simulated air quality indexes. However, the value of the exceedance days index is aspected not to respect the 2020 air quality standards in the more industrialized area of the domain.

Acknoledgments

The authors acknowledge Dr. Marco Bedogni (Agenzia Mobilitá e Ambiente, Italy), Dr. Guido Pirovano (CESI, Italy), for their valuable co-operation in the frame of CityDelta project. The work bas been partially supported by MIUR (Italian Ministry of University and Research) and by Agip Petroli.

References

1. Carnevale, C., Finzi, G., Volta, M.: Design and validation of a multiphase 3D model to simulate tropospheric pollution. In: Proc. 44th IEEE Conference on Decision and Control and European Control Conference, ECC-CDC (2005)
2. Volta, M., Finzi, G.: GAMES, a comprehensive Gas Aerosol Modelling Evaluation System. Environmental Modelling and Software 21, 578–594 (2006)
3. Carnevale, C., et al.: POEM-PM: An emission modelling for secondary pollution control scenarios. Environmental Modelling and Software 21, 320–329 (2006)
4. Scire, J.S., Insley, E.M., Yamartino, R.J.: Model formulation and user's guide for the CALMET meteorological model. Technical Report, California Air Resources Board, Sacramento, CA, A025-1 (1990)
5. Cuvelier, C., et al.: Citydelta: A model intercomparison study to explore the impact of emission reductions in european cities in 2010. Atmospheric Environment 41(1), 189–207 (2007)

6. Marchuk, G.I.: Methods of Numerical Mathematics. Springer, New York (1975)
7. Yamartino, R.J., et al.: The CALGRID mesoscale photochemical grid model — I. Model formulation. Atmospheric Environment 26A(8), 1493–1512 (1992)
8. Seinfeld, J.H., Pandis, S.N.: Atmospheric chemistry and physics. John Wiley & Sons, Chichester (1998)
9. Pepper, D.W., Kern, C.D., Long, P.E.: Modelling the dispersion of atmospheric pollution using cubic splines and chapeau functions. Atmospheric Environment 13, 223–237 (1979)
10. Forester, C.K.: Higher Order Monotonic Convection Difference Schemes. Journal of Computational Physics 23, 1–22 (1977)
11. Gery, M.W., Whitten, G.Z., Killus, J.P.: A Photochemical Kinetics Mechanism for Urban and Regional-Scale Computering Modeling. Journal of Geophysical Research 94, 12925–12956 (1989)
12. Carter, W.P.L.: A detailed mechanism for the gas-phase atmospheric reactions of organic compounds. Atmospheric Environment 24A, 481–518 (1990)
13. Carter, W.P.L., Luo, D.I.L., Malkina, I.L.: Environmental chamber studies for development of an updated photochemical mechanism for VOC reactivity assessment. Technical report, California Air Resources Board, Sacramento (CA) (1997)
14. Carter, W.P.L.: Documentation of the SAPRC-99 Chemical Mechanism for VOC Reactivity Assessment. Contract 92–329, 95–308, California Air Resources Board (2000)
15. Wexler, A.S., Seinfeld, J.H.: Second-Generation Inorganic Aerosol Model. Atmospheric Environment 25A, 2, 731–732, 748 (1991)
16. Chock, D.P., Winkler, S.L., Sun, P.: A Comparison of Stiff Chemistry Solvers for Air Quality Modeling. Air & Waste Management Association 87th Annual Meeting (1994)
17. Hindmarsh, A.C.: LSODE and LSODEI, Two New Initial Value Ordinary Differential Equation Solvers. ACM-SIGNUM Newsletter 15(4), 10–11 (1975)
18. Wille, D.R.: New Stepsize Estimators for Linear Multistep Methods. Numerical Analysis Report 247, inst-MCCM (1994)
19. Jaecker-Voirol, A., Mirabel, P.: Heteromolecular nucleation in the sulfuric acid-water system. Atmospheric Environment 23, 2053–2057 (1989)
20. Kim, Y.P., Seinfeld, J.H., Saxena, P.: Atmospheric Gas Aerosol Equilibrium I: Termodynamic Model. Aerosol Science and Technology 19, 157–187 (1993)
21. Wexler, A.S., Lurmann, F.W., Seinfeld, J.H.: Modelling Urban and Regional Aerosols-I. Model Development. Atmospheric Environment 28(3), 531–546 (1994)
22. Vestreng, V., Adams, M., Goodwin, J.: Inventory review 2004. emission data reportedd to CRLTAP and under the NEC Directive. Technical report, EMEP/EEA Joint Review Report (2004)
23. Carnevale, C., Volta, M.: Chemical and size characterization of PM road traffic emission: profile estimation for multiphase models. Journal of Aerosol Science — European Aerosol Conference 2, 1349–1350 (2003)

Modelling of Airborne Primary and Secondary Particulate Matter with the EUROS-Model

Felix Deutsch, Clemens Mensink, Jean Vankerkom, and Liliane Janssen

VITO, Centre for Integrated Environmental Studies, 2400 Mol, Belgium
felix.deutsch@vito.be
http://www.vito.be/english/environment/environmentalstudy9.htm

Abstract. We used the Eulerian Chemistry-Transport Model EUROS to simulate the concentrations of airborne fine particulate matter above Europe. Special attention was paid to both primary as well as secondary particulate matter in the respirable size range up to 10 μm diameter. Especially the small particles with diameters up to 2.5 μm are often formed in the atmosphere from gaseous precursor compounds. Comprehensive computer codes for the calculation of gas phase chemical reactions and thermodynamic equilibria between the compounds in the gas phase and those ones in the solid phase had been implemented into the EUROS-model. Obtained concentrations of PM_{10} for the year 2003 are compared to measurements. Additionally, calculations were carried out to assess the contribution of emissions derived from the sector agriculture in Flanders, the northern part of Belgium. The obtained results demonstrate the importance of ammonia emissions in the formation of secondary particulate matter. Hence, future abatement policy should consider more the role of ammonia in the formation of secondary particles.

1 Introduction

Many European countries currently face problems with episodes of high concentrations of fine particulate matter (PM) in the ambient air. These particles are associated with strong adverse health effects [1,2]. In 2003, there were more than ten episodes of high particle concentrations (PM_{10} > 100 μg/m^3) at several PM_{10} monitoring stations in Belgium. Advanced computer models including atmospheric transport and turbulent diffusion, but also atmospheric chemistry and microphysics, can help to understand the connections between emissions, chemical reactions and meteorological factors in the formation of PM. Not only primary emissions of particles contribute to high PM-concentrations, but also formation of secondary particulate matter from the emissions of precursor compounds contribute significantly to PM_{10}- and $PM_{2.5}$-concentrations. Especially the emissions of ammonia (NH_3), nitrogen oxides (NO_x) and sulphur dioxide (SO_2) are involved in these processes, leading preferentially to small secondary particles of the size fraction $PM_{2.5}$.

To investigate these processes, we extended the operational Eulerian air quality model EUROS with two special modular algorithms for atmospheric particles. The EUROS model is an Eulerian air quality model for the simulation of

I. Lirkov, S. Margenov, and J. Waśniewski (Eds.): LSSC 2007, LNCS 4818, pp. 385–392, 2008.

tropospheric ozone over Europe. It was originally developed at RIVM in the Netherlands and was implemented in Belgium in 2001. It is now an operational tool for policy support at the Belgian Interregional Environment Agency (IR-CEL) in Brussels [3,4]. The base grid of EUROS covers nearly whole Europe with a resolution of $60 * 60$ km. Several subgrids (e.g. around Belgium) with a resolution of down to $7.5 * 7.5$ km can be chosen.

A detailed emission module describes the emission of six pollutant categories (NO_x, non-methane volatile organic compounds (NMVOC), SO_2, NH_3, $PM_{2.5}$ and $PM_{10-2.5}$) and for 7 different emission sectors (traffic, residential emissions, refineries, solvents use, combustion, industry and agriculture). Both point sources and area sources are included. As far as the meteorology is concerned, the model uses three-dimensional input datasets derived from the ECMWF (European Centre for Medium-Range Weather Forecasts, Reading, UK) meteorological reanalysed datasets.

Following an extended literature study, the Caltech Atmospheric Chemistry Mechanism (CACM, [5]) and the Model of Aerosol Dynamics, Reaction, Ionization and Dissolution (MADRID 2, [6]) were selected and implemented into the EUROS model.

Currently, EUROS is able to model mass and chemical composition of aerosols in two size fractions ($PM_{2.5}$ and PM_{coarse}). The chemical composition is expressed in terms of seven components: ammonium, nitrate, sulphate, primary inorganic compounds, elementary carbon, primary organic compounds and secondary organic compounds (SOA).

In Section 2, we shortly discuss the implementation of CACM and MADRID 2 into the EUROS model. Section 3 presents and discusses the results of calculations with the EUROS model for PM_{10} and $PM_{2.5}$ for the year 2003. Additionally, the influence of emissions from the sector agriculture is investigated.

2 Methodology

For modelling aerosols, CACM is used as gas phase chemical mechanism in EU-ROS. This mechanism comprises 361 reactions among 122 components. With this, CACM contains a basic ozone chemistry plus the most important reactions of various generations of organic compounds during which condensable products are formed. Liquid phase chemistry and heterogeneous reactions are not yet integrated into the current implementation of the Chemistry-Aerosol-Module.

Emission data for the two size fractions of primary PM ($PM_{2.5}$ and $PM_{10-2.5}$) and for the precursor compounds of secondary particulate matter (NO_x, SO_2, NMVOC and NH_3) are derived from the EMEP-database for Europe [7] and for Flanders from recent emission inventories [8].

More information on the implementation of the aerosol module into the EUROS-model and results obtained with this model can be found in [9,10].

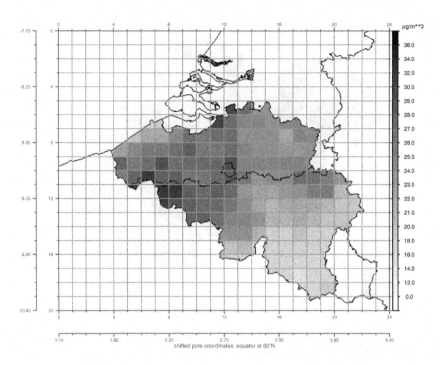

Fig. 1. Yearly averaged modeled PM_{10}-concentration in Belgium in 2003

3 Results

3.1 Comparison of Modeled and Observed PM_{10}-concentrations in Belgium

Figure 1 shows the yearly averaged PM_{10}-concentration map for Belgium for the year 2003. The three Belgian regions are shown on the map: the Flemish region in the northern part of Belgium, the capital Brussels in the centre and the Walloon region in the southern part of Belgium.

The highest PM_{10}-concentrations in Belgium were calculated for the western part of both the Flemish and the Walloon region. High concentrations of particulate matter were also calculated for the central part of Flanders. In contrast, lower PM_{10}-concentrations were calculated for the eastern part of Flanders. The lowest PM_{10}-concentrations were clearly modeled for the south-eastern part of the Walloon region.

In comparison, Figure 2 shows a map with the observed yearly averaged PM_{10}-concentrations in Belgium in 2003 [11]. In order to obtain PM_{10}-concen-trations for the whole territory of Belgium (and not only for the area around the measurement locations), concentrations were interpolated in between the measurement stations by means of the RIO interpolation technique, developed by VITO [12,13].

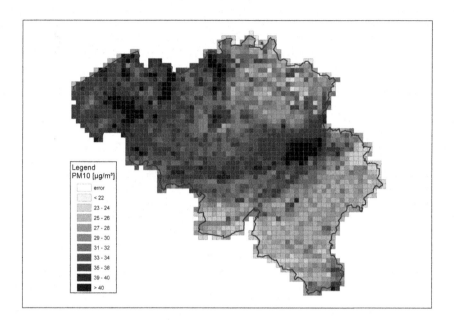

Fig. 2. Yearly averaged observed PM_{10}-concentration in Belgium in 2003 and interpolation with RIO (www.irceline.be)

The comparison shows that the EUROS-model underestimates the observed PM_{10}-concentrations. Note that the concentration scale used for Figure 1 and Figure 2 is not the same. However, this kind of underestimation is known from other models for fine particulate matter and can be traced back for an important part to the underestimation of diffusive emissions of primary particulate matter. However, the modeled geographical distribution of the PM_{10}-concentration across Belgium agrees well with the geographical distribution of the observed and interpolated concentrations. Especially the gradient within Flanders with higher concentrations in the west and in the centre and lower concentrations in the east is reproduced quite well by the model.

3.2   Contributions of Agricultural Emissions to PM_{10}- and $PM_{2.5}$-concentrations in Flanders

Figure 3 shows the contribution of emissions of the sector "agriculture" in Flanders to PM_{10}-concentrations in Belgium for the month of July, 2003. This calculation was carried out by performing a second model calculation, but leaving away the emissions from the sector "agriculture" in Flanders. These are mainly emissions of primary particulate matter ($PM_{10-2.5}$) and ammonia (NH_3). Afterwards, the difference was calculated between the obtained results with all emissions and the results obtained without the emissions from agricultural sources in Flanders. The calculation shows that the emissions of this sector contribute

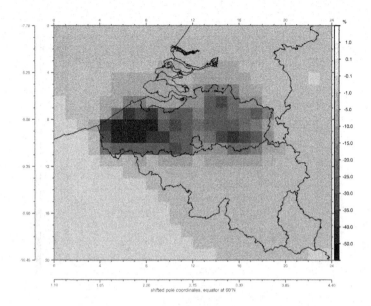

Fig. 3. Relative difference in PM_{10}-concentrations between a calculation with all emissions and a calculation without the Flemish agricultural emissions for July 2003

significantly to PM_{10}-concentrations in large parts of Flanders. Especially in the western part of Flanders, a high contribution of up to around 45 % is calculated. This high contribution can be explained by relatively high emissions of primary coarse particles ($PM_{10-2.5}$) due to agricultural sources in the Flemish emission inventory [8]. Figure 4 and Figure 5 show the contribution of the Flemish agricultural emissions to $PM_{2.5}$-concentrations in Belgium in January and in July, 2003, respectively. These results show that agricultural emissions contribute both locally and also on average in whole Flanders less to $PM_{2.5}$-concentrations than it is the case to PM_{10}-concentrations. As an average over whole Flanders, 7.5 % of $PM_{2.5}$ is derived from agricultural sources in January and 16.9 % in July, 2003. The maximum contribution in July in Flanders is 32 %. However, these are significant contributions also to the $PM_{2.5}$-fraction. This contribution is, in contrast to the one to PM_{10}-concentrations, only for a small part due to emissions of primary $PM_{2.5}$, as only 11 % of the Flemish primary $PM_{2.5}$-emissions are derived from agricultural sources [8].

Much more important is the contribution of agricultural emissions to the formation of secondary particulate matter, presumably mainly due to the formation of ammonium nitrate and ammonium sulphate in the size range $PM_{2.5}$. 95 % of all Flemish ammonia emissions are derived from agricultural sources, making this sector responsible for an important part of secondary formed $PM_{2.5}$ in Flanders.

Table 1 gives an overview of the obtained results concerning the calculation of the contribution of the Flemish agricultural emissions to the $PM_{2.5}$-concentrations in the three Belgian regions. The mean contribution to $PM_{2.5}$-concentrations is 12.2 % in Flanders.

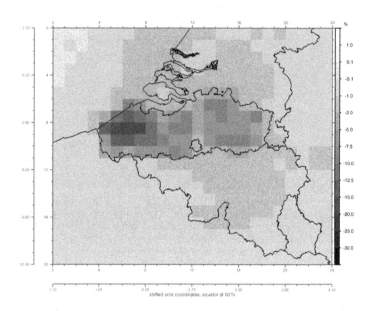

Fig. 4. Relative difference in $PM_{2.5}$-concentrations between a calculation with all emissions and a calculation without Flemish emissions from the sector agriculture for January, 2003

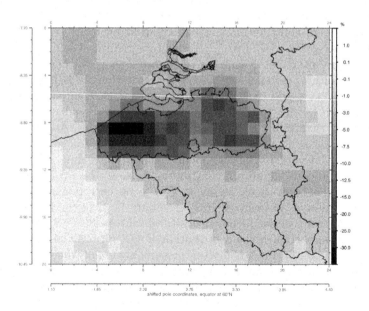

Fig. 5. Relative difference in $PM_{2.5}$-concentrations between a calculation with all emissions and a calculation without Flemish emissions from the sector agriculture for July, 2003

Table 1. Reduction of $PM_{2.5}$-concentrations when removing the emissions of the sector agriculture in Flanders for January and July, 2003 and mean contribution

$PM_{2.5}$	mean contribution [%]			max. contribution [%]		min. contribution [%]	
Location	January	July	av.	January	July	January	July
Flanders	7.5	16.9	12.2	17.1	32.3	0.7	2.3
Brussels	2.6	7.5	5.1	6.7	15.8	1.9	5.7
Wallonia	1.2	2.7	2.0	8.8	17.9	0.5	0.8
Belgium	4.0	9.0	6.5	17.1	32.3	0.5	0.8

4   Conclusions

The calculations carried out using the EUROS-model with implemented aerosol module show that PM_{10}-concentrations are underestimated by the model, mainly due to underestimations in the primary emissions of particles, but the modeled geographical pattern agrees well with that one obtained by observations. The contribution of emissions derived from the sector "agriculture" was found to be rather high, both to PM_{10}-concentrations as well as to $PM_{2.5}$-concentrations. The high contribution to the first size class of airborne particles is predominantly due to high emissions of primary particles in the size range $PM_{10-2.5}$, especially in the western part of Flanders. The high contributions to the size range of the fine particles have their origin mainly in the formation of ammonium salts such as ammonium nitrate and ammonium sulphate in the atmosphere. Approximately 12 % of $PM_{2.5}$-concentration in Flanders on average are due to Flemish agricultural emissions. For the small size range of particles, especially ammonia emissions are important. Hence, future emission reduction programs should consider more the importance of an abatement of ammonia emissions due to their potential in the formation of secondary $PM_{2.5}$.

References

1. Dockery, D.W., et al.: An association between air pollution and mortality in six US cities. N. Engl. J. Med. 329, 1753–1759 (1993)
2. Pope III, C.A., et al.: Particulate air pollution as predictor of mortality in a prospective study of US adults. Am. J. Resp. Crit. Care Med. 151, 669–674 (1995)
3. Delobbe, L., et al.: BelEUROS: Implementation and extension of the EUROS model for policy support in Belgium, Global Change and Sustainable Development, Federal Science Policy Office, Brussels (2002)
4. Mensink, C., Delobbe, L., Colles, A.: A policy oriented model system for the assessment of long-term effects of emission reductions on ozone. In: Borrego, C., Schayes, G. (eds.) Air Pollution Modelling and its Applications XV, pp. 3–11. Kluwer Academic/Plenum Publishers (2002)
5. Griffin, R.J., Dabdub, D., Seinfeld, J.H.: Secondary organic aerosol 1. Atmospheric chemical mechanism for production of molecular constituents. J. Geophys. Res. 107(D17), 4332 (2002)

6. Zhang, Y., et al.: Development and application of the Model of Aerosol Dynamics, Reaction, Ionization, and Dissolution (MADRID). J. Geophys. Res. 109 (2004)
7. Vestreng, V., et al.: Inventory Review 2004, Emission Data reported to CLRTAP and under the NEC Directive, EMEP/EEA Joint Review Report, EMEP/MSC-W Note 1/2004 (2004) ISSN 0804-2446
8. VMM: Lozingen in de lucht 1990-2003, Vlaamse Milieumaatschappij, Aalst, Belgium, p. 185 + appendix (in Dutch) (2004)
9. Deutsch, F., et al.: Modelling changes of aerosol compositions over Belgium and Europe. Int. J. Env. Poll. (in press, 2007)
10. Deutsch, F., et al.: Extension of the EUROS integrated air quality model to fine particulate matter by coupling to CACM/MADRID 2. Environmental Modeling and Assessment (in press, 2007)
11. IRCEL (2007), `http://www.irceline.be`
12. Hooyberghs, J., et al.: Spatial interpolation of ambient ozone concentrations from sparse monitoring points in Belgium. J. Environ. Monit. 8, 1129–1135 (2006)
13. Janssen, S., et al.: Spatial representativeness of AQ monitoring data and the relevance for model validation. In: Proceedings of the 6th International Conference on Urban Air Quality, Cyprus, March 27-29, 2007 (2007)

Comparative Study with Data Assimilation Experiments Using Proper Orthogonal Decomposition Method

Gabriel Dimitriu[1] and Narcisa Apreutesei[2]

[1] University of Medicine and Pharmacy "Gr. T. Popa",
Department of Mathematics and Informatics,
700115 Iaşi, Romania
dimitriu.gabriel@gmail.com
[2] Technical University "Gh. Asachi", Department of Mathematics,
700506 Iaşi, Romania
napreut@net89mail.dntis.ro

Abstract. In this work the POD approach to model reduction is used to construct a reduced-order control space for the simple one-dimensional transport equations. Several data assimilation experiments associated with these transport models are performed in the reduced control space. A numerical comparative study with data assimilation experiments in the full model space indicates that with an appropriate selection of the basis functions the optimization in the POD space is able to provide accurate results at a reduced computational cost.

1 Introduction

A major difficulty in the operational use of 4D-Var data assimilation for oceanographic and atmospheric global circulation models is the large dimension of the control space, which is the size of the discrete model initial conditions, typically in the range $10^6 - 10^8$. A way to significantly decrease the dimension of the control space without compromising the quality of the final solution for the 4D-Var data assimilation, motivates us to construct the control variable on a basis of characteristic vectors capturing most of the energy and the main features of variability of the model. We would then attempt to control the vector of initial conditions in the reduced space model.

In this paper the proper orthogonal decomposition (POD) technique is used to construct a reduced-order control space for the simple one-dimensional transport equations. The POD technique (it also goes by other names such as the Karhunen-Loève decomposition, principal component analysis and the Hotelling transform) is a method for deriving low order models of dynamical systems.

The paper is organized as follows. The numerical model under investigation is presented in Section 2. Section 3 is devoted to reviewing the POD method and 4D-Var formulation based on POD. Section 4 contains results from data assimilation experiments using 4D-Var and POD 4D-Var methods. Finally, Section 5 provides some conclusions and discussions of some related issues of this study.

I. Lirkov, S. Margenov, and J. Waśniewski (Eds.): LSSC 2007, LNCS 4818, pp. 393–400, 2008.

2 Numerical Model

Our model under study is a pure one-dimensional transport equation defined by the following partial differential equation ([5]):

$$\frac{\partial \mathbf{c}}{\partial t} = -\mathcal{V}\frac{\partial \mathbf{c}}{\partial x}, \quad x \in \Omega = [0, 2\pi], \quad t \in [0, T], \quad \mathbf{c}(x, 0) = f(x). \tag{1}$$

The initial condition f and the space distributed parameter $\mathcal{V}$ are chosen to be $f(x) = \sin(x)$ and $\mathcal{V}(x) = 6x(2\pi - x)/(4\pi^2)$. Then, the exact (analytical) solution of (1) is given by $\mathbf{c}_{exact}(x, t) = \sin(2\pi x/(x + (2\pi - x)\exp(3t/\pi)))$. Details about the numerical aspects and implementation of a data assimilation algorithm for this model are presented in [3,5].

3 POD Reduced Model and POD 4D-Var Assimilation

Basically, the idea of POD method is to begin with an ensemble of data, called *snapshots*, collected from an experiment or a numerical procedure of a physical system. The POD technique is then used to produce a set of basis functions which spans the snapshot collection. When these basis functions are used in a Galerkin procedure, one obtains a finite-dimensional dynamical system with the smallest possible degrees of freedom. For a successful application of the POD 4D-Var in data assimilation problems, it is of most importance to construct an accurate POD reduced model. In what follows, we only give a brief description of this procedure (see [1,2,4]).

For a temporal-spatial flow $\mathbf{c}(x, t)$, we denoted by $\mathbf{c}_1, \ldots, \mathbf{c}_n$ a set adequately chosen in a time interval $[0, T]$, that is $\mathbf{c}_i = \mathbf{c}(x, t_i)$. Defining the mean $\bar{\mathbf{c}} = \frac{1}{n}\sum_{i=1}^{n} \mathbf{c}_i$, we expand $\mathbf{c}(x, t)$ as

$$\mathbf{c}^{POD}(x, t) = \bar{\mathbf{c}}(x) + \sum_{i=1}^{M} \beta_i(t)\Phi_i(x), \tag{2}$$

where $\Phi_i(x)$ – the ith element of POD basis –, and M are appropriately chosen to capture the dynamics of the flow as follows:

1. Compute the mean $\bar{\mathbf{c}} = \frac{1}{n}\sum_{i=1}^{n} \mathbf{c}_i$;
2. Construct the correlation matrix $K = [k_{ij}]$, where $k_{ij} = \int_{\Omega}(\mathbf{c}_i - \bar{\mathbf{c}})(\mathbf{c}_j - \bar{\mathbf{c}})\, dx$;
3. Compute the eigenvalues $\lambda_1 \geq \lambda_2 \geq \cdots \geq \lambda_n \geq 0$ and the corresponding orthogonal eigenvectors $v_1, v_2, \ldots, v_n$ of K;
4. Set $\Phi_i := \sum_{j=1}^{n} v_j^i(\mathbf{c}_i - \bar{\mathbf{c}})$.

Now, we introduce a relative information content to select a low-dimensional basis of size $M \ll n$, by neglecting modes corresponding to the small eigenvalues. Thus, we define the index

$$I(k) = \frac{\sum_{i=1}^{k} \lambda_i}{\sum_{i=1}^{n} \lambda_i}$$

and choose M, such that $M = \text{argmin}\{I(m) : I(m) \geq \gamma\}$, where $0 \leq \gamma \leq 1$ is the percentage of total information captured by the reduced space $X^M = \text{span}\{\Phi_1, \Phi_2, \ldots, \Phi_M\}$. The tolerance parameter γ must be chosen to be near the unity in order to capture most of the energy of the snapshot basis. The reduced order model is then obtained by expanding the solution as in (2).

An atmospheric or oceanic flow $\mathbf{c}(x, t)$ is usually governed by the following dynamic model

$$\frac{d\mathbf{c}}{dt} = F(\mathbf{c}, t), \qquad \mathbf{c}(x, 0) = \mathbf{c}^0(x). \tag{3}$$

To obtain a reduced model of (3), we first solve (3) for a set of snapshots and follow above procedures, then use a Galerkin projection of the model equations onto the space X^M spanned by the POD basis elements (replacing $\mathbf{c}$ in (3) by the expansion (2), then multiplying Φ_i and integrating over spatial domain Ω):

$$\frac{d\beta_i}{dt} = < F\left(\overline{\mathbf{c}} + \sum_{i=1}^{M} \beta_i \Phi_i, t\right), \Phi_i >, \qquad \beta_i(0) = <\mathbf{c}(x, 0) - \overline{\mathbf{c}}(x), \Phi_i(x) > . \tag{4}$$

Equation (4) defines a reduced model of (3). In the following, we will analyze applying this model reduction to 4D-Var formulation. In this context, the forward model and the adjoint model for computing the cost function and its gradient are the reduced model and its corresponding adjoint, respectively.

At the assimilation time interval $[0, T]$, a prior estimate or 'background estimate', $\mathbf{c}_b$ of the initial state $\mathbf{c}_0$ is assumed to be known and the initial random errors $(\mathbf{c}_0 - \mathbf{c}_b)$ are assumed to be Gaussian with covariance matrix $\mathbf{B}$.

The aim of the data asimilation is to minimize the square error between the model predictions and the observed system states, weighted by the inverse of the covariance matrices, over the assimilation interval. The initial state $\mathbf{c}_0$ is treated as the required control variable in the optimization process. Thus, the objective function associated with the data assimilation for (3) is expressed by

$$J(\mathbf{c}_0) = (\mathbf{c}_0 - \mathbf{c}_b)^T \mathbf{B}^{-1}(\mathbf{c}_0 - \mathbf{c}_b) + (\mathbf{H}\mathbf{c} - \mathbf{y}^\circ)^T \mathbf{R}^{-1}(\mathbf{H}\mathbf{c} - \mathbf{y}^\circ). \tag{5}$$

Here, $\mathbf{H}$ is an observation operator, and $\mathbf{R}$ is the observation error covariance matrix.

In POD 4D-Var, we look for an optimal solution of (5) to minimize the cost function $J(\mathbf{c}_0^M) = J(\beta_1(0), \ldots, \beta_M(0))$ given by

$$J(\mathbf{c}_0^M) = (\mathbf{c}_0^{\text{POD}} - \mathbf{c}_b)\mathbf{B}^{-1}(\mathbf{c}_0^{\text{POD}} - \mathbf{c}_b) + (\mathbf{H}\mathbf{c}^{\text{POD}} - \mathbf{y}^\circ)\mathbf{R}^{-1}(\mathbf{H}\mathbf{c}^{\text{POD}} - \mathbf{y}^\circ), \tag{6}$$

where $\mathbf{c}_0^{\text{POD}}$ is the control vector.

In (6), $\mathbf{c}_0^{\text{POD}}(x) = \mathbf{c}_0^{\text{POD}}(x, 0)$ and $\mathbf{c}^{\text{POD}}(x) = \mathbf{c}^{\text{POD}}(x, t)$ are expressed by

$$\mathbf{c}_0^{\text{POD}}(x) = \overline{\mathbf{c}}(x) + \sum_{i=1}^{M} \beta_i(0)\Phi_i(x), \qquad \mathbf{c}^{\text{POD}}(x) = \overline{\mathbf{c}}(x) + \sum_{i=1}^{M} \beta_i(t)\Phi_i(x).$$

Therefore, in POD 4D-Var the control variables are $\beta_1(0), \ldots, \beta_M(0)$. As explained later, the dimension of the POD reduced space could be much smaller than that the original space. As a consequence, the forward model is the reduced model (4) which can be very efficiently solved. The adjoint model of (4) is then used to calculate the gradient of the cost function (6) and that will significantly reduce both the computational cost and the programming effort.

It is important to notice that the initial value of the cost function in the full model space is distinct from the initial value of the cost function in the POD space. Starting with the initial guess given by the background estimate $\mathbf{c}_0 = \mathbf{c}_b$, the value of the cost in the full model space is $J(\mathbf{c}_b)$. The corresponding initial guess in the reduced space is obtained by projecting the background on the POD space X^M.

The POD model in POD 4D-Var assimilation is established by construction of a set of snapshots, which is taken from the background trajectory, or integrate original model (3) with background initial conditions.

4   Computational Issues

This section presents the numerical results of six data assimilation experiments to examine the performance of POD 4D-Var by comparing them with the full 4D-Var. All the performed experiments have used as 'true' (exact) solution $\mathbf{c}_0$ of the assimilation problem, that one computed from the analytical solution $\mathbf{c}_{exact}$ of (1) given in Section 2.

We set in our approach $T = 1$, and used 19 discretization points in space and 60 points in time interval $[0, 1]$. By means of a perturbed initial condition we generated the observed state $\mathbf{y}^o$. We assumed that the correlation matrices in (5) and (6) are diagonal matrices, given by $\mathbf{B} = \sigma_b^2 I$ and $\mathbf{R} = \sigma_o^2 I$, with I denoting the identity matrix of appropriate order. We set $\sigma_b^2 = 0.05^2$ and $\sigma_o^2 = 0.06^2$ representing the variances for the background and observational errors, respectively.

As we already pointed out, the control variables in all these experiments are the initial conditions only. The numerical solution of the optimal control problem was obtained using *fminunc* – the Matlab unconstrained minimization routine. Its algorithm is based on the BFGS quasi-Newton method with a mixed quadratic and cubic line search procedure.

The first experiment is the standard 4D-Var assimilation problem. In the 4D-Var experiment, we applied a preconditioning by the inverse of square root of the background error covariance matrix.

In the second experiment we applied the POD technique with $n = 4$ snapshots. The POD model was constructed in the manner described in Section 3, and the snapshots were taken from the background model results. The eigenvalues λ_k and their corresponding values $I(k)$ are contained in Table 1. Figure 1 illustrates the reconstruction of the first snapshot using the three POD modes of

Table 1. The eigenvalues λ_k and their corresponding values $I(k)$ when using POD technique with 4 snapshots (Experiment 2)

k	1	2	3
Eigenvalue λ_k	2.09e+00	8.21e-01	6.14e-01
$I(k)$ (%)	59.28	82.58	> 99.99

Table 2. The eigenvalues λ_k and their corresponding values $I(k)$ when using POD technique with 10 snapshots (Experiment 3)

k	1	2	3	4	5	6	7
Eigenvalue λ_k	3.21e+00	1.53e+00	1.28e+00	9.97e-01	7.82e-01	7.61e-01	6.61e-01
$I(k)$ (%)	32.28	47.62	60.51	70.53	78.38	86.02	92.66
k	8	9	–	–	–	–	–
Eigenvalue λ_k	4.39e-01	2.93e-01					
$I(k)$ (%)	97.06	> 99.99					

Table 3. The eigenvalues λ_k and their corresponding values $I(k)$ when using POD technique with 15 snapshots (Experiment 4)

k	1	2	3	4	5	6	7
Eigenvalue λ_k	3.51e+00	1.42e+00	1.26e+00	1.23e–01	1.05e–01	8.70e–01	7.88e–01
$I(k)$ (%)	27.67	38.84	48.82	58.52	66.81	73.68	79.90
k	8	9	10	11	12	13	14
Eigenvalue λ_k	7.23e–01	5.33e–01	4.26e–01	3.20e–01	2.60e–01	1.74e–01	1.11e–01
$I(k)$ (%)	85.60	89.81	93.17	95.70	97.75	99.12	> 99.99

Table 4. The eigenvalues λ_k and their corresponding values $I(k)$ when using POD technique with 20 snapshots (Experiment 5)

k	1	2	3	4	5	6	7
Eigenvalue λ_k	4.46e+00	1.67e+00	1.63e+00	1.45e–00	1.35e–00	1.07e+00	9.50e–01
$I(k)$ (%)	26.08	35.85	45.38	53.85	61.76	68.02	73.57
k	8	9	10	11	12	13	14
Eigenvalue λ_k	8.68e-01	7.47e-01	6.63e-01	5.61e-01	5.45e-01	3.82e-01	3.32e-01
$I(k)$ (%)	78.65	83.02	86.89	90.17	93.35	95.59	97.53
k	15	16	17	18	19	–	–
Eigenvalue λ_k	2.45e-01	1.06e-01	7.17e-02	1.25e-08	9.20e-9		
$I(k)$ (%)	98.96	99.58	> 99.99	> 99.99	> 99.99		

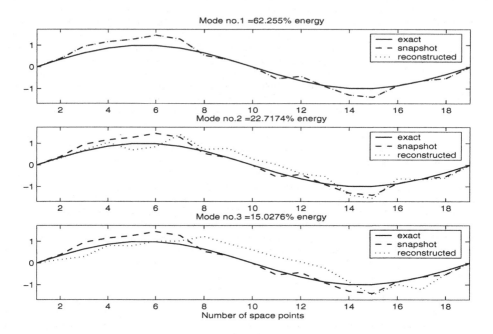

Fig. 1. Reconstruction of the first snapshot using the first three POD modes

this experiment. One can notice that the worst reconstruction is produced when using the third POD mode, since it only captures than 15.076% from the total energy, while the best reconstruction is performed with the first mode (their plots are overposed, becoming indistinguishable).

The experiments 3, 4, 5 and 6 apply the POD 4D-Var techniques, working with 10, 15, 20 and 30 snapshots, respectively. The eigenvalues λ_k and their corresponding values $I(k)$ associated with the experiments 3, 4, and 5 are presented in Tables 2, 3 and 4, respectively. Figure 2 also contains new information about these experiments, namely the relative energy spectrum for POD modes, corresponding to different sets of snapshots of the system state.

We can see from Table 3 that choosing $M = 5$ modes and $n = 15$ snapshots, the captured energy represents 66.81% of the total energy, while when $M = 11$, the captured energy percentage is 95.70%. We also notice from Table 4 that, when we use $M = 7$ modes and $n = 20$ snapshots in POD 4D-Var procedure, one obtains 73.57% captured energy and then, trying to increase the number of modes to $M = 16$ we get 99.58% energy of the dynamical model. The comparative assimilation results asssociated with experiments 4 and 6 and certain selections of M and n are depicted in Figure 3.

As a natural remark, we conclude that better results are obtained when one is using more POD modes and more snapshots.

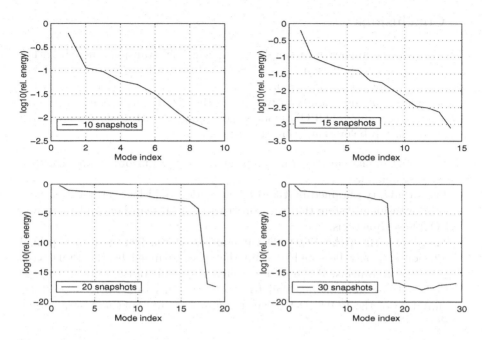

Fig. 2. Relative energy spectrum for POD modes, corresponding to different sets of snapshots for the system state

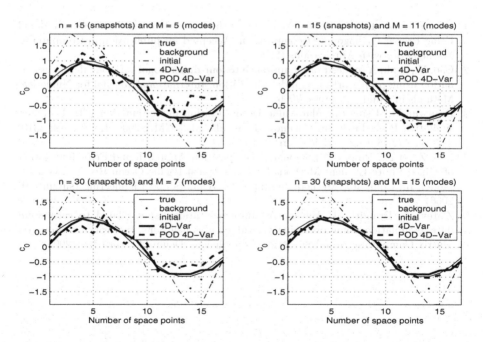

Fig. 3. Comparative results on the assimilation of c_0 using 4D-Var method with full order model and various reduced order models using POD 4D-Var technique

5 Conclusions

In this paper, we applied to a simple one-dimensional transport model, both a standard 4D-Var assimilation scheme and a reduced order approach to 4D-Var assimilation using POD. The approach not only reduces the dimension of the control space, but also significantly reduces the size of the dynamical model. The results from different POD models are compared with that of the original model. We sketch below some conclusions:

- The POD model can accurately approximate the full order model with a much smaller size.
- The variability obtained by the original model could be captured well by a low dimensional system that was constructed by 30 snapshots and 15 leading POD basis functions.
- The drawback of the POD 4D-Var consists of the fact that the optimal solution can only be sought within the space spanned by the POD basis of background fields. When observations lay outside of the POD space, the POD 4D-Var solution may fail to fit observations sufficiently. The above limitation of the POD 4D-Var can be improved by implementing adaptive POD 4D-Var.

References

1. Cao, Y., et al.: Reduced Order Modeling of the Upper Tropical Pacific Ocean Model Using Proper Orthogonal Decomposition. Accepted for publication in Computer & Mathematics with Applications (2006)
2. Cao, Y., et al.: A reduced-order approach to four-dimensional variational data assimilation using proper orthogonal decomposition. International Journal for Numerical Methods in Fluids (in press, 2006)
3. Dimitriu, G.: Using Singular Value Decomposition in Conjunction with Data Assimilation Procedures. In: Boyanov, T., et al. (eds.) NMA 2006. LNCS, vol. 4310, pp. 435–442. Springer, Heidelberg (2007)
4. Luo, Z., et al.: Proper Orthogonal Decomposition Approach and Error Estimation of Mixed Finite Element Methods for the Tropical Pacific Ocean Reduced Gravity Model. In: Computer Methods in Applied Mechanics and Engineering (submitted, 2006)
5. Zlatev, Z., Brandth, J., Havasi, À.: Implementation issues related to variational data assimilation: some preliminary results and conclusions. In: Working Group on Matrix Computations and Statistics, Copenhagen, Denmark, April 1–3 (2005)

Effective Indices for Emissions from Road Transport

Kostadin G. Ganev[1], Dimiter E. Syrakov[2], and Zahari Zlatev[3]

[1] National Institute of Meteorology and Hydrology, Bulgarian Academy of Sciences, 66 Tzarigradsko Chausee, Sofia 1784, Bulgaria
[2] Geophysical Institute, Bulgarian Academy of Sciences, Acad. G. Bonchev, Bl.3, Sofia 1113, Bulgaria
[3] National Environmental Research Institute Frederiksborgvej 399, P.O. Box 358 DK-4000 Roskilde, Denmark

Abstract. The local to regional processes of chemical transformations, washout and dry deposition cannot be directly resolved in global scale models, they rather need to be parameterized. A suitable way to account for the non-linearity, e.g., in chemical transformation processes, is the use of effective emission indices (EEI). EEI translate the actual (small scale) emissions into input for global scale models, partially accounting for unresolved processes occurring shortly after the release of the emissions.

The emissions from the road traffic have some specifics, because of which the concept of deriving EEI from the interaction of an instantaneous plume with the ambient air is perhaps not so convenient. A new parameterization scheme for the EEI from the road transport is suggested in the present paper. Based on few simplifying assumptions and introducing the adjoin equations approach this new scheme makes it possible to achieve unified, not depending on the specific emission pattern, procedure for calculating the EEI from road traffic.

1 Introduction

Transport emissions are released at relatively high concentrations at the exhaust of vehicles. They are then diluted to regional and then to global scales. The local to regional processes of chemical transformations, washout and dry deposition cannot be directly resolved in global scale models, they rather need to be parameterized. A suitable way to account for the non-linearity, e.g., in chemical transformation processes, is the use of effective emission indices (EEI). EEI translate the actual (small scale) emissions into input for global scale models, partially accounting for unresolved processes occurring shortly after the release of the emissions.

The emissions from the road traffic are in a way different from the ship and airplane emissions:

– The road network can be pretty dense in some cells of the large scale model grid;

I. Lirkov, S. Margenov, and J. Waśniewski (Eds.): LSSC 2007, LNCS 4818, pp. 401–409, 2008.

– The emissions are continuous with time;
– The road traffic sources are close to earth's surface, so the pollution deposition processes are important and should not be neglected.

That is why the concept of deriving effective emission indices from the interaction of an instantaneous plume with the ambient air (Paoli R., 2005) is perhaps not so convenient in the case of road transport emissions.

On the other hand, the vertical turbulent transport is a very important process near earth's surface, which means that it is relatively easy to parameterize the vertical structure of the pollution fields and so relegate the considerations to a two-dimensional problem within a layer where the emissions heterogeneity can be important for the nonlinear chemical reactions.

A new parameterization scheme for the EEI from the road transport is suggested in the present paper, based on the concept of small disturbances and introduction of the adjoin equations approach.

2   Vertical Parameterization

Let the road traffic emissions horizontal heterogeneity play significant role on the pollution transport and transformation in the layer $z_0 < z < h$, where z_0 is the roughness parameter and h is a kind of a blending height — a level at which the admixtures are already well mixed horizontally and mesoscale horizontal emission heterogeneity is "forgotten". The blending height h has still to be defined, but as an initial approximation it can be accepted that it coincides with the height of the surface layer (SL). For the sake of simplicity the road traffic emissions will be treated as sources at the earth surface and introduced by the lower boundary condition. In such a case, the transport and transformation equation, vertically integrated within this layer will transform into:

$$\frac{\partial \bar{c}}{\partial t} + L\bar{c} + A + F|_h - F|_0 = 0, \tag{1}$$

where $\bar{c} = \int_{z_0}^{h} c(z)dz$ is the total vertical pollution contents in the layer, $L = \frac{\partial}{\partial x}(u - k_x\frac{\partial}{\partial x}) + \frac{\partial}{\partial y}(v - k_y\frac{\partial}{\partial y})$ is the transport operator, (u, v, k_x, k_y) are the standard denotations of the meteorological variables, averaged in the layer $z_0 < z < h$, A — the term describing sources and sinks due to chemical transformations and possible wash-out by precipitation, $F|_h$ and $F|_0$ are the pollution fluxes trough the upper and lower boundaries.

In order to close the problem $F|_h$ and $F|_0$ have to be expressed as functions of $\bar{c}$. This is relatively easy to do, having in mind, that in the SL the vertical transport is a dominant process, and so the pollution concentration can be treated as locally horizontally homogeneous. Due to the limited volume of the present paper, however, the vertical parameterization will not be discussed in details. It will be mentioned only that it is based on the heavy particles dry deposition parameterization in the surface layer, suggested by Ganev and Yordanov (1981, 2005) and results in the following expressions for $F|_h$ and $F|_0$:

$$- F|_0 = k \frac{dc}{dz}\bigg|_{z_0} + w_g c_0 = (V_{d0} + w_g)c_0 - q \qquad (2)$$

$$= \frac{V_{d0} + w_g}{h - z_0 + V_{d0}J(h)} \bar{c} - \left(1 - J(h)\frac{V_{d0} + w_g}{h - z_0 + V_{d0}J(h)}\right), \qquad (3)$$

$$F|_h = \gamma\left(\bar{c} - \bar{c}^+\right) - w_g c^+, \qquad \gamma = \frac{\gamma'}{h - z_0}, \qquad \bar{c}^+ = (h - z_0)c^+, \qquad (4)$$

where V_{d0} is the dry deposition velocity (absorption coefficient) at earth's surface, $-w_g$, $(w_g > 0)$ is the gravity deposition velocity, q is the capacity of a flat (locally) homogeneous admixture source, c^+ is the admixture concentration at level h, obtained by the large-scale model, $J(h)$ is quite a complex function of the SL turbulent characteristics and the gravity deposition, γ' - a coefficient which has to be derived from the vertical turbulent exchange treatment in the large scale model.

3   On the Road Traffic Effective Emission Indices

3.1   The Concept of Small Disturbances

According to the suggested vertical parameterization the evolution of N interacting compounds in the near surface layer of a region D can described by the system of equations:

$$\frac{\partial \bar{c}_i}{\partial t} + L\bar{c}_i + A_i + B_{ij}\bar{c}_j + \gamma(\bar{c}_i - \bar{c}_i^+) - W_{ij}c_j^+ = E_i, \quad i = 1, \ldots, N, \qquad (5)$$

where $E_i(x, y, t) = \delta_{ij}\left(1 - J_j(h)\frac{V_{d0j} + w_{gj}}{h - z_0 + V_{d0j}J_j(h)}\right)q_j(x, y, t)$, $q_j(x, y, t)$ is the large scale pollution source, $\bar{c}_i(x, y, t)$ and $c_i(x, y, t)$ are the large scale pollution content and the respective concentration averaged in the layer, $c_i^+(x, y, t)$ is the concentration above the layer, $\bar{c}_i^+(x, y, t) = (h - z_0)c_i^+(x, y, t)$, $A_i(c_1, c_2, \ldots, c_N)$ — the term describing sources and sinks of the i-th admixture, due to chemical transformations, $\{B_{ij}\}$ and $\{W_{ij}\}$ are diagonal matrixes describing the large scale absorption by earth's surface and gravity deposition with elements along the main diagonal $B_{ii} = \frac{V_{d0i} + w_{gi}}{h - z_0 + V_{d0i}J_i(h)}$ and $W_{ii} = w_{gi}$ (no summing up for index i in this case), γ is a parameter describing the pollution exchange between the near surface layer and the upper atmosphere. Summing up for repeating indexes is assumed everywhere if anything else is not especially stated.

Accounting for smaller scale effects in the context of introducing some corrections into the large scale model results means mostly taking into account the more detailed pattern of emissions from road traffic and perhaps a more detailed description of the underlying surface (i.e. spatial specification of the absorption coefficients $\{B_{ij}\}$). A more detailed description of the meteorological fields means enhancing the spatial resolution of the model, which is not the task.

The set of parameters in (5) defines the solution:

$$A_i(c_1, c_2, \ldots, c_N), B_{ij}, E_i \rightarrow \bar{c}_1, \bar{c}_2, \ldots, \bar{c}_N.$$

Some small disturbances in the parameters lead to another solution:
$A_i(c_1 + \delta c_1, c_2 + \delta c_2, \ldots, c_N + \delta c_N), B_{ij} + \delta B_{ij}, E_i + \delta E_i \rightarrow \bar{c}_1 + \delta \bar{c}_1, \bar{c}_2 + \delta \bar{c}_2, \ldots, \bar{c}_N + \delta \bar{c}_N, \quad \delta \bar{c}_i = (h - z_0)\delta c_i.$

If the parameter disturbances are small enough, so that $\delta B_{ij}^\delta \bar{c}_j \ll B_{ij}^{\bar{c}_j}$ and

$$A_i(c_1 + \delta c_1, c_2 + \delta c_2, \ldots, c_N + \delta c_N) \approx A_i(c_1, c_2, \ldots, c_N) + \frac{\partial A_i}{\partial c_j}\delta c_j,$$

the problem of the model sensitivity to parameter disturbances can be defined:

$$\frac{\partial \delta \bar{c}_i}{\partial t} + L\delta \bar{c}_j + \alpha_{ij}\delta \bar{c}_j + B_{ij}\delta \bar{c}_j + \gamma \delta \bar{c}_i = \delta E_i - \delta B_{ij}\bar{c}_j, \quad i = 1, \ldots, N, \quad (6)$$

where $\delta c_i(x, y, t)$ and $\delta \bar{c}_i(x, y, t)$ are the concentration and vertical pollution contents disturbances; $\alpha_{ij} = \frac{1}{h-z_0}\frac{\partial A_i}{\partial c_j}$.

If equations (6) are solved for the interval $0 < t < \tau$, τ — the time step of the large scale model, under the initial conditions

$$\delta \bar{c}_i = 0 \text{ at } t = 0, i = 1, \ldots, N, \quad (7)$$

the solution $\delta \bar{c}_i(x, y, \tau)$ will give the one time step small scale corrections to the large scale model results due to the more detailed description of the emission fields and deposition processes.

Equation (6) needs also boundary conditions. From a point of view of the further considerations, the following boundary conditions are convenient:

$$\delta \bar{c}_i = 0 \text{ at } D \text{ boundaries}, i = 1, \ldots, N, \quad (8)$$

From a point of view of mass conservation it is obvious that the emission field disturbances $\delta q_i(x, y, t)$ have to fulfil the relation $\iint_{D^{ml}} \delta q_i dD = 0$, where $D^{ml} = \{x^m < x < x^m + \Delta x, y^l < y < y^l + \Delta y\}$ is the m, l cell of the large scale model grid. It is obvious that if the heterogeneity of the underlying surface within a grid cell is not accounted for (i.e. $\{\delta B_{ij}\} = 0$) the relation $\iint_{D^{ml}} \delta E_i dD = 0$ is also valid.

In the case when the heterogeneity of the underlying surface is accounted for ($\delta V_{d0i} \neq 0$) the large scale pollution sources and absorption coefficients, then their mesoscale disturbances can be defined by the following aggregation technique:

The detailed (accounting for mesoscale heterogeneities) of the source and absorption fields will obviously be:

$$E_i(x, y, t) + \delta E_i(x, y, t) = \delta_{ij}\left(1 - J_j(h)\frac{V_{d0j} + \delta V_{d0j} + w_{gj}}{h - z_0 + (V_{d0j} + \delta V_{d0j})J_j(h)}\right) \quad (9)$$

$$(q_j(x, y, t) + \delta q_j(x, y, t))$$

and

$$B_{ii}(y, y, t) + \delta B_{ii}(y, y, t) = \frac{V_{d0i} + \delta V_{d0i} + w_{gi}}{h - z_0 + (V_{d0i} + \delta V_{d0i})J_i(h)}. \quad (10)$$

Then the large scale fields $E_i(x, y, t)$ and $B_{ii}(y, y, t)$ can be presented in the form:

$$E_i(x, y, t) = E_i^{ml}(t) \cdot \chi^{ml}(x, y), \quad B_{ii}(x, y, t) = B_{ii}^{ml}(t) \cdot \chi^{ml}(x, y),$$

$$E_i^{ml}(t) = \iint\limits_{D^{ml}} (E_i + \delta E_i)\, dD = 0, \quad B_{ii}^{ml}(t) = \iint\limits_{D^{ml}} (B_{ii} + \delta B_{ii})\, dD = 0, \quad (11)$$

$$\chi^{ml}(x, y) = 1, \text{ when } (x, y) \in D^{ml}; \quad \chi^{ml}(x, y) = 0, \text{ when } (x, y) \notin D^{ml}, \quad (12)$$

i.e. calculated not by averaging the values of the dry deposition velocities and emissions, but by averaging the effect. Such a definition by "aggregation" grants that

$$\iint\limits_{D^{ml}} \delta E_i dD = 0 \text{ and } \iint\limits_{D^{ml}} \delta B_{ii} dD = 0. \tag{13}$$

If $\bar{C}_i^{ml} = (h - z_0)C_i^{ml}$ is the columnar pollution contents produced by the large scale model for the m, l grid cell for the moment $t = \tau$ and $\delta \bar{C}_i^{ml} = \frac{1}{\Delta x \Delta y} \iint\limits_{D^{ml}} \delta \bar{c}_i(x, y, \tau) dD$, then the corrected columnar pollution contents for the respective grid and $t = \tau$ will obviously be:

$$\tilde{\bar{C}}_i^{ml} = \bar{C}_i^{ml} + \delta \bar{C}_i^{ml}, \quad i = 1, \dots, N. \tag{14}$$

The same relation is obviously valid for the mesoscale corrections of the concentrations:

$$\tilde{C}_i^{ml} = C_i^{ml} + \delta C_i^{ml}, \quad i = 1, \dots, N, \tag{15}$$

where C_i^{ml} is the concentration produced by the large scale model for the m, l grid cell for the moment $t = \tau$ and $\delta C_i^{ml} = \frac{1}{\Delta x \Delta y} \iint\limits_{D^{ml}} \delta c_i(x, y, \tau) dD = \frac{\delta \bar{C}_i^{ml}}{h - z_0}$.

4   The Adjoin Problem

The functionals $\delta \bar{C}_i^{ml}$ can be calculated when the problem (6–8) is solved for the time period $[0, \tau]$, but there is also another way to obtain it. As the problem (6–8) is linear, the technique of functions of influence can be applied (Marchuck G.I., 1976, 1977, 1982, Penenko V. and A.Aloian, 1985, Ganev K. 2004). In such a case the problem adjoined to (6–8), concerning the i-th admixture and the D^{ml} cell is:

$$-\frac{\partial c_{(i)}{}_k^{ml^*}}{\partial t} + L^* c_{(i)}{}_k^{ml^*} + \alpha_{kj}^* c_{(i)}{}_j^{ml^*} + B_{kj}^* c_{(i)}{}_j^{ml^*} + \gamma c_{(i)}{}_k^{ml^*} = 0, \quad k = 1, \dots, N \tag{16}$$

$$c_{(i)}{}_k^{ml^*} = \frac{1}{\Delta x \Delta y} \chi^{ml}(x, y) \text{ for } k = i \text{ and } c_k^* = 0 \text{ for } k \neq i \text{ at } t = \tau, k = 1, \dots, N \tag{17}$$

$$c_{(i)}{}_k^{ml^*} = 0 \text{ at } D \text{ boundaries, } k = 1, \ldots, N, \tag{18}$$

where $c_{(i)}{}_k^{ml^*}(x, y, t)$ are the functions of influence, $L^* = -\left(u\frac{\partial}{\partial x} + \frac{\partial}{\partial x}k_x\frac{\partial}{\partial x}\right) - \left(v\frac{\partial}{\partial y} + \frac{\partial}{\partial y}k_y\frac{\partial}{\partial y}\right)$.

It is relatively easy to prove, that if $\alpha_{ij}^* = \alpha_{ji}$, $B_{ij}^* = B_{ji}$ the functional $\delta\bar{C}_i^{ml}$ can be also written in the form:

$$\delta\bar{C}_i^{ml} = \int_o^\tau dt \iint_D c_{(i)}{}_k^{ml^*}\left(\delta E_k - \delta B_{kj}\bar{c}_j\right) dD, \tag{19}$$

or if the dependence of $\delta E_k - \delta B_{kj}\bar{c}_j$ on time is not accounted for within a time step:

$$\delta\bar{C}_i^{ml} = \iint_D c_{(i)}{}_k^{\bar{m}l^*}\left(\delta E_k - \delta B_{kj}c_j\right) dD, \quad c_{(i)}{}_k^{\bar{m}l^*} = \int_o^\tau c_{(i)}{}_k^{ml^*} dt. \tag{20}$$

5 Further Simplifications

Even in the form (20) the expression for the mesoscale corrections is not suitable as a basis for a reasonable parameterization. $c_{(i)}{}_k^{\bar{m}l^*}(x, y)$ can be presented by superposition of "plumes", but still the description of the "plumes" in the large scale model code will be a tricky thing, requiring some geometrical considerations and will most probably be quite time consuming.

Therefore it seems reasonable (6), (7) to be replaced by:

$$\frac{\partial\delta\bar{c}_i}{\partial t} + L\delta\bar{c}_j + \alpha_{ij}\delta\bar{c}_j + B_{ij}\delta\bar{c}_j + \gamma\delta\bar{c}_i = 0, \quad i = 1, \ldots, N, \tag{21}$$

$$\delta\bar{c}_i = \tau.(\delta E_i - \delta B_{ij}\bar{c}_j) \text{ at } t = 0, i = 1, \ldots, N. \tag{22}$$

From mass conservation point of view this is an equivalent formulation, but all the disturbances are instantaneously introduced at the beginning of the time step. In such a case the conjugated functions approach leads to the following expression for $\delta\bar{C}_i^{ml}$:

$$\delta\bar{C}_i^{ml} = \tau \iint_D c_{(i)}{}_k^{ml^*}(x, y, 0).(\delta E_k - \delta B_{kj}\bar{c}_j) dD. \tag{23}$$

In a quite natural manner (it's all about processes within one time step τ) the functions of influence at the beginning of the time step can be presented by superposition of instantaneous "puffs", which originate from each point of the grid cell domain D^{ml} and move against the wind with velocity $(-U^{ml}, -V^{ml})$, where U^{ml} and V^{ml} are the (u, v) values of the large scale model for the cell D^{ml}.

Under such assumptions the conjugated functions $c_{(i)}{}_k^{ml*}(x,y,0)$ obtain the form:

$$c_{(i)}{}_k^{ml*}(x,y,t=0) = \int\limits_{x_m}^{x_m+\Delta x} \int\limits_{y_l}^{y_l+\Delta y} c_{(i)}{}_k^{\#}(x,y)\big|_{x_0,y_0}\, dx_0 dy_0, \qquad (24)$$

where

$$c_{(i)}{}_k^{\#}(x,y)\big|_{x_0,y_0} = c_{(i)0k}(0)\, c_{hor}(x,y)\big|_{x_0,y_0}, \qquad (25)$$

$$c_{hor}(x,y)\big|_{x_0,y_0} = \frac{1}{2\pi\sigma_x\sigma_y}\exp\left(-\frac{(x-(x_0-U^{ml}\tau))^2}{2\sigma_x^2} - \frac{(y-(x_0-V^{ml}\tau))^2}{2\sigma_y^2}\right), \qquad (26)$$

$$-\frac{\partial c_{(i)0k}}{\partial t} + \alpha_{kj}^* c_{(i)0j} + B_{kj}^* c_{(i)0j} + \gamma c_{(i)0k} = 0, \quad k=1,\dots,N \qquad (27)$$

$$c_{(i)0k} = \frac{1}{\Delta x \Delta y} \text{ for } k=i \text{ and } c_{(i)0k}=0 \text{ for } k\neq i \text{ at } t=\tau, k=1,\dots,N. \qquad (28)$$

As quantities $c_{(i)0k}$ do not depend on x_0,y_0, the explicit form of integral in (24) can be easily obtained, having in mind (26) and it is:

$$c_{(i)}{}_k^{ml*}(x,y,t=0) = c_{(i)0k}(0)c_{hor}^{ml}(x,y) \quad k=1,\dots,N, \qquad (29)$$

$$\begin{aligned}c_{hor}^{ml}(x,y) = \tfrac{1}{4}&\left(\Phi(\tfrac{x_m+\Delta x-U^{ml}\tau-x}{\sigma_x\sqrt{2}}) - \Phi(\tfrac{x_m-U^{ml}\tau-x}{\sigma_x\sqrt{2}})\right)\\ &\left(\Phi(\tfrac{y_l+\Delta y-V^{ml}\tau-y}{\sigma_y\sqrt{2}}) - \Phi(\tfrac{y_l-V^{ml}\tau-y}{\sigma_y\sqrt{2}})\right),\end{aligned} \qquad (30)$$

where $\Phi(z) = \frac{2}{\sqrt{\pi}}\int_0^z e^{-t^2}dt$ is the well known error function.

Thus, though $N\times N$ adjoined equations have to be solved in order to obtain the necessary set of functions of influence for each cell of the large model grid, the solutions (30) can be obtained in a rather "cheap" way, which makes the suggested approach suitable for making some mesoscale specifications of the large scale model simulations.

Finally, the effective emission corrections can also be calculated. From (21), (22) it is clear that the pollution outflows/inflows from cell D^{ml}, the mesoscale emission disturbances e_i^{ml} for a time step can be calculated by:

$$\begin{aligned}e_i^{ml} = \int_0^\tau dt &\left(\oint_{S^{ml}} (u_n\delta\bar{c}_i - k_n\tfrac{\partial\delta\bar{c}_i}{\partial n})dS + \gamma\iint_{D^{ml}}\delta\bar{c}_i dD\right) = \\ \Delta x \Delta y &\left(\delta\bar{C}_i^{ml}(0) - \delta\bar{C}_i^{ml}(\tau) - (\alpha_{ij}+B_{ij})\int_0^\tau \delta\bar{C}_j^{ml}(t)dt\right)\end{aligned} \quad i=1,\dots,N. \qquad (31)$$

The time integral can be roughly approximated by:

$$\int_0^\tau \delta\bar{C}_i^{ml}(t)dt = \tau\cdot\frac{\delta\bar{C}_i^{ml}(\tau) + \delta\bar{C}_i^{ml}(0)}{2} \quad i=1,\dots,N. \qquad (32)$$

$\sigma = 0.01\Delta$ $\sigma = 0.05\Delta$

$\sigma = 0.1\Delta$ $\sigma = 0.5\Delta$

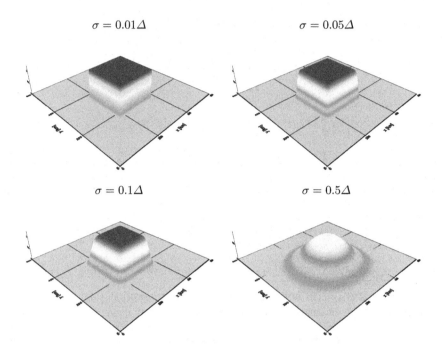

Fig. 1. Plots of c_{hor}^{ml} for different dependence of $\sigma_x = \sigma_y = \sigma$ on $\Delta, \tau|U| = \tau\sqrt{(U^{ml})^2 + (V^{ml})^2} = 0.1\Delta; \Delta x = \Delta y = \Delta = 200km, x_m = 200km, y_l = 200km, \frac{U^{ml}}{|U|} = \frac{V^{ml}}{|U|} = \frac{\sqrt{2}}{2}$

Having in mind that $\iint_{D^{ml}} \delta E_i dD = 0$ and $\iint_{D^{ml}} \delta B_{ij} dD = 0$ it is clear that $\delta \bar{C}_i^{ml}(0) = 0$ and so the expression (32) obtains the form:

$$e_i^{ml} = -\Delta x \Delta y \left(1 + 0.5\tau(\alpha_{ij} + B_{ij})\right) \delta \bar{C}_j^{ml}(\tau) \quad i = 1, \ldots, N. \tag{33}$$

The term $c_{(i)0k}(0)$ in (30) strongly depends on the chemical mechanism applied, but $c_{hor}^{ml}(x, y)$ is very easy to calculate. Some examples of $c_{hor}^{ml}(x, y)$ are given in Fig. 1. The calculations are made for the central cell of the grid shown, varying the horizontal dispersion and the large-scale model velocity.

6 Conclusions

Thus, though $N \times N$ adjoin equations have to be solved in order to obtain the necessary set of functions of influence for each cell of the large model grid, the solutions (31) can be obtained in a rather "cheap" way, which perhaps makes the suggested approach suitable for parameterization of the EEI for emissions from the road transport, thus introducing some mesoscale specifications of the large scale model simulations.

Acknowledgements

The present work is supported by EC through 6FP NoE ACCENT (GOCE-CT-2002-500337), IP QUANTIFY (GOGE-003893) and COST Action 728.

References

1. Ganev, K., Yordanov, D.: Parametrization of pollution from a heavy admixture source in the surface air layer. Compt. rend. Acad. bulg. Sci. 34(8), 1261–1264 (1981)
2. Ganev, K.: Functions of influence and air pollution models sensitivity. Compt. Rend. Acad. Bulg. Sci. 57(10), 23–28 (2004)
3. Ganev, K., Yordanov, D.: Parameterization of dry deposition processes in the surface layer for admixtures with gravity deposition. Int. J. Environment & Pollution 25(1–4), 60–70 (2005)
4. Marchuk, G.I.: The environment and some optimization problems (in Russian). Docl. AN USSR 226, 1056–1059 (1976)
5. Marchuk, G.I.: Methods of computational Mathematics. Science, Moscow (1977)
6. Marchuk, G.I.: Mathematical Modeling in Environmental Problems (in Russian). Science, Moscow (1982)
7. Paoli, R.: Review of computational methods for modelling Effective Emission Indices (2005)
8. Penenko, V., Aloian, A.: Models and methods for the environmental protection problems (in Russian). Science, Novosibirsk (1985)

On the Numerical Solution of the Heat Transfer Equation in the Process of Freeze Drying

K. Georgiev, N. Kosturski, and S. Margenov

Institute for Parallel Processing, Bulgarian Academy of Sciences
Acad. G. Bonchev, Bl. 25A, 1113 Sofia, Bulgaria
`georgiev@parallel.bas.bg`, `kosturski@parallel.bas.bg`
`margenov@parallel.bas.bg`

Abstract. The coupled process of vacuum freeze drying is modelled by a system of nonlinear partial differential equations. This article is focused on the submodel of heat and mass transfer in the absorption camera. The numerical treatment of the related parabolic partial differential equation is considered. The selected numerical results well illustrate the specific issues of the problem as well as some recently obtained results. Brief concluding remarks and prospectives for future investigations are given at the end.

Keywords: vacuum freeze drying, zeolites, heat and mass transfer, parabolic PDE, finite element method, MIC(0) preconditioning.

1 Introduction

The vacuum freeze drying is a process in which the water contents is removed from a frozen material by sublimation. The modelled dehydration phase starts from a frozen state. A strongly reduced pressure is maintained in the camera where the self-frozen substance is placed. Under certain technology conditions, the iced water is incrementally sublimated (directly transformed to vapor), leaving a fine porous structure favorable to re-hydration.

During the drying process a high level of vacuum is ensured and some (controlled) heat is supplied to keep a stable sublimation. The necessary amount of heat can be determined using the molecules' latent heat of sublimation.

The sublimated water molecules are absorbed in a s separate camera (absorption camera) which could be considered as dual (supporting) process to the primal process of freeze during. This part of the coupled process is governed by the material's absorption isotherms.

The freeze drying causes less damages of the substance than other (more classical) dehydration methods using higher temperatures. In particular, the freeze drying does not cause shrinkage or toughening of the material being dried. Complementary (which could be even more important), flavors and smells generally remain unchanged, making the process favorable for preserving of foods, spices, herbs, etc.

Zeolite's granules are used to absorb the sublimated water molecules. The zeolites are special type of silica–containing materials which have a strongly porous

I. Lirkov, S. Margenov, and J. Waśniewski (Eds.): LSSC 2007, LNCS 4818, pp. 410–416, 2008.

structure that makes them valuable as absorbents and catalysts. They are used in different environmental problems, e.g. for cleaning water, reducing methyl bromide emissions, etc. Let us mention, that not only natural zeolites but also synthetic ones can be used in the process of freeze drying. Both, natural and synthetic zeolites, have a substantially three–dimensional rigid crystalline structure (similar to honeycomb), consisting of a network of interconnected tunnels and cages [9].

The mathematical model of the coupled process of freeze drying is described by a system of time–depending partial differential equations. The model has a well established hierarchical structure. A splitting procedure according to the involved technological processes is applied. The objective of this article to study the process in the absorption camera which is modelled by a heat conduction equation. The nonlinearity of this submodel is represented by the right-hand-side depending on the unknown temperature.

The rest of the paper is organized as follows. In Section 2 the mathematical model is discussed. The numerical treatment of the related parabolic partial differential equation is presented in Section 3. Results of some representative numerical tests are given in Section 4. Finally, short concluding remarks and plans for future investigations are given in Section 5.

2   The Model and Its Numerical Treatment

Te process in the absorption camera is described by a two dimensional heat conduction equation [3]. It is a parabolic partial differential equation of the form:

$$c\rho \frac{\partial u}{\partial t} = Lu + f(u, x, t), \qquad x \in \Omega, \ t > 0, \tag{1}$$

where

$$Lu = \sum_{i=1}^{d} \frac{\partial}{\partial x_i} \left(k(x, t) \frac{\partial u}{\partial x_i} \right). \tag{2}$$

The following notations are used in (1) and (2):

- $u(x, t)$ – unknown distribution of the temperature;
- d – dimension of the space ($d = 2$ in this study);
- $\Omega \subset R^d$ – computational domain (see Fig. 3);
- $k = k(x, t) > 0$ – heat conductivity;
- $c = c(x, t) > 0$ – heat capacity;
- $\rho > 0$ – material density;
- $f(u, x, t)$ – right–hand side function, which represents the nonlinear (depending on the temperature) nature of heating.

The following initial (3) and boundary (4) conditions are assigned to the parabolic equation (1):

$$u(x, 0) = u_0(x), \qquad x \in \Omega, \tag{3}$$

$$u(x, t) = \mu(x, t), \qquad x \in \Gamma \equiv \partial\Omega, \quad t > 0. \tag{4}$$

The Crank–Nicolson time stepping scheme is used for numerical solution of the discussed initial boundary value problem. The Finite Element Method (FEM) with linear triangle elements (so called Courant linear finite elements) is applied for discretization in space of (1). The computational domain is divided in triangle elements (see Fig. 3). The triangulation is performed by the freely available computer mesh generator *Triangle* [6]. The choice of *Triangle* is based on the comparisons analysis of the following three mesh generators: *Triangle, NETGEN* [8] and *Gmsh* [7], see for more details in [5]. The resulting meshes are highly unstructured. The following mesh generating options are used: *minimal angle of a triangle* (the most often used value is 30°) and *maximal area of a triangle* (different for the different subdomains and depending on the geometry).

Let us denote with K and M the FEM stiffness and mass matrices. Then, the parabolic equation (1) is transferred in the following semi–discrete matrix form:

$$M\frac{du}{dt} + Ku = F(t). \tag{5}$$

If we denote with τ the time step, with u^{n+1} the unknown solution (temperature) at the current time level, with u^n the already computed solution at the previous time level, and if use central difference approximation of the time derivative in (5), we will obtain the following system of linear algebraic equations for the values of u^{n+1}:

$$\left(M + \frac{\tau}{2}K\right)u^{n+1} = \left(M - \frac{\tau}{2}K\right)u^n + \frac{\tau}{2}(F^{n+1} + F^n). \tag{6}$$

The system (6) is solved iteratively by the Preconditioned Conjugate Gradient (PCG) method [1]. The Modified Incomplete Cholesky (known as MIC(0)) preconditioner [2] is applied in the PCG framework.

3   Programming and Numerical Tests

The programming language **C++** is used for computer implementation of the above discussed numerical algorithm. The included numerical tests are performed under *Linux* operational system. Various experiments with preliminary known exact solutions were run in order to fix the mesh and time parameters of the computer model. A specially designed experimental laboratory apparatus was used (see Fig. 1) to evaluate the reliability and robustness of the developed computer model.

The product to be dried is placed in the left container. In the studied case this container is a *glass flask* (2 mm depth of the bottom and 1 mm depth of the walls), and the processed product is *grated carrots* (60 grams) (see Fig. 1a)). The right container is the absorption camera (in the laboratory apparatus it is a similar *glass flask* – 2 mm depth of the bottom and 1 mm depth of the walls, where 500 grams zeolites granules are put (see Fig.1b)).

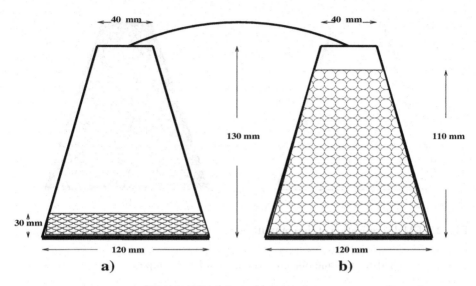

Fig. 1. The laboratory experiment

We study the temperature field in the absorption camera via numerical solution of the parabolic problem (1). The values of the related material coefficients are given in Table 1. The right-hand-side of the differential equation represents the intensity of the water molecules entering the absorption camera. The released heat depends on the current temperature (nonlinear nature of the process) and is strongly varying in space and time.

The computational domain consists of three very different subdomains concerning their physical characteristics: glass, zeolites granules and vacuum. The time to end the process of drying T is an input parameter (computed by a separate submodel of the integrated computer system). In the reported numerical experiments, $T = 67\ 930$ seconds, i.e. $t \in [0, T]$. The time step is set to $\tau = 10$ seconds corresponding to the relevant accuracy requirements. Let us note once again, that all parameters are tuned for the particular laboratory experiments. The computer tests were performed on Pentium IV, 1.5 GHz processor. Some basic characteristics of the computer model (mesh parameters) are given in Table 2.

Table 1. Coefficients of the parabolic problem

Subdomain	$k(x, y, t)$	$c(x, y, t)$	$\rho(x, y, t)$
glass	$0.75\ W\,m^{-1}\,K^{-1}$	$0.84\ kJ\,kg^{-1}\,K^{-1}$	$2600\ kg\,m^{-3}$
zeolite	$0.38\ W\,m^{-1}\,K^{-1}$	$1.226\ kJ\,kg^{-1}\,K^{-1}$	$760\ kg\,m^{-3}$
vacuum	$2.8\,10^{-6}\ W\,m^{-1}\,K^{-1}$	$1.012\ kJ\,kg^{-1}\,K^{-1}$	$1.2\,10^{-3}\ kg\,m^{-3}$

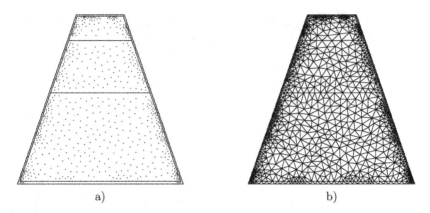

Fig. 2. FEM discretization of the computational domain. a) grid nodes; b) grid triangles.

Table 2. Some output parameters of the computer model

Number of elements	Number nodes	CPU time (in seconds)	No. iteration (total)	No. iteration (per time step)
2064	1173	123.87	61137	9
4244	2170	255.53	76867	11.3
6577	3365	393.82	84388	12.4

The number of PCG iterations indicates the robustness of the developed solver while the CPU time is representative for the total computational efficiency (scalability) of the code. The first row in the table corresponds to the initial triangulation of the computational domain (the mesh generator *Triangle* is used). The next two rows are related to two refined triangulations. The number of nodes (FEM degrees of freedom) N is consecutively increased by factors of of 1.84 and 1.55 at each refinement step.

It is easy to observe that the number of PCG iterations (per time step) behaves as $O(N^{1/4})$. This is the best result which could be expected based on the MIC(0) preconditioning theory. Moreover, the theory covers only some model cases on regular discretization grids while our meshes are unstructured.

The observed scalability of the CPU times is even better than the predicted by the theory (and/or corresponding to the number of iterations). The obtained favorable results could be additionally explained by the proper data locality as well as by the quality of the developed code.

The computed temperature field at the end of the time interval ($T = 67\,930$ seconds) is shown on Fig. 3.

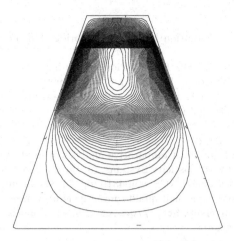

Fig. 3. The temperature field in the absorption camera

4   Concluding Remarks

The process of vacuum freeze drying is modelled by a coupled system of nonlinear partial differential equations. It contains: a) a time dependent non-linear balance equation describing the moving interface between dried and iced subregions of the product (to be dried) layer; b) a parabolic equation describing the heat and mass transfer in absorption camera. The model has a hierarchical structure, and therefore the natural way for solving the problem is to use a splitting according to the technological processes involved. The process in the absorption camera (box) is considered in the present paper. The Crank–Nicolson method is used for the numerical solution of the problem. The computational domain is discretized using the mesh generator *Triangle*. Some results from the computer test of the numerical algorithm for an experimental laboratory apparatus were presented and discussed. The behavior of the numerically computed temperature field well fits the behavior of the available experimental measurements.

The following improvements and future developments of the computer model are planned: a) further tuning of the boundary conditions and the related enlargement of the computational domain; b) development of accuracy preserving adaptive time stepping procedure; c) development of robust computer model in polar coordinates.

Acknowledgment

The research reported in this paper was partly supported by the Bulgarian IST Center of Competence in 21st century – BIS-21++funded by the European Commission in FP6 INCO via Grant 016639/2005.

References

1. Axelsson, O.: Iterative Solution Methods. Cambridge University Press, Cambridge (1996)
2. Gustafsson, I.: An incomplete factorization preconditioning method based on modification of element matrices. BIT 36(1), 86–100 (1996)
3. Jafar, F., Farid, M.: Analysis of heat and mass transfer in freeze–drying. Drying Technology 21(2), 249–263 (2003)
4. Kincaid, D., Cheney, W.: Numerical Analysis: Mathematics of Scientific Computing. Thomson Learning (2002)
5. Kosrurski, N., Margenov, S.: Comparative analysis of mesh generators and MIC(0) preconditioning of FEM Elasticity Systems. In: Boyanov, T., et al. (eds.) Numerical Methods and Applications. LNCS, vol. 4310, pp. 74–81. Springer, Heidelberg (2007)
6. TRIANGLE, a two dimensional quality mesh generator and Delainey triangulator, Carnegie Mellon University, http://www.cs.cmu.edu/~quake/triangle.html
7. http://geuz.org/gmsh/
8. http://www.hpfem.jku.at/netgen/
9. http://www.stcloudmining.com/what-is-zeolite.html

Results Obtained with a Semi-lagrangian Mass-Integrating Transport Algorithm by Using the GME Grid[*]

Wolfgang Joppich[1] and Sabine Pott[2]

[1] University of Applied Sciences Bonn-Rhein-Sieg, 53757 Sankt Augustin, Germany
[2] Fraunhofer Institute for Algorithms and Scientific Computing, Schloß Birlinghoven, 53754 Sankt Augustin, Germany

Abstract. The defect of mass by actual schemes within the GME has motivated the implementation of a conservative semi-Lagrangian scheme on an icosahedral mesh. This scheme for the GME-mesh is unique. The properties are demonstrated by applying the new algorithm to typical test cases as they can be derived from the shallow water test suite. The results using coarse grids show: this scheme is extremely attractive for a climate version of the GME. Even for periods longer than those as specified in the SWE test suite, a defect of mass in the range of accuracy of the machine is obtained. The relative error results show that the order of accuracy of the proposed algorithm is two.

1 Introduction

The global weather forecast model (GME) of the German weather service (DWD) uses an icosahedral mesh: created by recursively refining the initial icosahedron embedded into a sphere [1].

For the GME, the horizontal dynamics of the prognostic variables such as specific moisture, specific water content and ice content of clouds, and ozone mixing ratio is described by a semi-Langrangian advection scheme. This type of schemes is well-established for solving the equations of oceanic and atmospheric flow. One reason is that, compared to Eulerian approaches, there is no severe stability condition, such as the CFL condition [2]. Systematic mass deficiencies of the GME have caused the implementation of an alternative semi-Langrangian scheme on the icosahedral grid. This finally leads to a reduced defect of mass even for long periods of simulation and when using coarse meshes. Additional advantages of the method are shape preservation and almost no numerical dispersion.

The actually implemented method represents the transported quantities such as moisture or geopotential in shallow water equations by a quadratic Hermite polynomial for each grid cell, which are spherical triangles. The method does not only use values within points of the triangle but also requires values at the center position of the boundary arcs, calculated by cubic Hermite interpolation.

---

[*] This work has been funded by the German Research Foundation (DFG) under grant number Jo 554/1-1 and Jo 554/1-2 within SPP 1167.

I. Lirkov, S. Margenov, and J. Waśniewski (Eds.): LSSC 2007, LNCS 4818, pp. 417–424, 2008.

The mass-conserving modification of this interpolation scheme still uses interpolation for values in the corner points of the triangle. But interpolation at the centers of the edges is replaced by conditions which the polynomial has to satisfy in order to conserve both the prescribed mass integral within the spherical triangle and the prescribed variation of the mass integral along two of the edges.

The algorithm is described and applied to such test cases [3] where the exact solution is known, because this allows to determine the error. The advantages of the new method are demonstrated in Sect. 4.

2   Theoretical Basis of the Algorithm

A grid cell $\triangle$ of an icosahedral grid is a spherical triangle with corner points p_1, p_2, p_3. c_i, $i \epsilon \{1, 2, 3\}$ are the edges of $\triangle$. Let $V(t, x)$ describe the wind at time t at a point x on the sphere. We further assume that the wind field is a differentiable function. If $x(t)$ with $t, s \in I$ is the solution of the ordinary first order differential equation

$$\dot{x}(t) = V(t, x(t))$$
$$x(s) = x_0$$

with x_0 on the sphere, we call x the trajectory or integral curve of V. The mass of h over a measurable set U is given by

$$\iint_U h \, dM,$$

where dM is the area element of the sphere.

Now, select $\Delta t > 0$ such that for a grid point x_{grid} of an icosahedral grid the integral curve x with $I = [t, t + \Delta t]$ and

$$\dot{x}(\tilde{t}) = V(\tilde{t}, x(\tilde{t}))$$
$$x(t + \Delta t) = x_{grid}$$

does exist. The time sequence of points then leads to the description of $x(t)$ as departure point and x_{grid} as the arrival point of the integral curve.

Semi-Lagrangian methods are characterized by approximating such trajectory elements $x[t, t + \Delta t]$ which begin or end in a grid point [4]. In the following Δt is chosen such that all departure points are defined. Let $\Delta t > 0$, $\triangle$ is a typical grid cell, and $c_i, i \epsilon \{1, 2, 3\}$ are the edges of $\triangle$. We then refer to the inverse image of c_i with respect to the flux $\phi_{t,s}$ of V as the departure edge of c_i. They compose the departure triangle of $\triangle$. The orientation of the boundary of $\triangle$ canonically creates an orientation of the boundary of the departure triangle.

The fundamental equation upon which this semi-Langragian transport algorithm is based is the integral-differential form of the continuity equation [5,6]. At each time t_0 and for each measurable set U we have

$$\left. \frac{d}{dt} \right|_{t_0} \iint_{\phi_{t,s}(U)} h(t, x) \, dM = 0. \tag{1}$$

Here d/dt describes the total derivative with respect to time. Since

$$m(t, s, U) := \iint_{\phi_{t,s}(U)} h(t, x)\, dM$$

represents the mass at time t over the stream-invariant area $\phi_{t,s}(U)$, equation (1) represents an advective equation for the mass, and it states that the mass advected by the flux remains constant.

For $U = \triangle$ and $s = t + \Delta t$ it follows from (1) that

$$\iint_{\triangle} h(t + \Delta t, x)\, dM = \iint_{\phi_{t,t+\Delta t}(\triangle)} h(t, x)\, dM \qquad (2)$$

or $m(t + \Delta t, t + \Delta t, \triangle) = m(t, t + \Delta t, \triangle)$, which represents the equation of mass transport which has to be numerically solved for each grid cell at each discrete time step [7].

3   Implementation Issues for the GME Library

In this section we describe implementation aspects for the shallow water version of the GME.

3.1   Data Structure

The grid cells are numbered and local coordinates are introduced. The origin of this local coordinate system is the center of the respective triangle. This allows a parameterization of the geodesic arcs and numerical integration along these edges. For each triangle, the oriented boundary is a sequence of geodesic arcs. All the required variables, functions and functionals for each grid cell, edges, and corner points, respectively, are stored within a particularly introduced data type. Among these details are:

1. A real valued polynomial of second order in local coordinates

$$q(\eta, \chi) = a_0 + a_1 \chi + a_2 \eta + a_3 \chi^2 + a_4 \chi \eta + a_5 \eta^2,$$

   representing the distribution of transported quantities on the triangle.
2. The mass functional.
3. For each edge of the triangle the variation of mass.
4. The δ-functional of the corner points.

All conditions for these quantities finally result in a 6 x 6 system of equations. At the very beginning of all calculations, a highly accurate LU decomposition of this matrix is calculated. This LU decomposition is the left hand side of the equations to be solved for each time step for the coefficients of the interpolation polynomial. The system to be solved only depends on the geometry of the triangle and on the selected space of approximating functions. Therefore it remains constant for the entire prediction period.

3.2   Departure Grid

The departure grid consists of all the departure points of all icosahedral mesh points. Its edges are the departure edges, and its cells are the departure triangles. In general, departure edges do not represent great circle segments. Nevertheless, since numerically only the departure points are known, and departure grid and icosahedral grid are identical for $\Delta t = 0$, the departure edges are approximated by great circle segments. The intersection of an approximated departure triangle and a grid cell is a spherical polygon, unless empty. For each time step, the mass of departure triangles is calculated by summing up all mass integrals over such polygons.

3.3   Calculation of Mass and Its Variation in a Departure Triangle

Due to Stoke's theorem the integration over the above mentioned polygonial area is converted into line integrals of all boundary segments. The polygon boundary consists of departure edge segments and grid cell boundary segments. The segments are referred to as inside arc, and boundary arcs, respectively. Integration of mass and its variation is performed along the inside arcs using Gaussian quadrature formula.

3.4   Values of the Transported Property h at $t + \Delta t$

Given the advective form of the continuity equation

$$\frac{dh}{dt} + h\, divV = 0, \tag{3}$$

with dh/dt the total time derivative of h along the trajectory, it follows for a wind field with $divV = 0$ that the new value of $h(t + \Delta t)$ at a grid point is given by the value of $h(t)$ at the departure point. To calculate this quantity we have to locate the grid cell of the departure point and to evaluate the Hermitian polynomial $q(\xi, \eta)$ for h at time t at the departure point.

   If $divV \neq 0$ the new value of $h(t + \Delta t)$ at the grid point results from adding an approximate value for the integral

$$-\int h\, divV dt \tag{4}$$

along the trajectory to the interpolated value at the departure point (which is the right hand side of the δ-functional). After composing the right-hand sides of all equations, the LU solver delivers the coefficients of the quadratic polynomial at time $t + \Delta t$.

4   Two Test Cases for the New Scheme

The first test case of the Williamson Test Suite [3], a cosine bell advecting around the globe, has been the basis for our similar test cases. We consider the rotation of an inverted cosine bell across the poles by a non-divergent wind field (**A**) and the longitudinal transport of the cosine bell by a meridional wind field with $divV \neq 0$ (**B**).

The test cases have been applied for a time period of 10 to 18 days. With h_{num} the numerical solution and h_T the exact solution we are able to determine the relative global error in maximum-norm, the relative defect of mass, and the height error of the minimum. These quantities are given by

$$\frac{\| h_{num} - h_T \|_\infty}{\| h_T \|_\infty}, \quad \frac{mass(h_{num}) - mass(h_T)}{mass(h_T)}, \quad \frac{|\min h_{num} - \min h_T|}{|\Delta h(0)|},$$

respectively, where $\Delta h(0) = max\, h_T - \min h_T$ is the extension of the initial state at initial time $t = 0$. Similarly, the height error of the maximum is defined. For test case **A**, the last quantity describes the non-physical oscillations, see Fig. 3.

4.1   (A) Rotation of the Inverted Cosine Bell

In Cartesian coordinates (x, y, z), the wind field $V(x, y, z)$ is given by a tangential vector field of a single parameter rotational group around the y-axis. Its amplitude is chosen such that each point of the sphere has completed one revolution of 2π in 12 days. The geopotential (transported quantity) is described by a function h on the sphere, which takes the form of an inverted cosine bell. Except for only a few points h is infinitely differentiable. In these points, only the first derivative of h is continuous. This lack of smoothness is the reason for non-physical oscillations when using polynomial interpolation in an advection scheme. In Fig. 1 (left) the bell is plotted on a geographical grid. The coordinate lines at the left and right boundary represent the south pole and the north pole, respectively.

We have applied the above scheme to the horizontal transport of mass for icosahedral grids with different resolutions, specified by ni = 16, 32, and 48, which give the number of subintervals of an initial icosahedral arc. Although the wind field is non-divergent, we have performed a discretization of the full continuity equation at each time step. The simulated time period covers 18 days. The global error, Fig. 1 (right) is small. It can be recognized, Fig. 2 (left), that the implemented method for test case (**A**) is of second order at least. For an icosahedral grid with ni = 48, the implementation is mass-conserving up to machine accuracy, Fig. 2 (right). Using very coarse grids (for example ni = 16), we observe no mass conservation, but the loss of mass is extremely small. An explanation is the fact that, for test case (**A**), the departure edges are identical with edges of the icosahedral grid (boundary arcs), but departure arcs are numerically treated in a different way (see comment above).

Dispersive effects are not visible. The reduction of the non-physical oscillations is not yet of second order, but close to, Fig 3. Although the wiggles are low, we will pay special attention to a further reduction of them. Nakamura [7] emphasizes the nonoscillatory features of rational interpolation.

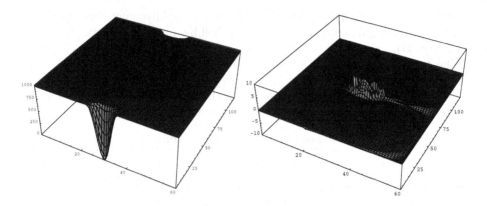

Fig. 1. (A) Initial state, left; global error, ni = 48, right

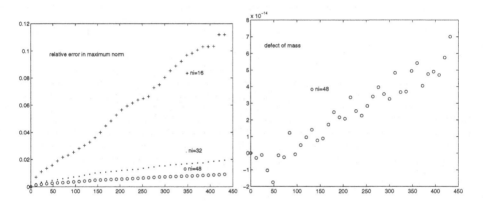

Fig. 2. (A) Relative error in maximum-norm, ni=16, 32, 48 (top-down), left; defect of mass, ni=48, right

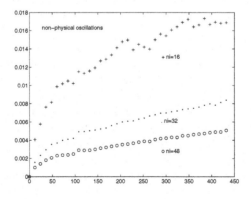

Fig. 3. (A) Measure of non-physical oscillations, ni = 16, 32, and 48 (top-down)

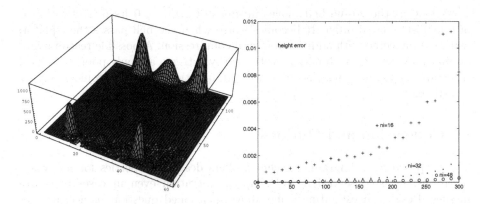

Fig. 4. (B) Deformation of the cosine bell, ni = 48, different positions, left; Height error, ni=16, 32, 48 (top-down), right

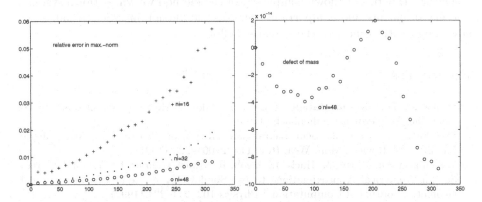

Fig. 5. (B) Relative error in maximum-norm, ni=16, 32, 48 (top-down), left; defect of mass, ni=48, right

4.2 (B) Transport of Mass by a Divergent Meridional Wind Field

This test case was generated to investigate the behavior of our numerical scheme in a wind field which alters the shape of the transported quantity. In a region of the globe with positive divergence of the wind, the shape of the cosine bell is flattened and broadened, whereas in regions with negative divergence, the cosine bell is narrowed and steepened. The relative error Fig. 5 (left), the non-physical oscillations, Fig. 4 (right), and the defect of mass, Fig. 5 (right), show the same quantitative behavior as for case (**A**). Figure 4 (left) shows the numerical solution at different time steps. The three states of the bell there represent the initial state, the bell close to the equator, and the bell approaching the south-pole (here on the same grid). Any transport of mass even closer to the north pole would force an increased concentration of the bell into a smaller region until the grid resolution would be no longer appropriate for a reasonable representation of the peak.

As long as the cosine bell moves in areas of $divV \geq 0$ its discretization is of (alomost) second order. It becomes worse when the bell passes the equator, which is connected with additional lateral compression. A possible reason might be that for coarse grids, functions with nearby extrema are only poorly modeled by quadratic polynomials. The answer to this question will be the topic of further investigation.

5   Conclusion and Outlook

The proposed Semi-Lagrangian scheme offers desired properties for many applications: mass-conservation and shape-preservation even in cases of coarse meshes. Besides investigating some above mentioned questions concerning algorithmic details we are implementing this method into a version of the GME which is close to the production code of the DWD. By this, a parallelization is obligatory. Due to the known complexity of the method we will perform systematic measurements and comparisons with the old version in order to judge the advantages with respect to the numerical effort.

References

1. Majewski, D.: Documentation of the new Global Model GME of the DWD, Deutscher Wetterdienst, Offenbach (1996)
2. Staniforth, A., Côté, J.: Semi-Lagrangian Integration Schemes for Atmospheric Models — A Review. Mon. Wea. Rev. 119, 2206–2223 (1991)
3. Williamson, D., Drake, J., Hack, J., Jacob, R., Swartztrauber, P.: A Standart Test Set for Numerical Approximations to the Shallow Water Equations in Spherical Geometry. Journal of Computational Physics 102, 211–224 (1992)
4. Temperton, C., Staniforth, A.: An efficient two-time level semi-Lagrangian semi-implicit Integration Scheme. Quart. J. Royal Meteorological Society 113, 1025–1039 (1987)
5. Laprise, J., Plante, A.: A Class of Semi-Lagrangian Integrated-Mass Numerical Transport Algorithms. Monthly Weather Review 123, 553–656 (1995)
6. Chorin, A., Marsden, J.E.: A Mathematical Introduction to Fluid Mechanics, 3rd edn. Springer, New York (1992)
7. Nakamura, T., Tanaka, R., Yabe, T., Takizawa, K.: Exactly Conservative Semi-Lagrangian Scheme for Multi-dimensional Hyperbolic Equations with Directional Splitting Technique. Journal of Computational Physics 174, 171–207 (2001)

The Evaluation of the Thermal Behaviour of an Underground Repository of the Spent Nuclear Fuel

Roman Kohut, Jiří Starý, and Alexej Kolcun

Institute of Geonics, Academy of Sciences of the Czech Republic,
Studentská 1768, 70800 Ostrava-Poruba, Czech Republic
kohut@ugn.cas.cz, stary@ugn.cas.cz, kolcun@ugn.cas.cz

Abstract. The paper concerns the evaluation of the thermal behaviour of an underground repository of the spent nuclear fuel where the canisters are disposed at a vertical position in the horizontal tunnels. The formulation of thermo-elastic problems should regard the basic steps of the construction of the repository. We tested the influence of the distance between the deposition places on the thermo-elastic response of the rock massif. The problems are solved by the in-house GEM-FEM finite element software. One sided coupling allows a separate solution of the temperature evolution and the computation of elastic responses only in predefined time points as a post-processing to the solution of the heat equations. A parallel solution of the arising linear systems by the conjugate gradient method with a preconditioning based on the additive Schwarz methods is used.

1 Introduction

Management of high-level, long-lived radioactive waste is an important issue today for all nuclear-power-generating countries. The deep geological disposal of these wastes is one of the promising options. The design of a safe underground depository of a spent nuclear fuel (SNF) from nuclear power stations requires careful study of the repository construction, reliability of the protecting barriers between SNF and the environment and study of all kinds of risks related to the behaviour of the whole repository system. For the assessment of the repository performance, it is fundamental to be able to do large-scale computer simulations in various coupled processes as heat transfer, mechanical behaviour, water and gas flow and chemical processes in rocks and water solutions. Generally, we speak about T-H-M-C processes and their modelling.

The T-H-M-C processes are generally coupled and a reliable mathematical modelling should respect at least some of the couplings. In this paper, we restrict to the modelling of T-M processes with one-directional T-M coupling via the thermal expansion term in the constitutive relations. Thus the problem can be divided in two parts. Firstly, the temperature distribution is determined by the solution of the nonstationary heat equation, secondly, at given time points the

I. Lirkov, S. Margenov, and J. Waśniewski (Eds.): LSSC 2007, LNCS 4818, pp. 425–432, 2008.
© Springer-Verlag Berlin Heidelberg 2008

linear elasticity problem is solved. The numerical solution of both the problems leads to a repeated solution of large systems of linear equations and our aim is to find efficient and parallelizable iterative solution methods.

Mathematically, the thermoelasticity problem is concerned with finding the temperature $\tau = \tau(x, t)$ and the displacement $u = u(x, t)$,

$$\tau : \Omega \times (0, T) \to R, \, u : \Omega \times (0, T) \to R^3$$

governed by the following equations

$$\kappa \rho \frac{\partial \tau}{\partial t} = k \sum_i \frac{\partial^2 \tau}{\partial x_i^2} + q(t) \quad in \quad \Omega \times (0, T), \tag{1}$$

$$-\sum_j \frac{\partial \sigma_{ij}}{\partial x_j} = f_i \quad (i = 1, \ldots, 3) \quad in \quad \Omega \times (0, T), \tag{2}$$

$$\sigma_{ij} = \sum_{kl} c_{ijkl} \left[\varepsilon_{kl}(u) - \alpha_{kl}(\tau - \tau_0) \right] \quad in \quad \Omega \times (0, T), \tag{3}$$

$$\varepsilon_{kl}(u) = \frac{1}{2} \left(\frac{\partial u_k}{\partial x_l} + \frac{\partial u_l}{\partial x_k} \right) \quad in \quad \Omega \times (0, T), \tag{4}$$

together with the corresponding boundary and initial conditions.

2   Numerical Methods

The initial-boundary value problem of thermo-elasticity (1)–(4) is discretized by finite elements in space and finite differences in time. Using the linear finite elements and the time discretization, it leads to the computation of vectors $\underline{\tau}^j$, $\underline{u}^j$ of nodal temperatures and displacements at the time levels $t_j, j = 1, N,$ with the time steps $\Delta t_j = t_j - t_{j-1}$. It gives the following time stepping algorithm:

> *find* $\underline{\tau}^0$: $M_h \underline{\tau}^0 = \underline{\tau}_0$, $\underline{u}^0$: $A_h \underline{u}^0 = b^0 = b_h(\underline{\tau}^0)$,
> *for j=1,...,N:*
>   *find* $\underline{\tau}^j$: $B_h^{(j)} \underline{\tau}^j = [M_h + \theta \Delta t_j K_h] \underline{\tau}^j = c^j$,
>   *find* $\underline{u}^j$: $A_h \underline{u}^j = b^j$.
> *end for*

Remark: The system

$$\underline{u}^j : A_h \underline{u}^j = b^j \tag{5}$$

we solve only in predefined time points.

Above, M_h is the capacitance matrix, K_h is the conductivity matrix, A_h is the stiffness matrix, $c_j = [M_h - (1 - \theta)\Delta t_j K_h]\underline{\tau}_{j-1} + \Delta t_j \phi_j$, $\phi_j = \theta q(t_j) + (1 - \theta)q(t_{j-1})$, $b^j = b_h(\underline{\tau}^j)$ and $\underline{\tau}_0$ is determined from the initial condition. Here parameter $\theta \in \{0, 0.5, 1\}$. It means that in each time level we have to solve the system of linear equations

$$[M_h + \theta \Delta t_j K_h]\underline{\tau}^j = [M_h - (1 - \theta)\Delta t_j K_h]\underline{\tau}^{j-1} + \Delta t_j \phi_j. \tag{6}$$

For $\theta = 0$ we obtain the explicit Euler scheme, for $\theta = 1$ we obtain the backward Euler (BE) scheme, $\theta = 0.5$ gives the Crank-Nicolson (CN) scheme. In our case we will use the BE scheme. If we substitute $\underline{\tau}^j = \underline{\tau}^{j-1} + \Delta\underline{\tau}^j$ into (6), we obtain the system of equations for the increment of temperature $\Delta\underline{\tau}^j$,

$$[M_h + \Delta t_j K_h]\Delta\underline{\tau}^j = \Delta t_j(\underline{q}_h^j - K_h\underline{\tau}^{j-1}), \tag{7}$$

where $\underline{q}_h^j = q(t_j)$.

To ensure accuracy and not waste the computational effort, it is important to adapt the time steps to the behaviour of the solution. We use the procedure based on local comparison of the backward Euler (BE) and the Crank-Nicolson (CN) scheme [1]. We solve the system (6) only using BE scheme. If this solution $\underline{\tau}^j = \underline{\tau}^{j-1} + \Delta\underline{\tau}^j$ is considered as the initial approximation for the solution of system (6) for $\theta = 0.5$ (CN scheme), then the first iteration of Richardson's method presents an approximation of the solution of the system (6) for $\theta = 0.5$. Thus $\underline{\tau}_{CN}^j \cong \underline{\tau}^j - r^j$, where

$$r^j = (M_h + 0.5\Delta t_j K_h)\underline{\tau}^j - (M_h - 0.5\Delta t_j K_h)\underline{\tau}^{j-1} - 0.5\underline{q}_h^j - 0.5\underline{q}_h^{j-1}. \tag{8}$$

The time steps can be controlled with the aid of the ratio $\eta = \frac{\|r^j\|}{\|\underline{\tau}^j\|}$. If $\eta < \varepsilon_{min}$ then we continue with time step $\Delta t = 2 \times \Delta t$, if $\eta > \varepsilon_{max}$ then we continue with time step $\Delta t = 0.5 \times \Delta t$, where $\varepsilon_{min}, \varepsilon_{max}$ are given values.

For the solution of the linear system $B_h\Delta\underline{\tau}^j = (M_h + \Delta_j K_h)\Delta\underline{\tau}^j = f_j$ (7) we shall use the preconditioned CG method where the preconditioning is given by the additive overlapping Schwarz method. In this case the domain is divided into m subdomains Ω_k. The nonoverlaping subdomains Ω_k are then extended to domains Ω'_k in the way that overlapping between the subdomains are given by two or more layers of elements. If B'_{kk} are the FE matrices corresponding to problems on Ω'_k, I'_k and $R'_k = (I'_k)^T$ are the interpolation and restriction matrices, respectively, then introduced matrices $B'_{kk} = R'_k B I'_k$ allow to define the one-level additive Schwarz preconditioner G,

$$g = Gr = \sum_{k=1}^{m} I'_k {B'_{kk}}^{-1} R'_k r.$$

Note that for the parabolic problems it is proved in [2] that under the assumption that Δ_j/H^2 is reasonably bounded, the algorithms based on one-level additive Schwarz preconditioning remain numerically scalable. Here Δ_j is in order of the time stepsize and H is the diameter of the largest subdomain.

3   Model Example

The model example comes out from the depository design proposed in [3] (see Figure 1). The whole depository is very large, but using symmetry we can solve the problem only on the part of the domain. The model domain contains three

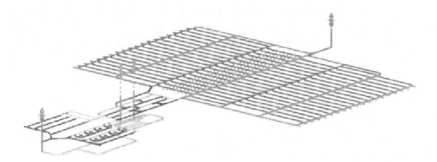

Fig. 1. The global design of depository

depository drifts (a half of the drift) , each with four 1.32 m diameter, 4.77 m deep deposition holes and one access drift. The heating canisters (0.67 m diameter, 3.67 m length) simulating the heating from the radioactive waste are emplaced in the holes. The highest allowable temperature on the surface between the canister and the bentonit is restricted to $100°C$. We solve six variants (A, B, C, D, E, F) which differ in the distance d_h between the holes (from 2.5 m for the variant A to 15 m for the variant F). The whole model domain is situated 800m under surface. A constructed $3D$ T-M model of repository is shown in Figure 2.

The computation domain is enlarged with increasing distance between the holes from dimensions $23.50 \times 58.00 \times 99.77$ m with FE grid $145 \times 105 \times 50$ nodes (614250 DOF for the heat problem, 1842750 DOF for the elasticity problem) for the variant A to dimensions $66.62 \times 58.00 \times 99.77$ m with FE grid $285 \times 105 \times$

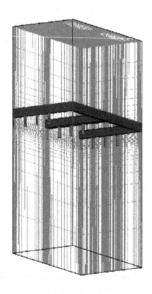

Fig. 2. Finite element mesh for repository model

Table 1. Material properties

	E (MPa)	ν	density (kg/m³)	conductivity (W/m°C)	capacity (J/kg°C)	expansion (1/°C)
granite	66500	0.23	2700	2.7	850	4.4 10⁻⁶
concrete	37000	0.30	2800	1.5	880	8.0 10⁻⁶
bentonite	30	0.30	2000	1.0	2500	3.0 10⁻⁶
steel	210000	0.30	7800	460.0	45	0.0
SNF	210000	0.30	7000	460.0	45	0.0

50 nodes (1496250 DOF for the heat problem, 4488750 DOF for the elasticity problem) for the variant F.

The thermal source is given by the radiactive waste. The power of SNF in the canister decays exponentialy in time according to formula determined from given data by MATLAB

$$q(t) = 14418.6(e^{-0.18444(t+t_c)} + 0.2193e^{-0.019993(t+t_c)} + 0.02376e^{-0.0006659(t+t_c)})$$

Here t_c presents the cooling time depending on the burn-up value of the fuel. In our case we suppose two possibilities for the canister power. In the first case the canister power C_p is 1500 W when disposed (this power is reached after $t_c = 50.0$ years pre-cooling time), in the second case the canister power C_p is 1600 W when disposed (this power is reached after $t_c = 42.16$ years pre-cooling time). Canister power is a very important parameter because the canister spacing can be reduced, if the power decreases. The materials are assumed to be isotropic, the mechanical properties do not change with the temperature variations.The thermal conductivity k and thermal expansion α of the rock are also assumed to be isotropic (see Table 1).

The boundary conditions for the mechanical parts consist of zero normal displacements and zero stresses on all outer faces except of the upper one. For the thermal part, we assume zero heat flux on all outer faces except of the bottom one, where the original rock temperature is given. On the faces of the drifts we suppose the heat transfer with the parameter $H = 7 \ W/m^2 \ {}^\circ C$, the temperature of air in the drifts is supposed to be constant in time and is equal to $27^\circ C$. The original temperature of rocks is determined by using geothermal gradient. This temperature also gives the initial condition.

The computations are done in four subsequent phases:

- the phase of virgin rocks — the initial stresses are determined from the weights of rocks, the initial temperature is determined using the geothermal gradient
- the drifts are excavated. The elasticity problem is solved using equivalent forces on the faces of drifts initiated by the excavation. The nonstationary heat problem is solved for period of 10 years with the initial condition determined in the phase 1 and with the heat transfer on the faces of drifts
- the deposition holes are excavated. The elasticity problem is solved using equivalent forces on the faces of holes

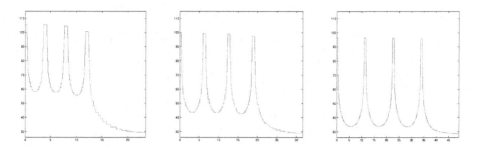

Fig. 3. The temperature on the line parallel with the drifts crossing the center of the canisters for $d_h = 2.5, 5.0,$ and 10.0 meters — the time is 1.6 year $(C_p = 1500$ W)

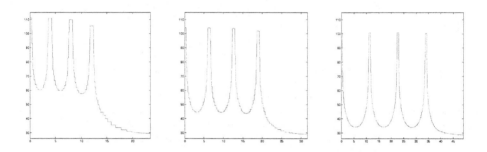

Fig. 4. The temperature on the line parallel with the drifts crossing the center of the canisters for $d_h = 2.5, 5.0,$ and 10.0 meters — the time is 1.6 year $(C_p = 1600$ W)

- the thermoelasticity problem is solved for the period of 200 years with the initial condition given by the temperature computed in the phase 2.

The highest temperature is encountered after about 1.6 years of deposition for both the cases $(C_p = 1500$ W, 1600 W). The results for the first case for the variants A, B, and D $(d_h = 2.5, 5.0,$ and 10.0) are shown in Figure 3. The results for the second case for the variants A, B and D are shown in Figure 4. Note that the figures present the behaviour of the temperature on the line parallel with the drifts crossing the center of the canisters. We can see that in the first case $(C_p = 1500$ W) the distance $d_h = 5.0$ m is sufficient to fulfil the restriction for the temperature on the surface of canister. In the second case $(C_p = 1600$ W) we can situate the holes in the distance $d_h = 10$ m.

Remark: The distance between drifts is supposed to be 25 metres. The results of our tests showed that the canisters deposition in one drift practically do not influence the temperature in the neighbouring drifts.

From the groundwater solute transport modelling point of view the knowledge of the stress field is very important. In Figure 5 we present the behaviour of the shear stress intensity for the first case $(C_p = 1500$ W).

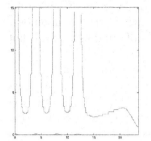 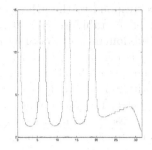 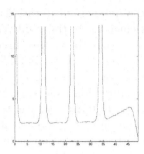

Fig. 5. The shear stress intensity on the line parallel with the drifts crossing the center of the canisters for $d_h = 2.5, 5.0$, and 10.0 meters — the time is 1.6 year ($C_p = 1500$ W)

Table 2. The numbers of iterations for the domain division in various directions

material	homogeneous			non-homogeneous		
direction	$x\ y\ z$	$x\ y\ z$	$x\ y\ z$	$x\ y\ z$	$x\ y\ z$	$x\ y\ z$
nbr of subdomains	1 1 3	1 3 1	3 1 1	1 1 3	1 3 1	3 1 1
nbr of iterations	153	162	267	354	354	394
nbr of subdomains	1 1 6	1 6 1	6 1 1	1 1 6	1 6 1	6 1 1
nbr of iterations	177	250	346	370	541	434

For the solution of the linear systems (5) and (6) we used the preconditioned CG methods with preconditioning given by the additive overlapping Schwarz method. The linear systems were solved in parallel. The parallel computations were performed on:

- the IBM xSeries 455 computer (symmetric multiprocessor (SMP), 8 processors) with Intel Itanium2 1.3 GHz 64bit Processor, 16 GB shared memory
- the PC cluster THEA with 8 AMD Athlon 1.4 GHz, 1.5 GB RAM computer nodes.

The parallel programming uses:

- OpenMP and MPI paradigms on SMP computer,
- MPI paradigm on the PC cluster.

The division of the domain to subdomains influences the efficiency of the preconditioning given by the additive overlapping Schwarz method if the materials are strong anisotropic or the material parameters have big jumps or the grid is anisotropic (narrow elements). We tested this efficiency in the variant B. In this case the averaged hexahedral element has dimensions $0.22 \times 0.55 \times 2.04$ m and the material parameters have big jumps on the canisters (see Table 1). Table 2 presents the numbers of PCG iterations for one timestep ($\Delta t_j = 10, \varepsilon = 10^{-6}$) , if the division to three or six subdomains in direction x, y or z is done. On the left part of the table we present results for the homogeneous case (we suppose that all materials have the same properties as granit). We can see that in this case the numbers of iterations correspond to the averaged dimensions of hexagonal elements. On the right part of the table we present results for the nonhomogenous case. If we use three subdomains, the boundaries of subdomains are not cutting the canisters and the numbers of iterations correspond to the averaged dimensions of elements. In the case of division to six subdomains the division in the direction x does not cut the canisters and the division in the direction y cuts the canisters directly in the centre. This fact distinctively influences the numbers of iterations. Therefore it's necessary to improve the code to enable the using of irregular division of the domain, which can guarantee that the boundaries of subdomains will not cut the areas with jumps of material parameters.

4   Conclusion

In the paper, the model problem of geological depository of the spent nuclear fuel is solved. We compare the results of the solution for various distances of the deposition holes. We tested the efficiency of the DD preconditioner from the point of the dependence on the division of the domain.

Acknowledgments

The work was supported by the Ministry of Education, Youth and Sports under the project 1M0554 and by the Academy of Sciences of the Czech Republic through the project No. 1ET400300415.

References

1. Blaheta, R., Byczanski, P., Kohut, R., Starý, J.: Algorithm for parallel FEM modelling of thermo–mechanical phenomena arising from the disposal of the spent nuclear fuel. In: Stephansson, O., Hudson, J.B., Jing, L. (eds.) Coupled Thermo-Hydro-Mechanical-Chemical processes in Geo-systems, Elsevier, Amsterdam (2004)
2. Cai, X.-C.: Additive Schwarz algorithms for parabolic convection-diffusion problems. Numer. Math. 60, 41–62 (1990)
3. Vavřina V.: Reference project of the underground and overground parts of the deep depository (in czech), SURAO 23–8024–51–001/EGPI444–990 009 (1999)

Study of the Pollution Exchange between Romania, Bulgaria, and Greece

Maria Prodanova[1], Dimiter Syrakov[1], Kostadin Ganev[2], and Nikolai Miloshev[2]

[1] National Institute of Meteorology and Hydrology, Bulgarian Academy of Sciences, 66 Tzarigradsko chausee, Sofia 1784, Bulgaria
[2] Geophysical Institute, Bulgarian Academy of Sciences, Acad. G. Bonchev, Bl.3, Sofia 1113, Bulgaria

Abstract. US EPA Models-3 system is used for calculating the exchange of ozone pollution between three countries in southeast Europe. For the purpose, three domains with resolution 90, 30, and 10 km are chosen in such a way that the most inner domain with dimensions 90×147 points covers entirely Romania, Bulgaria, and Greece.

The ozone pollution levels are studied on the base of three indexes given in the EU Ozone Directive, mainly AOT40c (Accumulated Over Threshold of 40 ppb for *crops*, period May-July), NOD60 (Number Of Days with 8-hour running average over 60 ppb), and ADM (Averaged Daily Maximum). These parameters are calculated for every scenario and the influence of each country emissions on the pollution of the region is estimated and commented.

Oxidized and Reduced Sulfur and Nitrogen loads over the territories of the three countries are also predicted. The application of all scenarios gave the possibility to estimate the contribution of every country to the S and N pollution of the others and detailed blame matrixes to be build.

Comparison of the ozone levels model estimates with data from the EMEP monitoring stations is made. The calculated data were use to draw several important conclusions.

1   Introduction

Regional studies of the air pollution over the Balkans, including country-to-country (CtC) pollution exchange, had been carried out for quite a long time [2,15,10,11,17,5,6]. These studies were focused on both studying some specific air pollution episodes and long-term simulations and produced valuable knowledge and experience about the regional to local processes that form the air pollution pattern over Southeast Europe.

The present paper will focus on some results which give an impression on the CtC regional scale pollution exchange (see also [13]). The simulations performed are oriented towards solving two tasks — CtC study of ozone pollution levels in the region and CtC study of sulphur and nitrogen loads in the region.

I. Lirkov, S. Margenov, and J. Waśniewski (Eds.): LSSC 2007, LNCS 4818, pp. 433–441, 2008.

2   Modeling Tools

The US EPA Model-3 system was chosen because it appears to be one of the most widely used modelling tools with proved simulation abilities. The system consists of three components: **MM5** — the 5th generation PSU/NCAR Meso-meteorological Model [7,12]; **CMAQ** — the Community Multiscale Air Quality System [3]; **SMOKE** — the Sparse Matrix Operator Kernel Emissions Modelling System [4].

3   Model Configuration and Brief Description of the Simulations

3.1   Model Domains

As far as the base meteorological data for 2000 is the **NCEP Global Analysis Data** with $1° \times 1°$ resolution, it was necessary to use MM5 and CMAQ nesting capabilities as to downscale to 10 km step for a domain over Balkans. The MM5 pre-processing program TERRAIN was used to define three domains with 90, 30, and 10 km horizontal resolution. They were chosen in such a way that the finest resolution domain contains Bulgaria, Romania, and Greece, entirely.

3.2   MM5 Simulations

First, MM5 was run on both outer grids (90 km and 30 km resolution) simultaneously with "two-way" nesting mode on. Then, after extracting the initial and boundary conditions, MM5 was run on the finer 10 km grid as a completely separate simulation with "one-way" nesting mode on. The MM5 model possesses four dimensional data assimilation option (FDDA) able to relax toward observed temperature, wind and humidity [14].

3.3   Emission Input to CMAQ

Two inventory files (for 30 and 10 km domains) were prepared exploiting the EMEP 50×50 km girded inventory [16] and its 16.67 km desaggregation described in [1]. The grid-values of the 30 km domain were obtained by bi-linear interpolation over the 50 km EMEP inventory. The values for the inner 10 km grid were obtained in the same way but over desaggregated inventory. Additional corrections were included for congruence between both inventories. These inventory files contain the annual emission rates of 5 generalized pollutants — SO_x, NO_x, VOC, NH_3 (ammonia), and CO for every grid cell of both domains.

CMAQ input emission files were prepared for the period May 1 – July 31, 2000 for the two inner domains. The inventory files were handled by a specially prepared computer code E_CMAQ that performs two main processes. First, the pollutant groups were speciated to the compounds, required by CB-IV chemical mechanism, following the way recommended in [18]. The next procedure in E_CMAQ is the over-posing of proper time profiles (annual, weekly, and daily).

The methodology developed in USA EPA Technology Transfer Network was adopted. As far as in the used gridded inventory the type of sources is not specified, some common enough area sources were chosen from the EPA SCC (Source Category Code) classification and their profiles were averaged, the resulting profiles implemented.

It must be stressed that the biogenic emissions of VOC were not estimated in E_CMAQ and this fact should be taken into consideration when interpreting model results.

3.4 CMAQ Simulations

From the MM5 output CMAQ meteorological input was created exploiting the CMAQ meteorology-chemistry interface — MCIP. The CB-4 chemical mechanism with Aqueous-Phase Chemistry was used. The CMAQ pre-defined (default) concentration profiles were used for initial conditions over both domains at the beginning of the simulation. The concentration fields obtained at the end of a day's run were used as initial condition for the next day. Default profiles were used as boundary conditions of the 30-km domain during all period. The boundary conditions for the 10-km domain were determined through the nesting capabilities of CMAQ.

Four emission scenarios were prepared: basic scenario with all emission sources (scenario **All**), scenario with Bulgarian emissions set to zero (**noBG**), scenario with Romanian emissions set to zero (**noRO**), and scenario with Greek emissions set to zero (**noGR**).

4 Comparison of the Ozone Simulations with Measurements

The number of background stations monitoring the ozone concentration is quite limited in the region. There are only 3 such stations belonging to EMEP monitoring network that used to operate all the year 2000. These are the Greek stations GR02-Finokalia and GR03-Livadi and the station K-puszta, located in the upper left corner of the 10-km modelling domain, in Hungary.

It occurs that during all the year 2000 two more stations were monitoring the ozone concentrations in Bulgaria in the frame of a research project . These stations are BG02-Rojen and BG03-Ahtopol.

Because of the lack of space the plots of measured and modelled hourly ozone values can not be fully demonstrated here (see Fig. 1 as an example). It can be stated that the agreement between measurements and simulations is as good (or as bad) as in many similar studies. The variations of ozone values as calculated by CMAQ are much more regular than the measurements. The possible reason is that one and the same mean temporal profile is over-posed on the anthropogenic emissions and that there are no biogenic emissions. In the future different temporal profiles would be used, at least, for the sources of different categories. And, of course, biogenic emissions must be accounted for.

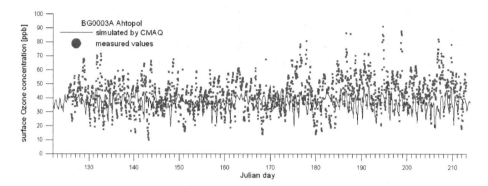

Fig. 1. CMAQ Ozone calculations *vs.* measurements — Ahtopol station

Generally speaking, CMAQ produces lower concentrations than the measurements almost everywhere with lower diurnal amplitude. The reason for these shortcomings is in the input data, mainly the emission data. Nevertheless, the results of calculation contain valuable information about space distribution of surface ozone and can be used for solving different tasks, as can be seen further.

5   Pollution Exchange in the Balkan Region

5.1   Ozone Pollution Levels and Analysis of the CtC Exchange

High ozone concentrations can cause damages on plants, animals, and human health. In fact, when the effects from high ozone levels were studied, one should look not at the ozone concentrations but on some related quantities. The following quantities are important: **AOT40** — accumulated over threshold of 40 ppb in the day-time hours during the period from May 1 to July 31 concentrations, which are damaging crops when they exceed 3000 ppb.hours; **NOD60** — number of days in which the running 8-hour average over ozone concentration exceeds at least once the critical value of 60 ppb. If the limit of 60 ppb is exceeded in at least one 8-hour period during a given day, then the day must be classified as "bad". People with asthmatic diseases have difficulties in "bad" days. Therefore, it is desirable not to have "bad" days at all. Removing all "bad" days is a too ambitious task. The requirement is often relaxed to the following: the number of "bad" days should not exceed 20. It turns out that in many European regions it is difficult to satisfy even this relaxed requirement.

The calculated AOT40 fields are given in Fig. 2. As already mentioned they are scaled by the threshold of 3000 ppb.hours and transformed in percents. It can be seen from the first plot in Fig. 2, where the scaled AOT40 field obtained by the **All**-scenario is presented, that this index is less than the threshold in almost all land territories. Only over western shore of the Balkan Peninsula values over 100 % can be seen. In the graph that shows the **noBG** scenario one can see that switching off Bulgarian sources leads to considerable decrease of this index not only over the territory of the country itself but over the European part of

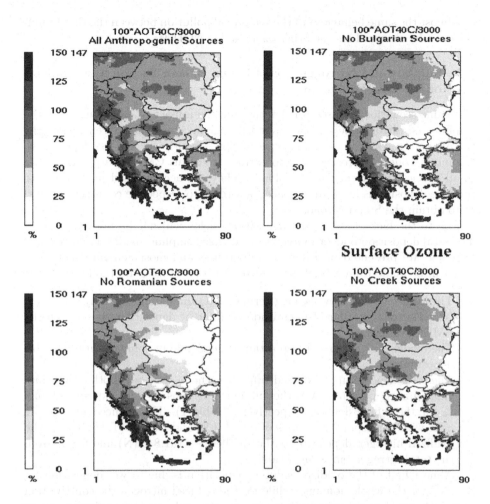

Fig. 2. AOT40C values normalized by the threshold of 3000 ppb.hours. for May-July 2000.

Turkey and northern Greece, i.e. 30–50 % of ozone pollution in these areas is due to ozone precursors (NO_x and VOC) emitted by Bulgarian sources. The fact that the AOT40 over Romania is almost not influenced by the elimination of Bulgarian emissions is due to the prevailing NW transport of air masses in the domain. From its side, Romania contribute essentially to the ozone pollution not only in Bulgaria and Moldova but even in Turkey and part of northern Greece, as can be seen from the lower left graph in Fig. 2 (Scenario **noRo**). The last graph in Fig. 2 shows the results of scenario **noGR**. Excluding the emissions over the Greek territory it decreases the ozone pollution mainly in the country itself. The decrease in some areas is 50–75 %. Only European Turkey is influenced to some extent by Greek NO_x and VOC pollution.

Almost the same behaviour of the reciprocal pollution between the three countries can be observed for the other ozone index NOD60.

5.2 Sulphur and Nitrogen Loads in the Balkan Region and Analysis of the CtC Exchange

The same scenarios are used for determining the impact of each country in the sulphur and nitrogen depositions over the region. Apart the concentration, the CMAQ output contains two main types of files — hourly dry and wet depositions. Computer code was created able to extract from this CMAQ output the fields of all species that form S and N loads. They are accumulated in time (as to produce depositions fields) and space (according to respective countries' masks) as to obtain the S and N loads.

Due to the lack of space, the depositions fields will not be presented, here. The total deposition budget matrixes for oxidized sulphur, oxidized and reduced nitrogen, are presented in Tables 1–3. The first and most general conclusions that can be made from a brief view of the fields and tables are the following:

- The loads calculated by the long-term CMAQ simulations are fully consistent (in terms of the order of magnitude of different deposition types) with the EMEP evaluations;
- The oxidized nitrogen wet deposition is negligible in comparison to the dry deposition;
- The total deposition of oxidized sulphur is the biggest one, both as absolute value and as percents from the sulphur emissions — 348.3 kt(S) and 46.6 % for the whole domain respectively (Table 1). The reduced nitrogen is next with corresponding values of 60.2 kt(N) and 34.5 % (Table 3) and the oxidized nitrogen deposition is the smallest — 101.8 kt(N) and 5.8 % from the total nitrogen emissions (Table 2).
- Almost half of the oxidized sulphur deposition is due to wet deposition ($\approx$ 43% for the whole domain); while for the reduced nitrogen the contribution of the wet deposition is much less — $\approx$ 30% of the total deposition.

The country to country pollution exchange can be followed from the tables. They present the emitter-receiver relations for 4 sub-domains in which the

Table 1. Blame matrix for oxidized sulphur, May-July 2005, 1000 t(S)

Emitter Receiver	BG	GR	RO	other	all sources
BG	14.772	0.381	3.021	5.838	31.717
GR	3.612	4.230	0.795	6.281	9.058
RO	3.278	0.200	18.203	16.652	28.251
other	10.055	4.246	6.232	64.605	93.376
deposited	31.717	9.058	28.251	93.376	162.401
total emission [S]	72.425	6.708	68.374	180.789	348.297
% Rec/Emit	43.792	33.913	41.318	51.649	46.627

Table 2. Blame matrix for oxidized nitrogen, May-July 2005, 1000 t(N)

Emitter Receiver	BG	GR	RO	other	all sources
BG	1.065	0.011	0.137	0.150	1.375
GR	0.077	0.890	0.000	0.458	1.015
RO	0.108	0.000	1.916	0.481	2.202
Other	0.125	0.114	0.149	4.496	5.584
Deposited	1.375	1.015	2.202	5.584	10.176
total emission [N]	18.317	20.406	44.210	91.430	174.363
% Rec/Emit	7.506	4.976	4.981	6.107	5.836

Table 3. Blame matrix for reduced nitrogen, May-July 2005, 1000 t(N)

Emitter Receiver	BG	GR	RO	other	all sources
BG	2.402	0.146	0.873	1.131	3.995
GR	0.351	2.314	0.230	1.549	3.672
RO	0.360	0.010	8.391	2.617	11.487
other	0.882	1.203	1.993	17.653	22.949
deposited	3.995	3.672	11.487	22.949	42.103
total emission [N]	18.317	20.406	44.210	91.430	174.363
% Rec/Emit	21.809	17.996	25.983	25.100	24.147

domain of integration is divided: Bulgaria (BG), Romania (RO), Greece (GR), and the other countries in the region (other). The impact of each country's sulphur and nitrogen emissions to the wet, dry and total depositions in these countries themselves and in the other countries is clearly demonstrated. The diagonal elements show the deposition quantity for each country due to its own sources.

The last three rows of the tables show respectively the total quantity deposited in the country in the column headers, the total quantity emitted by this country and the percentage of deposited quantities from the emitted. The last values in the tables can be treated as the relative part of sulphur/nitrogen that remains in the domain. The percents vary from 5–7.5 % for oxidized nitrogen trough 25–35 % for reduced nitrogen total deposition up to 34–52 % for oxidized sulphur total deposition.

The analysis of different scenarios of switching off the emissions from Bulgaria, Greece or Romania shows, that the impact is most prominent on the sulphur or nitrogen loads in the respective country. The impact on the neighbouring countries however can also be significant. For example it can be seen that the exclusion of Bulgarian sources leads to substantial decrease of oxidized sulphur loads in northeast Greece and European Turkey. The exclusion of Romanian sources also leads to substantial decrease of oxidized sulphur loads in northeast Bulgaria.

6   Conclusions

The main conclusions that can be made from this tentative study of the transboundary transport and transformation of air pollutants over the Balkans for the summer of year 2000 are the following:

1.) The comparison of the simulated results with measured data from the background stations in the region showed reasonable agreement and the loads calculated by the long-term CMAQ simulations are fully consistent (in terms of the order of magnitude of different deposition types) with the EMEP evaluations, which is quite an encouraging result.
2.) Enriching the number of background stations in the region will soundly contribute to understanding the mechanisms of regional scale transport and transformation of pollutants over Southeast Europe and to more reliable evaluation of CtC pollution exchange. The model results will be better verified by using such an enriched network of measurement stations.
3.) The emission inventories in the Balkan region need to be significantly more detailed both in spatial resolution and especially in temporal evolution — annual and diurnal course. Proper evaluation of the natural biogenic emissions is also very important for reliable ozone level simulations.

Acknowledgements

The present work is supported by EC through 6FP NoE ACCENT (GOCE-CT-2002–500337), IP QUANTIFY (GOGE–003893), COST Action 728, as well as by the Bulgarian National Science Council (ES-1002/00). Deep gratitude is due to US EPA, US NCEP and EMEP for providing free-of-charge data and software.

References

1. Ambelas Skjøth, C., Hertel, O., Ellermann, T.: Use of the ACDEP trajectory model in the Danish nation-wide Background Monitoring Programme. Physics and Chemistry of the Earth 27, 1469–1477 (2002) (Pergamon Press)
2. BG-EMEP 1994, 1995, 1996, 1997: Bulgarian contribution to EMEP, Annual reports for 1994, 1995, 1996, 1997, NIMH, EMEP/MSC-E, Sofia-Moscow
3. Byun, D., Ching, J.: Science Algorithms of the EPA Models-3 Community Multiscale Air Quality (CMAQ) Modeling System. EPA Report 600/R-99/030, Washington DC (1999)
4. CEP, Sparse Matrix Operator Kernel Emission (SMOKE) Modeling System, University of Carolina, Carolina Environmental Programs, Research Triangle Park, North Carolina (2003)
5. Chervenkov, H.D., Syrakov, M.: Prodanova, On the Sulphur Pollution over the Balkan Region. In: Lirkov, I., Margenov, S., Waśniewski, J. (eds.) LSSC 2005. LNCS, vol. 3743, pp. 481–489. Springer, Heidelberg (2006)
6. Chervenkov, H.: Estimation of the Exchange of Sulphur Pollution in the Southeast Europe. Journal of Environmental Protection and Ecology 7(1), 10–18 (2006)

7. Dudhia, J.: A non-hydrostatic version of the Penn State/NCAR Mesoscale Model: Validation tests and simulation of an Atlantic cyclone and cold front. Mon. Wea. Rev. 121, 1493–1513 (1993)
8. EC, Amended draft of the daughter directive for ozone, Directorate XI — Environment, Nuclear Safety, and Civil Protection, European Commission, Brussels (1998)
9. EC, Ozone position paper, Directorate XI — Environment, Nuclear Safety, and Civil Protection, European Commission, Brussels (1999)
10. Ganev, K., et al.: On some cases of extreme sulfur pollution in Bulgaria or Northern Greece. Bulg. Geoph. J. XXVIII, 1–4 (2002)
11. Ganev, K., et al.: Accounting for the mesoscale effects on the air pollution in some cases of large sulfur pollution in Bulgaria or Northern Greece. Environmental Fluid Mechanics 3, 41–53 (2003)
12. Grell, G.A., Dudhia, J., Stauffer, D.R.: A description of the Fifth Generation Penn State/NCAR Mesoscale Model (MM5). NCAR Technical Note, NCAR TN-398-STR, 138 (1994)
13. Prodanova, M., et al.: Preliminary estimates of US EPA Model-3 system capability for description of photochemical pollution in Southeast Europe. In: Proc. of the 26th ITM on Air Pollution Modelling and Applications, Leipzig, Germany, May 13–19 (2006)
14. Stauffer, D.R., Seaman, N.L.: Use of four-dimensional data assimilation in a limited area mesoscale model. Part I: Experiments with synoptic data 118, 1250–1277 (1990)
15. Syrakov, D., et al.: Exchange of sulfur pollution between Bulgaria and Greece. Environmental Science and Pollution Research 9(5), 321–326 (2002)
16. Vestreng, V.: Emission data reported to UNECE/EMEP: Evaluation of the spatial distribution of emissions. Meteorological Synthesizing Centre — West, The Norwegian Meteorological Institute, Oslo, Norway, Research Note 56, EMEP/MSC-W Note 1/2001 (2001)
17. Zerefos, C., et al.: Study of the pollution exchange between Bulgaria and Northern Greece. Int. J. Environment & Pollution 22(1/2), 163–185 (2004)
18. Zlatev, Z.: Computer Treatment of Large Air Pollution Models. Kluwer Academic Publishers, Dordrecht-Boston-London (1995)
19. Zlatev, Z., Syrakov, D.: A fine-resolution modelling study of pollution levels in Bulgaria. Part 2: high ozone levels. International Journal of Environment and Pollution 22(1/2), 203–222 (2004)

A Collaborative Working Environment for a Large Scale Environmental Model

Cihan Sahin, Christian Weihrauch, Ivan T. Dimov, and Vassil N. Alexandrov

Centre for Advanced Computing and Emerging Technologies,
University of Reading, Reading RG2 7HA
{c.sahin,c.weihrauch,i.t.dimov,v.n.alexandrov}@rdg.ac.uk

Abstract. Large scientific applications are usually developed, tested and used by a group of geographically dispersed scientists. The problems associated with the remote development and data sharing could be tackled by using collaborative working environments. There are various tools and software to create collaborative working environments. Some software frameworks, currently available, use these tools and software to enable remote job submission and file transfer on top of existing grid infrastructures. However, for many large scientific applications, further efforts need to be put to prepare a framework which offers application-centric facilities. Unified Air Pollution Model (UNI-DEM), developed by Danish Environmental Research Institute, is an example of a large scientific application which is in a continuous development and experimenting process by different institutes in Europe.

This paper intends to design a collaborative distributed computing environment for UNI-DEM in particular but the framework proposed may also fit to many large scientific applications as well.

Keywords: distributed computing, grid computing, grid services, modelling, large scale air pollution models.

1 Introduction

Large scientific applications are mostly developed, tested and used by the groups of scientists in geographically dispersed locations. It requires a framework for scientists to effectively communicate, develop and share data and experiments. E-science, as a newly emerging research area, proposes tools and software which people could use to create collaborative working environments by utilising grid infrastructures.

Web platforms (so called web portals), with their flexibility, accessibility, and platform independence, are the ultimate choice for many collaborative environments as the user interface. With the recent progress on grid computing, web services and grid services emerge to provide statefull web applications that could be used to offer a wide range of tools that are used to create such frameworks. However, current web portals are mostly written to provide generic services to submit jobs on remote resources or transfer data between different resources.

I. Lirkov, S. Margenov, and J. Waśniewski (Eds.): LSSC 2007, LNCS 4818, pp. 442–449, 2008.

Although it is relatively sufficient for some applications, some need a more application centric approach to create a collaborative working environment. Further efforts need to be put to facilitate the share of the data by offering standard data handling models, workflow creating abilities, a simple software versioning approach to keep track of the application, a facility to handle the modularisation of the application hence conceptually plug and play the modules of the application to improve the experiment development environment and many other application oriented functionality like visualisation and verification of the data for the case of UNI-DEM.

Large-scale air pollution models are used to simulate the variation and the concentration of emissions into the atmosphere and can be used to develop reliable policies and strategies to control the emissions [9,11]. The Unified Danish Eulerian Air Pollution Model (UNI-DEM model), developed by the Danish National Environmental Research Institute, simulates the transport and distribution of air pollutants in the atmosphere. UNI-DEM, as a complex simulation software, has high throughput and high performance computing requirements. As it uses and produces large sets of data, the establishment of a distributed, access controlled data server is essential as well. These requirements could be addressed with distributed technologies, specifically grid computing as certain toolkits define a set of tools on security, discovery, data transfer and distributed job submission for such complex systems. Grid computing also helps to target the collaborative issues as scientists on different domains would like to collaborate on the development of UNI-DEM or to carry out further studies on model code. Many portal systems already use an underlying grid infrastructure successfully to enable a collaborative working environment for geographically distributed scientists.

This paper looks at the overall design of such a collaborative working environment framework to further develop UNI-DEM by a group of geographically distributed scientists. Although the framework is designed for UNI-DEM, it could easily be generalised to many scientific applications. It takes into account the high throughput and high performance computing demands of UNI-DEM. As UNI-DEM is a data centric code, which uses and produces huge amount of data, its data control and access system sits at the centre of the design. A modular approach to the UNI-DEM code is also analysed to improve the experiment development environment by using the concepts of distributed computing technologies. The UNI-DEM specific requirements like visualisation and the verification of the data should correspondingly be part of the framework. The whole system design is put together for an experimental, collaborative working environment.

2   Requirement Analysis

There are a set of requirements that should be met for a collaborative working environment framework to provide the minimum facilities. Those requirements are as follows for UNI-DEM but could be as general as fitting to many other scientific applications:

- The framework should operate on top of a multi domain grid infrastructure of resources, to satisfy the high throughput and high performance computing requirements of UNI-DEM.
- Design of a data server to enable data sharing as well as listing, writing and reading of data. It should also have the data access control system to define and implement the sharing.
- A software versioning system to keep track of the changes on the source code providing the change information to the users.
- High level, easy to use interface for non-computational scientists.
- A simple workflow system to integrate job submissions, data access and other application centric functionality.
- Meta scheduler with an integrated resource broker to schedule the job on a remote resource automatically.
- An optional visualisation integration (embedded to the portal interface or suggesting a local visualisation).
- Definition of a verification schema for UNI-DEM data.

The requirements as mentioned above drive the need to use distributed computing technologies. Grid computing [4] is an ideal candidate to target all these requirements as it was coined to provide a heterogeneous but a unique infrastructure for loosely coupled systems.

2.1 Computational Model

The computational model looks at the details of how the computations are carried out. UNI-DEM is a parallel code with two versions; MPI and OpenMP. Many scientific applications like UNI-DEM have to run many times simultaneously (parametric studies) to carry out experiments on the code, initial data, or to assess the output data. Additionally, high throughput demand for UNI-DEM comes from the climatological studies where model runs should cover long periods of time. Also the development of an ensemble system requires a lot of compute power as several instances of the model have to run.

In addition to high throughput requirements, UNI-DEM is limited by high performance capability of the compute resources [1]. A typical high resolution, 3D version run of a UNI-DEM (over a period of a year) takes about 134878 seconds on 64 processor of a Blade Cluster. This run-time could be a problem when there are time constraints hence the model needs to be distributed over different domains.

The computational model of the framework consists of the heterogeneous platforms with grid computing capabilities. This requires the installation of a grid middleware on the resources. Although several different grid middleware are available, the one used in this study is Globus Toolkit 4.0 (gt4) [3,5]. It is proved to be working fine with the overlying interface, web portals, which provide the user interface to the virtual organisation.

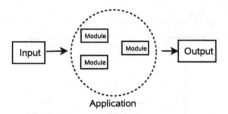

Fig. 1. A generic application prototype fitting to the framework

2.2 Application Model

Another important feature to improve the collaboration is to modularise the application. Two levels of modularisation could be defined; process based modularisation where pre-processing, computation and post-processing of the application could be wrapped up into services. Although this approach provides a good level of modularisation, it is often not sufficient to analyse the model physics in detail. The computational part of the application code is usually much more intensive then pre and post processing parts so a deeper level modularisation should be achieved.

This framework aims the type of the applications which falls on the category of Figure 1. For some scientific applications, this could be easily achieved by identifying the independent parts of the code and wrapping those independent parts into services. But for some applications like UNI-DEM, all physical modules are tightly coupled to each other and need to exchange messages on different levels of computations. At this level, a good understanding of the application code is required as certain computational pieces of the code could still be modularised. Modularised model code or pieces of model code could be made available to other scientists who may use these codes in their experiments. The modular approach provides a quick and efficient way of testing new segments of the code where the module could be replaced by a different piece of code.

Web and grid services could provide the technical foundation to achieve a modular model. As an example, the Linear Algebraic (LA) computations which are part of the computations in UNI-DEM, could be modularised into a grid service. Studies could be carried out by changing the method of the LA Grid service from standard LA methods to Monte Carlo LA methods [2]. A modular approach to application requires an integrated workflow system to be in place to create automated job sequences.

2.3 Data Model

UNI-DEM, like many other environmental models, use and produce a lot of data. Observations provide the initial conditions of the real situations and the accuracy of the model heavily relies on this observation data. The higher the quality of the data, the better the results of the model. As the model computations are suggested to be distributed over different domains, the data could not be

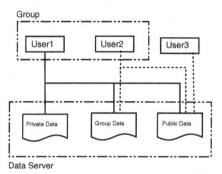

Fig. 2. A generic overview of the data model where data is organised into private, group and public segments on data server where access is controlled by user certificates

kept on a single domain any more. It should be accessible by all users (in a controlled manner) and all computational resources. This requires the setup of a data server, with a security and access control system.

The collaboration between the scientists could happen on different levels. The most apparent one is the collaboration on data level. Many different runs of the code means there is a lot of data produced and sharing this between the scientists requires strict definition of a sharing protocol. The data server should be segmented to keep private data, group data and public data, where access to each of them is controlled and granted with user credentials. This type of collaboration fits well with the security system of grid computing frameworks.

The data model lies in the core of the framework design. The organisation of the data reflects the sharing method and must be associated with metadata definitions. The data is categorised into 3 different groups; private, group and public data analogy to file permissions on Unix/Linux systems. Private data is only accessible by the owner of the data, group data is accessible by the group the owner belongs to and the public data is accessible by anyone on the virtual organisation as shown in Figure 2. Each piece of data is labelled with a metadata that contains the identifying information like the changes on application code (for example version control number) and input parameters of the run. This simple access control system provides an excellent database that logically identifies the accessible piece of data per user. Once integrated with the portal system, any user could check the previous runs of the application, which are accessible to this user not to duplicate an identical run. The data model suggested here could be applied to the application code as well where data is replaced with different versions of the application code or modules accessible by granted users.

UNI-DEM runs over a large geographical area. Even finer resolution of UNI-DEM may not be sufficient to analyse smaller scale areas especially on urban level. There are mathematical models suggested [10] to downscale the UNI-DEM for urban areas. Although UNI-DEM may not run for urban scale areas as it is now, but it could provide data for other air pollution models which are running at this scale [10]. Additionally, UNI-DEM uses wind data from atmospheric

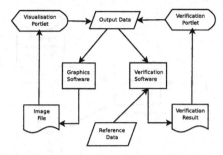

Fig. 3. A generic level representation of visualisation and verification functionality on user interface

models like MM5 Community Model [6]. The wind data from MM5 could be used as an input data in UNI-DEM. The data server suggested could solve many integration problems as all data could be transferred by using the data server. More decoding and encoding software need to be installed on data server to convert different data formats between the models.

2.4   Visualisation and Verification

Visualisation of UNI-DEM data gives a lot of information to scientists as it shows the distribution of the emissions on a geographical map. The verification helps to value the output results. From the computational point of view, the application must be rerun each time a change has been performed and the results should be verified by comparing to some reference data. The reference data could be collection of observations which are interpolated to the grid points or results of another run which proved to be correct enough for scientists. In the ideal case, visualisation and the verification software should be installed on data server as both need the output data and additionally, verification needs reference data. Portlets as part of the web portal of UNI-DEM take care of the invocation of the related software on the server and displays the results back as shown in Figure 3.

3   Technology and Software

Web portals, integrated with grid services, provide easy and flexible way to control and share information by utilising the distributed resources. The portal should have the following functionality:

- Portal login interface.
- Credential management interface to download, upload and renew certificates from a credential server.
- Resource management where users could select or monitor the state of the resources where application is installed or complied to be able to run.
- Workflow submission interface to configure the jobs.

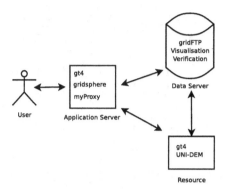

Fig. 4. Conceptual level design and interactions of the software components

- Job control interface to monitor and interact the running jobs.
- Data transfer functionality. Database access to download or upload data to local computer or to the database.
- Visualisation interface to run visualisation process and visualise the generated images.
- Verification interface to carry out verification of the output results by using reference data.

There are web portal frameworks which successfully use grid infrastructures and provide many functionality mentioned above on a generic level. GridSphere portal framework [7] is one of them and it offers generic portlets to submit and control jobs, transfer data between resources and resource monitoring. P-GRADE web portal [8] is built on top of GridSphere framework and provides additional functionality like workflow management. The GridSphere portal framework is used and additional functionality specific to the application like collaboration, visualisation and verification functionality will be added by using self contained portlets. The overall design has the following software components and relations which are shown in Figure 4. Note that only one resource is shown on Figure 4 just for illustration. In real life cases, there are many resources distributed over different domains.

4   Conclusion

Many large scientific applications are products of collaborative multi-institutional efforts. Collaboration between geographically dispersed scientists are not easy due to problems associated with application and data sharing. However, collaborative working environments built on distributed resources and servers could help a lot on the development, testing and using of scientific applications.

In this paper, we show a framework of a collaborative working environment for a large scale unified air pollution model (UNI-DEM) by using distributed computing concepts and technologies. We implemented the relevant infrastructure containing the application and data servers, and also enabled the resources on the grid infrastructure. Furthermore, UNI-DEM has been deployed on the resources. The fundamental job submission portlet has been developed and customised to the requirements of UNI-DEM application. The methods of the collaboration on the data level have been defined to enable a grid aware sharing mechanism. The job submission parameters and UNI-DEM job configuration file provide the necessary meta-data to access and use the model output data for granted users. Automatic transfer of files between resources and the data server have been setup however the implementation of sharing policies are currently in progress. Data collaboration is also crucial for the integration of UNI-DEM with different models.

In the future, we are looking at the further development of the UNI-DEM portal by introducing application versioning system and portlet. This requires the definition of the metadata associated to each version of the application whenever the application is modified. Data sharing rules apply to application versions to grant user access. Visualisation, verification, application based grid services are the possible research and development areas to achieve a complete collaborative UNI-DEM application.

References

1. Alexandrov, V., et al.: Parallel runs of a large air pollution model on a grid of sun computers. Mathematics and Computers in Simulation 65, 557–577 (2004)
2. Branford, S., et al.: Monte Carlo Methods for Matrix Computations on the Grid. Future Generation Computer Systems (to be published, 2007)
3. Foster, I.: Globus Toolkit Version 4: Software for Service-Oriented Systems. In: Jin, H., Reed, D., Jiang, W. (eds.) NPC 2005. LNCS, vol. 3779, pp. 2–13. Springer, Heidelberg (2005)
4. Foster, I., Kesselman, C., Tuecke, S.: The Anatomy of the Grid: Enabling Scalable Virtual Organizations. International Journal of High Performance Computing Applications 15(3), 200–222 (2001)
5. Globus Toolkit Website, http://www.globus.org
6. MM5 Community Model Web Site, http://www.mmm.ucar.edu/mm5
7. Novotny, J., Russell, M., Wehrens, O.: GridSphere: A Portal Framework for Building Collaboration, GridSphere Project Website, http://www.gridsphere.org
8. Sipos, G., Kacsuk, P.: Multi-Grid, Multi-User Workflows in the P-GRADE Portal. Journal of Grid Computing 3(3-4), 221–238 (2006)
9. Zlatev, Z.: Computer treatment of large air pollution models. Kluwer Academic Publishers, Dorsrecht-Boston-London (1995)
10. Zlatev, Z., Dimov, I.: Computational and Numerical Challenges in Environmental Modeling. Elsevier, Amsterdam (2006)
11. Zlatev, Z., Dimov, I., Georgiev, K.: Three-dimensional version of the Danish Eulerian Model. Zeitschrift für Angewandte Mathematik und Mechanik 76(S4), 473–476 (1996)

Advances on Real-Time Air Quality Forecasting Systems for Industrial Plants and Urban Areas by Using the MM5-CMAQ-EMIMO

Roberto San José[1], Juan L. Pérez[1], José L. Morant[1], and Rosa M. González[2]

[1] Environmental Software and Modelling Group, Computer Science School, Technical University of Madrid — UPM, Campus de Montegancedo, Boadilla del Monte 28660 Madrid, Spain
roberto@fi.upm.es
http://artico.lma.fi.upm.es

[2] Department of Meteorology and Geophysics, Faculty of Physics, Complutense University of Madrid — UCM, Ciudad Universitaria, 28040 Madrid, Spain

Abstract. The system MM5-CMAQ-EMIMO-MICROSYS produces reliable air quality forecasts for urban areas with street level detail over the Internet. In this contribution we will show the special example applied to Las Palmas (Canary Islands, Spain). Additionally, the MM5-CMAQ-EMIMO has been used to know the air quality impact of several industrial plants such as combined cycle power and cement plants. Additional runs are performed in parallel without the emissions of the different chimneys forming the industrial complex (up to 5 scenarios in our experiences for electric companies and cement industrial plants). The differences ON1-OFF and so on, show the impact in time and space of the different industrial sources. The system uses sophisticated cluster technology to take advantage of distributed and shared memory machines in order to perform the parallel runs in an efficient and optimal way since the process should operate under daily basis. We will show the methodology and results of these applications also in two industrial complex in the surrounding area of Madrid City.

Keywords: CFD, air quality modeling, MM5, CMAQ, turbulence, fluid dynamics, software tools.

1 Introduction

New advances on air quality modeling have been produced during the last years. In particular applications for operational services by using state-of-the-art models over industrial plants and urban areas. Up to now, these applications were limited by the highest spatial resolution of the air quality mesoscale models which received boundary and initial data from the meteorological global models. Nowadays, we have made a step forward going down up to 1-10 m spatial resolution at urban scale to understand, visualize and analyze the detailed concentrations found in the urban canopy with the complex building structure. Urban canopy is dominated by turbulence on very local scales and the turbulence

I. Lirkov, S. Margenov, and J. Waśniewski (Eds.): LSSC 2007, LNCS 4818, pp. 450–457, 2008.
© Springer-Verlag Berlin Heidelberg 2008

parameterization in our classical RANS (Reynolds-Averaged Numerical Simulations) simulations can include aspects which can be visualized by using the LES (Large-eddy simulations, where large eddies are producing the differences with RANS or those models with much more computational cost, such as the Direct Numerical Simulations (DNS) — no turbulence model is included — and the whole range of spatial and temporal scales of the turbulence must be resolved. We are still far from operational implementations of DNS models and even for LES models. Since our objective is tom provide some information related to the existence of so called "hot spots" in CFD modeling and by considering that the boundary and initial conditions of these models are essential aspects on such a matter, we are implementing in diagnostic mode an adapted application of the MIMO model (University of Karlsruhe, 2002) with a cellular automata model (UPM, 2003) which receives boundary and initial conditions from the MM5-CMAQ-EMIMO (NCEP/EPA/UPM) mesoscale air quality modeling systems. So that we are covering and linking global modeling systems, operating under daily basis, and microsimulations — represented by the so called MICROSYS CFD modeling tool.

In addition to these aspects new operational applications in industrial plants are running and producing operational services since 2005, July and January, 2007. These complex systems could evaluate the impact of several urban strategic emission reduction measures such as reduction of private traffic, increase of public transportation, impact on introduction of new fuel cell vehicles, etc. Also, they could be used for analysis of pollution concentrations at different heights (buildings) and on different areas of urban neighborhoods. Air dispersion in urban areas is affected by atmospheric flow changes produced by building-street geometry and aerodynamic effects. The traffic flow, emissions and meteorology are playing also an important role. Microscale air pollution simulations are a complex task since the time scales are compared to the spatial scales (micro) for such a type of simulations. Boundary and initial conditions for such a simulations are also critical and essential quantities to influence fundamentally the air dispersion results. Microscale Computational Fluid Dynamical Models (CFDM) are playing an increasing role on air quality impact studies for local applications such as new road and building constructions, emergency toxic dispersion gases at urban and local scale, etc. Microscale air dispersion simulations are applied to predict air-flow and pollution dispersion in urban areas [8]. Different combinations and applications appear in the literature as in Reference [9] by integrating a Lagrangian model and a traffic dynamical model into a commercial CFD code, Star-CD to simulate the traffic-induced flow field and turbulence.

In this contribution we have applied the microscale dispersion model MIMO [10] to simulate different emission reduction scenarios in Madrid (Spain) related to the vehicle traffic conditions. The MIMO CFD code has been adapted and incorporated into a mesoscale air quality modeling system (MM5-CMAQ-EMIMO) to fit into the one-way nesting structure. MM5 is a meteorological mesoscale model developed by Pennsylvania State University (USA) and NCAR (National Center for Atmospheric Research, USA) [4]. The CMAQ model is the

Community Multiscale Air Quality Modeling System developed by EPA (USA) [1] and EMIMO is the Emission Model [6]. MM5 is a well recognized non-hydrostatic mesoscale meteorological models which uses global meteorological data produced by global models such as GFS model (NCEP, USA) to produce high resolution detailed three dimensional fields of wind, temperature and humidity which are used in our case as input for the photochemical dispersion model CMAQ [5]. In addition of MM5 output data, EMIMO model produces for the specific required spatial resolution, hourly emission data for different inorganic pollutants such as particulate matter, sulphur dioxide, nitrogen oxides, carbon monoxide and total volatile organic compounds VOC's. The VOC's are split according to SMOKE (Sparse Matrix Operator Kernel Emissions) [2,3,7]. The CFD and mesoscale models solve the Navier-Stokes equations by using different numerical techniques to obtain fluxes and concentrations at different scales. Mesoscale air quality models cover a wide range of spatial scales from several thousands of kilometers to 1 km or so. In this contribution we have applied the MM5-CMAQ-EMIMO models over Madrid domain to obtain detailed and accurate results of the pollutant concentrations at this spatial resolution and the MIMO CFD model over a 1 × 1 km domain with several spatial resolutions (2–10 m) and different vertical resolutions. MM5-CMAQ-EMIMO data serves as initial and boundary conditions for MIMO modeling run.

The MM5-CMAQ-EMIMO modeling system has been used to provide detailed initial and boundary conditions to a system called MICROSYS which is composed by the MIMO CFD microscale dispersion model and CAMO which is a cellular automata traffic model. The results show that the air quality modeling system offers realistic results although no comparison with eddy-correlation measurement system has been performed in the area. The tool can be used for many air quality impact studies but in particular for traffic emission reduction strategies. In Figure 1 we observe the spatial architecture for the application of the MM5-CMAQ-EMIMO mesoscale air quality modeling system. In Figure 2 we show a detailed diagram of the EMIMO modeling system. EMIMO is currently operating with the so called Version 2 which includes the CLCL2000 with 44 different landuse types with 100 m spatial resolution. EMIMO 2.0 also uses the CIESIN 30″ (CIESIN, 2004), population database and the Digital Chart of the World 1 km land use database to produce adequate emission data per 1 km grid cell per hour and per pollutant. In order to apply the EMIMO CFD model, we need detailed information related to the building structure in the 1 km grid cell. Figure 3 shows an scheme showing the nesting between global models and urban applications. Figure 4 shows a GIS vector file for Las Palmas de Gran Canaria (Cary Islands, Spain) with an area of 1 × 1 km where one of our applications has been implemented operationally since January, 2007. The height of the buildings is included in this file. A cellular automata traffic model (CAMO) has been developed. CAMO — which has been included into the EMIMO modeling system — is based on transitional functions defined in a discrete interval t as follows:

$$s(t+1) = p(s(t), a(t))$$
$$u(t) = v(s(t))$$

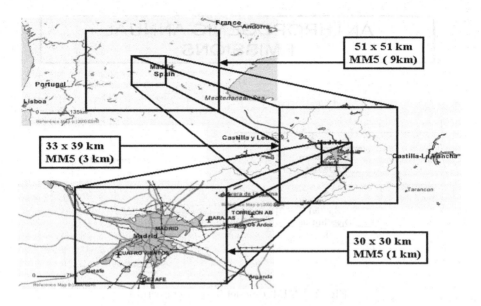

Fig. 1. MM5-CMAQ-EMIMO architecture for this application

where $s(t)$ is a defined state, $s(t + 1)$ is the defined state after one step, a(t) is an input symbol and u(t) is an output symbol. We have used the Moore neighborhood with 8 different surrounding cells where each cell — representative of a vehicle — can move on. The whole system is focusing on Las Palmas de Gran Canaria (Canary Islands, Spain) with 7 areas with 1×1 km each one. The structure of the system is shown in Figure 5. We have a regular grid and a variable spacing grid area shown in the Figure 5. In this particular case the horizontal spatial resolution is set to 10 m and the vertical spatial resolution varies from 0 to 167 m with 10 m the regular vertical spatial resolution up to 100 m in height.

2 Results

The system has been implemented to produce operational air quality forecasts for seven areas with 1×1 km spatial resolution each one for 72 hours. The operational mesoscale air quality system — MM5-CMAQ-EMIMO, OPANA V3 — which was operating since 2005 had a urban spatial resolution of 1 km over an area of 16×16 km and two mesoscale domains covering up to 180×180 km over the Canary Islands with 9 km spatial resolution and a second domain over the Gran Canaria island with 3 km spatial resolution, was used to provide boundary conditions to the new seven 1×1 km domains where MICROSYS was applied with the CAMO model. The full new system is called OPANA V4. The system should run on a daily basis over the 72 hours. The MICROSYS model is running on diagnostic mode for 2 minutes each hour and then using the boundary conditions produced

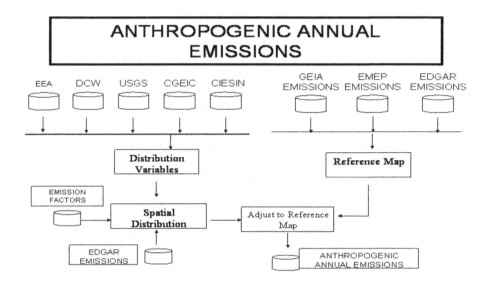

Fig. 2. EMIMO model basic architecture

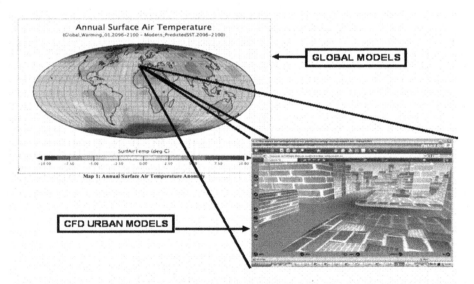

Fig. 3. From global models to urban applications

in forecasting mode by OPANA V3. The full system performs under a daily basis on a unique PC computer platform for 16–20 hours CPU time. Figure 6 shows an example of the operational web site where the system is producing daily data. The user can access to all the information, including different GIS capabilities taken from the city data base, time series, zooming capabilities, etc. Figure 7 shows an example in a point (10 m resolution) of a street in Las Palmas where the O3 values can be access in a forecasting mode (72 hours).

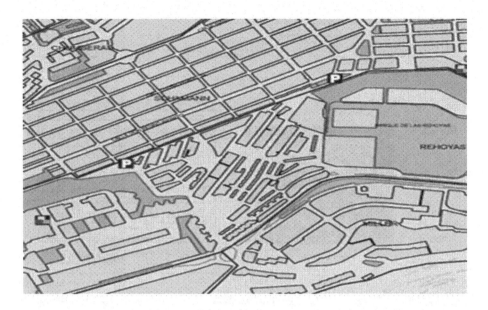

Fig. 4. One of the sub-domains with 1×1 km in Las Palmas de Gran Canaria (Canary Islands, Spain) where the MM5-CMAQ-EMIMO-MICROSYS air quality modeling system has been implemented for this experiment

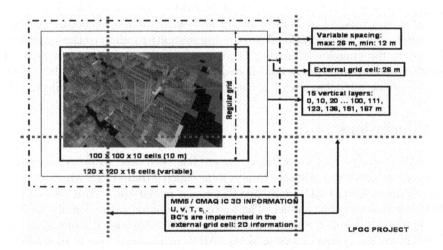

Fig. 5.

In addition to this application, we have installed an operational system in a cluster platform to obtain the impact of two industrial plants in forecasting mode. ACECA (South of Madrid Community) — a thermal and combined power plant with 4 emitters in the area — and Portland Valderrivas with five scenarios

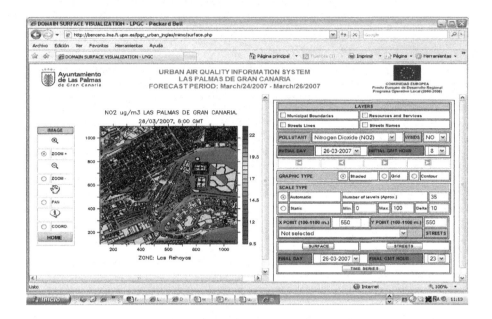

Fig. 6. OPANA V4 Las Palmas web site

also in forecasting mode in the south of Madrid Community. The system has been also implemented in Madrid urban area where it has been used as part of research tool (not operational forecasting system) to analyze the impact of different traffic emission scenarios. Similar implementation has been done in Helsinki (Finland) and Marylebone (London, UK). These cases can be illustrated in http://www.eu-oscar.org/ as results of the OSCAR EU FP5 project.

3   Conclusions

The MM5-CMAQ-EMIMO modeling system has been used to provide detailed initial and boundary conditions to a system called MICROSYS which is composed by the MIMO CFD microscale dispersion model and CAMO which is a cellular automata traffic model. The system is running operationally for Las Palmas (Canary Islands, Spain) since January, 2007. Additional a mesoscale air quality modeling system, MM5-CMAQ-EMIMO, has been put in operational service to produce air quality forecasts for several scenarios produced by switching on and off the emissions of different chimneys on ACECA (thermal and combined power plant located in the South of Madrid Community) and Portland Valderrivas (Cement company located in the south of Madrid Community). Both systems are in operation over cluster parallel platforms since July, 2005 and January, 2007, respectively.

Acknowledgement.

We would like to thank Prof. N. Moussiopoulos from Aristotle University for providing in 1996 and 2003 several versions of the MIMO model. We also would like to than to EPA and PSU/NCAR for the CMAQ and MM5 models. We would like to thank also to the OSCAR EU project. We also would like to thank to Las Palmas environmental authorities for the full support to implement this forecasting system with the EU structural funds support.

References

1. Byun, D.W., et al.: Description of the Models-3 Community Multiscale Air Quality (CMAQ) model. In: Proceedings of the American Meteorological Society 78th Annual Meeting Phoenix, January 11–16, 1998, pp. 264–268 (1998)
2. Center for International Earth Science Information Network (CIESIN). Global Rural-Urban Mapping Project (GRUMP): Urban/Rural population grids. Palisades, NY: CIESIN, Columbia University (2004),
   http://sedac.ciesin.columbia.edu/gpw/
3. Coats Jr., C.J.: High Performance Algorithms in the Sparse Matrix Operator Kernel Emissions (SMOKE) Modelling System, Microelectronics Center of North Carolina, Environmental Systems Division, Research Triangle Park, NC, p. 6 (1995)
4. Grell, G., Dudhia, J., Stauffer, D.: A Description of the Fifty Generation Penn State/NCAR Mesoscale Model (MM5). NCAR Tech. Note, TN-398+STR, p. 117 (1994)
5. San José, R., et al.: Sensititivity study of dry deposition fluxes in ANA air quality model over Madrid mesoscale area. In: José, S., Brebbia (eds.) Measurements and Modelling in Environmental Pollution, pp. 119–130 (1997)
6. San José, R., et al.: EMIMO: An emission model, pp. 292–298. Springer, Heidelberg (2003)
7. Williams, A., et al.: Comparison of emissions processing by EM-S95 and SMOKE over the Midwestern U.S. In: Preprint of International Emission Inventory Conference: One Atmosphere, One Inventory, Many Challenges, Denver, Colorado, May 1-3, 2001, pp. 1–13 (2001)
8. Pullen, J., et al.: A comparison of contaminant plume statistics from a Gaussian puff and urban CFD model for two large cities. Atmospheric Environment 39, 1049–1068 (2005)
9. Pospisil, J., Katolicky, J., Jicha, M.: A comparison of measurements and CFD model predictions for pollutant dispersion in cities. In: Science of the total Environment, pp. 334–335, 185-195 (2004)
10. Ehrhard, J., et al.: The microscale model MIMO: Development and assessment. Journal of Wind Engineering and Industrial Aerodynamics 85, 163–176 (2000)

Part VIII

Computational Grid and Large-Scale Problems

Ultra-fast Semiconductor Carrier Transport Simulation on the Grid

Emanouil Atanassov, Todor Gurov, and Aneta Karaivanova

Institute for Parallel Processing, Bulgarian Academy of Sciences,
Acad. G. Bonchev, Bl. 25A, 1113 Sofia, Bulgaria
{emanouil,gurov,anet}@parallel.bas.bg

Abstract. We consider the problem of computer simulation of ultra-fast semiconductor carrier transport. The mathematical description of this problem includes quantum kinetic equations whose approximate solving is a computationally very intensive problem. In order to reduce the computational cost we use recently developed Monte Carlo methods as a numerical approach. We study intra-collision field effect, i.e. effective change of phonon energy, which depends on the field direction and the evolution time. In order to obtain results for different evolution times in a reasonable time-frame, we implement simulation on the computational grid. We split the task into thousands of subtasks (jobs) which are sent to different grid sites to be executed. In this paper we present new results for inhomogeneous case in the presence of electric field, and we describe our grid implementation scheme.

Keywords: Grid computing, Ultra-fast carrier transport, Monte Carlo methods, Wigner function, Grid performance.

1 Introduction

The Monte Carlo Methods for quantum transport in semiconductors and semiconductor devices have been actively developed during the last decade [4,12,9,10]. These Monte Carlo calculations need large amount of computational power and the reason is as follows: if temporal or spatial scales become short, the evolution of the semiconductor carriers cannot be described in terms of the Boltzmann transport and therefore a quantum description is needed. Let us note that in contrast to the semiclassical transport when the kernel is positive, the kernel in quantum transport can have negative values. The arising problem, sometimes referred to as the "negative sign problem", leads to additional computational efforts for obtaining the desired solution. That is why the quantum problems are very computationally intensive and require parallel and Grid implementations.

On the other hand, the intrinsically parallel aspect of Monte Carlo applications makes them an ideal fit for the grid-computing paradigm. For more information about Grid as a new computational infrastructure see [5].

Development of the grid infrastructure motives development of different grid implementation schemes for these algorithms. This paper presents our grid implementation scheme which uses not only the computational capacity of the grid

I. Lirkov, S. Margenov, and J. Waśniewski (Eds.): LSSC 2007, LNCS 4818, pp. 461–469, 2008.

but also the available grid services in a very efficient way. With this scheme
we were able to obtain new estimates about important physical quantities. The
paper is organised as follows: section 2 describes very briefly the problem and
the algorithms, section 3 presents the grid implementation scheme, section 4
contains the new estimates and performance analysis.

2   Background (Algorithms for Ultra-Fast Carrier Transport)

We consider algorithms for solving Wigner equation for the nanometer and
femtosecond transport regime. In the homogeneous case we solve a version of
the Wigner equation called Levinson (with finite lifetime evolution) or Barker-
Ferry equation (with infinite lifetime evolution) [8,7]. Another formulation of
the Wigner equation considers inhomogeneous case when the electron evolution
depends on the energy and space coordinates [11]. The problem is relevant e.g.
for description of the ultra-fast dynamics of confined carriers. Particularly we
consider a quantum wire, where the carriers are confined in the plane normal to
the wire by infinite potentials. The initial condition is assumed both in energy
and space coordinates.

The mathematical models that we consider include:

- Homogeneous case (one-band semiconductors):
  - Levinson and Barker-Ferry equations without an eclectic field;
  - Levinson and Barker-Ferry equations with an applied electric field;
  - Both equations with finite temperature and an applied electric field.
- Inhomogeneous case (quantum wire — more realistic case): we solve the
  above mentioned quantum equations with additional terms.

A set of Monte Carlo and Quasi-Monte Carlo algorithms was developed to solve
the problems arising from the mathematical models under consideration [8,6,1].
All algorithms were integrated in a Grid application named SALUTE (Stochastic
ALgorithms for Ultra-fast Transport in sEmiconductors). The description of the
first version of SALUTE can be found in [2,3].

The numerical results that we present in this paper are for the inhomogeneous
case with applied electric field (see figures in the Numerical tests section). We
recall the integral form of the quantum-kinetic equation, [6]:

$$f_w(z, k_z, t) = f_w(z - \frac{\hbar k_z}{m}t + \frac{\hbar \mathbf{F}}{2m}t^2, k_z, 0) + \int_0^t \partial t'' \int_{t''}^t \partial t' \int d\mathbf{q}'_\perp \int dk'_z \times \quad (1)$$

$$\left[S(k'_z, k_z, t', t'', \mathbf{q}'_\perp) f_w \left(z - \frac{\hbar k_z}{m}(t - t'') + \frac{\hbar \mathbf{F}}{2m}(t^2 - t''^2) + \frac{\hbar q'_z}{2m}(t' - t''), k'_z, t'' \right) \right.$$

$$\left. - S(k_z, k'_z, t', t'', \mathbf{q}'_\perp) f_w \left(z - \frac{\hbar k_z}{m}(t - t'') + \frac{\hbar \mathbf{F}}{2m}(t^2 - t''^2) - \frac{\hbar q'_z}{2m}(t' - t''), k_z, t'' \right) \right],$$

$$S(k_z', k_z, t', t'', \mathbf{q}_\perp') = \frac{2V}{(2\pi)^3} |G(\mathbf{q}_\perp') \mathcal{F}(\mathbf{q}_\perp', k_z - k_z')|^2 \times$$

$$\left[(n(\mathbf{q}') + 1) cos \left(\frac{\epsilon(k_z) - \epsilon(k_z') + \hbar\omega_{\mathbf{q}'}}{\hbar}(t' - t'') + \frac{\hbar}{2m}\mathbf{F}.q_z'(t'^2 - t''^2) \right) \right.$$

$$\left. + n(\mathbf{q}') cos \left(\frac{\epsilon(k_z) - \epsilon(k_z') - \hbar\omega_{\mathbf{q}'}}{\hbar}(t' - t'') + \frac{\hbar}{2m}\mathbf{F}.q_z'(t'^2 - t''^2) \right) \right]$$

Here, $f_w(z, k_z, t)$ is the Wigner function described in the $2D$ phase space of the carrier wave vector k_z and the position z, and t is the evolution time.

$F = e\mathbf{E}/\hbar$, where $\mathbf{E}$ is a homogeneous electric field along the direction of the wire z, e being the electron charge and $\hbar$ — the Plank's constant.

$n_{\mathbf{q}'} = 1/(\exp(\hbar\omega_{\mathbf{q}'}/\mathcal{K}T) - 1)$ is the Bose function, where $\mathcal{K}$ is the Boltzmann constant and T is the temperature of the crystal, corresponds to an equilibrium distributed phonon bath.

$\hbar\omega_{\mathbf{q}'}$ is the phonon energy which generally depends on $\mathbf{q}' = \mathbf{q}_\perp' + q_z' = \mathbf{q}_\perp' + (k_z - k_z')$, and $\varepsilon(k_z) = (\hbar^2 k_z^2)/2m$ is the electron energy.

$\mathcal{F}$ is obtained from the Fröhlich electron-phonon coupling by recalling the factor $i\hbar$ in the interaction Hamiltonian, Part I:

$$\mathcal{F}(\mathbf{q}_\perp', k_z - k_z') = -\left[\frac{2\pi e^2 \omega_{\mathbf{q}'}}{\hbar V} \left(\frac{1}{\varepsilon_\infty} - \frac{1}{\varepsilon_s} \right) \frac{1}{(\mathbf{q}')^2} \right]^{\frac{1}{2}},$$

where (ε_∞) and (ε_s) are the optical and static dielectric constants. The shape of the wire affects the electron-phonon coupling through the factor

$$G(\mathbf{q}_\perp') = \int d\mathbf{r}_\perp e^{i\mathbf{q}_\perp' \mathbf{r}_\perp} |\Psi(\mathbf{r}_\perp)|^2,$$

where Ψ is the ground state of the electron system in the plane normal to the wire.

In the inhomogeneous case the wave vector (and respectively the energy) and the density distributions are given by the integrals

$$f(k_z, t) = \int \frac{dz}{2\pi} f_w(z, k_z, t); \qquad n(z, t) = \int \frac{dk_z}{2\pi} f_w(z, k_z, t). \qquad (2)$$

Our aim is to estimate these quantities, as well as the Wigner function (1) by MC approach.

3   Grid Implementation

The evolvement of the grid implementation scheme of SALUTE was motivated by the development of the SEE-GRID[1] infrastructure which we use. In this section we first describe the grid infrastructure, then we show the preliminary and present implementation schemes.

---

[1] SEE-GRID2 (South Eastern European Grid enabled e-Infrastructure Development-2) initiative is co-funded by the European Commission under the FP6 Research

3.1 SEE-GRID Infrastructure

The SEE-GRID infrastructure integrates computational and storage resources, provided by the project partners, into a pilot grid infrastructure for South Eastern Europe. Currently there are more than 30 clusters with a total of more than 500 CPUs and more than 10 TB of storage. The peculiarities of the region are that the network connectivity of many of these clusters is insufficient, which implies the necessity to avoid network-hungry applications and emphasize computationally intensive applications, that make efficient use of the available resources. It also imposes the need of fault-tolerant implementations.

The first phase of the deployment of the SEE-GRID infrastructure used the LCG middleware. The services and APIs, available in LCG middleware, motivated the development of the first simpler scheme of SALUTE.

The second phase of the deployment of the SEE-GRID infrastructure was built using the gLite middleware. The gLite middleware provides Web Service APIs for most of its services, and provides new types of services, like the gLite WMS, gLite FTS, AMGA, etc. It also improved the reliability and scalability of the other services.

- Each of the SEE-GRID clusters has the mandatory services:
  - Computing Element (CE) — provides user acces to the Grid resources;
  - Worker Nodes (WN) — execute the jobs, perform calculations;
  - Storage Element (dCache, DPM or classic SE) — reliable data storage;
  - MON box (Monitoring and accounting) — monitors the current Grid status and reports complete jobs and resources used.
- The Worker Nodes provide the computational resource of the site, and the Storage Element provides the storage resources.
- The set of services, that are not tied to the specific site are called core services. In SEE-GRID the core services are distributed among partners. They include
  - VOMS (Virtual organisation management system)
  - MyProxy
  - R-GMA registry/schema server (distributed data-base)
  - BDII (provides comprehensive information about the resources)
  - WMS (distributes and manages the jobs among the different grid sites)
  - FTS (file transfer service)
  - AMGA (metadata catalog

Infrastructures contract # 031775 towards sustainable grid-related activities in the SEE region. The SEE-GRID2 consortium consists of thirteen contractors from SE European countries and CERN: GRNET (Greece), CERN, IPP-BAS (Bulgaria), ICI (Romania), SZTAKI (Hungary), TUBITAK (Turkey), INIMA (Albania), UOB (Serbia), UKIM (FYROM), RENAM (Moldova), RBI (Croatia), FEE-UoBL (Bosnia-Herzegovina), UOM (Montenegro). More information is availed on http://www.see-grid.eu.

3.2   Grid Implementation Scheme

Using the experience gained with the first scheme and the new services and APIs, available in SEE-GRID 2, we developed a new grid implementation scheme outlined below.

In the new scheme we incorporated the use of the FTS and AMGA services, available in the gLite, and we were able to include the estimation of several new physical quantities, which increased to total amount of data to be generated, stored, processed and visualized. We increased the spacial resolution of all graphs, since we had more computational resources available. Here are the details:

1. On the User Interface (UI) computer the scientist launches the Graphical User Interface (GUI) of the application. The job submission, monitoring and analysis of the results is controlled from there. The GUI is written using PyQt and pyopengl for the 3D visualization, and the python bindings for the grid functions (mainly the wmproxymethods module provided by gLite).
2. The *Web service* computer (WS) provides a grid-enabled secure gateway to the MySQL database, so that no direct mysql commands are run by the user or from inside the grid jobs.
3. The AMGA (ARDA Metadata Catalog) is used to hold information about the results obtained so far by the user — for example input parameters, number of jobs executed, execution date etc.
4. The user selects input parameters and queries the AMGA server to find if data for these parameters is already present or not.
5. If a new run is necessary, the user submits request to the WS computer for calculation.
6. From the GUI the jobs are sent to the Workload Management System (WMS) and information about them is stored at the MySQL database computer via WS invocation.
7. The WMS sends the job to the selected sites.
8. When the job starts, it downloads the executable from the dCache storage element. The executable is responsible for obtaining the input parameters from the WS, performing the computations, and storing the results in the local Storage Element. After finishing the store operation, it calls the WS computer in order to register the output.
9. The jobs are monitored from a monitoring thread, started from the GUI, and information about their progress is displayed to the user.
10. Another thread run from the GUI is responsible for collecting the output results from the various Storage Elements to the local dCache server. For each output file a request for transfer is sent to the File Transfer Service.
11. The FTS is used in order to limit the number of files that are transferred simultaneously, because of the limited bandwidth. In this way we also avoid scalability limitations of the middleware and we do not overload the Storage Elements.
12. After a file has been transferred to the dCache, it is registered in the MySQL database (by WS invocation).

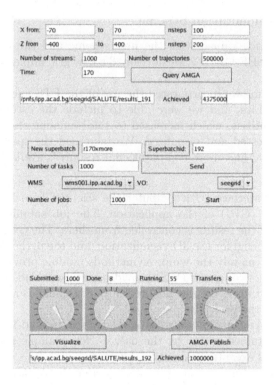

Fig. 1. Graphical User Interface (GUI) for submission and monitoring of SALUTE jobs, and accumulation and visualization of their results

13. A special computational job is run at a local WN and it is responsible for gradual accumulation of the outputs of all jobs into one final result. It checks the MySQL database for new results and if they are available at the dCache server, it retrieves them locally and performs the accumulation. At regular intervals the accumulated results are registered to the dCache and made available for the user.

We utilized several programming languages and technologies: Java and tomcat for the web services, mysql as a database back end, python with the Qt and OpenGL bindings for the GUI, MPI for parallel execution, SPRNG for the random number generation, etc.

The new GUI (see Figure 1) allows the user to control the whole process of job submission, monitoring, and collection of results. Partial results can be viewed in 3D view, and from there the user can see if the accuracy is going to be enough, or more jobs need to be submitted. One can also search for older results and compare them, taking the information from the metadata catalogue. This system has been used to submit up to 5000 jobs, and since it offloads most of the work to the gLite service nodes in SEE-GRID infrastructure, it has not shown performance problems. This gives us the advantage that the GUI does not have to be running all the time while the simulation is going on. By using Grid authentication and via the metadata catalogue we can have several scientists working jointly on similar problems, without duplicating computations.

4   Numerical Tests and Grid Performance Analysis

The problems arising when we solve the Wigner equation using Monte Carlo approach are due to the large statistical error. This error is a product of two factors: standard deviation and sample size on the power one half. The standard deviation increases exponentially with time, so in order to achieve reasonable accuracy, we must increase considerably the sample size. This implies the need of computational resources.

Using the grid and above described grid implementation scheme, we were able to obtain new results about important physical quantities: Wigner function, wave vector, electron density, and energy density. The results presented here are for inhomogenious case with applied electric field. The new Graphical user interface gives additional possibilities for analysing the results.   On the Figures 2, 3, and 4 one can see how Wigner function changes when the number of jobs successfully

Table 1. The approximate average CPU time per job, number of jobs and total CPU time for different evolution times. The number of the trajectories is fixed on $N = 2000000$ for each job.

t	Number of Jobs	CPU time per job	Total CPU time
20 fs	5	1 h	5 h
100 fs	100	1 h 20 min	133 h 20 min
180 fs	2000	2 h 20 min	4666 h 40 min

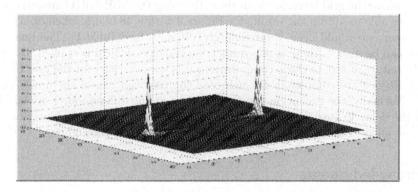

Fig. 2. The Wigner function solution at 20 fs presented in the plane $z \times k_z$. The electric field is 15[kW/cm] along to the nanowire.

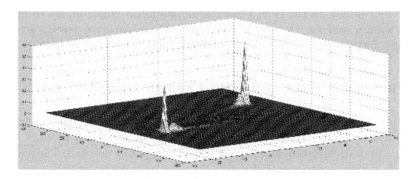

Fig. 3. The Wigner function solution at 100 fs presented in the plane $z \times k_z$. The electric field is 15[kW/cm] along to the nanowire.

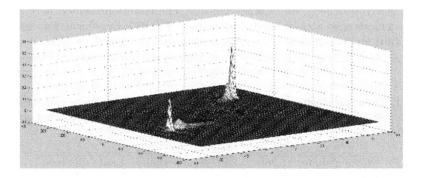

Fig. 4. The Wigner function solution at 180 fs presented in the plane $z \times k_z$. The electric field is 15[kW/cm] along to the nanowire.

executed on the grid increases with time. By using the SEE-GRID infrastructure we were able to obtain the results for Figure 4 within 48 hours, when on a single computer one would need approximately 195 days (see Table 1). The quantum effect can be seen — there is no symmetry when electric field is applied. Because of the nature of Monte Carlo computations it is not necessary for the user to wait until all jobs are completed. In such case the user can cancel the unnecessary jobs. Normally, the execution times of the jobs at the different sites are similar, and the delay in starting is caused by lack of free Worker Nodes. Thus our new scheme allows the user to achieve the maximum possible throughput.

Acknowledgments

Supported by the Ministry of Education and Science of Bulgaria under Grant No: I-1405/04 and by the EC FP6 under Grant No: INCO-CT-2005-016639 of the project BIS-21++.

References

1. Atanassov, A., et al.: New algorithms in the Grid Application SALUTE. In: Proceedings of MIPRO convention, Opatija, Croatia, May 21–25, 2007, pp. 217–222 (2007)
2. Atanassov, E., et al.: SALUTE — an MPI-GRID Application. In: Proceeding of the 28th International convention, MIPRO 2005, Opatija, Croatia, May 30 – June 3, 2005, pp. 259–262 (2005)
3. Atanassov, E., et al.: Monte Carlo Grid Application for Electron Transport. In: Alexandrov, V., et al. (eds.) ICCS 2006. LNCS, vol. 3993, pp. 616–623. Springer, Heidelberg (2006)
4. Fischetti, M.V., Laux, S.E.: Monte Carlo Analysis of Electron Transport in Small Semiconductor Devices Including Band-Structure and Space-Charge Effects. Phys. Rev. B 38, 9721–9745 (1988)
5. Foster, J., Kesselmann, C.: The Grid: Blueprint for a New Computing Infrastructure. Morgan Kaufmann, San Francisco (1998)
6. Gurov, T., et al.: Femtosecond Evolution of Spatially Inhomogeneous Carrier Excitations: Part II: Stochastic Approach and GRID Implementation. In: Lirkov, I., Margenov, S., Waśniewski, J. (eds.) LSSC 2005. LNCS, vol. 3743, pp. 157–163. Springer, Heidelberg (2006)
7. Gurov, T.V., et al.: Femtosecond relaxation of hot electrons by phonon emission in presence of electric field. Physica B 314, 301–304 (2002)
8. Gurov, T.V., Whitlock, P.A.: An efficient backward Monte Carlo estimator for solving of a quantum kinetic equation with memory kernel. Math. and Comp. in Sim. 60, 85–105 (2002)
9. Nedjalkov, M., et al.: Convergence of the Monte Carlo Algorithm for the Solution of the Wigner Quantum-Transport Equation. J. Math. Comput. Model 23(8/9), 159–166 (1996)
10. Nedjalkov, M., et al.: Unified particle approach to Wigner-Boltzmann transport in small semiconductor devices. Physical Review B 70, 115319–115335 (2004)
11. Nedjalkov, M., et al.: Femtosecond Evolution of Spatially Inhomogeneous Carrier Excitations: Part I: Kinetic Approach. In: Lirkov, I., Margenov, S., Waśniewski, J. (eds.) LSSC 2005. LNCS, vol. 3743, pp. 149–156. Springer, Heidelberg (2006)
12. Schmidt, T.C., Moehring, K.: Stochastic Path-Integral Simulation of Quantum Scattering. Physical Review A 48(5), R3418–R3420 (1993)

Simple Grid Access for Parameter Study Applications

Péter Dóbé, Richárd Kápolnai, and Imre Szeberényi

Budapest University of Technology and Economics, Hungary
dobe@iit.bme.hu, kapolnai@iit.bme.hu, szebi@iit.bme.hu

Abstract. The problem domain of Parameter Study covers a wide range
of engineering and scientific tasks which can be easily processed parallel.
However, the research community users, who would like to benefit from
the supercomputing power of a Grid are not familiar with the required
distributed computing techniques. For this reason, the presented *Salève*
development framework aims to free this limit by providing a general
solution for PS tasks by inserting an abstraction layer between the user
application and the Grid middleware. We design a new Salève plugin for
the gLite middleware of the EGEE Grid and discuss the authentication
issues.

Keywords: Grid Application Development, gLite, Parameter Study.

1 Introduction

The increasing demand for computing resources in various research and engineering activities makes the use of distributed computing technologies inevitable. Although this field can be proud of important results risen in the past decade (for a survey see [3]), the real application of high performance computing paradigms in everyday work can be highly challenging. In addition, the research community users who would like to benefit from the supercomputing power of a Grid are not familiar with the required techniques.

For this reason, the presented *Salève* development framework [6] aims to free this limit by hiding the technological details from the end-user and providing a general solution for *parameter study (PS) tasks*. The problem domain of PS covers a wide range of the engineering and scientific tasks such as data analysis in high energy physics or in astronomy, Monte Carlo methods etc. An independent PS computation generally defines a parameter space which has to be fully traversed to obtain the desired result. We suppose that the parameter space can be partitioned into *subdomains* where the final result is simply calculated from the subresults given by covering the subdomains, respectively. The independence property is essential to our case: we need to be able to process the subdomains independently, therefore the execution of the task can be easily paralleled.

After developing a computationally intensive but paralleled application, a researcher probably intends to submit it to a Grid. Salève now supports the submission of jobs to the EU supported EGEE Grid. The *Enabling Grid for*

I. Lirkov, S. Margenov, and J. Waśniewski (Eds.): LSSC 2007, LNCS 4818, pp. 470–475, 2008.
© Springer-Verlag Berlin Heidelberg 2008

E-sciencE (EGEE) project was initiated by CERN and aims to provide a production quality Grid infrastructure for academic and industrial researchers [2]. The EGEE users can access Grid resources only through *Virtual Organizations (VO)* which is a group of Grid users who e.g. plan to share some data or simply have similar interests. In order to offer a transparent access to the EGEE infrastructure, Salève hides the communication with the *gLite* middleware, the engine making the whole Grid run [4].

Salève's plugin-based architecture allows easy adaptation of a new Grid middleware or a local scheduler. In this paper we present the plugin mechanism of Salève through the design of a new plugin for gLite, and we also propose a new method for authentication between the Salève server and the client application. In Section 2 we introduce the working of the Salève system from the user aspect, and Section 3 gives an outline of Salève's client-server architecture and the security issues. In Section 4 we show the gLite plugin for Salève, and in Section 5 we present our goals to reach in the future.

2   Capabilities of Salève

In this section we give a brief summary of the operation of the Salève system through the phases of a PS task. First, the PS application written in C programming language has to be transformed to the Salève client. One of our main goal is to make this step as simple as possible. At present, the user needs to modify the source code slightly, and then link the Salève client library to the program producing a single executable application. The partition method of parameter space has to be implemented arbitrarily by the user, however it is aided by the logic of the Salève client library.

The second step is starting the client program which, by default, performs the calculations locally (without a server), just like the original program. When provided with an URL of a Salève server, the client submits its own binary code along with the required input data to the server, and waits for the subresults. In the third phase the server creates a job for every subdomain, and dispatches these jobs to a Grid or to a local scheduler. When the jobs are done, the server returns the subresults to the client as the fourth phase. Finally, after receiving the subresults, the client computes locally and then returns the final result.

There are several typical scenarios of Salève usage that we illustrate in Figure 1. The most simple involves no Salève server but local resources of the user's desktop machine. In this case, the client code acts almost exactly like the original sequential code, except that it takes advantage of multiple processors for parallel computation of subdomains if possible.

In the second approach (II–IV. cases on Figure 1), the processing is carried out by a Salève server. By default, the client submits itself and its input data and waits for the result from the server. During this phase what we called the third phase above, the client can be interrupted and resumed later to reattach to the server from another machine, and eventually retrieve the result data on the other client machine. Thus we are able to exploit the high availability and

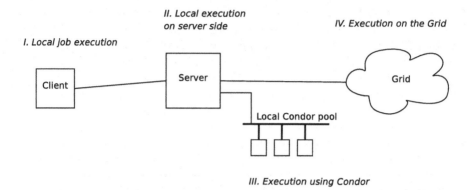

Fig. 1. Salève architecture and usage scenarios

location-independence of the Salève services, and eliminate the risks of job check-pointing on the desktop computer.

The server can execute the job locally, which can be useful if it is a multi-processor machine with much higher computing performance than that of the desktop computer running the client program. In addition, the server can pass on the task to an advanced distributed computing resource, e.g. a batch system or a Grid middleware. We emphasize that the way the Salève server uses these resources is fully transparent to the client, i.e. no recompilation nor reconfiguration is required if the server switches from one batch system or middleware to another. Furthermore, no thorough knowledge of distributed technology is necessary from the user.

3   The Internals of the Salève Server

The main components of the server are shown in Figure 2 which are the communication component and the plugins. The component responsible for the communication with the clients is a collection of SOAP web services [7]. Providing interfaces based on web services might be a commonly accepted practice in Grid environments in the near future. For the SOAP protocol implementation the highly portable gSOAP [5] toolkit has been used. We mention that the input and output data transfers are realized via HTTP messages.

We call the other important group of components *the plugins*. Salève's plugin mechanism has been designed in order to handle the diverse types of schedulers and middlewares (see Figure 2). For each distributed technology a lightweight plugin has to be developed, thus avoiding a radical redesign of the server architecture. The core Salève server already contains support for server-side local execution, and for Condor [1], which is one of the most widespread batch systems in the research community. The functionality of the server can be extended by adding new plugins to keep up with the new technologies. For the time being only one plugin can be handled by a server instance thus deploying a different plugin needs replacing the previous one.

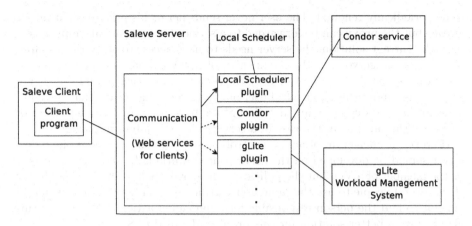

Fig. 2. The components of the Salève server

In a multi-user environment the security issues are crucial, particularly the authentication of the Salève client. We propose to use HTTP over SSL instead of the HTTP plain text version, so the client can authenticate herself or himself with a certificate signed by a trusted authority (e.g. a server-side certificate authority). Moreover the user can avoid typing in a password each time starting the client application.

On the other hand, server has to authenticate itself towards the Grid as well. Currently this is the responsibility of the corresponding plugin, without involving the client or the user in this process. If we follow this approach, no modification is required on the client side whenever a new server plugin is deployed.

4   The gLite Plugin

Writing a new plugin for the Salève server is relatively simple. The core of the server is implemented in C++ with high level object-oriented design. Developing a plugin consists of extending a base class and implementing functions for submitting a job and checking its status in the corresponding middleware. There is also a generic, plugin-implemented method of handling input and output data.

Besides the core server, a plugin extension for the gLite middleware has been developed in order to help the researchers to exploit the potential of the European Grid. For the most part, the design of the plugin was straightforward, but, since this is a pilot plugin, we have left the most important question open: the problem of authentication. In the following part, we outline the actual working mechanism.

The gLite middleware uses a certificate-based authentication mechanism, with the users and resources grouped into *Virtual Organizations* (VOs) built on the necessity of collaboration. So as to submit jobs, a *proxy certificate* must be generated using the user's private key (certificate). The lifetime of this proxy certificate is short to prevent damages in case of being compromised, therefore the proxy has to be periodically renewed. The user never sends her or his private certificate to anyone but the proxy can be forwarded to a service for authentication purposes.

In the current solution, the server needs to have access to the Grid resources directly, i.e. the server must be the member of a VO and have its own certificate. The proxy renewal is accomplished by the server automatically, and the clients' jobs are run under this single Grid user of the server. This means that the client does not have to deal with any detail of gLite: from its own view, the execution is exactly the same as in the case of a server-side local job or a Condor job.

However, a finer grain of access control among different clients of one Salève server should be developed on the server side. Additionally, the server needs to be the member of a VO, but the policies of the currently existing VOs might not allow several people to use the same user certificate, mainly because it would be difficult to find the person responsible for a damaging action. We have a future plan to give a better solution for this problem detailed in Section 5.

5   Future Work

As mentioned in Section 4, integrating the authentication in the server plugin independently from the client leads to difficulties. It might be better to release the server from playing a role in the authentication, and then the client would send with the input data a valid certificate. The certificate sent could be a proxy, but the client side proxy generation could harm the client's lightweight property.

Another important point is to make even less work to adapt the traditional sequential programs to Salève. A major step could be in this direction the creation of the *webstream* interface. The present state of bulk file transfer implementation is not always satisfactory. It is only possible to retrieve complete files after the communicating partner has finished writing the whole data. There are situations however, where the files should be streamed on the fly with minimal delay. The future implementations of Salève will contain the support for webstreams to support on the fly streaming, which can be used in the same manner as other input/output stream objects of C++. Webstreams would allow better integration of the Salève client into the native C++ code, and the migrator would need less work and knowledge to make the program Salève-compatible.

Acknowledgements

Part of this work was funded by the Péter Pázmány program (RET-06/2005) of the National Office for Research and Technology, and the authors would like to thank EGEE project (EU INFSO-RI-031688).

References

1. Condor Project homepage, http://www.cs.wisc.edu/condor/
2. EGEE Information Sheets, http://www.eu-egee.org/information_sheets/

3. Foster, I., Kesselman, C.: The Grid 2: Blueprint for a New Computing Infrastructure. Morgan Kaufmann Publishers Inc., San Francisco (2003)
4. gLite 3 User Guide, http://glite.web.cern.ch/
5. gSOAP homepage, http://gsoap2.sourceforge.net/
6. Zs. Molnár and I. Szeberényi. Saleve: simple web-services based environment for parameter study applications. In: The 6th IEEE/ACM International Workshop on Grid Computing (2005)
7. SOAP Version 1.2. W3C Recommendation (June 24, 2003), http://www.w3.org/TR/soap12/

A Report on the Effect of Heterogeneity of the Grid Environment on a Grid Job

Ioannis Kouvakis[1] and Fotis Georgatos[2]

[1] Univerity of Aegean, Department of Mathematics,
gkouvakis@hep.ntua.gr
[2] University of Cyprus, Department of Computer Science,
fotis@mail.cern.ch

Abstract. Grids include heterogeneous resources, which are based on different hardware and software architectures or components. In correspondence with this diversity of the infrastructure, the execution time of any single job, as well as the total grid performance can both be affected substantially, which can be demonstrated by measurements.

In need to effectively explore this issue, we decided to apply micro benchmarking tools on a subset of sites on the EGEE infrastructure, employing in particular the lmbench suite, for it includes latency, bandwidth and timing measurements.

Furthermore we retrieved and report information about sites characteristics, such as kernel version, middleware, memory size, cpu threads and more. Our preliminary conclusion is that any typical grid can largely benefit from even trivial resource characterization and match-making techniques, if we take advantage of this information upon job scheduling.

These metrics, which in this case were taken from the South Eastern Europe VO of EGEE, can provide a handle to compare and select the more suitable site(s), so that we can drive the grid towards maximum capacity and optimal performance.

1 Introduction and Outline

Grid computing emphasizes on the sharing heterogeneous resources. Those can be based on varying hardware/software architectures and computer components and may be located at different sites belonging to multiple administrative domains. They are interconnected by a network and a middleware which both have open standards and this is where their commonality ends.

The objective of this work is to document how the grid enviroment heterogeneity can affect a job running on a grid environment, so we send the same benchmarking program over different grid sites of the South Eastern Europe Virtual Organization. We contemplate that if those metrics are consistent over time we could reach better performance of the grid, simply by profiling the current infrastructure and tools.

The effects of Grid Heterogeneity were been measured from a user point of view. Various aspects of the WN enviroment can influence benchmarking measurements, including hyperthreading, deamon and rogue processes, other users'

I. Lirkov, S. Margenov, and J. Waśniewski (Eds.): LSSC 2007, LNCS 4818, pp. 476–483, 2008.

jobs, etc. We understand that this is an effect we can't isolate and the measurements are indeed objective and representative for any given grid job.

A glossary concerning Grid can be found on the EGEE-II Technical Pages on egee-technical.web.cern.ch/egee-technical/documents/glossary.htm.

2   Related Work

Lmbench has been extensively used for profiling Linux kernels by its own developers [3]. Indeed, such benchmarking techniques have already been demonstrated to be of interest [1]. The current -static- Information System-based practices should be replaced by -dynamic- characterization of grid resources and can be greatly advanced in at least some cases [6].

A similar approach was made by the developers of the Gridbench [7], which is a tool for evaluating the performance of Grids and Grid resources through benchmarking. Measurements were taken from the CrossGrid testbed for resources characterization. Lately more tests have been done on the EGEE infrastructure, with more conclusive results [8].

3   Issues and Methodology

The current grid provides insufficient information for sites' characteristics, which results in longer queue and job execution times and, indirectly, to more failures. The information available currently on the Information System is total memory of a node, OS distribution name and version, processor model and total cpus per site. This information is not always complete and the data that provide technical specifications, such as processor model, total memory and total cpus per site are far from optimal in selecting among sites.

Benchmarks are standardized programs or detailed specifications of programs designed to investigate well-defined performance properties of computer systems according to a widely demonstrated set of methods and procedures. For many years, benchmarks have been used to characterize a large variety of systems ranging from CPU architectures and caches to file-systems, databases, parallel systems, Internet infrastructures and middleware. Computer benchmarking provides a commonly accepted basis for comparing the performance of different computer systems in a fair manner, so it appears appealing to use them for resource metrics.

Operating Systems, like Linux, include tools that provide us with technical information regarding cpu speed, detailed model, vendor and threads, total and available memory for system and swap space. In addition to that, we are able to collect and supplement information on total and available size for disk and in some cases the model of the hard disk, linux distribution, kernel version and middleware version and also the number of users for the current VO per node. This information collectively can both document the heterogeneity of the grid, as well as provide input for processes that have particular needs for their environment.

In this paper we see real results of production sites, by use of the lmbench benchmarking suite which is a set of microbenchmark tools, and a collection of system information based on linux tools, mostly driven by customized scripts for the purpose of this work, which provide extra data and a detailed study of interactions between the operating system and the hardware architecture, e.g. in disk throughput.

LMbench [5], is a suite of micro benchmarking tools that measure a variety of important aspects of system performance. It is written in a portable ANSI-C code using POSIX interfaces that makes it possible to run on a wide range of systems without modification. Those micro-benchmarks report either latency or bandwidth of an operation or data pathway which is the most significant performance issues. We used lmbench-3.0-a4 released [4] for its better benchmarking techniques and more micro-benchmarking tools.

4   The Results

A script based on the python language has been developed, in order to manage the tools, and get the results from the sites in an organized manner; the results were exported in csv format and, combined with data from the Information System, the outcome helped to generate the charts and tables below. More charts and tables can be found in the technical report which is posted on arxiv [2]. Full result measurements can be found on the university's site http://www.fme.aegean.gr/research/physics/index.html.

In the sites ce101.grid.ucy.ac.cy, wipp-ce.weizmann.ac.il and ce.ipp.acad.bg we observed an internal heterogeneity. Those sites were using two different types of hardware for nodes and the results as being an average of those are less valid. We don't know, also, the exact number of machines from each type, so our uncertainty is even greater. In the future, this aspect could be addressed by the feature of Subclusters, which is supported in the GLUE schema, but unfortunately not by the current gLite-based systems.

Furthermore, even for the rest of the sites that have coherent results, we can hardly have any guarantee for their internal homogeneity or consistency; for, we have obtained no information on systems' structure and their evolution over time.

It is possible to observe that a substancial percentage of sites use the latest linux distributions 1(a) and kernel versions 1(b) and a take advantage of smp-capable kernel versions which can manage multiple execution queues and process threads in a satisfying manner. Most sites have upgraded to the latest middleware version as of March 2007.

The amount that we counted were about 1900 cpus on about 600 nodes. Cpus are mostly Intel with four cpus 1(e) and a small percentage of AMD processors. Each cpu, hyperthreaded or not, on each node can use full system memory or share it with other cpus on the same node.

Nodes are based on Intel Xeon 1.6GHz, Intel Xeon 3.4GHz and Intel P4 2.66GHz processors and there are few with AMD and other Intel models 1(f).

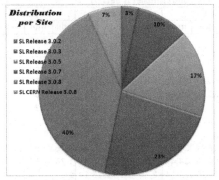

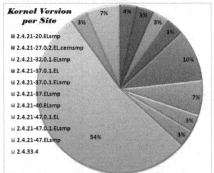

(a) Distribution Per Site

(b) Kernel Version Per Site

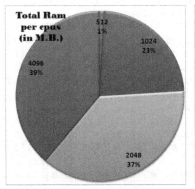

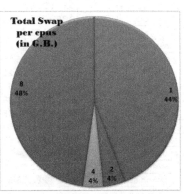

(c) Total Ram per Node*Cpu

(d) Total Swap per Node*Cpu

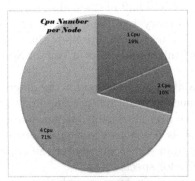

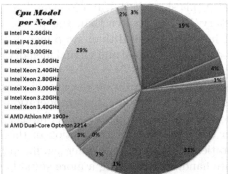

(e) CPUs Number per Site

(f) CPU Models

Basic Number Operations - in nanoseconds (smaller is better)											
SiteName	Integer					Float			Double		
	bit	add	mul	div	mod	add	mul	div	add	mul	div
ce.hep.ntua.gr	0,38	0,36	3,59	23,26	30,07	2,26	3,65	16,48	2,25	3,67	16,42
ce.phy.bg.ac.yu (lcg)	0,42	0,28	5,26	21,48	25,51	3,00	3,41	16,56	3,96	2,85	15,93
ce.ulakbim.gov.tr	0,19	0,19	5,30	21,79	24,26	1,88	2,64	16,26	1,88	2,63	16,26
ce001.imbm.bas.bg	0,18	0,18	5,03	20,68	23,01	1,78	2,50	15,43	1,78	2,50	15,43
ce001.ipp.acad.bg (glite)	0,33	0,33	3,34	20,72	27,80	2,00	2,67	15,11	2,01	2,68	15,14
ce002.ipp.acad.bg (lcg)	0,25	0,25	5,05	23,08	27,36	2,06	2,84	17,10	2,06	2,84	17,10
ce01.afroditi.hellasgrid.gr	0,29	0,29	2,94	18,24	24,63	1,76	2,35	13,30	1,76	2,36	13,31
ce01.ariagni.hellasgrid.gr	0,29	0,29	2,94	18,24	24,64	1,76	2,35	13,30	1,76	2,35	13,30
ce01.athena.hellasgrid.gr (long)	0,29	0,29	2,98	18,24	24,52	1,76	2,35	13,31	1,76	2,35	13,31
ce01.athena.hellasgrid.gr (see)	0,29	0,29	2,94	18,24	24,47	1,76	2,35	13,31	1,76	2,35	13,30
ce01.isabella.grnet.gr	0,27	0,22	5,18	22,19	25,14	2,54	10,89	15,54	2,85	5,14	15,52
ce01.kallisto.hellasgrid.gr	0,32	0,31	2,96	18,59	24,64	1,76	2,35	13,31	1,76	2,39	13,31
ce01.marie.hellasgrid.gr	0,29	0,31	2,94	18,24	24,54	1,77	2,36	13,30	1,76	2,35	13,30
ce02.athena.hellasgrid.gr (long)	0,29	0,29	2,94	18,24	24,48	1,76	2,35	13,30	1,76	2,35	13,30
ce02.athena.hellasgrid.gr (see)	0,30	0,29	2,95	18,24	24,67	1,76	2,35	13,30	1,76	2,35	13,30
ce02.marie.hellasgrid.gr	0,63	0,63	2,50	25,67	26,90	2,51	2,50	15,05	2,50	2,50	15,06
ce101.grid.ucy.ac.cy	0,39	0,39	1,16	15,83	16,21	1,54	1,54	9,30	1,55	1,54	9,29
ctb31.gridctb.uoa.gr	0,33	0,33	3,33	20,67	27,72	2,00	2,67	15,07	2,00	2,67	15,07
cox01.grid.metu.edu.tr	0,63	0,63	1,88	20,88	12,71	1,88	3,13	23,89	1,88	3,13	23,90
g02.phy.bg.ac.yu (glite)	0,43	0,27	5,13	21,66	24,18	4,68	5,79	16,64	4,69	5,62	16,69
grid-ce.ii.edu.mk	0,51	0,29	5,25	22,75	24,57	4,79	5,46	15,52	4,23	5,81	15,54
grid001.ics.forth.gr	0,36	0,36	3,58	22,20	29,99	2,15	2,87	16,21	2,15	2,87	16,21
grid01.cu.edu.tr	0,63	0,63	1,88	20,85	12,71	1,88	3,13	23,85	1,88	3,13	23,85
grid01.erciyes.edu.tr	0,63	0,63	1,88	20,84	12,71	1,88	3,13	23,85	1,88	3,14	23,85
kalkan1.ulakbim.gov.tr (lcg)	0,64	0,63	1,88	20,85	12,72	1,88	3,22	23,85	1,88	3,13	23,85
kalkan2.ulakbim.gov.tr (glite)	0,63	0,63	1,88	20,85	12,71	1,88	3,13	23,85	1,88	3,13	23,86
node001.grid.auth.gr	0,24	0,23	4,39	20,87	24,25	1,80	2,66	15,75	1,81	2,63	15,74
testbed001.grid.ici.ro	0,27	0,22	5,56	23,06	25,64	1,96	4,96	18,03	5,97	4,39	18,02
wipp-ce.weizmann.ac.il	0,82	0,82	3,91	34,44	38,94	2,78	4,42	32,02	2,78	4,42	32,00

Fig. 1. Basic Number Operations

The majority of the nodes are based on 4096 or 2048 MB of system Ram 1(c) making this share better and job running faster. On half of the nodes, the swap memory 1(d) is greater than the physical memory but there are many sites that have less physical than swap memory. The overall picture hints that some standardization in real and virtual memory organization aspects of Worker Nodes is missing.

The variety of hardware resources is certainly a good reason for the different job timings. If a job is based on operations such as addition, division, multiplication or modulo for float, double, integer (see Fig. 1) or 64bit integer numbers then the total time for completion of the job is based on the timings of those operations. Those timings differ significantly from site to site and are an appropriate handle for selecting a more suitable site for specific jobs.

In fact, it appears that performance differences are huge enough that a continuous and extremely exhaustive grid benchmarking system imposing a total 10% overhead on the total resources of the grid, even if it could only provide an

File & VM system latencies in microseconds (smaller is better)							
Site	OK File	OK File	10K	10K	Mmap	Prot	100fd
ce.hep.ntua.gr	27,17	22,27	317,27	491,57	9641	3,68	15,83
ce.phy.bg.ac.yu (lcg)	45,90	31,80	147,10	218,70	3364	5,18	22,50
ce.ulakbim.gov.tr	13,40	7,40	45,43	13,00	8920	2,74	2,84
ce001.imbm.bas.bg	15,10	10,74	73,37	31,87	3517	2,15	6,11
ce001.ipp.acad.bg (glite)	21,20	18,60	68,00	34,20	8809	2,83	12,40
ce002.ipp.acad.bg (lcg)	25,60	28,90	249,50	85,57	9702	2,95	6,11
ce01.afroditi.hellasgrid.gr	19,30	16,47	66,87	30,73	17267		11,57
ce01.ariagni.hellasgrid.gr	18,97	16,20	64,80	30,57	7990	2,54	11,37
ce01.athena.hellasgrid.gr (long)	18,23	15,60	63,10	29,47	16600	2,50	10,87
ce01.athena.hellasgrid.gr (see)	18,27	15,63	61,40	29,50	16567	2,50	10,87
ce01.isabella.grnet.gr	33,37	29,33	114,70	53,53	11222		20,20
ce01.kallisto.hellasgrid.gr	21,73	19,33	71,37	61,10	16600	2,74	12,17
ce01.marie.hellasgrid.gr	18,90	16,17	64,07	31,77	17000	2,58	11,40
ce02.athena.hellasgrid.gr (long)	18,20	15,70	60,50	29,50	16600	2,51	10,90
ce02.athena.hellasgrid.gr (see)	18,60	15,60	68,70	29,90	16600	2,51	10,80
ce02.marie.hellasgrid.gr	22,97	19,53	98,53	38,43	8323	3,15	6,74
ce101.grid.ucy.ac.cy	11,87	9,43	41,73	18,73	11414	2,18	4,17
ctb31.gridctb.uoa.gr	21,40	18,80	65,20	32,80	17000	2,68	12,30
cox01.grid.metu.edu.tr	19,10	15,10	60,30	28,10	17000	3,88	11,80
g02.phy.bg.ac.yu (glite)	35,00	31,00	139,70	60,30	17800	5,57	22,00
grid-ce.ii.edu.mk	35,40	32,50	91,60	46,50	5581	4,31	18,60
grid001.ics.forth.gr	22,80	19,70	74,20	36,00	3070	1,99	13,40
grid01.cu.edu.tr	19,00	15,10	58,40	27,50	16900	2,82	11,90
grid01.erciyes.edu.tr	19,30	15,30	89,70	27,90	16900	2,86	11,90
kalkan1.ulakbim.gov.tr (lcg)	17,73	15,10	56,73	26,53	7515		11,93
kalkan2.ulakbim.gov.tr (glite)	17,87	15,10	57,40	26,10	8504		11,93
node001.grid.auth.gr	26,40	24,07	87,43	42,80	11067	3,31	18,30
testbed001.grid.ici.ro	82,60	80,33	274,07	951,53	17876	3,98	22,10
wipp-ce.weizmann.ac.il	26,60	21,93	95,17	41,30	21633	3,79	14,63

Fig. 2. File And VM System Latencies Integer Operations

average 20% improvement upon job scheduling -on the RB or WMS - it would still be benefial for the grid system as a whole, as well as nearly any individual grid job.

Still, it should be possible to rip most the benefits of the technique with only 0.1% overhead, by doing selective benchmark scheduling at the moments that interesting results could occur, i.e. reconfiguration of clusters, and only do some rare regular runs for validation purposes.

Also, there are jobs that use local storage for temporary data manipulation and those need better latencies in creating / deleting files. It is possible to find examples of sites with low-end processors or low-ram that still have the best timings for those jobs. For reference, we suggest comparing Fig. 1 and 2.

5   Conclusions

We observed that there are important differences in sites' characteristics, and provided concrete output of some typical benchmark results.

Most sites do typically have internal homogeneity, i.e. most large sites include the same kind of Worker Nodes, which helps us categorize them depending on the measurements. Some percentage of the sites still present internal heterogeneity, which we suggest as a reseach topic for future work.

Timings of the various microbenchmarks are presented in nanoseconds, and even when they appear very small according to the measurements, once they occupy a repetitive part of a grid job's process they can have an important and highly-impacting factor: A job that could be executed in a site in a time period T, in some other site could be executed in the half time, T/2; ignoring at this point the communication overheads.

Also, in the special case that a grid job can be further parallelized, each part could be sent in the most suitable site depending on the nature of the subprocess, and then the time of individual subjobs will be also decreased, and consequently the total time as well.

Finally, it is impossible for each user to know or measure the characteristics of each site. Therefore some mechanism must exist that allows the matchmaking to happen in an automatic way. There is some ongoing discussion if the best way to implement this would be through a job description technique (i.e. in the .jdl file), or at the global scheduling stage (i.e. RB or WMS), or both. The latter of course is advantageous, if it is combined with an Information System that can provide such benchmarking results; then it becomes possible for the middleware to identify the sites that are best for a specific job, assuming all other issues equal. For once thing, resource ranking is deemed necessary [8].

We hope that this information will be used for further cluster performance research and that it will help future system administrators choose better hardware and/or software components during the deployment of new clusters.

In fact, a new era begins where instead of brute-force usage of resources, we will be able to load-balance grids according to their true capabilities, just as is envisaged in power, transportation and communication systems.

Acknowledgements

The authors of this paper would like to thank Prof. Ioannis Gialas for his useful remarks and valuable support, as well as NTUA and GRNET for providing the infrastructure necessary for this experiment.

References

1. Kenny, E., et al.: Heterogeneous Grid Computing: Issues and Early Benchmarks. In: Sunderam, V.S., et al. (eds.) ICCS 2005. LNCS, vol. 3516, pp. 870–874. Springer, Heidelberg (2005)
2. Kouvakis, J., Georgatos, F.: A Technical Report On Grid Benchmarking using SEE V.O., TR-MOD07-001. University of Aegean (March 2007),
   http://arxiv.org/pdf/cs/0703086
3. Mackerras, P.: Low-level Optimizations in the PowerPC Linux Kernels. IBM Linux Technology Center OzLabs (July 2003)

4. Staelin, C.: lmbench3: Measuring Scalability (November 2002)
5. Staelin, C., McVoy, L.: LMBench: Portable tools for performance analysis. In: USENIX Winter Conference (January 1996)
6. Tirado-Ramos, A., et al.: Grid Resource Selection by Application Benchmarking for Computational Haemodynamics Applications. In: Sunderam, V.S., et al. (eds.) ICCS 2005. LNCS, vol. 3514, pp. 534–543. Springer, Heidelberg (2005)
7. Tsouloupas, G., Dikaiakos, M.: GridBench: A Tool for Benchmarking Grids. In: Proceedings of the 4th International Workshop on Grid Computing (November 2003)
8. Tsouloupas, G., Dikaiakos, M.: Grid Resource Ranking using Low-level Performance Measurements, TR-07-02 Dept. of Computer Science, University of Cyprus (February 2007)

Agents as Resource Brokers in Grids — Forming Agent Teams

Wojciech Kuranowski[1], Marcin Paprzycki[2], Maria Ganzha[3],
Maciej Gawinecki[3], Ivan Lirkov[4], and Svetozar Margenov[4]

[1] Software Development Department, Wirtualna Polska
ul. Traugutta 115C, 80–226 Gdansk, Poland
wkuranowski@wp-sa.pl
[2] Institute of Computer Science, Warsaw School of Social Psychology, ul.
Chodakowska 19/31, 03–815 Warszawa, Poland
marcin.paprzycki@swps.edu.pl
[3] Systems Research Institute, Polish Academy of Science,
ul. Newelska 6, 01-447 Warszawa, Poland
{maria.ganzha,maciej.gawinecki}@ibspan.waw.pl
[4] Institute for Parallel Processing, Bulgarian Academy of Sciences,
Acad. G. Bonchev, Bl. 25A, 1113 Sofia, Bulgaria
ivan@parallel.bas.bg, margenov@parallel.bas.bg

Abstract. Recently we have proposed an approach to utilizing agent
teams as resource brokers and managers in the Grid. Thus far we have
discussed the general overview of the proposed system, how to efficiently
implement matchmaking services, as well as proposed a way by which
agents select a team that will execute their job. In this paper we focus
our attention on processes involved in agents joining a team.

1 Introduction

In our recent work we have discussed how teams of software agents can be utilized
as resource brokers and managers in the Grid. Thus far we have presented an
initial overview of the proposed approach [7], studied the most effective way
of implementing yellow-page-based matchmaking services [6], and considered
processes involved in agents seeking teams to execute their jobs [5]. The aim of
this paper is to start addressing the question: how agent teams are formed?

To this effect, we start with an overview of the proposed system, consisting
of the basic assumptions that underline our approach, followed by a UML Use
Case Diagram. In the next section we discuss issues involved in agent to agent-
team matchmaking. The paper is completed with UML-based formalization of
the main process involved in agent joining an existing team, and report on the
status of the implementation.

2 System Overview

Let us start by making it explicit that in our work we follow these who claim
that software agents will play an important role in design, implementation and

I. Lirkov, S. Margenov, and J. Waśniewski (Eds.): LSSC 2007, LNCS 4818, pp. 484–491, 2008.

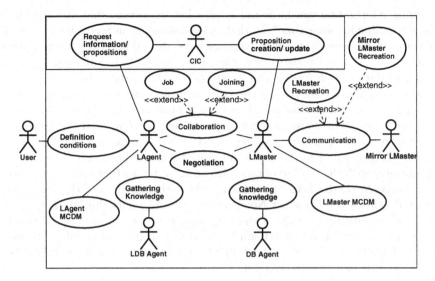

Fig. 1. Use Case diagram of the proposed system

long-term upkeep of large-scale software systems (see e.g. [9]). Second, our work assumes that software agents will be crucially involved in the future development of the Grid. While these two assumptions are not uncontroversial, arguments supporting them can be found, among others, in [8,10]. The latter assumption is further supported by the body of research devoted to combining software agents and the Grid; summarized in [5]. Finally, we view the Grid as a global infrastructure (rather than a local / laboratory-based Grid). As a result, we deal with a situation similar to the P2P environment, where no centralized control over individual Grid nodes is exerted.

As a result of these assumptions we have functionalized the Grid as an environment in which workers (in our case *agent workers*) that want to contribute their resources (and be paid for their usage), meet and interact with users (in our case *agent users*) that want to utilize offered services to complete their tasks and (in [7]) proposed a system based on the following tenets:

- agents work in teams (groups of agents)
- each team has a single leader—*LMaster agent*
- each *LMaster* has a mirror *LMirror agent* that can take over its job
- incoming workers (*worker agents*) join teams based on individual criteria
- teams (represented by *LMasters*) accept workers based on individual criteria
- decisions about joining and accepting involve multicriterial analysis
- each *worker agent* can (if needed) play role of an *LMaster*
- matchmaking is yellow page based [11] and facilitated by the *CIC agent* [3]

Combining these propositions resulted in the system represented in Figure 1 as a Use Case diagram. Let us now focus our attention on interactions between the *User* and its representative: *LAgent* and agent teams residing in the system (remaining information can be found in [7]). Let us assume that the system is already "running for some time", so that at least some agent teams have been already formed. As a result, team "advertisements" describing: (1) what resources they offer, and (2) characteristics of workers they would like to join their team are posted with the *Client Information Center* (*CIC*). Let us also note that the *User*, can either contribute resources to the Grid, or utilize resources available there. Interestingly, both situations are "Use Case symmetric" and involve the same pattern of interactions between agents representing the *User* and the system.

User who wants to utilize resources in the Grid communicates with its local agent (*LAgent*) and formulates conditions for executing a job. The *LAgent* communicates with the *CIC* to obtain a list of agent teams that satisfy its predefined criteria. Next, the *LAgent* communicates with *LMasters* of the remaining teams and utilizes the Contract Net Protocol [1] and multicriterial analysis [4] to evaluate obtained proposals. If the *LAgent* selects a team to execute its job, a contract is formed. If no such team is found (e.g. if nobody is willing to execute a 10 hour job for 5 cents), the *LAgent* informs its *User* and awaits further instructions (for more details see [5]).

The remaining part of the text will be devoted to the situation when *User* requests that its *LAgent* joins a team and works within it (e.g. to earn extra income for the *User*).

3   Selecting Team to Join

The general schema of interactions involved in *LAgent* selecting the team to join is very similar to that described above. First, the *User* specifies the conditions of joining, e.g. minimum payment for job execution, times of availability etc. Then she provides its *LAgent* with the description of resources offered as a service, e.g. processor power, memory, disk space etc. The *LAgent* queries the *CIC* which agent teams seek workers with specified characteristics. Upon receiving the list of such teams, it prunes teams deemed untrustworthy (e.g. teams that did not deliver on promised payment) and contacts *LMasters* of the remaining teams (if no team is left on the list, the *LAgent* informs its *User* and awaits further instructions). Negotiations between the *LAgent* and the *LMasters* take form of the FIPA Contract Net Protocol [1]. The summary of this process is depicted as a sequence diagram in Figure 2. For clarity, this sequence diagram is simplified and does not include possible "negative responses" and/or errors. Note that registering with the *CIC* takes place only once — when a new *LAgent* joins the system (or when it wants to start anew, i.e. to erase bad reputation). All subsequent interactions between the *CIC* and a given *LAgent* involve only checking credentials. The sequence diagram includes also processes involved in "mirroring". In our system we assume that the *LMaster* has its mirror, the *LMirror* agent. The role of this agent is to become the *LMaster* in the case when the current *LMaster* "disappears". Let us note that it is only the *LMaster* that has complete information about team members, jobs that are executed (and by whom), etc. Therefore, disappearance of the *LMaster* would imemdiately "destroy the

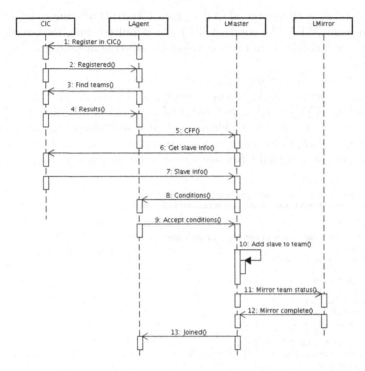

Fig. 2. Sequence diagram of interactions when an agent is seeking a team to join

team". In an attempt to avoid such a situation the *LMaster* shares all vital information with the *LMirror*. Obviously, it is possible that both the *LMaster* and the *LMirror* "go down" simultaneously, but our goal is only to introduce some degree of resilience (not to build a fault tolerant environment). Since this subject is out of scope of this paper it is omitted from further considerations.

3.1   Representing Conditions of Joining

Let us now discuss representation of (1) resources that the *LAgent* brings to the team, and (2) its conditions of joining. Before we proceed, let note that in an ideal situation, an all-agreed "ontology of the Grid" (that would include both the resources and the economical model) would exist. Unfortunately, while there exist separate and incompatible attempts at designing such an ontology, currently they are only "work in progress". Therefore, we focused our work on designing and implementing agent system skeleton, while using simplistic ontologies (and thus all proposals presented below should be viewed with this fact in mind). Obviously, when the Grid ontology will be agreed on, our system can be easily adapted to utilize it. In [7] we presented our ontological representation of computational resources. Here, we describe parameters used to negotiate conditions of joining.

Currently we utilize three parameters of joining: (1) price per work-hour, (2) work time—specific times of the day when the resource is to be available, and (3) length of contract—time interval that a given *LAgent* is offering to be a member of a given team. While the contract holds for a limited time, we assume that if both sides are satisfied, it can be extended for subsequent (and possibly longer) time periods.

What follows is an instance of joining conditions that ontologically depicts a computer: (1) with an Intel processor running at 3 GHz, (2) that offers to users 256 Mbytes of RAM and (3) 20 Gbytes of disk space, and that is offered to the team under the following conditions: (4) it is available every night between 23:50 and 8:15, and (5) wants to sign a contract for 7 days. Note that payment conditions are not specified (they are a part of the response of the *LMaster*).

```
(cfp
  :sender (agent-identifier :name proteus@bach:1099/JADE)
  :receiver (agent-identifier :name zerg@chopin:1099/JADE)
  :content
    ((action
      (agent-identifier :name zerg@chopin:1099/JADE)
      (take-me
        :configuration (hardware
          :cpu 3.0
          :memory 256
          :quota 20)
        :conditions (condition
          :availability (every-day
            :when (period
              :from 00000000T23500000
              :to 00000000T08150000))
        :contract-duration +00000007T000000000))
  :language fipa-sl0
  :ontology joining-ontology
  :protocol fipa-contract-net
)
```

This type of an information is used in two situations. First, each team looking for members advertises the resources it is looking for. Such an advertisement is an instance of an ontology, where parameters with numerical values (e.g. processor speed or available disk space) are treated as minimal requirements, while parameters that describe necessary software are hard constraints that have to be satisfied. Note that descriptions of sought workers include only resource parameters, but they do not include specific offers related to, for instance, payments for working for the team. In this way, when the *LAgent* requests list of teams that look for members, information about its own resources is used as a filter. For querying ontologically demarcated information we use SPARQL query language [2]. Therefore when the *LAgent* representing the above described computer communicated with the *CIC*, the following SPARQL query is executed.

```
PREFIX Grid: <http://Gridagents.sourceforge.net/Grid#>
SELECT ?team
WHERE {
  ?team Grid:needs ?machine .
  ?machine Grid:hasCPU ?cpu ;
           Grid:hasMemory ?mem ;
           Grid:hasQuota ?quota .
  FILTER ( ?cpu <= "3.0"^xsd:float ) .
  FILTER ( ?mem <= "256"^xsd:integer ) .
  FILTER ( ?quota <= "20480"^xsd:integer ) .
}
```

Second, when the *LAgent* issues a CFP (arrow 5 in Figure 2), the complete information describing resources and conditions of joining is included in the CFP and is used by the *LMaster* to prepare an offer. Let us note that specific offers are based on: available resources, overall availability to do the job, etc. Furthermore, note that each time a given *LAgent* issues a CFP it may specify different resource as: (1) the same *LAgent* may represent *User*'s different machines, or (2) for a single machine at one time available disk space may be 5 Gbytes, while at another time 25 Gbytes (e.g. depending on the number of stored MP3 files).

3.2   Negotiations

Let us now focus our attention on negotiations. The first step is the *LAgent* sending a CFP (arrow 5 in Figure 2) containing resource description and conditions of joining (see ontology snippet above). Upon receiving the CFP each *LMaster* contacts the *CIC* to make sure that this particular *LAgent* is registered with the system (arrows 6 and 7 in Figure 2). To somewhat improve safety of the system we assume that only *LAgents* that are registered with the *CIC* can join agent teams.

On the basis of the CFP, *LMasters* prepare their response. First, CFPs that do not satisfy hardware / software requirement are refused (e.g. worker that does not have Maple, cannot join a team that requires Maple). Second, each *LMaster* utilizes its knowledge about past jobs to establish base price per hour and base system that matches it. Currently, this price is split between the three components that exist in our ontology (processor speed: P_b, memory: M_b, disk space: D_b). As a result we obtain processor cost P_c, memory cost M_c and disk cost D_c (such that the base cost $B_c = P_c + M_c + D_c$). This information is used to estimate the "value" of the new worker in the following way, (assume that the new worker has processors speed P, memory M and disk space D):

$$Cost = \alpha\Big(\frac{P}{P_b}P_c + \frac{M}{M_b}B_c + \frac{D}{D_b}D_c\Big), \qquad (1)$$

Where $\alpha \in [0,1]$ denotes the overhead charged by the *LMaster*. Obviously, this model is extremely simplistic, but our goal was not to build a complete economical model of the Grid (for this, one would need a Grid ontology), but to specify a replaceable function that can be used in our system skeleton.

Responses from *LMasters* can have the following forms: (1) refusal (an ACL REFUSE message), (2) lack of response in a predefined by the *LAgent* time, (3) a specific offer (an ACL PROPOSE message). The *LAgent* awaits a specific time for responses and then finds the best of them (currently the response contains only the proposed price; as soon as a more complicated response is to be used a multicriterial analysis has to be applied). If the best available offer is above its own private valuation an agent team is selected to be joined (arrow 9 in Figure 2). If no acceptable offer is received, *User* is informed and *LAgent* awaits further instructions. Note that, the final confirmation is depicted as arrow number 11 in Figure 2. According to the Contract Net Protocol, since the *LAgent* was the originator of the negotiations, it has to be the receiver of the final confirmation.

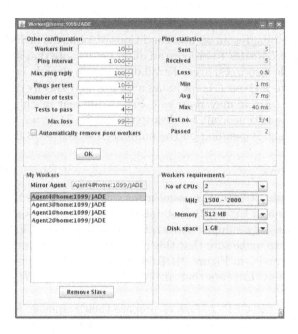

Fig. 3. GUI of the *LMaster* agent

3.3   Implementation

Currently we are implementing the above described processes. Note that they cannot be implemented without additional mechanisms involved in agent team management (that were omitted here due to the lack of space). To illustrate the state of our implementation, in Figure 3, we present the GUI of the *LMaster* agent. Most important informations, in the context of this paper, are (1) the *Workers requirements* box and (2) the *My Workers* box. The first one specifies that this *LMaster* is interested in workers that have 2 processors running at between 1.5 and 2.0 GHz, minimal memory of 512 Mbytes and disk space of 1 Gbyte. At the same time we can see that this *LMaster* is currently managing a team of 4 workers.

The *Other configuration* box represents options related to agent team management. We can see there that this *LMaster* will accept no more that 10 workers, as well as a number of parameters used to monitor which worker agents are down and thus will not continue executing their jobs. Finally, the *Ping statistics* box provides statistical results of monitoring sessions. Describing these (already working) mechanisms is outside of scope of this paper.

4   Concluding Remarks

The aim of this paper was to discuss processes involved in an agent joining a team, conceptualized within the framework of the proposed earlier agent-team-based

Grid resource brokering and management system. Processes described in this paper, while relatively simplistic, can be easily augmented to a more robust version. Currently we are proceeding with implementation of the above described processes. This involves also development of agent team management tools that have been briefly mentioned in Section 3.3.

Acknowledgments

Work presented here is a part of the Poland-Bulgaria collaborative grant: "Parallel and distributed computing practices".

References

1. Fipa contract net protocol specification,
   http://www.fipa.org/specs/fipa00029/SC00029H.html
2. Sparql query language for rdf, http://www.w3.org/TR/rdf-sparql-query
3. Bădică, C., et al.: Developing a model agent-based e-commerce system. In: Lu, J., et al. (eds.) E-Service Intelligence - Methodologies, Technologies and Applications, pp. 555–578. Springer, Berlin (2007)
4. Dodgson, J., et al.: DTLR multi-criteria analysis manual. In: UK: National Economic Research Associates (2001)
5. Dominiak, M., Ganzha, M., Paprzycki, M.: Selecting grid-agent-team to execute user-job—initial solution. In: Proceedings of the Conference on Complex, Intelligent and Software Intensive Systems, pp. 249–256. IEEE Computer Society Press, Los Alamitos (2007)
6. Dominiak, M., et al.: Efficient matchmaking in an agent-based grid resource brokering system. In: Proceedings of the International Multiconference on Computer Science and Information Technology, pp. 327–335. PTI Press (2006)
7. Dominiak, M., et al.: Utilizing agent teams in grid resource management—preliminary considerations. In: Proceedings of the IEEE J. V. Atanasoff Conference, pp. 46–51. IEEE Computer Society Press, Los Alamitos (2006)
8. Foster, I., Jennings, N.R., Kesselman, C.: Brain meets brawn: Why grid and agents need each other. In: Kudenko, D., Kazakov, D., Alonso, E. (eds.) AAMAS 2004. LNCS (LNAI), vol. 3394, pp. 8–15. Springer, Heidelberg (2005)
9. Jennings, N.R.: An agent-based approach for building complex software systems. CACM 44(4), 35–41
10. Tianfield, H., Unland, R.: Towards self-organization in multi-agent systems and grid computing. Multiagent and Grid Systems 1(2), 89–95 (2005)
11. Trastour, D., Bartolini, C., Preist, C.: Semantic web support for the business-to-business e-commerce lifecycle. In: WWW 2002. Proceedings of the 11th international conference on World Wide Web, pp. 89–98. ACM Press, New York (2002)

Parallel Dictionary Compression Using Grid Technologies

Dénes Németh

Budapest University of Technology,
Magyar Tudósok körútja 2, H-1117 Budapest, Hungary
nemeth.denes@iit.bme.hu
http://dictionary.homelinux.com

Abstract. This paper introduces a novel algorithm which approaches dictionary compression without the preliminary knowledge of the grammatical rules. Any type of languages except for incorporating ones can be processed by this solution in an effective way. The algorithm cuts words derived from the same stem into base word, prefix and suffix groups from which a hierarchical dictionary is constructed allowing spell checking, possible stem determination, and efficient distributed parallel pattern matching. By eliminating the severe redundancy in the word's simple treerepresentation, the compression ratio can be significantly better than by using conventional techniques.

1 Introduction

Nowadays, with the spread of different embedded systems, the need of an efficient, transparent dictionary compression is becoming more intense. The development of input methods is evolving from the unaccustomed formal commands to the natural human language. This is mainly caused by the fact that the amount of digitally exchanged information is accelerating in a tremendous rate. This information mainly consists of three parts: audio, video and text. In most cases the problems of audio and video compression have been extensively analyzed and partially solved by the industry due to the demanding public need. Since the demand for natural language support in electronic equipments is also increasing, it is indispensable to develop an effective method to compress and store languages.

Describing languages has several difficulties [2,3]. First, every language has its own specialities, which means that the structure of the languages is diverse, meanwhile the grammar varies a lot too. Second, the words derived from a stem can not be determined by the grammatical rules and the grammatical category of the stem only, the meaning has to be taken into account too [4]. This renders a grammatical rule based generative algorithm nearly useless. Third, the size of the uncompressed dictionary is extreme(5–40 GB), and it would be desirable to use the dictionary in an environment where the resources are limited. This means that the dictionary has to be compact enough to fit into the device, and has to be accessible through a low-cost methods, since the available computation resources

I. Lirkov, S. Margenov, and J. Waśniewski (Eds.): LSSC 2007, LNCS 4818, pp. 492–499, 2008.
© Springer-Verlag Berlin Heidelberg 2008

are limited too. These contradictionary requirements have to be simultaneously met in order to create a viable and widely usable system.

The main aim of this paper is to solve the problems of currently used dictionary compression methods, and provide a distributed multi-language dictionary which in the first step facilitates word level storage, but can be extended to support sentence level rules. The algorithm is able to extract the grammatical or meaning based rules from the input. This eliminate the dictionary's direct dependency form the generative grammatical rules. These extracted rules does not have to be consistent, but the more consistent they are, the better the compression is. This construction allows high quality spellchecking, possible stem determination, and efficient parallel pattern matching which is difficult in traditional generative language compression.

The report can be divided into four parts. In the first section, the general problems and requirements are introduced, while the second gives a brief overview of the grammatical environment. The third part explains the compression and matchmaking algorithm, and how it can be applied in the described environment. In the last part we analyze the results of the measurements in case of the Hungarian language, which has one of the most complex grammatical structures among the languages of the world.

2   Environment

Before introducing the compression algorithm, I would like to describe the different features of the languages, which are relevant to dictionary compression. In typology, there are four categories: isolating, flecting, agglutinating and incorporating. None of the languages belong strictly to one of these categories. For example, English is mainly isolating, but partly shows agglutinating characteristics due to French influence.

In isolating languages, grammatical forms are expressed by separate words in the neighborhood of the original word. For example in *"I jump over"*, both *I* and *over* are connected to the verb *jump*. In contradiction to this, agglutinating languages like Hungarian attach affixes to the words like *"ÁtUgrOm"*, in which *"Át"* means *over* and *"Om"* means that I perform the jumping *"Ugr"* action. Flecting behavior means that the stem changes, for example *"I sing, I sang and I have sung."*, and incorporating behavior means the incorporation of the affix into the stem.

In every language, when words are altered, the allowed extensions are regulated not only by the grammatical category of the original word, but the meaning also. This causes that the compression algorithm can not depend only on the grammatical rules of a given language.

So far, most of the dictionary compression programs are based on the English language which is mainly isolating, with minimal stem alterations. However, these algorithms cannot give an optimal solution in languages showing different characteristics. For example, in agglutinating languages, the number of words derived from a stem can be 10^5 times more than in isolating languages.

Now that we have a basic understanding of the different features of the languages, I am going to describe how this novel algorithm is able to cope with all these problems and requirements.

3   The Algorithm

This section is divided into three parts: the first describes how a group of words can be converted into a compact structure. The second outlines how this structure can be added to the database, while the third depicts how the information can be retrieved from the database.

3.1   Chopping the Words

Let W denote a subset of derived words from one stem. Let us assume that the size of this set is N, and the words in this set are noted by $\omega_1, \omega_2, \cdots, \omega_N$. Let $\omega_{i,j}$ be the i-th word's j-th character, and let $|\omega_i|$ be the length of the i-th word. Let us define $*$ as the operation of concatenation, so $\omega_i = \omega_{i,1} * \omega_{i,2} * \cdots * \omega_{i,|\omega_i|}$. Let us define the base word (from this point forward B) the longest character sequence which is included in all the ω_i words. If this is epsilon, then word chopping stops and all of the original words should be considered as separate words derived from a different stem. If B is not ϵ, then the decomposition of the i-th word is $\omega_i = p_i * B * s_i$.

Let us call the parts before the base word prefixes (from this point onward P), and the parts after the base word suffixes (from this point onward S). Let $|P|$ be the number of different prefixes, and $|S|$ the number of different suffixes. Let us define $\overline{P}$, $\overline{S}$ and the $\overline{W}$ vectors as follows:

$$\overline{P} = \left[p_1, p_2, \cdots, p_{|P|} \right], \forall i \neq j : p_i \neq p_j \tag{1}$$

$$\overline{S} = \left[s_1, s_2, \cdots, s_{|S|} \right], \forall i \neq j : s_i \neq s_j \tag{2}$$

$$\overline{W} = \left[\omega_1, \omega_2, \cdots, \omega_N \right] \tag{3}$$

Decomposition Theory: Let us find the minimal $\overline{\overline{F_{2 \times M}}}$ size matrix, which is defined in the following way:

$$\overline{\overline{F_{2 \times M}}} = \begin{bmatrix} \overline{\overline{F}} \, [1, i] \subseteq \{1, 2, \cdots, |P|\} \\ \overline{\overline{F}} \, [1, i] \subseteq \{1, 2, \cdots, |S|\} \end{bmatrix} \tag{4}$$

and met the following two requirements:

$$- \text{ for } \quad \forall i, j \in \overline{\overline{F}} \, [1, i] \, , k \in \overline{\overline{F}} \, [2, i] \, \exists m : \overline{P} \, [i] * B * \overline{S} \, [k] = \overline{W} \, [m] \tag{5}$$

$$- \text{ for } \quad \forall m \quad \exists i, j \in \overline{\overline{F}} \, [1, i] \, , k \in \overline{\overline{F}} \, [2, i] : \overline{P} \, [i] * B * \overline{S} \, [k] = \overline{W} \, [m] \tag{6}$$

This informally means that all original and no other words can be generated by concatenating the appropriate (defined by the $\overline{\overline{F}}$ matrix) prefix, base word and suffix. The following example illustrates these matrices.

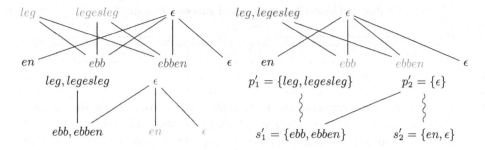

Fig. 1. Prefix and suffix combinations using graphs

$$\overline{W} = \left\{ \begin{array}{l} \mathbf{k\acute{e}k}, \mathbf{k\acute{e}k}en, \mathbf{k\acute{e}k}ebb, \mathbf{k\acute{e}k}ebben, leg\mathbf{k\acute{e}k}ebb, \\ leg\mathbf{k\acute{e}k}ebben, legesleg\mathbf{k\acute{e}k}ebb, legesleg\mathbf{k\acute{e}k}ebben \end{array} \right\}$$

$$B = \mathbf{k\acute{e}k} \qquad \overline{P} = \begin{bmatrix} \epsilon\ leg\ legesleg \end{bmatrix} \qquad \overline{S} = \begin{bmatrix} \epsilon\ ebb\ ebben\ en \end{bmatrix} \tag{7}$$

$$\overline{\overline{F}} = \begin{bmatrix} \{2,3\} & \{1\} \\ \{3,4\} & \{1,2,3,4\} \end{bmatrix} \tag{8}$$

It is clear that neither the $\overline{\overline{F}}[1,i] \cap \overline{\overline{F}}[1,j]$, nor the $\overline{\overline{F}}[2,i] \cap \overline{\overline{F}}[2,j]$ set has to be empty, but the compression rate becomes higher if the intersections of these sets are minimized. In the following section we go through the different steps needed to produce these decomposed sets in $\overline{\overline{F}}$.

Word decomposition

1. *Finding the longest base word:* We denote the length of the shortest (L) and longest (R) word in W by L an R respectively. Let us find one of (I chose the one to head of L) the longest character sequence of which is included by all words of W. The cost of this is $\mathcal{O}(N) + \mathcal{O}\left(N\frac{(L+1)}{2}R\right) = \mathcal{O}(NLR)$

2. *Finding the minimal number* $\overline{\overline{F}}[1,i]$ *and* $\overline{\overline{F}}[2,i]$ *set pairs:* Let us define the $G(P,S,E)$ bipartite graph, where $P = \{p_1, p_2, \cdots, p_{|P|}\}$, $S = \{s_1, s_2, \cdots, s_{|S|}\}$, and let $(p_i, s_i) \in E$ if $\exists w \in W$ that $p_i * B * s_i = w$. This means that the two set of vertexes should be the prefixes and suffixes, and only those should be connected which form a valid word by adding the prefix before, and the suffix after the base word. Let us define $N(X)$ as the neighbors of the X set of vertexes, where $X \subseteq P$ or $X \subseteq S$. Let us define x_1 and x_2 vertexes as combinable if $(x_1, x_2) \subseteq P$ or $(x_1, x_2) \subseteq S$ and $N(x_1) = N(x_2)$. Let us combine the P and S vertexes, that $\nexists x_1, x_2$, that $\{x_1, x_2\} \subseteq P$ or $\{x_1, x_2\} \subseteq S$ and $N(x_1) = N(x_2)$. This procedure is illustrated on Fig. 1.

These procedures can be effectively executed if we create a $\overline{\overline{K}}_{|P| \times |S|}$ matrix, in which the rows are the prefixes and the columns are the suffixes. It should contain 1 on the $[i,j]$ element, if $(p_i, s_j) \in E$, otherwise 0. Let us combine the identical columns and rows in the matrix, which is illustrated on table 1. The cost of this combination is $\mathcal{O}\left(|P|^2 + |S|^2\right)$.

Table 1. Combining prefixes and suffixes using matrices

	en	ebb	ebben	ε
leg		1	1	
legesleg		1	1	
ε	1	1	1	1

	en	ebb	ebben	ε
leg,legesleg		1	1	
ε	1	1	1	1

	en,ε	ebb	ebben
leg,legesleg		1	1
ε	1	1	1

	en,ε	ebb,ebben
leg,legesleg		1
ε	1	1

As there are many zeros in the matrix, this solution could be further accelerated if a list of edges is used. As a result we get a graph, the vertexes of which can not be combined any more. In the next step, we determine the size of the $\overline{\overline{F}}$ matrix and the elements itself.

Let us define a new $G'(P', S', E')$ bipartite graph, where P' is the combined prefixes and S' is the combined suffixes. Let $T(X')$ denote the original uncombined vertexes of G, where $X' \in P'$ or $X' \in S'$. Let us find the maximal vertex matching using the Hungarian method. P_M is defined as those vertexes in P', which are included in the maximal matching, and S_M is defined as those vertexes in S', which are included in the maximal matching. Let $P_{\overline{M}} = P' \setminus P_M$ and $S_{\overline{M}} = S' \setminus S_M$, and let $U(X)$ denote a set of vertexes, which can be reached through odd length alternating paths from the X set of vertexes. Let P_{opc} be those vertexes in P', which can not be reached from $S_{\overline{M}}$, and let S_{opc} be those vertexes in S', which cannot be reached from $P_{\overline{M}}$. These definitions are illustrated on Fig. 2.

Since edges represent words, our original goal was to determine the minimal number of vertexes needed to cover all edges. This minimal set can be constructed if we add either P_{opc} or S_{opc} vertexes to the $U(P_{\overline{M}}) \cup U(S_{\overline{M}})$ vertexes. The size of this set determines the second dimension's size of the $\overline{\overline{F}}$. Assuming that the minimal points are $v_1, v_2, \cdots, v_M$ we can determine the elements of the $\overline{\overline{F}}$ matrix.

$$\text{if } v_i \in P', \text{ then } F[1, i] = T(v_i) \text{ and } F[2, i] = N(T(v_i)) \qquad (9)$$

$$\text{if } v_i \in S', \text{ then } F[1, i] = N(T(v_i)) \text{ and } F[2, i] = T(v_i) \qquad (10)$$

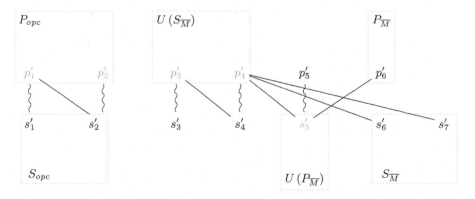

Fig. 2. Minimal vertex coverage in bipartite graphs

The determination of the $\overline{\overline{F}}$ matrix in our original example is illustrated on the last graph on Fig. 1. If we choose the P_{opc} set, then the two coverage points are p'_1 and p'_2. Since $v_1 = p'_1$ and the only neighbor of p'_1 is s'_1, then $\overline{\overline{F}}[1,1] = \{2,3\}$ and $\overline{\overline{F}}[1,2] = \{3,4\}$. The second coverage point is $v_2 = p'_2$ whose neighbors are s'_1 and s'_2, causing that $\overline{\overline{F}}[2,1] = \{1\}$ and $\overline{\overline{F}}[2,2] = \{1,2,3,4\}$.

The full cost of this procedure is: $\mathcal{O}\left(\mathbf{L^2 N}\right) + \mathcal{O}\left(|\mathbf{P}|^2 + |\mathbf{P}|^2\right) + \mathcal{O}\left(|\mathbf{E}|\,|\mathbf{P}|\right)$.

The dictionary database: The database consists of prefixes(P_i), suffixes(S_i) and base words(B) which are represented with a variable branching tree. A prefix is a path from a leaf to a root in P_i, while a suffix is constructed from a root to a leaf in S_i. Let PRE be the forest of the P_i trees, and let SUF be the forest of the S_i trees. Let PRE_i be the i-th tree in PRE, and SUF_i be the i-th tree in SUF. Let DIC be the words stored in the dictionary, and let $B_{b_1,b_2,\cdots,b_j}$ denote the last node in the B tree on the $b_1, b_2, \cdots, b_j$ path. Let $B_{b_1,b_2,\cdots,b_j}(END)$ be true if the $b_1 * \cdots * b_j$ word is valid, then $\omega \in DIC \Leftrightarrow \exists i,j,k$, that $\omega = p_1 * \cdots * p_i * b_1 * \cdots * b_j * s_1 * \cdots * s_k$ and $\exists B_{b_1,b_2,\cdots,b_j}$ and $B_{b_1,b_2,\cdots,b_j}(END)$ or $\exists P_{p_i,p_{i-1},\cdots,p_1}$ and $\exists S_{s_i,s_{i-1},\cdots,s_1}$.

Adding the created compact structure to the database: After we have executed the word decomposition, we have a set of P_i, S_i and a B variable branching trees. Let us merge the P_i with PRE and S_i with SUF, which means that if PRE already contains one of the P_i prefix then the tree in PRE is used instead of P_i. After the merging the affixes, B should be combined with $BASE$ and the already merged prefixes and suffixes should be connected to the last node of B. By this procedure the compact structure is integrated into the database.

4   Applying Grid Technology

In the last section we have introduced the algorithm. Since the data needed to be processed is enormous, this algorithm should be applied in a parallel way. The set of data can be cut into word groups, in which every word is derived from the same stem. These datasets are transformed into the compact structure using separate gLite[1] grid jobs. After a certain number of these results are processed over the grid, they are inserted into an empty database. These databases are recursively combined into each other resulting in one single database. It is clear that since the number of these word groups is huge, the problem can be interpreted as a parameter sweep problem. Meaning that the problem can be fully parallelized, and the outputs produced by the parallel processing can be easily merged together.

5   Measurement Results

This section presents our compression measurement results in the Hungarian language. We have processed 78000 word groups which had the total uncompressed size of 19.8 GB. After the compression, we have created two different

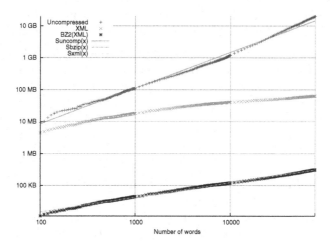

Fig. 3. The size of the processed data

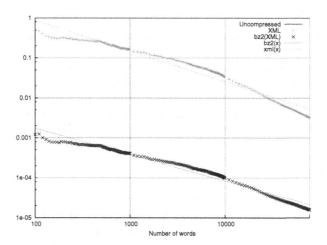

Fig. 4. The compression ratio

output formats: the first is XML, the second is a zipped format of this XML. It is trivial that the XML with its long labels contains a lot of simple redundancy, which can be eliminated with a run length based compression like LZW. The XML output was 65 MB, but after compression, it is only 318 KB, which is extremely low. This measurement is illustrated on Fig. 3. The compression ratio is approximately $1 : 10^5$ if only 78000 word groups (40% of all the word groups) are processed. The compression ratio in the case of XML and the binary output is illustrated on Fig. 4.

It is visible that the compression ratio decreases linearly on a double logarithmic scale graph. If we fit a power law using the least squares method on the compression dataset, we get $xml(x) = 2.8 \cdot 10^{2}1x^{-0.77}$ and $bz2(x) = 3.7 \cdot 10^{-2}x^{-0.66}$, which means that if the language contains 250000 words, then the size of the dictionary will be less than 1 MB. We also measured the cost of searches and pattern mattchings in the database. We found that the cost of the search only depends on the depth of the tree, which is allways less than the longest word, and in pattern matching the cost mainly depends on the number of result returned by the search.

6   Conclusions

By exploiting the structural redundancy of the data, any language can be efficiently compressed. The algorithm provides both a compact and searchable structure, and takes the needs of most embedded systems into account. By eliminating the dictionary's direct dependency from the generative grammatical rules, the algorithm is able to cope with the most serious problems of the currently used dictionary compression methods, which is that the derived words can only be determined if the meaning of the word is also considered. It also provides a powerful tool in the hand of linguists to analyze or compare the temporal changes in languages or individual texts.

Acknowledgments

Part of this work was funded by the Pázmány Péter program (RET-06/2005) of the National Office for Research and Technology, and the author would like to thank the EU EGEE (EU INFSO-RI-031688) program.

References

1. gLite Webpage, http://glite.web.cern.ch/glite
2. Revuzl, D., Zipstein, M.: DZ A text compression algorithm for natural languages. Combinatorial Pattern Matching, 315–333 (1992)
3. Dolev, D., Harari, Y., Parnas, M.: Finding the neighborhood of a query in a dictionary. Theory and Computing Systems, 33–42 (1993)
4. Nemeth, L., et al.: Leveraging the open-source ispell codebase for minority language analysis. In: Proceedings of SALTMIL (2004)

A Gradient Hybrid Parallel Algorithm to One-Parameter Nonlinear Boundary Value Problems

Dániel Pasztuhov and János Török

Budapest University of Technology and Economics, Budapest, Hungary
dani@iit.bme.hu

Abstract. We present a novel global algorithm for parallel computers, suitable to solve nonlinear boundary value problems depending on one parameter. The existing scanning and solution following algorithms are extended by a gradient method which is performed on an artificial potential created from the equation system. All three components of the algorithm can be parallelized and thus used in a GRID network. We validate our algorithms on a few small examples.

1 Introduction

Engineers need a lot of computing power to solve problems about the behaviour of the reinforced concrete beams on the influence of several forces [1]. These problems can often be described as boundary value problems (BVP). In this paper we present a novel method to handle these problems.

The boundary value problems can be traced back to finding the solutions of a non-linear equation system in a multidimensional space. These solutions are always a collection one dimensional objects (lines). Once a small part of one of them is found it can be followed in both directions which is the first component of our algorithm [4]. We use two methods to find pieces of the solution: a stochastic and a gradient one [5]. The latter is a new method in this field and is performed on a non-negative potential obtained by the transformation of the equation system, where the solutions are the minima of the potential with zero values.

The gradient extension does not make the algorithm scan the entire space much faster instead it can deliver the solutions much faster than the scanning algorithm. However, in principle we have to scan the whole space to find *all* solutions.

The aim of this work is to present the gradient algorithm. We implement the simplest possible solution to demonstrate the power of the new algorithm. As the dimension of the GRS [1] gets higher and the scanning of the whole GRS would need exponentially large times. The gradient algorithm with some extra calculation need helps to deliver the solutions earlier.

2 Boundary Value Problem

If we get the deformations of the beam as an integration along the length, so position, forces and moments are known at one end of the beam, we are talking

I. Lirkov, S. Margenov, and J. Waśniewski (Eds.): LSSC 2007, LNCS 4818, pp. 500–507, 2008.

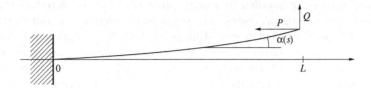

Fig. 1. Euler problem

about an *initial value problem (IVP)*. Most of the cases, the position and/or forces and/or moments are given in both end of the beam, which transfers it to a *boundary value problem (BVP)*.

We will illustrate our method on the example of the axially compressed, uniform, elastic cantilever beam, illustrated in Fig. 1. The ordinary differential equation describing the shape of the beam in terms of the slope α as a function of the arclength s was first described by Euler:

$$EI\alpha'' + P\sin\alpha + Q\cos\alpha = 0. \tag{1}$$

The vertical force Q will be used as a small imperfection parameter which is *constant* during the loading process. The trajectories of this equation are uniquely determined by the three scalars $\alpha(0)$, $\alpha'(0)$, and P (the former ones being 'true' initial conditions, the latter one a parameter, Q is treated as a constant). However, we are not interested in all trajectories, only the ones which meet the boundary conditions $\alpha(0) = 0$ and $\alpha'(L) = 0$, which express zero slope at the left end and zero curvature at the right end of the beam respectively.

If we denote $x_1 \equiv \alpha'(0)$, $x_2 \equiv P$, the scalars x_i are called *global coordinates*, the space spanned by $[x_1, x_2]$ will be called *Global Representation Space* (GRS) of the BVP and we denote its dimension by D. In these problems there are $D-1$ equations to be solved which results in a line (one dimensional object) as the set of solutions. The function we have to solve is $f_1 : \alpha'(L) = 0$. This example is a simple one, in case of more complex problems GRS can have more dimensions, some of them exceed even the 20th dimension!

In order to solve globally a BVP in *moderate dimensions*, our algorithm discretizes the GRS into hypercubes and splits up the cubes into simplices (that are triangles in two dimensions and tetrahedron in three dimensions) [2]. The original IVP (Eq. 1) is solved in each apex of the simplex, and the solution, if any is determined for each side by a linear interpolation algorithm. Solution points of all the simplex sides are taken as the global results of the BVP.

An acceleration for simplex algorithm is the path-continuation extension: If we find a solution anywhere in the GRS, we suppose that this solution continues in two ways (as it is said, solutions are one-dimensional objects), so we examine the neighbouring cubes of the solution provider cube to find the solutions sooner. Simplex algorithm extended by path-continuation method is called *hybrid* algorithm.

The parallel version of the hybrid algorithm [3] is implemented in the following way: The space is divided into large *primary hypercubes* (consisting of l^D

hypercubes) which are handled by a slave processes in order to reduce the communication need among processes. The primary hypercubes are not distributed arbitrarily to the slaves but a weighting is applied. If a solution was found to leave a primary hypercube on a specific side the neighbouring primary hypercube is marked with the maximum weight. On the other hand primary hypercubes with the most unchecked neighbours have higher weights in order to scan more distant hypercubes earlier and thus find the solutions earlier.

As number of dimensions increase, size of GRS increases *exponentially*. Hybrid algorithm is good enough for problems with dimensions not higher than about 6. Problems with higher dimensions cannot be handled by this algorithm.

3   Gradient Method

There is more information in the functions than it is used by hybrid algorithm. The main new idea is that we construct a potential which has minima at the solutions:

$$U(p, x_1, x_2, \ldots, x_n) = \sum_{i=1}^{n} c_i f_i^2(p, x_1, x_2, \ldots, x_n), \tag{2}$$

where c_i denote positive constants. The above construction ensures that the value of U is always non-negative and zero values indicate the solution. Thus a gradient method may be used to find these points.

The gradient algorithm is implemented to step from every point to the neighbouring hypercube in the direction of the largest gradient. It stops if it would leave the examined parameter space or if it found a local minimum, where the U is smaller than in the neighbouring hypercubes. The primary hypercube with the found local minima is marked with high weight for the hybrid algorithm irrespect of the value of U.

4   Implementation and Problems

The aim of the present work is to justify the effectiveness of the gradient algorithm without any further optimizations. There are many aspects that may render this algorithm useless. Since we have no *a priori* knowledge about the above potential it may contain too many local minima making the gradient algorithm useless. The different components of f_i are in general of different unit and thus can be of different magnitude which might introduce anomalies. In the followings we note other optimization or implementation possibilities in parallel with the present simplest choice of realization.

We have a parallel algorithm where the balance between different slave types must be synchronized. This requires an elaborate weighting of the unchecked primary hypercubes as well as a fine tuning of the slave types. We show here that even the dumbest choice can deliver a considerable performance increase. At the beginning we let the gradient algorithm run a few times ($1 - 10$ in our simple examples) and then we mark with high weight the primary hypercubes found as local minima by the gradient algorithm to be the first candidates for the hybrid algorithm then we switch back to the old algorithm.

The same simplicity is followed in choosing the initial points for the gradient algorithms which is done randomly. Since many gradient runs may find the same solution in the future it would be important that the starting points are well distributed in the GRS. In higher dimensions methods of determining attraction zone by some negative gradient algorithm might be also helpful.

The role of the c_i coefficients is to bring the values of the f_i of different units in the same magnitude. This can be done e.g. by an initial scan of the space where we calculate f in m points and set

$$c_i = \left(\frac{1}{m} \sum_{j=1}^{m} \left(f_i^{(j)} \right)^2 \right)^{-1} \tag{3}$$

Since there might be huge differences among quarters of the GRS the coefficients should be calculated for the perimeter of each gradient run. An other possibility is to let c_i evolve in time as the gradient algorithm advances but in this case one has to care about the algorithm making cycles. In our case the best choice was to take the simplest possibility: $c_i \equiv 1$.

We chose that the gradient method is stepping into the neighbouring hypercube in the direction of the largest gradient. This might be inappropriate in more complex problems where a conventional gradeint method should be used.

In summary the only difference compared to the hybrid algorithm is to run a few gradient algorithm slaves at the beginning and then switch back to the old algorithm with the found local minima being the first to be scanned by the simplex method.

The extra calculation of the gradient algorithm increases the overall computation need of the algorithm but the solutions may be found earlier. We also note that the gradient algorithm has much less computation need than the scanning one. The gradient algorithm generates the function values in the neighbouring hypercubes of the actual point. There are $2D$ such points. On the other hand to test whether a hypercube contains a solution the simplex algorithm has to calculate the functions at the corners which means 2^D function calls. It is also important to note that in the case of the gradient algorithm we do not check all the hypercubes of the primary hypercube but we follow the gradient which means again in average l^{D-1} factor advantage for the gradient method, where l is the linear size of the primary hypercube. This means that the gradient method can be considered instantaneous compared to the scanning algorithm in high dimensions.

5   Results

As we already mentioned the aim of the gradient algorithm is not to scan the GRS in less time but to find the solutions earlier than the stochastic algorithm. We chose to measure the time by calculating the number of function calls needed to find a primary hypercube with a solution in it. Measuring the time would be

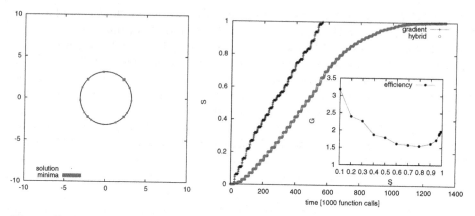

Fig. 2. Circle problem (a) The continuous line shows the solution curve the squares indicate the local minima found by the gradient algorithm, the area of the squares is proportional of its occurrence in the 100 runs. (b) The variation of S in time for both the gradient and hybrid algorithm. The inset shows the variation of the efficiency with the percentage of the found solutions.

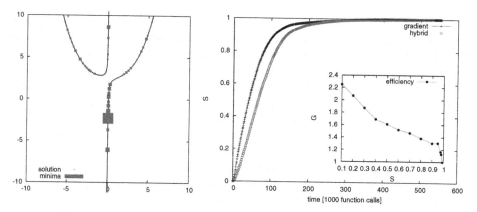

Fig. 3. Cantilever beam with $Q = 0.1$. Graphs are same as in Fig. 2.

misleading as more complicated problems have functions with heavy computation need which requires most of the evaluation time while the simple examples presented here spend relatively more time for communication etc. Therefore we measure the time in 1000 function calls and denote it by τ. The time is measured independently for each slave.

In all test cases we did the following procedure: Two series of runs were done, one with the gradient algorithm and the other one with the hybrid algorithm. Each series consisted of 100 runs with different random seeds. We present the averaged results.

We note by $S(\tau)$ the ratio of the found primary hypercubes with solutions compared to the total one. The efficiency of the gradient algorithm is defined

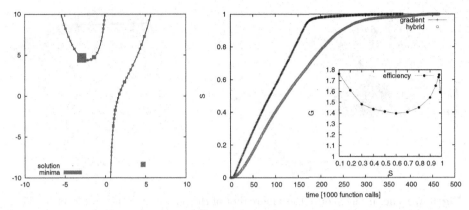

Fig. 4. Cantilever beam with $Q = 2.0$. Graphs are same as in Fig. 2.

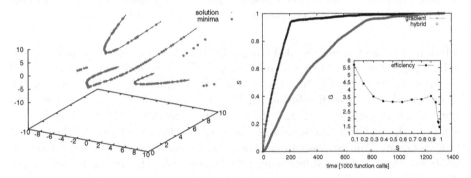

Fig. 5. Cantilever beam with $|Q|=P$ (three-dimensional GRS). (a) The continuous line shows the solution curve the points indicate the local minima found by the gradient algorithm (occurrence is not shown!). (b) Graph is same as in Fig. 2. (b).

by the ratio of the time needed to find S part of the solutions with the hybrid algorithm and with the gradient: $G \equiv \tau_{\mathrm{hybr}}(S)/\tau_{\mathrm{grad}}(S)$.

The first test we performed is not a real problem. The function was chosen to be $x^2 + y^2 = 10$ which corresponds to no real BVP. The GRS space was set to the $[-10 : 10] \times [-10 : 10]$ space the size of the primary hypercube is 1.5. The solution is a circle with $\sqrt{10}$ radius (see Fig. 2. (a)). A single gradient algorithm was run at the beginning and only one slave was working. The efficiency is $1.5 - 2$. The time evolution of S for the gradient algorithm is quite linear indicating a scenario where the gradient algorithm found a solution and then the path following algorithm completed it. The hybrid algorithm has a curved shape indicating an exponential distribution of the time when the first point of the solution was found. The time gain very well corresponds to the average time needed for the hybrid algorithm to find the first solution.

The second example consist of the already illustrated cantilever problem with a very small imperfection parameter $Q = 0.1$. The resulting GRS with the

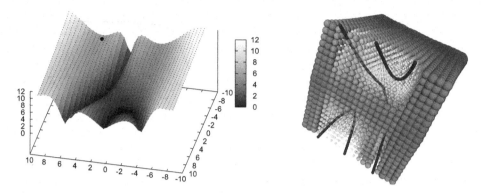

Fig. 6. (a) The surface plot of the square root of the potential of the cantilever problem with $Q=2$. The monotonic square root function was taken to visually enhance the view of the potential. The local minimum is shown with a black point. (b) Visualization of the potential in the 3 dimensional problem. The size of the balls is proportional to the square root of the potential. The solution is shown with a thick black line.

solution is presented on Fig. 3. Five gradient algorithms were run prior to the hybrid algorithm and both ran on 5 slaves simultaneously. Due to the small imperfection the solution line is cut into two distinct parts. The ratio of the primary hypercubes with solution is lower than before (5.7% instead of 8%). In spite of this the efficiency [Fig. 3 (b)] is less than in the previous example. This is due to the fact that the solutions are relatively long lines which takes a long time to follow. This process can only be done by one or two slaves and the others are free to look for new solution in a stochastic way. This example emphasizes the importance of the well planned parallelization of the algorithm.

We present the third example on Fig. 4 which is different from the previous one only in the imperfection parameter $Q = 2$ and the number of slaves which was set to 2. The efficiency changes only little: It gets worse for small S but gets better for large S. Where the low number of slaves does not let for a free scanning of GRS while the others are following a solution. On Fig. 4 (a) we can see that a local minimum with no solution at the point $(4.7, -8.4)$. It is also visible on the surface map of the potential on Fig. 6 (a). It has a considerable attraction range but with sufficient gradient runs the solutions are found with very high probability.

The last example we analyze here is a three dimensional one. The imperfection parameter is no longer a constant but may change on condition that its absolute value equals to $|Q|=P$. The GRS is 3 dimensional in this example with Q being the third dimension. The problem was run on two slaves. The solution probability is 0.6%. The position of the solution is shown on Fig. 5 and 6 (b). It consists of three disjunct branches. The figure 5 also shows the local minima found by the gradient algorithm where it is obvious that it nicely finds the solution as well as other local minima lines. In spite of these false local minima the algorithm is very efficient and needs about 3 times less function calls than the hybrid algorithm to find the majority of the solution, which shows that the efficiency of the gradient algorithm compared to the hybrid one increases rapidly with the dimension of the GRS.

6   Conclusion

In this paper we introduced a gradient algorithm to find the solution of boundary value problems faster than by the existing scanning and solution following hybrid algorithm. We showed that even the easiest implementation of this algorithm brings a considerable time gain. This is achieved despite the fact that the gradient runs do not deliver solutions but just alter the weighting of the primary hypercubes of the discretized GRS. We also showed that in high dimensions the computation need of the gradient algorithm is negligible.

On the other hand we showed that this dump implementation lack many feature that could make the algorithm run faster. We expect the most efficiency gain by developing a starting point choosing mechanism and a much better weighting of the primary hypercubes in parallel with an elaborate selection of slave types.

Acknowledgements

Part of this work was funded by the Péter Pázmány program (RET-06/2005) of the National Office for Research and Technology, and the authors would like to thank EU INFSO-RI-031688 program.

References

1. Domokos, G.: Global Description of Elastic Bars ZAMM. 74, T289–T291 (1994)
2. Domokos, G., Gáspár, Z.: A global, direct algorithm for path-following and active static control of elastic bar structures. Mech. of Struct. and Mach. 23, 549–571 (1995)
3. Gáspár, Z., Domokos, G., Szeberényi, I.: A parallel algorithm for the global computation of elastic bar structures. Comp. Ass. Mech. Eng. Sci. 4, 55–68 (1997)
4. Domokos, G., Szeberényi, I.: A hybrid parallel approach to nonlinear boundary value problems. Comp. Ass. Mech. Eng. Sci. 11, 15–34 (2004)
5. Várkonyi, P., Sipos, A.Á., Domokos, G.: Structural design on the internet: options to accelerate a computing algorithm (in Hungarian). Épités és Építészettudomány 34, 271–291 (2006)

Quantum Random Bit Generator Service for Monte Carlo and Other Stochastic Simulations

Radomir Stevanović, Goran Topić, Karolj Skala, Mario Stipčević,
and Branka Medved Rogina

Rudjer Bošković Institute, Bijenička c. 54, HR–10000 Zagreb, Croatia
`radomir.stevanovic@irb.hr`

Abstract. The work presented in this paper has been motivated by scientific necessity (primarily of the local scientific community) of running various (stochastic) simulations (in cluster/Grid environments), whose results often depend on the quality (distribution, nondeterminism, entropy, etc.) of used random numbers. Since *true random numbers* are impossible to generate with a finite state machine (such as today's computers), scientists are forced either to use specialized expensive hardware number generators, or, more frequently, to content themselves with suboptimal solutions (like pseudorandom numbers generators). *Quantum Random Bit Generator Service* has begun as a result of an attempt to fulfill the scientists' needs for quality random numbers, but has now grown to a global (public) high-quality true random numbers service.

1 Introduction

1.1 On Random Number Generation

The random numbers, which are (by their definition) nondeterministic and ruled by some prescribed probability distribution, can only be (as it is generally accepted) extracted from observation of some physical process that is believed to exhibit nondeterministic behavior. Various randomness extraction methods have been used in the past and different processes were being observed, but these usually could be grouped into either (1) measurements of macroscopic effects of an underlying noise ruled by statistical mechanics (e.g. quantum noise manifested as electronic shot noise or quantum effects in optics [13,8], thermal noise [1], avalanche noise, radioactive decay [16], atmospheric noise [4], etc.), or (2) sampling of a strictly nonlinear process (or iterated function system) that inherently exhibits chaotic behavior and intrinsical sensitivity to initial conditions (which are generally unknown or unmeasurable) and as such is considered to be random for practical purposes (e.g. chaotic electronic circuits like phase–locked loops [3], or chaotic mechanical systems [10], etc.). The only scientifically provable randomness (nondeterminism) sources, at the present state of the art, are quantum systems. Contrary to those, classical physics systems (including chaotic ones) only hide determinism behind complexity.

A cheap alternative to complex true random number generators (in terms of speed of generation or resource requirements, but on account of randomness) are

I. Lirkov, S. Margenov, and J. Waśniewski (Eds.): LSSC 2007, LNCS 4818, pp. 508–515, 2008.

algorithmic generators implemented on a digital finite state machines outputting pseudorandom numbers, i.e. deterministic, periodic sequences of numbers that are completely determined by the initial state (seed). Entropy of an infinite pseudorandom sequence is finite, and upward limited with entropies of the generating algorithm and the seed. This is *not* the case with infinite true random numbers sequences.

Regardless of the randomness acquisition method used, random number sequences are often post-processed [7] to remove bias, shape probability distribution, remove correlation, etc.

1.2   The Need for Random Numbers

Random numbers seem to be of an ever increasing importance — in cryptography, various stochastic numerical simulations and calculations (e.g. Monte Carlo methods), statistical research, various randomized or stochastic algorithms, etc. and the need for them is spanning a wide range of fields — from engineering to physics to bioinformatics. The applications usually put constraints on properties of input random numbers (probability distribution, bias, correlation, entropy, determinism, sequence repeatability, etc.). Consequently, these constraints dictate the choice of a random number generator.

If the quality of simulation or calculation results would be substantially affected, or dominated, by the (lack of) randomness of input random number sequences, then the true random number hardware generator should be used. And that imposes additional project costs, of both financial and time resources.

2   The Random Numbers Service

To ease and simplify the acquisition of high quality true random numbers for our local scientific community, we have developed the *Quantum Random Bit Generator Service* (QRBG Service for short), which is based on the *Quantum Random Bit Generator* [13]. This service is now publicly available on the Internet, from [11]. Design requirements for the Service were:

- *true randomness* of data served (nondeterminism and high entropy),
- *high speed* of data generation and serving,
- *high accessibility* of the service (easy and transparent access),
- *great robustness* of the service, and
- *high security* for users that require it.

The development of the QRBG Service is still a work in progress, but all requirements except the last one have been hitherto implemented and tested. If we exclude the security requirement from the list above, it can be said that QRBG Service tops currently available random number acquisition methods (including existing Internet services like [9,4,16]) in at least one of the remaining categories (to the best knowledge of the authors).

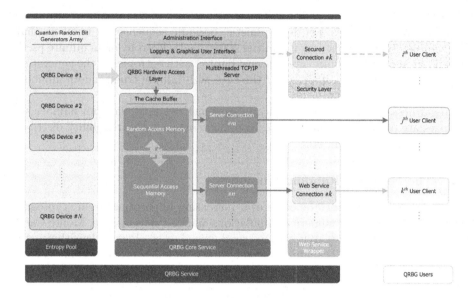

Fig. 1. The structure of the Service

To ensure the high quality of supplied random numbers (true randomness) and the high speed of serving, we have used the Quantum Random Bit Generator (described in Section 2.1) and a carefully designed server (described in Section 2.2). The server design also gives the Service its robustness, which, in turn, acts as an important factor in service accessibility, namely its temporal availability. Transparent access to random data, or *access modes availability* is achieved through multiple client connectors developed (described in Section 2.3 and Section 2.4), including "black-box" C/C++ libraries and web service (SOAP) access, but also more user-friendly Mathematica and MATLAB add-ons. To facilitate high security, an SSL wrapper is being implemented and tested which will enable the encryption of transferred random data with user certificates.

The structural overview of the Service is depicted in Fig. 1. Implementation details of the Service components are given in the following sections.

2.1 Randomness Source

As a source of random numbers, an array of *Quantum Random Bit Generators* (QRBGs) is being used. The QRBG device was designed and developed at the Rudjer Bošković Institute, in the Laboratory for Stochastic Signals and Process Research, and is still in its prototype phase [13].

QRBG is a fast, nondeterministic and novel random number generator whose randomness relies on intrinsic randomness of the quantum physical process of photonic emission in semiconductors and subsequent detection by the photoelectric effect. The timing information of detected photons is used to generate binary random digits — bits, with efficiency of nearly 0.5 bits per detected random event. Device consists of a light source (LED), one single–photon detector and fast electronics for the timing analysis of detected photons providing random output numbers (bits) at (currently) 16 Mbit/sec. By using only one photodetector (in contrast to other similar solutions) there is no need to perform any fine-tuning of the generator, moreover, the method used is immune to detector instability problems, which fosters the autonomous work of the service (without the usually required periodic calibrations of the generator). For the purpose of eliminating correlations, a restartable clock method is used for time interval measurement.

The collection of statistical tests (including NIST's "Statistical Test Suite for Random and Pseudorandom Number Generators for Cryptographic Applications" and DIEHARD battery of strong statistical randomness tests) applied to random numbers sequences longer than 1 Gb produced with this quantum random number generator presents results which demonstrate the high quality of randomness resulting in bias[1] less than 10^{-4}, autocorrelation[2] consistent with zero, near maximal binary entropy and measured min–entropy near theoretical maximum. For much more details on these and other performed tests results, see [12].

2.2   QRBG Core Service

The core of the QRBG Service has been written as a native Microsoft Windows NT (XP / 2003 Server) stand-alone multithreaded server application with a graphical user interface, in the C++ programming language with the support of the standard Microsoft Foundation Classes library.

All application components (servers, database and device access layers, user interface and logging, client authentication and quota management) communicate using a thread-safe notification message based queue-like structure which enables decoupling and asynchronous work of independent components (refer to Fig. 1). The flow of random data from randomness generator(s) to end–users could be seen as standard FIFO buffer–centric producer(s) – buffer – consumer(s) problem, and solved accordingly. However, to maximize aggregate throughput, transition from a FIFO to randomly accessed buffer has to be made. This enables asynchronous buffer *writers* and *readers*, and thus not only greatly improves the overall performance, but also transfers the scheduling problem of equally prioritized readers (end–users) from the application to the operating system.

Hardware Access Layer. The QRBG device connects to a computer via USB 2.0 interface. This enables connecting several QRBG devices and achieving random numbers acquisition speeds much higher than those of a single device (16 Mbps), since the USB standard allows connecting of up to 127 devices onto one host controller and upstream speed of up to 480 Mbps.

---

[1] Defined as $b = p_1 - 0.5$, where p_1 is probability of ones.
[2] Defined in [5].

QRBG hardware access (layer) is implemented as a dynamic link library that communicates directly with the QRBG device driver. Random data from all the input devices in the array[3] are being constantly read (at maximum speed) and stored into the application cache buffer. All status messages are dispatched into the application notify queue.

Application Cache Buffer. To optimize data transfer for speed, a Buffer class has been implemented with the behavior similar to that of `malloc` (the standard C library memory allocation routine). Namely, the complete buffer storage space is split into blocks of variable size. Each block can be either: *loaded* or *empty* (state flag), and at the same time: *ready* or *busy* (access flag). The state flag specifies whether the block contains random data or it's empty, and the access flag tells if some reader or writer thread is allowed to access the block. Interface of the Buffer component features two sets of grasp/release methods — for loaded and empty block: `graspEmpty(size,...)`, `releaseEmpty(...)`, `graspLoaded(size,...)` and `releaseLoaded(...)`. All of these methods work with block descriptors only. When a loader thread wants to copy data from QRBG device(s) into the buffer, it "grasps" an empty block in the buffer, copies the data, and then "releases" the block. Whenever a reader thread needs random data, it similarly "grasps" an loaded block (with random data), serves the data, and then "releases" the block (which is after that considered empty). The Buffer component takes care of joining consecutive free or loaded blocks. Also, it transfers data between randomly accessed (and faster) and sequentially accessed (and slower) parts of the buffer, when needed.

TCP/IP Core Server. When a user (directly, or indirectly through some of the QRBG extension services) requests random data from the QRBG Service, it is served (over TCP/IP network protocol) by the QRBG Core Server. Upon connection attempt, if a client is allowed to connect (IP is allowed and server isn't overloaded), communication begins. The communication protocol is inspired by the Simple Authentication and Security Layer protocol [14] but is extremely simplified — only two binary messages are exchanged. First, the client sends a request packet (with requested operation, login credentials and other data specific for requested operation — usually number of requested random bytes). Server responds with status message followed by a requested amount of random data (if the user was authenticated and his download quotas weren't exceeded) and closes the connection. Users' login credentials and all usage statistics are stored in a MySQL database through the QRBG Database Access Layer.

Due to random data caching, the data transfer rates achieved with QRBG Core Server (on a mid-range Microsoft Windows XP computer), exceed 45 MiB/s in loopback configuration and 11 MiB/s on a 100 Mbps local network. With empty cache, transfer rate falls bellow QRBG Device theoretical speed limit, adding some 20–30% overhead.

---

[3] Due to a high price of a single QRBG device, we currently have only one device in the array.

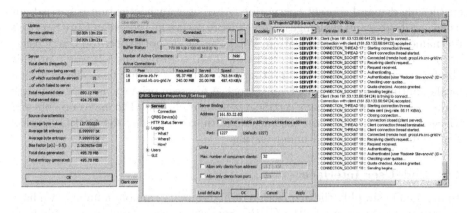

Fig. 2. The application's user interface (from left to right: statistics box, main window, settings dialog and log output.)

HTTP Status Server. While the Service is running, various statistics are collected concerning the clients connected, quantity and quality of data generated and served, etc. Simple HTTP server is running on the same address as the Core Server is and it enables users to inspect the status of the Service from their web browsers.

Administration User Interface. Configuration and administration of the Service is performed locally, through its graphical user interface. Screen capture of application windows is given in Fig. 2.

2.3 Extension Services

Additional features, new protocol support and other extensions of the QRBG Core Service can be easily implemented as QRBG Extension Services. Two such extensions are developed.

Web Service Wrapper. Provides SOAP protocol extension [15] of the basic QRBG Service. The web service is implemented in a standard fashion and executes in a stand-alone web server. Various random data fetching methods (getByte, getInt, getFloat, etc.) simply relay user requests to the QRBG Core Service and return SOAP–encoded results to the user. Slowdown due to relaying is irrelevant (non-significant), since web services are inherently slow. The only purpose of implementing the web service wrapper was to simplify connectivity to the Service from environments that have natural support for the web services (this could extend the QRBG usage to high-level web and similar scripting applications).

Secure Access Wrapper. Provides SSL protocol extension of the QRBG Core Service. Like the web service wrapper, it relays user requests, but in addition enables on–demand content encryption. Also, user authentication is carried out using certificates and challenges, a much more secure method than Core Service's username/password authentication. We, however, do not expect too heavy usage of the secure access, since it (1) will slow down transfer, and (2) won't provide absolute security for sensitive cryptography (which can only be achieved with a local quantum random number generator). A foreseen primary usage is on behalf of the users that already have certificates in their cluster/Grid environments and whose Virtual Organization becomes an authorized user of the QRBG Service. The secure wrapper is still under testing.

2.4 End–User Interface

To maximally simplify the acquisition of high-quality random numbers, we have tried to make the access to the QRBG Service as transparent as possible from as many platforms/environments we could. The work in this segment is in no way over, and many more clients (service connectors) are still to be written.

Basic Access. A simple C++ class for transparent access to the QRBG Service has been developed. It features the acquisition of standard data types and a local cache of user–defined size. Since it is written in a standard, widespread language, it compiles cross–platform (Windows/Linux, 32/64–bit) and makes a good starting point for both users and developers. To illustrate the simplicity of its usage, we quote here a complete code segment that acquires a 100 double precision floating point numbers uniformly distributed on interval $[0, 1\rangle$.

```
QRBG random;
random.defineUser("username", "password");
double x[100];
random.getDoubles(x, 100);
```

Command-line Utility. Based on the C++ client, a powerful and option–rich cross–platform command–line tool has been written. It is intended primarily, although not exclusively, for Linux users. A GUI–enabled version was written for Windows users.

QRBG Toolbox for MathWorks MATLAB. For the users of this powerful engineering platform we have also developed seamlessly integrateable QRBG extension. Its main function, **qrand**, has a syntax similar to the MATLAB built–in **rand** function, with notable semantic difference — it returns a matrix of *true* random numbers (64-bit floats from $[0, 1\rangle$), and not pseudorandom numbers.

```
> qinit('username', 'password');
> m = qrand(5, 5);
```

QRBG Add-on for Wolfram Mathematica. Similar to the client examples above, we also developed a Mathematica add-on based on C/C++ MathLink Software Developer Kit [17].

3   Conclusions and Future Work

We have presented a solution for the problem of simple acquisition of high quality true random numbers — an online random number service. While using a fast nondeterministic quantum random number generator and writing a robust and scalable, performance–tuned application around it, we were driven by requirements of true randomness delivery, fast serving, high access transparency and high service availability.

The development of the QRBG Service is still a work in progress, and future work will include: extending access transparency by creating more client access modes (connectors), testing and opening a secure wrapper of the Service, and opening the Service to a wider public.

References

1. Intel 80802 Firmware Hub chip with included thermic noise based RNG, http://www.intel.com/design/software/drivers/platform/security.htm
2. VIA C3 CPU with included chaotic electronic system based RNG, http://www.via.com.tw/en/initiatives/padlock/hardware.jsp
3. Bernstein, G.M., Lieberman, M.A.: Secure random number generation using chaotic circuits. IEEE Trans. Circuits Syst. 37, 1157–1164 (1990)
4. Haahr, M.: Random.org — An atmospheric noise based online true random numbers service, http://random.org/
5. Knuth, D.E.: The Art of Computer Programming, 3rd edn. Semi-numerical Algorithms, vol. 2. Addison-Wesley, Reading (1997)
6. MathWorks MATLAB Documentation: MATLAB's interface to DLLs
7. Proykova, A.: How to improve a random number generator. Computer Physics Comm 124, 125–131 (2000)
8. Quantis: Quantum random number generator, http://www.idquantique.com/
9. Quantis: Quantum RNG online service, http://www.randomnumbers.info
10. Silicon Graphics: Method for seeding a pseudo-random number generator with a cryptographic hash of a digitization of a chaotic system (also known as Lavarand). U.S. Patent 5732138
11. Stevanović, R.: QRBG Service Online, http://random.irb.hr/
12. Stipčević, M.: Quantum Random Bit Generator (QRBG), http://qrbg.irb.hr
13. Stipčević, M., Medved Rogina, B.: Quantum random number generator. Review of Scientific Instruments (to appear)
14. The Internet Engineering Task Force: Simple Authentication and Security Layer (SASL). RFC 2222, http://www.ietf.org/rfc/rfc2222.txt
15. W3C: Web Services Architecture, http://www.w3.org/TR/ws-arch/
16. Walker, J.: Hotbits — An radioactive decay based online true random numbers service, http://www.fourmilab.ch/hotbits/
17. Wolfram Mathematica Documentation: MathLink

A Hierarchical Approach in Distributed Evolutionary Algorithms for Multiobjective Optimization

Daniela Zaharie, Dana Petcu, and Silviu Panica

Department of Computer Science, West University of Timişoara,
and Institute e-Austria Timişoara
bv. V. Pârvan, no. 4, 300223 Timişoara, Romania
dzaharie@info.uvt.ro, petcu@info.uvt.ro, silviu@info.uvt.ro

Abstract. This paper presents a hierarchical and easy configurable framework for the implementation of distributed evolutionary algorithms for multiobjective optimization problems. The proposed approach is based on a layered structure corresponding to different execution environments like single computers, computing clusters and grid infrastructures. Two case studies, one based on a classical test suite in multiobjective optimization and one based on a data mining task, are presented and the results obtained both on a local cluster of computers and in a grid environment illustrates the characteristics of the proposed implementation framework.

1 Introduction

Evolutionary algorithms proved to be adequate metaheuristics in solving multiobjective optimization problems. However, for complex problems characterized by a large number of decision variables and/or objective functions they need large populations and a lot of iterations in order to obtain a good approximation of the Pareto optimal set. In order to solve this problem, different variants for parallelizing and distributing multiobjective evolutionary algorithms (MOEAs) have been proposed in the last years [3,4,7]. Choosing the appropriate variant for a particular problem is a difficult task, thus simultaneously applying different variants and combining their results could be beneficial. The huge computational power offered today by grid infrastructures allows the use of such strategies which could be beneficial especially when the human knowledge on the problem to be solved or on the method to be applied is lacunar.

The approach proposed in this paper is developed in order to be used either on a cluster or in a grid environment and is based on the idea of using one or several colonies of populations. Each colony consists of a set of populations and can be characterized by its own strategies for assigning a search subspace to each population and for ensuring the communication between populations. The results obtained by all colonies are to be collected and combined in order to obtain the global approximation of the Pareto optimal set and/or Pareto front.

I. Lirkov, S. Margenov, and J. Waśniewski (Eds.): LSSC 2007, LNCS 4818, pp. 516–523, 2008.

The paper is organized as follows. Section 2 presents a brief overview of existing distributed variants of MOEAs. The hierarchical approach and the particularities of the cluster and grid layers are presented in Section 3. In Section 4 two case studies are presented, one involving a classical test suite in MOEAs analysis, and the other one related to a data mining task, the problem of attributes' selection.

2 Distributed Versions of Multiobjective Evolutionary Algorithms

Most evolutionary algorithms for multi-objective optimization use a population of elements which are transformed by recombination and mutation during a given number of generations. At each generation all objective functions are evaluated for all elements of the population and the non-dominance relationship between them is analyzed. In the case of a minimization problem involving r objective functions, $f_1,...,f_r$ an element $x \in D \subset R^n$ is considered non-dominated if there does not exist another element $y \in D$ such that $f_i(y) \leq f_i(x)$ for all $i \in \{1,...,r\}$ and the inequality is strict for at least one function. The non-dominated elements of the population represent an approximation of the Pareto optimal set. Both the evaluation of elements and the analysis of the nondominance relationship are high cost operations. These costs can be reduced by dividing the population in subpopulations and by evolving them in parallel. In order to design such a distributed variant of a MOEA some key issues should be addressed: the division of the search space, the communication between subpopulations and the combination of the results obtained by all subpopulations.

The division of the search space can be made before starting the evolution (apriori division rule) or dynamically during the evolution (dynamic division rule). In the last years different strategies have been proposed, most of them being based on dynamic division rules [4,7]. Dynamic division rules usually involve some operations aiming to periodically reorganize the structure of the search space. In some cases this operations could be costly by themselves, as for instance in [7], where a clustering step involving the elements of all subpopulations is executed, or in [4], where all subpopulations are gathered and their elements are sorted. On the other hand, applying apriori division rules (e.g. dividing the decision variables space in disjoint or overlapping regions) does not usually involve supplementary costs.

The communication between subpopulations plays a critical role in the behavior of a distributed MOEA. The main components of a communication process are: the communication topology, the communication policy and the communication parameters. The communication topology defines the relationship between subpopulations and the most common variants are: fully connected topology (each subpopulation is allowed to communicate with any other subpopulation) and neighborhood based topologies like the ring topology (a subpopulation S_i can communicate only with subpopulations S_{i-1} and S_{i+1}, in a circular manner).

The communication policy refers to the manner the migrants are selected from the source subpopulation and the way they are assimilated into the destination subpopulation. The migrants can be randomly selected from the entire population or just from the non-dominated subset (elitist selection). The assimilation of the migrants can be realized by just replacing an element of the target subpopulation with the immigrant (the so-called pollination [3]) or by sending back an element to the source subpopulation in order to replace the emigrant (plain migration [9]). The communication topologies and policies can be combined in different manners leading to a large number of strategies.

The parameters influencing the behavior of the distributed variant are: the communication frequency (number of generations between consecutive migration steps) and the migration probability (the probability of an element to be selected for migration).

Besides the large number of parallel implementations of MOEAs, grid implementations have been also recently reported. In [6] is proposed a Globus based implementation of a Pareto archived evolution strategy characterized by remotely executing a number of sequential algorithms on different grid machines and storing the approximated fronts which satisfy some quality criteria. In [5] is presented a grid-enabled framework which allows the design and deployment of parallel hybrid meta-heuristic, including evolutionary algorithms.

3   The Hierarchical Approach

The existence of different MOEAs distribution strategies on one hand and of different architectures on which such algorithms can be executed, on the other hand, motivated us to search for an easy configurable framework for the execution of distributed MOEAs. The approach we propose is based on a layered structure, as illustrated in Figure 1, allowing the execution either on a single computer, on a cluster of computers or in a grid infrastructure.

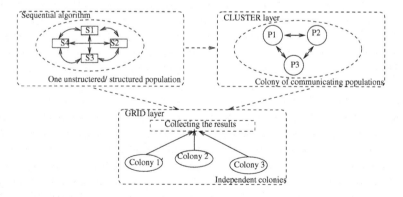

Fig. 1. The layered structure corresponding to the hierarchical approach

The first layer corresponds to the evolution of either a single unstructured population which explores the entire decision space or to a structured population consisting of communicating subpopulations. From an implementation point of view this would correspond to a sequential implementation on one processor. Using a structured population could be beneficial by increasing the population diversity and reducing the cost of the selection step even in the case of a sequential implementation.

The second layer corresponds to a colony of populations which evolve independently but which periodically change some information during a migration step. From an implementation point of view this layer would correspond to a parallel variant executed on a cluster of computers (each processor deals with the evolution of one or several populations from the colony).

The third layer corresponds to the evolution of several independent colonies of populations each one being executed in a location of the grid environment. Unlike the populations in a colony, the colonies are loosely coupled in order to deal with the heterogeneous character of the grid infrastructure. In the current implementation the only communication between the colonies is in the end of the evolution process where the results are collected from all of them. Since the same problem is solved on each colony this induce a certain level of redundancy. In order to reduce the redundancy, each colony can be based on a different MOEA and on different distribution strategies.

The cost of communication between (sub)populations is highly dependent on the layer, thus the communication strategy should be adequately chosen for each layer. There are two main communication processes involved: a periodical communication corresponding to migration stages and a final communication step corresponding to the collection of the partial results obtained by all processes.

Since the aim of the final communication step is just to collect the results, a natural way to implement it is: all processes send their results to a master process which collect them and construct the final result. These partial results are sets of non-dominated elements having almost the same number of elements as the population. If the final results produced by p populations of size m are collected through a message passing interface, the cost of this communication is $\mathcal{O}(pmL)$, L being the size of each element (which depends on the number of decision variables and objective functions).

In the first layer an intensive periodical communication between subpopulations can be applied, including strategies based on gathering and redistributing the entire population. If the evolution of a colony of populations is executed in parallel on a cluster of computers there could be different approaches in implementing the periodical communication when using a message passing interface. Let us consider the case of a colony consisting of p populations. For instance, in the case of random pollination (each population sends some elements to other randomly selected populations), a direct implementation strategy could lead to a number of messages of $\mathcal{O}(p^2)$ to be transferred between p processes. In the case of plain migration, when for each immigrant a replacing element is sent back to the source population the number of messages is twice as in the case of pollination.

The number of messages can be significantly reduced ($\mathcal{O}(p)$) if all processes send their migrants to a master process which distribute them to the target processes. If each message containing migrants is preceded by a short message containing their number, the total number of messages is at most $4(p-1)$. The number of messages transmitted between processors is even smaller in the case of ring topologies ($2p$). The length of messages containing migrants depends on the size of each element, L, on the population size, m, and on the migration probability, p_m. The averaged length of the messages containing migrants is $mp_mL(p-1)/p$.

For the grid layer at least two scenarios can be identified: (i) in each grid location a sequential job corresponding to one colony is executed; (ii) a parallel job involving a message passing interface in the grid infrastructure is executed. In the first case there is no direct communication between colonies, the jobs launched in the grid environment are independently executed (as in [6]) and they send their results through files transfer to the location which initiated the jobs. In the second case low frequency periodical migration should be applied between colonies in order to limit the communication between different sites.

4   Experimental Results and Discussion

The experiments were conducted for the first from the above mentioned scenarios and the behavior of the proposed approach was tested in two distinct contexts: (i) one problem — several strategies; (ii) one strategy — multiple subproblems (e.g. data subsets).

4.1   Case Studies

In the first case we used the test suite from [10], characterized by two objective functions, and we applied different MOEAs (e.g. Nondominated Sorting Genetic Algorithm [1] and Pareto Differential Evolution [9]) with different parameters and distribution strategies for different colonies. Both algorithms use similar selection operators based on computing non-domination ranks and crowding factors.

The second case study is related to the problem of attributes subset selection which consists in identifying, starting from a training set, the most relevant attributes. Such a problem can be solved by assigning to attributes some weights which optimize three criteria [8]: intra-class distance (to be minimized), inter-class distance (to be maximized) and an attribute-class correlation measure (to be maximized). Interpreting this optimization problem as a multiobjective one, the result will be a set of attributes weights, each one leading to a ranking of attributes, where the first attributes are the most relevant ones. The final ranking can be obtained by averaging the rankings corresponding to all elements of the approximated Pareto set. Since the estimation of the attribute-class correlation measure is quadratic with respect to the number of elements in the training set, the evaluation of each element of the population is costly. A natural approach is to split, by using a proportional sampling strategy, the data set in smaller subsets, and to apply a MOEA independently for each subset. The results obtained for all subsets are combined in order to construct the final ranking.

4.2 Results in a Cluster Environment

The tests corresponding to the cluster layer were based on a parallel implementation using mpiJava and were conducted on a local heterogeneous cluster of 8 nodes (Intel P4, 6 CPUs at 3.0 GHz and 2 CPUs at 2.4 GHz) connected through optical fiber and a Myricom switch at 2 Gb/s. The evolutionary process involved a colony of c populations to be executed on p processors. Since $c \geq p$ each processor deals with a subcolony of c/p populations. Therefore, different communication strategies can be applied between the populations in the subcolony assigned to one processor and between populations assigned to different processors. Figure 2 illustrates the influence of the communication strategy and that of the problem complexity on the speedup ratio in two cases: when the time needed for collecting the results is ignored and when this final communication time is taken into account. The reported results were obtained in the case of 24 populations each one having 20 elements which evolve for 250 generations by using a NSGA-II algorithm [1] and communicate every 25 generations. It follows that for simple test problems (e.g. ZDT2 from [10] with $n = 100$) the cost of the final communication step (involving long messages) is significant with respect to the cost of other steps, leading to low speedup ratios, while for real problems (e.g. attribute selection in the case of a set of real medical data consisting of 177 instances, each one with $n = 14$ attributes) the final communication cost does not significantly alter the speedup ratio. The decrease in the speedup ratio when 8 processors were used is generated by the heterogeneous character of the processors.

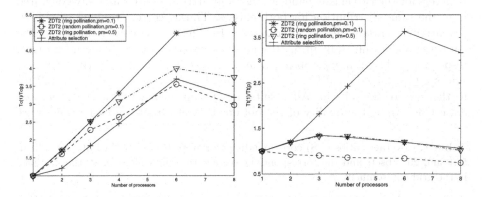

Fig. 2. Speedup ratios when the final communication time is ignored (left) and when the final communication time is taken into consideration (right)

4.3   Results in a Grid Environment

The experiments concerning the grid layer were conducted on the European SEE-GRID infrastructure by using up to 24 nodes with Intel P4 at 3.0 GHz, 1 GB RAM and 100 GB HDD. The tests consisted in launching several sequential and parallel jobs corresponding to different instances of MOEAs. The code was ported on remote sites using gLite. Each MOEA instance is described in a user configuration file specified in the job description. The results generated by different jobs at different sites are transferred through files back to the site which launched the jobs.

The first case study involved 24 sequential jobs corresponding to 24 variants based on two MOEAs, four communication strategies, four variants of search space division and some different values of the specific parameters. All Pareto fronts were compared by using the coverage ratio measure [2] which allows identifying the best result (in our example it was the NSGA-II with one population of 200 elements and a recombination probability of 0.9; the worst behavior corresponds to the same strategy but for a recombination probability of 0.2). Besides the tests involving sequential jobs, experiments with parallel codes executed on clusters from the SEE-GRID virtual organization were also conducted. The possibility of using a larger number of processors than that in the local cluster (e.g. 24 instead of 8) led to a significant decrease of the running time of the evolutionary process. This simple case study illustrates the opportunity offered by the computational grid to efficiently conduct experimental designs when we are looking for appropriate strategies for a given problem.

The second case study was related to the attribute selection problem and involved a set of 2000 synthetic data corresponding to two classes and having 10 attributes. First attribute is just the class label, the next five attributes are randomly generated starting from different distributions for the two classes (e.g. random values generated according to the normal distribution with different parameters for the two classes) and the last four attributes are randomly generated from the same distribution for both classes. Thus a correct ranking would be: first attribute, attributes 2–6, attributes 7–10. Three variants were analyzed: the data were uniformly split in 5, 10 and 20 subsets leading to 5,10 and 20 jobs, respectively. The rankings obtained are: (1,3,5,6,2,4,7,8,9,10) in the case of 5 subsets, (1,6,3,5,2,4,8,7,9,10) in the case of 10 subsets and (1,3,6,5,4,2,7,8,9,10) in the case of 20 subsets. All results are in concordance with the generated data. Concerning the quality of the obtained Pareto front the best results were

Table 1. Coverage ratios (CS) corresponding to Pareto fronts for the data set split in 5,10 and 20 subsets and the corresponding average running time of executing the jobs in the grid environment

CS	5 jobs	10 jobs	20 jobs	Average time (s)
5 jobs	0.0	0.0322	0.002	4485.15± 64.77
10 jobs	0.563	0.0	0.031	1041.95± 194.42
20 jobs	0.912	0.764	0.0	374.71± 70.12

obtained by the variant using 20 subsets. This is illustrated by the coverage ratio measures presented in Table 1($CS(F_1, F_2)$ denotes the ratio of the elements in F_2 which are dominated by elements in F_1) .

5 Conclusions

The hierarchical approach in distributing MOEAs leads to an easy configurable framework allowing the execution either on computational clusters or in a grid infrastructure. Two situations when the grid infrastructure can be efficiently exploited were identified: experimental design of evolutionary algorithms when a large set of strategies should be applied to the same problem and distributed attributes selection for large sets of data when one method can be applied to different data subsets.

Acknowledgment. This work is supported by the projects RO-CEEX 95/ 03.10.2005 (GridMOSI — UVT) and RO-CEEX 65/31.07.2006 (SIAPOM — IeAT).

References

1. Deb, K., et al.: A fast and elitist multi-objective genetic algorithm: NSGA-II. IEEE Trans. on Evolutionary Computation 6(2), 181–197 (2002)
2. Coello, C.A., van Veldhuizen, D.A., Lamont, G.B.: Evolutionary algorithms for solving multi-objective problems. Kluwer Academic Publishers, Dordrecht (2002)
3. Deb, K., Zope, P., Jain, A.: Distributed computing of Pareto-optimal solutions using multi-objective evolutionary algorithms. In: Fonseca, C.M., et al. (eds.) EMO 2003. LNCS, vol. 2632, pp. 535–549. Springer, Heidelberg (2003)
4. Hiroyasu, T., Miki, M., Watanabe, S.: The new model of parallel genetic algorithm in multi-objective optimization problems — divided range multi-objective genetic algorithm. In: Proc. of IEEE Congress on Evolutionary Computation (CEC 2000), vol. 1, pp. 333–340. IEEE Computer Society, Los Alamitos (2000)
5. Melab, N., Cahon, S., Talbi, E.G.: Grid computing for parallel bioinspired algorithms. J. Parallel Distrib. Comput. 66, 1052–1061 (2006)
6. Nebro, A.J., Alba, E., Luna, F.: Observations in using grid technologies for multi-objective optimization. In: Di Martino, B., et al. (eds.) Engineering the Grid, pp. 27–39 (2006)
7. Streichert, F., Ulmer, H., Zell, A.: Parallelization of multi-objective evolutionary algorithms using clustering algorithms. In: Coello Coello, C.A., Hernández Aguirre, A., Zitzler, E. (eds.) EMO 2005. LNCS, vol. 3410, pp. 92–107. Springer, Heidelberg (2005)
8. Wang, L., Fu, X.: Data Mining with Computational Intelligence. Springer, Berlin (2005)
9. Zaharie, D., Petcu, D.: Adaptive Pareto differential evolution and its parallelization. In: Wyrzykowski, R., et al. (eds.) PPAM 2004. LNCS, vol. 3019, pp. 261–268. Springer, Heidelberg (2004)
10. Zitzler, E., Deb, K., Thiele, L.: Comparison of multiobjective evolutionary algorithms. IEEE Transactions on Evolutionary Computation 8(2), 125–148 (2000)

Part IX

Application of Metaheuristics to Large-Scale Problems

Optimal Wireless Sensor Network Layout with Metaheuristics: Solving a Large Scale Instance

Enrique Alba and Guillermo Molina

Departamento de Lenguajes y Ciencias de la Computación
University of Málaga, 29071 Málaga, Spain
eat@lcc.uma.es, guillermo@lcc.uma.es

Abstract. When a WSN is deployed in a terrain (known as the *sensor field*), the sensors form a wireless ad-hoc network to send their sensing results to a special station called the *High Energy Communication Node* (HECN). The WSN is formed by establishing all possible links between any two nodes separated by at most R_{COMM}, then keeping only those nodes for which a path to the HECN exists. The sensing area of the WSN is the union of the individual sensing areas (circles of radius R_{SENS}) of these kept nodes. The objective of this problem is to maximize the sensing area of the network while minimizing the number of sensors deployed. The solutions are evaluated using a geometric fitness function. In this article we will solve a very large instance with 1000 preselected available locations for placing sensors (ALS). The terrain is modelled with a 287×287 point grid and both R_{SENS} and R_{COMM} are set to 22 points. The problem is solved using simulated annealing (SA) and CHC. Every experiment is performed 30 times independently and the results are averaged to assure statistical confidence. The influence of the allowed number of evaluations will be studied. In our experiments, CHC has outperformed SA for any number of evaluations. CHC with 100000 and 200000 evaluations outperforms SA with 500000 and 1,000,000 evaluations respectively. The average fitness obtained by the two algorithms grows following a logarithmic law on the number of evaluations.

1 Introduction

Nowadays, the trend in telecommunication networks is having highly decentralized, multinode networks. From small, geographically close, size-limited local area networks the evolution has led to the huge worldwide Internet. This same path is being followed by wireless communications, where we can already see wireless telephony reaching virtually any city in the world.

Wireless networks started as being composed by a small number of devices connected to a central node. Recent technological developments have enabled smaller devices with computing capabilities to communicate in the absence of any infrastructure by forming ad-hoc networks. The next step in wireless communications begins with ad-hoc networks and goes towards a new paradigm: Wireless Sensor Networks (WSN) [1].

I. Lirkov, S. Margenov, and J. Waśniewski (Eds.): LSSC 2007, LNCS 4818, pp. 527–535, 2008.
© Springer-Verlag Berlin Heidelberg 2008

A WSN allows an administrator to automatically and remotely monitor almost any phenomenon with a precision unseen to the date. The use of multiple small cooperative devices yields a brand new horizon of possibilities yet offers a great amount of new problems to be solved.

We discuss in this paper an optimization problem existing in WSN: the layout (or coverage) problem [6,4]. This problem consists in placing sensors so as to get the best possible coverage while saving the number of sensors as low as possible. A genetic algorithm has already been used to solve an instance of this problem in [4]. In this paper we define a new instance for this problem, and tackle it using some metaheuristic techniques [7,5,3] and solve a large dimension instance.

This work is structured as follows. After this introduction, the WSN layout problem (WSN problem for short) will be presented, and its formulation described in Section 2. Section 3 explains the optimization techniques employed for solving this problem. Then in Section 4 the experiments performed and the results obtained are analyzed. Finally, Section 5 shows the conclusions and future work.

2   WSN Problem

In this section we describe the layout problem for WSN, then present the formulation employed for its resolution.

2.1   Problem Description

A Wireless Sensor Network allows to monitor some physical set of parameters in a region known as the *sensor field*. When a WSN is placed in the sensor field, every sensor monitors a region of the field; ideally the complete network is able to monitor all the field by adding all the pieces of information together. It is the duty of the designer to establish what is the sensor field that the WSN has to monitor.

A node sensing area (the area that a single sensor can sense) can be modelled with a circle whose radius R_{SENS} -or *sensing radius*- indicates the sensing range of the sensor. The value of this range is determined by both the magnitude that is sensed and the sensor itself (hardware employed). Similarly, R_{COMM}, the *communication radius* of a sensor, defines the circle where any other sensor can establish a direct communication link with it. The value of this range depends on the environment, the radio hardware, the power employed and other factors.

When a WSN is deployed in the sensor field, the sensors form a wireless ad-hoc network in order to communicate their sensing results to a special station called the *High Energy Communication Node* (HECN). The data can then be analyzed by the HECN processor, or be accessed by the network administrator. Any sensor unable to transmit its sensing data to the HECN is useless. The sensing information is not sent through a direct link to the HECN, but rather a hop by hop communication is employed. Thus, for any node to be useful, it has to be within communication range of another useful node.

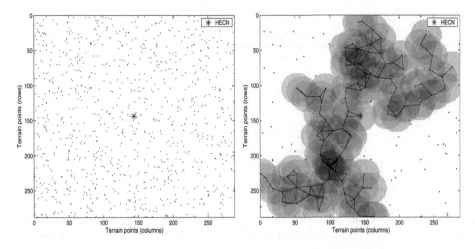

Fig. 1. Available sites in the problem instance (left), random solution (right)

The sensing area of the WSN is the union of the individual sensing areas of all the useful nodes. The designer wants the network to cover the complete sensing area, or, if this is unfeasible, to cover as much of it as possible. On the other hand, the number of sensor nodes must be kept as low as possible, since using many nodes represents a high cost of the network, possibly influences the environment, and also provokes a high probability of detection (when stealth monitoring is desired).

The problem of designing the layout for a WSN can be defined as an extension of an existing problem: the radio network design problem (RND) [2]. The objective of this problem is to maximize the sensing area of the network while minimizing the number of sensors deployed.

2.2 Problem Formulation

For this work we employ a square terrain as the sensor field, and use a discrete model to represent it. This model is a 287×287 point grid as in [2], where every point can be either monitored or not.

Sensor nodes can only be placed in some of those field points. If a sensor can communicate with the HECN, then a discretized circular area around its location is considered to be monitored. The available field points for placing the sensors are given as an ordered list (the *Available Location Sites*, ALS for short) that constitutes the specific *problem instance*. Figure 1 shows a graphical example of a WSN instance (left) and a solution layout with its underlying topology (right).

The WSN problem can be reduced to selecting from the list of available points a subset of locations that form the optimal sensor network. The list is ordered so that any bit string of the same length a the ALS represents a solution attempt to the problem (the '1's in the string indicating the chosen locations).

```
t:= 0;
Initialize(T,Sa);
while not end_condition(t,Sa) do
        while not cooling_condition(t)
                Sn := Choose_neighbor(Sa);
                Evaluate(Sa,Sn);
                if Accept(Sa,Sn,T) then
                        Sa := Sn;
                end if
                t := t+1;
        end while
        Cooldown(T);
end while
```

Fig. 2. Pseudocode for SA

From the previous definition of the problem, a fitness function that combines both objectives is employed [2] (Equation 1). The objective is to **maximize** the fitness value of the solution.

$$f(\boldsymbol{x}) = \frac{Coverage(\boldsymbol{x})^2}{Nb.\ of\ sensors(\boldsymbol{x})}, \qquad Coverage(\boldsymbol{x}) = 100 \cdot \frac{Covered\ points}{Total\ points} \qquad (1)$$

3   Optimization Techniques

In this section, we describe the two techniques used to solve the problem: simulated annealing and CHC.

3.1   SA Algorithm

Simulated annealing is a trajectory based optimization technique. It was first proposed by Kirkpatrick et al. in [5]. SA is a fairly commonly used algorithm that provides good results and constitutes an interesting method for comparing results and test other optimizing methods. The pseudocode for this algorithm is shown in Fig. 2.

The algorithm works iteratively and keeps a single tentative solution S_a at any time. In every iteration, a new solution S_n is generated from the old one, S_a, and depending on some acceptance criterion, it might replace it.

The acceptance criterion is the true core of the algorithm. It works as follows: both the old (S_a) and the new (S_n) solutions have an associated quality value — determined with a *fitness* function. If the new solution is better than the old one, then it will replace it. If it is worse there is still some chance that it will replace it. The replacing probability is calculated using the quality difference between both solutions and a special control parameter T named *temperature*.

The acceptance criterion ensures a way of escaping local optima by choosing solutions that are actually worse than the previous one with some probability. That probability is calculated using Boltzmann's distribution function:

$$P = \frac{2}{1 + e^{\frac{fitness(S_a) - fitness(S_n)}{T}}} \qquad (2)$$

```
t:=0;
Initialize(Pa,convergence_count);
while not ending_condition(t,Pa) do
        Parents := Selection_parents(Pa);
        Offspring := HUX(Parents);
        Evaluate(Offspring);
        Pn := Elitist_selection(Offspring,Pa);
        if not modified(Pa,Pn) then
                convergence_count := convergence_count-1;
                if (convergence_count == 0) then
                        Pn := Restart(Pa);
                        Initialize(convergence_count);
                end if
        end if
        t := t+1;
        Pa := Pn;
end while
```

Fig. 3. Pseudocode for CHC

As iterations go on, the value of the temperature parameter is progressively reduced following a cooling schedule, thus reducing the probability of choosing worse solutions and increasing the biasing of SA towards good solutions. In this work we employ a geometric rule, such that every k (*Markov chain length*) iterations the temperature is updated as $T(n+1) = \alpha \cdot T(n)$, where $0 < \alpha < 1$ is called the temperature decay.

3.2   CHC Algorithm

The second algorithm we propose for solving the RND problem is Eshelman's CHC (*Cross generational elitist selection, Heterogenous recombination, and Cataclysmic mutation*), a kind of Evolutionary Algorithm (EA) surprisingly not used in many studies despite it has unique operations usually leading to very efficient and accurate results [3]. Like all EAs, it works with a set of solutions (*population*) at any time. The algorithm proceeds iteratively, producing new solutions at each iteration, some of which will be placed into the population replacing others that were previously included. The pseudocode for this algorithm is shown in Fig. 3.

The algorithm CHC works with a population of individuals (solutions) that we will refer to as P_a. In every step, a new set of solutions is produced by selecting pairs of solutions from the population (the parents) and recombining them. This selection is made in such a way that individuals that are too similar can not mate each other, and recombination is made using a special procedure known as HUX (*Half Uniform crossover*). This procedure copies first the common information for both parents into both offspring, then it translates half the diverging information from each parent to each of the offspring. This is done in order to preserve the maximum amount of diversity in the population, as no new diversity is introduced during the iteration (there is no mutation operator). The next population is formed by selecting the best individuals among the old population and the new set of solutions (elitist criterion).

As a result of this, at some point of the execution, population convergence is achieved, so the normal behavior of the algorithm should be to stall on it.

A special mechanism is used to generate new diversity when this happens: the *restart* mechanism. When restarting, all of the solutions except the very best ones are significantly modified. This way, the best results of the previous phase of evolution are maintained and the algorithm can proceed again.

4   Tests and Results

In this section we describe the experiments and present the results obtained using the two algorithms described in Section 3. The results are then analyzed rigorously in order to determine the statistical confidence of the observed differences.

The instance solved in this work is a very large instance (1000 available locations), specially if compared with the previously existing work [2] where the biggest instance had only 349 available locations. The sensor field is modelled by a 287×287 point grid. All sensors behave equally and both their sensing and communication radii are set to 22 terrain points. The ALS is formed by 1000 locations randomly distributed over the sensor field following a uniform distribution. Figure 1 illustrates the instance of the problem, and shows a random solution for this instance using 167 sensors and covering 56.76% of the sensor field. The low quality achieved by random search, the NP nature of the problem, and its high dimensionality clearly suggest the utilization of metaheuristics.

The models and parameters employed in our problem instance are summed up in Table 1.

The problem is solved using simulated annealing (SA) and CHC. The same instance of the problem is used for both algorithms, and a parameter tuning is made to get good results from them (the values of the parameters can be seen in Table 2). We will analyze the algorithm's effectiveness for solving the problem by inspecting the fitness obtained. The influence of the number of solution evaluations will also be studied by running several experiments with both algorithms using increasingly higher number of allowed evaluations. The number of evaluations will range from $100,000$ up to $1,000,000$.

For every experiment the results are obtained by performing 30 independent runs, then averaging the fitness values obtained in order to ensure statistical confidence. Table 3 summarizes the results obtained for this study. Analysis of the data using Matlab's ANOVA/Kruskal-Wallis test plus Multcompare function has been used to get statistical confidence on the results with a confidence level

Table 1. Models and parameters

Concept	Model
Sensor Field	287×287 point grid
ALS	1000 points, uniform distribution
Solution	Bit string (1000 bits)
R_{SENS}	22 points
R_{COMM}	22 points

Table 2. Parameters of the algorithms

Algorithm	CHC
Population size	100
Crossover	*HUX*
Cataclysmic mutation	*Bit flip with prob.* 35%
Incest threshold	25% *of instance size*
Selection of parents	*Random*
Selection of next generation	*Elitist*
Algorithm	SA
Mutation	*Bit flip prob.* 1/*Length*
Markov-Chain length	50
Temperature decay	0.99
Initial temperature	1.05

Table 3. Fitness results

Evals.	50,000	100,000	200,000	300,000	400,000	500,000	1,000,000
SA	74.793	76.781	78.827	79.836	80.745	81.602	84.217
CHC	75.855	83.106	87.726	89.357	90.147	90.974	**92.107**

of 95%. A minimum mean square error approximation function is calculated (from a list of standard functions) to estimate the relation between the average fitness value and the allowed number of evaluations, for both SA and CHC.

From the results in Table 3 we can state that the average fitness obtained with either SA or CHC improves when the number of evaluations is increased. In the first case (SA) the average fitness goes from 74.793 for 50,000 evaluations to 84.217 for 1,000,000 evaluations. In the second case (CHC) it goes from 75.855 to 92.107. Analysis of the data shows that the increment of the fitness values is meaningful for both algorithms when the difference in number of evaluations is bigger than 100,000.

When it comes to comparing the two algorithms, CHC outperforms SA. The average fitness value obtained for any number of evaluations is greater using CHC than using SA. The analysis of the data confirms that CHC's results are significantly better than SA's for any number of evaluations except 50,000, for which they are equivalent. Furthermore, the executions using CHC have outperformed the ones using SA that performed *five times* more solution evaluations. CHC with 100,000 and 200,000 evaluations has outperformed SA with 500,000 and 1,000,000 evaluations respectively (though analysis couldn't show the significance at 95% confidence). CHC with 200,000 and 500,000 evaluations is significantly better than SA with 500,000 and 1,000,000 evaluations respectively.

The improvement obtained augmenting the number of evaluations is sublineal and is best modelled for this range of values using a logarithmic function for both SA and CHC. Figure 4 shows the average fitness obtained by both algorithms in the different experiments as well as the mathematical models calculated for

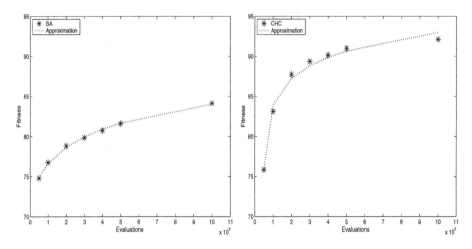

Fig. 4. Results obtained with SA and approximation (left), results obtained with CHC and approximation (right)

them. Equations 3 and 4 show the mathematical models for the fitness values obtained using SA and CHC respectively.

$$SA_{fitness}(evals) = 3.556 \cdot \log(evals/100,000 + 0.287) + 75.733 \qquad (3)$$
$$CHC_{fitness}(evals) = 3.155 \cdot \log(evals/100,000 - 0.459) + 85.867 \qquad (4)$$

5   Conclusions

We have defined a coverage problem for wireless sensor networks with its innate connectivity constraint. A very large instance containing $1,000$ available locations has been solved for this problem using two different metaheuristic techniques: simulated annealing and CHC.

CHC has been able to solve the problem more efficiently than SA. In our experiments CHC has been able to reach high fitness values with an effort (number of performed solution evaluations) less than five times smaller than the effort required by SA to reach that same fitness. The average fitness obtained by any of the algorithms improves if the allowed number of evaluations per execution is increased within the range employed for our experiments ($50,000$ to $1,000,000$ evaluations), however their growths are sublineal. Mathematical models for this dependence have been calculated for both algorithms, resulting in logarithmic functions modelling SA's and CHC's fitness growth.

In future work the effect of the relation between sensing and communication radii will be studied. We also plan to redefine the problem so as to be able to place the sensors anywhere in the sensor field (instead of only in the available positions), and also take into account the power constraints existing in WSN (much harder than in other systems).

Acknowledgements

This paper has been partially funded by the Spanish Ministry of Education and Science and by European FEDER under contract TIN2005-08818-C04-01 (The OPLINK project, http://oplink.lcc.uma.es). Guillermo Molina is supported by grant AP2005-0914 from the Spanish government.

References

1. Akyildiz, I., et al.: A survey on sensor networks. IEEE Communications Magazine (2002)
2. Alba, E., Molina, G., Chicano, F.: Optimal placement of antennae using meta-heuristics. In: Boyanov, T., et al. (eds.) NMA 2006. LNCS, vol. 4310, pp. 214–222. Springer, Heidelberg (2007)
3. Eshelman, L.J.: The CHC Adaptive Search Algorithm: How to Have Safe Search When Engaging in Nontraditional Genetic Recombination. In: Foundations of Genetic Algorithms, pp. 265–283. Morgan Kaufmann, San Francisco (1991)
4. Jourdan, D., de Weck, O.: Layout optimization for a wireless sensor network using a multi-objective genetic algorithm. In: Proceedings of the IEEE Semiannual Vehicular Technology Conference, vol. 5, pp. 2466–2470 (2004)
5. Kirkpatrick, S., Gelatt, C.D., Vecchi, M.P.: Optimization by simulated annealing. Science 4598(220), 671–680 (1983)
6. Meguerdichian, S., et al.: Coverage problems in wireless ad-hoc sensor networks. In: INFOCOM, pp. 1380–1387 (2001)
7. Michalewicz, Z., Fogel, D.: How to Solve It: Modern Heuristics. Springer, Heidelberg (1998)

Semi-dynamic Demand in a Non-permutation Flowshop with Constrained Resequencing Buffers

Gerrit Färber[1], Said Salhi[2], and Anna M. Coves Moreno[1]

[1] Institut d'Organització i Control de Sistemes Industrials,
Universitat Politècnica de Catalunya, Barcelona, Spain
[2] The Center for Heuristic Optimisation, Kent Business School,
University of Kent, Canterbury, UK
Gerrit_Faerber@gmx.de, s.salhi@kent.ac.uk, Anna.Maria.Coves@upc.edu

Abstract. This work presents the performance comparison of two conceptually different approaches for a mixed model non-permutation flowshop production line. The demand is a semi-dynamic demand with a fixed job sequence for the first station. Resequencing is permitted where stations have access to intermediate or centralized resequencing buffers. The access to the buffers is restricted by the number of available buffer places and the physical size of the products. An exact approach, using Constraint Logic Programming (CLP), and a heuristic approach, a Genetic Algorithm (GA), were applied.

Keywords: Semi-dynamic demand, Constraint Logic Programming, Genetic Algorithm, Non-Permutation Flowshop, Mixed model assembly line.

1 Introduction

Mixed model production lines consider more than one model being processed on the same production line in an arbitrary sequence. Nevertheless, the majority of publications are limited to solutions which determine the job sequence before the jobs enter the line and maintain it without interchanging jobs until the end of the production line, known as permutation flowshop. In the case of more than three stations and with the objective function to minimize the makespan, a unique permutation for all stations is no longer optimal. In [23] and [16] studies of the benefits of using non-permutation flowshops are presented.

Various designs of production lines, which permit resequencing of jobs, exist [13,12,6,24]. Resequencing of jobs on the line is even more relevant with the existence of an additional cost or time, occurring when at a station the succeeding job is of another model, known as setup-cost and setup-time [1].

The case of infinite buffers is basically a theoretical case in which no limitation exists with respect to the number of jobs that may be buffered between two stations. Surveys on heuristics treating the case of infinite buffers are presented in [17] and [22]. Approaches which consider a limited number of buffer places for

I. Lirkov, S. Margenov, and J. Waśniewski (Eds.): LSSC 2007, LNCS 4818, pp. 536–544, 2008.
© Springer-Verlag Berlin Heidelberg 2008

the flowshop problem are studied in [5,25,21,20,28,14,3]. In [4] and [18] limited resequencing possibilities are considered for jobshop problems.

The introduction of resequencing possibilities generally leads to additional costs, caused by additional equipment to be mounted, like buffers, but also extra efforts in terms of logistics complexity may arise. In the case in which there exist jobs with large and small physical size, the investment for additional resequencing equipment can be reduced by, e.g., only giving small jobs the possibility to resequence. Consequently, only small resequencing buffer places are installed. Following this concept, in a chemical production line where the demand of customers is different, only resequencing tanks that permit to resequence the request of a small customer order are used.

In what follows, the problem is formulated in more detail, and the exact and the heuristic approaches are explained. Thereafter promising results are presented for medium and large sized problems, which demonstrate the relevance of the proposed concept, followed by the conclusions.

2   Problem Definition

This paper considers a mixed model non-permutation flowshop with the possibility of resequencing jobs between consecutive stations. The jobs $(J_1, J_2, ..., J_j,$ $..., J_n)$ pass consecutively through the stations $(I_1, I_2, ..., I_i, ..., I_m)$ and after determined stations, off-line buffers B_i permit to resequence jobs. The buffer provides various buffer places $(B_{i,1}, B_{i,2}, ...)$ and each buffer place is restricted by the physical size of the jobs to be stored. As can be seen in figure 1a, job J_2 can be stored in buffer place $B_{i,1}$ as well as in $B_{i,2}$. Whereas, the next job J_3 can be stored only in buffer place $B_{i,2}$, because of the physical size of the job exceeding the physical size of buffer place $B_{i,1}$, see figure 1b.

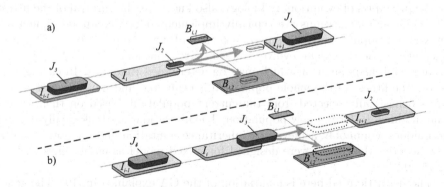

Fig. 1. Scheme of the considered flowshop. The jobs J_j pass consecutively through the stations I_i. The buffer B_i permits to temporally store a job with the objective of reinserting it at a later position in the sequence. a) Job J_2 can pass through any of the two buffer places $B_{i,1}$ or $B_{i,2}$ of buffer B_i. b) Job J_3 can pass only through buffer place $B_{i,2}$, due to its physical size.

The buffers are located off-line, in a first step accessible from a single station (intermediate case). Then, for an additional benefit, a single resequencing buffer is used, accessible from various stations (centralized case). The objective function is the weighted sum of the makespan and the setup-cost, where there is no weight associated with the setup-time though this is indirectly included in the calculation of the makespan. An exact approach, using Constraint Logic Programming (CLP) and a heuristic approach, using a Genetic Algorithm (GA), were applied to the problem under study.

3   Approaches

3.1   Exact Approach: CLP

The concept of CLP can be described as a powerful extension of conventional logic programming [11]. It involves the incorporation of constraint languages and constraint solving methods into logic programming languages [26].

The formulation used here is explained in more detail in [7] and was implemented in OPL Studio version 3.7. Apart from job and station precedences, the CLP formulation determines the jobs which are to be taken off the line for the purpose of resequencing, given that a free buffer place is available and that the physical size of the job does not exceed the physical size of the buffer place. The formulation also includes computational enhancements like imposing the start time of jobs and the reduction of the size of the variables and considers the intermediate as well as the centralized location of the resequencing buffers.

3.2   Heuristic Approach: GA

The concept of Genetic Algorithms (GA) can be understood as the application of the principles of evolutionary biology, also known as the survival of the fittest [9,10]. Genetic algorithms are typically implemented as a computer simulation in which a population of chromosomes, each of which represents a solution of the optimization problem, evolves toward better solutions. The evolution starts from an initial population which may be determined randomly. In each generation, the fitness of the whole population is evaluated and multiple individuals are stochastically selected from the current population, based on their fitness and modified to form a new population. The alterations are biologically-derived techniques, commonly achieved by inheritance, mutation and crossover. Multiple Genetic Algorithms were designed for mixed model assembly lines such as [2,15,29,27].

The heuristic used here is a variation of the GA explained in [19]. The genes represent the jobs which are to be sequenced. The chromosomes v, determined by a series of genes, represent a sequence of jobs. A generation is formed by R chromosomes and the total number of generations is G. In the permutation case, the size of a chromosome is determined by the number of jobs, the fraction Π. In the non-permutation case, the chromosomes are $L + 1$ times larger, resulting in the fractions $\Pi_1', ..., \Pi_{L+1}'$, being L the number of resequencing possibilities. In both cases, special attention is required when forming the chromosomes, because of the fact that for each part of the production line every job has to be sequenced exactly one time.

The relevant information for each chromosome is its fitness value (objective function), the number of job changes and the indicator specifying if the chromosome represents a feasible solution. A chromosome is marked infeasible and is imposed with a penalty. This situation arises if a job has to be taken off the line and no free resequencing buffer place is available or the physical size of the job exceeds the size limitation of the available resequencing buffer places. When two solutions result in the same fitness, the one with fewer job changes is preferred. In [8] the detailed formulation can be found.

4   Performance Study

The performance study considers a medium sized problem with 10 stations and up to 10 jobs. This is applied to the exact (CLP) as well as to the heuristic approach (GA). The second instance uses a large problem with 5 stations and up to 100 jobs where the heuristic approach (GA) is applied only.

4.1   Instance-1: Medium Sized Problem (CLP Versus GA)

A flowshop which consists of 10 stations is considered. After station 3, 5 and 8 a single intermediate buffer place is located. The range of the production time is [1...100], for the setup cost [2...8] and for the setup time [1...5]. The number of jobs is varied in the range of 4 to 10 and the objective function is the weighted sum of the makespan (factor 1.0) and the setup cost (factor 0.3), where the setup time is not concerned with a weight but is indirectly included in the calculation of the makespan.

Three differently sized buffer places (large, medium, small) are available and the ratio of jobs is $\frac{3}{10}$, $\frac{3}{10}$ and $\frac{4}{10}$ for large, medium and small, respectively. The allocation of the buffer places to the buffers considers five scenarios for the intermediate case ("I111", "I231", "I132", "I222", "I333") and three scenarios for the centralized case ("C1", "C2", "C3"). "I132" represents 1 small, 1 large and 1 medium buffer place, located as intermediate resequencing buffer places after stations 3, 5 and 8, respectively. "C2" represents 1 medium buffer place, located as a centralized buffer place, accessible from stations 3, 5 and 8. "I333"

Table 1. Semi-dynamic demand using the exact approach (CLP)

Jobs	Perm	I111	I231	I132	I222	I333	C1	C2	C3
4	483,1	480,9	480,7	480,7	480,9	480,7	480,9	480,7	480,7
5	552,0	490,7	490,7	490,7	490,7	490,7	490,7	490,7	490,7
6	647,5	627,0	620,1	620,1	620,1	620,1	628,5	622,2	622,2
7	636,5	627,7	627,7	625,0	625,0	609,5	628,0	627,7	616,1
8	673,6	646,6	646,6	644,8	644,8	644,8	646,6	644,8	632,3
9	744,3	719,4	719,4	716,1	716,1	716,1	719,4	716,1	712,1
10	813,7	786,5	763,2	762,5	785,9	791,2	786,5	786,5	764,0

Table 2. Semi-dynamic demand using the heuristic approach (GA)

Jobs	Perm	l111	l231	l132	l222	l333	C1	C2	C3
4	483,1	480,9	480,7	480,7	480,7	480,7	480,9	480,7	480,7
5	552,0	490,7	490,7	490,7	490,7	490,7	490,7	490,7	490,7
6	647,5	627,0	620,1	620,1	620,1	620,1	628,5	622,7	622,2
7	636,5	627,7	628,8	626,1	625,0	609,5	629,1	628,5	616,3
8	673,6	669,2	651,3	646,0	647,2	638,2	672,4	653,4	637,0
9	744,3	736,5	724,2	728,6	721,8	709,4	736,5	729,5	714,7
10	813,7	808,3	788,0	781,7	805,1	757,5	809,2	805,9	772,2

and "C3" are the two cases which provide the largest flexibility in terms of physical size restrictions.

The results of the CLP are shown in table 1. In all cases, when offline re-sequencing buffers are considered, the results are improved compared to the permutation sequence. In the studied flowshop, an average of 4.3% is achieved for the CLP, whereas, in the case of the GA, see table 2, the average is 3.7%.

In the case of the exact approach, as well as in the GA, the semi-dynamic demand with a fixed job sequence for the first station, leads to a considerable improvement. In table 3 the improvement of the CLP with respect to the GA is shown. For up to 5 jobs, both methods achieve the same solutions. When 6 or more jobs are to be sequenced, in general, the CLP outperforms the GA when smaller buffer places are used.

The execution time in the vast majority of the cases was inferior to 600 seconds for the case of the CLP. The execution time of the GA, limited to 1000 iterations, required up to 55 seconds; increasing the number of iterations did not result in a major improvement. In general, the solutions of the CLP show better results in the more restricted problem. Nevertheless, the GA has to process a large number of infeasible solutions when the resequencing possibilities are heavily restricted. The GA consequently performs better in the less restricted the problem. This behavior was also observed when the demand is not a semi-dynamic demand, considering that the jobs can be resequenced before they enter the production line.

4.2   Instance-2: Large Problem (GA)

A flowshop which consists of 5 stations is considered. The range of the production time is [0...20] such that for some jobs zero-processing time at some stations exists, for the setup cost [2...8] and for the setup time [1...5]. The objective

Table 3. Comparison of the GA and the Constrained Logic Programming for semi-dynamic demand. The values show the improvement of the CLP with respect to the GA.

Jobs	Perm	l111	l231	l132	l222	l333	C1	C2	C3
4	483,1	0,0%	0,0%	0,0%	0,0%	0,0%	0,0%	0,0%	0,0%
5	552,0	0,0%	0,0%	0,0%	0,0%	0,0%	0,0%	0,0%	0,0%
6	647,5	0,0%	0,0%	0,0%	0,0%	0,0%	0,0%	0,1%	0,0%
7	636,5	0,0%	0,2%	0,2%	0,0%	0,0%	0,2%	0,1%	0,0%
8	673,6	3,4%	0,7%	0,2%	0,4%	-1,0%	3,8%	1,3%	0,7%
9	744,3	2,3%	0,7%	1,7%	0,8%	-0,9%	2,3%	1,8%	0,4%
10	813,7	2,7%	3,1%	2,5%	2,4%	-4,5%	2,8%	2,4%	1,1%

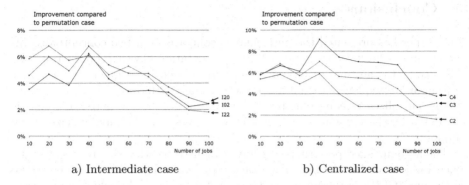

a) Intermediate case

b) Centralized case

Fig. 2. Semi-dynamic demand without sequencing before the first station: a) In I22, both buffers are provided with two buffer places each; in I20, only station 2 has access to a resequencing buffer with two places; and in I02, only station 3 has access to a resequencing buffer with two places. b) Station 2 and station 3 have access to a centralized buffer with two buffer places (C2), three buffer places (C3) and four buffer places (C4).

function is the weighted sum of the makespan (factor of 1.0) and the setup cost (factor of 0.3). The setup time has no weight associated with but is indirectly included in the calculation of the makespan. For the intermediate case, three scenarios are considered: in I22, both buffers are provided with two buffer places each; in I20, only station 2 has access to a resequencing buffer with two places; and in I02, only station 3 has access to a resequencing buffer with two places. For the centralized case, station 2 and station 3 have access to a centralized buffer with two places (C2), three places (C3) and four places (C4).

Figure 2 shows the amounts of improvement which are achieved by the GA, when resequencing of a randomly generated sequence within the production line (semi-dynamic case) is permitted, compared to the case without resequencing. In the intermediate case, see figure 2a, the use of the two resequencing possibilities (I22) achieves best results up to 30 jobs, then, the case of one resequencing possibility at station 2 (I20) outperforms in nearly all cases until 100 jobs. Even though I22 provides more flexibility in terms of resequencing, the GA performs better for instances with fewer resequencing possibilities when more than 30 jobs are to be sequenced. This comes from the fact that the length of the chromosomes is dependent on the number of resequencing possibilities (L).

In the centralized case, see figure 2b, the variable parameter is the number of buffer places. The use of two buffer places (C2) in all of the considered cases is inferior compared to the case of three and four buffer places (C3, C4). Until 30 jobs, the cases C3 and C4 achive nearly equivalent results which means that the fourth buffer place is not required. Then, for 40 jobs and more, the fourth buffer place shows a considerable impact on the possible improvements, compared to the case without resequencing.

5   Conclusions

This paper has presented the performance comparison of two conceptually different approaches for a mixed model non-permutation flowshop production line. The demand is a semi-dynamic demand with a fixed job sequence for the first station and resequencing is permitted where stations have access to intermediate or centralized resequencing buffers. Furthermore, the access to the buffers is restricted by the number of available buffer places and the physical size of the products.

The accomplished performance study demonstrated the effectiveness of resequencing jobs within the line. The exact approach, using Constraint Logic Programming (CLP), outperforms the Genetic Algorithm (GA), when the physical size of the resequencing buffer places is limited. Due to the limited applicability of the exact approach, the performance study for larger problems of up to 100 jobs was performed using the GA. The chromosome size is dependent on the number of resequencing possibilities and therefore the performance of the GA for more than 30 jobs shows better results when fewer resequencing possibilities are present.

The results revealed the benefits that come with a centralized buffer location, compared to the intermediate buffer location. It either improves the solution or leads to the use of fewer resequencing buffer places. An increased number of large buffer places clearly improves the objective function and including buffers, constrained by the physical size of jobs to be stored, on one side limits the solutions but on the other side minimizes the necessary buffer area.

In order to take full advantage of the possibilities of resequencing jobs in a mixed model flowshop, additional installations may be necessary to mount, like buffers, but also extra efforts in terms of logistics complexity may arise. The additional effort is reasonable if it pays off the necessary investment.

Acknowledgement

This work is partially supported by the Ministry of Science and Technology, and the funding for regional research DPI2004-03472.

References

1. Bolat, A.: Sequencing jobs on an automobile assembly line: Objectives and procedures. International Journal of Production Research 32(5), 1219–1236 (1994)
2. Bolat, A., Al-Harkan, I., Al-Harbi, B.: Flow-shop scheduling for three serial stations with the last two duplicate. Computers & Operations Research 32(3), 647–667 (2005)
3. Brucker, P., Heitmann, S., Hurink, J.: Flow-shop problems with intermediate buffers. OR Spektrum 25, 549–574 (2003)
4. Brucker, P., et al.: Job-shop scheduling with limited capacity buffers. OR Spektrum 28, 151–176 (2006)

5. Dutta, S., Cunningham, A.: Sequencing two-machine flow-shops with finite intermediate storage. Management Science 21, 989–996 (1975)
6. Engström, T., Jonsson, D., Johansson, B.: Alternatives to line assembly: Some swedish examples. International Journal of Industrial Ergonomics 17(3), 235–245 (1996)
7. Färber, G.: Sequencing in mixed model non-permutation flowshop production line using constrained buffers. PhD Thesis, Universitat Politècnica de Catalunya, Spain (2006)
8. Färber, G., Coves, A.: Performance study of a genetic algorithm for sequencing in mixed model non-permutation flowshops using constrained buffers. In: Gavrilova, M.L., et al. (eds.) ICCSA 2006. LNCS, vol. 3982, pp. 638–648. Springer, Heidelberg (2006)
9. Holland, J.: Genetic algorithms and the optimal allocation of trials. SIAM J. Comput. 2(2), 88–105 (1973)
10. Holland, J.: Adaptation in natural and artificial systems. University of Michigan Press, Ann Arbor (1975)
11. Jaffar, J., Lassez, J.: Constraint logic programming. Technical Report 74, Department of Computer Science, Monash University 74 (1986)
12. Lahmar, M., Ergan, H., Benjaafar, S.: Resequencing and feature assignment on an automated assembly line. IEEE Transactions on Robotics and Automation 19(1), 89–102 (2003)
13. Lee, H., Schaefer, S.: Sequencing methods for automated storage and retrieval systems with dedicated storage. Computers and Industrial Engineering 32(2), 351–362 (1997)
14. Leisten, R.: Flowshop sequencing problems with limited buffer storage. International Journal of Production Research 28(11), 2085–2100 (1990)
15. Levitin, G., Rubinovitz, J., Shnits, B.: A genetic algorithm for robotic assembly line balancing. European Journal of Operational Res. 168, 811–825 (2006)
16. Liao, C., Liao, L., Tseng, C.: A performance evaluation of permutation vs. non-permutation schedules in a flowshop 44(20), 4297–4309 (2006)
17. Liesegang, G., Schirmer, A.: Heuristische verfahren zur maschinenbelegungsplanung bei reihenfertigung. Zeitschrift für Operations Research 19, 195–211 (1975)
18. Mascic, A., Pacciarelli, D.: Job-shop scheduling with blocking and no-wait constraints. European Journal of Operational Research 143, 498–517 (2002)
19. Michaelewicz, Z.: Genetic Algorithms + Data Structures = Evolution Programs, 3rd edn. Springer, Heidelberg (1996)
20. Nowicki, E.: The permutation flow shop with buffers: A tabu search approach. European Journal of Operational Research 116, 205–219 (1999)
21. Papadimitriou, C., Kanellakis, P.: Flowshop scheduling with limited temporary storage. Journal of the Association for Computing Machinery 27, 533–554 (1980)
22. Park, Y., Pegden, C., Enscore, E.: A survey and evaluation of static flowshop scheduling heuristics. International Journal of Production Research 22, 127–141 (1984)
23. Potts, C., Shmoys, D., Williamson, D.: Permutation vs. non-permutation flow shop schedules 10(5), 281–284 (1991)
24. Rachakonda, P., Nagane, S.: Simulation study of paint batching problem in automobile industry (2004), (consulted 14.07.2004),
   http://sweb.uky.edy/~pkrach0/Projects/MFS605Project.pdf
25. Reddi, S.: Sequencing with finite intermediate storage. Management Science 23, 19–25 (1976)

26. Riezler, S.: Probabilistic constraint logic programming. AIMS Arbeitspapiere des Instituts für Maschinelle Sprachverarbeitung Lehrstuhl für Theoretische Computerlinguistik, Universität Stuttgart 5(1) (1999)
27. Ruiz, R., Maroto, C.: A genetic algorithm for hybrid flowshops with sequence dependent setup times and machine eligibility. European Journal of Operational Research 169(3), 781–800 (2006)
28. Smutnicki, C.: A two-machine permutation flow shop scheduling problem with buffers. OR Spektrum 20, 229–235 (1998)
29. Wang, L., Zhang, L., Zheng, D.: An effective hybrid genetic algorithm for flow shop scheduling with limited buffers. In: Computers & Operations Research (Article in Press, 2006)

Probabilistic Model of Ant Colony Optimization for Multiple Knapsack Problem

Stefka Fidanova

Institute for Parallel Processing, Bulgarian Academy of Science, Acad. G. Bonchev
bl. 25A, 1113 Sofia, Bulgaria
`stefka@parallel.bas.bg`

Abstract. The Ant Colony Optimization (ACO) algorithms are being applied successfully to a wide range of problems. ACO algorithms could be good alternatives to existing algorithms for hard combinatorial optimization problems (COPs). In this paper we investigate the influence of the probabilistic model in model-based search as ACO. We present the effect of four different probabilistic models for ACO algorithms to tackle the Multiple Knapsack Problem (MKP). The MKP is a subset problem and can be seen as a general model for any kind of binary problems with positive coefficients. The results show the importance of the probabilistic model to quality of the solutions.

1 Introduction

There are many NP-hard combinatorial optimization problems for which it is impractical to find an optimal solution. Among them is the MKP. For such problems the reasonable way is to look for algorithms that quickly produce near-optimal solutions. ACO [2,4,3] is a meta-heuristic procedure for quickly and efficiently obtaining high quality solutions of complex optimization problems [11]. The ACO algorithms were inspired by the observation of real ant colonies. Ants are social insects, that is, insects that live in colonies and whose behavior is directed more to the survival of the colony as a whole than to that of a single individual component of the colony. An important and interesting aspect of ant colonies is how ants can find the shortest path between food sources and their nest. ACO is the recently developed, population-based approach which has been successfully applied to several NP-hard COPs [6]. One of its main ideas is the indirect communication among the individuals of a colony of agents, called "artificial" ants, based on an analogy with trails of a chemical substance, called pheromones which real ants use for communication. The "artificial" pheromone trails are a kind of distributed numerical information which is modified by the ants to reflect their experience accumulated while solving a particular problem. When constructing a solution, at each step ants compute a set of feasible moves and select the best according to some probabilistic rules. The transition probability is based on the heuristic information and pheromone trail level of the move (how much the movement is used in the past). When we apply ACO algorithm to

I. Lirkov, S. Margenov, and J. Waśniewski (Eds.): LSSC 2007, LNCS 4818, pp. 545–552, 2008.

MKP various probabilistic models are possible and the influence on the results is shown.

The rest of the paper is organized as follows: Section 2 describes the general framework for MKP as a COP. Section 3 outlines the implemented ACO algorithm applied to MKP. In section 4 four probabilistic models are described. In Section 5 experimental results over test problems are shown. Finally some conclusions are drawn.

2   Formulation of the Problem

The Multiple Knapsack Problem has numerous applications in theory as well as in practice. It also arise as a subproblem in several algorithms for more complex COPs and these algorithms will benefit from any improvement in the field of MKP. We can mention the following major applications: problems in cargo loading, cutting stock, bin-packing, budget control and financial management may be formulated as MKP. In [12] there is proposed to use the MKP in fault tolerance problem and in [1] there is designed a public cryptography scheme whose security realize on the difficulty of solving the MKP. Martello and Toth [10] mention that two-processor scheduling problems may be solved as a MKP. Other applications are industrial management, naval, aerospace, computational complexity theory.

Most of theoretical applications either appear where a general problem is transformed to a MKP or where a MKP appears as a subproblem. We should mention that MKP appears as a subproblem when solving the generalized assignment problem, which again is used when solving vehicle routing problems. In addition, MKP can be seen as a general model for any kind of binary problems with positive coefficients [7].

The MKP can be thought as a resource allocation problem, where we have m resources (the knapsacks) and n objects. The object j has a profit p_j, each resource has its own budget c_i (knapsack capacity) and consumption r_{ij} of resource i by object j. We are interested in maximizing the sum of the profits, while working with a limited budget.

The MKP can be formulated as follows:

$$\max \sum_{j=1}^{n} p_j x_j$$

$$\text{subject to} \sum_{j=1}^{n} r_{ij} x_j \leq c_i \quad i = 1, \ldots, m \tag{1}$$

$$x_j \in \{0, 1\} \quad j = 1, \ldots, n$$

x_j is 1 if the object j is chosen and 0 otherwise.

There are m constraints in this problem, so MKP is also called m-dimensional knapsack problem. Let $I = \{1, \ldots, m\}$ and $J = \{1, \ldots, n\}$, with $c_i \geq 0$ for all $i \in I$. A well-stated MKP assumes that $p_j > 0$ and $r_{ij} \leq c_i \leq \sum_{j=1}^{n} r_{ij}$ for all $i \in I$ and $j \in J$. Note that the matrix $[r_{ij}]_{m \times n}$ and the vector $[c_i]_m$ are both non-negative.

In the MKP we are not interested in solutions giving a particular order. Therefore a partial solution is represented by $S = \{i_1, i_2, \ldots, i_j\}$ and the most recent
elements incorporated to S, i_j need not to be involved in the process for selecting
the next element. Moreover, solutions of ordering problems have a fixed length
as we search for a permutation of a known number of elements. Solutions of
MKP, however, do not have a fixed length. We define the graph of the problem
as follows: the nodes correspond to the items, the arcs fully connect nodes. A
fully connected graph means that after the object i we can choose the object j
for every i and j if there are enough resources and object j is not chosen yet.

3   ACO Algorithm for MKP

Real ants foraging for food lay down quantities of pheromone (chemical clues)
marking the path that they follow. An isolated ant moves essentially at random
but an ant encountering a previously laid pheromone will detect it and decide to
follow it with high probability and thereby reinforce it with a further quantity of
pheromone. The repetition of the above mechanism represents the auto catalytic
behavior of real ant colony where the more the ants follow a trail, the more
attractive that trail becomes.

The above behavior of real ants has inspired ACO algorithm. This technique,
which is a population-based approach, has been successfully applied to many NP-
hard optimization problems [2,4]. The ACO algorithm uses a colony of artificial
ants that behave as co-operative agents in a mathematical space where they are
allowed to search and reinforce pathways (solutions) in order to find the optimal
ones. A solution satisfying the constraints is said to be feasible.

```
procedure ACO
begin
    Initialize
    while stopping criterion not satisfied do
        Position each ant in a starting node
        repeat
            for each ant do
                Chose next node by applying the state transition rate
                Apply step-by-step pheromone update
            end for
        until every ant has build a solution
        Update best solution
        Apply offline pheromone updating
    end while
end
```

After initialization of the pheromone trails, ants construct feasible solutions,
starting from random nodes, then the pheromone trails are updated. At each step
ants compute a set of feasible moves and select the best one (according to some

probabilistic rules) to carry out the rest of the tour. The transition probability is based on the heuristic information and pheromone trail level of the move. The higher the value of the pheromone and the heuristic information, the more profitable it is to select this move and resume the search. In the beginning, the initial pheromone level is set to a small positive constant value τ_0 and then ants update this value after completing the construction stage.

ACO algorithms adopt different criteria to update the pheromone level. In our implementation we use the Ant Colony System (ACS) [4] approach.

In ACS the pheromone updating stage consists of local update stage and global update stage.

3.1 Local Update Stage

While ants build their solution, at the same time they locally update the pheromone level of the visited paths by applying the local update rule as follows:

$$\tau_{ij} \leftarrow (1 - \rho)\tau_{ij} + \rho\tau_0, \tag{2}$$

where ρ is a persistence of the trail and the term $(1 - \rho)$ can be interpreted as trail evaporation.

The aim of the local updating rule is to make better use of the pheromone information by dynamically changing the desirability of edges. Using this rule, ants will search in wide neighborhood around the best previous solution. As shown in the formula, the pheromone level on the paths is highly related to the value of evaporation parameter ρ. The pheromone level will be reduced and this will reduce the chance that the other ants will select the same solution and consequently the search will be more diversified.

3.2 Global Updating Stage

When all ants have completed their solution, the pheromone level is updated by applying the global updating rule only on the paths that belong to the best solution since the beginning of the trail as follows:

$$\tau_{ij} \leftarrow (1 - \rho)\tau_{ij} + \Delta\tau_{ij} \tag{3}$$

where $\Delta\tau_{ij} = \begin{cases} \rho L_{gb} \text{ if } (i, j) \in \text{best solution} \\ 0 \quad \text{otherwise} \end{cases}$,

L_{gb} is the cost of the best solution from the beginning. This global updating rule is intended to provide a greater amount of pheromone on the paths of the best solution, thus the search is intensified around this solution.

Let $s_j = \sum_{i=1}^{m} r_{ij}$. For heuristic information we use:

$$\eta_{ij} = \begin{cases} p_j^{d_1}/s_j^{d_2} \text{ if } s_j \neq 0 \\ p_j^{d_1} \quad \text{if } s_j = 0 \end{cases} \tag{4}$$

Hence the objects with greater profit and less average expenses will be more desirable.

The MKP solution can be represented by string with 0 for objects that are not chosen and 1 for chosen objects. The new solution is accepted if it is better than current solution.

4 Transition Probability

In this section we describe four possibilities for transition probability model. For ant k, the probability p_{ij}^k of moving from a state i to a state j depends on the combination of two values:

- The attractiveness η_{ij} of the move as computed by some heuristic.
- The pheromone trail level of the move.

The pheromone τ_{ij} is associated with the arc between nodes i and j.

4.1 Proportional Transition Probability

The quantity of the pheromone on the arcs between two nodes is proportional to the experience of having the two nodes in the solution. Thus the node j is more desirable if the quantity of the pheromone on arc (i, j) is high. For ant k which moves from node i to node j the rule is:

$$
p_{ij}^k(t) = \begin{cases} \dfrac{\tau_{ij}\eta_{ij}(S_k(t))}{\sum_{q \in allowed_k(t)} \tau_{iq}\eta_{iq}(S_k(t))} & \text{if } j \in allowed_k(t) \\ 0 & \text{otherwise} \end{cases}, \qquad (5)
$$

where $allowed_k$ is the set of remaining feasible states, $S_k(t)$ is the partial solution at step t from ant k.

4.2 Transition Probability with Sum

This probability takes into account how desirable in the past has been the node j, independently how many ants have reached it from the node i or from some other. Thus the node j is more desirable if the average quantity of the pheromone on the arcs which entry in the node j is high. In this case the transition probability becomes:

$$
p_{ij}^k(t) = \begin{cases} \dfrac{(\sum_{i=1}^n \tau_{ij})\eta_{ij}(S_k(t))}{\sum_{q \in allowed_k(t)}(\sum_{l=1}^n \tau_{lq})\eta_{iq}(S_k(t))} & \text{if } j \in allowed_k(t) \\ 0 & \text{otherwise} \end{cases}. \qquad (6)
$$

4.3   Maximal Transition Probability

This probability is proportional to the maximal pheromone on the arcs which entry in the node j. Thus the node j can be more desirable independently of the quantity of the pheromone on the arc (i,j) if there is some other arc with high quantity of the pheromone which entry in the node j. In this case the transition probability is changed as follows:

$$p_{ij}^k(t) = \begin{cases} \dfrac{(\max_l \tau_{lj})\eta_{ij}(S_k(t))}{\sum_{q \in allowed_k(t)}(\max_l \tau_{lq})\eta_{iq}(S_k(t))} & \text{if } j \in allowed_k(t) \\ 0 & \text{otherwise} \end{cases}. \qquad (7)$$

4.4   Minimal Transition Probability

This probability is proportional to the minimal pheromone on the arcs which entry in the node j. Thus the node j will be more desirable if the quantity of the pheromone on all arcs which entry in the node j is high. In this case the transition probability is as follows:

$$p_{ij}^k(t) = \begin{cases} \dfrac{(\min_l \tau_{lj})\eta_{ij}(S_k(t))}{\sum_{q \in allowed_k(t)}(\min_l \tau_{lq})\eta_{iq}(S_k(t))} & \text{if } j \in allowed_k(t) \\ 0 & \text{otherwise} \end{cases}. \qquad (8)$$

5   Experimental Results

In this section we describe the experimental analysis on the performance of MKP as a function of the transition probability. We show the computational experience of the ACS using 10 MKP instances from "OR-Library" available at http://people.brunel.ac.uk/~{}mastjjb/jeb/orlib, with 100 objects and 10 constraints. To provide a fair comparison for the above implemented ACS algorithm, a predefined number of iterations, $k = 400$, is fixed for all the runs. The developed technique has been coded in C++ language and implemented on a Pentium 4 (2.8 GHz).

Because of the random start of the ants in every iteration, we can use fewer ants than the number of the nodes. After the tests we found that 10 ants are enough to achieve good results. Thus we decrease the running time of the program. We run the same instance using different transition probability models on the same random sequences for starting nodes and we find different results. Thus we are sure that the difference comes from the transition probability. For all 10 instances we were running experiments for a range of evaporation rates and the parameters d_1 and d_2 in order to find the best parameters for every instance. We fixed the initial pheromone value to be $\tau_0 = 0.5$. After choosing for every problem instance the best rate for the parameters we could compare the different transition probabilities. In Figure 1 we show the average results over all 10 problem instances and every instance is run 20 times with the same parameter settings.

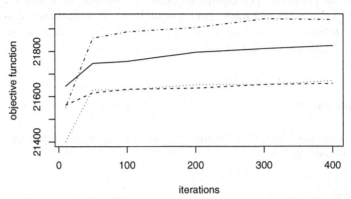

Fig. 1. The graphics shows the average solution quality (value of the total cost of the objects in the knapsack) over 20 runs. Dash dot line represents proportional probability, dash line — probability with sum, dots line — maximal probability and thick line — minimal probability.

Our first observation is that the proportional and the minimal transition probabilities show advantage over the sum and maximal transition probabilities. In a small number of iterations (less than 50), probability with sum and minimal probability achieve better results, but after that the proportional probability outperforms them. The MKP is not ordered problem. It means that the quality of the solution is not related to the order we choose the elements. Using maximal transition probability it is enough only one arc to have high quantity of the pheromone and the node will be more desirable. This kind of probability is more suitable to ordered problems: the node j is more desirable by node i than by node q. Using transition probability with sum the node is more desirable if the average quantity of the pheromone is high, but for some of the arcs this quantity can be very high and for other arc it can be very low and the average pheromone to be high. Thus we can explain the worst results with this two models of the transition probability. If the minimal probability is high, the quantity of pheromone for all arcs which entry the node is high. If the proportional probability is high it means that after node i is good to chose node j. The last two models of transition probability are better related to the unordered problems and thus we can achieve better results using them.

6   Conclusion

The design of a meta-heuristic is a difficult task and highly dependent on the structure of the optimized problem. In this paper four models of the transition probability have been proposed. The comparison of the performance of the ACS coupled with these probability models applied to different MKP problems are reported. The goal is to find probability model which is more relevant to the structure of the problem. The obtained results are encouraging and the ability of the developed models to rapidly generate high-quality solutions for MKP can be seen. For future work another important direction for current research is to try different strategies to explore the search space more effectively and provide good results.

Acknowledgments

Stefka Fidanova was supported by the European Community program "Center of Excellence" BIS 21++.

References

1. Diffe, W., Hellman, M.E.: New direction in cryptography. IEEE Trans. Inf. Theory IT-36, 644–654 (1976)
2. Dorigo, M., Di Caro, G.: The Ant Colony Optimization metaheuristic. In: Corne, D., Dorigo, M., Glover, F. (eds.) New Idea in Optimization, pp. 11–32. McGrow-Hill (1999)
3. Dorigo, M., Di Caro, G., Gambardella, L.M.: Ant Algorithms for Distributed Discrete Optimization. J. of Artificial Life 5, 137–172 (1999)
4. Dorigo, M., Gambardella, L.M.: Ant Colony System: A cooperative Learning Approach to the Traveling Salesman Problem. IEEE Transaction on Evolutionary Computation 1, 53–66 (1999)
5. Ferreira, C.E., Martin, A., Weismantel, R.: Solving Multiple Knapsack Problems by Cutting Planes. SIAM Journal on Optimization 6(3), 858–877 (1996)
6. Gambardella, M.L., Taillard, E.D., Dorigo, M.: Ant Colonies for the QAP. J. of Oper. Res. Soc. 50, 167–176 (1999)
7. Kochenberger, G., McCarl, G., Wymann, F.: A Heuristic for General Integer Programming. J. of Decision Sciences 5, 34–44 (1974)
8. Leguizamon, G., Michalevich, Z.: A New Version of Ant System for Subset Problems. In: Proceedings of Int. Conf. on Evolutionary Computations, Washington, pp. 1459–1464 (1999)
9. Marchetti-Spaccamela, A., Vercellis, C.: Stochastic on-line Knapsack Problems. J. of Mathematical Programming 68(1), 73–104 (1995)
10. Martello, S., Toth, P.: A mixtures of dynamic programming and branch-and-bound for the subset-sum proble. Management Science 30, 756–771 (1984)
11. Osman, I.H., Kelley, J.P.: Metaheuristic: An Overview. In: Osman, I.H., Kelley, J.P. (eds.) Metaheuristic: Theory and Applications, pp. 1–21. Kluwer Academic Publishers, Dordrecht (1996)
12. Sinha, A., Zoltner, A.A.: The multiple-choice knapsack problem. J. of Operational Research 27, 503–515 (1979)

An Ant-Based Model for Multiple Sequence Alignment

Frédéric Guinand and Yoann Pigné*

LITIS laboratory, Le Havre University, France
www.litislab.eu

Abstract. Multiple sequence alignment is a key process in today's biology, and finding a relevant alignment of several sequences is much more challenging than just optimizing some improbable evaluation functions. Our approach for addressing multiple sequence alignment focuses on the building of structures in a new graph model: the factor graph model. This model relies on block-based formulation of the original problem, formulation that seems to be one of the most suitable ways for capturing evolutionary aspects of alignment. The structures are implicitly built by a colony of ants laying down pheromones in the factor graphs, according to relations between blocks belonging to the different sequences.

1   Introduction

For years, manipulation and study of biological sequences have been added to the set of common tasks performed by biologists in their daily activities. Among the numerous analysis methods, multiple sequence alignment (MSA) is probably one of the most used. Biological sequences come from actual living beings, and the role of MSA consists in exhibiting the similarities and differences between them. Considering sets of homologous sequences, differences may be used to assess the evolutionary distance between species in the context of phylogeny. The results of this analysis may also be used to determine conservation of protein domains or structures. While most of the time the process is performed for aligning a limited number of thousands bp-long sequences, it can also be used at the genome level allowing biologists to discover new features that could not be exhibited at a lower level of study [5]. In all cases, one of the major difficulties is the determination of a biologically relevant alignment, performed without relying explicitly on evolutionary information like a phylogenetic tree.

Among existing approaches for determining such relevant alignments, one of them rests on the notion of block. A block is a set of factors present in several sequences. Each factor belonging to one block is an almost identical substring. It may correspond to a highly conserved zone from an evolutionary point of view. Starting from the set of factors for each sequence, the problem we address is the building of blocks. It consists in choosing and gathering almost identical factors

---

* Authors are alphabetically sorted. The work of Y. Pigné is partially supported by French Ministry of Research and Higher Education.

I. Lirkov, S. Margenov, and J. Waśniewski (Eds.): LSSC 2007, LNCS 4818, pp. 553–560, 2008.

common to several sequences in the most appropriate way, given that one block cannot contain more than one factor per sequence, that each factor can belong to only one block and that two blocks cannot cross each other. For building such blocks, we propose an approach based on ant colonies. This problem is very close to some classical optimization issues except that the process does not use any evaluation function since it seems unlikely to find a biologically relevant one. As such, it also differs notably from other works in the domain setting up ant colonies for computing alignments for a set of biological sequences [6,2].

Next section details the proposed graph model. Section 3 goes deeper into the ant algorithm details. Finally, Section 4 studies the behavior of the algorithm with examples.

2 Model

There exist many different families of algorithms for determining multiple sequence alignments, dynamic programming, progressive or iterative methods, motif-based approaches... However, if the number of methods is important, the number of models on which these methods operate is much more limited. Indeed, most algorithms use to consider nucleotide sequences either as strings or as graphs. In any case however, the problem is formulated as an optimization problem and an evaluation function is given. Within this paper, we propose another approach based on a graph of factors, where the factors are sub-sequences present in, at least, two sequences. Instead of considering these factors individually, the formulation considers that they interact with each other when they are neighbors in different sequences, such that our *factor graph* may be understood as a factor/pattern interaction network. Considering such a graph, a multiple sequence alignment corresponds to a set of structures representing highly interacting sets of factors. The original goal may be now expressed as the detection of such structures and we propose to perform such a task with the help of artificial ants.

2.1 Graph Model

An alignment is usually displayed sequence by sequence, with the nucleotides or amino acids that compose it. Here the interest is given to the factors that compose each sequence. So each sequence of the alignment is displayed as a list of the factors it is composed of. Fig. 1 illustrates such a representation, where sequences are displayed as series of factors.

There exists a relation between factors (named 1, 2 and 3 in Fig. 1) as soon as there are almost identical. Indeed, two identical factors on different sequences may be aligned. Such an alignment aims at creating blocks. Together with factors, these relations can be represented by a graph $G = (V, E)$. The set V of nodes represents all the factors appearing in the sequences, and edges of E link factors that may be aligned. These graphs are called *factor graphs*. A *factor graph* is a complete graph where edges linking factors attending on the same sequence are removed. Indeed, a given factor f may align with any other identical

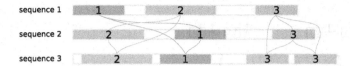

sequence 1

sequence 2

sequence 3

Fig. 1. This is a set of three sequences. Common subsequences of these sequences which are repeated are labeled. After the conversion, each sequence of the alignment is displayed as a list of factors. Here sequence $1 = [1,2,3]$, sequence $2 = [2,1,3]$ and sequence $3 = [2,1,3,3]$. Thin lines link factors that may be aligned together.

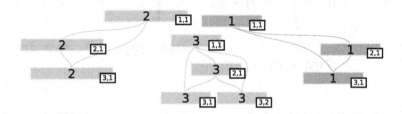

Fig. 2. The alignment seen in Fig. 1 displayed as a set of *factor graphs*

factor f' provided f' does not belong to the same sequence. In Fig. 1, thin links between factors illustrate the possible alignments between them.

From a graph point of view the sequential order "sequence by sequence" has no sense. The alignment problem is modeled as a set of *factor graphs*. So as to differentiate the different factors, a unique identifier is given to each of them. Each factor is assigned a triplet $[x, y, z]$ where x is an identifier for the pattern, y is the identifier of the sequence the factor is located on and z is the occurrence of this pattern on the given sequence. For instance, on Fig. 1, the bottom right factor of the third sequence is identified by the triplet $[3, 3, 2]$. Namely, it is the pattern "3", located on the sequence 3 and it occurs for the second time on this sequence, given the sequences are read from left to right.

Using that model the sequences are not ordered as it is the case when considering progressive alignment methods. Fig. 2 illustrates such graphs according to the representation of Fig. 1.

From each *factor graph* a subset of factors may be selected to create a block. If the block is composed of one factor per sequence, it is a complete block, but if one sequence is missing the block is said partial. Not all block constructions are possible since blocks crossing is not allowed and the selection of one block may prevent the construction of another one. For instance, from Fig. 1 one can observe that a block made of factors "1" may be created. Another block with factor "2" may also be created. However, both blocks cannot be present together in the alignment. These blocks are said *incompatible*. A group of blocks is said to be *compatible* if all couples of blocks are compatible.

Another relation between potential blocks can be observed in their neighborhood. Indeed, a strong relevance has to be accorded to potential blocks that are closed to one another and that do not cross. If two factors are neighbors

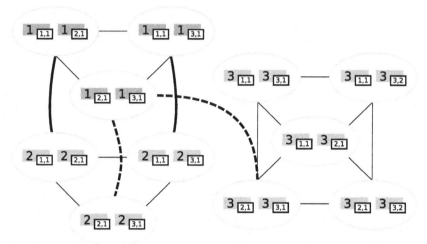

Fig. 3. The relation graph G_r based on the dual graph of G with additional edges corresponding to compatibility constraints, represented by thick plain edges, and friendly relations, represented by dashed links

in many sequences, there is a high probability for these factors to be part of a bigger factor with little differences. Such relations are taken into account within our approach and are called friendly relations. An example of friendly relation in the sample alignment can be observed sequences 2 and 3 between factors "1" and "2".

The *factor graphs* are intended to represent the search space of blocks according to the set of considered factors. However, they do not capture compatibility constraints and friendly relations between blocks. For that purpose, we first consider $G' = (V', E')$ dual graph of $G = (V, E)$, G being the graph composed of the entire set of *factor graphs*. The set V' of G' corresponds to the set E of G. Each couple of adjacent edges $e_1, e_2 \in E$ corresponds to one edge in E'. Moreover, in order to represent compatibility constraints and friendly relations two new kinds of edges have to be added to this graph: namely E_c for compatibility constraints and E_f for friendly relations. We call *relation graph* (Fig. 3) the graph $G_r = (V', E' \cup E_f \cup E_c)$.

Our approach makes use of both graphs. Ants move within the factor graphs, but their actions may also produce some effects in remote parts of the factor graphs, according to relation graph topology as explained in Section 3.

2.2   Ants for Multiple Sequence Alignment

Ant based approaches have shown their efficiency for single or multiple criteria optimization problems. The central methodology being known as Ant Colony Optimization [3]. Ants in these algorithms usually evolve in a discrete space modeled by a graph. This graph represents the search space of the considered problem. Ants collectively build and maintain a solution. Actually, this construction takes place thanks to pheromone trails laid down the graph. The search is led both by local information in the graph and by the global evaluation of the produced solutions.

The model issued in the previous section proposed a graph model that raises local conflicts and attractions that may exist in the neighborhood of the factors in the alignment. For this, the model may handle the local search needed by classical ACOs. However, providing a global evaluation function for this problem is unlikely. Indeed, defining a relevant evaluation function for MSA is in itself a problem since the evaluation should be aware of the evolutionary history of the underlying species, which is part of the MSA problem. Molecular biologists themselves do not all agree whether or not one given alignment is a good one. Popular evaluation functions like the classical *sum of pairs* [1] are still debated. As a consequence, instead of focusing on such a function, our approach concentrates on building structures in the factor graphs according to the relation graph. These structures correspond to compatible blocks. The building of blocks is made by ants which behavior is directly constrained by the pheromones they laid down in the graph and indirectly by the relation graph since this graph has a crucial impact on pheromone deposit location.

The global process can be further refined by taking into account the size of the selected factors, as well as the number of nucleotides or amino acids located between the factors of two neighbor blocks. These numbers are called *relative distances* between factors in the sequel. Thus, the factor graphs carry local information necessary to the ant system and acts like the environment for ants. Communication via the environment also known as stigmergic communication [4] takes place in that graph; pheromones trails are laid down on the edges according to ants move, but also according to the relation graph. A solution of the original problem is obtained by listing the set of compatible blocks that have been bring to the fore by ants and revealed by pheromones.

3 Algorithm

The proposed ant-based system does not evaluate the produced solutions, in this way it is not an ACO. However, the local search process remains widely inspired from ACOs. The general scheme of the behavior of the ant based system follows these rules:

- Ants perform walks into the factor graphs.
- During these walks each ant lay constant quantities of pheromones down on the edges they cross.
- This deposit entails a change in pheromone quantities of some remote edges according to the relation graph G_r as described in Section 2.
- Ants are attracted by the pheromone trails already laid down in the environment.

Finally a solution to the problem is a set of the most pheromone loaded edges of the *factor graph* that are free from conflicts. In the following section pheromone management is more formally detailled.

3.1   Pheromone Trails

Let τ_{ij} be the quantity of pheromone present on edge (i, j) of graph G which links nodes i and j. If $\tau_{ij}(t)$ is the quantity of pheromone present on the edge (i, j) at time t, then $\Delta\tau_{ij}$ is the quantity of pheromone to be added to the total quantity on the edge at the current step (time $t + 1$). So:

$$\tau_{ij}(t + 1) = (1 - \rho).\tau_{ij}(t) + \Delta\tau_{ij} \tag{1}$$

Note that the initial amount of pheromone in the graph is close to zero and that ρ represents the evaporation rate of the pheromones. Indeed, the modeling of the evaporation (like natural pheromones) is useful because it makes it possible to control the importance of the produced effect. In practice, the control of this evaporation makes it possible to limit the risks of premature convergence of the process.

The quantity of pheromone $\Delta\tau_{ij}$ added on the edge (i, j) is the sum of the pheromone deposited by all the ants crossing the edge (i, j) with the new step of time. The volume of pheromone deposited on each passage of an ant is a constant value Q. If m ants use the edge (i, j) during the current step, then:

$$\Delta\tau_{ij} = mQ \tag{2}$$

3.2   Constraints and Feedback Loops

Generally speaking, feedback loops rule self-organized systems. Positive feedback loops increase the system tendencies while negative feedback loops prevent the system from continually increasing or decreasing to critical limits. In this case, pheromone trails play the role of positive feedback loop attracting ants that will deposit pheromones on the same paths getting them more desirable. Friendly relationship found in the graph G_r may also play a positive feedback role. Indeed, more pheromones are laid down around friendly linked blocks. On the other side, conflicts between blocks act as negative feedback loops laying down 'negative' quantities of pheromone.

Let us consider an edge (i, j) that has conflict links with some other edges. During the current step, the c ants that cross edges in conflict with edge (i, j) define the amount of negative pheromones to assign to (i, j): $\Delta\tau_{ij}^{conflict} = cQ$.

Besides, an edge (i, j) with some friendly relations will be assigned positive pheromones according to the f ants that cross edges in friendly relation with (i, j) during the current step : $\Delta\tau_{ij}^{friendly} = fQ$.

Finally, the overall quantity of pheromone on one given edge (i, j) for the current step defined in (2) is modified as follow:

$$\Delta\tau_{ij} = \Delta\tau_{ij} + \Delta\tau_{i,j}^{friendly} - \Delta\tau_{i,j}^{conflict} \tag{3}$$

3.3   Transition Rule

When an ant is on vertex i, the choice of the next vertex to be visited must be carried out according to a defined probability rule.

According to the method classically proposed in ant algorithms, the choice of a next vertex to visit is influenced by 2 terms. The first is a local heuristic based on local information, namely the *relative distance* between the factors (d). The second term is representative of the stigmergic behavior of the system. It is the quantity of pheromone deposited (τ).

Remark 1. Interaction between positive and negative pheromones can lead on some edges to an overall negative value of pheromone. Thus, pheromones quantities need normalization before the random draw is made. Let max be an upper bound value computed from the largest quantity of pheromones on the neighborhood of the current vertex. The quantity of pheromone τ_{ij} between edge i and j is normalized as $max - \tau_{ij}$.

The function $N(i)$ returns the list of vertices adjacent to i (its neighbors). The next vertex will be chosen in this list. The probability for an ant being on vertex i, to go on j (j belonging to $N(i)$) is:

$$P(ij) = \frac{\left[\frac{1}{max - \tau_{ij}}^{\alpha} \cdot \frac{1}{d_{ij}}^{\beta} \right]}{\sum_{s \in N(i)} \left[\frac{1}{max - \tau_{is}}^{\alpha} \cdot \frac{1}{d_{is}}^{\beta} \right]} \tag{4}$$

The parameters α and β make it possible to balance the impact of pheromone trails relatively to the *relative distances*.

4   Analysis

First experiments on real sequences have been performed. Fig. 4 compares some results obtained by the proposed structural approach and the alignment provided by ClustalW. This sample shows that the visible aligned blocks can be regained in the results given by ClustalW. However, results are still to be evaluated. Indeed, it may happen that some blocks are found by our method while they are not detected by ClustalW. Anyway, we are aware of the necessity of additional analyses and discussions with biologists in order to validate this approach.

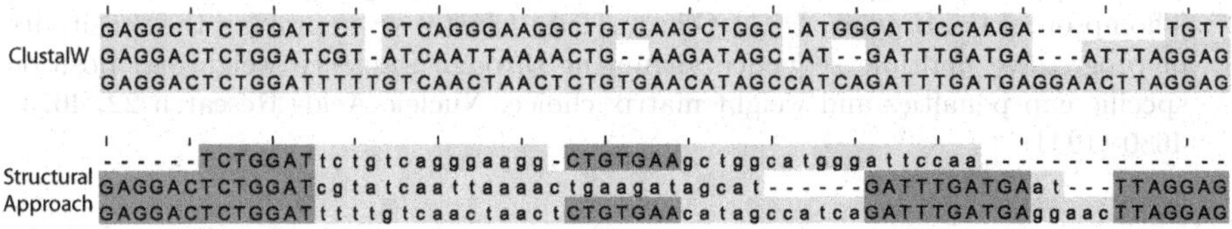

Fig. 4. Alignment of 3 sequences of TP53 regulated inhibitor of apoptosis 1 for Homo sapiens, Bos taurus and Mus musculus. Comparison of the alignment given by ClustalW and our structural approach. The red uppercase nucleotides on the structural approach are the blocks.

5   Conclusion

In this paper there was proposed a different approach, for the problem of multiple sequence alignment. The key idea was to consider a problem of building and maintaining a structure in a set of biological sequences instead of considering an optimization problem. Preliminary results show that the structures built by our ant-based algorithm can be informally compared, on a pattern basis, with the results given by ClustalW.

The outlook for the project is now to prove the efficiency and the relevance of the method, in particular, an important chunk of future work will concern the comparison of the differences between conserved regions provided by ClustalW and other well-known multiple sequence alignment methods and our approach. The second perspective focuses on the performance of the method. Indeed, the way blocks are built and intermediate results allow us to consider a kind of divide-and-conquer parallel version of this tool. Most recent results and advances will be made available on www.litislab.eu.

References

1. Altschul, S.F.: Gap costs for multiple sequence alignment. Journal of Theoretical Biology 138, 297–309 (1989)
2. Chen, Y., et al.: Partitioned optimization algorithms for multiple sequence alignment. In: Proceedings of the 20th International Conference on Advanced Information Networking and Applications (AINA 2006), vol. 2, pp. 618–622. IEEE Computer Society Press, Los Alamitos (2006)
3. Dorigo, M., Di Caro, G.: New Ideas in Optimization. In: Corne, D., Dorigo, M., Glover, F. (eds.) The Ant Colony Optimization Meta-Heuristic, pp. 11–32. McGraw-Hill, New York (1997)
4. Grassé, P.-P.: La reconstruction du nid et les coordinations inter-individuelles chez belicositermes natalensis et cubitermes s.p. la théorie de la stigmergie: Essai d'interprétation du comportement des termites constructeurs. Insectes sociaux 6, 41–80 (1959)
5. Kurtz, S., et al.: Versatile and open software for comparing large genomes. Genome Biology 5(2), 12 (2004)
6. Moss, J.D., Johnson, C.G.: An ant colony algorithm for multiple sequence alignment in bioinformatics. In: Pearson, D.W., Steele, N.C., Albrecht, R.F. (eds.) Artificial Neural Networks and Genetic Algorithms, pp. 182–186. Springer, Heidelberg (2003)
7. Thompson, J.D., Higgins, D.G., Gibson, T.J.: Clustal w: improving the sensitivity of progressive multiple sequence alignment through sequence weighting, position specific gap penalties and weight matrix choice. Nucleic Acids Research 22, 4673–4680 (1994)

An Algorithm for the Frequency Assignment Problem in the Case of DVB-T Allotments

D.A. Kateros[1], P.G. Georgallis[1], C.I. Katsigiannis[1],
G.N. Prezerakos[1,2], and I.S. Venieris[1]

[1] National Technical University of Athens,
School of Electrical and Computer Engineering,
Intelligent Communications and Broadband Networks Laboratory,
Heroon Polytechniou 9, 15773 Athens, Greece
{dkateros,chkatsig,prezerak}@telecom.ntua.gr,
pgeorgal@mail.ntua.gr, venieris@cs.ntua.gr
[2] Technological Education Institute of Pireaus,
Dpt. of Electronic Computing Systems,
Petrou Ralli & Thivon 250, 12244 Athens, Greece
prezerak@teipir.gr

Abstract. In this paper, we investigate the allocation of frequency channels to DVB-T allotments for frequency Bands IV/V (UHF). The problem is modeled with a constraint graph represented by an interference matrix. The aim is to optimize the spectrum usage, so that the maximum number of possible channels is assigned to each allotment area. The problem is a variation of the Frequency Assignment Problem (FAP). We have developed an algorithm that uses metaheuristic techniques, in order to obtain possible solutions. Three versions of the algorithm, a Tabu Search, a Simulated Annealing, and a Genetic Algorithm version, are evaluated. We present and discuss comparative results of these versions for generated networks, as well as for real allotment networks that are included in the GE06 digital frequency plan produced by ITU during the RRC-06 conference.

1 Introduction

The developments in standardization of digital radio broadcasting systems have introduced new possibilities regarding the methodology of planning networks of such systems. The introduction of single frequency networks (SFNs), allows the synchronization of the transmitters of a relatively small geographic area, so that they may operate at the same multiplex frequency channel without any destructive interference occurring at the receiver. Therefore, the planning of digital radio broadcasting networks need not to be conducted at transmitter level, but using the notion of an allotment area instead. In this paper, we investigate the allocation of frequency channels to DVB-T allotments for frequency bands IV/V (UHF) by means of an algorithm that employs metaheuristic techniques.

The paper is organized as follows. The frequency assignment problem in the case of DVB-T allotments is defined in the next section. In section 3 we present

I. Lirkov, S. Margenov, and J. Waśniewski (Eds.): LSSC 2007, LNCS 4818, pp. 561–568, 2008.
© Springer-Verlag Berlin Heidelberg 2008

the proposed algorithm and the three metaheuristic techniques, Tabu Search, Simulated Annealing, and the Genetic Algorithm in relation to their application on it. Comparative results of the three versions of the algorithm for generated networks, as well as a real DVB-T allotment network involving allotment areas of countries located in Southeastern Europe, are presented and discussed in section 4. Finally, in section 5 we conclude the paper.

2   The FAP for DVB-T Allotments

The allotment planning approach has been used during the past decade in order to establish T-DAB and DVB-T frequency plans. The planning procedure can be separated into two distinct stages; in the first one the compatibility analysis is conducted, while in the second the plan synthesis takes place, i.e. the allocation of frequency channels to the allotment areas. The compatibility analysis determines whether two allotment areas can share the same frequency channel without causing harmful interference to each other. In order to make this assessment, the knowledge of the acceptable interference levels on the borders of each allotment area and the interference potential of each allotment area must be known. The acceptable interference levels can be calculated based on the desired reception type (mobile, portable, indoor), the desired coverage location percentage and the modulation type of the wanted radio signal. On the other hand, the interference potential of each allotment area can be modeled with generic network structures which are defined in order to be used as representatives of real network implementations (reference networks) [1]. Technical criteria for the compatibility analysis can be found in [2].

Let $\mathbf{A}$ be the set of allotment areas in the planning area and $\mathbf{F}$ the set of available frequency channels, in our case UHF channels 21-69. $\mathbf{C}$ denotes the constraints matrix, a symmetrical $N \times N$ matrix, where N is the number of allotment areas. The elements $c_{i,j}$ of $\mathbf{C}$ are defined as follows:

$$c_{i,j} = \begin{cases} 1, \text{ if the same channel cannot be reused between allotments } i, j \in \mathbf{A} \\ 0, \text{ otherwise} \end{cases}$$

$\mathbf{C}$ models co-channel constraints. Adjacent channel constraints are not taken into account due to the low protection ratios [3] that apply. The objective is to optimize the spectrum usage while satisfying the constraints included in $\mathbf{C}$. Since the multiplicity $m(a)$, i.e. the number of channels required for allotment $a \in \mathbf{A}$ is not fixed or known, we try to assign as many channels as possible to each allotment area. The problem is often referred to as Maximum Service FAP or Max-FAP. To model the problem let $n(a)$ be the number of channels allocated to allotment $a \in \mathbf{A}$. Then we aim to maximize the total number of channels allocated to all allotment areas or equivalently the mean number of channels per allotment area ($MCpA$):

$$\max \sum_{a \in \mathbf{A}} n(a) \Longleftrightarrow \max \frac{1}{N} \sum_{a \in \mathbf{A}} n(a)$$

For a detailed overview and classification of FAP problems the reader can refer to [4].

3   The Proposed Algorithm

Due to the difficulty of FAP, heuristic approaches have been widely employed. Max-FAP, discussed here, exhibits the additional difficulty, that the multiplicity is not fixed, which leads to the vast increase of the solution space. We have developed an algorithm that utilizes known metaheuristic techniques, in order to obtain possible solutions. The objective of the algorithm is to maximize the $MCpA$, while disallowing frequency allocations that lead to unacceptable interference, as indicated by the constraints matrix $\mathbf{C}$. Additionally, the algorithm aims to balance the distribution of frequency channels between allotment areas, as it was observed that the maximization of $MCpA$ tends to produce solutions that assign very few frequency channels to some allotments. This issue was also observed in [9], where the authors set lower and upper bounds on the channel demands, in order to avoid strongly unbalanced solutions. We have implemented three versions of the algorithm, a Tabu Search, a Simulated Annealing, and a Genetic Algorithm version. The reader can find useful details for the implementation of methods that deal with Fixed Spectrum FAP (FS-FAP) and Minimum Span FAP (MS-FAP) in [11,8].

The proposed algorithm includes two phases. In the first, we employ the meta-heuristic technique in order to solve the subproblem of the allocation of exactly L channels in every allotment area (fixed and known $m(a)$). We increment L gradually and repeat this procedure, until the metaheuristic fails to provide an acceptable solution. The final value of L, for which an acceptable solution is obtained, is the lower bound for the number of frequency channels allocated to each allotment area. In the second phase, the algorithm attempts iteratively to allocate one additional channel to each allotment area using a full execution of the corresponding metaheuristic to generate a candidate additional frequency for each allotment area. In the end of every iteration we examine the violations of the constraints and remove the necessary frequencies in order to obtain an acceptable solution. In order to determine the allotment from which a channel will be removed for a given constraint violation, we take into account the following criteria:

1. If the two allotments have a different number of allocated channels, remove the channel from the one that has more channels. This ensures that in the final channel allocation L will be the minimum number of channels allocated to an allotment area.
2. If the two allotment areas have the same number of allocated channels, remove the channel from the one that participates in the highest number of constraint violations at the current stage.
3. If the two allotment areas have the same number of allocated channels and additionally participate in the same number of constraint violations at the current stage remove the channel added more recently.

4. If the previous criteria do not apply choose the allotment from which the channel will be removed randomly.

The algorithm terminates in case N or more channels need to be removed in order to obtain an acceptable solution for two consecutive iterations or when a user defined calculation time is reached.

3.1   Simulated Annealing Version

Simulated Annealing (SA) [10] is a stochastic computational technique for finding good approximations of a given function in a large search space. The method searches for the best solution in a neighborhood. This new solution is accepted if it is better than the current one, or with an acceptance probability. The acceptance probability is controlled by the difference between the current and the new value of the cost function and an additional parameter T, the temperature, which is gradually decreased with the number of executed iterations, based on an annealing schedule.

For the implementation of SA in the framework of our algorithm, we define the cost function as follows: $E = 2v + va$, where v is the number of violated constraints and va the number of allotments that participate in the violations. The annealing schedule is a simple geometric cooling scheme: $T_{k+1} = 0.9 \cdot T_k$ and the initial temperature is calculated so that the acceptance ratio exceeds 0.9. The transition to new frequency allocations is performed with a *Single Move* generator for the first phase of the algorithm, while a *Double Move* generator was used for the second phase. This leads to a more thorough search of the solution space for the first phase in order to obtain the best possible basis for the final solution. Lastly, the acceptance probability is calculated using the following formula:

$$h\left(\Delta E, T\right) = e^{\frac{-\Delta E}{T}}$$

The procedure terminates when the cost function is zeroed or when no transition to a new frequency allocation is accepted after ten attempts for a given temperature or, lastly, after a user defined minimum temperature (T_{min}) is reached.

3.2   Tabu Search Version

Tabu Search (TS) [6] is a local search method. The method attempts to explore the solution space in a sequence of moves. In order to try to avoid cycling and to escape from local optimal solutions, for each iteration, some moves are declared tabu and cannot be selected. Tabu moves are based on the long- and short-term history of moves. Additionally, aspiration criteria might make a move that is included in the tabu list possible, when it is deemed favorable.

For the implementation of TS in the framework of our algorithm, the cost function is defined as in the SA implementation. The transition to new frequency allocations is performed inside a *Restricted Random Neighborhood* of the current allocation. Specifically, for every allotment that is involved in a constraint violation we change the violating channel to a new one $\in \mathbf{F}$ randomly.

Transitions inside a *Full Neighborhood* or a *Random Neighborhood* were also examined, but produced inferior performance. We defined two Tabu criteria, one based on short- and one based on long-term memory. The short-term memory criterion involved moves that lead to the allocation of a channel to an allotment that had been also allocated to that particular allotment in the last R iterations, while the long-term memory criterion involves moves for allotments that have an increased percentage of accepted new allocations in relation with the total number of accepted new allocations for all allotments. The defined aspiration criteria involve the acceptance of a tabu transition if the resulting allocation leads to the minimum value of the cost function recorded.

The procedure terminates when the cost function is zeroed or when no improvement of the cost function has been observed for $a = 150$ iterations or the total number of iterations reaches $b = 500$.

3.3 Genetic Algortihm

Genetic Algorithms (GA) [7] are inspired by the natural process of reproduction and mimic natural evolution and selection. An initial population of n possible solutions (chromosomes) is created. Each chromosome is evaluated against a fitness function that represents the quality of the corresponding solution. There is a small probability P_m that a chromosome may suffer mutation in order to introduce random alteration of genes. Pairs of chromosomes are then selected for reproduction based on the relative value of their fitness function in the current population. These pairs produce offspring based on a crossover probability P_c. This procedure takes place until the new population contains n chromosomes, which may consist of offspring and original chromosomes that have survived and may or may not have suffered mutation. New generations are produced until a satisfactory solution is found.

For the implementation of GA in the framework of our algorithm, the fitness function is based on a fitness remapping [5] of the cost function used in TS and SA implementations. Namely, the fitness function for chromosome i is derived from the equation:

$$ff_i = J_i - (\bar{J} - 2 \cdot \sigma)$$

where $J_i = -E_i$, $\bar{J}$ and σ are respectively the arithmetic mean and the standard deviation of J_i. If ff_i evaluates to < 0 then it is set to 0. We have chosen a P_c starting from a value of 0.85 and gradually decreasing to 0.55, in order to allow the fitter chromosomes to survive to the new generations as the algorithm proceeds. On a similar manner, we have chosen P_m starting from a value of 0.001 and gradually increasing to 0.02, in order to allow mutations of the chromosomes of the latter generations that increase the variety within the generations and can therefore lead to better solutions. We have used *Uniform Crossover* and the mutation takes place by replacing a random gene of the chromosome with an allowable frequency. The size of the population is 100 chromosomes.

The GA procedure terminates if the cost function for a chromosome is evaluated to 0 or if no improvement has been observed for $a = 150$ generations or the number of generations has reached $b = 300$.

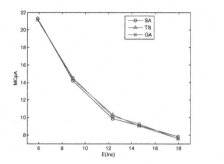 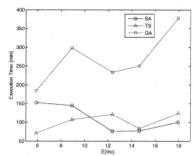

Fig. 1. $MCpA$ and Execution Time (minutes) versus corresponding mean number of incompatibilities. $N = 50$ and $Var(Inc) \cong 29$.

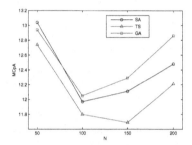 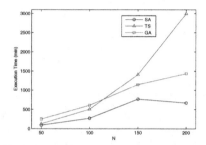

Fig. 2. $MCpA$ and Execution Time (minutes) versus corresponding size of benchmark constraint matrices. $E(Inc) \cong 9$ and $Var(Inc) \cong 16$.

4   Results and Discussion

In order to analyze the performance of the three versions of the proposed algorithm, we have generated benchmark constraints matrices using as input the mean and the standard deviation of the number of incompatibilities (Inc) per allotment area. The benchmark problems include sets of 50, 100, 150, and 200 allotment areas. For every instance we performed three executions of each version of the algorithm and present the average values of the results obtained.

In Fig. 1 we observe that all versions of the algorithm exhibit similar performance in terms of $MCpA$ values, which can be attributed to the small size of N. $MCpA$ decreases as $E(Inc)$ increases, which is normal, as the problem becomes harder. What is more interesting is that the calculation times show that the GA version is slower and additionally becomes generally slower as $E(Inc)$ increases, while the SA and TS versions are faster and do not seem to be effected much by the difficulty of the problem in terms of execution time. In Fig. 2 we notice that the largest values of $MCpA$ are obtained using the GA version, which becomes more evident as N increases. This leads to the conclusion that the GA version examines the solution space more thoroughly, which explains also the

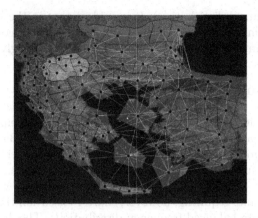

Fig. 3. Constraint graph of the allotment network used for the evaluation of the algorithm

Table 1. Results of frequency allocation for the allotment network illustrated in Fig. 3

	SA_L	TS_L	GA_L	SA	TS	GA
MCpA	11.1	11.2	11.5	12.2	12.1	12.4
Execution Time (min)	151	317	434	183	444	562
Minimum Number of Channels	6	6	5	2	1	1

fact that it demands greater calculation times for small values of N. We also note that the execution time required by the TS version increases exponentially as N increases, which can be attributed to the fact that higher values of N (and therefore higher numbers of allotments violating constraints) increase the *Restricted Random Neighborhood* of the current allocation, while the $MCpA$ values produced are the smallest. The SA version produces inferior results to the GA version, but requires the least amount of time, which moreover exhibits a smooth increase with higher values of N.

Lastly, we would like to make some observations concerning the fairness of the frequency channel allocations produced, which can be quantified using the minimum number of channels L that has been allocated to an allotment area. In nearly all cases, we observed that the results produced by the SA version are the most fair, while the results produced by the TS version were the same or closely inferior. The GA version produced the most unbalanced results, which explains the superior performance in terms of $MCpA$ values, as there is an obvious tradeoff between $MCpA$ and L.

Fig. 3 shows the constraint graph of the allotment network used. It consists of 95 allotments included in the GE-06 plan produced by the Regional Radiocommunication Conference 2006 (RRC-06). Table 1 displays the obtained results, both when the algorithm operates exactly as described in section 3 (SA_L,TS_L,GA_L columns), as well as when criterion 1 is not taken into account during the removal of constraint violations (i.e. no provision is taken for the fairness of the channel distribution). It is evident that in the latter case the algorithm leads to more optimal solutions as far as MCpA is concerned, however the results have limited application value, as they are strongly unbalanced.

5   Conclusion

In this paper we described an algorithm to solve the problem of frequency assignment in the case of DVB-T allotments for frequency Bands IV/V. Three versions of the algorithm, each utilizing a different metaheuristic technique, are evaluated. The results show that the optimal results in terms of the maximization of MCpA are obtained by the GA version, while the SA version produces closely inferior results requiring, however, shorter execution times and additionally leading to more balanced frequency distributions. The TS version exhibits the worst performance in terms of both result quality and time efficiency.

References

1. ERO, ECC REPORT 49: Technical Criteria of DVB-T and T-DAB Allotment Planning (2004)
2. ITU: Report of the Regional Radiocommunication Conference 2004 (2004)
3. ITU-R: BT.1368-6, Planning criteria for digital terrestrial television services in the VHF/UHF bands (2006)
4. Aardal, K.I., et al.: Models and Solution Techniques for Frequency Assignment Problems, ZIB Report (2001)
5. Beasley, D., Bull, D.R., Martin, R.R.: An overview of genetic algorithms: Part I, fundamentals. University Computing 15(2), 58–69 (1993)
6. Glover, F.: Tabu Search — Part I. ORSA Journal on Computing 1(3), 190–206 (1989)
7. Holland, J.H.: Adaptation in Natural and Artificial Systems. University of Michigan Press (1975)
8. Hurley, S., Smith, D.H., Thiel, S.U.: FASoft: A System for Discrete Channel Frequency Assignment. Radio Science 32(5), 1921–1939 (1997)
9. Jaumard, B., et al.: Comparison of column generation models for channel assignment in cellular networks. Discrete Applied Mathematics 112, 217–240 (2001)
10. Kirkpatrick, S., Gelatt, C.D., Vecchi, M.P.: Optimization by Simulated Annealing. Science 220(4598), 671–680 (1983)
11. Thiel, S.U., Hurley, S., Smith, D.H.: Frequency Assignment Algorithms, project report (1997), http://data.research.glam.ac.uk/media/files/documents/2006-10-24/FASoft.pdf

Optimizing the Broadcast in MANETs Using a Team of Evolutionary Algorithms*

Coromoto León, Gara Miranda, and Carlos Segura

Dpto. Estadística, I.O. y Computación
Universidad de La Laguna
38271 La Laguna, Tenerife, Spain

Abstract. This work presents a new approach to optimize the broadcast operation in MANETs based on a team of evolutionary algorithms. A library of parallel algorithmic skeleton for the resolution of multi-objective optimization problems has been applied. This tool provides a C++ implementation of a selection of the literature best-known evolutionary multi-objective algorithms and introduces the novelty of the algorithms cooperation for the resolution of a given problem. The algorithms used in the implementation are: SPEA, SPEA2, and NSGA2. The computational results obtained on a cluster of PCs are presented.

1 Introduction

A *Mobile Ad-Hoc Network* (MANET) [9] is a set of autonomous mobile nodes, connected by wireless links. This kind of networks has numerous applications because of its capacity of auto-configuration and its possibilities of working autonomously or connected to a larger network. Broadcasting is a common operation at the application level. Hence, having a well-tuned broadcasting strategy will result in a major impact in network performance. This work focuses on the study of a particular kind of MANETs called, *metropolitan* MANETs. These MANETs have some specific features that make difficult the testing in real environments: the network density is heterogeneous and it is continuously changing because devices in a metropolitan area move and/or appear/disappear from the environment. Therefore we need to have a simulation tool to easily test the quality of broadcasting strategies. Fixed the DFCN protocol and the mall scenario in the simulator, the optimization implies satisfying different objectives simultaneously: the number of devices to reach (*coverage*) must be maximized, a minimum usage of the network (*bandwidth*) is desirable and the process must take a time as short as possible (*duration*).

The improvement of the broadcast in MANETs is based on the optimization of more than one objective function. The multiple objectives are conflicting and

* This work has been supported by the EC (FEDER) and by the Spanish Ministry of Education inside the 'Plan Nacional de I+D+i' with contract number TIN2005-08818-C04-04. The work of G. Miranda has been developed under the grant FPU-AP2004-2290.

I. Lirkov, S. Margenov, and J. Waśniewski (Eds.): LSSC 2007, LNCS 4818, pp. 569–576, 2008.

must be simultaneously satisfied. In this kind of *multi-objective optimization problems*, a solution optimizing every objective might not exist. Since exact approaches are practically unaffordable, a wide variety of evolutionary algorithms has been designed. The goal of such algorithms is to obtain an approximated *Pareto front* as closer as possible to the optimal one.

The first issue when using an evolutionary technique in the resolution of a problem is to select the particular algorithm to apply. There are many of them available in the literature but it is important to notice that every alternative has some particular properties that make it appropriated or not for a certain kind of problems. So, if the chosen algorithm is not suitable for the problem to solve, poor quality solutions will be found. Generally, this forces the user to test several algorithms before making a final decision. Usually the users do not have a prior knowledge about the behaviour of the algorithm applied to a particular problem, so if they have to try many alternatives, the process could take too much user and computational effort.

One possibility to decrease the time invested in such task consists in the application of parallel schemes. An example of these strategies is the island model [7]. In particular, the use of heterogeneous island-based parallel techniques can avoid the inconvenient of selecting an algorithm. In this scheme, on each island (or processor) a different algorithm is executed for the resolution of the same problem. Taking into account that certain algorithms will obtain quite better results than others, a promising approach is found on the technique called *team algorithms* [3]. This approximation makes possible to weigh up the algorithms depending on their obtained results and also allows to assign more computational resources to the algorithms with better expectations.

This work presents a team of evolutionary algorithms which work parallelly and cooperatively in the resolution of the broadcast operation in a fixed scenario of MANET. The implementation uses a library of parallel algorithmic skeleton for the resolution of multi-objective optimization problems [8]. This tool provides a C++ implementation of a selection of the literature best-known evolutionary multi-objective algorithms and introduces the novelty of the algorithms cooperation for the resolution of a given problem. For MANETs, the particular properties of the problem have been programmed. Also, the configuration of the evolutionary algorithms that will participate in the problem resolution has been customized.

The remaining content of the article is structured in the following way: Section 2 presents the broadcast optimization problem in MANETs. The implementation using the team algorithm skeleton is described in section 3. The computational study is presented in section 4. Finally, the conclusions and some lines of future work are given in section 5.

2   Broadcast Operation in MANETS

As stated in previous section, the optimization of the broadcast operation in MANETs involves a multi-objective optimization problem. The objectives are: minimize the broadcast duration, maximize the percentage of reached devices and minimize the bandwidth used in the operation. MANETs have several features that hinders the testing in real environments. There are different simulation tools [5] and the choice for this work was *Madhoc* [4]. This tool provides a simulation environment for several levels of services based on different types of MANETs technologies and for a wide range of MANET real environments. It also provides implementations of several broadcasting algorithms. For the simulation of a broadcast operation, the user must specify the characteristics of the environment to simulate, a mall in this case, and the type of broadcasting algorithm to be used.

The chosen broadcast protocol was DFCN (*Delayed Flooding with Cumulative Neighbourhood*) [6]: a deterministic and totally localized algorithm. It uses heuristics based on the information from one hop. Thus, it achieves to get a high scalability. The DFCN behaviour is determined by a set of configuration parameters that must be appropriately tuned to obtain an optimum behaviour:

- **minGain** is the minimum gain for forwarding a message. It ranges from 0.0 to 1.0.
- [**lowerBoundRAD**, **upperBoundRAD**] define the *Random Delay for rebroadcasting* values in milliseconds. Values must be in the range [0.0, 10.0].
- **proD** is the maximum density for which it is still needed using proactive behaviour for complementing the reactive behaviour. It ranges in [0, 100].
- **safeDensity** is the maximum density below which DFCN always rebroadcasts. It ranges from 0 up to 100 devices.

Given the values for the five DFCN configuration parameters, the *Madhoc* tool will do the corresponding simulation and it will provide the average values obtained during the simulation for each of the three objectives: the broadcast duration, the percentage of reached devices and the bandwidth. The goal is to find the five parameter configuration that optimizes the DFCN behaviour regarding to duration, coverage and bandwidth features.

One possibility of optimizing the strategy behaviour is to systematically vary each of the five DFCN parameters. The drawback is that the number of possible parameter combinations is too large. Moreover, evaluations in the simulator need certain time with the used processors, so an enumerative solution is unaffordable. Furthermore, since the broadcast algorithm presents an important complexity and the simulator introduces many random factors, doing a deep analysis of the problem to extract information in order to define an heuristic strategy is very difficult. For these reasons, one usual way of affording this problem is through evolutionary techniques [1]. Here the proposed strategy is based on the merging of some of this techniques to build a team of algorithms.

3   Team Algorithms

A team algorithm scheme is based on the cooperation of different algorithms to solve a problem. In the used scheme [8] the team consists of evolutionary algorithms.Specifically, the algorithms available are: SPEA (*Strength Pareto Evolutionary Algorithm*) [11], SPEA2 (*Strength Pareto Evolutionary Algorithm 2*)

1 *initConfiguration (fichConfig);*	1 *recvConfigMigration ();*
2 *initAllIsland ();*	2 *while (1) {*
3 *while (1) {*	3 *data = recvDataConfig ();*
4 *islandIdle = recvParetoFront();*	4 *if (data is messageFinalize ())*
5 *if (stop()) {*	5 *break;*
6 *for (i = 0; i < numIsland-1; i++)*	6 *initParamAlgorithm (data);*
7 *recvParetoFront();*	7 *recvInitialPopulation ();*
8 *sendMessageFinalize();*	8 *while (!finishAlgorihtm ()) {*
9 *printOptimalParetoFront();*	9 *runGeneration ();*
10 *}*	10 *sendMigration ();*
11 *config = selectionConfigAlgorithm();*	11 *recvMigration ();*
12 *initIsland(config);*	12 *}*
13 *}*	13 *sendParetoFront ();*
	14 *}*
(a) Coordinator	(b) Island

Fig. 1. Team Algorithm Pseudocode

[10] and NSGA2 (*Nondominated Sorting Genetic Algorithm 2*) [2]. The team algorithm model consists of a *coordinator* process and as many *islands* or *slaves* processes (parallel and asynchronous) as specified by the user.

The aim of the team coordinator is the initialization of an algorithm configuration on each of the slave processes and the management of the global solution. Figure 1(a) shows the pseudocode for the coordinator process. Before initiating the slave processes, the coordinator reads and stores the tool setup which has been specified by the user through a configuration file (line 1). Then, the coordinator initiates all the slave processes and assigns to each of them an algorithm instance (line 2). When all the configurations have been executed at least once, the coordinator begins to apply the selection criterion given in the tool customization for deciding, every time a slave gets idle, which is the next configuration to be executed. However, when beginning the execution of a new evolutionary algorithm it is not worth enough to restart the search without considering the solutions obtained until that moment. Every time an algorithm execution is going to begin, the corresponding slave will take some of the initial individuals from the current global solution. After each algorithm completion, the slaves send the set of obtained solutions to the coordinator (line 4), so the coordinator is able to update its Pareto front. The size of this global Pareto front is fixed by the user. This solution size is maintained by applying the NSGA2 crowding operator among slave completions. Until the global stop condition is not reached (line 5), the coordinator selects the next algorithm configuration to execute (line 11) and allocates it to the just finalized slave (line 12). If the stop condition is verified, the coordinator waits until all the slaves finish and collects all the local solutions in order to build the final global Pareto front (lines 6-7). To conclude, the coordinator process sends a finalization message to the slaves (line 8) giving as result the global Pareto front and the execution statistics for each of the algorithm configurations.

Each slave process represents an execution island. The aim of an execution island is to search problem solutions from an initial set of individuals applying an evolutionary algorithm configuration. Figure 1(b) shows the pseudocode for the slave processes. At the beginning, all the slaves receive the migration parameters to be applied: migration probability and number of individuals to migrate (line 1). Next, each slave receives a message with the configuration and the algorithm to execute (line 3). The slave initiates the algorithm instance (line 6) and afterwards receives from the coordinator process some individuals randomly selected from the global Pareto front. This set of individuals is used to fill a part of the algorithm initial population (line 7). In the case where the island must continue executing the same algorithm configuration, the population of the last execution generation is taken as the new initial population. Once the algorithm instance is initialized in the island process and the local front is cleared, the instance execution is able to be started (lines 8-12). Every time the algorithm finishes a generation (line 9) the slave determines, considering the established migration probability, if it has to do a migration or not (line 10). In affirmative case, a certain number of randomly selected individuals from the local Pareto front is sent to a randomly chosen slave. Also it is necessary to check for received individuals migrated from others islands (line 11). When the algorithm execution finishes, the slave process sends its local Pareto front to the coordinator (line 13). This event indicates to the coordinator that now the slave is idle and available for doing more work. If the final stop condition has not been reached yet, the coordinator will send to the slave the new algorithm configuration to execute. These steps are successively repeated until the coordinator indicates the end of the program (line 4).

The implementation of the MANETs problem consists in defining a C++ class representing an individual (Figure 2) and define a configuration file with the skeleton operation characteristics: algorithms and configurations to use, the algorithm selection method and the execution stop condition.

4   Computational Results

The executions have been done over a Debian GNU/Linux cluster of 8 bi-processor nodes. Each processor is an Intel Xeon 2.66 GHz with 512 MB RAM and a Gigabit Ethernet interconnection network. The MPI implementation used is MPICH version 1.2.6 and the C++ compiler version is 3.3.5. Due to the random component inherent to the evolutionary algorithms, all the executions were repeated 30 times and average values have been considered.

In a first experiment, the broadcast problem was solved with each one of the single evolutionary algorithms, fixing the number of evaluations to 10,000. The approximated Pareto fronts obtained with each of the single algorithms were merged obtaining a new solution called MERGE. Afterwards, each test problem was solved using a team algorithm constituted by three islands fixing the maximum Pareto front size to 100 elements. The number of evaluations with this method was fixed to 30,000, that is, the sum of the evaluations executed by the

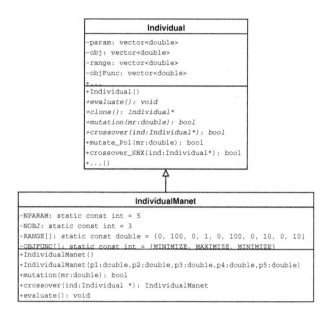

Fig. 2. Skeleton classes for the broadcasting problem in MANETs

individual algorithms. The overhead of the team algorithm for this problem is negligible due to the high computational effort required by each single individual evaluation. In average, the team algorithm execution time is similar to the single execution time. The solutions obtained by this method were named TEAM. With these configurations, the execution time of the parallel algorithm varies between the lowest and highest sequential times. The mutation and crossover rates used in every execution were fixed to 0.01 and 0.8 respectively. The population and archive size used in the individual executions of the evolutionary algorithms were fixed to 100 and 60, respectively. In the team algorithm executions both values were fixed to 60. A smaller population was used in the team algorithm because there is a population of that size on each one of the execution islands. The migration rate was fixed to 0.01, using 4 individuals in each migration. The 100% of the elements of the initial population of each execution in the islands were filled from the current global Pareto front.

Left side of Table 1 presents the average, maximum and minimum generational distances using as reference front, for each of the 30 executions, the join of the solutions given by all the methods. The ratio of contribution of each one of the methods is also presented. The average generational distance obtained with the team algorithm execution is the lowest one. Results for the team algorithm are clearly better than the results obtained with the individual executions, and it is also better than the merge of the individual executions. The contribution of points to the join of the Pareto fronts is less than 50%, but it is important to note that the team algorithm Pareto front is limited to 100 points, that is, in the join 400 points are considered and only 100 points come from the team algorithm, so a value of 44% represents a

Table 1. Single Algorithms and Team Algorithm Results

	GD_{avg}	GD_{max}	GD_{min}	Ratio	GD_{avg}	GD_{max}	GD_{min}	Ratio
NSGA2	0.38	1.17	0.19	13%	0.46	0.73	0.25	11%
SPEA	0.3	0.55	0.04	21%	0.43	0.71	0.28	17%
SPEA2	0.29	0.87	0.17	22%	0.40	0.84	0.27	21%
MERGED	0.16	0.29	0.01	56%	0.36	0.49	0.25	49%
TEAM	0.15	0.33	0.05	44%	0.34	0.49	0.22	51%

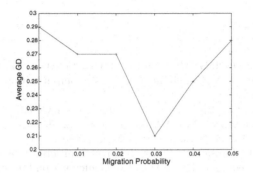

Fig. 3. Effects of Migration Probability

high ratio. Right side of Table 1 presents a similar comparative but using as reference front the join of all the 30 executions. Similar results are obtained in this case. It is important to note that the contribution of points to the final Pareto front given by the team algorithm method is greater than 50%.

A second experiment checks the effect of the migration scheme used in the team algorithm. The parameters that can be fixed are: the migration probability and the number of individuals to send when a migration is done. The team algorithm was executed using the same configuration as in the first experiment, but variating the value of the migration probability between 0.0 and 0.05 with increases of 0.01 among executions. Figure 3 represents the average generational distance reached for each migration rate using as reference front the join of all the executions. The error reached with low migration probability is lower than the error reached without migration, proving the positive effect of the migration scheme. However, high values of migration probability are not positive because it can produce that several slaves find the same solutions or very similar solutions, causing an increase in the average generational distance achieved.

5   Conclusions and Future Work

In this work an optimization of the broadcast operation in MANETs based on a team of evolutionary algorithms has been presented. The implementation uses a library of parallel algorithmic skeleton which implements a set of the best-known multi-objective evolutionary algorithms and proposes a parallel interaction and cooperation between them for the resolution of the problem. The results presented here prove that an improvement of the results quality can be obtained using this approach.

The main line of future work is to extend the set of available multi-objective evolutionary algorithms. Experiments have been developed over a single metropolitan scenario: a mall. It would be worth to verify that results can be generalized to other type of metropolitan MANETs and even to other kind of multi-objective real-world optimization problems.

References

1. Alba, E., et al.: A Cellular Multi-Objective Genetic Algorithm for Optimal Broadcasting Strategy in Metropolitan MANETs. Computer Communications 30(4), 685–697 (2007)
2. Deb, K., et al.: A Fast Elitist Non-Dominated Sorting Genetic Algorithm for Multi-Objective Optimization: NSGA-II. In: Deb, K., et al. (eds.) PPSN 2000. LNCS, vol. 1917, pp. 849–858. Springer, Heidelberg (2000)
3. Fernández, J.M., Barán, B.: Elitist Team of Multiobjective Evolutionary Algorithms. In: Proceedings of Latin-American Conference on Informatics, Cali, Colombia (2005)
4. Hogie, L.: Mobile Ad Hoc networks: modelling, simulation and broadcast-based applications. PhD thesis, Le Havre University and Luxembourg University (2007)
5. Hogie, L., Bouvry, P., Guinand, F.: An Overview of MANETs Simulation. Electronics Notes in Theorical Computer Science 150(1), 81–101 (2006)
6. Hogie, L., et al.: A Bandwidth-Efficient Broadcasting Protocol for Mobile Multi-hop Ad hoc Networks. IEEE, Los Alamitos (2006)
7. Horii, H., et al.: Asynchronous migration of island parallel ga for multi-objective optimization problem. In: Asia-Pacific Conference on Simulated Evolution and Learning, Singapore, pp. 86–90 (2002)
8. León, C., Miranda, G., Segura, C.: Parallel Skeleton for Multi-Objective Optimization. In: Genetic and Evolutionary Computation Conference, London, England (to appear, 2007)
9. Macker, J., Corson, M.: Mobile Ad Hoc Networking and the IETF. ACM Mobile Computing and Communications Review 2(1) (1998)
10. Zitzler, E., Laumanns, M., Thiele, L.: SPEA2: Improving the Strength Pareto Evolutionary Algorithm for MultiObjective Optimization. In: Evolutionary Methods for Design, Optimization and Control (2002)
11. Zitzler, E., Thiele, L.: An Evolutionary Algorithm for Multiobjective Optimization: The Strength Pareto Approach. Technical Report 43, Gloriastrasse 35, CH-8092 Zurich, Switzerland (1998)

Ant Colony Models for a Virtual Educational Environment Based on a Multi-Agent System

Ioana Moisil[1], Iulian Pah[2], Dana Simian[1], and Corina Simian[2]

[1] University "Lucian Blaga" of Sibiu, Romania
[2] University Babeş Bolyai of Cluj-Napoca, Romania

Abstract. We have designed a virtual learning environment where students interact through their computers and with the software agents in order to achieve a common educational goal. The Multi-Agent System (MAS) consisting of autonomous, cognitive and social agents communicating by messages is used to provide a group decision support system for the learning environment. Learning objects are distributed in a network and have different weights in function of their relevance to a specific educational goal. The relevance of a learning object can change in time; it is affected by students', agents' and teachers' evaluation. We have used an ant colony behavior model for the agents that play the role of a tutor and organizing the group-work activities for the students.

Keywords: learning objects, stigmergic collaboration, multi-agent system, e-learning, ant colony optimization.

1   Introduction

In one of our recent papers [13] we had presented a socio-cultural model of the student as the main actor of a virtual learning environment, as part of a larger project — DANTE — Socio-Cultural Models implemented through multi-agent architecture for e-learning. DANTE has as main objective the development of a global model for the virtual education system, student centered, that facilitates the learning through collaboration as a form of social interaction. In our vision, the global model requires its own universe in which the human agents interact with software agents. The global model is considered the core of an e-learning system. The proposed e-Learning system has a general architecture with three levels: user, intermediary, supplier educational space, on each level heterogeneous families of human and software agents are interacting. The main human actors are: the student, the teacher, and the tutor. In the virtual learning environment we have the corresponding agents. The human actors are interacting with the e-learning system via several agentified environments. The teacher (human agent) is assisted by two types of software agents: personal assistant (classic interface agent) and didactic assistant. The SOCIAL agentified environment has social agents and a database with group models (profiles of social behavior). The agentified DIDACTIC environment assists the cognitive activities of the student and/or of the teachers. The student (human agent) evolves in an

I. Lirkov, S. Margenov, and J. Waśniewski (Eds.): LSSC 2007, LNCS 4818, pp. 577–584, 2008.

agentified environment with three types of agents. He/she has a personal assistant (software interface agent) who monitors all the student's actions and communicates (interacts) with all the other agents, with the agentified environments of other students and with the teacher's agentified environment. The student has at his/her disposal two more agents: the TUTOR and the mediating agent. The TUTOR assistant evaluates the educational objectives of the student and recommends her/him some kind of activities. The decisions are based on the knowledge of the students' cognitive profile (which takes into account the social component). The TUTOR agent interacts with the personal assistant of the student, with the mediating agent and with the social agentified environment. As the system is conceived, the accent is put on collaboration activities between students, which consist in knowledge exchange, realization of common projects, tasks' negotiation, sharing resources, common effort for the understanding of a subject, problem-solving in-group. The TUTOR is mainly evolving in a network populated with learning objects. A learning object is a resource with three components: a learning objective, a learning activity, and a learning assessment. From a technical point of view, a learning object is, according to IEEE- Learning Technology Standards Committee (LTSC), "any entity, digital, or non-digital, which can be used, reused, or referenced during technology supported learning".

In this paper we are presenting a possible model for one of the components of the TUTOR sub-system, i.e. the one responsible for organizing the group-work activities for the students. A group-work activity can be, for example, a project that will be developed by a team of students. In a team, students are organized according to their skills and preferences. For example, if the work assigned to a team is from the computer science class and the goal is to develop an online art exhibition, a member of the team can be "specialized" in web design, another in PHP programming or MySQL, another in .NET, etc. As someone has to know what an art exhibition is, this task will be assigned to a student from the art class, a student that can be in another team. At one moment, a student can be asked by different teams to perform a certain task.

An activity consists of several tasks. A member of a team will perform in a time-interval the task for which she has (or intent to obtain) the required knowledge and skills. The goal is to minimize the maximum of the completion times for all tasks with the constraint that a student cannot participate in two group-work activities at the same time.

So, in our learning environment we have a set of students' teams and a set of group work activities consisting of several tasks. There are task-classes and at this stage of the modelling process the potential assignment of one student to a specific task is based on the concept of response threshold combined with a function of the background knowledge of the student. As we can see, this problem is scheduling one, belonging to the NP-class of problems. Works developed in the last ten years have shown that social insects provide us with a powerful metaphor to create decentralized systems of interacting agents. Swarm intelligence — the emergent collective intelligence of social insects — in networks of interactions among individuals and between individuals and their environment. Several models of division of labor in colonies of insect have been used to solve task allocation problems. In our paper we are using for the TUTOR agent an Ant System [5] combined with a simple reinforcement of response thresholds.

2 Ant Social Behavior

Insects with social behavior, and among them several species of ant colonies, have been studied for more than hundred years. Their organizing capacity, the way they communicate and interact and their performances have become models especially for highly distributed artificial multi-agent system.

Fifty years ago, the French entomologist Pierre-Paul Grass [1] observed that some species of termites react to what he called "significant stimuli". He observed something very interesting: that the effects of these reactions can act as new significant stimuli for both the insect that produced them and for the other insects in the colony. Grass described this particular type of communication, in which the "workers are stimulated by the performance they have achieved", by the term stigmergy.

Almost thirty years ago, in Douglas R. Hofstadter's fantastic book Godel, Escher, Bach: an Eternal Golden Braid (1979), one of the characters, Anteater, is telling Achilles about Aunt Hillary, his friend, who is in fact an ant colony, viewed as a single intelligent entity. The dialog is evolving around the question if there is the possibility for collective intelligence to emerge from the interaction of hundreds of simple not-so-intelligent agents and if we can learn from studying these natural multi-agents at work. Hofstadter was asking the question: "Could there be an Artificial Ant Colony?", and the answer, today, is Yes!

Today we have many methods and techniques "ants inspired", one of the most successful being the general purpose optimization technique known as ant colony optimization — ACO. Ants, in their quest for food (foraging behavior) lay pheromone on the ground in order to mark some favorable path that should be followed by other members of the colony. Ant colony optimization exploits a similar mechanism for solving optimization problems [1,2]. ACO can be applied to discrete optimization problems with a finite set of interconnected components and an associated cost function (objective function). There are constrains on what components and connection can be part of a feasible solution. Each feasible solution has a global "quality" determined from a function of all costs components.

ACO algorithms use a population of agents-artificial ants in order to build a solution to a discrete optimization problem. Solutions' information (individual connections of the problem) are kept in a global memory — the pheromone mapping. Specific heuristics can be considered on a-priori information.

In an ACO there are basically three processes: ants' generation and activity, pheromone trail evaporation and daemon actions. In the problem graph, an artificial ant searches for a minimum or maximum cost solution. Each ant has a memory, can be assigned a starting position and can move to any feasible vertex. An ant decides to move by a probabilistic manner, taking into account a pheromone value and a heuristic value. When an ant is moving from one location to another pheromone values are altered (step-by-step pheromone update). The set of allowed nodes (vertexes) is kept in a tabu-list. After completing the solution, the artificial ant dies, freeing allocated resources. The amount of pheromone on all connections is regulated through pheromone evaporation and by daemon actions that are centralized processes that cannot be performed to an individual ant solution (frequently this are local optimization procedures).

In order to apply ACO to a given combinatorial optimization, Dorigo et al. [4,7,8,5] have formalized Ant Colony Optimization (ACO) into a metaheuristic. The formalization starts with a model of the combinatorial problem.
A model

$$P = (S, \Omega, f)$$

of a combinatorial optimization problem consists of:

- a search space S defined over a finite set of discrete decision variables: X_i, $i = 1, \ldots, n$
- a set Ω of constraints among the variables
- an objective function (or cost function)

$$f : S \to R_0^+$$

to be minimized.

X_i is a generic variable that takes values in $D_i = \{v_i^1, \ldots, v_i^{|D_i|}\}$.
A feasible solution $s \in S$ is a complete assignment of values to variables that satisfies all constraints in Ω.
A solution $s^* \in S$ is called a global optimum if and only if:

$$f(s^*) \le f(s), \ \forall \ s \in S.$$

The model of a combinatorial optimization problem is used to define the pheromone model of ACO. To each possible solution component we associate a pheromone value. That means that the pheromone value τ_{ij} is associated with the solution component c_{ij} , by the assignment $X_i = v_i^j$. The set of all possible solution components is denoted by C. In ACO, an artificial ant builds a solution by traversing the fully connected construction graph $G_C(V, E)$, where V is a set of vertices and E is a set of edges. This graph can be obtained from the set of solution components C in two ways: components may be represented either by vertices or by edges. Artificial ants move from vertex to vertex along the edges of the graph incrementally building a partial solution. Additionally, ants deposit a certain amount of pheromone on the components; that is, either on the vertices or on the edges that they traverse. The amount of pheromone deposited may depend on the quality of the solution found. Subsequent ants use the pheromone information as a guide toward promising regions of the search space [6].

3   TUTOR Group-Work Activities Sub-system

In our virtual learning environment we have a set G of students' teams and a set A of group work activities consisting of several tasks. Let

$$a_j \in A \text{ with } j = 1, \ldots, J \text{ be an activity from the set } A \tag{1}$$

$$g_i \in G \text{ with } i = 1, \ldots, I \text{ be a team from the set } G \tag{2}$$

$$t_{ji}^k \in TK \text{ with } i = 1, \ldots, I, \ j = 1, \ldots, J \tag{3}$$

be a task of the activity a_j that has to be performed by the team g_i.

An activity a_j consists of an ordered sequence of tasks from a set $TK = \{t_{ji}^k\}$. A task t_{ji}^k has to be performed by the team g_i in a number d_{ji} of time units. $N = |TK|$ is then the total number of tasks. As in a classical Job-Shop Scheduling Problem [12] the goal is to assign tasks to time intervals in such a way that no two activities are performed at the same time by the same team and that the maximum completion time of all tasks is minimized.

For the convenience of the validation process, at the first stage, we will consider that all teams have the same number of members. A team $g_i \in G$ is supposed to have M members (student-agents):

$$g_i = \bigcup_{m=1}^{M} g_{im} \tag{4}$$

This is almost the real situation, when students are assigned in teams with the same size, based on an arbitrary criterion. There are task-classes and at this stage of the modelling process the potential assignment of one student to a specific task is based on the concept of response threshold combined with a function of the background knowledge of the student and her decision making behavior. A student-agent g_{im}, is attracted to a task t_{ji}^k with a probability P depending on her background knowledge, beliefs (cognitive), affects (emotive) and intentions [13] and on response threshold θ_{imj}:

$$P(\theta_{imj}, s_{imj}) = \frac{s_{imj}^2}{s_{imj}^2 + \theta_{imj}^2} \tag{5}$$

where $s_{imj} = f(\text{beliefs, affects, intention, qualification-knowledge and skills})$.

The response threshold θ_{imj} of the ant-agent g_{im} to the task t_{ji}^k is decreased when the agent has performed the task t_{ji}^k; at the same time, thresholds for the other tasks are increasing proportional to the time t to perform the task.

$$\theta_{imj}^{new} = \theta_{imj}^{old} - x_{ij}\xi\Delta_t + (1 - x_{ij})\varphi\Delta_t \tag{6}$$

where x_{ij} is the time spent by the agent i for the task j, ξ is a learning coefficient, φ is a forgetting coefficient, and Δ_t is a time unit.

If a task is performed, the response threshold is decreased with a quantity depending on the learning coefficient, in the opposite situation; the threshold is increased with a function of the forgetting coefficient. The state of an ant-agent is evolving from "active" — performing a task, to "inactive" — idle, and vice versa. An inactive agent starts to perform a task with a probability P given by (5). An active agent completes the task, or abandons it, with a probability p per time unit:

$$p = P(state = \text{``active''} \rightarrow state = \text{``inactive''}) \tag{7}$$

the average time spent by an agent in task performing before giving up the task is $1/p$.

The problem is represented as a weighted graph $Q = (TK', L)$, where $TK' = TK\{tk_0\}$ and L is the set of edges that connect node t_0 with the first task of each activity. The vertexes of T are completely connected, with exception of the nodes of the tasks from the same activity that are connected sequentially, each such node being linked only to its direct successor. There are $N(N-1)/2 + |A|$ edges. Each edge (i,j) is weighted by two numbers: τ_{ij} — the pheromone level (trail level) and η_{ij} — the so called visibility and represents the desirability of a transition from node i to node j. Each ant-agent has an associated data structure — the tabu-list, that memorizes the tasks of an activity that have been performed at the time moment t.

A transition probability function from node i to node j for the k-th ant-agent was defined as:

$$p_{ij}^k(t) = \begin{cases} \dfrac{[\tau_{ij}(t)]^\alpha \cdot [\eta_{ij}]^\beta}{\displaystyle\sum_{k \in \ allowed_k} [\tau_{ij}(t)]^\alpha \cdot [\eta_{ij}]^\beta}, & \text{if } j \in allowed_k \\ \\ 0 & \text{otherwise} \end{cases} \tag{8}$$

where $allowed_k = (TK \backslash tabu_k)$, $tabu_k$ is a vector that changes dynamically and contains the tabu list of the k-th ant, and α and β are parameters used to control the relative importance of pheromone level and visibility.

Considering NR the number of potential active agents, NR_{act} the number of active agents at the time moment t, we have the following formula for the variation of the attraction of a task (pheromone deposit) in a discrete time situation:

$$s_{imj}(t+1) = s_{imj}(t) + \beta - (\alpha \cdot NR_{act})/NR \tag{9}$$

The order in which the nodes are visited by each ant-agent specifies the proposed solution.

4   Discussion

We have presented a complex model of one of the components of a virtual learning environment — the TUTOR sub-system, i.e. the one responsible for organizing the group-work activities for the students. We have used an ant colony behavior model for the agents that play the role of a tutor and are organizing the group-work activities for the students. We have chosen an ant social behavior model because it is natural and it is efficient and effective. Studies on the students' learning behavior have shown that in choosing the tasks to perform they are heavily influenced by colleagues and friends. This behavior is similar to the foraging behavior of ant colonies. The validation of the model was made having in mind that, at this stage of development, we are interested more in the good fitness of the model with the real problem than in the performance of the algorithms compared to other approaches. We have collected data from groups of students in high schools and from the tutors (human agents) in order to determine s_{imj} — the student attraction towards a task and the probabilities involved in the model. We used the STUDENT model presented in [13]. For testing the tutor-agent behavior we have used special designed questionnaires. The first results were quite promising. However, there was a problem concerning s_{imj}. For two of the data samples our model was weak in the sense that the parameter reflecting the student's knowledge and skills, based on a combination of scores obtained at different tests, was almost irrelevant; students with a low score, managed to perform very well, and students with an average score had a high rate of task rejection. That is why one of the future research directions will be focused in improving the STUDENT behavioral learning model. Another will consist in building a small pilot and testing the whole system, but with a reduced number of variables and parameters.

Acknowledgement

This paper benefits from funding by the Romanian Ministry of Education and Research (INFOSOC — CEEX 73/31.07.2006).

References

1. Bonabeau, E., Dorigo, M., Theraulaz, G.: Swarm intelligence: From natural to artificial systems. In: Santa Fe Institute Studies of Complexity, Oxford University Press, Oxford (1999)
2. Bonabeau, E., Dorigo, M., Theraulaz, G.: Inspiration for optimization from social insect behaviour. Nature 406, 39–42 (2000)
3. Deneubourg, J.L., et al.: The Self-Organizing Exploratory Pattern of the Argentine Ant. Journal of Insect Behaviour 3, 159–168 (1990)
4. Dorigo, M.: Optimization, Learning and Natural Algorithms, Ph.D. Thesis, Politecnico di Milano, Italy (1992)
5. Dorigo, M., Birattari, M., Stützle, T.: Ant Colony Optimization, Artificial Ants as a Computational Intelligence Technique. IEEE Computational Intelligence Magazine (2006)
6. Dorigo, M., Birattari, M., Stützle, T.: Ant Colony Optimization, Artificial Ants as a Computational Intelligence Technique, IRIDIA — Technical Report Series, Technical Report No.TR/IRIDIA/2006-023 (2006)
7. Dorigo, M., Di Caro, G.: The Ant Colony Optimization Meta-Heuristic. In: Crone, D., Dorigo, M., Glover, F. (eds.) New Ideas in Optimization, pp. 11–32. McGraw-Hill, New York (1999)
8. Dorigo, M., Gambardella, L.M.: Ant Colonies for the Traveling Salesman Problem. BioSystems 43, 73–81 (1997)

9. Goldberg, D.E., Deb, K.: A comparative analysis of selection schemes used in genetic algorithms. In: FOGA 1991, vol. 1, pp. 69–93 (1991)
10. Goss, S., et al.: Self-Organized Shortcuts in the Argentine Ant. Naturwissenschaften 76, 579–581 (1989)
11. Jennings, N.R., Wooldridge, M.: Applications of Intelligent Agents. In: Jennings, N., Wooldridge, M. (eds.) Agent Technology: Foundations, Applications, and Markets, Springer, Heidelberg (1998)
12. Lenstra, J.K., Rinnooy Kan, A.H.G.: Optimization and Approximation in Deterministic Sequencing and Scheduling: A Survey. Annals of Discrete Mathematics 5, 287–326 (1979)
13. Moisil, I., et al.: Socio-cultural modelling of student as the main actor of a virtual learning environment. WSEAS Transaction on Information Science and Applications 4, 30–36 (2007)
14. Nouyan, Sh., Dorigo, M.: Path Formation in a Robot Swarm, IRIDIA — Technical Report Series, Technical Report No.TR/IRIDIA/2007-002 (2007)
15. Resnick, M.: Turtles, Termites and Traffic Jams. Explorations in Massively Parallel Microworlds. In: Complex Adaptive Systems, MIT Press, Cambridge (1994)
16. Stützle, T., Dorigo, M.: ACO Algorithms for the Travelling Salesman Problem. In: Makela, M., et al. (eds.) Proceedings of the EUROGEN conference, pp. 163–184. John Wiley & Sons, Chichester (1999)
17. Tsui, K.C., Liu, J.: Multiagent Diffusion and Distributed Optimization. In: Proceedings of the Second International Joint Conference on Autonomous Agents & Multiagent Systems, pp. 169–176. ACM, New York (2003)

Simulated Annealing Optimization of Multi-element Synthetic Aperture Imaging Systems

Milen Nikolov and Vera Behar

Institute for Parallel Processing, Bulgarian Academy of Sciences
Acad. G. Bonchev, bl. 25A, 1113 Sofia, Bulgaria
`milenik@bas.bg, behar@bas.bg`

Abstract. In conventional ultrasound imaging systems with phased arrays, the improvement of lateral resolution of images requires enlarging of the number of array elements that in turn increases both, the complexity and the cost, of imaging systems. Multi-element synthetic aperture focusing (MSAF) systems are a very good alternative to conventional systems with phased arrays. The benefit of the synthetic aperture is in reduction of the system complexity, cost, and acquisition time.

A general technique for parameter optimization of an MSAF system is described and evaluated in this paper. The locations of all "transmit-receive" subaperture centers over a virtual linear array are optimized using the simulated annealing algorithm. The optimization criterion is expressed in terms of the beam characteristics — beam width and side lobe level.

The comparison analysis between an optimized MSAF system and an equivalent conventional MSAF system shows that the optimized system acquires images of equivalent quality much faster.

1 Introduction

Images produced by ultrasound imaging systems must be of sufficient quality in order to provide accurate clinical interpretation. The image quality (lateral resolution and contrast) is primarily determined by the beam characteristics of a transducer used in an imaging system. In conventional ultrasound imaging systems with phased array (PA), all transducer elements transmit signals and receive the echoes, reflected from the tissue. Thus the modern conventional PA systems produce high-resolution images at high cost because the system complexity and thus the system cost depends on the number of transducer elements [1]. The further improvement of lateral resolution in a conventional PA imaging system requires enlarging of the number of transducer elements. It is often not possible because of physical constrains or too high cost. The same effect of high lateral resolution and contrast can be accomplished by using various synthetic aperture techniques. The benefit of the synthetic aperture is the reduction of system complexity and cost. Moreover, in a synthetic aperture imaging system, the acquisition

I. Lirkov, S. Margenov, and J. Waśniewski (Eds.): LSSC 2007, LNCS 4818, pp. 585–592, 2008.

time can be drastically reduced and dynamical steering and focusing can be applied in both transmit and receive. There are different methods for synthetic aperture imaging — *Synthetic Receive Aperture* (SRA) technique, *Synthetic Transmit Aperture* (STA) technique, *Synthetic Aperture Focusing technique* (SAFT), and *Multi-Element Synthetic Aperture Focusing* (MSAF) technique [2].

In this paper the MSAF technique for image formation is considered and optimized. In an MSAF system, a group of elements transmits and receives signals simultaneously, and the transmit beam is defocused to emulate a single element response [3]. The acoustic power and the signal-to-noise ratio (SNR) are increased compared to the classical SAFT technique where a single element is used in both transmit and receive [5]. A disadvantage is that the method requires more memory for data recordings.

The relation between the effective aperture function and the corresponding beam pattern of the imaging system can be used as a tool for analysis and optimization of an MSAF imaging system. In MSAF imaging, the aperture function depends on the number of transmit elements, the number of subapertures and their locations within the array, and the weight coefficients applied to the array elements. Hence it appears that the shape of the effective aperture function of a system and, as consequence, the shape of the two-way beam pattern can be optimized depending on the positions and the weights, applied to the transmit/receive subapertures.

2   MSAF Imaging

Consider a conventional MSAF system with N-element transducer, where a K_t-element transmit subaperture sends an ultrasound pulse and the reflected echo signals are received at a K_r-element receive subaperture. Usually $K_t = K_r = k$. At the next step, one element is dropped, a new element is added at the end of the subaperture and the transmit/receive process is repeated again (Fig. 1). The process of transmit/receive is repeated for all positions of transmit/receive subapertures. The number of transmissions needed to create a synthetic aperture equivalent to a physical array with N elements is:

$$M = N - k + 1 \tag{1}$$

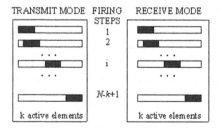

Fig. 1. Conventional MSAF system

All the data recordings associated with each corresponding pair transmit/receive subaperture are then focused synthetically by a computer producing low-resolution images. The final high-resolution image is formed as a sum of all low-resolution images.

The image acquisition time is an important quality parameter of the imaging systems, which are used for real time applications. It defines the ability of a system to resolve movements of structures in the image. In an MSAF imaging system, the image acquisition time is evaluated as:

$$T_{MSAF} = M.T_{REC} \tag{2}$$

where M is the number of transmissions, and T_{REC} is the time needed to acquire RF-signals at each transmission.

In this paper we investigate a variant of the MSAF system and the possibilities to optimize it. In contrast to the classical MSAF system mentioned above where one element is dropped between neighboring transmissions, in our MSAF system a random number of elements is dropped between neighboring transmissions. This approach makes it possible to reduce the image acquisition time T_{MSAF}, compared to the classical MSAF system, by reducing the number of emissions, needed to form an image.

3   Description of the Optimization Problem

The image quality parameters, lateral resolution and contrast, are determined by the main lobe beam width (W) and the side lobe peak (SL) of the beam pattern of an imaging system. The resolution is improved by decreasing the main lobe beam width, and the contrast is improved by lowering the level of the side lobe peak. The two-way beam pattern of an MSAF system that employs a transducer with N elements is evaluated as the Fourier Transform of the effective aperture function e_{MSAF}, defined as:

$$e_{MSAF} = \sum_{m=1}^{N-k+1} w_R(m) \otimes w_T(m) \tag{3}$$

where $w_R(m) = w_T(m) = [0, 0, \ldots, i_m, i_{m+1}, \ldots, i_{m+k-1}, 0, \ldots, 0]$, k is the number of elements in each subaperture, i_m to i_{m+k-1} are the weighting coefficients applied to each transmit/receive subaperture and $\otimes$ is the convolution operator.

The effective aperture for the modified MSAF system is changed to:

$$e_{MSAF} = \sum_{m=1}^{N_{SUB}} w_R(m) \otimes w_T(m) \tag{4}$$

where

$$w_R(m) = w_T(m) = [0, 0, \ldots, i_n, i_{n+1}, \ldots, i_{n+k-1}, 0, \ldots, 0]; \quad n = 1 + \sum_{j=1}^{m} S(j);$$

S is an array which shows the number of elements dropped after each transmission, $S(1) = 0$ because the first subaperture starts from the first element of the transducer array; $N_{SUB} = size(S)$ is the number of used subapertures.

The optimization problem of an MSAF system can be formulated as an optimization of both the positions of the transmit/receive subapertures, which depend on the number of elements which are dropped after each transmit/receive step, and the weights applied to the subaperture elements.

The set of the subaperture positions is found by using the simulated annealing algorithm. It is done for a set of known weighting functions $\{B\}_L$. Such a set weighting functions may include several well-known window-functions (Hamming, Hann, Kaiser, Chebyshev, etc.). The optimization criterion can be written as follows:

$$Given \quad k, \quad N \quad and \quad \{B\}_L, \quad choose \quad S(j_1, j_2, \ldots, j_{N_{SUB}}) \qquad (5)$$
$$to \quad minimize \quad the \quad cost \quad C(S, B_l)$$

where B_l is the currently used weighting function.

The cost function $C(S, B_l)$ is defined as:

$$C(S, B_l) = \min W(I, B_l) \ subject \ to \ SL < Q \qquad (6)$$

where $W(I, B_l)$ is the main lobe width corresponding to the current positions of the subapertures and the weighting functions, applied to them, SL is the side lobe peak and Q is a threshold of acceptable level of the side lobe peak.

4   The Simulated Annealing Algorithm

The simulated annealing algorithm was suggested by Kirkpatrick et al. [4]. It realizes an iterative procedure that is determined by simulation of the arrays with variable number of elements dropped after each transmit/receive step. The process is initiated with a random layout of the positions of all transmit/receive subapertures (the first subaperture starts with the first element of the transducer array and the last subaperture ends with the last element of the transducer array). These positions are found by generating the step array, where each step is a random number, smaller than the number of elements in each subaperture. The step array S should satisfy the following expression:

$$N = k + \sum_{j=1}^{N_{SUB}} S(j) \qquad (7)$$

where N is the total number of elements in the aperture and k is the number of elements in each subaperture. The number of subapertures N_{SUB} (expressed by the size of the step array) depends on the total number of dropped elements, which is random, hence in each run of the algorithm the step array is of different size.

The simulated annealing algorithm is composed of two loops (Fig. 2). At each perturbation is chosen a new neighboring layout, by shifting the current aperture with one element to the left or to the right with equal probability of 0.5.

begin
set the weighting function B_l
compute a random initial state I_0
set the initial temperature T_0
 for i=1 **to** *number of iterations*
  for j=1 **to** *number of perturbations*
   compute at random a neighboring
   layout $I_p = perturbate(I_{j-1})$
   if I_p is better than I_{j-1} **then**
    set $I_j = I_p$
   else
    set $I_j = I_p$ with probability
    $p(T)$ depending on the cost function
   endif
  endfor
  $T_i = T_{i-1}.\alpha$
 endfor
end

Fig. 2. The simulated annealing algorithm

The algorithm accepts or rejects the new layout according to a certain criterion. The number of perturbations is equal to the number of subapertures.

The acceptance is described in terms of probability $p(T)$ that depends on the cost function $C(S, B_l)$. For the cost function defined by (6), the expression for the acceptance probability $p(T)$ takes the form:

$$p(T) = \begin{cases} 1 & \text{if} \quad \Delta W < 0 \quad \& \quad \Delta SL < 0 \\ exp(-\Delta W/T_{1,k}) & \text{if} \quad \Delta W > 0 \quad \& \quad \Delta SL < 0 \\ exp(-\Delta SL/T_{2,k}) & \text{if} \quad \Delta W < 0 \quad \& \quad \Delta SL > 0 \\ exp(-\Delta W/T_{1,k}) \times exp(-\Delta SL/T_{2,k}) & \text{if} \quad \Delta W > 0 \quad \& \quad \Delta SL > 0 \end{cases}$$
(8)

where ΔW is the difference of width of the main lobe, ΔSL is the difference of the height of the peak of the side lobe between the current locations of transmit/receive subapertures and the threshold of acceptable level Q. The choice of Q depends on the used weighting function. If no weighting is used $Q = 40$ and $Q = 90$ otherwise. T_k is the current value of "temperature", evaluated as $T_k = 0.99T_{k-1}$. Since in our case two parameters are evaluated (W and SL), we use two "temperatures" T_1 and T_2 defined as 10% of the expected value of the corresponding parameter. The algorithm proceeds until 1000 iterations are executed.

5 Simulation Results

The physical array utilized in computer simulations is of 64 elements with half wavelength spacing. Three different systems are evaluated with 40, 45, and 50 active elements in each subaperture. The properties of the MSAF system are

Table 1. Numerical results obtained by employing the optimization algorithm

No of elements in subaperture	Step Array	Weighting	$W[°]$		SLB [dB]		Average number of iterations
			Optimal	Average	Optimal	Average	
40	0,1,3,2,5, 3,3,4,3	No weighting	1.78	1.73	-42.07	-38.54	43.30
45	0,1,5, 3,4,6	No weighting	1.83	1.82	-40.09	-33.50	66.57
50	0,2,3, 2,3,4	No weighting	1.83	1.82	-52.84	-45.19	352.10
40	0,1,1,1,1, 2,15,1,1,1	Hamming Window	1.75	1.92	-100.59	-96.85	650.00
45	0,1,1,5, 4,6,1,1	Hamming Window	1.97	1.99	-94.09	-97.11	580.20
50	0,1,1, 10,1,1	Hamming Window	2.10	2.18	-90.13	-94.05	549.90
40	0,1,1,3, 16,1,1,1	Chebishev Window	1.75	1.88	-109.97	-107.79	558.57
45	0,1,5, 6,6,1	Chebishev Window	1.97	1.98	-103.39	-103.38	568.26
50	0,1,1, 11,1	Chebishev Window	2.11	2.14	-104.74	-103.87	432.43

Table 2. Beam characteristics of a conventional MSAF system

No of elements in subaperture	No weighting		Hamming Window		Chebishev Window	
	$W[°]$	SLB [dB]	$W[°]$	SLB [dB]	$W[°]$	SLB [dB]
40	1.87	-43.44	2.11	-102.65	2.12	-105.81
45	1.88	-41.15	2.29	-102.16	2.32	-103.12
50	1.84	-49.58	2.41	-103.39	2.45	-102.47

optimized using the algorithm described in Section 3. The optimal positions of the subapertures are found for three weighting functions. For each weighting function, the positions of the subapertures are shifted until optimal performance is obtained.

The number of elements, dropped between neighboring transmissions and found to optimize the performance of the system, according the optimization criterion, together with the optimal and the average quality parameters (width of the main beam lobe and peak of the sidelobes) are presented in Table 1. The results are achieved in 30 runs of the algorithm. In the last column is shown the average number of iterations, needed to produce the best result in each run. For comparison the numerical results obtained for a conventional MSAF system with one element dropped between neighboring transmissions are presented in Table 2.

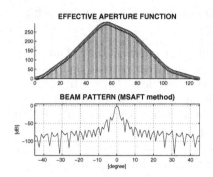

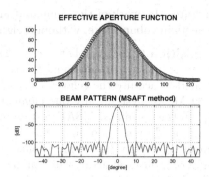

Fig. 3. The optimized effective aperture function and two-way beam pattern; 50/64 elements; No weighting

Fig. 4. The optimized effective aperture function and two-way beam pattern; 50/64 elements; Hamming window

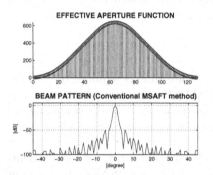

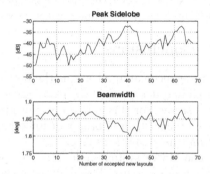

Fig. 5. Effective aperture function and two-way beam pattern of a conventional MSAF system; 50/64 elements; No weighting

Fig. 6. Illustration of the work of the optimization algorithm

It can be seen that the weighting, applied to the subapertures reduces the peaks of the side lobes from -40 dB to -95 dB for Hamming window and -104 dB for Chebishev window. This is done at the cost of widening the main lobe width (hence reducing the lateral resolution of the system) from 1.73 degrees for a system without weighting, to more than 2 degrees for a system with Chebishev window. A comparison of the optimized system with a conventional MSAF system shows that the optimized system has slightly better main lobe width, but the side lobe peaks of the conventional system are a little bit lower. The differences are small so practically the systems have similar beam parameters.

Both optimized functions, the effective aperture function and the corresponding two-way beam pattern, are plotted for optimized system with 50 elements in subaperture, without weighting (Fig. 3); with 50 elements in subaperture and hamming weighting (Fig. 4). For comparison a conventional MSAF system with 50 elements in subaperture and without weighting is shown on Fig. 5. Fig. 6

illustrates the work of the simulated annealing algorithm for a system with 45 elements in subaperture, without weighting.

Acquisition time. As follows from (2) the image acquisition time of an MSAF system is linearly proportional to the number of emissions (M) and the time required to acquire the echoes from one direction of view (T_{REC}). In a conventional MSAF system only one element is dropped after each transmit/receive step. In the optimized MSAF system on the other hand, after each step usually are dropped more than one elements. This significantly reduces the number of transmissions and, as a consequence, the image acquisition time for the same length of a physical array, while maintaining the same image quality.

Analysis of Table 1 shows that the best compromise between the desired main beam lobe width and the side lobe peaks is achieved by the system with 40 elements in each subaperture, Chebishev weighting and step array: 0, 1, 1, 3, 16, 1, 1, 1. Such a system employs a 64-element physical array and produces an image in 8 transmissions. In comparison the equivalent conventional MSAF system produces an image with the same quality in 25 transmissions, which is more than 3 times slower.

6   Conclusions

In most medical applications the image acquisition time is required to be minimized in order to avoid the phase errors caused by tissue motion during the data acquisition. For this aim, an optimization algorithm based on the simulated annealing algorithm is proposed for optimizing the locations of the subapertures within the array transducer.

The MSAF system under study employs a 64-element array, with 40 elements in each subaperture. The analysis shows that an optimized MSAF system obtains images of similar quality and acquires data 3 times faster than a conventional MSAF system.

Acknowledgments. This work is partially supported by the Bulgarian Foundation for Scientific Investigations: MI-1506/05 and by Center of Competence BIS21++ (FP6-2004-ACC-SSA-2).

References

1. Angelsen, B.: Ultrasound imaging: Waves, signals, and signal processing, Emantec, Norway (2000)
2. Holm, S., Yao, H.: Method and apparatus for synthetic transmit aperture imaging. In: US patent No 5.951.479, September 14 (1999)
3. Karaman, M., Bilge, H., O'Donnell, M.: Adaptive multi-element synthetic aperture imaging with motion and phase aberration correction. IEEE Trans. Ultrason. Ferroelec. Freq. Contr. 45(4), 1077–1087 (1998)
4. Kirkpatrik, S., Gelatt, C., Vecchi, M.: Optimization by simulated annealing. Science 220(4598), 671–680 (1988)
5. Ylitalo, J.: On the signal-to-noise ratio of a synthetic aperture ultrasound imaging method. European J. Ultrasound 3, 277–281 (1996)

Adaptive Heuristic Applied to Large Constraint Optimisation Problem

Kalin Penev

Southampton Solent University, Southhampton, UK
Kalin.Penev@solent.ac.uk

Abstract. The article presents experimental results achieved by Free Search on optimization of 100 dimensional version of so called bump test problem. Free Search is adaptive heuristic algorithm. It operates on a set of solutions called population and it can be classified as population-based method. It gradually modifies a set of solutions according to the prior defined objective function. The aim of the study is to identify how Free Search can diverge from one starting location in the middle of the search space in comparison to start from random locations in the middle of the search space and start from stochastic locations uniformly generated within the whole search space. The results achieved from the experiments with above initialization strategies are presented. A discussion focuses on the ability of Free Search to diverge from one location if the process stagnates in local trap during the search. The article presents, also, the values of the variables for the best achieved results, which could be used for comparison to other methods and further investigation.

1 Introduction

In this study Free Search [10,12,13] is applied to 100 dimensional variant of so-called bump problem [6,7]. The test problem is hard constraint non-linear optimization problem generalized for multidimensional search space. It is widely discussed in the literature. Large research efforts have been directed towards its exploration. The optimum is unknown, and the best-achieved results are published [2,5,6,7,8,9,11,13,15,16,17,18]. "It's a difficult problem that has been studied in the scientific literature and no traditional optimization method has given a satisfactory result. The function is non-linear and the global maximum is unknown." [9] An earlier investigation of the bump problem, published in the literature applies Genetic Algorithm [6] and some evolutionary algorithms [2,7] to twenty and fifty dimensional variants of the problem. These investigations accept 20000 and 50000 iterations limit respectively for 20 and 50 dimensions. Better results, published in the literature, for 20 and 50 dimensional variants are achieved by evolutionary algorithms modified for precise exploration of the space near to the constraint boundaries [8,9,18]. The best results, published in the literature [5], for the bump problem from 2 up to 50 dimensions achieved by asynchronous parallel evolutionary algorithm APEMA on distributed MIMD computational system indicates best value 0.80361910 for 20 dimensions and

I. Lirkov, S. Margenov, and J. Waśniewski (Eds.): LSSC 2007, LNCS 4818, pp. 593–600, 2008.

0.83523753 for 50 dimensions. The best results achieved by Free Search are for 20 dimensions 0.80361910412558 and for 50 dimensions 0.83526234835811175. This investigation has no available publications for other algorithms for 100 dimensional variant of the bump problem. The search space of this test is continuous. It has many peaks. An essential condition of this test is start from single location in the middle of the search space. This condition guarantees start from location, which is relatively far from the maximal hill. In contrast to start from multiple start locations uniformly distributed within the search space, it eliminates possibility for starting from initial locations accidentally generated near to the best value. Start from one location facilitates a measurement of the divergence across the whole space and then convergence to the best value. The work presented in this article is a continuation of the experiments on this problem published earlier [11,17]. The experiments with the bump optimization problem for $n = 50$, $x_i \in (0, 10)$, $i = 1, \ldots, 50$ require, for clarification of the results with precision of seven decimal digits, exploration of 10^{400} solutions ... Free Search achieves the maximum with such precision after exploration of less than 10^9 solutions... The relation between the number of all possible values with a certain level of precision and the number of explored locations deserves attention. These results can be considered from two points of view. The first point of view is when these results are accepted as accidental. The algorithm is "lucky" to "guess" the results. If that point of view is accepted, it follows that the algorithm "guesses" the appropriate solution from 10^{400} possible for 50-dimensional space. Another point of view is to accept the results as an outcome of the intelligent behavior modeled by the algorithm. Free Search abstracts from the search space essential knowledge. That knowledge leads to the particular behavior and adaptation to the problem. In that case the relation between explored and all possible locations can be a quantitative measure of the level of abstraction. The second point of view that the algorithm models intelligent behavior is accepted. Abstraction knowledge from the explored data space; learning, implemented as an improvement of the sensibility; and then individual decision-making for action, implemented as selection of the area for next exploration; can be considered as a model of artificial thinking [11]. The experiments with 100 dimensional variant in this study are harder and Free Search confirms its excellent exploration abilities on large-scale optimization problems. In the article and in the tables the following notation is accepted: n is the number of dimensions, i is dimensions indicator, $i = 1, \ldots n$, x_i are initial start locations, X^{max} and X^{min} are search space boundaries, x^{max} are variables of an achieved local maximum, F_{max100} is a maximal achieved value for the objective function, r_i is random value, $r_i \in (0, 1)$.

2   Test Problem

The objective function and the conditions are:

$$\text{Maximize:} \left| \sum_{i=1}^{n} \cos^4(x_i) - 2 \prod_{i=1}^{n} \cos^2(x_i) \right| \Big/ \sqrt{\sum_{i=1}^{n} i x_i^2}$$

$$\text{subject to: } \prod_{i=1}^{n} x_i > 0.75 \tag{1}$$

and

$$\sum_{i=1}^{n} x_i < 15\frac{n}{2} \tag{2}$$

for $0 < x_i < 10$, $i = 1, \ldots n$, starting from $x_i = 5$, $i = 1, \ldots, n$, where x_i are the variables (expressed in radians) and n is the number of dimensions [7].

3 Methodology

Free Search is applied to the 100 dimensional variant of the bump problem as follows: The population size is 10 (ten) individuals for all experiments. Four series of 320 experiments with four different start conditions are made:

- start from one location in the middle of the search space $x_i = 5$, $i = 1, \ldots, n$; (This is an original condition of the test, ignored from majority authors due to inability of other methods to diverge successfully from one location [8,9,18]
- start from random locations in the middle of the search space $x_i = 4 + 2r_i$, $r_i \in (0, 1)$;
- start from locations stochastically generated within the whole search space $x_i = X_i^{\min} + (X_i^{\max} - X_i^{\min})r_i$, $r_i \in (0, 1)$;
- additional experiments with start from the best achieved location $x_i = x_i^{\max}$ are made. The last result of these experiments is presented also in Table 1.

For the first three series maximal neighbor space per iteration is restricted to 5% of the whole search space and sensibility is enhanced to 99.999% from the maximal. For the fourth series, in order to distinguish very near locations with very similar high quality, maximal neighbor space per iteration is restricted to 0.0005% from the whole search space and sensibility is enhanced to 99.9999999% from the maximal.

4 Experimental Results

The maximal values achieved from 320 experiments for 100-dimensional search space for the four different start conditions are presented in Table 1. In Table 1: $x_i = 5$ start from one location; $x_i = 4 + 2r_i$ start from random locations in the middle of the search space; $x_i = X_i^{\min} + (X_i^{\max} - X_i^{\min})r_i$ start from random locations uniformly distributed within the whole search space; $x_i = x_i^{\max}$ start from a local sub-optimum.

The best results from 320 experiments with start from currently achieved best local sub-optimum $x_i = x_i^{\max}$ are presented in Table 2. The variables' values for the best achieved objective function value for $n = 100$ are presented respectively in Table 3. The constraint parameter values are an indicator whether the found

Table 1. Maximal values achieved for n = 100 bump problem

n	start from	iterations	$\prod x_i$	F_{max100}
100	$x_i = 5$	100000	0.750305761739250	0.838376006873
100	$x_i = 4 + 2r_i$	100000	0.7750016251034640	0.836919742344
100	$x_i = X_i^{\min} + (X_i^{\max} - X_i^{min})r_i$	100000	0.750125157302112	0.838987184912
100	$x_i = x_i^{\max}$	100000	0.750000000039071	0.845685456012

Table 2. Best achieved results from 320 experiments start from $x_i = x_i^{max}$

$F_{\max} = 0.84568545600962819$	$\prod x_i = 0.75000000007771150$
$F_{\max} = 0.8456854560029361$	$\prod x_i = 0.75000000002732459$
$F_{\max} = 0.84568545601228962$	$\prod x_i = 0.75000000003907141$
$F_{\max} = 0.84568545600160894$	$\prod x_i = 0.75000000003325862$
$F_{\max} = 0.84568545601179412$	$\prod x_i = 0.75000000001648370$
$F_{\max} = 0.84568545600045464$	$\prod x_i = 0.75000000009027890$
$F_{\max} = 0.84568545600126788$	$\prod x_i = 0.75000000002984646$
$F_{\max} = 0.84568545600471334$	$\prod x_i = 0.75000000013973589$
$F_{\max} = 0.84568545600142975$	$\prod x_i = 0.75000000000654199$
$F_{\max} = 0.84568545600266842$	$\prod x_i = 0.75000000002129508$
$F_{\max} = 0.84568545600390022$	$\prod x_i = 0.75000000011346923$
$F_{\max} = 0.84568545600150213$	$\prod x_i = 0.75000000021696134$
$F_{\max} = 0.84568545600245093$	$\prod x_i = 0.75000000004234102$
$F_{\max} = 0.84568545600018008$	$\prod x_i = 0.75000000002819023$
$F_{\max} = 0.84568545600770795$	$\prod x_i = 0.75000000006311085$
$F_{\max} = 0.84568545600403033$	$\prod x_i = 0.75000000001039713$
$F_{\max} = 0.84568545600069034$	$\prod x_i = 0.75000000001963218$

Table 3. Variables values for the best achieved F_{max100}=0.84568545601228962

x[0]=9.4220107126347745	x[1]=6.2826322016103955	x[2]=6.2683831748189975
x[3]=3.1685544118834987	x[4]=3.1614825164328604	x[5]=3.1544574053305716
x[6]=3.1474495832290019	x[7]=3.1404950479045355	x[8]=3.1335624853236599
x[9]=3.1266689747394079	x[10]=3.1197805974560233	x[11]=3.1129253611190442
x[12]=3.1061003015872641	x[13]=3.0992700218084379	x[14]=3.0924554966175832
x[15]=3.0856546345558589	x[16]=3.0788589079539115	x[17]=3.0720627263662514
x[18]=3.0652657000077048	x[19]=3.0584746707061194	x[20]=3.0516637034865339
x[21]=3.0448584949106863	x[22]=3.0380286236964169	x[23]=3.0311906892219178
x[24]=3.024325642570814	x[25]=3.0174564447484493	x[26]=3.0105540636060151
x[27]=3.0036206348205221	x[28]=2.996654984477289	x[29]=2.9896634779600619
x[30]=2.9612192303388043	x[31]=2.9755508892857203	x[32]=2.9684029022824512
x[33]=2.9612192303388043	x[34]=2.9539969359393266	x[35]=2.9467002904105652
x[36]=2.9393411868452524	x[37]=2.9319091645017501	x[38]=2.9243754870396326
x[39]=2.916779048836506	x[40]=0.48215961508911009	x[41]=0.48103824318067195
x[42]=0.47987816774664849	x[43]=0.47878167955313988	x[44]=0.47768451249416577
x[45]=0.47661282330477983	x[46]=0.47553403486022883	x[47]=0.47446785125492774
x[48]=0.47342756022045934	x[49]=0.47239007924579712	x[50]=0.47137147513069294
x[51]=0.47032878010013335	x[52]=0.46930402745591204	x[53]=0.46831098721684394
x[54]=0.46735104878920491	x[55]=0.46636407600760849	x[56]=0.46539891729486565
x[57]=0.46441864961851576	x[58]=0.46347276946009547	x[59]=0.46251323773794134
x[60]=0.4616002801440135	x[61]=0.46065790486354485	x[62]=0.45975509243021634
x[63]=0.45882641352502623	x[64]=0.45794463889695297	x[65]=0.45703335381561577
x[66]=0.45616083797292917	x[67]=0.45530545866090766	x[68]=0.4544181045335004
x[69]=0.45356136754388732	x[70]=0.45270186592373279	x[71]=0.45185207109498432
x[72]=0.45101805630688263	x[73]=0.45019912098968307	x[74]=0.44936498811072595
x[75]=0.4485406848660335	x[76]=0.44772742155880246	x[77]=0.44690550854857802
x[78]=0.44610180301058927	x[79]=0.44532932347961096	x[80]=0.44452998407145683
x[81]=0.4437461852789148	x[82]=0.44294433416028023	x[83]=0.44217581932603778
x[84]=0.44144485059511346	x[85]=0.44065197401737571	x[86]=0.43992044125167573
x[87]=0.43915615308675215	x[88]=0.43841742447730969	x[89]=0.43766920812505228
x[90]=0.43695640149660347	x[91]=0.43620589323988396	x[92]=0.43550809702824705
x[93]=0.43477896755895717	x[94]=0.43406590088092134	x[95]=0.43335315514811895
x[96]=0.43265970793420877	x[97]=0.43194897101473584	x[98]=0.43126789176092778
x[99]=0.43057416262765469		

maximum belongs to the feasible region. They indicate also expected possible improvement and can be valuable for further research.

The best results presented in the Table 2 suggest that for 100-dimensional search space n = 100 with the precision of four decimal digits (0.0001) the optimum is $F_{opt100} = 0.8456$. Considering constraint value with high probability could be accepted that the optimal value for 100-dimensional variant of the bump problem is between 0.845685 and 0.845686. Clarification of this result could be a subject of future research. From previous experiments Free Search achieves: "For $n = 50$, $x_i \in (0, 10)$, $i = 1, \ldots, 50$ there are 10^{400} solutions with a precision of seven decimal digits. Free Search achieves the maximum with such precision after exploration of less than 10^9 solutions" [11].

For $n = 100$, $x_i \in (0, 10)$, $i = 1, \ldots, 100$ with a precision of three decimal digits there are 10^{400} solutions. To reach the result with this precision current version of Free Search needs exploration of more than 10^{12} solutions. These results suggest that perhaps 100-dimensional space is more complex than 50-dimensional space with the same size.

Let us note that: (1) these results are achieved on a probabilistic principle;(2) the search space is continuous and the results can be clarified to an arbitrary precision; (3) precision could be restricted from the hardware platform but not from the algorithm [15,16].

5   Discussion

Comparison between the best achieved result and the results achieved within 100000 iterations with start from (1) single location in the middle of the search space, (2) random locations in the middle of the search space, and (3) random locations stochastically distributed within the whole search space suggests difference of around 1%. Consequently Free Search can diverge successfully starting from one location and then can reach the optimal hill with the same speed as start from random locations. This in high extent is a confirmation of the independence of the algorithm from the initial population published earlier [12,15,16]. Theoretically these results can be interpreted as an ability of the algorithm to abstract knowledge during the process of search and to utilize this knowledge for self-improvement and successful, satisfactory completion of the search process. Rational value of the results is a confirmation of the high adaptivity of Free Search, which can support scientific search and engineering design in large complex tasks. Ability to continue the search process starting from the best achieved from previous experiments location brings additional value to the method. By means of the engineering design practice Free Search can overcome stagnation and can continue the search process until reaching an acceptable value and an arbitrary precision.

Another aspect of the results is what is the overall computational cost for exploration of this multidimensional task. A product between the number of iterations and the number of individuals in the algorithm population could be considered as a good quantitative measure for overall computational cost. For all experiments population size is 10 (then) individuals. In comparison to other publications [7,8,9] this is low number of individuals, which lead to low computational cost. To guarantee diversification and high probability of variation of the dimensions values within the population other methods operate on higher number of individuals, pay high computational cost and require large or distributed computational systems and extensive redundant calculations [5,18]. For his task Free Search minimizes required computational resources to a single processor PC.

6 Conclusion

In summary Free Search demonstrates good exploration abilities on 100–dimensional variant of the bump problem. Implemented novel concepts lead to an excellent performance. The results suggest that: (1) FS is highly independent from the initial population; (2) the individuals in FS adapt effectively their behavior during the optimization process taking into account the constraints on the search space, (3) FS requires low computational resources and pays low computational cost keeping better exploration and search abilities than the methods tested with the bump problem and discussed in the literature [2,5,7,9,18] (4) FS can be reliable in solving real-world non-linear constraint optimization problems. The results achieved on the bump optimization problem illustrate the ability of Free Search for unlimited exploration. With Free Search, clarification of the desired results can continue until reaching an acceptable level of precision. Therefore, Free Search can contribute to the investigation of continuous, large (or hypothetically infinite), constrained search tasks. A capability for orientation and operation within multidimensional search space can contribute also to the studying of multidimensional spaces and to a better understanding of real space. Presented experimental results can be valuable for evaluation of other methods. The algorithm is a contribution to the research efforts in the domain of population-based search methods, and can contribute, also, in general to the Computer Science in exploration and investigation of large search and optimization problems.

References

1. Eiben, A.E., Smith, J.E.: Introduction to Evolutionary Computing. Natural Computing Series, pp. 15–35. Springer, Heidelberg (2007)
2. El-Beltagy, M.A., Keane, A.I.: Optimisation for Multilevel Problems: A Comparison of Various Algorithms. In: Parmee, I.C. (ed.) Adaptive computing in design and manufacture, pp. 111–120. Springer, Heidelberg (1998)
3. Fogel, G.: Evolutionary Computation: Towards a New Philosophy of Machine Inteligence, 2nd edn. IEEE Press, ISBN (2000)
4. Holland, J.: Adaptation In Natural and Artificial Systems. University of Michigan Press (1975)
5. Li, J., et al.: Automatic Data Mining by Asynchronous Parallel Evolutionary Algorithms, tools. In: 39th International Conference and Exhibition on Technology of Object-Oriented Languages and Systems (TOOLS39), p. 99 (2001)

6. Keane, A.J.: Genetic algorithm optimization of multi-peak problems: studies in convergence and robustness. Artificial Intelligence in Engineering 9(2), 75–83 (1995)

7. Keane, A.J.: A Brief Comparison of Some Evolutionary Optimization Methods. In: Rayward-Smith, V., et al. (eds.) Modern Heuristic Search Methods, pp. 255–272. J. Wiley, Chichester (1996)

8. Michalewicz, Z., Schoenauer, M.: Evolutionary Algorithms for Constrained Parameter Optimization Problems. Evolutionary Computation 4(1), 1–32 (1996)

9. Michalewicz, Z., Fogel, D.: How to Solve It: Modern Heuristics. Springer, Heidelberg (2002)

10. Penev, K., Littlefair, G.: Free Search — A Novel Heuristic Method. In: Proceedings of PREP 2003, Exeter, UK, April 14-16, 2003, pp. 133–134 (2003)

11. Penev, K.: Adaptive Computing in Support of Traffic Management. In: Adaptive Computing in Design and Manufacturing VI (ACDM 2004), pp. 295–306. Springer, Heidelberg (2004)

12. Penev, K., Littlefair, G.: Free Search — A Comparative Analysis. Information Sciences Journal 172(1-2), 173–193 (2005)

13. Penev, K.: Adaptive Search Heuristics Applied to Numerical optimisation, PhD thesis, Nottingham Trent University, UK (2005)

14. Penev, K.: Heuristic Optimisation of Numerical Functions. In: 20th International Conference Systems for Automation of Engineering and Research, Varna, Bulgaria, 124–129 (September 2006) ISBN-10: 954-438-575-4

15. Penev, K.: Novel Adaptive Heuristic for Search and Optimisation. In: IEEE John Vincent Atanasoff International Symposium on Modern Computing, pp. 149–154. IEEE Computer Society, Los Alamitos (2006)

16. Penev, K.: Free Search Towards Multidimensional Optimisation Problems. In: Guerrero-Bote, V.P. (ed.) Current Research in Information Science and Technologies, Multidisciplinary Approaches to Global Information Systems, vol. 2, pp. 233–237 (2006) ISBN-10: 84-611-3105-3

17. Penev, K.: Free Search and Differential Evolution Towards Dimensions Number Change. In: International Conference Artificial Neural Networks and Intelligent Engineering 2006 (ANNIE 2006), St. Louis, Missouri, USA (2006)

18. Schoenauer, M., Michalewicz, Z.: Evolutionary Computation at the Edge of Feasibility. In: Voigt, H.M., et al. (eds.) PPSN 1996. LNCS, vol. 1141, pp. 245–254. Springer, Heidelberg (1996)

Parameter Estimation of a Monod-Type Model Based on Genetic Algorithms and Sensitivity Analysis

Olympia Roeva

Centre of Biomedical Engineering Prof. Ivan Daskalov,
Bulgarian Academy of Sciences
105 Acad. G. Bonchev Str., 1113 Sofia, Bulgaria,
olympia@clbme.bas.bg

Abstract. Mathematical models and their parameters used to describe cell behavior constitute the key problem of bioprocess modelling, in practical, in parameter estimation. The model building leads to an information deficiency and to non unique parameter identification. While searching for new, more adequate modeling concepts, methods which draw their initial inspiration from nature have received the early attention. One of the most common direct methods for global search is genetic algorithm. A system of six ordinary differential equations is proposed to model the variables of the regarded cultivation process. Parameter estimation is carried out using real experimental data set from an *E. coli MC4110* fed-batch cultivation process. In order to study and evaluate the links and magnitudes existing between the model parameters and variables sensitivity analysis is carried out. A procedure for consecutive estimation of four definite groups of model parameters based on sensitivity analysis is proposed. The application of that procedure and genetic algorithms leads to a successful parameter identification.

1 Introduction

The costs of developing mathematical models for bioprocesses improvement are often too high and the benefits too low. The main reason for this is related to the intrinsic complexity and non-linearity of biological systems. In general, mathematical descriptions of growth kinetics assume hard simplifications. These models are often not accurate enough at describing the underlying mechanisms. Another critical issue is related to the nature of bioprocess models. Often the parameters involved are not identifiable. Additionally, from the practical point of view, such identification would require data from specific experiments which are themselves difficult to design and to realize. The estimation of model parameters with high parameter accuracy is essential for successful model development. All parameter estimation problems involve minimization and the choice of minimization algorithm is problem-dependent. There are many possible variants such as numerical methods [7,14]. During the last decade evolutionary techniques have been applied in a variety of areas. A concept that promises a lot is the genetic

I. Lirkov, S. Margenov, and J. Waśniewski (Eds.): LSSC 2007, LNCS 4818, pp. 601–608, 2008.

technique. Genetic algorithm (GA) is global, parallel, stochastic search method, founded on Darwinian evolutionary principles. Since its introduction, and subsequent popularization [6], the GA has been frequently utilized as an alternative optimization tool to conventional methods. Specific particularities of the considered processes lead to estimation of a large-scale problem and as a successful tool for solving this problem are examined GA. The GA effectiveness and robustness have been already demonstrated for identification of fed-batch cultivation processes [2,15,16,17,18,19].

Even if experimental curves are successfully matched by fermentation process model outputs, it does not imply that the estimates of parameters are unique. Sensitivity analysis is an efficient tool in parameter estimation in fermentation processes models and throws light on the conditions that make parameters identifiable. For mathematical models that involve a large number of parameters and comparatively few responses, sensitivity analysis can be performed very efficiently by using methods based on sensitivity functions [10,11]. The output sensitivity functions (partial derivates of the measured states with respect to the parameters) are central to the evaluation of practical identifiability.

The aim of this paper is to investigate the sensitivity of the *E. coli* model parameters thus to propose particular identification procedure using genetic algorithms. The main purpose is to cope with the problem of adequate estimation of large number of parameters in such complex and non-linear cultivation processes models.

The paper is organized as follows. A dynamic model of an *E. coli* cultivation process using Monod kinetics is described in Section 2. In Section 3 a sensitivity analysis of the model parameters concerning process variables are presented and an identification procedure is proposed. The genetic algorithm performance for parameter estimation is discussed in Section 4. The results and discussion are presented in Section 5. Conclusion remarks are done in Section 6.

2   Mathematical Model

Application of the general state space dynamical model [1] to the *E. coli* cultivation fed-batch process leads to the following system:

$$\frac{dX}{dt} = \mu_{max}\frac{S}{k_S + S}X - \frac{F_{in}}{V}X \tag{1}$$

$$\frac{dS}{dt} = -\frac{1}{Y_{S/X}}\mu_{max}\frac{S}{k_S + S}X + \frac{F_{in}}{V}(S_{in} - S) \tag{2}$$

$$\frac{dA}{dt} = \frac{1}{Y_{A/X}}\mu_{max}\frac{A}{k_A + A}X - \frac{F_{in}}{V}A \tag{3}$$

$$\frac{dpO_2}{dt} = -\frac{1}{Y_{pO_2/X}}\mu_{max}\frac{pO_2}{k_{pO_2} + pO_2}X + k_La^{pO_2}(pO_2^* - pO_2) - \frac{F_{in}}{V}pO_2 \tag{4}$$

$$\frac{dCO_2}{dt} = \frac{1}{Y_{CO_2/X}}\mu_{max}\frac{CO_2}{k_{CO_2} + CO_2}X + k_La^{CO_2}(CO_2^* - CO_2) - \frac{F_{in}}{V}CO_2 \tag{5}$$

$$\frac{dV}{dt} = F_{in} \tag{6}$$

where: X is biomass concentration, [g/l]; S — substrate concentration, [g/l]; A — acetate concentration, [g/l]; pO_2 — dissolved oxygen concentration, [%]; CO_2 — carbon dioxide concentration, [%]; pO_2^* — saturation concentration of dissolved oxygen, [%]; CO_2^* — saturation concentration of carbon dioxide, [%]; F_{in} — feeding rate, [1/h]; V — bioreactor volume, [l]; S_{in} — substrate concentration in the feeding solution, [g/l]; μ_{max} — maximum value of the specific growth rate, $[h^{-1}]$; k_S and k_A, [g/l]; k_{pO_2} and k_{CO_2}, [%] — saturation constants; $k_L a^{pO_2}$, $k_L a^{CO_2}$ — volumetric oxygen transfer coefficients, $[h^{-1}]$; $Y_{S/X}$, $Y_{A/X}$, $Y_{pO_2/X}$ and $Y_{CO_2/X}$ — yield coefficients, [-].

3   Sensitivity Analysis

In order to provide a precisely and accurate model parameter identification a sensitivity analysis using sensitivity functions is carried out [10,11]. If mathematical model (Eqs. (1)–(6)) is presented as follows:

$$\frac{dx_j}{dt} = f_j(x_1, ..., x_m, t, p_1, ..., p_n) \tag{7}$$

Accordingly [10,11] the sensitivity functions are defined as:

$$s_{ji} = \left. \frac{\partial x_j(p, t)}{\partial p_i} \right|_{p=p_0} \tag{8}$$

where s_{ji} are the sensitivity functions of ith parameter according j^{th} variable, x_j — state variables, p_i — model parameters.

The derivatives $\frac{\partial x_j}{\partial p_i}$ are obtained by:

$$\frac{d}{dt} \frac{\partial x_j}{\partial p_i} = \sum_{j=1}^{m} \frac{\partial f_j}{\partial x_j} \frac{\partial x_j}{\partial p_i} + \frac{\partial f_j}{\partial p_i} \tag{9}$$

Mathematical model (Eq. (7)) and sensitivity equations (Eq. (9)) together formed the sensitivity model of considered system. In the considered case $x = [X \quad S \quad A \quad pO_2 \quad CO_2]$ and $p = [\mu_{max} \quad k_S \quad k_A \quad k_{pO_2} \quad k_{CO_2} \quad Y_{S/X} \quad Y_{A/X} \quad Y_{pO_2/X} \quad Y_{CO_2/X} \quad k_L a^{pO_2} \quad k_L a^{CO_2}]$. In solving the sensitivity model of the system the following parameter values are used: $p = [0.46 \ 0.014 \ 0.012 \ 0.01 \ 0.012 \ 0.49 \ 0.015 \ 0.04 \ 0.03 \ 250 \ 200]$. Based on these parameter values two parameter groups, respectively +15% variation and -15% variation are formed. The two sensitivity models are analytically worked out and the sensitivity function are calculated. The results of the sensitivity analysis of the both sensitivity models can be summarized as follows: (i) The biggest sensitivity has the parameter μ_{max}; (ii) After that are the following parameters: $Y_{S/X}$, $Y_{A/X}$, $Y_{pO_2/X}$ and $Y_{CO_2/X}$; (iii) The influence of the rest of parameters is less in comparison with sensitivity of parameters mentioned in (i) and (ii).

Based on these results an particular identification procedure is proposed. The model parameters are divided into four groups according to their sensitivities. First parameters that will be estimated are μ_{max}, k_S, and $Y_{S/X}$. At this step only Eqs. (1), (2), and (3) of the model are used. Second parameters are $Y_{A/X}$ and k_A, using Eqs. (1), (2), (3), and (6). Third parameters that will be estimated are $Y_{pO_2/X}$, k_{pO_2}, and $k_L a^{pO_2}$, using Eqs. (1), (2), (4), and (6). Finally, based on Eqs. (1), (2), (5), and (6), the parameters $Y_{CO_2/X}$, k_{CO_2}, and $k_L a^{CO_2}$ will be estimated. All identification procedures are carried out by applying genetic algorithm tuned for the considered problem.

4 Genetic Algorithm Performance

Difficulties in parameter estimation of cultivation processes arise when the estimation involves many parameters that interact in highly non-linear ways. Objective functions characterized by many local optima, expansive flat planes in multi-dimensional space, points at which gradients are undefined, or when the objective function is discontinuous, pose difficulty for traditional mathematical techniques. Due to the complex non-linear structures of cultivation models, often the parameters involved are not identifiable. In these situations, heuristic methods like GA offer a powerful alternative, and can greatly enhance the set of tools available to researchers. Outline of the used here genetic algorithm could be presented as:

1. [**Start**] Generate random population of n chromosome (suitable solution of the problem).
2. [**Fitness**] Evaluate the fitness $f(x)$ of each chromosome x in the population.
3. [**New population**] Create a new population by repeating the following steps:
   (a) [**Selection**] Select two parent chromosomes from a population according to their fitness (the better fitness, the bigger chance to be selected).
   (b) [**Crossover**] With a crossover probability cross over the parents to form new offspring (children).
   (c) [**Mutation**] With a mutation probability mutate new offspring.
   (d) [**Accepting**] Place new offspring in the new population.
4. [**Replace**] Use new generates population for a further run of the algorithm.
5. [**Test**] If the condition is satisfied, **stop**, and return the best solution in current population.
6. [**Loop**] Go to step 2.

In the GA, there are many operators, functions, parameters, and settings that can be implemented differently in various problems. The adjustment of the GA is based on results in [2,13,15,16,17,18,19]. The first decision to be taken is how to create chromosomes and what type of encoding to be chosen. Binary representation is the most common one, mainly because of its relative simplicity. A binary 20 bit representation is considered here. The next question is how to select parents for crossover. There are many methods for selection of the best chromosomes [3,6,9]. The selection method used here is the roulette wheel selection. Crossover can be quite complicated and depends (as well as the technique of mutation) mainly on the encoding of chromosomes. A double point crossover is used here [13]. In accepted encoding here a bit inversion mutation is used. In proposed genetic algorithm fitness-based reinsertion (selection of offspring) is used.

There are two basic parameters of genetic algorithms — crossover probability and mutation probability. Crossover rate should be high generally, about 65–95%, here — 75%. Mutation is randomly applied with low probability — 0.01 [12,13]. The rate of individuals to be selected — generation gap — should be defined as well. In proposed genetic algorithm generation gap is 0.97 [12,13]. Particularly important parameters of GA are the population size and number of generations. If there is too low number of chromosomes, GA has a few possibilities to perform crossover and only a small part of search space is explored. On the other hand, if there are too many chromosomes, GA slows down. For considered algorithm a number of generations of 200 and a population size of 100 are chosen.

5   Results and Discussion

In practical view, modelling studies are performed to identify simple and easy-to-use models that are suitable to support the engineering tasks of process optimization and especially of control. The most appropriate model must satisfy the following conditions: (i) the model structure should be able to represent the measured data in a proper manner and (ii) the model structure should be as simple as possible compatible with the first requirement. On account of that the cultivation process dynamic is described using simple Monod-type model, the most common kinetics applied for modelling of cultivation processes [1].

For the parameter estimation problem real experimental data of the *E. coli MC4110* fed-batch cultivation process are used. The cultivation condition and the experimental data have been already published [5]. As it is shown, the model consists of six differential equations (Eqs. (1)–(6)) thus represented five dependent state variables (x) and eleven parameters (p). The model parameters are estimated in conformity with the proposed in Section 3 identification procedure. Genetic algorithm using *Genetic Algorithm Toolbox* [3,9] is applied. The optimization criterion is presented as a minimization of a distance measure J between experimental and model predicted values of state variables, represented by the vector x:

$$J = \sum_{i=1}^{n}\sum_{j=1}^{m}\{[x_{exp}(i) - x_{mod}(i)]_j\}^2 \rightarrow \min \qquad (10)$$

In the case of identification of all model parameters most of the numerical methods can not find the correct decision [15,19] whereas the GA is able to find the correct decision but for too much computation time. The estimation of all

Table 1. Identified model parameters

Parameter	Value	Parameter	Value
$\mu_{max}, [h^{-1}]$	0.52	$Y_{A/X}, [\text{-}]$	0.013
$k_S, [g/l]$	0.023	$Y_{pO_2/X}, [\text{-}]$	0.20
$k_A, [g/l]$	0.59	$Y_{pO_2/X}, [\text{-}]$	0.10
$k_{pO_2}, [\%]$	0.023	$k_L a^{pO_2}, [h^{-1}]$	155.87
$k_{CO_2}, [\%]$	0.020	$k_L a^{CO_2}, [h^{-1}]$	53.41
$Y_{S/X}, [\text{-}]$	0.50		

eleven parameters takes more than four hours. Using different parameters groups in identification allows estimation of two or three parameters simultaneously instead of estimation of eleven parameters simultaneously. Thus the computation time is decreased — for proposed GA and identification scheme the computation time is about 10 minutes.

The numerical results from the identification are presented in Table 1. The presented estimates are mean values of 25 runs of the genetic algorithm. The algorithm produce the same estimations with more than 85% coincidence. The resulting parameters values are in admissible range according to [5,8,20].

The model predictions of the state variables are compared to the experimental data points of the real *E. coli MC4110* cultivation. The results are depicted in Fig. 1. Model predicted data are presented with solid line. Presented figure indicates that the model predicts successfully the process variables dynamics during the fed-batch cultivation of *E. coli MC4110*. However, graphical comparisons can clearly show only the existence or absence of systematic deviations between model predictions and measurements. It is evident that a quantitative measure of the differences between calculated and measured values is an important criterion for the adequacy of a model. The most important criterion for the valuation of models is that the deviations between measurements and model calculations (J) should be as small as possible. This criterion cannot be used alone, because it favors the use of complex models with many parameters which are difficult to identify uniquely. For this reason, this criterion has to be complemented by a criterion of 'parsimony' leading to a preference for simple model structures, as example — Fisher criterion (FC) and minimum description length (MDL) criterion [4]. The numerical results for values of these criteria are as follows: $J = 1.23$, $FC = 0.47$ and $MDL = 3.34$.

The obtained criteria values show that the developed model is adequate and has a high degree of accuracy. The proposed estimation procedure leads to simplification of the identification and as a result accurate estimations are obtained. The presented results are a confirmation of successfully application of the proposed identification procedure and of the choice of genetic algorithms.

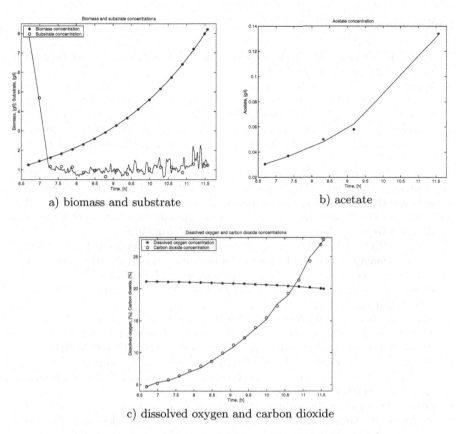

a) biomass and substrate   b) acetate

c) dissolved oxygen and carbon dioxide

Fig. 1. Time profiles of the process variables

6   Conclusion

The use of genetic algorithms in the parameter estimation of nonlinear dynamical model of *E. coli* cultivation process has been investigated in this paper. The identification problem is formulated as an optimization problem. The mathematical model is presented by a system of six ordinary differential equations, describing the regarded process variables. Proposed particular identification procedure, based on an sensitivity analysis, leads to the simplification of the parameter estimation process and to the evaluating of accurate estimates. Numerical and simulation results reveal that correct and consistent results can be obtained using considered procedure and genetic algorithms. The results confirm that the genetic algorithm is powerful and efficient tool for identification of the parameters in the non-linear dynamic model of cultivation processes.

Acknowledgement

This work is supported from National Science Fund Project No. MI–1505/2005.

References

1. Bastin, G., Dochain, D.: On-line Estimation and Adaptive Control of Bioreactors. Els. Sc. Publ., Amsterdam (1991)
2. Carrillo-Ureta, G.E., Roberts, P.D., Becerra, V.M.: Genetic Algorithms for Optimal Control of Beer Fermentation. In: Proc. of the 2001 IEEE International Symposium on Intelligent Control, Mexico City, Mexico, pp. 391–396 (2001)
3. Chipperfield, A.J., Fleming, P.J.: The Matlab Genetic Algorithm Toolbox, IEE Colloquium Applied Control Techniques Using MATLAB, Sheffield, UK, 10/1–10/4 (1995)
4. Garipov, E.: Systems Identification, Technical University, Sofia, Bulgaria (2004)
5. Georgieva, O., Arndt, M., Hitzmann, B.: Modelling of Escherichia coli Fed-Batch Fermentation, International Symposium "Bioprocess Systems", I.61–I.64 (2001)
6. Goldberg, D.: Genetic algorithms in search, optimization and machine learning. Addison-Wesley Publishing Company, Massachusetts (1989)
7. Lagarias, J.C., et al.: Convergence Properties of the Nelder-Mead Simplex Method in Low Dimensions. SIAM Journal of Optimization 9(1), 112–147 (1998)
8. Levisauskas, D., et al.: Model-based Optimization of Viral Capsid Protein Production in Fed-batch Culture of recombinant. Escherichia coli Bioprocess and Biosystems Engineering 25, 255–262 (2003)
9. MatWorks Inc., Genetic Algorithms Toolbox, User's Guide (1999)
10. Müller, T.G., et al.: Parameter Identification in Dynamical Models of Anaerobic Waste Water Treatment. Mathematical Biosciences 177, 147–160 (2002)
11. Noykova, N., Gyllenberg, M.: Sensitivity Analysis and Parameter Estimation in a Model of Anaerobic Waste Water Treatment Process with Substrate Inhibition. Bioprocess Engineering 23, 343–349 (2000)
12. Obitko, M.: Genetic Algorithms (2005), http://cs.felk.cvut.cz/~xobitko/ga
13. Pohlheim, H.: Genetic and Evolutionary Algorithms: Principles, Methods and Algorithms, Technical Report, Technical University Ilmenau (1994-2007), http://www.geatbx.com/docu/algindex.html
14. Press, W.H., et al.: Numerical Recipes — The Art of Scientific Computing. Cambridge University Press, Cambridge (1986)
15. Roeva, O.: Multipopulation genetic algorithm: A tool for parameter optimization of cultivation processes models. In: Boyanov, T., et al. (eds.) NMA 2006. LNCS, vol. 4310, pp. 255–262. Springer, Heidelberg (2007)
16. Roeva, O.: A Modified Genetic Algorithm for a Parameter Identification of Fermentation Processes. Biotechnology and Biotechnological Equipment 20(1), 202–209 (2006)
17. Roeva, O.: Application of Genetic Algorithms in Fermentation Process Identification. Journal of the Bulgarian Academy of Sciences CXVI(3), 39–43 (2003)
18. Roeva, O.: Genetic Algorithms for a Parameter Estimation of a Fermentation Process Model: A Comparison. Bioautomation 3, 19–28 (2005)
19. Roeva, O., St., T.: Parameter Identification of Fermentation Processes using Multipopulation Genetic Algorithms. Technical Ideas XL(3-4), 18–26 (2003)
20. Zelic, B., et al.: Modeling of the Pyruvate Production with Escherichia coli in a Fed-batch Bioreactor. Bioprocess and Biosystems Engineering 26, 249–258 (2004)

Analysis of Distributed Genetic Algorithms for Solving a Strip Packing Problem

Carolina Salto[1], Enrique Alba[2], and Juan M. Molina[2]

[1] Universidad Nacional de La Pampa, La Pampa, Argentina
saltoc@ing.unlpam.edu.ar
[2] Universidad de Málaga, E.T.S.I. Informátic, Grupo GISUM (NEO)
Málaga, España
{eat,jmmb}@lcc.uma.es

Abstract. This paper presents a solution of a constrained two dimensional strip packing problem using genetic algorithms. The constraint consists of considering three-stage guillotine patterns. This is quite a real constraint motivated by technological considerations in some industries. An analysis of including distributed population ideas and parallelism into the basic genetic algorithm is carried out to solve the problem accurately and efficiently. Experimental evidence in this work shows that the proposed parallel versions of the distributed algorithms outperform their sequential counterparts in time, although there are no significant differences either in the mean best values obtained or in the effort.

1 Introduction

The two-dimensional strip packing problem (2SPP) is present in many real-world applications such as in the paper or textile industries, and each of them could impose different constraints and objectives to its basic formulation. The 2SPP can be described as having to pack a set of small rectangular pieces onto a larger rectangle with a fixed width W of unlimited length, designated as the *strip*. There are M different pieces, each piece $j \in \{1, \ldots, M\}$ is characterized by its length and its width, and all pieces have fixed orientation. The search is for a layout of all the pieces in the strip that minimizes the required strip length and, where necessary, takes additional constraints into account.

In some cases, a problem constraint consists in n-stage guillotine packing, where the corresponding packing pattern is built as a series of *levels*: each piece is placed so that its bottom rests on one of these levels. The first level is simply the bottom of the strip. Each subsequent level is defined by a horizontal line drawn through the top of the tallest piece on the previous level. Particularly, in this work, we are focusing on a three-stage guillotine pattern. In the first stage, horizontal cuts (parallel to horizontal edge of the strip) are performed to the strip, producing an arbitrary number of *levels (stripes)*. In the second stage, those levels are processed by vertical cuts generating an arbitrary number of so-called *stacks*. The third stage produces the final elements (and waste) from the stacks by performing only horizontal cuts. Many real application of

I. Lirkov, S. Margenov, and J. Waśniewski (Eds.): LSSC 2007, LNCS 4818, pp. 609–617, 2008.

cutting and packing in the glass, wood, and paper industries consider n-stage guillotine packing (or cutting) patterns, hence the importance of incorporating this restrictions in the problem formulation. Few authors restrict to problems involving guillotine patterns [5,9] and n-stage guillotine patterns [10,11,13,14,15].

Some existing approaches for solving the 2SPP include the utilization of a genetic algorithm (GA) [6,9], among others. But, as it happens frequently in practice, the high complexity of this task poses problems and results in time-consuming scenarios for industrial problems. This gives rise to the application of parallel algorithms not to only reduce the resolution time but also to improve the quality to the provided solutions. Few works on packing address this problem with parallelism techniques [12].

This paper presents a GA with an order-based representation of tentative solutions, problem-dependent genetic operators, and a layout algorithm. In order to reduce the trim loss in each level, an additional final operation, denominated as adjustment operator, is always applied to each generated child. Moreover, the initial population is seeded using a set of rules including information of the problem (such as the piece's width, pieces area, etc.), resulting in a more specialized initial population. In particular, we will analyze the advantages of using a single population (panmixia) versus an algorithm having multiple distributed populations. The main goal of this paper is to present how to build an improved GA to solve larger problems than the ones found in the literature at present, and to quantify the effects of including these operations into the algorithms.

The paper is organized as follows. In Section 2 we will briefly describe parallel GAs for the 2SPP. In Section 3 we present the parameterization used. Then we analyze the results in Section 4. Finally, we summarize the conclusions and discuss the future research in Section 5.

2   Parallel Genetic Algorithms for the 2SPP

In this work we use a distributed GA (dGA) [17], which is a multi-population (island) model performing sparse exchanges of individuals (migration) among the elementary subpopulations P_i. The migration policy must define the island topology, when migration occurs, which individuals are being exchanged, the synchronization among the subpopulations, and the kind of integration of exchanged individuals within the target subpopulations.

In Algorithm 1 we can see the structure of an elementary genetic subalgorithm (dGA_i) in which we will now explain the steps for solving our packing tasks. Each dGA_i creates an initial subpopulation P_i of μ solutions to the 2SPP in a random way, and then evaluates these solutions. The evaluation uses a placement (ad hoc or heuristic) algorithm to arrange the pieces into the strip to construct a feasible packing pattern. After that, the population goes into a cycle where it undertakes evolution, which means the application of genetic operators, to create λ offspring. This cycle also includes an additional phase of individual exchange with a set of neighboring subalgorithms, denoted as dGA_j. Finally, each iteration ends by selecting μ individuals to build up the new subpopulation from the set

Algorithm 1. Distributed Genetic subalgorithm

dGA_i
$t = 0$; {current evaluation}
initialize($P_i(t)$);
evaluate($P_i(t)$);
while (**not** $max_{evaluations}$) **do**
    $P_i'(t)$ = evolve($P_i(t)$); {recombination, mutation and adjustment}
    evaluate ($P_i'(t)$);
    $P_i'(t)$ = send/receive individuals from dGA_j; {interaction with neighbors}
    $P_i(t+1)$ = select new population from $P_i'(t) \cup P_i(t)$;
    $t = t + 1$;
end while

of $\mu + \lambda$ existing ones. The best solution is identified as the best individual ever found which minimizes the strip length needed.

In what follows we will discuss some design issues of the GAs proposed to solve the 2SPP. Issues such as encoding, fitness function, genetic operators and the generation of the initial population must be taken care of in a methodological way in any application of GAs to a given problem.

Representation. We encode a packing pattern into a chromosome as a sequence of pieces that defines the input for the layout algorithm. Therefore, a chromosome will be a permutation $\pi = (\pi_1, \pi_2, ..., \pi_M)$ of M natural numbers (piece identifiers).

In order to generate three-stage guillotine patterns, a modified *next-fit decreasing height* heuristic is used here —in the following referred as *modified next-fit*, or *MNF*— which proved to be very efficient in [11,12]. This heuristic gets a sequence of pieces as its input, not necessarily decreasing height sorted, and constructs the packing pattern by placing pieces into stacks, and then stacks into levels in a greedy way, i.e., once a new stack or a new level is started, previous ones are never reconsidered. Deeper explanation of the MNF procedure can be found in [15].

Fitness Function. In our problem, the objective is to minimize the strip length needed to build the layout corresponding to a given solution π. An important consideration is that two packing patterns could have the same length —so their fitness will be equal— although, from the point of view of reusing the trim loss, one of them can be actually better because the trim loss in the last level (which still connects with the remainder of the strip) is greater than the one present in the last level of the other layout. In order to distinguish these situations we are using the following fitness function:

$$F(\pi) = strip.length - \frac{l.waste}{l.area} \qquad (1)$$

where *strip.length* is the length of the packing pattern corresponding to the permutation π, and *l.area* and *l.waste* are the areas of the last level and of the reusable trim loss in the last level, respectively. Hence, $F(\pi)$ is both simple and accurate.

Recombination Operator. The Best Inherited Level Recombination (BIL) [15] transmits the best levels of the parent to the child, i.e. those with the highest filling rate (fr) or, equivalently, with the least trim loss. This rate is calculated as follows, for a given level l:

$$fr(l) = \sum_{i=1}^{n} \frac{width(\pi_i) \times length(\pi_i)}{W \times l.length} \tag{2}$$

where $\pi_1, ..., \pi_n$ are the pieces in l, $width(\pi_i)$ and $length(\pi_i)$ are the piece dimensions, and W and $l.length$ the level dimensions. Actually, BIL recombination works as follows. In the first step the filling rates of all levels from one parent, $parent_1$, are calculated. After that, a selection probability for each level l, proportional to the filling rate, is determined. A number k of levels are selected from $parent_1$ by proportional selection according to their filling rate. The pieces π_i belonging to the inherited levels are placed in the first positions of the child. Meanwhile, the remaining positions are filled with the pieces which do not belong to that levels, in the order they appear in the other parent $parent_2$.

Mutation Operator. Best and Worst Stripe Exchange (BW_SE) [15] mutation changes the location of the best and the worst level, so that the final cost is reduced. The pieces of the best level (the one with highest filling rate) are allocated in the first positions of the new packing pattern while the pieces of the worst level are assigned to the last positions. The middle positions are filled with the remaining pieces in the order they appeared in the original packing pattern. In BW_SE, the movements can help to the involved levels or their neighbors to accommodate pieces from neighboring levels, thus improving their trim loss.

Adjustment Operator. Given a solution, the operator *MFF_Adj* [15] consists of the application of a modified *first-fit decreasing height* (FFDH) heuristics, with the aim of improving the filling rate of all levels. The possible new layout obtained in this way has to be transmitted to the chromosome in such a way that we can obtain the same layout by applying MNF to the chromosome. MFF_Adj works as follows. It considers the pieces in the order given by the permutation π. The piece π_i is packed into the first stack in the first level it fits, as in MNF. If piece π_i does not fit into any existing stack and there is room enough in that level, a new stack is created, as in MNF. Otherwise the following levels are considered and checked in the previous order. If no space were found, a new stack containing π_i is created and packed into a new level in the remaining length of the strip. The above process is repeated until no piece remains in π.

Initial Seeding. The search process is started from a specialized initial population, created by following some building rules, hopefully allowing to reach good solutions in early stages of the search. The rules (see Table 1) will include some characteristics from the problem such as piece sizes, and also incorporate ideas from the *best fit* (BF) and *first fit* (FF) heuristics [7]. These rules are proposed with the aim of producing individuals with improved fitness values and also for introducing diversity in the initial population. Individuals are generated in two steps. In the first step, the packing patterns are randomly sampled from the

Table 1. Rules to generate the initial population

#	Rule Description	#	Rule Description
1	sorts pieces by decreasing width.	2	sorts pieces by increasing width.
3	sorts pieces by decreasing length.	4	sorts pieces by increasing length.
5	sorts pieces by decreasing area.	6	sorts pieces by increasing area.

#	Rule Description
7	sorts pieces by alternating between decreasing width and height.
8	sorts pieces by alternating between decreasing width and increasing height.
9	sorts pieces by alternating between increasing width and height.
10	sorts pieces by alternating between increasing width and decreasing height.
11	the pieces are reorganized following the BFDH heuristic.
12	the pieces are reorganized following the FFDH heuristic.
13	The packing pattern remains without modifications, so here the rule preserves the original piece position (random generation).

search space with a uniform distribution. After that, each of them is modified by one rule, randomly selected, with the aim of improving the piece location inside the random packing pattern. Each application of a rule yields a (possibly) different solution because of the randomization used in the first step.

3 Implementation

The specific GA we have implemented is a steady state, or a $(\mu+1)$-GA, where only one new solution is built in each step, with a binary tournament selection for each parent. The new generated individual in every step replaces the worst individual in the population only if it is fitter. We wanted to compare the following approaches: a panmitic (sequential) GA (seqGA) and some distributed GAs or dGAn, where n indicates the number of islands of the model. In this test we used two, four and eight islands. The dGAs were run on a single processor system and on a cluster of workstations, where each island was assigned to a different processor.

In our GAs, the whole population was composed of 512 individuals and each island had a population of $512/n$ individuals, where n is the number of islands. The stopping criterion was based on the quality of the final solution or when the maximum number 2^{16} of evaluations were reached because we wanted to measure the time to find equivalent solutions among the GAs versions. The probability of recombination was 0.8, the mutation probability 0.1, and the adjustment probability was set to 1.0. The method of seeding the initial population consists of an uniform decision of what rule from a set of problem aware rules should be used for seeding. In dGAs, one migrant is randomly selected to be exchanged to the neighboring subpopulation, while the target island selects the worst individual to be replaced with the incoming one (only if it is better). The migration frequency was set to 1,024 evaluations. The subalgorithms were disposed in a unidirectional ring with asynchronous communications (individuals are integrated into the population whenever they arrive) for efficiency. The above parameters had been previously tuned [15,14], but we do not include the tuning details in the article due to room restrictions. These algorithms were run in MALLBA [3], a C++ software library fostering rapid prototyping of hybrid and parallel algorithms, and the platform was an Intel Pentium 4 at 2.4 GHz and 512 MB RAM, linked by Fast Ethernet, under SuSE Linux with 2.4.19-4GB kernel version.

We have considered five randomly generated problem instances with M equal to 100, 150, 200, 250 and 300 pieces and a known global optimum equal to 200 (the minimal length of the strip). These instances belong to the 2SPP only and were generated as guillotine patterns by an own implementation of a data set generator, following the ideas proposed in [18] with the length-to-width ratio of all rectangles in the range $1/3 \leq l/w \leq 3$ (publicly available at http://mdk.ing.unlpam.edu.ar/~lisi/2spp.htm). As the optimum value does not correspond to the three-stage guillotine pattern constraint, then the target fitness to reach for instance corresponds to the average of the best fitness found in previous works for a maximum of 2^{16} evaluations. These values were 217.99, 216.23, 213.12, 213.59 and 213.90 for instances 100, 150, 200, 250 and 300 respectively.

4   Computational Analysis

Let us proceed with the analysis of the results. For each algorithm we have performed 50 independent runs per instance. Table 2 shows a summary of the results for the seqGA and each dGAn running both in sequential (1 processor) and in parallel (n processors). The most relevant aspects that were measured in this comparison are the following ones: the number of times each algorithm reached the target value for each instance (column $hits$), the average values of the best found feasible solutions along with their standard deviations (column avg), the average number of evaluations (column $eval$) and the average run time expressed in seconds (column t/s). Also in this table we report the speedup (column s), the efficiency (column e), and the serial fraction (column sf) using the orthodox definition of speedup of [1] (comparing the same algorithm, dGAn, both in sequential and in parallel).

From this table we can infer that there are a significant difference between dGAn and seqGA regarding the numerical effort to solve the problem (the ANOVA test is always significant: p-values well below 0.05) corroborating that this two algorithms performed a different search process. For example, seqGA samples near three times more the number of points in the search space than dGA2 before locating good solutions. Also, a decrease in the number of evaluations is observed as the number of islands n increases, independently of whether the dGAn were run on 1 or n processors. Overall, it seems that the two versions needed a similar effort to solve all the instances (the t-test for this column gave p-values greater than 0.05). Hence we cannot conclude anything about the superiority of any of the two dGAn versions.

With respect to the number of hits: dGA4 and dGA8 running in parallel reached higher numbers than their counterparts running on 1 processor, dGA2

Table 2. Experimental results

Inst	Alg	1 processor				n processors				s	e	sf
		hits	avg±σ	eval	t[s]	hits	avg±σ	eval	t[s]			
100	seqGA	50	218.03 ± 0.19	1325.96	426.32							
	dGA2	60	217.99 ± 0.00	588.88	441.85	54	217.99 ± 0.00	470.56	225.01	1.96	0.98	0.02
	dGA4	36	217.99 ± 0.00	270.00	419.61	46	217.99 ± 0.00	277.44	109.22	3.84	0.96	0.01
	dGA8	44	218.01 ± 0.25	138.04	398.86	62	218.01 ± 0.14	267.08	52.92	7.54	0.94	0.01
150	seqGA	64	216.18 ± 0.68	1333.30	563.42							
	dGA2	72	216.09 ± 0.68	753.70	476.74	68	216.11 ± 0.72	696.02	274.05	1.74	0.87	0.15
	dGA4	50	216.37 ± 0.70	326.74	833.41	72	216.09 ± 0.68	325.94	120.83	6.90	1.72	-0.14
	dGA8	62	216.25 ± 0.72	191.60	631.60	64	216.27 ± 0.61	201.98	100.73	6.27	0.78	0.04
200	seqGA	78	213.22 ± 0.43	1091.23	824.13							
	dGA2	80	213.19 ± 0.40	525.26	709.60	70	213.27 ± 0.49	570.54	559.55	1.27	0.63	0.58
	dGA4	76	213.23 ± 0.43	291.68	872.25	76	213.19 ± 0.49	267.14	222.69	3.92	0.98	0.01
	dGA8	58	213.20 ± 0.42	139.22	800.72	62	213.37 ± 0.49	143.30	189.42	4.23	0.53	0.13
250	seqGA	38	213.68 ± 0.55	2288.15	4001.02							
	dGA2	28	213.72 ± 0.49	869.86	4433.81	34	213.73 ± 0.60	856.44	2049.38	2.16	1.08	-0.08
	dGA4	18	214.03 ± 0.64	403.54	5242.06	26	213.83 ± 0.59	510.02	1274.57	4.11	1.03	-0.01
	dGA8	24	213.83 ± 0.55	203.88	4851.44	36	213.63 ± 0.49	263.20	539.87	8.99	1.12	-0.02
300	seqGA	6	213.94 ± 0.19	1171.35	11014.84							
	dGA2	4	213.90 ± 0.34	426.68	11256.02	0	214.01 ± 0.20	424.89	5958.72	1.89	0.94	0.06
	dGA4	2	213.95 ± 0.11	291.85	11369.00	4	213.93 ± 0.20	236.71	2836.35	4.10	1.03	-0.01
	dGA8	4	213.93 ± 0.22	125.14	11010.61	4	213.91 ± 0.31	278.86	1420.12	8.16	1.02	0.00

running on 1 processor reached a higher number of hits than the corresponding algorithm running on 2 processors —except for the instance 250—, and dGA4 had the lowest number of hits for each instance. As we expected, there are no significative differences in mean best values in running each dGAn in sequential or in parallel, since the two versions of dGAs correspond to the same algorithm, and only the execution time should be affected by the different number of processors used. These conclusions are supported by a t-test.

The speedup is quite high. In instances 250 and 300, with $n = 4, 8$, the speedup is slightly superlinear. This results indicate that we are using a good parallel implementation. There are a reduction of efficiency for the instance 200. As expected in a well-parallelized algorithm, the serial fraction is quite stable, although we can notice a reduction of this value as the number of processors increases (except in the instance 200).

5   Conclusions

In this work we have shown how distributed genetic algorithms can be used to give a solution to the strip packing problem using three-stage guillotine patterns. The characteristics of the distributed search have been shown to lead to fast techniques computing accurate results, which represents a promising advance in this area. The distributed algorithms were capable of a higher numerical performance (lower efforts) with similar levels of accuracy with respect to the sequential panmitic algorithm. Also we have shown the high speedup of the distributed GAs proposed running in parallel, since we firmly believe that time is very important in the research in this area which is actually aimed at a practical utilization.

As future work we plan to investigate non-permutation representations with a direct mapping to the final layout of the pieces and to use longer problem instances.

Acknowledgements

This work has been partially funded by the Spanish Ministry of Education and the European FEDER under contract TIN2005-08818-C04-01 (the OPLINK project, http://oplink.lcc.uma.es). We also acknowledge the Universidad Nacional de La Pampa, and the ANPCYT in Argentina from which we received continuous support.

References

1. Alba, E.: Parallel evolutionary algorithms can achieve super-linear performance. Information Processing Letters 82(1), 7–13 (2002)
2. Alba, E.: Parallel Metaheuristics: A New Class of Algorithms. Wiley, Chichester (2005)
3. Alba, E., et al.: MALLBA: A Library of Skeletons for Combinatorial Optimisation. In: Monien, B., Feldmann, R.L. (eds.) Euro-Par 2002. LNCS, vol. 2400, pp. 63–73. Springer, Heidelberg (2002)
4. Bäck, T., Fogel, D., Michalewicz, Z.: Handbook of Evolutionary Computation. Oxford University Press, New York (1997)
5. Bortfeldt, A.: A genetic algorithm for the two-dimensional strip packing problem with rectangular pieces. EJOR 172(3), 814–837 (2006)
6. Hopper, E., Turton, B.: An empirical investigation of meta-heuristic and heuristic algorithms for a 2D packing problem. EJOR 128(1), 4–57 (2000)
7. Lodi, A., Martello, S., Monaci, M.: Recent advances on two-dimensional bin packing problems. Discrete Applied Mathematics 123, 379–396 (2002)
8. Michalewicz, M.: Genetic Algorithms + Data Structures = Evolution Programs, 3rd edn. Springer, Heidelberg (1996)
9. Mumford-Valenzuela, C.L., Vick, J., Wang, P.Y.: Metaheuristics: Computer Decision-Making. In: Chapter Heuristics for large strip packing problems with guillotine patterns: An empirical study, pp. 501–522 (2003)
10. Puchinger, J., Raidl, G.R.: Models and algorithms for three-stage two-dimensional bin packing. Technical Report TR-186-04-04, Technische Universität Wien, Institut für Computergraphik und Algorithmen (2004)
11. Puchinger, J., Raidl, G.R., Koller, G.: Solving a Real-World Glass Cutting Problem. In: Gottlieb, J., Raidl, G.R. (eds.) EvoCOP 2004. LNCS, vol. 3004, pp. 165–176. Springer, Heidelberg (2004)
12. Salto, C., Molina, J.M., Alba, E.: Sequential versus distributed evolutionary approaches for the two-dimensional guillotine cutting problem. In: Proc. of International Conference on Industrial Logistics, pp. 291–300 (2005)
13. Salto, C., Molina, J.M., Alba, E.: Analysis of distributed genetic algorithms for solving cutting problems. ITOR 13(5), 403–423 (2006)
14. Salto, C., Molina, J.M., Alba, E.: A comparison of different recombination operators for the 2-dimensional strip packing problem. In: Proc. of the XII Congreso Argentino de Ciencias de la Computación, pp. 1126–1138 (2006)

15. Salto, C., Molina, J.M., Alba, E.: Evolutionary algorithms for the level strip packing problem. In: Proc. of the Workshop on Nature Inspired Cooperative Strategies for Optimization, pp. 137–148 (2006)
16. Spiessens, P., Manderick, B.: A massevily parallel genetic algorithm. In: Proc. of the 4th. International Conference on Genetic Algorithms, pp. 279–286 (1991)
17. Tanese, R.: Distributed genetic algorithms. In: Proc. of the 3rd. International Conference on Genetic Algorithms, pp. 434–439 (1989)
18. Wang, P.Y., Valenzuela, C.L.: Data set generation for rectangular placement problems. EJOR 134, 378–391 (2001)

Computer Mediated Communication and Collaboration in a Virtual Learning Environment Based on a Multi-agent System with Wasp-Like Behavior

Dana Simian[1], Corina Simian[2], Ioana Moisil[1], and Iulian Pah[2]

[1] University "Lucian Blaga" of Sibiu, Romania
[2] University Babeş Bolyai of Cluj-Napoca, Romania

Abstract. In this paper is presented a model for an adaptive multi-agent system for dynamic routing of the grants' activities from a learning environment, based on the adaptive wasp colonies behavior. The agents use wasp task allocation behavior, combined with a model of wasp dominance hierarchy formation. The model we introduced allows the assignment of activities in a grant, taking into account the specialization of students, their experience and the complexity of activities already taken. An adaptive method allows students to enter in the Grant system for the first time. The system is changing dynamic, because both the type of activities and the students involved in the system change. Our approach depends on many system's parameters. For the implementation these parameters were tuned by hand. The Grant-system we built is integrated in a virtual education system, student centered, that facilitates the learning through collaboration as a form of social interaction.

Keywords: Multi-Agent System, E-learning, Wasp Models.

1 Introduction

In [6] was developed a model for the virtual education system, student centered, that facilitates the learning through collaboration as a form of social interaction. The general architecture of the e-Learning system proposed there, is one with three levels (user, intermediary, supplier educational space), to each corresponding heterogeneous families of human agents and software. The teacher (human agent) is assisted by two types of software agents: personal assistant (classic interface agent), with role of secretary and didactic assistant, which is the assistant from the classical educational system. The SOCIAL agentified environment has a social agent and a database with group models (profiles of social behavior). The social agent has as main aim the construction of models for the groups of students who socialize in the virtual educational environment. The agentified DIDACTIC environment assists the cognitive activities of the student and/or of the teachers. Within this environment a Web searching agent evolves together with a semiotic agent who stimulates the interceding agent of the student sending him pictogram type stimuli, text, numbers. The environment is endowed

I. Lirkov, S. Margenov, and J. Waśniewski (Eds.): LSSC 2007, LNCS 4818, pp. 618–625, 2008.

with a collection of instruments and signs recorded in a knowledge base. The student (human agent) evolves in an agentified environment with three types of agents. He also has a personal assistant (software interface agent) who monitors all the students' actions and communicates (interacts) with all the other agents, with the agentified environments of other students and the TEACHER agentified environment. The student has at his disposal two more agents: TUTOR and the mediating agent. The TUTOR assistant evaluates the educational objectives of the student and recommends her/him some kind of activities. The decisions are based on the knowledge of the students' cognitive profile (which takes into account the social component). The TUTOR agent interacts with the personal assistant of the student, with the mediating agent and with the social agentified environment. Student population is considered a closed one and individuals are separated into groups, called classes. Students from a class communicate one with another and also with students from other classes. We will have intra-class and inter-class communication models and a different student-software agent communication model. A class consists of several teams. Students interact through their computers and with the software agents in order to achieve a common educational goal.

We want to enlarge this educational system adding a component named GRANTS, which allows students to participate to some projects or grants, depending on their qualification. The qualification of a student, on a certain area is given by the tests he had passed. A test is passed if the associated score is situated between two values, the minimum and the maximum value. When a student chooses the courses he wants to take in a period (a week, a month, a semester, a year) a zero qualification variable is assigned for this student at the chosen course. Then, the qualification of the student i for the course j is computed as:

$$q_{i,j} = \frac{p_{i,j} - c_{j,min}}{c_{j,max} - c_{j,min}}, \tag{1}$$

with $p_{i,j}$ being the score obtained by the student i to the test associated to the course j. Every course has specified a minimum and a maximum score: $c_{j,min}$, $c_{j,max}$.

The complexity of a grant is given in complexity points associated to each activity. Every activity requires a qualification in one or many courses, case in which an average qualification for all the required courses is used.

A student can be involved in many activities of many grants, such that the total number of complexity points for these activities don't exceed a given maximum value. That is, many activities may exists in a student's grant queue.

The time period for every grant is strictly determinated.

The aim of this paper is to build a multi-agent virtual environment where agents use wasp task allocation behavior, combined with a model of wasp dominance hierarchy formation, to determine which activity of a set of grants should be accepted into a student's queue, such that the execution time of every grant be respected and the number of students involved in these grants be maximized. Wasp-like computational agents that we call learning routing wasps act as overall student proxies. The policies that the learning routing wasps adapt for their student are the policies for deciding when to bid or when not to bid for arriving activities.

Our environment is a dynamic one, because both new grants and new qualified students appears in time.

Effective coordination of multiple agents interacting in dynamic environments is an important part of many practical problems. Few of these problems are presented in [5]. Our system has many common characteristics with a distributed manufacturing system. The desire for a more robust basis for coordination has motivated research into agent-based approaches to manufacturing scheduling and control [4,7]. Looking to our system from this point of view we can associate grants' activities with the factory commands and the students with the factory machines. The main difference is that in our system not only the commands have a dynamic behavior but also the number and the type of machines. That is why the coordination policies in our system must be viewed as an adaptive process.

The adaptive behavior in many natural multi-agents systems has served as inspiration for artificial multi-agents systems. A survey of adaptive multi-agent systems that have been inspired by social insect behavior, can be found in [1].

The paper is organized as follows. In the part 2 we present the model of wasp behavior and a brief survey of the papers in which this model is used. The section 3 contains the main results: our model for an adaptive multi-agent system which makes a dynamic allocation of grants' activities in a learning environment. In the section 4 we present the conclusions.

2   Wasp Behavior Model

Our approach for the Grant System is based on the natural multi-agent system of a wasp colony. Theraulaz et al. present a model for self-organization within a colony of wasps [9]. In a colony of wasps, individual wasp interacts with its local environment in the form of a stimulus-response mechanism, which governs distributed task allocation. An individual wasp has a response threshold for each zone of the nest. Based on a wasp's threshold for a given zone and the amount of stimulus from brood located in this zone, a wasp may or may not become engaged in the task of foraging for this zone. A lowest response threshold for a given zone amounts to a higher likelihood of engaging in activity given a stimulus.

In [3] is discussed a model in which these thresholds remain fixed over time. Later, in [8] is considered that a threshold for a given task decreases during time periods when that task is performed and increases otherwise. In [5], Cicirello and Smith, present a system which incorporates aspects of the wasp model which have been ignored by others authors. They consider three ways in which the response thresholds are updated. The first two ways are analogous to that of real wasp model. The third is included to encourage a wasp associated with an idle machine to take whatever jobs rather than remaining idle.

The model of wasp behavior also describes the nature of wasp-to-wasp interaction that takes place within the nest. When two individuals of the colony encounter each other, they may with some probability interact in a dominance contest. The wasp with the higher social rank will have a higher probability of dominating in the interaction. Wasps within the colony self-organize themselves into a dominance hierarchy. In [5] is incorporated this aspect of the behavior model, that is when two or more of the wasp-like agents bid for a given job, the winner is chosen through a tournament of dominance contests.

3   Main Results. Learning Routing Wasps

In this section we present our approach to the problem of allocating dynamically the activities from many grants, to qualified students, such that the time period allocated to every grant be respected and the number of students involved, be maximized. We next define the problem's terms.

The student i has one or more course qualifications $q_{i,j}$ given by the equality (1) and can be involved in various activities from grants such that the sum of the complexity points for these activities must not exceed a limit value:

$$MCPS = \text{Maximum Complexity Points /student} \qquad (2)$$

Each activity of a grant has associated a number of complexity points:

$$ncp_{j,k} = \text{number of complexity points for the activity } j \text{ in the grant } k \qquad (3)$$

and a set of courses which define the area of each activity.

$$A_{j,k} = \{i_{1,j,k}, \ldots, i_{n_{j,k},j,k}\} \qquad (4)$$

is the set of indexes of courses required by the activity j from the grant k.
The minimum and maximum score that allow a student to realize the activity j, from the grant k are:

$$min_{j,k} = \sum_{l \in A_{j,k}} c_{l,min}/N_{j,k}, \qquad (5)$$

with $N_{j,k} = \#A_{j,k}$.

$$max_{j,k} = \sum_{l \in A_{j,k}} c_{l,max}/N_{j,k} \qquad (6)$$

The activities are classified using the type of activities. First, in the system are introduced a number of activity types, characterized by the sets T_m, which contain the courses that were required by these types.

$$T_m = \{c_{i_1}, \ldots, c_{i_{k_m}}\} \qquad (7)$$

The intersection of two sets of this kind has the following property:

$$\#(T_m \bigcap T_n) \leq p\% \cdot \min(\#T_m, \#T_n) \qquad (8)$$

Each activity belongs to an unique type. The value p is a dynamical system parameter, that is, it is modified in a dynamical way such that every activity which is in the system in every moment, belongs to an unique activity type set. We will denote by

$$NT(t) = \#T = \#\{T_1, \ldots\} \text{ at the moment } t \qquad (9)$$

and by

$$ta_{j,k} = m, \qquad (10)$$

the type of activity j from the grant k (the activity j, from the grant k is of type T_m). To each student i we associate two sets of indexes:

$$M_i = \bigcup_{student\ i} ta_{j,k} \qquad (11)$$

is the set of all types of activities in which he is or was involved.

$$M_{i,f} \subseteq M_i \qquad (12)$$

is the set of all types of activities he had already finished. The qualification of a student for the activity j, from the grant k is

$$qa_{i,j,k} = average\{q_{i,l} | l \in A_{j,k}\} \qquad (13)$$

If the incoming flow of new activities allows, then, ideally each of the students should specialize to one or more types of activities among the ones he is capable to do. To model this requirement, we introduce the activity specialization of a student. It takes into account the qualification of the student for the courses required by this activity and the participation to other likewise activities.

$$s_{i,j,k} = \omega \cdot qa(i,j,k) + \alpha \cdot \sum_{j \in M_i \bigcap \{ta_{j,k}\}} qa(i,j,k) + \beta \sum_{j \in M_{i,f} \bigcap \{ta_{j,k}\}} qa(i,j,k), \qquad (14)$$

where ω, α, β are parameters of the system and have a major role in modelling of what "specialization" must represent. For the first step of our system modelling, these parameters will be tuned by hand. If we choose $\omega = 1$, α, $\beta > 1$ it means that the experience of student is more important than the initial score from different courses required from the activities. If we choose $\alpha = 0$, $\beta = 0, \omega = 1$ it means that only the initial qualification is taking into account.

We denote by

$$sq_i = \text{number of activities in the queue of student } i \qquad (15)$$

This number satisfies the restriction

$$\sum_{l=1}^{sq(i)} ncp_{l,k_l} \leq MCPS \qquad (16)$$

Each student in our system has an associated *learning routing wasp*. Each routing wasp is in charge of choosing which activity to bid for possible assignment to the queue of its associate student. Each learning routing wasp has a set of response thresholds, as like in the wasp behavior model. The response thresholds, are associated to every type of activity

$$W_i = \{w_{i,j,k}\} \tag{17}$$

where $w_{i,j,k}$ is the response threshold of wasp associated to student i to activity j from the grant k. Every activity from the system has associated for each wasp such a response thresholds. The threshold value $w_{i,j,k}$ may vary in the interval $[w_{min}, w_{max}]$.

Activities in the system that have not been assigned yet to a student and that are awaiting assignment, broadcast to all of the learning routing wasp a stimulus $S_{j,k}$, which is proportional to the length of time the activity has been waiting for assignment to a student. The learning routing wasp i will bid for an activity k only if

$$\sum_{l \in A_{j,k}} q_{i,l}/N_{j,k} \geq min_{j,k} \tag{18}$$

In this case the learning routing wasp i will bid for this activity with probability

$$P(i,j,k) = \frac{S_{j,k}^{\gamma}}{S_{j,k}^{\gamma} + w_{i,j,k}^{\gamma}} \tag{19}$$

The exponent γ is a system parameter. In [8] such a rule for task allocation is used with $\gamma = 2$. If, in this rule, $\gamma \geq 1$, then the lower the response thresholds is, the bigger the probability of binding an activity is. But, using this rule, a wasp can bid for an activity if a hight enough stimulus is emitted.

Each learning routing wasp, at all times, knows what its student is doing: the status of the queue, the characteristics of the activity that is realized (all the variables associated to activity, that is $A_{j,k}$, $ncp_{j,k}$, $qa_{i,j,k}$, $ta_{j,k}$, and if the student is idle). This knowledge is necessary in order to adjust the response thresholds for the various activities. This update occurs at each time step. If the student i is currently realizing an activity of the same type of the activity $a_{j,k}$, or is in process of starting up this activity, then

$$w_{i,j,k} = w_{i,j,k} - \delta_1 \tag{20}$$

If the student is involved in other type of activity, then

$$w_{i,j,k} = w_{i,j,k} + \delta_2 \tag{21}$$

If the student is currently idle and has empty queue then

$$w_{i,j,k} = w_{i,j,k} - \delta_3^{\tau}, \tag{22}$$

where τ is the length of time the student has been idle and is an exponent.

The δ_1, δ_2 and δ_3 are positive system constants.

Therefore, wasp stochastically decides whether or not to bid for the activity, according to the type of activity, the length of time the activity has been waiting and the response threshold. The response thresholds for the activity type currently being realized are reinforced as to encourage the learning routing wasp to bid on activity of the same type. The equation (22) encourages a wasp associated to an idle student to take whatever activity it can get, rather than remaining idle. This equation makes easier the integration of a student just entered in the GRANT system, and helps him to get his first activity.

The main characteristics of our system, which differentiates it from the system in [5], is that the number of activity's type dynamically changes in time and that the number of activities from a student queue depends on the restriction (16).

If two or more learning routing wasps respond positively to the same stimulus, that is bid for the same activity, these learning routing wasps enter in a dominance contest. We introduced a method for deciding which learning routing wasp from a group of competing wasps gets the activity. We take into account the student specialization for the activity $a_{j,k}$, computed in (14), with an adaptive choice of parameters ω, α and β. If $M_i = \emptyset$, then it will be chosen $\omega > 1$, to encourage students to enter in GRANT system, otherwise, the parameters ω, α and β will have constant values, depending on the specialization policy selected for the GRANT system. We define for a learning routing wasp the force F_i, as:

$$F_i = 1 + \sum_{a_{j,k} \in \; queue(i)} \left(ncp_{j,k} + \frac{1}{s_{i,j,k}} \right) \tag{23}$$

where, $s_{i,j,k}$ is the student specialization given in (14).

Let i and l be the learning routing wasps in a dominance contest. Learning routing wasp i will get the activity with probability

$$P_c(i,l) = P(Wasp \; i \; win \; |F_i, F_l) = \frac{F_l^2}{F_i^2 + F_l^2} \tag{24}$$

In this way, learning routing wasps associated with students of equivalent specializations and equivalent complexity of the activity in their queue, will have equal probabilities of getting the activity. For the same specialization, if the complexity of the activities in the queue is different, then the wasp with the smaller complexity has a higher probability of taking the new activity. For the same complexity of the activities in the queue the wasp with the higher specialization has a higher probability of taking the new activity. If the specialization of a wasp is lower but the complexity of the activity in its queue is also lower, the probability for this wasp may increase.

4   Conclusions

In this paper a model for an adaptive multi-agent system for dynamic routing of the grants' activities from a learning environment, based on the adaptive wasp colonies behavior, is presented. The model we introduced allows the assignment of activities taking into account the specialization of students, their experience and the complexity of activities already taken. An adaptive method allows students to enter in the Grant system for the first time. The system is changing dynamically, because both the type of activities and the students involved in the system change. Our approach depends on many system parameters.

For the implementation these parameters were tuned by hand. The next direction of our studies is to compare the results obtained for different sets of parameters and then, to use meta-level optimization of the control parameters. Another paper will be dedicated only to the implementation aspects and to the analysis of the results for different sets of system parameters.

Acknowledgement

This work benefits from founding from the research grant of the Romanian Ministry of Education and Research, code CNCSIS 33/2007.

References

1. Bonabeau, E., Dorigo, M., Theraulaz, G.: Swarm intelligence: From natural to artificial systems. Santa Fe Institute Studies of Complexity. Oxford University Press, Oxford (1999)
2. Bonabeau, E., Dorigo, M., Theraulaz, G.: Inspiration for optimization from social insect behaviour. Nature 406, 39–42 (2000)
3. Bonabeau, E., Theraulaz, G., Demeubourg, J.I.: Fixed response thresholds and the regulation of division of labor in insect societies. Bull. Math. Biol. 60, 753–807 (1998)
4. Bussemann, S.: Agent — oriented programming of manufacturing control tasks. In: Proceeding of the 3rd International Conference on Multi-Agent Systems, ICMAS 1998, pp. 57–63 (1998)
5. Cicirelo, V.A., Smith, S.F.: Wasp-like Agents for Distributed Factory coordination. Autonomous Agents and Multi-Agent Systems 8(3), 237–267 (2004)
6. Moisil, I., et al.: Socio-cultural modelling of student as the main actor of a virtual learning environment. WSEAS Transaction on Information Science and Applications 4, 30–36 (2007)
7. Parunak, V., Baker, A., Clark, S.: The AARIA agent architecture. From manufacturind requirements to agent-based system design. In: Proceedings of the ICAA 1998 Workshop on Agent based Manufacturing (1998)
8. Theraulaz, G., Bonabeau, E., Demeubourg, J.I.: Response threshold reinforcement and division of labour in insects societies. Proc. R Spob London B. 265(1393), 327–335 (1998)
9. Theraulaz, G., et al.: Task differention in policies waspcolonies. A model for self-organizing groups of robots. In: From animals to Animats: Proceedings of the First International Conference on Simulation of Adaptive behavior, pp. 346–355 (1991)

Design of 2-D Approximately Zero-Phase Separable IIR Filters Using Genetic Algorithms

F. Wysocka-Schillak

University of Technology and Life Sciences,
Institute of Telecommunications, al. Prof. S. Kaliskiego 7,
85-796 Bydgoszcz, Poland
felicja@mail.atr.bydgoszcz.pl

Abstract. The paper presents a method for designing approximately zero-phase 2-D IIR filters with a quadrantally symmetric magnitude response. The method is based on two error criteria: equiripple error criterion in the passband and least-squared error criterion in the stopband. The filter design problem is transformed into an equivalent bicriterion optimization problem which is converted into a single criterion optimization problem using the weighted sum strategy. The stability constraints are explicitly included into this problem. A two-step solution procedure of the considered problem is proposed. In the first step, a genetic algorithm is applied. The final point from the genetic algorithm is used as the starting point for a local optimization method. A design example is given to illustrate the proposed technique.

1 Introduction

In recent years, design and implementation of two-dimensional (2-D) digital filters have been extensively investigated. There are two types of 2-D digital filters: finite impulse response (FIR) and infinite impulse response (IIR). IIR filters can have considerably lower order than FIR filters with similar performance, but in case of causal IIR filters, zero-phase or linear-phase response can be achieved only approximately [4]. Approximately zero-phase IIR filters are useful in wide range of applications where the phase of 2-D signals needs to be preserved.

The design of 2-D IIR filters is more complicated than the design of 2-D FIR filters. The transfer functions of IIR filters are rational functions and the resulting approximation problems are highly nonlinear. As IIR filters can be unstable, the stability conditions must also be included into IIR filter design problems. Several optimization-based methods have been developed for designing 2-D IIR filters that approximate desired magnitude and phase specifications. In these methods, either the least-square (LS) or the minimax approximation is applied [1,3,4,5,7].

In the paper, a new approach for the design of 2-D approximately zero-phase IIR filters with a separable denominator is proposed. This approach is based on two error criteria: equiripple error criterion in the passband and least-squared error criterion in the stopband. Two objective functions are introduced and the filter design problem is transformed into an equivalent bicriterion optimization

I. Lirkov, S. Margenov, and J. Waśniewski (Eds.): LSSC 2007, LNCS 4818, pp. 626–633, 2008.

problem. The obtained problem is converted into a single criterion one using the weighted sum strategy.

The objective function in the considered constrained optimization problem is highly nonlinear, and may have many local minima. Such difficult optimization problems may be solved using local optimization methods in certain cases, but generally global optimization methods are more suitable. Global methods, such as a genetic algorithm (GA), are particularly effective when the goal is to find an approximate global minimum in case of high-dimensional, difficult optimization problems and multiobjective optimizations.

In the paper, a two-step procedure for solving the considered problem is proposed. At the first step, a GA is applied. The final point from the genetic algorithm is used as the starting point for a local optimization method.

The paper is organized as follows. In Section 2, a frequency-domain filter design problem is formulated. In Section 3, the design problem is transformed into an equivalent bicriterion optimization problem. Section 4 deals with the two-step solution procedure. Section 5 comprises an illustrative design example. Section 6 concludes the paper.

2   Formulation of the Design Problem

The transfer function $H(z_1, z_2)$ of a 2-D IIR filter is given by

$$H(z_1, z_2) = \frac{A(z_1, z_2)}{B(z_1, z_2)} \tag{1}$$

where $A(z_1, z_2)$ and $B(z_1, z_2)$ are finite order polynomials in z_1, z_2.

Let $B(z_1, z_2) = D_1(z_1)D_2(z_2)$, where $D_1(z_1)$ and $D_2(z_2)$ are 1-D polynomials. The assumption that $B(z_1, z_2)$ is separable restricts the filter being designed to the class of quadrantally symmetric 2-D filters. It is known that the transfer function of 2-D IIR filters with quadrantally symmetric frequency response has the separable denominator [5,7]. The class of quadrantally symmetric 2-D IIR filters covers practically all types of 2-D IIR filters that have been found useful in 2-D signal processing applications [5].

A two-variable function $F(\omega_1, \omega_2)$ possesses quadrantal symmetry if it satisfies the following condition [7]:

$$F(\omega_1, \omega_2) = F(-\omega_1, \omega_2) = F(\omega_1, -\omega_2) = F(-\omega_1, -\omega_2) \tag{2}$$

If $F(\omega_1, \omega_2)$, in addition, satisfies

$$F(\omega_1, \omega_2) = F(\omega_2, \omega_1) \tag{3}$$

then it has octagonal symmetry. The presence of these symmetries results in certain relations among the filter coefficients. These relations can be used to reduce the number of independent parameters in filter design procedures.

Using the symmetry conditions, the transfer function of a quadrantally symmetric 2-D IIR filter can be expressed in the form [7]:

$$H(z_1, z_2) = \frac{Q_1(z_1, z_2 + z_2^{-1})Q_2(z_1 + z_1^{-1}, z_2)}{D(z_1)D(z_2)}$$

$$= \frac{(\sum_{m=0}^{M} \sum_{n=0}^{N} a_{mn} z_1^m (z_2 + z_2^{-1})^n)(\sum_{m=0}^{M} \sum_{n=0}^{N} b_{mn} (z_1 + z_1^{-1})^m z_2^n)}{(z_1^K + \sum_{i=0}^{K-1} d_i z_1^i)(z_2^K + \sum_{i=0}^{K-1} d_i z_2^i)} \quad (4)$$

In case of an octagonally symmetric 2-D IIR digital filter, the number of independent coefficients to optimize is still reduced and the transfer function can be written as [7]:

$$H(z_1, z_2) = \frac{Q(z_1 + z_1^{-1}, z_2)Q(z_2 + z_2^{-1}, z_1)}{D(z_1)D(z_2)}$$

$$= \frac{(\sum_{m=0}^{M} \sum_{n=0}^{N} a_{mn} (z_1 + z_1^{-1})^m z_2^n)(\sum_{m=0}^{M} \sum_{n=0}^{N} a_{mn} (z_2 + z_2^{-1})^m z_1^n)}{(z_1^K + \sum_{i=0}^{K-1} d_i z_1^i)(z_2^K + \sum_{i=0}^{K-1} d_i z_2^i)} \quad (5)$$

A 2-D IIR filter design problem is to determine the coefficients of the stable transfer function $H(z_1, z_2)$ such that the resulting frequency response is the best approximation of the desired frequency response in the given sense.

The 2-D IIR filter with the separable denominator is stable if and only if both $D_1(z_1)$ and $D_2(z_2)$ are stable 1-D polynomials [7]. An 1-D polynomial is stable if its zeros are strictly inside the unit circle.

Let $\mathbf{Y}$ be a vector of the transfer function coefficients. In case of a quadrantally symmetric filter, $\mathbf{Y}$ is defined as follows:

$$\mathbf{Y} = [a_{00}, a_{01}, \ldots, a_{NM}, b_{00}, b_{01}, \ldots, b_{NM}, d_0, d_1, \ldots, d_{K-1}]^T \quad (6)$$

In case of an octagonally symmetric filter, $\mathbf{Y}$ is given by:

$$\mathbf{Y} = [a_{00}, a_{01}, \ldots, a_{NM}, d_0, d_1, \ldots, d_{K-1}]^T \quad (7)$$

Assume that the continuous (ω_1, ω_2) - plane is discretized by using a $K_1 \times K_2$ rectangular grid $(\omega_{1k}, \omega_{2l})$, $k = 0, 1, \ldots, K_1 - 1$, $l = 0, 1, \ldots, K_2 - 1$.

The desired zero-phase frequency response $H_d(\omega_{1k}, \omega_{2l})$ of the 2-D filter is:

$$H_d(\omega_{1k}, \omega_{2l}) = \begin{cases} 1 & \text{for } (\omega_{1k}, \omega_{2l}) \text{ in the passband } P, \\ 0 & \text{for } (\omega_{1k}, \omega_{2l}) \text{ in the stopband } S. \end{cases} \quad (8)$$

Let $H(e^{j\omega_1}, e^{j\omega_2}, \mathbf{Y})$ denote the frequency response of the filter obtained using the coefficients given by vector $\mathbf{Y}$.

In case of the proposed method, the approximation error is defined differently in the passband and in the stopband. In the passband P, the approximation is to be equiripple. The error function $E(\omega_{1k}, \omega_{2l}, \mathbf{Y})$ is given by:

$$E(\omega_{1k}, \omega_{2l}, \mathbf{Y}) = |H(e^{j\omega_{1k}}, e^{j\omega_{2l}}, \mathbf{Y}) - H_d(\omega_{1k}, \omega_{2l})|, \quad \omega_{1k}, \omega_{2l} \in P. \quad (9)$$

Note that the function $E(\omega_{1k}, \omega_{2l}, \mathbf{Y})$ is real.

In the stopband S, the LS error $E_2(\mathbf{Y})$ to be minimized is:

$$E_2(\mathbf{Y}) = \sum_{(\omega_{1k}, \omega_{2l}) \in S} |H(e^{j\omega_{1k}}, e^{j\omega_{2l}}, \mathbf{Y}) - H_d(\omega_{1k}, \omega_{2l})|^2 \quad (10)$$

The considered 2-D IIR filter design problem can be formulated as follows: For desired zero-phase response $H_d(\omega_{1k}, \omega_{2l})$ defined on a rectangular grid $K_1 \times K_2$, and given degrees of numerator and denominator find a vector $\mathbf{Y}$ for which the designed filter is stable and the error function $E(\omega_{1k}, \omega_{2l}, \mathbf{Y})$ is equiripple in the passband and, simultaneously, the LS error $E_2(\mathbf{Y})$ is minimized in the stopband. Optionally, the following condition on the maximum allowable approximation error $\delta > 0$ in the passband can be additionally imposed:

$$\forall \omega_{1k}, \omega_{2l} \in P \quad \left| H(e^{j\omega_{1k}}, e^{j\omega_{2l}}, \mathbf{Y}) - H_d(\omega_{1k}, \omega_{2l}) \right| \leq \delta \tag{11}$$

Adding the above condition results in obtaining the magnitude ripple equal or smaller than δ.

3   Transformation of the Problem

We solve the considered filter design problem by transforming it into an equivalent bicriterion optimization problem. We introduce two objective functions $X_1(\mathbf{Y})$ and $X_2(\mathbf{Y})$. Let us assume that the function $X_1(\mathbf{Y})$ possesses the property that it has the minimum equal to zero when the error function $E(\omega_{1k}, \omega_{2l}, \mathbf{Y})$ is equiripple in the passband. The error function $E(\omega_{1k}, \omega_{2l}, \mathbf{Y})$ is equiripple in the passband when the absolute values $\Delta E_i(\mathbf{Y})$, $i = 1, 2, \ldots, J$, of all the local extrema of the function $E(\omega_{1k}, \omega_{2l}, \mathbf{Y})$ in the passband, as well as the maximum value $\Delta E_{J+1}(\mathbf{Y})$ of $E(\omega_{1k}, \omega_{2l}, \mathbf{Y})$ at the passband edge are equal, i.e.:

$$\Delta E_i(\mathbf{Y}) = \Delta E_k(\mathbf{Y}), \quad k, i = 1, 2, \ldots, J + 1. \tag{12}$$

Let the objective function $X_1(\Delta E_1, \Delta E_2, \ldots, \Delta E_{J+1})$ be defined as follows:

$$X_1(\Delta E_1, \Delta E_2, \ldots, \Delta E_{J+1}) = \sum_{i=1}^{J+1} (\Delta E_i - R)^2, \tag{13}$$

where:

$$R = \frac{1}{J+1} \sum_{k=1}^{J+1} \Delta E_k. \tag{14}$$

is the arithmetic mean of all ΔE_k, $k = 1, 2, \ldots, J + 1$.

Note that X_1 is non-negative function of $\Delta E_1, \Delta E_2, \ldots, \Delta E_{J+1}$ and it is equal to zero if and only if $\Delta E_1 = \Delta E_2 = \cdots = \Delta E_{J+1}$. As $\Delta E_1, \Delta E_2, \ldots, \Delta E_{J+1}$ are the functions of the vector $\mathbf{Y}$, the function X_1 can be used as the first objective function in our bicriterion optimization problem. As the second objective function, we apply the LS error $E_2(\mathbf{Y})$, so $X_2(\mathbf{Y}) = E_2(\mathbf{Y})$. The weighted combination of the two objective functions $X_1(\mathbf{Y})$ and $X_2(\mathbf{Y})$ allows simultaneous control of both the equiripple error in the passband and the LS error in the stopband.

The equivalent optimization problem can be stated as follows: For given filter specifications and a weighting coefficient α, find a vector $\mathbf{Y}$ such that the function

$$X(\mathbf{Y}, \alpha, \beta) = \alpha X_1(\mathbf{Y}) + (1 - \alpha)\beta X_2(\mathbf{Y}) \tag{15}$$

is minimized, when the following stability constraints are given:

$$D(z_1) \neq 0 \qquad \text{for } |z_1| \geq 1 \tag{16}$$

$$D(z_2) \neq 0 \qquad \text{for } |z_2| \geq 1 \tag{17}$$

Note that $X(\mathbf{Y}, \alpha, \beta)$ is a convex combination of two objective functions $X_1(\mathbf{Y})$ and $X_2(\mathbf{Y})$. Introducing the parameter β enables obtaining comparable initial values of the two terms in (15).

4   Hybrid Solution Procedure

In the previous section, we have formulated a 2-D IIR filter design problem as a constrained bicriterion optimization problem. This optimization problem is highly nonlinear, may be multimodal and has high dimensionality. The stability constraints on the filter coefficients are also included into the problem. Local optimization methods may work well in case of some 2-D filter design problems, but they generally are less suited for solving such difficult optimization problems. Global methods, such as GAs, are more likely to obtain better solutions in case of high-dimensional, difficult optimization problems, multimodal problems and multi-objective optimizations. GAs are also largely independent of the initial conditions. Because of these reasons, GAs are well suited for solving the optimization problem formulated in the previous section.

GAs are stochastic search and optimization techniques based on the mechanism of natural selection. GAs operate on a population of individuals (chromosomes) in each generation. A chromosome represents one possible solution to a given optimization problem. Chromosome coding is the way of representing the design variables. GAs use various coding schemes and the choice of coding scheme depends on the kind of the optimization problem.

To start implementing a GA, an initial population is considered. Successive generations are produced by manipulating the solutions in the current populations. Each solution has a fitness (an objective function) that measures its competence. New solutions are formed using crossover and mutation operations. The crossover mechanism exchanges portions of strings between the chromosomes. Mutation operation causes random alternations of the strings introducing new genetic material to the population. According to the fitness value, a new generation is formed by selecting the better chromosomes from the parents and offspring, and rejecting other so as to keep the population size constant. The algorithm converges to the best chromosome, which represents the solution of the considered optimization problem. The detailed description of a simple GA is presented by Goldberg in [2].

In order to solve the optimization problem formulated in the previous section, we propose a hybrid procedure, i.e., a combination of the GA and a local optimization method [2]. Such hybrid approach is useful in our case because GAs are slow in convergence, especially when the solution is close to the optimum. In order to improve the speed of convergence, after a specified number of generations in the GA has been reached, a local optimization method — the Davidon, Fletcher, and Powell (DFP) method is applied to solve the considered problem.

In the implementation of the GA used in the first step of the solution procedure, the elements of chromosome vectors are double precision floating point numbers. The GA terminates when a predefined maximum number of generations is exceeded. The final point from the GA (the best solution) is used as the starting point for the DFP method. The DFP method is a quasi-Newton method which approximates the inverse Hessian matrix [6]. Note that the proposed hybrid approach combines the advantages of a GA with the fast convergence and accuracy of a quasi-Newton method.

Numerical calculations have shown that it is possible to achieve better convergence if, instead of the minimization problem formulated in the previous section, we apply the GA to the following least square approximation problem

$$E_2(\mathbf{Y}) = \sum_{(\omega_{1k}, \omega_{2l}) \in P \cup S} |H(e^{\omega_{1k}}, e^{\omega_{2l}}, \mathbf{Y}) - H_d(\omega_{1k}, \omega_{2l})|^2 \qquad (18)$$

Then, the solution of this problem is used as a starting point for solving the problem of minimizing $X(\mathbf{Y})$ using the DFP method. As the final result, we get a vector $\mathbf{Y}$ of the transfer function coefficients for which the error function $E(\omega_{1k}, \omega_{2l}, \mathbf{Y})$ is equiripple in the passband and LS error is minimized in the stopband. The local extrema of the error function $E(\omega_{1k}, \omega_{2l}, \mathbf{Y})$ are determined by searching the grid.

In applying the GA, the choice of the probability of crossover, the probability of mutation as well as the choice of the population size are very important. Their settings are dependent on the form of objective function. In the developed program, the population size is 30, the probability of crossover is 0.8, and the probability of mutation is 0.01.

The constrained optimization problem has been transformed into an unconstrained problem using penalty function technique.

5   Design Example

In this section, we apply the proposed approach to the design of a circularly symmetric, lowpass 2-D IIR filter. The passband of the filter is a circular region centered at $(0,0)$ with a radius $r_p = 0.45\pi$. The stopband corresponds to the region outside the circle with a radius of $r_s = 0.7\pi$. The desired magnitude response is 1 in the passband P, 0 in the stopband S and varies linearly in the transition band Tr. The filter is designed with $M = 1$, $N = 9$, $K = 8$, $\delta = 0.05$, and $\beta = 4 \times 10^4$. The weighting coefficient is $\alpha = 0.5$. A square grid of 101×101 points is used for discretizing the (ω_1, ω_2)-plane. The magnitude and phase responses of the resulting filter are shown in Fig. 1 and 2, respectively. Note that the resulting phase response is close approximation of the zero-phase response in the passband. The designed filter is stable. The maximum pole magnitude is 0.824. It should be pointed out that the order of the designed filter is relatively high and the considered design problem is quite difficult.

As GAs are stochastic methods, several runs (at least 30 executions) should be performed to analyze the accuracy of the method. We have performed 30

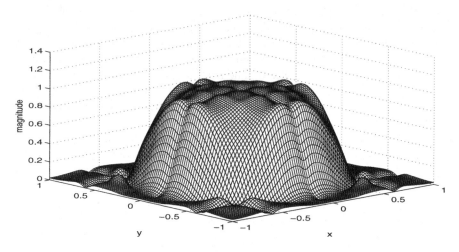

Fig. 1. Magnitude response of the filter designed in the example $(x = \omega_1/\pi,\ x = \omega_2/\pi)$

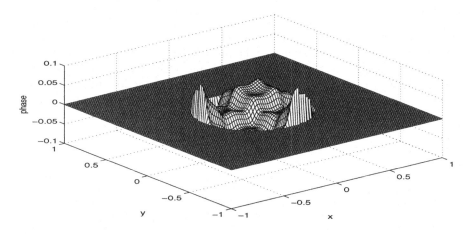

Fig. 2. Phase response of the filter designed in the example $(x = \omega_1/\pi,\ x = \omega_2/\pi)$

executions of the program for the filter specifications as in the above example and it has turned out that the coefficients of the vector $\mathbf{Y}$ obtained after solving the problem of minimizing $X(\mathbf{Y})$ using the DFP method have been differing quite unsignificantly.

In order to compare the resulting filter with the filter obtained using the LS approach, the LS filter was designed for the same filter specifications. In case of the proposed approach, the maximum values of the error function $E(\omega_{1k}, \omega_{2l}, \mathbf{Y})$ are: in the passband $\delta_p = 0.050$ and in the stopband $\delta_s = 0.051$. For the LS filter, $\delta_p = 0.170$ and $\delta_s = 0.130$. Note that in case of the proposed approach, the maximum values of the error function are smaller than in case of the LS approach, both in the passband and in the stopband.

6   Conclusions

A new technique for the design of 2-D approximately zero-phase IIR filters with separable denominator has been presented. Using this technique, phase responses that are close approximations of the zero-phase response in the passband can be achieved. The application of symmetry conditions reduces the number of design parameters. The filter design problem is transformed into a bicriterion optimization problem. Stability constraints are explicitly included into this problem. Additional linear and/or nonlinear constraints, such as e.g., the impulse response decay that characterizes the width of the boundary effect layer in the filtered signal domain [3], can be also included. In the proposed approach, a standard GA along with a local optimization method is used to solve the considered optimization problem. The proposed technique can also be applied for solving other 2-D filter design problems in which a compromise between the equiripple and LS errors is required [8]. It should be possible to extend the proposed technique to the design of 2-D IIR filters with separable denominator and approximately constant group delay. This topic will be studied in the future.

References

1. Dumitrescu, B.: Optimization of Two-Dimensional IIR Filters with Nonseparable and Separable Denominator. IEEE Trans. on Signal Processing 53, 1768–1777 (2005)
2. Goldberg, D.E.: Genetic Algorithms in Search, Optimization, and Machine Learning. Addison Wesley, New York (1989)
3. Gorinevsky, D., Boyd, S.: Optimization-Based Design and Implementation of Multi-Dimensional Zero-Phase IIR filters. IEEE Trans. on Circuits and Syst. I 52, 1–12 (2005)
4. Lim, J.S.: Two-Dimensional Signal and Image Processing. Prentice Hall, Englewood Cliffs (1990)
5. Lu, W.-S., Hinamoto, T.: Optimal Design of IIR Digital Filters with Robust Stability Using Conic-Quadratic-Programming Updates. IEEE Trans. on Signal Processing 51, 1581–1592 (2003)
6. Nocedal, J., Wright, S.J.: Numerical Optimization. Springer, Heidelberg (1999)
7. Reddy, H.C., Khoo, I.-H., Rajan, P.K.: 2-D Symmetry: Theory and Filter Design Applications. IEEE Circuits and Syst. Mag. 3, 4–32 (2003)
8. Wysocka-Schillak, F.: Design of 2-D FIR Filters with Real and Complex Coefficients Using Two Approximation Criteria. Electronics and Telecommunications Quarterly 53, 83–95 (2007)

Part X
Contributed Talks

Optimal Order Finite Element Method for a Coupled Eigenvalue Problem on Overlapping Domains

A.B. Andreev[1] and M.R. Racheva[2]

[1] Department of Informatics, Technical University of Gabrovo
5300 Gabrovo, Bulgaria
[2] Department of Mathematics, Technical University of Gabrovo
5300 Gabrovo, Bulgaria

Abstract. The aim of this paper is to present a new finite element approach applied to a nonstandard second order elliptic eigenvalue problem, defined on two overlapping domains. We derive optimal error estimates as distinguished from [1], where they are suboptimal. For this purpose we introduce a suitable modified degrees of freedom and a corresponding interpolation operator. In order to fix the ideas and to avoid technical difficulties, we consider an one-dimensional case. The conclusive part presents numerical results.

1 Introduction

Let Ω_1 and Ω_2 be the overlapping intervals (a_1, b_1) and (b_2, a_2) respectively, i.e. $a_1 < b_2 < b_1 < a_2$. Let also $H^m(\Omega_i)$ be the usual m-th order Sobolev space on Ω_i, $i = 1, 2$ with norm $\| \cdot \|_{m,\Omega_i}$.

Consider the one-dimensional elliptic operators

$$L^{(i)} = -\frac{d}{dx}\left(\alpha^{(i)}(x)\frac{d}{dx}\right) + \alpha_0^{(i)}(x),$$

where $\alpha^{(i)}(x) > 0$ and $\alpha_0^{(i)}(x) \geq 0$ are bounded functions on Ω_i, $i = 1, 2$. For notational convenience we shall often drop the argument x.

The eigenvalue problem is defined by:
Find $(\lambda, u_1, u_2) \in \mathbf{R} \times H^2(\Omega_1) \times H^2(\Omega_2)$ which obey the differential equation

$$L^{(i)}u_i + (-1)^i \chi_{\Omega_1 \cap \Omega_2}.K = \lambda u_i \quad \text{in } \Omega_i, \; i = 1, 2, \tag{1}$$

and the boundary conditions

$$\alpha^{(i)}u_i'(b_i) - (-1)^i \sigma^{(i)}u_i(b_i) = 0,$$

$$u_i(a_i) = 0, \quad i = 1, 2, \tag{2}$$

as well as the following nonlocal coupling condition:

$$\int_{\Omega_1 \cap \Omega_2} [u_1(x) - u_2(x)] \, dx = 0. \tag{3}$$

I. Lirkov, S. Margenov, and J. Waśniewski (Eds.): LSSC 2007, LNCS 4818, pp. 637–644, 2008.

Herein $\sigma^{(i)} \geq 0$ and $\chi_{\Omega_1 \cap \Omega_2}$ denotes the characteristic function of $\Omega_1 \cap \Omega_2$.

Remark 1. K is a real number depending on unknown function $u = (u_1, u_2)$. It is easy to obtain its explicit representation:

$$K = \frac{1}{2\text{meas}(\Omega_1 \cap \Omega_2)} \sum_{i=1}^{2} (-1)^i \left\{ \left[\alpha^{(i)}(b_1) u_i'(b_1) - \alpha^{(i)}(b_2) u_i'(b_2) \right] \right.$$
$$\left. - \int_{\Omega_1 \cap \Omega_2} \alpha_0^{(i)}(x) u_i(x) \, dx \right\}.$$

The study of problem (1)–(3) is motivated by its applications in many engineering disciplines. Such kind of "contact problems" appear in heat conduction, soil airing, and semiconductors.

We introduce the spaces

$$V_i = \left\{ v_i \in H^1(\Omega_i) : \ v_i(a_i) = 0 \right\}, \ i = 1, 2 \quad \text{and} \quad \widetilde{V} = V_1 \times V_2.$$

Let the space, which incorporates the nonlocal coupling condition (3) on $\Omega_1 \cap \Omega_2$, be defined by

$$V = \left\{ v \in \widetilde{V} : \ \int_{\Omega_1 \cap \Omega_2} [v_1(x) - v_2(x)] \, dx = 0 \right\}.$$

Obviously, V is a closed subspace of $\widetilde{V}$.

Consider the following variational eigenvalue problem: Find $(\lambda, u) \in \mathbf{R} \times V$ such that for all $v \in V$

$$a(u, v) = \lambda(u, v), \tag{4}$$

where

$$a(u, v) = \sum_{i=1}^{2} \left[\int_{\Omega_i} \left(\alpha^{(i)}(x) u_i'(x) v_i'(x) \right. \right.$$
$$\left. \left. + \alpha_0^{(i)}(x) u_i(x) v_i(x) \right) \, dx + \sigma^{(i)} u_i(b_i) v_i(b_i) \right],$$

$$(u, v) = \sum_{i=1}^{2} \int_{\Omega_i} u_i(x) v_i(x) \, dx.$$

Using the definition of trial and test functions of V as well as the properties of coefficient functions, it is easy to see that:

- $a(\cdot, \cdot)$ is bounded, symmetric and strongly coersive on $V \times V$;
- V is a closed subspace of $H^1(\Omega_1) \times H^1(\Omega_2)$.

Thus, for the problem (4) we can refer to the theory of abstract elliptic eigenvalue problems in Hilbert space [2]. We shall use the following results proved by De Shepper (see [1], Theorem 6):

Theorem 1. *The problems (1)–(3) and (4) are formally equivalent. Both problems have a countable infinite set of eigenvalues λ_l, all being strictly positive and having finite multiplicity, without a finite accumulation point. The corresponding eigenfunctions u_l can be chosen to be a Hilbert basis of V, orthonormal with respect to $(\cdot, \cdot)$.*

2   Finite Element Approximations

We are interested in the approximation of the eigenpairs of (4) by the finite element method (FEM). Consider families of regular finite element partitions $\tau_{h_i}^{(i)}$ of Ω_i, $i = 1, 2$, which fulfill standard assumptions [3]. Herein h_1 and h_2 are mesh parameters. Nodal points on the intervals Ω_i, $i = 1, 2$ are chosen in a following way:

$$a_1 = s_0^{(1)} < s_1^{(1)} < \ldots < s_{k_1}^{(1)} = b_1, \quad a_2 = s_0^{(2)} > s_1^{(2)} > \ldots > s_{k_2}^{(2)} = b_2,$$

$$h_j^{(i)} = |s_j^{(i)} - s_{j-1}^{(i)}|, \quad h_i = \max_j h_j^{(i)}, \quad j = 1, \ldots, k_i, \quad h = \max_i h_i, \quad i = 1, 2.$$

Then the partitions $\tau_{h_i}^{(i)}$ consist of intervals T_j^i with endpoints $s_{j-1}^{(i)}$ and $s_j^{(i)}$ such that $\tau_{h_i}^{(i)} = \bigcup_{j=1}^{k_i} T_j^i$. Let us note that b_1 and b_2 are nodes for both partitions $\tau_{h_1}^{(1)}$ and $\tau_{h_2}^{(2)}$, i.e. $T_j^i \in \Omega_1 \cap \Omega_2$ or $T_j^i \in \Omega_i \setminus \Omega_1 \cap \Omega_2$.

We introduce the following finite element spaces related to the partitions $\tau_{h_i}^{(i)}$:

$$X_{h_i}^i = \left\{ v_i \in C(\overline{\Omega}_i) : \; v_{i|_{T^i}} \in \mathcal{P}_2(T^i) \; \forall T^i \in \tau_{h_i}^{(i)} \right\}, \quad i = 1, 2,$$

where $\mathcal{P}_2(T^i)$ is the set of polynomials of degree ≤ 2.

Remark 2. Our presentation is restricted to polynomials of second degree. Nevertheless, the approach we will present could be generalized when $v_{i|_{T^i}} \in \mathcal{P}_k(T^i)$, $k > 2$. As it will be clarified later, the case $k = 1$ is not applicable to our method.

On the base of $X_{h_i}^i$, $i = 1, 2$ we define $X_h = X_{h_1}^1 \times X_{h_2}^2$ and

$$X_{h,0} = \left\{ v = (v_1, v_2) \in X_h : \; v_i(a_i) = 0 \right\}, \quad i = 1, 2,$$

Then, the finite element space related to the nonlocal boundary condition on the intersection of the domains is:

$$V_h = \left\{ v \in X_{h,0} : \; \int_{\Omega_1 \cap \Omega_2} [v_1(x) - v_2(x)] \, dx = 0 \right\}, \quad V_h \subset V.$$

Now we will present a method which gives an optimal order FE approximation applied to the problem (4).

For any element T_j^i we choose its degree of freedom in such a way that every polynomial $p(x) \in \mathcal{P}_2(T_j^i)$ is determined by the values at the endpoints of T_j^i and the integral value $\int_{T_j^i} p(x) \, dx$, $j = 1, \ldots, k_i$, $i = 1, 2$.

We define the interpolation operator $\pi_h : C(\Omega_1) \times C(\Omega_2) \to X_{h,0}$, where $\pi_h = (\pi_{h_1}, \pi_{h_2})$ by means of following conditions:

$$\pi_h v = (\pi_{h_1} v_1, \pi_{h_2} v_2) \in X_{h,0};$$

$$\pi_{h_i} v_i(s_j^{(i)}) = v_i(s_j^{(i)}), \quad j = 0, \ldots, k_i, \quad i = 1, 2,$$

$$\int_{T_j^i} \pi_{h_i} v_i(x)\, dx = \int_{T_j^i} v_i(x)\, dx, \quad j = 1, \ldots, k_i, \quad i = 1, 2.$$

In view of integral condition, it is evident that $\pi_h v \in V_h$ for any $v \in V$.

Likewise, let Π_{h_i}, $i = 1, 2$ be the Lagrange quadratic interpolation operators on Ω_i, $i = 1, 2$ respectively. Then we denote $\Pi_h = (\Pi_{h_1}, \Pi_{h_2}) \in X_{h,0}$.

First, we estimate the difference between both interpolants Π_h and π_h. This is a crucial point of our approach. In contrast to [1] (cf. Proposition 8) an optimal order error estimate is obtained. Considerations are restricted to polynomials of degree two, but this result could be proved for polynomials of higher degree. Also, the results could be extended to two-dimensional case for second-order problems. It is relevant to remark that, in account of (3), one can use integral degrees of freedom for fourth-order problems on overlapping domains (see e.g. [5]).

For the case we consider the next theorem contains the main result:

Theorem 2. *Let the function* $v = (v_1, v_2)$ *belong to* $V \cap H^3(\Omega)$, $\overline{\Omega} = \overline{\Omega}_1 \cup \overline{\Omega}_2$. *Then there exists a constant* $C = C(\Omega) > 0$, *independent of* h, *such that*

$$\|v - \pi_h v\|_{m,\Omega} \le C h^{3-m} \|v\|_{3,\Omega}, \quad m = 0, 1. \tag{5}$$

Proof. We shall estimate $\Pi_{h_i} v_i - \pi_{h_i} v_i$ on each finite element T_j^i, $j = 1, \cdots, k_i$, $i = 1, 2$.

For this purpose, we denote $t = x - s_{j-1}^{(i)}$, $x \in T_j^i$.

Evidently $t \in [0, h_j^{(i)}]$. The basis functions of the Lagrange interpolant, related to this interval, are:

$$\psi_1(t) = \frac{2}{h_j^{(i)^2}} t^2 - \frac{3}{h_j^{(i)}} t + 1; \quad \psi_2(t) = -\frac{4}{h_j^{(i)^2}} t^2 + \frac{4}{h_j^{(i)}} t; \quad \psi_3(t) = \frac{2}{h_j^{(i)^2}} t^2 - \frac{1}{h_j^{(i)}} t.$$

Analogously, for π_{h_i} the basis functions, corresponding to the degrees of freedom that we have chosen, are:

$$\varphi_1(t) = \frac{3}{h_j^{(i)^2}} t^2 - \frac{4}{h_j^{(i)}} t + 1; \quad \varphi_2(t) = -\frac{6}{h_j^{(i)^3}} t^2 + \frac{6}{h_j^{(i)^2}} t; \quad \varphi_3(t) = \frac{3}{h_j^{(i)^2}} t^2 - \frac{2}{h_j^{(i)}} t.$$

Then

$$(\Pi_{h_i} v_i - \pi_{h_i} v_i)_{|_{T_j^i}}$$

$$= \left[\psi_1(t) v_i(s_{j-1}^{(i)}) + \psi_2(t) v_i \left(\frac{s_{j-1}^{(i)} + s_j^{(i)}}{2} \right) + \psi_3(t) v_i(s_j^{(i)}) \right]$$

$$- \left[\varphi_1(t) v_i(s_{j-1}^{(i)}) + \varphi_2(t) \int_{T_j^i} v_i(x)\, dx + \varphi_3(t) v_i(s_j^{(i)}) \right] \tag{6}$$

$$= \left(-\frac{1}{h_j^{(i)^2}} t^2 + \frac{1}{h_j^{(i)}} t \right) \left[v_i \left(s_{j-1}^{(i)} \right) + 4 v_i \left(\frac{s_{j-1}^{(i)} + s_j^{(i)}}{2} \right) + v_i \left(s_j^{(i)} \right) \right.$$

$$\left. - \frac{6}{h_j^{(i)}} \int_{T_j^i} v_i(x)\, dx \right]$$

$$= \frac{6}{h_j^{(i)}} \left(-\frac{1}{h_j^{(i)^2}} t^2 + \frac{1}{h_j^{(i)}} t \right) \left[h_j^{(i)} \left(\frac{1}{6} v_i(s_{j-1}^{(i)}) \right. \right.$$

$$\left. \left. + \frac{4}{6} v_i \left(\frac{s_{j-1}^{(i)} + s_j^{(i)}}{2} \right) + \frac{1}{6} v_i(s_j^{(i)}) \right) - \int_{T_j^i} v_i(x)\, dx \right].$$

Expression in square brackets represents error functional of a quadrature formula, i.e.

$$E_{T_j^i}(v_i) = h_j^{(i)} \left(\frac{1}{6} v_i(s_{j-1}^{(i)}) + \frac{4}{6} v_i \left(\frac{s_{j-1}^{(i)} + s_j^{(i)}}{2} \right) + \frac{1}{6} v_i(s_j^{(i)}) \right) - \int_{T_j^i} v_i(x)\, dx.$$

It is easy to verify, that $E_{T_j^i}(v_i) = 0$ for all $v_i \in \mathcal{P}_2(T_j^i)$. However, the quadrature formula is more accurate, i.e. $E_{T_j^i}(v_i) = 0$ for $v_i \in \mathcal{P}_3(T_j^i)$. But, to make use of this property a higher order of regularity of the function v_i is needed.

Using the Bramble-Hilbert lemma [4], we have

$$|E_{T_j^i}(v_i)| \le C h_j^{(i)^4} |v|_{3,T_j^i}, \tag{7}$$

where $|\cdot|_{3,T_j^i}$ is third order Sobolev seminorm.

On the other hand

$$\left| \frac{6}{h_j^{(i)}} \left(-\frac{1}{h_j^{(i)^2}} t^2 + \frac{1}{h_j^{(i)}} t \right) \right| \le \frac{3}{2} h_j^{(i)^{-1}}, \quad t \in [0, h_j^{(i)}].$$

From this inequality and (7) it follows that

$$|\Pi_{h_i} v_i - \pi_{h_i} v_i|_{T_j^i} \le C h_j^{(i)^3} |v_i|_{3,T_j^i}.$$

Then, finally we obtain the following L_2-norm error estimate:

$$\|\Pi_h v - \pi_h v\|_{0,\Omega} = \left(\sum_{i=1}^{2} \sum_{T_j^i \in \tau_{h_i}} \int_{T_j^i} |\Pi_{h_i} v_i - \pi_{h_i} v_i|^2\, dx \right)^{1/2} \le C h^3 \|v\|_{3,\Omega}. \tag{8}$$

By the same way as (6), we calculate

$$(\Pi_{h_i} v_i - \pi_{h_i} v_i)'|_{T_j^i} = \left(-\frac{2}{h_j^{(i)^2}} t + \frac{1}{h_j^{(i)}} \right) \left[v_i(s_{j-1}^{(i)}) + 4 v_i \left(\frac{s_{j-1}^{(i)} + s_j^{(i)}}{2} \right) + v_i(s_j^{(i)}) \right.$$

$$\left. -\frac{6}{h_j^{(i)}} \int_{T_j^i} v_i(x)\, dx \right] = \frac{6}{h_j^{(i)}} \left(-\frac{2}{h_j^{(i)^2}} t + \frac{1}{h_j^{(i)}} \right) E_{T_j^i}(v_i).$$

We estimate

$$\left| \frac{6}{h_j^{(i)}} \left(-\frac{2}{h_j^{(i)2}} t + \frac{1}{h_j^{(i)}} \right) \right| \leq 6 h_j^{(i)-2}, \quad t \in [0, h_j^{(i)}].$$

Applying again (7) we obtain $|(\Pi_{h_i} v_i - \pi_{h_i} v_i)'|_{T_j^i} \leq C h_j^{(i)2} |v_i|_{3, T_j^i}$. Then, the H^1-norm error estimate is

$$\|\Pi_h v - \pi_h v\|_{1, \Omega} \leq C h^2 \|v\|_{3, \Omega}.$$

This inequality and (8) give

$$\|\Pi_h v - \pi_h v\|_{m, \Omega} \leq C h^{3-m} \|v\|_{3, \Omega}, \quad m = 0, 1. \tag{9}$$

Finally, using classical interpolation theory [3], inequality (9) and applying

$$\|v - \pi_h v\|_{m, \Omega} \leq \|v - \Pi_h v\|_{m, \Omega} + \|\Pi_h v - \pi_h v\|_{m, \Omega}$$

we complete the proof.

As a consequence of this theorem one can prove

Proposition 1. *The finite element space $V_h \subset V$ satisfies the following approximation property:*

$$\inf_{v_h \in V_h} \{ \|v - v_h\|_{0, \Omega} + h|v - v_h|_{1, \Omega} \} \leq C h^3 \|v\|_{3, \Omega},$$

$$\|v - \mathcal{R}_h v\|_{1, \Omega} \leq C h^3 \|v\|_{3, \Omega}, \quad \forall v \in V \cap H^3(\Omega), \tag{10}$$

where $\mathcal{R}_h : V \to V_h$ is the elliptic projector defined by

$$a(u - \mathcal{R}_h u, v_h) = 0, \quad \forall u \in V, \ v_h \in V_h.$$

Let us define finite element approximation of the eigenvalue problem (4): Find $(\lambda_h, u_h) \in \mathbf{R} \times V_h$ such that

$$a(u_h, v_h) = \lambda_h (u_h, v_h) \quad \forall v_h \in V_h. \tag{11}$$

The estimates (10) enable us to adapt the theory of the error analysis [2] to the case of one-dimensional problem on overlapping domains. Namely, using quadratic finite elements to solve (11), we get optimal order error estimate. If (λ, u) is an exact eigenpair of (4) and (λ_h, u_h) is the corresponding approximate solution of (11), then

$$\|u - u_h\|_{1, \Omega} \leq C h^2 \|u\|_{3, \Omega},$$

$$|\lambda - \lambda_h| \leq C h^4 \|u\|_{3, \Omega}^2,$$

where $C = C(\Omega)$ is independent of the mesh parameters.

3 Numerical Results

We apply the theoretical results obtained in previous section taking an example which gives a good illustration of the proposed approach and at the same time its exact eigenpairs could be determined.

Let $\Omega_1 = [0, 2\pi]$ and $\Omega_2 = [\pi, 3\pi]$. The model problem is:

$$-u_1'' - K\chi_{(\pi,2\pi)} = \lambda u_1 \quad \text{on } (0, 2\pi),$$

$$-u_2'' + K\chi_{(\pi,2\pi)} = \lambda u_2 \quad \text{on } (\pi, 3\pi),$$

$$u_1(0) = 0, \quad u_2(3\pi) = 0,$$

$$u_1'(2\pi) = 0, \quad u_2'(\pi) = 0,$$

$$\int_{\pi}^{2\pi} [u_1(x) - u_2(x)] \, dx = 0.$$

The exact eigenvalues are $\lambda_{2j+1} = \lambda_{2j+2} = \left(\dfrac{2j+1}{4}\right)^2$, $j = 0, 1, \ldots$.

Table 1. The eigenvalues computed by the quadratic mesh

λ_h / N	4	8	16	32
$\lambda_{1,h}$	0.0625001288	0.0625000081	0.0625000005	0.0625000000
$\lambda_{2,h}$	0.0625001288	0.0625000081	0.0625000005	0.0625000000
$\lambda_{3,h}$	0.5625923908	0.5625058521	0.5625003670	0.5625000230
$\lambda_{4,h}$	0.5926287762	0.5625058568	0.5625003670	0.5625000230
$\lambda_{5,h}$	1.5644208630	1.56262443651	1.5625078502	1.5625004918

Table 2. The relative error R

R_h / N	4	8	16	32
$R_{1,h}$	2.06×10^{-6}	1.29×10^{-7}	$8 \times 10-9$	0
$R_{2,h}$	2.06×10^{-6}	1.29×10^{-7}	$8 \times 10-9$	0
$R_{3,h}$	1.64×10^{-4}	1.04×10^{-5}	6.52×10^{-7}	4.09×10^{-8}
$R_{4,h}$	5.36×10^{-2}	1.04×10^{-5}	6.52×10^{-7}	4.09×10^{-8}
$R_{5,h}$	1.23×10^{-3}	7.96×10^{-5}	5.02×10^{-6}	3.15×10^{-7}

For the constant K it is easy to obtain $K = \dfrac{u_1'(\pi) + u_2'(2\pi)}{2\pi}$.

We use a quadratic mesh partitions of both domains Ω_i, $i = 1, 2$ consisting of N identical subintervals. It means that h is taken to be equal to $2\pi/N$. The numerical results for the first five eigenvalues computed by finite element method in terms of integral degrees of freedom are given in Table 1. Table 2 shows the relative error $R_{i,h} = \lambda_{i,h}/\lambda_i - 1$.

Acknowledgement

This work is supported by the Bulgarian Ministry of Science under grant VU-MI 202/2006.

References

1. De Shepper, H.: Finite element analysis of a coupling eigenvalue problem on overlapping domains. JCAM 132, 141–153 (2001)
2. Raviart, P.A., Thomas, J.M.: Introduction a l'Analyse Numerique des Equations aux Derivees Partielles, Masson Paris (1983)
3. Ciarlet, P.G.: The Finite Element Method for Elliptic Problems. In: The Finite Element Method for Elliptic Problems, North-Holland, Amsterdam (1978)
4. Bramble, J.H., Hilbert, S.: Bounds for the class of linear functionals with application to Hermite interpolatio. Numer. Math. 16, 362–369 (1971)
5. Andreev, A.B., Dimov, T.T., Racheva, M.R.: One-dimensional patch-recovery finite element method for fourth-order elliptic problems. In: Li, Z., Vulkov, L.G., Waśniewski, J. (eds.) NAA 2004. LNCS, vol. 3401, pp. 108–115. Springer, Heidelberg (2005)

Superconvergent Finite Element Postprocessing for Eigenvalue Problems with Nonlocal Boundary Conditions

A.B. Andreev[1] and M.R. Racheva[2]

[1] Department of Informatics
Technical University of Gabrovo
5300 Gabrovo, Bulgaria
[2] Department of Mathematics
Technical University of Gabrovo
5300 Gabrovo, Bulgaria

Abstract. We present a postprocessing technique applied to a class of eigenvalue problems on a convex polygonal domain Ω in the plane, with nonlocal Dirichlet or Neumann boundary conditions on $\Gamma_1 \subset \partial\Omega$. Such kind of problems arise for example from magnetic field computations in electric machines. The postprocessing strategy accelerates the convergence rate for the approximate eigenpair. By introducing suitable finite element space as well as solving a simple additional problem, we obtain good approximations on a coarse mesh. Numerical results illustrate the efficiency of the proposed method.

1 Introduction

This study deals with second order eigenvalue problems with nonlocal boundary conditions. Here, we propose a procedure for accelerating the convergence of finite element approximations of the eigenpairs.

Let $\Omega \subset \mathbf{R}^2$ be a bounded polygonal domain with boundary $\partial\Omega = \overline{\Gamma}_1 \cup \overline{\Gamma}_2$. Here Γ_1 and Γ_2 are disjoint parts of $\partial\Omega$, each consisting of an integer number of sides of Ω. We consider the following two model problems:

$(\mathcal{P}_1)$: Find $u(x) \in H^2(\Omega)$, $u(x) \neq 0$ and $\lambda \in \mathbf{R}$ satisfying the differential equation

$$-\sum_{i,j=1}^{2} \frac{\partial}{\partial x_i}\left(a_{ij}\frac{\partial u}{\partial x_j}\right) + a_0 u = \lambda u, \quad x \in \Omega, \tag{1}$$

subject to nonlocal Dirichlet boundary condition

$$\int_{\Gamma_1} u\, ds = 0, \tag{2}$$

$$\frac{\partial u}{\partial \nu} \equiv \sum_{i,j=1}^{2} a_{ij}\frac{\partial u}{\partial x_j}\nu_i = K = \text{const}, \quad x \in \Gamma_1, \tag{3}$$

I. Lirkov, S. Margenov, and J. Waśniewski (Eds.): LSSC 2007, LNCS 4818, pp. 645–653, 2008.

and usual Robin boundary condition on Γ_2

$$\frac{\partial u}{\partial \nu} + \sigma u = 0, \quad x \in \Gamma_2. \tag{4}$$

$(\mathcal{P}_2)$: Find $u(x) \in H^2(\Omega)$, $u(x) \neq 0$ and $\lambda \in \mathbf{R}$ satisfying (1) as well as the nonlocal Neumann boundary condition

$$\int_{\Gamma_1} \sum_{i,j=1}^{2} \left(a_{ij} \frac{\partial u}{\partial x_j} \nu_i \right) ds = 0, \tag{5}$$

$$u = K = \text{const}, \quad x \in \Gamma_1, \tag{6}$$

and the eigenfunctions $u(x)$ obey the Robin boundary condition (4) on Γ_2.

The constant value K for both problems is unknown and it must be determined as a part of the solution.

The data used in (1)–(6) are as follows:

$$\exists \alpha > 0, \ \forall \xi \in \mathbf{R}^2 \quad \sum_{i,j=1}^{2} a_{ij}(x)\xi_i\xi_j \geq \alpha|\xi|^2, \quad \text{a.e. in } \Omega,$$

$$a_{ij}(x) \in L_\infty(\Omega); \quad a_{ij} = a_{ji}, \ i,j = 1,2, \quad \text{a.e. in } \Omega,$$

$$a_0(x) \in L_\infty(\Omega); \quad \exists a > 0: \ a_0(x) \geq a, \quad \text{a.e. in } \Omega,$$

$$\sigma(x) \in L_\infty(\Omega); \quad \exists \sigma_0 \geq 0: \ \sigma(x) \geq \sigma_0, \quad \text{a.e. in } \Gamma_2.$$

Moreover, ν_i is the i-th component of the outward unit normal vector $\bar{\nu}$ to $\partial\Omega$.

Eigenvalue problems for linear elliptic differential equations in one or more dimensions are in themselves important in various physical and engineering contexts. In addition they form a link between linear elliptic boundary value problems on one hand and some initial problems for evolution equations on the other. The considered problems $(\mathcal{P}_1)$ and $(\mathcal{P}_2)$ could be referred to some type of Helmholz equations applied to electro-magnetic field computations (see, for example [7] or [6]).

In engineering practice the variational eigenvalue problem is the starting point for an internal approximation method as in the standard finite element method [6]. Our aim is, using the ideas developed in [4] (see also [1]), to extend the results applied to the problems with nonlocal boundary conditions. We derive a postprocessing algorithm that allows to get higher order convergence for the postprocessed eigenpairs.

2   Some Preliminaries

For positive integer k we shall use the conventional notations for the Sobolev spaces $H^k(\Omega)$ and $H_0^k(\Omega)$ provided with the norm $\|\cdot\|_{k,\Omega}$ and seminorm $|\cdot|_{k,\Omega}$

[2]. We denote the L_2-inner product by $(\cdot, \cdot)$ and by $\| \cdot \|_{0,\Omega}$ — the L_2-norm on Ω. We use C as a generic positive constant which is not necessarily the same at each occurrence.

Further on, we shall also use some results obtained by De Shepper and Van Keer [7,3]. First, let us note that the differential equation (1) and the Robin boundary condition (4) coincide in Ω and on Γ_2 respectively for both problems (1)–(4) and (1), (4)–(6). Thus we have one and the same presentation of the variational a-form

$$a(u, v) = \int_{\Omega} \left(\sum_{i,j=1}^{2} a_{ij} \frac{\partial u}{\partial x_j} \frac{\partial v}{\partial x_i} + a_0 uv \right) dx + \int_{\Gamma_2} \sigma uv \, ds \quad \forall u, v \in V, \qquad (7)$$

where V is a subspace of $H^1(\Omega)$.

We shall present our postprocessing procedure using only the problem with nonlocal Neumann boundary condition. The space of trial functions is defined by

$$V = \left\{ v \in H^1(\Omega) : v \text{ is constant on } \Gamma_1 \right\}.$$

The variational problem corresponding to $(\mathcal{P}_2)$ is: Find $(\lambda, u) \in \mathbf{R} \times V$ such that

$$a(u, v) = \lambda(u, v) \quad \forall v \in V. \qquad (8)$$

Using the definition of V, it is evident from (7) that:

- $a(\cdot, \cdot)$ is bounded, symmetric and strongly coersive on $V \times V$;
- V is a closed subspace of $H^1(\Omega)$. Also, V is densely and compactly embedded in $L_2(\Omega)$.

Thus, the following theorem is valid [7,5]:

Theorem 1. *Problem (8) has a countable infinite set of eigenvalues λ_i, all being strictly positive and having finite multiplicity, without a finite accumulation point. The corresponding eigenfunctions u_i can be chosen to be orthonormal in $L_2(\Omega)$. They constitute a Hilbert basis for V.*

For any function $f \in L_2(\Omega)$ let us consider the following elliptic problem [1]:

$$a(u, v) = (f, v) \quad \forall v \in V.$$

Then the operator $T : L_2(\Omega) \to V$ defined by $u = Tf$, $u \in V$ is the solution operator for boundary value (source) problem. Evidently:

$$a(Tu, v) = a(u, Tv) \quad \forall u, v \in H^1(\Omega),$$

$$(Tu, v) = (u, Tv) \quad \forall u, v \in L_2(\Omega).$$

Accordingly, the operator T is symmetric and positive. It follows by the Riesz representation theorem ($a(\cdot, \cdot)$ is an inner product on V), that T is bounded. As it is shown in [5], λ is an eigenvalue and u is the corresponding eigenfunction if and only if $u - \lambda Tu = 0$, $u \neq 0$.

We are interested in the approximation of the eigenpairs of (8) by the finite element method. Let τ_h be a regular family of triangulations of Ω which fulfill the standard assumptions (see [2], Chapter 3). The partitions τ_h consist of triangular or quadrilateral elements e.

Using τ_h we associate the finite-dimensional subspaces V_h of $V \cap C(\overline{\Omega})$ such that the restriction of every function of these spaces over every finite element $e \in \tau_h$ is a polynomial of $P_k(e)$ or $Q_k(e)$ if e is a triangle or a rectangle, respectively. Here $P_k(e)$ is the set of polynomials of degree which is less than or equal to k and $Q_k(e)$ is the set of polynomials of degree which is less than or equal to k in each variable. Moreover, $h = \max_{e \in \tau_h} h_e$, where h_e denotes the diameter of e.

Let us introduce the following finite element space related to the partition τ_h:

$$X_h = \left\{ v \in C_0(\overline{\Omega}) : \ v|_e \in P_k(e) \ (\text{or} \ Q_k(e)), \ \forall e \in \tau_h \right\} \subset H^1(\Omega).$$

Also, we will use the space

$$X_{0,h} = \left\{ v \in X_h : v = 0 \ \text{on} \ \Gamma_1 \right\}.$$

Let $\{a_i\}_{i=1}^N$, where a_i and N depend on h, be the set of nodes associated with X_h and $\{\varphi_i\}_{i=1}^N$ be the canonical basis for X_h. The nodes are numbered in such a way that the first N_0 of them belong to Γ_1.

Defining the function $\psi = \sum_{i=1}^{N_0} \varphi_i$ it is easy to see, that $\psi(a_i) = 1$, $i = 1, \ldots, N_0$ and $\psi(a_i) = 0$, $i = N_0 + 1, \ldots, N$.

Then the finite element space V_h can be represented as $V_h = X_{0,h} \oplus \operatorname{span} \psi$.

Clearly, $\dim X_{0,h} = N - N_0$. Also, it is important to note that, constructing the mass and stiffness matrix for the corresponding problem, the first N_0 nodes should be treated as a single node and the functions ψ and $\{\varphi\}_{i=N_0+1}^N$ form a basis for the finite element space V_h.

The approximate eigenpairs (λ_h, u_h) obtained by the finite element method corresponding to (8) are determined by: Find $\lambda_h \in \mathbf{R}$, $u_h \in V_h$, $u_h \neq 0$ such that

$$a(u_h, v) = \lambda_h(u_h, v) \quad \forall v \in V_h. \tag{9}$$

Crucial point in the finite element analysis is to construct an appropriate space V_h. Some computational aspects related to V_h will be discussed later. The construction of V_h proposed by De Shepper and Van Keer (see [7], Lemmas 3.1 and 3.2) fulfills the standard approximation property:

$$\inf_{v_h \in V_h} \left\{ \|v - v_h\|_{0,\Omega} + h\|v - v_h\|_{1,\Omega} \right\} \leq Ch^{r+1}\|v\|_{r+1,\Omega} \quad \forall v \in V \cap H^{r+1}(\Omega), \tag{10}$$

where $1 \leq r \leq k$.

On the base of this result the rate of convergence of finite element approximation to the eigenvalues and eigenfunctions can be given by the following estimates [5]:

$$\|u - u_h\|_{m,\Omega} \leq Ch^{k+1-m}\|u\|_{k+1,\Omega}, \quad m = 0, 1, \tag{11}$$

$$|\lambda - \lambda_h| \leq Ch^{2k}\|u\|_{k+1,\Omega}. \tag{12}$$

3   Postprocessing Technique

Our aim is to prove that the ideas, developed in [4] and [1] could be applied to the eigenvalue problem with nonlocal boundary conditions. Namely, the post-processig method is reduced to the solving of more simple linear elliptic problem on a higher-order space.

Let u_h be any approximate eigenfunction of (9) with $(u_h, u_h) = 1$. Using finite element solution we consider the following elliptic problem:

$$a(\tilde{u}, v) = (u_h, v) \quad \forall v \in V. \tag{13}$$

Let us define the number

$$\tilde{\lambda} = \frac{1}{(\tilde{u}, u_h)},$$

where $\tilde{u}$ and u_h are solutions of (13) and (9) respectively.

We now consider the approximate elliptic problem corresponding to (13). Using the same partition τ_h, we define the finite-dimensional subspace $\tilde{V}_h \subset V \cap C(\overline{\Omega})$ such that the restriction of every function of $\tilde{V}_h$ over every finite element $e \in \tau_h$ is a polynomial of higher degree. More precisely, since V_h contains polynomials from $P_k(e)$ $(Q_k(e))$, it is sufficient to choose $\tilde{V}_h$ in such a manner that it contains polynomials from $P_{k+1}(e)$ $(Q_{k+1}(e))$. The finite element solution, which corresponds to (13) is:

$$a(\tilde{u}_h, v) = (u_h, v) \quad \forall v \in \tilde{V}_h. \tag{14}$$

Then we define

$$\tilde{\lambda}_h = \frac{1}{(\tilde{u}_h, u_h)},$$

where u_h and $\tilde{u}_h$ are solutions of (9) and (14) respectively.

Theorem 2. *Let the finite element subspaces V_h and $\tilde{V}_h$ contain piece-wise poly-nomials of degree k and $k+1$ respectively. If (λ, u) is an eigenpair of problem (8) with nonlocal boundary condition, $u \in H^{k+1}(\Omega)$ and (λ_h, u_h) is the corresponding solution of (9). Let also eigenfunctions be normalized, i.e. $(u, u) = (u_h, u_h) = 1$. Then the following superconvergent estimate holds:*

$$|\lambda - \tilde{\lambda}_h| \leq Ch^{2k+2}\|u\|_{k+1,\Omega}^2. \tag{15}$$

Proof. Because of the fact, that T is a solution operator, it is evident that $a(Tu, v) = (u, v)$, $\forall v \in V$. Then $a(Tu, u) = 1$ and $a(u, Tu) = \lambda(u, Tu)$. Consequently

$$\lambda = \frac{1}{(Tu, u)}.$$

Thus we obtain

$$\frac{1}{\lambda} - \frac{1}{\tilde{\lambda}} = (Tu, u) - (Tu_h, u_h) + (T(u - u_h); u - u_h) - (T(u - u_h), u - u_h)$$

$$= 2(Tu, u) - 2(Tu, u_h) - (T(u - u_h), u - u_h).$$

Finally,

$$\frac{1}{\lambda} - \frac{1}{\widetilde{\lambda}} = 2(Tu, u - u_h) - (T(u - u_h), u - u_h). \tag{16}$$

Using the boundness of the operator T, we get

$$|(T(u - u_h), u - u_h)| \leq C\|u - u_h\|_{0,\Omega}^2.$$

The first term in the right-hand side of (16) we estimate as follows:

$$2(Tu, u - u_h) = \frac{2}{\lambda}(1 - (u - u_h)) = \frac{1}{\lambda}((u, u) - 2(u, u_h) + (u_h, u_h))$$

$$= \frac{1}{\lambda}(u - u_h, u - u_h) \leq \frac{1}{\lambda}\|u - u_h\|_{0,\Omega}^2.$$

From (16), it follows

$$|\lambda - \widetilde{\lambda}| \leq C\|u - u_h\|_{0,\Omega}^2. \tag{17}$$

On the other hand,

$$\frac{1}{\widetilde{\lambda}} - \frac{1}{\widetilde{\lambda}_h} = (\widetilde{u}, u_h) - (\widetilde{u}_h, u_h) = a(\widetilde{u} - \widetilde{u}_h, \widetilde{u}) + a(\widetilde{u}_h, \widetilde{u}) - a(\widetilde{u}_h, \widetilde{u}_h)$$

$$= a(\widetilde{u} - \widetilde{u}_h, \widetilde{u}) - a(\widetilde{u} - \widetilde{u}_h, \widetilde{u}_h) - a(\widetilde{u} - \widetilde{u}_h, \widetilde{u} - \widetilde{u}_h).$$

The continuity of the a-form leads to the inequality

$$|\widetilde{\lambda} - \widetilde{\lambda}_h| \leq C\|\widetilde{u} - \widetilde{u}_h\|_{1,\Omega}^2. \tag{18}$$

From (17) and (18) we obtain

$$|\lambda - \widetilde{\lambda}_h| \leq C\left(\|u - u_h\|_{0,\Omega}^2 + \|\widetilde{u} - \widetilde{u}_h\|_{1,\Omega}^2\right).$$

Finally, applying the estimates (10) and (11), we complete the proof.

The estimate (15) shows that the postprocessing procedure gives two order higher accuracy compared to the consistent mass error estimate (12).

We can improve the estimate (11) by the same postprocessing argument (see [1]).

Introduce an elliptic projection operator $\widetilde{R}_h : V \to \widetilde{V}_h$ defined by (see [7], Lemma 3.2): $\forall u \in V, \ \forall v \in \widetilde{V}_h, \ a(u - \widetilde{R}_h u, v) = 0.$

For any exact eigenfunction u and its finite element approximation u_h we define:

$$\widetilde{w} = \widetilde{\lambda}_h \widetilde{u} = \widetilde{\lambda}_h T u_h \quad \text{and} \quad \widetilde{w}_h = \widetilde{R}_h \widetilde{w} = \widetilde{\lambda}_h \widetilde{R}_h \circ T u_h = \widetilde{\lambda}_h \widetilde{u}_h.$$

Theorem 3. *Let the conditions of Theorem 2 be fulfilled. Then the following superconvergence estimate is valid:*

$$\|u - \widetilde{w}_h\|_{1,\Omega} \leq Ch^{k+1}\|u\|_{k+1,\Omega}. \tag{19}$$

Proof. In order to estimate $\|u - \widetilde{w}_h\|_{1,\Omega}$ we shall use the following equalities:

$$a(u, u) = \lambda^2 a(Tu, Tu), \quad a(u, \widetilde{w}) = \lambda \widetilde{\lambda}_h a(Tu, Tu_h), \quad a(\widetilde{w}, \widetilde{w}) = \widetilde{\lambda}_h^2 a(Tu_h, Tu_h).$$

Then, since the eigenfunctions u and u_h are normalized in $L_2(\Omega)$, we obtain:

$$a(u - \widetilde{w}, u - \widetilde{w}) = \lambda^2 (Tu, u) - 2\lambda \widetilde{\lambda}_h (u, Tu_h) + \widetilde{\lambda}_h^2 a(Tu_h, Tu_h)$$

$$= \lambda a(Tu, u) - 2\widetilde{\lambda}_h a(Tu_h, u) + \widetilde{\lambda}_h^2 a(Tu_h, Tu_h)$$

$$= \lambda - 2\widetilde{\lambda}_h (u_h, u) + \frac{\widetilde{\lambda}_h^2}{\lambda} = 2\widetilde{\lambda}_h - 2\widetilde{\lambda}_h (u, u_h) + \lambda - \widetilde{\lambda}_h + \frac{\widetilde{\lambda}_h^2}{\lambda} - \widetilde{\lambda}_h$$

$$= \widetilde{\lambda}_h \left[(u, u) - 2(u, u_h) + (u_h, u_h) \right] + \lambda - \widetilde{\lambda}_h + \frac{\widetilde{\lambda}_h}{\lambda} (\widetilde{\lambda}_h - \lambda)$$

$$= \widetilde{\lambda}_h \|u - u_h\|_{0,\Omega}^2 + (\lambda - \widetilde{\lambda}_h) + \frac{\widetilde{\lambda}_h}{\lambda} (\widetilde{\lambda}_h - \lambda).$$

Evidently, the bilinear form $a(\cdot, \cdot)$ defined by (7) is V-elliptic. Moreover, to estimate $\|u - u_h\|_{0,\Omega}^2$ we apply the regularity of the a-form on $V \times V$. From (11) and (12) it follows that

$$\|u - \widetilde{w}\|_{1,\Omega}^2 \leq C h^{2(k+1)} \|u\|_{k+1,\Omega}^2. \tag{20}$$

The approximation property of the operator $\widetilde{R}_h$ and the standard assumptions of the smoothness of $\widetilde{w}$ imply

$$\|\widetilde{w} - \widetilde{w}_h\|_{1,\Omega}^2 \leq C(\lambda) h^{2(k+1)}.$$

Combining this result with (20) we arrive at the estimate (19).

4   Numerical Results

To illustrate our theoretical results we shall refer to the example on related two-dimensional eigenvalue problem. Let Ω be a square domain: $\{(x_1, x_2) : 0 < x_i < 1, \ i = 1, 2\}$, $\Gamma_1 = \{(x_1, x_2) : 0 < x_1 < 1, \ x_2 = 1\}$ and $\Gamma_2 = \partial\Omega \setminus \Gamma_1$.

Consider the following model problem: Find a pair $(\lambda, u) \in \mathbf{R} \times H^2(\Omega)$ which obeys the differential equation

$$-\Delta u = \lambda u \quad \text{in} \quad \Omega,$$

with nonlocal Neumann boundary condition

$$\int_{\Gamma_1} \left(\frac{\partial u}{\partial x_1} \nu_1 + \frac{\partial u}{\partial x_2} \nu_2 \right) ds = 0,$$

where u is a constant on Γ_1 and satisfies the Robin boundary condition

$$\frac{\partial u}{\partial x_1} \nu_1 + \frac{\partial u}{\partial x_2} \nu_2 + 0.21u = 0 \quad \text{on} \quad \Gamma_2.$$

Table 1. The eigenvalues computed by the finite element method

N	$\lambda_{1,h}$	$\lambda_{2,h}$	$\lambda_{3,h}$	$\lambda_{4,h}$
16	0.808932859	11.63731126	14.21535688	36.70151076
64	0.804947034	11.21636601	13.72260280	35.34580736
256	0.797117258	11.12108553	13.60201416	33.52161620
1024	0.781913261	11.10150221	13.57203629	33.36752919

Table 2. The eigenvalues obtained after applying the postprocessing

N	$\widetilde{\lambda}_{1,h}$	$\widetilde{\lambda}_{2,h}$	$\widetilde{\lambda}_{3,h}$	$\widetilde{\lambda}_{4,h}$
16	0.796993204	11.11973533	13.60842419	33.50785135
64	0.780697718	11.10580126	13.58073619	33.30999380
256	0.780111186	11.10090389	13.56982857	33.28591706

The numerical results for the first four eigenvalues computed by the standard finite element method with N bilinear elements are given in Table 1. Further, in Table 2 we present the results of the same four eigenvalues computed by our postprocessing method, applied on a finite element space which uses polynomials of degree 2, i.e. by means of biquadratic Serendipity finite elements.

The exact eigenvalues for this problem are not known. The best we can do is to make a comparison between the eigenvalues computed by the finite element method and those obtained after applying postprocessing. For instance, according to the theoretical results, the accuracy obtained as a result of postprocessing when $N = 16$, should be similar to the obtained accuracy from FEM implementation with 256 elements.

On the base of our numerical experiments it may be concluded that the global postprocessing method presented and studied here gives an effective and accurate algorithm for calculating the eigenvalues for problems with nonlocal boundary conditions using the lowest order (bilinear) finite elements on a coarse mesh instead of usual finite element method on a fine mesh or using a finite element space with approximating polynomials of higher degree.

Acknowledgement

This work is supported by the Bulgarian Ministry of Science under grant VU-MI 202/2006.

References

1. Andreev, A.B., Racheva, M.R.: On the postprocessing technique for eigenvalue problems. In: Dimov, I.T., et al. (eds.) NMA 2002. LNCS, vol. 2542, pp. 363–371. Springer, Heidelberg (2003)
2. Ciarlet, P.G.: The Finite Element Method for Elliptic Problems. North-Holland, Amsterdam (1978)

3. De Shepper, H., Van Keer, R.: On a finite element method for second order elliptic eigenvalue problems with nonlocal Dirichlet boundary conditions. Numer. Funct. Anal. and Optimiz 18(384), 283–295 (1997)

4. Racheva, M.R., Andreev, A.B.: Superconvergence postprocessing for eigenvalues. Comp. Methods in Appl. Math. 2(2), 171–185 (2002)

5. Raviart, P.H., Thomas, J.M.: Introduction a l'Analyse Numerique des Equation aux Derivees Partielles, Masson Paris (1983)

6. Reece, A.B., Presfon, T.W.: Finite Element Methods in Electric Power Engineering. Oxford University Press, Oxford (2004)

7. Van Keer, R., De Shepper, H.: Finite element approximation for second order elliptic eigenvalue problems with nonlocal boundary or transition conditions. Applied Math. and Computation 82, 1–16 (1997)

Uniform Convergence of Finite-Difference Schemes for Reaction-Diffusion Interface Problems

Ivanka T. Angelova and Lubin G. Vulkov

University of Rousse, Department of Mathematics, 7017 Rousse, Bulgaria
iangelova@ru.acad.bg, vulkov@ami.ru.acad.bg

Abstract. We consider a singularly perturbed reaction-diffusion equation in two dimensions (x, y) with concentrated source on a segment parallel to axis Oy. By means of an appropriate (including corner layer functions) decomposition, we describe the asymptotic behavior of the solution. Finite difference schemes for this problem of second and fourth order of local approximation on Shishkin mesh are constructed. We prove that the first scheme is almost second order uniformly convergent in the maximal norm. Numerical experiments illustrate the theoretical order of convergence of the first scheme and almost fourth order of convergence of the second scheme.

2000 Mathematics Subject Classification: 65M06, 65M12.

Keywords: reaction-diffusion, interface problems, singular perturbation, uniform convergence, Shishkin mesh.

1 Introduction

There is an extensive literature on analytical and numerical methods for singularly perturbed problems, cf. [1,2,5,6,7,8,9,10,12,13] and the references therein. Our interest lies in examine parameter-uniform numerical methods [13] of high order for singularly perturbed interface problems. That is, we are interested in numerical methods for which the following error bound can be theoretically established

$$\|u - \bar{U}_N\|_\infty \leq CN^{-p}, \ p > 0,$$

where N is the number of mesh elements employed in each coordinate direction, $\bar{U}_N$ is a polynomial interpolant generated by the numerical method, $\|.\|_\infty$ is the global pointwise maximum norm and C is a constant independent of ε and N. In general, the gradients of the solution [2,3,13] become unbounded in the boundary/interface and corner layers as $\varepsilon \to 0$; however, parameter-uniform numerical methods guarantee that the error in the numerical approximation is controlled solely by the size of N.

Let us consider the boundary value problems for the reaction-diffusion model:

$$-div\,(a(x,y)\nabla u) + b(x,y)u = f(x,y) + \varepsilon\delta_\Gamma K(y), \ (x,y) \in \Omega, \tag{1}$$

I. Lirkov, S. Margenov, and J. Waśniewski (Eds.): LSSC 2007, LNCS 4818, pp. 654–660, 2008.
© Springer-Verlag Berlin Heidelberg 2008

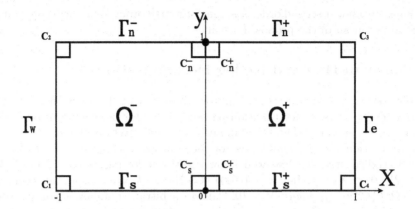

Fig. 1. The domain

where $\Omega = (-1, 1) \times (0, 1)$, the diagonal matrix

$$(i)\ a(x, y) = \begin{pmatrix} \varepsilon^2 & 0 \\ 0 & \varepsilon^2 \end{pmatrix}\ or\ (ii)\ a(x, y) = \begin{pmatrix} \varepsilon^2 & 0 \\ 0 & 1 \end{pmatrix}$$

and δ_Γ is the Dirac-delta distribution concentrated on the segment $\{(x, y)| x = 0,\ 0 < y < 1\}$, $b(x, y) \geq \beta^2 > 0$ and the diffusion parameter can be arbitrary small, but $0 < \varepsilon \leq 1$. In [5] the case (ii) is considered, in which case only boundary interface layers can arise when $\varepsilon \to 0$. In this paper our attention is concentrated on (i).

This type of problem (1), (i) is characterized by the presence of regular exponential layers in a neighborhoods of $\partial \Omega$ and Γ and corner layer at the intersections of the external and internal boundaries, of width $O(\varepsilon)$. We define:

$$\Gamma_s^- = \{(x, 0)|\ -1 \leq x \leq 0\},\ \Gamma_s^+ = \{(x, 0)|\ 0 \leq x \leq 1\},\ \Gamma_s = \Gamma_s^- \cup \Gamma_s^+,$$
$$\Gamma_n^- = \{(x, 1)|\ -1 \leq x \leq 0\},\ \Gamma_n^+ = \{(x, 1)|\ 0 \leq x \leq 1\},\ \Gamma_n = \Gamma_n^- \cup \Gamma_n^+,.$$
$$\Gamma_w = \{(-1, y)|\ 0 \leq y \leq 1\},\quad \Gamma_e = \{(1, y)|\ 0 \leq y \leq 1\},$$

Fig. 1 shows the edges, the corners $c_1 = \Gamma_s^- \cap \Gamma_w$, $c_2 = \Gamma_w \cap \Gamma_n^-$, $c_3 = \Gamma_n^+ \cap \Gamma_e$, $c_4 = \Gamma_e \cap \Gamma_s^+$, and the interface corners of the domain $c_s^- = \Gamma_s^- \cap \Gamma$, $c_s^+ = \Gamma_s^+ \cap \Gamma$, $c_n^- = \Gamma_n^- \cap \Gamma$, $c_n^+ = \Gamma_n^+ \cap \Gamma$. We adopt the following notation for the boundary conditions:

$$g(x, y) = g_i(x),\ (x, y) \in \Gamma_i,\ i = s, n;\ g(x, y) = g_i(y),\ (x, y) \in \Gamma_i,\ i = e, w.\ (2)$$

In the next section by means of an appropriate Shishkin-Han&Kellog decomposition we describe the asymptotic behavior of the solution of the problem. In section 3 we propose two difference schemes on piecewise-uniform Shishkin mesh: one of second-order convergent at fixed ε and the other one — of fourth order. For the first scheme on Shishkin mesh we proved rigorously almost second-order of uniform (with respect to ε) convergence. Numerical experiments in the last section show almost second order and almost fourth order of uniform convergence on Shishkin mesh of the proposed schemes.

2   Shishkin-Han and Kellog Decomposition

In this section we examine the asymptotic behavior of the solution of (1) with respect to the singular perturbation parameter ε. This behavior will be used in the next section at the theoretical analysis of the discrete problems.

The question for the smoothness of the solution without interface ($K \equiv 0$) is well studied [1,9,10]. The well known is the following fact [1]. Let f, $b \in C^{4,\lambda}(\bar{\Omega})$, $0 < \lambda < 1$. If $g \in C(\partial\Omega) \bigcap C^{4,\lambda}(\Gamma_k)$, $k = s$, e, n, w, then $u \in C^{1,\lambda'}(\bar{\Omega}) \bigcap C^{6,\lambda}(\Omega)$, where $\lambda' \in (0,1)$ is an arbitrary number. In order to obtain a higher smoothness compatibility conditions in the corners c_i for $i = 1,...,4$ and $c_s^{\mp}$, $c_n^{\mp}$ must be imposed. Such compatibility conditions for the case without interface ($K \equiv 0$) are derived in [10], and for the interface case in [5,11]. Further, we suppose that all necessary compatibility conditions are fulfilled.

Using results for smoothness of solution [11], maximum principle and stretching arguments [5,7,6,10,12,13] one can establish crude bounds on the derivatives of the solution of the form

$$\|u^{(k,j)}\| \le C\varepsilon^{-k/2-j/2}, \; 0 \le k+j \le 4, \; u^{(k,j)} = \frac{\partial^{k+j}u}{\partial x^k \partial y^j}.$$

These bounds are not sufficient to analyze the uniform convergence of the difference schemes studied in this paper, because they do not explicitly show the presence of boundary/interface and corner layers in Ω^-, Ω^+. In the next theorem we present a decomposition of u and appropriate bounds of its derivatives with respect to ε, which is used in the error analysis in Section 3.

Theorem 1. *Let* $u \in C^{5+\lambda}\left(\bar{\Omega}^-\right) \bigcap C^{5+\lambda}\left(\bar{\Omega}^+\right) \bigcap C\left(\bar{\Omega}\right)$ *and* $\frac{\partial^i u}{\partial y^i} \in C\left(\bar{\Omega}\right)$, $i = 0,1,2,3,4$. *The solution* u *may be written as a sum*

$$u = v + \sum_{i=1}^{4}\left(w_i^- + w_i^+\right) + \sum_{i=1}^{4}\left(z_i^- + z_i^+\right),$$

where

$$Lv \equiv f + \delta_\Gamma\left(x\right)K\left(y\right), \; Lw_i^{\mp} \equiv 0, \; Lz_i^{\mp} \equiv 0, \; i = 1,2,3,4.$$

Boundary conditions for v, w_i, z_i, $i = 1,2,3,4$, *can be specified so that the following bounds on the derivatives of the components hold:*

$$\|v^{(k,j)}\| \le C\left(1 + \varepsilon^{2-k-j}\right), \qquad 0 \le k+j \le 4$$

$$|w_1^{\mp}\left(x,y\right)| \le C\exp\left(-\frac{\sqrt{b^{\mp}}}{\varepsilon}y\right), \qquad |w_3^{\mp}\left(x,y\right)| \le C\exp\left(-\frac{\sqrt{b^{\mp}}}{\varepsilon}(1-y)\right)$$

$$|w_2^-\left(x,y\right)| \le C\exp\left(-\frac{\sqrt{b^-}}{\varepsilon}(1+x)\right), |w_4^-\left(x,y\right)| \le C\exp\left(-\frac{\sqrt{b^-}}{\varepsilon}x\right)$$

$$|w_2^+\left(x,y\right)| \le C\exp\left(-\frac{\sqrt{b^+}}{\varepsilon}x\right), \qquad |w_4^+\left(x,y\right)| \le C\exp\left(-\frac{\sqrt{b^+}}{\varepsilon}(1-x)\right)$$

$$\max\left\{\|w_i^{(k,j)}\|, \|z_i^{(k,j)}\|\right\} \le C\varepsilon^{-k-j}, \; 0 \le k+j \le 4,$$

$$\|w_i^{(k,0)}\| \le C, \; i = 1,3, \; \|w_i^{(0,j)}\| \le C, \; j = 2,4,$$

$$|z_1^-(x,y)| \le C\exp\left(-\tfrac{\sqrt{b^-}}{\varepsilon}(1+x)\right)\exp\left(-\tfrac{\sqrt{b^-}}{\varepsilon}y\right),$$

$$|z_2^-(x,y)| \le C\exp\left(-\tfrac{\sqrt{b^-}}{\varepsilon}(1+x)\right)\exp\left(-\tfrac{\sqrt{b^-}}{\varepsilon}(1-y)\right),$$

$$|z_3^-(x,y)| \le C\exp\left(-\tfrac{\sqrt{b^-}}{\varepsilon}x\right)\exp\left(-\tfrac{\sqrt{b^-}}{\varepsilon}(1-y)\right),$$

$$|z_4^-(x,y)| \le C\exp\left(-\tfrac{\sqrt{b^-}}{\varepsilon}x\right)\exp\left(-\tfrac{\sqrt{b^-}}{\varepsilon}y\right),$$

where $b^{\mp} = \max b(x,y)$ on $\bar{\Omega}^-$ and $\bar{\Omega}^+$, respectively.

3   The Discrete Problems

On Ω we introduce the mesh ω_h that is the tensor product of two one dimensional piecewise uniform Shishkin meshes, i.e. $\omega_h = \omega_h^x \times \omega_h^y$, ω_h^x splits the interval $[-1,1]$ into six subintervals $[-1,-1+\sigma_x]$, $[-1+\sigma_x,-\sigma_x]$, $[-\sigma_x,0]$, $[0,\sigma_x]$, $[\sigma_x,1-\sigma_x]$ and $[1-\sigma_x,1]$. The mesh distributes $N/4$ points uniformly within each of the subintervals $[-1,-1+\sigma_x]$, $[-\sigma_x,0]$, $[0,\sigma_x]$, $[1-\sigma_x,1]$ and the remaining N mesh points uniformly in the interior subintervals $[-1+\sigma_x,-\sigma_x]$ and $[\sigma_x,1-\sigma_x]$. ω_h^y splits the in interval $[0,1]$ into three intervals $[0,\sigma_y]$, $[\sigma_y,1-\sigma_x]$ and $[1-\sigma_y,1]$. The mesh distributes uniformly within each of the subintervals $[0,\sigma_y]$ and $[1-\sigma_y,1]$ and the remaining $N/2$ mesh points uniformly in the interior subinterval $[\sigma_y,1-\sigma_y]$. To simplify our discussion we take $\sigma_x = \sigma_y$; these boundary interface transition points are defined as $\sigma = \sigma_x = \sigma_y = \min\left[\frac{1}{4}, \sigma_0 \varepsilon \frac{\ln N}{\beta}\right]$. Below we denote by $h = \frac{4\sigma}{N}$, $H = \frac{2(1-2\sigma)}{N}$; $h_i = x_i - x_{i-1}$, $0 < i \le 2N$, $[h] = [h]_{x_i} = h_{i+1} - h_i$. Similarly $k_j = y_j - y_{j-1}$, $0 < j \le N$, $[k] = [k]_{y_j} = k_{j+1} - k_j$. In the numerical experiments we take $\sigma_0 = 2$ for the scheme (3) and $\sigma_0 = 4$ for the scheme (4). For a mesh function $U = U_{ij} = U(x_i, y_j)$ on ω_h, we define:

$$U_{\bar{x}} = U_{\bar{x},i} = (U(x_i,y_j) - U(x_{i-1},y_j))/h_i, \; U_x = U_{x,i} = U_{\bar{x},i+1},$$

$$U_{\bar{y}} = U_{\bar{y},j} = (U(x_i,y_j) - U(x_i,y_{j-1}))/k_j, \; U_y = U_{y,j} = U_{\bar{y},j+1},$$

$$U_{\hat{x}} = U_{\hat{x},i} = \frac{U(x_{i+1},y_j) - U(x_i,y_j)}{\hbar_i}, \; U_{\hat{y}} = U_{\hat{y},j} = \frac{U(x_i,y_{j+1}) - U(x_i,y_j)}{\bar{k}_j},$$

$$\hbar_i = \frac{h_i + h_{i+1}}{2}, \hbar_0 = \frac{h_1}{2}, \hbar_{2N} = \frac{h_{2N}}{2}, \bar{k}_j = \frac{k_j + k_{j+1}}{2}, \bar{k}_0 = \frac{k_1}{2}, \bar{k}_N = \frac{k_N}{2},$$

$$U_{\bar{x}\hat{x}} = U_{\bar{x}\hat{x},i} = (U_{x,i} - U_{\bar{x},i})/\hbar_i, \; U_{\bar{y}\hat{y}} = U_{\bar{y}\hat{y},j} = (U_{y,j} - U_{\bar{y},j})/\bar{k}_j,$$

$$U_{\overset{\circ}{x}} = U_{\overset{\circ}{x},i} = \frac{h_{i+1}U_{\bar{x}} + h_iU_x}{h_i + h_{i+1}}, U_{\overset{\circ}{y}} = U_{\overset{\circ}{y},j} = \frac{k_{j+1}U_{\bar{y}} + k_jU_y}{k_j + k_{j+1}}, U_{\overset{\circ\circ}{xy}} = (U_{\overset{\circ}{x}})_{\overset{\circ}{y}}.$$

To discretize problem (1)–(2), first we use the standard central difference scheme

$$\Lambda U = -\varepsilon^2(U_{\bar{x}})_{\hat{x}} - \varepsilon^2(U_{\bar{y}})_{\hat{y}\hat{x}} + b_{\hat{x}}U = \begin{cases} f, & x_i \ne 0, (x_i,y_j) \in \omega_h \\ f_{\hat{x}} - \frac{1}{\hbar}K, & x_i = 0, (x_i,y_j) \in \omega_h \end{cases}$$

$$U = g, \text{ on the boundary } \partial\omega_h,$$

where

$$f_{\bar{x}} = f_{\bar{x}_i}(y) = \frac{h_i f(x_i-, y) + h_{i+1} f(x_i+, y)}{h_i + h_{i+1}}.$$

Theorem 2. Let $u \in C^{5+\lambda}(\bar{\Omega}^-) \bigcap C^{5+\lambda}(\bar{\Omega}^+) \bigcap C(\bar{\Omega})$ and $\frac{\partial^i u}{\partial y^i} \in C(\bar{\Omega})$, $i = 0, 1, 2, 3, 4$. Then the error at the mesh points satisfies

$$\| (u - U)(x_i, y_j) \| \leq C \left(N^{-1} \ln N \right)^2, \quad (x_i, y_j) \in \omega_h. \tag{3}$$

In [3] the following 9-points scheme is derived:

$$\Lambda' U = -\varepsilon^2 U_{\bar{x}\hat{x}} - \varepsilon^2 U_{\bar{y}\hat{y}} + b_{\check{x}} U - \frac{\varepsilon^2}{12} \left(\left(k^2 U_{\bar{x}\hat{x}\bar{y}} \right)_{\hat{y}} + \left(h^2 U_{\bar{y}\hat{y}\bar{x}} \right)_{\hat{x}} \right)$$

$$- \frac{\varepsilon^2}{6} \left([k] \, U_{\bar{x}\hat{x}\overset{\circ}{y}} + [h] \, U_{\bar{y}\hat{y}\overset{\circ}{x}} \right) + \frac{1}{12} \left(\left(h^2 (bU)_{\bar{x}} \right)_{\hat{x}} + \left(k^2 (bU)_{\bar{y}} \right)_{\hat{y}} \right)$$

$$+ \frac{1}{6} \left([h] \, (bU)_{\overset{\circ}{x}} + [k] \, (bU)_{\overset{\circ}{y}} + \frac{2}{3} [h] \, [k] \, (bU)_{\overset{\circ}{x}\overset{\circ}{y}} \right) \tag{4}$$

$$= \begin{cases} f + \frac{1}{12} \left(\left(h^2 f_{\bar{x}} \right)_{\hat{x}} + \left(k^2 f_{\bar{y}} \right)_{\hat{y}} \right) \\ \quad + \frac{1}{6} \left([h] \, f_{\overset{\circ}{x}} + [k] \, f_{\overset{\circ}{y}} + \frac{2}{3} [h] \, [k] \, f_{\overset{\circ}{x}\overset{\circ}{y}} \right), & x_i \neq 0, \\ f + \frac{1}{12} \left(h^2 f_{\bar{x}\hat{x}} + \left(k^2 f_{\bar{y}} \right)_{\hat{y}\check{x}} \right) + \frac{1}{6} [k] \, f_{\overset{\circ}{y}\check{x}} + \frac{h}{12} \left[\frac{\partial f}{\partial x} \right]_{x=0} \\ \quad - \left(\frac{h}{12\varepsilon} b_{\check{x}} + \frac{\varepsilon}{h} \right) K + \frac{h\varepsilon}{12} K_{\bar{y}\hat{y}} - \frac{\varepsilon}{12h} \left(k^2 K_{\bar{y}} \right)_{\hat{y}} \\ \quad - \frac{[k]}{18} \left(\frac{3\varepsilon}{h} K_{\overset{\circ}{y}} + \frac{h}{\varepsilon} b_{\check{x}} K' - h \left[\frac{\partial^2 f}{\partial x \partial y} \right]_{x=0} \right), & x_i = 0. \end{cases}$$

Table 1. The maximum error and the numerical order of convergence for $\varepsilon = 2^{-2}, ..., 2^{-12}$, 10^{-4} for the scheme (3) illustrate the estimate (3)

$\varepsilon \backslash N$	8	16	32	64	128	256	512
2^{-2}	5,1200E-2	1,4900E-2	3,9000E-3	9,8717E-4	2,4750E-4	6,1920E-5	1,5483E-5
	1,7808	1,9338	1,9821	1,9959	1,9990	1,9997	
2^{-4}	5,4780E-1	2,9160E-1	1,0530E-1	2,9800E-2	7,7000E-3	1,9000E-3	4,8667E-4
	0,9097	1,4695	1,8211	1,9524	2,0189	1,9650	
2^{-6}	5,5480E-1	4,4530E-1	2,9290E-1	1,5200E-1	6,3700E-2	2,3100E-2	7,7000E-3
	0,3172	0,6044	0,9463	1,2547	1,4634	1,5850	
2^{-8}	5,5550E-1	4,4540E-1	2,9290E-1	1,5200E-1	6,3700E-2	2,3100E-2	7,7000E-3
	0,3187	0,6047	0,9463	1,2547	1,4634	1,5850	
2^{-10}	5,5550E-1	4,4540E-1	2,9290E-1	1,5200E-1	6,3700E-2	2,3100E-2	7,7000E-3
	0,3187	0,6047	0,9463	1,2547	1,4634	1,5850	
2^{-12}	5,5550E-1	4,4540E-1	2,9290E-1	1,5200E-1	6,3700E-2	2,3100E-2	7,7000E-3
	0,3187	0,6047	0,9463	1,2547	1,4634	1,5850	
10^{-4}	5,5550E-1	4,4540E-1	2,9290E-1	1,5200E-1	6,3700E-2	2,3100E-2	7,7000E-3
	0,3187	0,6047	0,9463	1,2547	1,4634	1,5850	

4   Numerical Experiments

In this section we present numerical experiments obtained by applying the difference schemes (3), (4). The test problem is

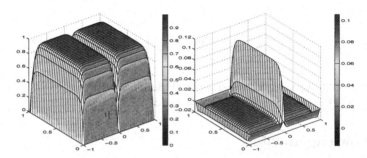

Fig. 2. Numerical solution and the local error for the scheme (3), $\varepsilon = 2^{-5}$, $N = 64$

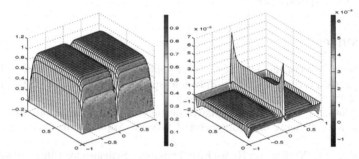

Fig. 3. Numerical solution and the local error for the scheme (4), $\varepsilon = 2^{-5}$, $N = 64$. The right picture shows well the increase of the error near the interface corner layers.

Table 2. The maximum error and the numerical order of convergence for $\varepsilon = 2^{-2}, ..., 2^{-12}$, 10^{-4} for the scheme (4)

$\varepsilon \backslash N$	8	16	32	64	128	256
2^{-2}	1.3402E-4	8.4042E-6	5.2576E-7	3.2868E-8	2.0544E-9	1.2852E-10
	3.9952	3.9986	3.9996	3.9999	3.9987	
2^{-4}	1.7900E-2	1.4000E-3	8.9588E-5	5.6273E-6	3.5460E-7	2.2168E-8
	3.6765	3.9660	3.9928	3.9882	3.9996	
2^{-6}	1.4580E-1	9.0100E-2	1.7900E-2	1.4000E-3	8.9588E-5	5.6273E-6
	0.6944	2.3316	3.6765	3.9660	3.9928	
2^{-8}	1.3900E-1	9.0100E-2	3.3700E-2	6.9000E-3	8.1896E-4	8.9588E-5
	0.6255	1.4188	2.2881	3.0747	3.1924	
2^{-10}	1.3630E-1	9.0100E-2	3.3700E-2	6.9000E-3	8.1896E-4	8.9588E-5
	0.5972	1.4188	2.2881	3.0747	3.1924	
2^{-12}	1.3630E-1	9.0100E-2	3.3700E-2	6.9000E-3	8.1896E-4	8.9588E-5
	0.5898	1.4188	2.2881	3.0747	3.1924	
10^{-4}	1.3550E-1	9.0100E-2	3.3700E-2	6.9000E-3	8.1896E-4	8.9588E-5
	0.5887	1.4188	2.2881	3.0747	3.1924	

$$-\varepsilon^2 \triangle u + u = f, \; x \in (-1,0) \cup (0,1), \; y \in (0,1), \; \text{where}$$

$$u_\varepsilon(x,y) = \left(1 - \frac{\exp(-\frac{1+x}{\varepsilon}) + \exp(\frac{x}{\varepsilon})}{1 + \exp(-\frac{1}{\varepsilon})}\right)\left(1 - \frac{\exp(-\frac{y}{\varepsilon}) + \exp(-\frac{1-y}{\varepsilon})}{1 + \exp(-\frac{1}{\varepsilon})}\right)$$

for $(x,y) \in (-1,0) \times (0,1)$,

$$u_\varepsilon(x,y) = \left(1 - \frac{\exp(-\frac{x}{\varepsilon}) + \exp(-\frac{1-x}{\varepsilon})}{1 + \exp(-\frac{1}{\varepsilon})}\right)\left(1 - \frac{\exp(-\frac{y}{\varepsilon}) + \exp(-\frac{1-y}{\varepsilon})}{1 + \exp(-\frac{1}{\varepsilon})}\right)$$

for $(x,y) \in (0,1) \times (0,1)$, $u(-1,y) = 0$, $u(1,y) = 0$, $u(x,0) = 0$, $u(x,1) = 0$, and f, $K(y) = \varepsilon[\frac{\partial u}{\partial x}]_{x=0}(y)$ are calculated from the exact solution.

Acknowledgements

This research was supported by the Bulgarian National Fund of Science under Project VU-MI-106/2005.

References

1. Andreev, V.: On the accuracy of grid approximation to nonsmooth solutions of a singularly perturbed reaction-diffusion equation in the square. Differential Equations 42, 954–966 (2006) (in Russian)
2. Angelova, I., Vulkov, L.: Singularly perturbed differential equations with discontinuous coefficients and concentrated factors. Appl. Math. Comp. 158, 683–701 (2004)
3. Angelova, I., Vulkov, L.: High-Order Difference Schemes for Elliptic Problems with Intersecting Interfaces. Appl. Math. Comp. 158(3), 683–701 (2007)
4. Angelova, I., Vulkov, L.: Marchuk identity-type second order difference schemes of 2-D and 3-D elliptic problems with intersected interfaces. In: Krag. J. Math (2007)
5. Braianov, I., Vulkov, L.: Numerical solution of a reaction-diffusion elliptic interface problem with strong anisotropy. Computing 71(2), 153–173 (2003)
6. Clavero, C., Gracia, J.: A compact finite difference scheme for 2D reaction-diffusion singularly perturbed problems. JCAM 192, 152–167 (2006)
7. Clavero, C., Gracia, J., O'Riordan, E.: A parameter robust numerical method for a two dimensionsl reaction-diffusion problem. Math. Comp. 74, 1743–1758 (2005)
8. Dunne, R., O'Riordan, E.: Interior layers arising in linear singularly perturbed differential equations with discontinuous coefficients, MS-06-09, 1–23 (2006)
9. Grisvard, P.: Boundary value Problems in Non-smooth Domains, Pitman, London (1985)
10. Han, H., Kellog, R.: Differentiability properties of solutions of the equation $-\varepsilon^2 \triangle u + ru = f(x,y)$ in a square. SIAM J. Math. Anal. 21, 394–408 (1990)
11. Jovanović, B., Vulkov, L.: Regularity and a priori estimates for solutions of an elliptic problem with a singular source. In: J. of Diff. Eqns. (submitted)
12. Li, J., Wheller, M.: Uniform superconvergence of mixed finite element methods on anisotropically refined grids. SINUM 38(3), 770–798 (2000)
13. Miller, J., O'Riordan, E., Shishkin, G.: Fitted numerical methods for singular perturbation problems. World-Scientific, Singapore (1996)

Immersed Interface Difference Schemes for a Parabolic-Elliptic Interface Problem

Ilia A. Brayanov, Juri D. Kandilarov, and Miglena N. Koleva

University of Rousse, Department of Mathematics, 7017 Rousse, Bulgaria
brayanov@ru.acad.bg, ukandilarov@ru.acad.bg, mkoleva@ru.acad.bg

Abstract. Second order immersed interface difference schemes for a parabolic-elliptic interface problem arising in electromagnetism is presented. The numerical method uses uniform Cartesian meshes. The standard schemes are modified near the interface curve taking into account the specific jump conditions for the solution and the flux. Convergence of the method is discussed and numerical experiments, confirming second order of accuracy are shown.

1 Introduction

Let Ω be a bounded domain in $\mathbb{R}^2$ with a piecewise smooth boundary $\partial\Omega$ and $\Gamma \subset \Omega$ be a smooth curve, which splits the domain Ω into two separate regions Ω^+, Ω^-, $\Omega = \Omega^+ \cup \Omega^- \cup \Gamma$. We consider the following parabolic-elliptic interface problem

$$u_t = \Delta u + f(x,y,t) \quad (x,y,t) \in \Omega^- \times (0,T], \tag{1}$$

$$0 = \Delta u + f(x,y,t) \quad (x,y,t) \in \Omega^+ \times (0,T], \tag{2}$$

$$u(x,y,0) = u_0(x,y), \quad (x,y) \in \bar{\Omega}^-, \tag{3}$$

$$u(x,y,t) = u_B(x,y,t), \quad (x,y,t) \in \partial\Omega \times [0,T], \tag{4}$$

with conjugation conditions on the interface $\Gamma_T = \Gamma \times [0,T]$

$$[u]_\Gamma := u^+(x(s),y(s),t) - u^-(x(s),y(s),t) = \varphi(x(s),y(s),t), \tag{5}$$

$$\left[\frac{\partial u}{\partial \mathbf{n}}\right]_\Gamma = \psi(x(s),y(s),t), \tag{6}$$

where s is a parameter of Γ, the superscripts $+$ and $-$ denote the limiting values of a function from one side in Ω^+ and another side in Ω^- respectively, and $\mathbf{n}$ is the normal vector at the point $(x(s),y(s)) \in \Gamma$, directed from Ω^- to Ω^+. For the initial data f, u_0, u_B, φ, ψ, and the solution u we assume to be sufficiently smooth with the exception of f and u, that may have discontinuity on Γ.

The problem (1)–(6) arises, when we study the production of eddy currents in a metallic cylinder due to particular type of external electromagnetic field as a special case, see [1]. Existence and uniqueness of the solution has been studied by MacCamy and Suri [11] and Al-Droubi [2]. The basic strategy is to solve the

I. Lirkov, S. Margenov, and J. Waśniewski (Eds.): LSSC 2007, LNCS 4818, pp. 661–669, 2008.

exterior problem by use of potential theory and then to reduce the problem to an interior problem.

Numerous methods have been developed for interface problems [8,10,15]. The IIM proposed by LeVeque and Li [7] solves elliptic equations with jump relations on Γ, which are known functions, defined on the interface. It has been successfully implemented for 1D and 2D linear and nonlinear elliptic and parabolic equations [8,9]. Some 2D problems with jump conditions, that depend on the solution on the interface are considered by Kandilarov and Vulkov [4,5,6,13]. The main goal of this work is the application of the IIM to the proposed parabolic-elliptic problem and theoretical validation of its implementation.

We present an algorithm for numerical solution of problem (1)–(6) which consists in two parts. First, using Shortley-Weller approximation [3] of Δu, we solve an elliptic problem on Ω^+ for $t = 0$. Next, a finite difference scheme is constructed using IIM for the problem on the whole domain Ω. Convergence properties of the proposed schemes are discussed. A comparison of the numerical results against the exact solution shows that the method is near second order accurate.

2   The Numerical Method

Let $\overline{\Omega}$ be a unit square and let us introduce on $\overline{\Omega}_T = \overline{\Omega} \times [0,T]$ the uniform mesh $\overline{\omega}_{h,\tau} = \overline{\omega}_h \times \overline{\omega}_\tau$, where $\overline{\omega}_h = \overline{\omega}_{h_1} \times \overline{\omega}_{h_2}$ and

$$\overline{\omega}_{h_1} = \{x_i = ih_1, \ i = 0, 1, ..., M_1, \ h_1 = 1/M_1\},$$
$$\overline{\omega}_{h_2} = \{y_j = jh_2, \ j = 0, 1, ..., M_2, \ h_2 = 1/M_2\},$$
$$\overline{\omega}_\tau = \{t_n = n\tau, \ n = 0, 1, ..., N, \ \tau = T/N\}.$$

Let ω_h and $\partial\omega_h$ be the sets of mesh points of Ω and $\partial\Omega$ respectively. Let also ω_h^+ and ω_h^- be the sets of mesh points of Ω^+ and Ω^-. With γ_h we denote the points of ω_h, that lie on the interface curve Γ. Then $\omega_h = \omega_h^+ \cup \omega_h^- \cup \gamma_h$.

Let us introduce the level set function $\phi(x,y)$ for the curve Γ, such that $\phi(x,y) = 0$ when $(x,y) \in \Gamma$, $\phi(x,y) < 0$ for $(x,y) \in \Omega^-$ and $\phi(x,y) > 0$ for $(x,y) \in \Omega^+$. The outward normal $\mathbf{n}(n_1, n_2)$ of the curve Γ is directed from Ω^- to Ω^+. We call the node (x_i, y_j) **regular**, if $\phi(x_i, y_j)$, $\phi(x_{i-1}, y_j)$, $\phi(x_{i+1}, y_j)$, $\phi(x_i, y_{j+1})$ and $\phi(x_i, y_{j-1})$ are together positive (negative), i.e. the curve Γ doesn't intersect the stencil. The rest of the nodes we call **irregular**. The set of irregular points is divided into three subsets: ω_{ir}^+ if the point $(x_i, y_j) \in \Omega^+$, ω_{ir}^- if $(x_i, y_j) \in \Omega^-$ and γ_h if $(x_i, y_j) \in \Gamma$.

First stage of the algorithm. The function $u_0(x,y)$ is definite on Ω^-, therefore to have initial data on the whole domain Ω for $t = 0$ we must solve numerically the elliptic problem for $w(x,y) = u(x,y,0)$ with Dirihlet boundary conditions:

$$-\Delta w = f(x,y,0), \qquad\qquad (x,y) \in \Omega^+, \qquad\qquad (7)$$
$$w(x,y) = u_B(x,y,0), \qquad\quad (x,y) \in \partial\Omega, \qquad\qquad (8)$$
$$w(x,y) = u_0(x,y) + \varphi(x,y,0), \ (x,y) \in \Gamma. \qquad\qquad (9)$$

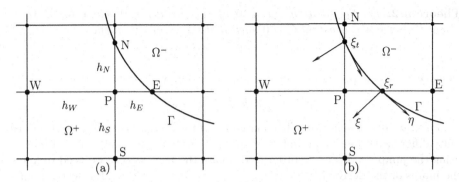

Fig. 1. (a) The stencil at irregular point $P = (x_i, y_j)$ for the Shortley-Weller approximation; (b) The stencil for the IIM and local coordinate system at points (ξ_r, y_j), (x_i, ξ_t), when the interface curve Γ intersects the right and top arm of the standard 5-point stencil

The interior boundary Γ is a curve. For the discretization of (7)–(9) we use the well known Shortley-Weller approximation of the discrete Laplace operator $-\Delta_h \overline{U}$, defined in [3,12]

$$
\begin{aligned}
-\Delta_h \overline{U}(P) &= -f(P), & P &\in \omega_h^+, \\
\overline{U}(P) &= u_B(P, 0), & P &\in \partial \omega_h, \\
\overline{U}(P) &= u_0(P) + \varphi(P, 0), & P &\in \gamma_h,
\end{aligned}
\tag{10}
$$

where

$$
-\Delta_h \overline{U}(P) = \left(\frac{2}{h_E h_W} + \frac{2}{h_N h_S} \right) \overline{U}(P) - \frac{2}{h_E(h_E + h_W)} \overline{U}(E)
$$
$$
- \frac{2}{h_W(h_E + h_W)} \overline{U}(W) - \frac{2}{h_N(h_N + h_S)} \overline{U}(N) - \frac{2}{h_S(h_N + h_S)} \overline{U}(S).
$$

This approximation is standard at regular grid points and then $h_E = h_W = h_1$, $h_N = h_S = h_2$. At irregular grid points, which are closed to the interior boundary Γ we must know the intersection points of the interface Γ with the grid lines through the point P, see for an example the points E and N in Fig. 1(a), when Γ crosses the right and top arm of the stencil.

Let $h = \min\{h_1, h_2\}$, $Tr^0(P)$ be the local truncation error (LTE) and $e^0(P) = \overline{U}(P) - u(P, 0)$ be the error. Then the following results hold.

Theorem 1. *Let* $u(x, y, 0)$, *the solution of (1)–(6) at* $t = 0$ *belongs to* $C^4(\Omega^+)$. *Then there exists a constant* $C > 0$ *such that the local truncation error* $Tr^0(P)$ *of the scheme (10) at point* P *satisfies*

$$
\begin{aligned}
|Tr^0(P)| &\le Ch^2, & P &\in \omega_h^+ \backslash \omega_{ir}^+, \\
|Tr^0(P)| &\le Ch, & P &\in \omega_{ir}^+.
\end{aligned}
$$

Theorem 2. *Let the solution $u(x, y, 0)$ of (1)–(6) at $t = 0$ belong to $C^4(\Omega^+)$. Then there exists a constant $C > 0$ such that the error e^0 satisfies*

$$\|e^0\|_\infty \leq Ch^2, \qquad P \in \omega_h^+.$$

Remark 1. At the points $P \in \omega_{ir}^+$ the super convergence property is satisfied [3]:

$$|e^0(P)| \leq Ch^3, \qquad P \in \omega_{ir}^+.$$

Second stage of the algorithm. The main idea of the IIM is to modify the standard finite difference schemes on uniform Cartesian grids at the irregular grid points, using the jump conditions in order to decrease the LTE [9]. So, we need to know the jumps of the derivatives $[u_x]$, $[u_y]$, $[u_{xx}]$, and $[u_{yy}]$ at the intersection points of Γ by the grid lines. For this goal we use the idea of Z. Li from [7]. We introduce local coordinate system at each intersection point of the interface curve with the standard 5-point stencil, for example (ξ_r, y_j) of the right arm, see Fig. 1(b),

$$\xi = (x - \xi_r)\cos\theta_r + (y - y_j)\sin\theta_r, \quad \eta = -(x - \xi_r)\sin\theta_r + (y - y_j)\cos\theta_r.$$

Here θ_r is the angle between the axis Ox and the normal vector $\mathbf{n} = (\cos\theta_r, \sin\theta_r)$ at the point (ξ_r, y_j). Then Γ can be locally parameterized by $\xi = \chi(\eta)$ and $\eta = \eta$. Note that $\chi(0) = 0$ and for a smooth curve $\chi'(0) = 0$.

For the first derivatives in the new directions we get

$$u_\xi = u_x \cos\theta_r + u_y \sin\theta_r, \qquad u_\eta = -u_x \sin\theta_r + u_y \cos\theta_r. \tag{11}$$

In a similar way we find the derivatives of second order and the jumps up to second order in the new coordinates. After an inverse transformation we have at the point (ξ_r, y_j) (see also [6,9] for more details):

$$[u_x] = \psi\cos\theta_r - \varphi_\eta\sin\theta_r, \qquad [u_y] = \psi\sin\theta_r + \varphi_\eta\cos\theta_r, \tag{12}$$

$$[u_{xx}] = (\chi''\psi - \varphi_{\eta\eta} - [f] - u_t^-)\cos^2\theta_r - 2(\chi''\varphi_\eta + \psi_\eta)\cos\theta_r\sin\theta_r$$
$$+(-\chi''\psi + \varphi_{\eta\eta})\sin^2\theta_r, \tag{13}$$

$$[u_{yy}] = (\chi''\psi - \varphi_{\eta\eta} - [f] - u_t^-)\sin^2\theta_r + 2(\chi''\varphi_\eta + \psi_\eta)\cos\theta_r\sin\theta_r$$
$$+(-\chi''\psi + \varphi_{\eta\eta})\cos^2\theta_r,$$

where $(.)_\eta$ is the derivative in tangential direction, $[f]$ is the jump of f, χ'' is the curvature of Γ at the intersection point (ξ_r, y_j) and u_t^- is the limiting value of the derivative in time at the same point.

At every irregular point we use the Taylor expansion. As example, for the situation on Fig. 1(b) we have:

$$\frac{\partial^2 u}{\partial x^2}(x_i, y_j) = \frac{u_{i+1,j} - 2u_{ij} + u_{i-1,j}}{h_1^2} \tag{14}$$

$$+ \frac{[u]}{h_1^2} + \frac{(x_{i+1} - \xi_r)}{h_1^2}[u_x] + \frac{(x_{i+1} - \xi_r)^2}{2h_1^2}[u_{xx}] + O(h_1),$$

$$\frac{\partial^2 u}{\partial y^2}(x_i, y_j) = \frac{u_{i,j+1} - 2u_{ij} + u_{i,j-1}}{h_2^2} \tag{15}$$

$$+ \frac{[u]}{h_2^2} + \frac{(y_{j+1} - \xi_t)}{h_2^2}[u_y] + \frac{(y_{j+1} - \xi_t)^2}{2h_2^2}[u_{yy}] + O(h_2),$$

where the jumps are evaluated at the points (ξ_r, y_j) and (x_i, ξ_t) respectively.

Using this expressions and (5), (12), (13), the implicit difference scheme with the IIM can be written in the form:

$$\sigma \frac{U_{ij}^{n+1} - U_{ij}^n}{\tau} = \frac{U_{i+1,j}^{n+1} - 2U_{ij}^{n+1} + U_{i-1,j}^{n+1}}{h_1^2} + D_{x,ij}^{n+1} \tag{16}$$

$$+ \frac{U_{i,j+1}^{n+1} - 2U_{ij}^{n+1} + U_{i,j-1}^{n+1}}{h_2^2} + D_{y,ij}^{n+1} + f_{i,j}^{n+1}, \quad (x_i, y_j) \in \omega_h,$$

$$U_{i,j}^n = u_B(x_i, y_j, t_n), \qquad (x_i, y_j) \in \partial\omega_h, \tag{17}$$

$$U_{i,j}^0 = u_0(x_i, y_j), \qquad (x_i, y_j) \in \omega_h^- \cup \gamma_h, \tag{18}$$

$$U_{i,j}^0 = \overline{U}^0(x_i, y_j), \qquad (x_i, y_j) \in \omega_h^+, \tag{19}$$

where U_{ij}^n approximates $u(x_i, y_j, t_n)$ and $f_{i,j}^{n+1} = f(x_i, y_j, t_{n+1})$. Here $D_{x,ij}^{n+1} = D_{xl,ij}^{n+1} + D_{xr,ij}^{n+1}$ and $D_{y,ij}^{n+1} = D_{yt,ij}^{n+1} + D_{yb,ij}^{n+1}$ are additional terms chosen in order to improve the LTE to the first order at the irregular points. By l, r, t, b, we show the intersection of the interface curve, respectively, with the left, right, top or bottom arm of the standard 5-point stencil for the discrete elliptic operator at (x_i, y_j). The coefficient σ is: $\sigma = 1$ if $(x_i, y_j) \in \omega_h^- \backslash \omega_{ir}^-$; $\sigma = 0$ if $(x_i, y_j) \in \omega_h^+ \backslash \omega_{ir}^+$; $\sigma = 1 - \sum_{k=l,r,t,b} \rho_k$ if $(x_i, y_j) \in \omega_{ir}^- \cup \gamma_h$; $\sigma = \sum_{k=l,r,t,b} \rho_k$ if $(x_i, y_j) \in \omega_{ir}^+$ and ρ_k are the coefficients of the terms, including u_t^- (see (13)). At irregular points of ω_{ir}^+ one must use also the relation $u_t^+ = u_t^- - \varphi_t$. If $(x_i, y_j) \in \gamma_h$, for definiteness we choose $U_{ij}^n \approx u^+(x_i, y_j, t_n)$, see [14].

The convergence results are based on the maximum principle, described in [12], so we will also use the notation $U^n(P)$ for the numerical solution at the node (x_i, y_j, t_n).

Theorem 3. *Let the interface curve $\Gamma \in C^2$ and the solution $u(x, y, t)$ of (1)– (6) belongs to $C^{4,2}((\Omega^+ \cup \Omega^-) \times (0, T])$. Then there exists a constant $C > 0$ such that the local truncation error $Tr^n(P)$ of the scheme (16)–(19) satisfies*

$$\begin{aligned} |Tr^n(P)| &\leq C(\tau + h^2), & P &\in \omega_h^- \backslash \omega_{ir}^-, \\ |Tr^n(P)| &\leq C(\sigma\tau + h), & P &\in \omega_{ir}^- \cup \omega_{ir}^+ \cup \gamma_h, \\ |Tr^n(P)| &\leq Ch^2, & P &\in \omega_h^+ \backslash \omega_{ir}^+. \end{aligned}$$

Proof. The first and third estimates are standard. The estimate for the LTE at irregular grid nodes follows by construction, see (14)–(15), the jump conditions (5), (12), (13), and the terms therein. □

Remark 2. If the mesh parameter h and the curvature χ'' of Γ satisfy the inequality $h \leq 1/\max|\chi''|$, then for the parameter σ we have: $1/2 < \sigma < 1$ if $P \in \omega_{ir}^- \cup \gamma_h$ and $0 \leq \sigma < 1/2$ if $P \in \omega_{ir}^+$, and the constant C does not depend on h and τ.

Theorem 4. *Let the interface curve $\Gamma \in C^2$ and the solution $u(x, y, t)$ of (1)–(6) satisfy $u \in C^{4,2}((\Omega^+ \cup \Omega^-) \times (0, T])$. Then the numerical solution U of the scheme (16)–(19) converges to the solution of (1)–(6) and*

$$\|U^n(P) - u(x_i, y_j, t_n)\|_\infty \leq C(\tau + h), \qquad P \in \omega_{h,\tau}. \tag{20}$$

Proof. We outline the proof. Let for simplicity $\Omega = [-1, 1]^2$ and $h_1 = h_2 = h$. We define the difference operator $L_h U^{n+1}(P)$ in the form

$$L_h U^{n+1}(P) := -\sigma \frac{U_{ij}^{n+1} - U_{ij}^n}{\tau} + \frac{U_{i+1,j}^{n+1} + U_{i-1,j}^{n+1} + U_{i,j+1}^{n+1} + U_{i,j-1}^{n+1} - 4U_{ij}^{n+1}}{h^2}.$$

Then from (16) we have:

$$L_h U^{n+1}(P) + f^{n+1}(P) + g^{n+1}(P) = 0,$$

where $g^{n+1}(P) = D_{x,ij}^{n+1} + D_{y,ij}^{n+1}$ at irregular grid points. For the error $e^n(P) = U^n(P) - u(x_i, y_j, t_n)$ we obtain:

$$L_h e^n(P) = Tr^n(P), \qquad P \in \omega_h.$$

Let us define a barrier function $\Phi^n(P)$ of the form

$$\Phi^n(P) := At_n + E_1(2 - x_i^2 - y_j^2)/4 + E_2, \qquad P = (x_i, y_j) \in \omega_h,$$

Then we apply the difference operator to $e^n(P) - \Phi^n(P)$. If the nonnegative constants A and E_1 satisfy $A + E_1 \geq Tr_{max} := \max_{n,\omega_h} |Tr^n(P)|$ it follows

$$L_h(e^n(P) - \Phi^n(P)) = A + E_1 + Tr^n(P) \geq 0, \qquad P \in \omega_h.$$

On the boundary $e^n(P) = 0$ and $\Phi^n(P) = At_n + E_1(1 - x^2)/4 + E_2$ or $\Phi^n(P) = At_n + E_1(1 - y^2)/4 + E_2$, $P \in \partial\omega_h$. Similarly, on the zero time layer $t_0 = 0$ $\Phi^0(P) = E_1(2 - x_i^2 - y_j^2)/4 + E_2$. If $E_2 = \|e^0\|_\infty$ (see Theorem 2), then $e^n(P) - \Phi^n(P) \leq 0$ on the boundary. The conditions of Theorem 6.1 of [12] are fulfilled and therefore $e^n(P) - \Phi^n(P) \leq 0$ at every mesh point of $\overline{\omega}_{h,\tau}$. The same procedure for $-e^n(P) - \Phi^n(P)$ leads to $-e^n(P) - \Phi^n(P) \leq 0$ and hence $|e^n(P)| \leq \Phi^n(P)$. Taking $A = E_1 = Tr_{max}$ we obtain (20). $\qquad \square$

3   Numerical Experiments

Example 1. On the region $\Omega = (-1, 1)^2 \setminus \Gamma = \Omega^- \cup \Omega^+$ ($\Omega^\pm = x^2 + y^2 - r_0^2 \gtrless 0$) and interface curve $\Gamma : x^2 + y^2 = r_0^2$ we consider the problem (1)–(2) with exact solution

$$u(x, y, t) = \exp(-t) \begin{cases} k J_0(r), & (x, y) \in \Omega^-, \\ J_0(r_0) Y_0(r)/Y_0(r_0), & (x, y) \in \Omega^+, \end{cases}$$

where J_0 and Y_0 are the Bessel functions of order zero, $r = \sqrt{x^2 + y^2}$. The boundary condition, the function $u_0(x, y)$, $(x, y) \in \bar{\Omega}^-$ and $f(x, y, t)$ one obtains

Table 1. Mesh refinement analysis for *Example 1*, where Γ is a circle with $r_0 = 0.505$

$M_1 = M_2$	$\|Tr^1\|_\infty$	m	$\|e^1\|_\infty$	m	$\|e^N\|_\infty$	m
10	2.7103e-01	-	8.4659e-03	-	8.9430e-03	-
20	3.4306e-01	0.34	1.3527e-03	2.64	2.6852e-03	1.74
40	1.4364e-01	1.26	2.8032e-04	2.27	5.2057e-04	2.37
80	8.1532e-02	0.82	7.2053e-05	1.96	1.4589e-04	1.83
160	4.6742e-02	0.80	1.7409e-05	2.05	4.2135e-05	1.79

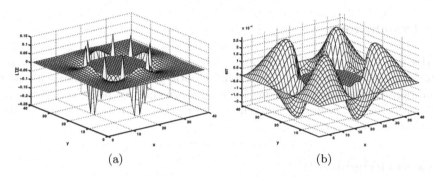

(a)   (b)

Fig. 2. (a) The local truncation error Tr^0 and (b) the error e^0 of the numerical solution for *Example 1*, when $r_0 = .505$, $k = 3$, $t = 0$, $M_1 = M_2 = 40$

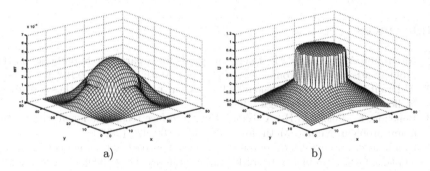

a)   b)

Fig. 3. (a) The error e^N and (b) the numerical solution U at final time $T = 1$ for *Example 1*, $M_1 = M_2 = 40$, $k = 3$, $\tau/h^2 = 10$

from the exact solution. The solution is discontinuous (if $k \neq 1$), as well as the jumps of the normal derivative on the interface, of the function f and u_t:

$$[u] = \exp(-t)J_0(r_0)(1 - k), \quad \left[\frac{\partial u}{\partial \mathbf{n}}\right] = \exp(-t)\left(\frac{J_0(r_0)}{Y_0(r_0)}Y_0'(r) - kY_0'(r)\right)$$
$$[f] = -\exp(-t)J_0(r_0), \quad [u_t] = -\exp(-t)J_0(r_0)(1 - k).$$

We choose $k = 3$ and circle interface with radius $r_0 = .505$. Since the method is expected to be of order $O(\tau + h^2)$, to prevent the dominant influence of the parameter τ into the error we keep the ratio τ/h^2 constant (in our example it is equal to 10). The mesh refinement analysis is presented in Table 1. We control the LTE $\|Tr\|_\infty$ on the first time layer, and also the error on the first $(t_1 = \tau)$ and on the last time layer $(t_N = T)$. The rate of convergence we denote by m. The results show that the LTE is $O(h)$. The second order of accuracy of the method is also confirmed. In Fig. 2(a) the LTE and in Fig. 2(b) the error of the solution at $t_0 = 0$ are plotted, when $M_1 = M_2 = 40$. In Fig. 3(a) the error and in Fig. 3(b) the numerical solution at final time $T = 1$ are shown with the same data.

4   Conclusions

In this paper we have developed a finite difference method based on the immersed interface method for a two-dimensional elliptic-parabolic problem with jump conditions on the interface. Using maximum principle we prove convergence of first order of the proposed difference scheme. The numerical results indicate second order of accuracy.

Acknowledgements

This work was supported in part by National Science Fund of Bulgaria under contract HS-MI-106/2005.

References

1. Al-Droubi, A., Renardy, M.: Energy methods for a parabolic-hyperbolic interface problem arising in electromagnetism. J. Appl. Math. Phys. 39, 931–936 (1988)
2. Al-Droubi, A.: A two-dimensional eddy current problem. Ph. D. thesis, Carnegie-Mellon University, Pittsburgh (1987)
3. Bouchon, F., Peichl, G.H.: A second order interface technique for an elliptic Neumann problem. Numer. Meth. for PDE 23(2), 400–420 (2007)
4. Kandilarov, J.: Immersed-boundary level set approach for numerical solution of elliptic interface problems. In: Lirkov, I., et al. (eds.) LSSC 2003. LNCS, vol. 2907, pp. 456–464. Springer, Heidelberg (2004)
5. Kandilarov, J., Vulkov, L.: The immersed interface method for a nonlinear chemical diffusion equation with local sites of reactions. Numer. Algor. 36, 285–307 (2004)
6. Kandilarov, J., Vulkov, L.: The immersed interface method for two-dimensional heat-diffusion equations with singular own sources. Appl. Num. Math. 57, 486–497 (2007)
7. LeVeque, R., Li, Z.: The immersed interface method for elliptic equations with discontinuous coefficients and singular sources. SIAM J. Num. Anal. 31, 1019–1044 (1994)
8. Li, Z.: An overview of the immersed interface method and its applications. Taiwanese J. of Mathematics 7(1), 1–49 (2003)
9. Li, Z., Ito, K.: The Immersed Interface Method: Numerical Solutions of PDEs Involving Interfaces and Irregular Domains. SIAM, Philadelphia (2006)

10. Liu, X., Fedkiw, R., Kang, M.: A boundary condition capturing method for Poisson's equation on irregular domain. J. Comput. Phys. 160, 151–178 (2000)
11. MacCamy, R., Suri, M.: A time-dependent interface problem for two-dimentional eddy currents. Quart. Appl. Math. 44, 675–690 (1987)
12. Morton, K.W., Mayers, D.F.: Numerical Solution of Partial Differential Equations. Cambridge University Press, Cambridge (1995)
13. Vulkov, L., Kandilarov, J.: Construction and implementation of finite-difference schemes for systems of diffusion equations with localized nonlinear chemical reactions. Comp. Math. Math. Phys. 40, 705–717 (2000)
14. Wiegmann, A., Bube, K.: The explicit jump immersed interface method: Finite difference methods for PDE with piecewise smooth solutions. SIAM J. Numer. Anal. 37, 827–862 (2000)
15. Zhou, Y.C., et al.: High order matched interface and boundary method for elliptic equations with discontinuous coefficients and singular sources. J. Comput. Phys. 213, 1–30 (2005)

Surface Reconstruction and Lagrange Basis Polynomials

Irina Georgieva[1] and Rumen Uluchev[2]

[1] Institute of Mathematics and Informatics, Bulgarian Academy of Sciences,
Acad. G. Bonchev, Bl. 8, Sofia 1113, Bulgaria
irina@math.bas.bg
[2] Department of Mathematics and Informatics, University of Transport,
158 Geo Milev Str., Sofia 1574, Bulgaria
rumenu@vtu.bg

Abstract. Surface reconstruction, based on line integrals along segments of the unit disk is studied. Various methods concerning with this problem are known. We consider here interpolation over regular schemes of chords by polynomials. We find the interpolant in Lagrange form and investigate some properties of Lagrange basis polynomials. Numerical experiments for both surface and image reconstruction are presented.

1 Introduction and Preliminaries

There are a lot of practical problems, e.g. in tomography, electronic microscopy, etc., in which information about the relevant function comes as values of its Radon projections. Because of the importance of such non-destructive methods for applications in medicine and technics they have been intensively investigated by many mathematicians [2,5,10,11,12,13], and others. Part of the algorithms are based on the inverse Radon transform (see [11,12] and the bibliography therein) and others, like interpolation and smoothing (see [1,4,6,7,8,10]) are direct methods. It turns out that the smoothing problem and the interpolation problem are closely related. In [7] it was shown that existence and uniqueness of the best smoothing polynomial relies on a regularity property of the scheme of chords.

Here we express the interpolant of Radon type of data in Lagrange form and investigate some properties of the corresponding Lagrange basis polynomials. Similarly to the univariate case it is convenient to use Lagrange form of the interpolant if one and the same configuration of chords should be used for a large amount of calculations.

We denote by Π_n^2 the set of all algebraic polynomials in two variables of total degree at most n and real coefficients. Then, Π_n^2 is a linear space of dimension $\binom{n+2}{2}$ and $P \in \Pi_n^2$ if and only if

$$P(x,y) = \sum_{i+j \leq n} \alpha_{ij} x^i y^j, \qquad \alpha_{ij} \in \mathbb{R}.$$

Let $\mathbf{B} := \{\mathbf{x} = (x,y) \in \mathbb{R}^2 : \|\mathbf{x}\| \leq 1\}$ be the unit disk in the plane, where $\|\mathbf{x}\| = \sqrt{x^2 + y^2}$. Given $t \in [-1,1]$ and an angle of measure $\theta \in [0,\pi)$,

I. Lirkov, S. Margenov, and J. Waśniewski (Eds.): LSSC 2007, LNCS 4818, pp. 670–678, 2008.

the equation $x \cos \theta + y \sin \theta - t = 0$ defines a line ℓ perpendicular to the vector $\langle \cos \theta, \sin \theta \rangle$ and passing through the point $(t \cos \theta, t \sin \theta)$. The set $I(\theta, t) := \ell \cap \mathbf{B}$ is a *chord* of the unit disk $\mathbf{B}$ which can be parameterized in the manner

$$\begin{cases} x = t \cos \theta - s \sin \theta, \\ y = t \sin \theta + s \cos \theta, \end{cases} \quad s \in [-\sqrt{1 - t^2}, \sqrt{1 - t^2}],$$

where the quantity θ is the direction of $I(\theta, t)$ and t is the distance of the chord from the origin. Suppose that for a given function $f : \mathbb{R}^2 \to \mathbb{R}$ the integrals of f exist along all line segments on the unit disk $\mathbf{B}$. *Radon projection* (or X-*ray*) of the function f over the segment $I(\theta, t)$ is defined by

$$\mathcal{R}_\theta(f; t) := \int_{I(\theta, t)} f(\mathbf{x}) \, d\mathbf{x} = \int_{-\sqrt{1 - t^2}}^{\sqrt{1 - t^2}} f(t \cos \theta - s \sin \theta, t \sin \theta + s \cos \theta) \, ds.$$

Clearly, $\mathcal{R}_\theta(\cdot; t)$ is a linear functional. Since $I(\theta, t) \equiv I(\theta + \pi, -t)$ it follows that $\mathcal{R}_\theta(f; t) = \mathcal{R}_{\theta + \pi}(f; -t)$. Thus, the assumption above for the direction of the chords $0 \le \theta < \pi$ is not loss of generality.

It is well-known that the set of Radon projections

$$\left\{ \mathcal{R}_\theta(f; t) \ : \ -1 \le t \le 1, \ 0 \le \theta < \pi \right\}$$

determines f uniquely (see [9,13]). According to a more recent result in [14], an arbitrary function $f \in L^1(\mathbb{R}^2)$ with compact support in $\mathbf{B}$ is uniquely determined by any infinite set of X-rays. Since the function $f \equiv 0$ has all its projections equal to zero, it follows that the only function which has the zero Radon transform is the constant zero function. It was shown by Marr [10] that every polynomial $P \in \Pi_n^2$ can be reconstructed uniquely by its projections only on a finite number of directions.

Another important property (see [10,3]) is the following: if $P \in \Pi_n^2$ then for each fixed θ there exists a univariate polynomial p of degree n such that $\mathcal{R}_\theta(P; t) = \sqrt{1 - t^2} \, p(t)$, $-1 \le t \le 1$.

The space Π_n^2 has a standard basis of the power functions $\{x^i y^j\}$. Studying various problems for functions on the unit disk it is often helpful to use some orthonormal basis. In [2] the following orthonormal basis was constructed. Denote the Chebyshev polynomial of second kind of degree m as usual by

$$U_m(t) := \frac{1}{\sqrt{\pi}} \frac{\sin(m + 1)\psi}{\sin \psi}, \quad t = \cos \psi.$$

If set $\psi_{mj} := \frac{j\pi}{m+1}$, $m = 0, \ldots, n$, $j = 0, \ldots, m$, then the ridge polynomials

$$U_{mj}(\mathbf{x}) := U_m(x \cos \psi_{mj} + y \sin \psi_{mj}), \quad m = 0, \ldots, n, \ j = 0, \ldots, m,$$

form an orthonormal basis in Π_n^2.

The interpolation problem. For a given scheme of chords I_k, $k = 1, \ldots, \binom{n+2}{2}$, of the unit circle $\partial \mathbf{B}$, find a polynomial $P \in \Pi_n^2$ satisfying the conditions:

$$\int_{I_k} P(\mathbf{x}) \, d\mathbf{x} = \gamma_k, \quad k = 1, \ldots, \binom{n+2}{2}. \tag{1}$$

If (1) has a unique solution for every given set of values $\{\gamma_k\}$ the interpolation problem is called *poised* and the scheme of chords — *regular*.

The first known scheme which is regular for every degree n of the interpolating polynomial was found by Hakopian [8]. Hakopian's scheme consists of all $\binom{n+2}{2}$ chords, connecting given $n+2$ points on the unit circle $\partial\mathbf{B}$. Bojanov and Xu [4] proposed a regular scheme consisting of $2\lfloor\frac{n+1}{2}\rfloor + 1$ equally spaced directions with $\lfloor\frac{n}{2}\rfloor + 1$ chords, associated with the zeros of the Chebyshev polynomials of certain degree, in each direction.

Another family of regular schemes was provided by Bojanov and Georgieva [1]. In this case the Radon projections are taken along a set of $\binom{n+2}{2}$ chords $\{I(\theta,t)\}$ of the unit circle, partitioned into $n+1$ subsets, such that the k-th subset consists of $k+1$ parallel chords. More precisely, consider the scheme (Θ, T), where $\Theta := \{\theta_0, \theta_1, \ldots, \theta_n\}$, $\theta_0 < \cdots < \theta_n$ are in $[0, \pi)$, and $T := \{t_{ki}\}$ is a triangular matrix of points with the distances $t_{kk} > \cdots > t_{kn}$, associated with the angle measures θ_k, $k = 0, \ldots, n$:

$$\begin{array}{ll}
\theta_0 \to & t_{00} \; t_{01} \; \cdots \; t_{0n} \\
\theta_1 \to & \quad\;\; t_{11} \; \cdots \; t_{1n} \\
\vdots \;\; \vdots & \qquad\quad \ddots \;\; \vdots \\
\theta_n \to & \qquad\qquad\quad t_{nn}
\end{array}$$

The problem is to find a polynomial $P \in \Pi_n^2$ satisfying the interpolation conditions

$$\int_{I(\theta_k, t_{ki})} P(\mathbf{x})\,d\mathbf{x} = \gamma_{ki}, \qquad k = 0, \ldots, n, \quad i = k, \ldots, n, \tag{2}$$

A necessary and sufficient condition for regularity of the schemes of this type is proved by Bojanov and Georgieva [1]. Nevertheless, given a set of points T, it is often difficult to determine if the problem (2) has a unique solution.

Several regular schemes of this type were suggested by Georgieva and Ismail [6] and by Georgieva and Uluchev [7]. We summarize these results in the following theorem.

Theorem A. *Let n be given positive integer, $\Theta = \{\theta_0, \theta_1, \ldots, \theta_n\}$, $\theta_0 < \cdots < \theta_n$ be arbitrary in $[0, \pi)$, and let $T = \{t_{ki}\}_{k=0, i=k}^{n}$ be one of the following:*

(a) $t_{ki} = \xi_{i+1} = \cos\frac{(2i+1)\pi}{2(n+1)}$, $i = k, \ldots, n$, *are the zeros of the Chebyshev polynomial of first kind* $T_{n+1}(x)$;

(b) $t_{ki} = \eta_{i+1} = \cos\frac{(i+1)\pi}{n+2}$, $i = k, \ldots, n$, *are the zeros of the Chebyshev polynomial of second kind* $U_{n+1}(x)$;

(c) $t_{ki} = \cos\frac{(2i+1)\pi}{2n+3}$, $i = k, \ldots, n$, *are the zeros of the Jacobi polynomial* $P_{n+1}^{(1/2,-1/2)}(x)$;

(d) $t_{ki} = \cos\frac{2(i+1)\pi}{2n+3}$, $i = k, \ldots, n$, *are the zeros of the Jacobi polynomial* $P_{n+1}^{(-1/2,1/2)}(x)$.

Then the interpolation problem (2) is poised, i.e., the interpolatory scheme (Θ, T) is regular.

2   Lagrange Form of the Interpolant

Let us consider interpolation problem (2) for a given regular scheme (Θ, T). We define the Lagrange basis polynomials $\{\ell_{ki}(x, y)\}_{k=0, i=k}^{n \quad n}$ for this configuration of chords as follows: $\ell_{ki}(x, y)$ is the unique bivariate polynomial of degree at most n such that

$$\mathcal{R}_{\theta_j}(\ell_{ki}; t_{mj}) = \delta_{km}\delta_{ij}, \qquad m = 0, \ldots, n, \quad j = m, \ldots, n, \tag{3}$$

δ_{ki} being the Kroneker symbol. Then the interpolant can be written in the form

$$L_n(f; x, y) = \sum_{k=0}^{n} \sum_{i=k}^{n} \ell_{ki}(x, y)\mathcal{R}_{\theta_i}(f; t_{ki}).$$

It is a well-known property of the univariate Lagrange basis polynomials that they sum up to 1. For some special choices of the distances $\{t_{ki}\}_{k=0, i=k}^{n \quad n}$ of the chords from the origin we prove a bivariate analogue of this result.

Theorem 1. *Let (Θ, T) be any of the regular schemes of chords specified in Theorem A. Then the sum of all Lagrange basis polynomials $\{\ell_{ki}(x, y)\}_{k=0, i=k}^{n \quad n}$ for this configuration of chords is a radial polynomial of degree $2\lfloor \frac{n}{2} \rfloor$.*

Proof. Let us set $r = \lfloor \frac{n}{2} \rfloor$, i.e., $2r = n$ if n is even and $2r = n - 1$ for odd n. We shall need the non-negative zeros $0 \le \tau_r < \tau_{r-1} < \cdots < \tau_0 < 1$ of the corresponding orthogonal polynomial of degree $n+1$. For example, for the scheme from Theorem A, case (a), $\{\tau_q\}_{q=0}^{r}$ are the non-negative zeros of the Chebyshev polynomial of first kind $T_{n+1}(x)$; for the scheme of case (b) we take the non-negative zeros of the Chebyshev polynomial of second kind $U_{n+1}(x)$, etc.

Now we shall prove that there exists a unique radial polynomial

$$\varphi(x, y) = \sum_{\nu=0}^{r} \alpha_\nu (x^2 + y^2)^\nu \tag{4}$$

such that

$$\mathcal{R}_0(\varphi(x, y); \tau_q) = 1, \qquad q = 0, 1, \ldots, r. \tag{5}$$

Note that $\varphi \in \Pi_n^2$ since $2r \le n$. Moreover, $\theta = 0$ in (5) means that we take all chords perpendicular to the x-axis.

By the linearity of the functional $\mathcal{R}_\theta(\,\cdot\,; t)$ conditions (5) are equivalent to the linear system

$$\sum_{\nu=0}^{r} \alpha_\nu \mathcal{R}_0\big((x^2 + y^2)^\nu; \tau_q\big) = 1, \qquad q = 0, 1, \ldots, r, \tag{6}$$

with respect to the coefficients $\{\alpha_\nu\}$. Along a chord perpendicular to the x-axis and crossing the abscissa at $x = \tau_q$ we have

$$\mathcal{R}_0\big((x^2+y^2)^\nu; \tau_q\big) = \int_{-\sqrt{1-\tau_q^2}}^{\sqrt{1-\tau_q^2}} (\tau_q^2+s^2)^\nu \, ds = 2\sqrt{1-\tau_q^2} \sum_{j=0}^{\nu} \binom{\nu}{j} \frac{\tau_q^{2j}(1-\tau_q^2)^{\nu-j}}{2\nu - 2j + 1}.$$

Denote for convenience $p_{2\nu}(t) := 2 \sum_{j=0}^{\nu} \binom{\nu}{j} \dfrac{t^{2j}(1-t^2)^{\nu-j}}{2\nu - 2j + 1}$. Observe that $p_{2\nu}(t)$ is an even univariate polynomial of degree exactly 2ν (see also [3]) and the coefficient of $t^{2\nu}$ is

$$\frac{2(2\nu)!!}{(2\nu+1)!!} > 0. \tag{7}$$

Then

$$\mathcal{R}_0\big((x^2+y^2)^\nu; \tau_q\big) = \sqrt{1 - \tau_q^2}\, p_{2\nu}(\tau_q)$$

and the determinant of the system (6) is

$$\Delta = \begin{vmatrix} \sqrt{1-\tau_0^2}\, p_0(\tau_0) & \sqrt{1-\tau_0^2}\, p_2(\tau_0) & \cdots & \sqrt{1-\tau_0^2}\, p_{2r}(\tau_0) \\ \sqrt{1-\tau_1^2}\, p_0(\tau_1) & \sqrt{1-\tau_1^2}\, p_2(\tau_1) & \cdots & \sqrt{1-\tau_1^2}\, p_{2r}(\tau_1) \\ \cdots\cdots\cdots\cdots\cdots\cdots\cdots\cdots\cdots\cdots\cdots\cdots\cdots \\ \sqrt{1-\tau_r^2}\, p_0(\tau_r) & \sqrt{1-\tau_r^2}\, p_2(\tau_r) & \cdots & \sqrt{1-\tau_r^2}\, p_{2r}(\tau_r) \end{vmatrix}$$

$$= \begin{vmatrix} p_0(\tau_0) & p_2(\tau_0) & \cdots & p_{2r}(\tau_0) \\ p_0(\tau_1) & p_2(\tau_1) & \cdots & p_{2r}(\tau_1) \\ \cdots\cdots\cdots\cdots\cdots\cdots\cdots \\ p_0(\tau_r) & p_2(\tau_r) & \cdots & p_{2r}(\tau_r) \end{vmatrix} \cdot \prod_{q=0}^{r} \sqrt{1-\tau_q^2}.$$

Having in mind that all polynomials $p_{2\nu}(t)$ are even it follows from (7) that

$$\begin{vmatrix} p_0(\tau_0) & p_2(\tau_0) & \cdots & p_{2r}(\tau_0) \\ p_0(\tau_1) & p_2(\tau_1) & \cdots & p_{2r}(\tau_1) \\ \cdots\cdots\cdots\cdots\cdots\cdots\cdots \\ p_0(\tau_r) & p_2(\tau_r) & \cdots & p_{2r}(\tau_r) \end{vmatrix} = 2^{r+1} \prod_{\nu=0}^{r} \frac{(2\nu)!!}{(2\nu+1)!!} \begin{vmatrix} 1 & \tau_0^2 & \cdots & \tau_0^{2r} \\ 1 & \tau_1^2 & \cdots & \tau_1^{2r} \\ \cdots\cdots\cdots\cdots \\ 1 & \tau_r^2 & \cdots & \tau_r^{2r} \end{vmatrix}.$$

The last determinant is Vandermondeian, since $0 \le \tau_r^2 < \tau_{r-1}^2 < \cdots < \tau_0^2 < 1$. Therefore

$$\Delta = 2^{r+1} \prod_{q=0}^{r} \sqrt{1-\tau_q^2} \cdot \prod_{\nu=0}^{r} \frac{(2\nu)!!}{(2\nu+1)!!} \cdot \prod_{0 \le l < m < r} (\tau_l^2 - \tau_m^2) > 0.$$

Consider now the sum of all Lagrange basis polynomials of degree n constructed for the given scheme of chords

$$\Phi(x,y) = \sum_{k=0}^{n} \sum_{i=k}^{n} \ell_{ki}(x,y).$$

Clearly $\Phi(x,y) \in \Pi_n^2$ and

$$\mathcal{R}_{\theta_k}\big(\Phi; t_{ki}\big) = \mathcal{R}_{\theta_k}\big(\ell_{ki}; t_{ki}\big) = 1, \qquad k = 0, \ldots, n, \quad i = k, \ldots, n, \tag{8}$$

Since $\varphi(x,y)$ is a radial polynomial, from (5) it follows that

$$\mathcal{R}_{\theta_k}\big(\varphi; t_{ki}\big) = \mathcal{R}_0\big(\varphi; t_{ki}\big) = \begin{cases} \mathcal{R}_0\big(\varphi; \tau_i\big) = 1, & i \le \lfloor \frac{n}{2} \rfloor, \\ \mathcal{R}_0\big(\varphi; -\tau_{n-i}\big) = \mathcal{R}_0\big(\varphi; \tau_{n-i}\big) = 1, & i > \lfloor \frac{n}{2} \rfloor. \end{cases}$$

Therefore $\varphi(x,y)$ satisfies the interpolatory conditions (8). Moreover $\varphi(x,y)$ is a polynomial of degree $2r \leq n$, hence $\varphi \in \Pi_n^2$. The regularity of the scheme of chords (Θ, T) implies uniqueness of the interpolatory polynomial of degree n for (8), hence $\Phi(x,y) \equiv \varphi(x,y)$. The proof is complete.

3   Numerical Experiments

Here we give results from some numerical experiments made with the *Mathematica* software developed by Wolfram Research Inc.

A regular scheme of chords based on the zeros of U_{n+1} with $\theta_k = \frac{(k+1)\pi}{n+2}$, $k = 0, \ldots, n$ and $\{t_{ki} = \cos\frac{(i+1)\pi}{n+2}\}_{i=k}^{n}$, $k = 0, \ldots, n$, is used in all examples.

Example 1. We give the Lagrange basis polynomials $\{\ell_{ki}(x,y)\}_{k=0,i=k}^{n}$ of degree $n = 2$ defined with (3). We have $\theta_0 = \frac{\pi}{4}$, $t_{00} = \frac{1}{\sqrt{2}}$, $t_{01} = 0$, $t_{02} = \frac{-1}{\sqrt{2}}$, $\theta_1 = \frac{\pi}{2}$, $t_{11} = 0$, $t_{12} = \frac{-1}{\sqrt{2}}$, $\theta_2 = \frac{3\pi}{4}$, $t_{22} = \frac{-1}{\sqrt{2}}$ for the chords and

$$\ell_{00}(x,y) = -0.042893 + 0.207107x + 0.12868x^2 + 0.5y + 0.62132xy + 0.75y^2,$$
$$\ell_{01}(x,y) = -0.06066 - 0.707107x + 0.181981x^2 + 0.707107y - 2.12132xy$$
$$+ 1.06066y^2,$$
$$\ell_{02}(x,y) = -1.103553 - 1.207107x + 3.31066x^2 + 0.5y - 1.5xy + 1.81066y^2,$$
$$\ell_{11}(x,y) = 1.06066 + 0.707107x - 1.681981x^2 - 0.707107y + 2.12132xy$$
$$- 2.56066y^2,$$
$$\ell_{12}(x,y) = 1.5 + x - 4.5x^2 - y + 3xy - 1.5y^2,$$
$$\ell_{22}(x,y) = -1.06066 + 3.181981x^2 - 2.12132xy + 1.06066y^2.$$

E.g., for the function $f(x,y) = \sin xy$ the Radon projections along the above chords are: $\gamma_{00} = 0.234023$, $\gamma_{01} = -0.32748$, $\gamma_{02} = 0.234023$, $\gamma_{11} = \gamma_{12} = 0$, $\gamma_{22} = -0.234023$, $\gamma_{ki} = \mathcal{R}_{\theta_k}(f; t_{ki})$. Now we can write the interpolation polynomial of degree $n = 2$ for the problem (2) in Lagrange form:

$$P(x,y) = 0.234023\ell_{00}(x,y) - 0.3274\ell_{01}(x,y) + 0.234023\ell_{02}(x,y) + 0\,\ell_{11}(x,y)$$
$$+ 0\,\ell_{12}(x,y) - 0.234023\ell_{22}(x,y).$$

Hence,

$$P(x,y) = -0.00021 - 0.0025x + 0.00064x^2 + 0.0025y + 0.98539xy + 0.00374y^2.$$

The relative L_2-norm of the error on the unit disk is $\|f - P\|_2 / \|f\|_2 = 0.01579$.

Example 2. We interpolate the function

$$f(x,y) = \ln(x^2 + y + 1.5)\cos(5x + y)$$

using $N = \binom{42}{2} = 861$ pieces of Radon projections by polynomial of degree $n = 40$. The surface $z = f(x,y)$, its reconstruction by the interpolant $P(x,y)$,

Original surface   Reconstructed surface   The error
$z = f(x,y)$ $z = P(x,y)$ $z = f(x,y) - P(x,y)$

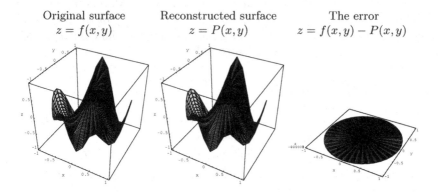

Fig. 1. $f(x,y) = \ln(x^2 + y + 1.5)\cos(5x + y)$

and the error function $f(x,y) - P(x,y)$ are presented in Figure 1. The relative L_2-norm of the error function on the unit disk is $\|f - P\|_2/\|f\|_2 = 0.000110356$.

Example 3. A gray-scale image is recovered by interpolation method. The original image and its reconstruction by polynomial of degree $n = 40$ are shown in Figure 2.

Example 4. Here we reconstruct the surface $z = f(x,y)$, where

$$f(x,y) = \sin(6x)\cos(4y + 2x),$$

using $N = \binom{52}{2} = 1326$ pieces of Radon projections by polynomial $P(x,y)$ of degree $n = 50$. The surface, its grey-scale image, and their reconstructions are presented in Figure 3. The relative L_2-norm of the error function on the unit disk is $\|f - P\|_2/\|f\|_2 = 0.000156029$.

Original image   Reconstructed image

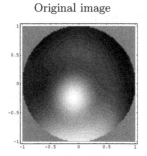 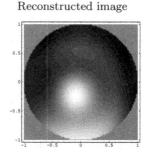

Fig. 2.

Original surface $z = f(x, y)$

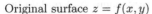

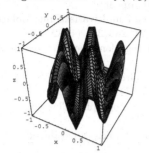

Original grey-scale image

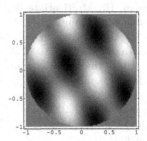

Reconstructed surface $z = P(x, y)$

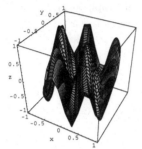

Reconstructed grey-scale image

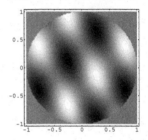

Fig. 3. $f(x, y) = \sin(6x)\cos(4y + 2x)$

Acknowledgements

This research was partially supported by the Bulgarian Ministry of Education and Science under Grant No. MM-1402/04 and by Sofia University under Grant No. 83/2006.

References

1. Bojanov, B., Georgieva, I.: Interpolation by bivariate polynomials based on Radon projections. Studia Math. 162, 141–160 (2004)
2. Bojanov, B., Petrova, G.: Numerical integration over a disc. A new Gaussian cubature formula. Numer. Math. 80, 39–59 (1998)
3. Bojanov, B., Petrova, G.: Uniqueness of the Gaussian cubature for a ball. J. Approx. Theory 104, 21–44 (2000)
4. Bojanov, B., Xu, Y.: Reconstruction of a bivariate polynomials from its Radon projections. SIAM J. Math. Anal. 37, 238–250 (2005)
5. Davison, M.E., Grunbaum, F.A.: Tomographic reconstruction with arbitrary directions. Comm. Pure Appl. Math. 34, 77–120 (1981)
6. Georgieva, I., Ismail, S.: On recovering of a bivariate polynomial from its Radon projections. In: Constructive Theory of Functions, pp. 127–134. Marin Drinov Academic Publishing House, Sofia (2006)

7. Georgieva, I., Uluchev, R.: Smoothing of Radon projections type of data by bivariate polynomials. In: J. Comput. Appl. Math. (to appear)

8. Hakopian, H.: Multivariate divided differences and multivariate interpolation of Lagrange and Hermite type. J. Approx. Theory 34, 286–305 (1982)

9. John, F.: Abhängigkeiten zwischen den Flächenintegralen einer stetigen Funktion. Math. Anal. 111, 541–559 (1935)

10. Marr, R.: On the reconstruction of a function on a circular domain from a sampling of its line integrals. J. Math. Anal. Appl. 45, 357–374 (1974)

11. Natterer, F.: The Mathematics of Computerized Tomography. Classics in Applied Mathematics 32 (2001)

12. Pickalov, V., Melnikova, T.: Plasma Tomography, Nauka, Novosibirsk (1995) (in Russian)

13. Radon, J.: Über die Bestimmung von Funktionen durch ihre Integralwerte längs gewisser Mannigfaltigkeiten. Ber. Verch. Sächs. Akad. 69, 262–277 (1917)

14. Solmon, D.C.: The X-ray transform. J. Math. Anal. Appl. 56(1), 61–83 (1976)

A Second-Order Cartesian Grid Finite Volume Technique for Elliptic Interface Problems

Juri D. Kandilarov, Miglena N. Koleva, and Lubin G. Vulkov

Faculty of Natural Science and Education, University of Rousse, 8 Studentska str.,
7017 Rousse, Bulgaria
ukandilarov@ru.acad.bg, mkoleva@ru.acad.bg, vulkov@ami.ru.acad.bg

Abstract. A second-order finite volume method (FVM) difference scheme for elliptic interface problems is discussed. The method uses bi-linear functions on Cartesian grid for the solution resulting in a compact nine-point stencil. Numerical experiments show second order of accuracy.

2000 Mathematics Subject Classification: 65M06, 65M12.

Keywords: parabolic equations, dynamical boundary conditions.

1 Introduction

In this paper we construct a finite-difference scheme that has second order of accuracy for an elliptic equation of the form

$$\nabla(\beta(x,y)\nabla u) - k(x,y)u = f(x,y), \quad (x,y) \in \Omega \backslash \Gamma \qquad (1)$$

with an embedded interface Γ. For simplicity we assume $\overline{\Omega}$ to be a rectangle and impose Dirichlet boundary conditions. The curve Γ separates two disjoint sub-domains $\overline{\Omega}^+$ and $\overline{\Omega}^-$ with $\overline{\Omega} = (\overline{\Omega}^+ \cup \overline{\Omega}^-)\backslash \Gamma$, see Figure 1 (a) for an illustration. Along the interface Γ we prescribe jump conditions of a generalized proper lumped source:

$$r^-(x,y)u_n^- + r^+(x,y)u_n^+ = [u] + g_1(x,y), \quad (x,y) \in \Gamma \qquad (2)$$

$$[\beta(x,y)u_n]_\Gamma = \delta(\alpha^+(x,y)u^+ - \alpha^-(x,y)u^-) + g_2(x,y), \quad (x,y) \in \Gamma, \qquad (3)$$

where the symbol $[v]$ stands for the jump of the function v across Γ, i.e.,

$$[v] = v^+ - v^-, \quad v^+(x,y) = \lim_{\zeta \to (x,y), \zeta \in \Omega^+} v(\zeta), v^-(x,y) = \lim_{\zeta \to (x,y), \zeta \in \Omega^-} v(\zeta), \qquad (4)$$

$r^-(x,y)$, $r^+(x,y)$ and δ are given nonnegative functions. Moreover, r^- and r^+ do not simultaneously vanish, $\alpha^+ = r^-/(r^- + r^+)$, $\alpha^- = r^+/(r^- + r^+)$, $u_n = (\nabla u.n)$ and $n = (n_1, n_2)$ is the unit normal vector on Γ, pointing from Ω^+ to Ω^-. Note that, for $g_1 = 0$ and $\delta = 0$, the transmission conditions (2) and (3) become the nonhomogeneous conditions of nonperfect contact [5,6].

I. Lirkov, S. Margenov, and J. Waśniewski (Eds.): LSSC 2007, LNCS 4818, pp. 679–687, 2008.

In the literature one can find a great number of different approaches for the numerical solution of elliptic interface problems. We limit our discussion here to the immersed interface method (IIM). It is a second order finite difference method on Cartesian grids for second order elliptic and parabolic equations with variable coefficients, see [2]. The finite element IIM, based on Cartesian triangulations, is developed in [2,3]. Starting with an idea of P. Vabischevich [6], for piecewise linear approximation of the interface curve in the integrointerpolation method ($\equiv$ FVM), we use as in [4] piecewise bilinear functions on Cartesian grid, which makes our method similar to the finite element method FEM.

To obtain finite volume formulation of (1)-(4), we integrate the equation (1) over an arbitrary control volume $e \in \bar{\Omega}$ and for irregular control volumes (intersected by the interface) we have

$$\int_{\partial e} \beta \nabla u \cdot \boldsymbol{n} dS - \int_e k u dV = \int_{e^+} f dV + \int_{e^-} f dV - \int_{\Gamma_e} [\beta u_n] dS, \qquad (5)$$

where Γ_e is the part of the embedded interface Γ, lying inside e and $\partial e = (\partial e^+ \bigcup \partial e^-) \setminus \Gamma_e$.

The finite volume method (FVM) is based on a 'balance' approach and originates from the integrointerpolation difference scheme method of Samarskii [5], designed first of all to be **locally conservative**. Some of the important features of the FVM are similar to those of the FEM: it may be used on arbitrary geometries, using structured or unstructured meshes and it leads to robust schemes. The survey paper [1], is devoted to a review of principles of the FVM and to the analysis tools for the mathematical study of cell centered finite volume schemes in the past years.

In the next section we describe the numerical method. Numerical results are discussed in Section 3. Finally, some conclusions are formulated.

2   Numerical Method

We develop some ideas, presented in [4,6] for (1)–(4) in the case $k = r^- = r^+ = \delta \equiv 0$. The discretization of (5) is on uniform grid with h_1 and h_2 grid spacing in x and y-directions (see Figure 1) and the numerical solution is denoted by u_{ij} at point (x_i, y_j). The control volumes e_{ij} are centered around the corresponding grid nodes (i, j), having edges of length h_1 and h_2. Let $\mathcal{M}_{ij}$ be the set of rectangles, called cells in this work, adjacent to the node (i, j) (I–IV, Figure 1 (a)). The discrete form of (5) for the control volume e_{ij} now reads as

$$\sum_{N \in \mathcal{M}_{ij}} \sum_{k=1}^2 \int_{l_k^N} \beta \nabla u \cdot \boldsymbol{n} dS - \int_{e_{ij}} k u dV = \int_{e_{ij}} f dV - \int_{\Gamma_{e_{ij}}} [\beta u_n] dS, \qquad (6)$$

where l_k^N, $k = 1, 2$ are the two boundary edges with normals $\boldsymbol{n}_1$ and $\boldsymbol{n}_2$ of ∂e_{ij}, lying inside N.

To evaluate the left hand side of (6) we apply a finite element approach with piecewise bilinear functions for u on each rectangular cell $N \in \mathcal{M}_{ij}$. The local coordinate systems $((\xi, \eta)$ and $(\widetilde{\xi}, \widetilde{\eta}))$ are involved:

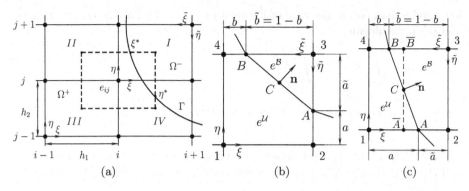

Fig. 1. (a): Control volume e_{ij}. Regular cells II, III, irregular cells I, IV; (b): Irregular cell type A; (c): Irregular cell type B in local $\xi - \eta$ coordinate system.

$$\xi = (x - x_0^N)/h_1 = 1 - \widetilde{\xi}, \quad \eta = (y - y_0^N)/h_2 = 1 - \widetilde{\eta}, \quad \text{with } \xi, \eta \in [0, 1]. \quad (7)$$

Here (x_0^N, y_0^N) denotes the origin of the local (ξ, η)-coordinate system in global (x, y) space. Irregular cells can always be mapped onto one of the two unit-square cells, shown on Figure 1 (b) and (c). A cell with an interface cutting the two adjacent edges of the upper right corner, we call type A (Figure 1 (b)) and type B (Figure 1 (c)), otherwise. The interface curve is assumed to be a straight line within the cell and is given by the zero level set of a signed normal distance function $\Phi(x)$. For any cell $N \in \mathcal{M}_{ij}$ we apply a bilinear local ansatz ($e^B \equiv 0$ in regular cell):

$$u^{\mathcal{U}}(\xi, \eta) = a_0 + a_1\xi + a_2\eta + a_3\xi\eta, \quad \xi, \eta \in e^{\mathcal{U}}, \quad (8)$$

$$u^B(\widetilde{\xi}, \widetilde{\eta}) = b_0 + b_1\widetilde{\xi} + b_2\widetilde{\eta} + b_3\widetilde{\xi}\widetilde{\eta}, \quad \widetilde{\xi}, \widetilde{\eta} \in e^B. \quad (9)$$

The procedure of obtaining the eight (only first four in regular case) unknown coefficients $\hat{u} = [a_0, a_1, a_2, a_3, b_0, b_1, b_2, b_3]^T$ is given below. **The idea is to write $\hat{u}$ as a linear combination of the four unknown corner values u_i, i = 1...4**, i.e. $\hat{u} = Mb$, $b = [u_1, u_2, u_3, u_4, \bar{b}]^T$, where $\bar{b} = [-g_1(x_A, y_A), -g_1(x_B, y_B), g_2(x_A, y_A), g_2(x_B, y_B)]$ for irregular case and $\bar{b} \equiv 0$ for regular case or irregular case with $g_1(x, y) = g_2(x, y) = 0$.

Now, using (8)–(9) we can evaluate any one of the two integrals on the left hand size of (6) analytically on each irregular (and regular) cell. As an example (for the first integral), consider cell I (Figure 1 (a)), which is irregular of type B. For boundary edge l_1^I with unit normal vector $\boldsymbol{n} = [n_x, n_y]^T = [0, 1]^T$ we have

$$\int_{l_1^I} \beta u_y dx = \frac{h_1}{h_2} \left(\int_0^{\xi^*} \beta^{\mathcal{U}} u_\eta^{\mathcal{U}} d\xi + \int_{\widetilde{\xi}^*}^{1/2} \beta^B u_{\widetilde{\eta}}^B d\widetilde{\xi} \right)$$

$$= \frac{\beta^{\mathcal{U}} h_1}{h_2} \left(a_2\xi^* + a_3\frac{\xi^{*2}}{2} \right) + \frac{\beta^B h_1}{h_2} \left(b_2 \left(\frac{1}{2} - \widetilde{\xi}^* \right) + b_3 \left(\frac{1}{8} - \frac{\widetilde{\xi}^{*2}}{2} \right) \right).$$

For regular case the same integral is valid with $\xi^* = 0.5$ and $\widetilde{\xi}^* = 0.5$, $\widetilde{\xi}^*$ is the corresponding value of ξ^* in $(\widetilde{\xi}, \widetilde{\eta})$ coordinates. For the second integral in the left hand size of (6), in the case of Figure 1 (a)($e^{\mathcal{U}}$ and $e^{\mathcal{B}}$ are as on Figure 1 (b) and (c)), we obtain

$$\int_{e_{ij}} ku \, dV = \int_{e_{ij}^{\mathcal{U}}} ku \, dV + \int_{e_{ij}^{\mathcal{B}}} ku \, dV$$

$$= h_1 h_2 \left(k(\xi_c^{\mathcal{U}}, \eta_c^{\mathcal{U}}) \int_{e_{ij}^{\mathcal{U}}} u^{\mathcal{U}}(\xi, \eta) d\xi d\eta + k(\widetilde{\xi}_c^{\mathcal{B}}, \widetilde{\eta}_c^{\mathcal{B}}) \int_{e_{ij}^{\mathcal{B}}} u^{\mathcal{B}}(\widetilde{\xi}, \widetilde{\eta}) d\widetilde{\xi} d\widetilde{\eta} \right),$$

where $(\widetilde{\xi}_c^{\mathcal{B}}, \widetilde{\eta}_c^{\mathcal{B}})$ and $(\xi_c^{\mathcal{U}}, \eta_c^{\mathcal{U}})$ denote the barycenters in local coordinates of $e_{ij}^{\mathcal{B}}$ and $e_{ij}^{\mathcal{U}}$, respectively. The remainder integrals in (6) are calculated as follows:

$$\int_{\Gamma_{e_{ij}}} [\beta u_n] dS = \sum_{N \in \mathcal{M}_{ij}} [\beta u_n] l_{\Gamma_{e_{ij}^N}}, \tag{10}$$

where $l_{\Gamma_{e_{ij}^N}}$ is the part of the interface $\Gamma_{e_{ij}^N}$ in cell N, which belongs to the control volume e_{ij}^N. The flux jump in (10) we substitute, using (3), (7)-(9). Next,

$$\int_{e_{ij}} f \, dV = |e_{ij}^{\mathcal{U}}| f^{\mathcal{U}}(x_c^{\mathcal{U}}, y_c^{\mathcal{U}}) + |e_{ij}^{\mathcal{B}}| f^{\mathcal{B}}(x_c^{\mathcal{B}}, y_c^{\mathcal{B}}),$$

where $(x_c^{\mathcal{U} \text{ or } \mathcal{B}}, y_c^{\mathcal{U} \text{ or } \mathcal{B}})$ is the barycenter of $e_{ij}^{\mathcal{U} \text{ or } \mathcal{B}}$, respectively. Now, we shall discuss the technique of obtaining $\hat{u}$.

- In **regular case**, the four unknown coefficients (a_i) are uniquely determined by the four corner values of u, matrix $M = \bar{M}$, $\bar{M}$ is given in the Appendix.
- In **irregular case**: To avoid the singularities (along a line parallel to any of the two coordinate axes, i. e. $\xi = const$ or $\eta = const$ and along a line with $n_\xi = \pm n_\eta$, see [4] for details), instead of (8)–(9), we propose a two-step asymptotic approach, [4].

$$u^{\mathcal{U}}(\xi, \eta) = u^{(\mathcal{U},0)}(\xi, \eta) + \varepsilon u^{(\mathcal{U},1)}(\xi, \eta),$$
$$u^{\mathcal{B}}(\widetilde{\xi}, \widetilde{\eta}) = u^{(\mathcal{B},0)}(\widetilde{\xi}, \widetilde{\eta}) + \varepsilon u^{(\mathcal{B},1)}(\widetilde{\xi}, \widetilde{\eta}),$$

where ε is a properly defined small parameter. We also present the jump conditions $[u]_C = ([u]_A + [u]_B)/2$ and $[\beta u_n]_C = ([\beta u_n]_A + [\beta u_n]_B)/2$. The point C is in the middle of the segment AB and is involved in order to avoid the above mentioned singularities.

$\rightarrow$ **Irregular cell of type A**: Let $\varepsilon = \min(\widetilde{a}, \widetilde{b})$ where $\widetilde{a} = 1 - a$ and $\widetilde{b} = 1 - b$. We define $[u]_A = [u]_B$, if $\varepsilon = 0$ and $\widetilde{b} = 0$, also $[u]_B = [u]_A$, if $\varepsilon = 0$ and $\widetilde{a} = 0$. For $\hat{u}$, [4], $\hat{u} = P\hat{b}$, where $P = \left[M_0^A + \varepsilon (M_1^A)^{-1} \left(B_1^A M_0^A + B_2^A \right) \right]$, see Appendix.

Now, for problem (1)–(4), substituting (2) and (3) in $\hat{b}$, we obtain $M = (S)^{-1} \widetilde{P}$, where for $i = 1, \ldots, 8$ and E — 8×8 unit matrix:

$$\widetilde{P}(i,1) = P(i,1) - \mu_1 P(i,5) - \bar{\mu}_1 P(i,6) + \rho_1 P(i,7) + \bar{\rho}_1 P(i,8),$$

$$\widetilde{P}(i,2) = P(i,2) - \mu_2 P(i,5) - \bar{\mu}_2 P(i,6) + \rho_2 P(i,7) + \bar{\rho}_2 P(i,8),$$

$$\widetilde{P}(i,3) = P(i,3) + \rho_3 P(i,7) + \bar{\rho}_3 P(i,8),$$

$$\widetilde{P}(i,4) = P(i,4) - \mu_4 P(i,5) - \bar{\mu}_4 P(i,6) + \rho_4 P(i,7) + \bar{\rho}_4 P(i,8),$$

$$\widetilde{P}(i,j) = P(i,j), \quad j = 5,\ldots,8,$$

$$S(i,j) = E + \begin{cases} \mu_3 P(i,5) + \bar{\mu}_3 P(i,6) - \rho_5 P(i,7) - \bar{\rho}_5 P(i,8), & j = 4, \\ -\nu_1 P(i,5) - \bar{\nu}_1 P(i,6) + \rho_6 P(i,7) + \bar{\rho}_6 P(i,8), & j = 6, \\ -\nu_2 P(i,5) - \bar{\nu}_2 P(i,6) + \rho_7 P(i,7) + \bar{\rho}_7 P(i,8), & j = 7, \\ -\nu_3 P(i,5) - \bar{\nu}_3 P(i,6) & j = 8, \\ 0, & \text{otherwise.} \end{cases}$$

The following notations are used (variables with bars are computed at point B, $\bar{\rho} = \bar{\rho}(B)$, variables without bars are computed at point A, $\rho = \rho(A)$):
$\nu_1 = r^B \beta^B n_{\bar{\xi}}/\tilde{a} h_1$, $\nu_2 = r^B \beta^B n_{\bar{\eta}}/\tilde{b} h_2$, $\nu_3 = r^B \beta^B n_{\bar{\xi}}/\tilde{b} h_2$, $\bar{\nu}_1 = r^B \beta^B n_{\bar{\xi}}/\tilde{a} h_1$, $\bar{\nu}_2 = r^B \beta^B n_{\bar{\eta}}/\tilde{b} h_2$, $\bar{\nu}_3 = r^B \beta^B \varepsilon n_{\bar{\xi}}/\tilde{a} h_2$, $\mu_1 = \mu_2 + \mu_4$, $\mu_2 = r^U \beta^U n_{\xi}/h_1$, $\mu_3 = r^U \beta^U(an_{\xi}/h_1 + n_{\eta}/h_2)$, $\mu_4 = r^U \beta^U n_{\eta}/h_2$, $\bar{\mu}_1 = \bar{\mu}_2 + \bar{\mu}_4$, $\bar{\mu}_2 = r^U \beta^U n_{\xi}/h_1$, $\bar{\mu}_3 = r^U \beta^U(n_{\xi}/h_1 + bn_{\eta}/h_2)$, $\bar{\mu}_4 = r^U \beta^U n_{\eta}/h_2$, $\rho_1 = \pm\delta\alpha^U(1 - \xi - \eta)$, $\rho_2 = \pm\delta\alpha^U\xi$, $\rho_3 = \pm\delta\alpha^U\eta$, $\rho_4 = \pm\delta\alpha^B$, $\rho_5 = \pm\delta\alpha^U\xi\eta$, $\rho_6 = \pm\delta\alpha^B\bar{\xi}/\tilde{b}$, $\rho_7 = \pm\delta\alpha^B\bar{\eta}/\tilde{a}$, $\bar{\rho}_1 = \pm\delta\alpha^U(1 - \xi - \eta)$, $\bar{\rho}_2 = \pm\delta\alpha^U\xi$, $\bar{\rho}_3 = \pm\delta\alpha^U\eta$, $\bar{\rho}_4 = \pm\delta\alpha^B$, $\bar{\rho}_5 = \pm\delta\alpha^U\xi\eta$, $\bar{\rho}_6 = \pm\delta\alpha^B\bar{\xi}/\tilde{b}$, $\bar{\rho}_7 = \pm\delta\alpha^B\bar{\eta}/\tilde{a}$, where the sign is '+', if $e_{ij}^+ \equiv e_{ij}^U$ (then $e_{ij}^- \equiv e_{ij}^B$) and '$-$', otherwise.

$\rightarrow$ **Irregular cell of type B**: Now $\varepsilon = (a - b)$. The leading order solutions $u^{(U,0)}$ and $u^{(B,0)}$ are determined by the four corner values of u and the jump conditions $[u]_{\bar{A}}$, $[u]_{\bar{B}}$, $[\beta u_n]_{\bar{A}}$, and $[\beta u_n]_{\bar{B}}$. The points $\bar{A}$ and $\bar{B}$ are defined to have the same ξ-coordinate as point C, i.e. $\xi_{\bar{A}} = \xi_{\bar{B}} = \xi_{\bar{C}} = (a + b)/2$, so that $\bar{A} = A$ and $\bar{B} = B$ in the limit $\varepsilon = 0$, see Figure 2 (a). The matrix equation $\hat{u} = P\hat{b}$ is valid, with $P = \left[M_0^B + (M_1^B)^{-1}\left(B_1^B M_0^B + B_2^B\right)\right]$, see Appendix. Now, for $M = (S)^{-1}\widetilde{P}$, $i = 1,\ldots,8$ we obtain

$$\widetilde{P}(i,1) = P(i,1) - \mu_4 P(i,5) - \bar{\mu}_4 P(i,6) + \rho_{11} P(i,7) + \bar{\rho}_{11} P(i,8),$$

$$\widetilde{P}(i,2) = P(i,2) + \nu_2 P(i,5) + \bar{\nu}_2 P(i,6) - \rho_9 P(i,7) - \bar{\rho}_9 P(i,8),$$

$$\widetilde{P}(i,3) = P(i,3) - \nu_2 P(i,7) - \bar{\nu}_2 P(i,8) - \rho_{10} P(i,7) - \bar{\rho}_{10} P(i,8),$$

$$\widetilde{P}(i,4) = P(i,4) + \mu_4 P(i,5) + \bar{\mu}_4 P(i,6) + \rho_{11} P(i,7) + \bar{\rho}_{11} P(i,8),$$

$$\widetilde{P}(i,j) = P(i,j), \quad j = 5,\ldots,8,$$

$$S(i,j) = E - \begin{cases} \mu_2 P(i,5) + \bar{\mu}_2 P(i,6) + \rho_2 P(i,7) + \bar{\rho}_2 P(i,8), & j = 2, \\ \mu_5 P(i,5) + (\bar{\mu}_2 + \bar{\mu}_5)P(i,6) + \rho_5 P(i,7) + \bar{\rho}_5 P(i,8), & j = 4, \\ \nu_1 P(i,5) + \bar{\nu}_1 P(i,6) - \rho_6 P(i,7) - \bar{\rho}_6 P(i,8), & j = 6, \\ (\nu_1 + \nu_3)P(i,5) + \bar{\nu}_3 P(i,6) - \rho_8 P(i,7) - \bar{\rho}_8 P(i,8), & j = 7, \\ 0, & \text{otherwise.} \end{cases}$$

The following notations are involved or redefined: $\mu_5 = a\mu_4$, $\bar{\mu}_5 = b\bar{\mu}_4$, $\rho_6 = \pm\delta\alpha^{\mathcal{B}}\tilde{\xi}$, $\rho_8 = \pm\delta\alpha^{\mathcal{B}}\tilde{\xi}\tilde{\eta}$, $\rho_9 = \pm\delta\alpha^{\mathcal{B}}\tilde{\eta}$, $\rho_{10} = 1-\rho_9$, $\rho_{11} = 1-\rho_3$, $\bar{\rho}_6 = \pm\delta\alpha^{\mathcal{B}}\tilde{\xi}$, $\bar{\rho}_8 = \pm\delta\alpha^{\mathcal{B}}\tilde{\xi}\tilde{\eta}$, $\bar{\rho}_9 = \pm\delta\alpha^{\mathcal{B}}\tilde{\eta}$, $\bar{\rho}_{10} = 1 - \bar{\rho}_9$, $\bar{\rho}_{11} = 1 - \bar{\rho}_3$, $\nu_1 = r^{\mathcal{B}}\beta^{\mathcal{B}}n_{\tilde{\xi}}/h_1$, $\nu_2 = r^{\mathcal{B}}\beta^{\mathcal{B}}n_{\tilde{\eta}}/h_2$, $\nu_3 = \tilde{a}\nu_2$, $\bar{\nu}_1 = r^{\mathcal{B}}\beta^{\mathcal{B}}n_{\tilde{\xi}}/h_1$, $\bar{\nu}_2 = r^{\mathcal{B}}\beta^{\mathcal{B}}n_{\tilde{\eta}}/h_2$, $\bar{\nu}_3 = \tilde{ab}\bar{\nu}_2$.

3 Numerical Results

From physical point of view, the problem (1)–(4), where $\delta = g_1(x,y) = g_2(x,y) = r^-(x,y) \equiv 0$ and $r^+(x,y) = g(x,y)\beta^+$ is of interest, see [5,6]. This will be our test example. Let $f^{\pm} = (4 - k\beta^{\pm}(x^2 + y^2))/\beta^+\beta^-$, $g = 0.5(\beta^- - \beta^+)(x^2 + y^2)/(xn_1 + yn_2)$, $k(x,y) \equiv 1$, $-1 \leq x, y \leq 1$ and Dirichlet boundary conditions are defined by the exact solution: $u^- = (x^2 + y^2)/\beta^+$, $u^+ = (x^2 + y^2)/\beta^-$. The interface is a simple circle with radius $\sqrt{0.23}$ and midpoint $(0,0)$. The level set

Table 1. Convergence results and error in the L_2 and L_∞-norm

N	$\beta^- = 1000, \beta^+ = 1$				$\beta^- = 1, \beta^+ = 0.005$			
	L_∞	CR	L_2	CR	L_∞	CR	L_2	CR
40	3.08591e-4	*2.1078*	2.74498e-4	*2.1205*	6.27347e-2	*2.1244*	5.18968e-2	*2.1302*
80	8.34144e-5	*1.8873*	7.29559e-5	*1.9117*	1.67619e-2	*1.9040*	1.36504e-2	*1.9267*
160	2.01086e-5	*2.0524*	1.74572e-5	*2.0632*	4.57867e-3	*1.8721*	3.68324e-3	*1.8899*
320	5.40968e-6	*1.8942*	4.58033e-6	*1.9303*	1.17416e-3	*1.9633*	9.27085e-4	*1.9902*
640	1.35148e-6	*2.0010*	1.13379e-6	*2.0143*	3.15569e-4	*1.8956*	2.47770e-4	*1.9037*

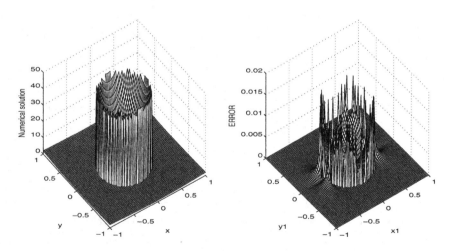

Fig. 2. Numerical solution (left) and absolute error (right), $N = 80$, $\beta^- = 1$, $\beta^+ = 0.005$

function is $\Phi = \sqrt{x^2 + y^2} - \sqrt{0.23}$, $\Phi(x) > 0$ in Ω^+ and $\Phi(x) \leq 0$ in Ω^-. Figure 2 plots the numerical solution and absolute error. In Table 1 convergence results and error in the L_2 and L_∞-norm are given on grid set: $h_1 = h_2 = h$, $N = 1/h$. The rate of convergence (CR) is computed, using double mesh principle.

4   Conclusions

We used FVM instead of finite difference method or FEM. Our algorithm always automatically achieves second order accuracy on compact 9-point stencil. We used piecewise bilinear functions on the Cartesian grid instead of piecewise linear functions on triangles, as is in IIM-FEM of [2]. The method can handle the problems when the solution and/or interfaces are weaker than C^2. For example, $u \in H^2(\Omega^{\mp})$, Γ is Lipschitz continuous.

To date, we have not been able to treat the theoretic framework of the method. Matrix theoretic considerations must be done, in order to obtain a variant of the discrete maximum principle.

Acknowledgments

The authors thank to the referee for the critical remarks that improved the exposition. This research is supported by the Bulgarian National Fund of Science under Project VU-MI-106/2005.

References

1. Eymard, R., et al.: Analysis tools for finite volume schemes. Acta Math. Univ. Comenciane LXXVI(1), 111–136 (2007)
2. Li, Z., Ito, K.: The Immersed Interface Method: Numerical Solutions of PDEs Involving Interfaces and Irregular Domains. SIAM Frontiers in Applied Mathematics 33 (2006)
3. Li, Z., Lin, T., Lin, Y., Rogers, R.: Error estimates of an immersed finite element method for interface problems, *Numerical Methods of PDEs* **20** (3) (2004) 338–367.
4. Oevermann, M., Klein, R.: A cartesian grid finite volume method for elliptic equations with variable coefficients amd embedded interfaces. J. Comp. Phys. 219(2), 749–769 (2007)
5. Samarskii, A.A.: The Theory of Difference Schemes. Marcel Dekker Inc. (2001)
6. Vabishchevich, P.N.: Numerical methods for solution of problems with free boundary. Moscow SU Publisher (1987) (in Russian)

Appendix

$$
\bar{M} = \begin{bmatrix} 1 & 0 & 0 & 0 \\ -1 & 1 & 0 & 0 \\ -1 & 0 & 0 & 1 \\ 1 & -1 & 1 & -1 \end{bmatrix}, \quad
B_1^A = \begin{bmatrix}
0 & 0 & \dots & & & \dots 0 \\
0 & 0 & \dots & & & \dots 0 \\
0 & 0 & \dots & & & \dots 0 \\
0 & 0 & \dots & & & \dots 0 \\
\frac{1}{\varepsilon} & \frac{1}{\varepsilon} & \frac{a}{\varepsilon} & \frac{a}{\varepsilon} & -\frac{1}{\varepsilon} & 0 & 0 & 0 \\
\frac{1}{\varepsilon} & \frac{b}{\varepsilon} & \frac{1}{\varepsilon} & \frac{b}{\varepsilon} & -\frac{1}{\varepsilon} & 0 & 0 & 0 \\
\frac{1}{\varepsilon} & \frac{1+b}{2\varepsilon} & \frac{1+a}{2\varepsilon} & \frac{(1+a)(1+b)}{4\varepsilon} & -\frac{1}{\varepsilon} & 0 & 0 & 0 \\
0 & \frac{\beta^{\mathcal{U}} n_\xi}{h_1} & \frac{\beta^{\mathcal{U}} n_\eta}{h_2} & \frac{\beta^{\mathcal{U}} n_\xi(1+a)}{2h_1} + \frac{\beta^{\mathcal{U}} n_\eta(1+b)}{2h_2} & 0 & 0 & 0 & 0
\end{bmatrix},
$$

$$
B_1^B = \begin{bmatrix}
0 & 0 & \dots & & & & \dots & 0 \\
0 & 0 & \dots & & & & \dots & 0 \\
0 & 0 & \dots & & & & \dots & 0 \\
0 & 0 & \dots & & & & \dots & 0 \\
-1 & -a & 0 & 0 & 1 & \widetilde{a} & 1 & \widetilde{a} \\
-1 & -b & -1 & -b & 1 & \widetilde{b} & 0 & 0 \\
0 & 0 & 0 & -\frac{1}{4} & 0 & 0 & 0 & \frac{1}{4} \\
0 & \frac{-\beta^{\mathcal{U}} n_\xi}{h_1} & \frac{-\beta^{\mathcal{U}} n_\eta}{h_2} & t^{\mathcal{U}} & 0 & \frac{-\beta^{\mathcal{B}} n_\xi}{h_1} & \frac{-\beta^{\mathcal{B}} n_\eta}{h_2} & t^{\mathcal{B}}
\end{bmatrix}, \quad
\begin{aligned}
t^{\mathcal{U}} &= \frac{-\beta^{\mathcal{U}} n_\xi}{2h_1} - \frac{\beta^{\mathcal{U}} n_\eta(a+b)}{2h_2}, \\
t^{\mathcal{B}} &= \frac{-\beta^{\mathcal{B}} n_\xi}{2h_1} - \frac{\beta^{\mathcal{B}} n_\eta(a+b)}{2h_2},
\end{aligned}
$$

$$
B_2^A = \begin{bmatrix}
0 & 0 & \dots & & \dots 0 \\
0 & 0 & \dots & & \dots 0 \\
0 & 0 & \dots & & \dots 0 \\
0 & 0 & \dots & & \dots 0 \\
0 & 0 & \dots & \frac{1}{\varepsilon} & 0 & 0 & 0 \\
0 & 0 & \dots & 0 & \frac{1}{\varepsilon} & 0 & 0 \\
0 & 0 & \dots & \frac{1}{2\varepsilon} & \frac{1}{2\varepsilon} & 0 & 0 \\
0 & 0 & \dots & 0 & 0 & \frac{1}{2} & \frac{1}{2}
\end{bmatrix}, \quad
B_2^B = \begin{bmatrix}
0 & 0 & \dots & & \dots 0 \\
0 & 0 & \dots & & \dots 0 \\
0 & 0 & \dots & & \dots 0 \\
0 & 0 & \dots & & \dots 0 \\
0 & 0 & \dots & 1 & 0 & 0 & 0 \\
0 & 0 & \dots & 0 & 1 & 0 & 0 \\
0 & 0 & \dots & 0 & 0 & 0 & 0 \\
0 & 0 & \dots & 0 & 0 & \frac{1}{2} & \frac{1}{2}
\end{bmatrix},
$$

$$
M_0^A = \begin{bmatrix}
1 & 0 & 0 & 0 & 0 & 0 & 0 & 0 \\
-1 & 1 & 0 & 0 & 0 & 0 & 0 & 0 \\
-1 & 0 & 0 & 1 & 0 & 0 & 0 & 0 \\
1 & -1 & 1 & -1 & \frac{1}{2} & \frac{1}{2} & 0 & 0 \\
0 & 0 & 1 & 0 & 0 & 0 & 0 & 0 \\
0 & 0 & \dots & & & \dots 0 \\
0 & 0 & \dots & & & \dots 0 \\
0 & 0 & \dots & & & \dots 0 \\
0 & 0 & \dots & & & \dots 0
\end{bmatrix}, \quad
\hat{b} = \begin{bmatrix}
u_1 \\ u_2 \\ u_3 \\ u_4 \\ [u]_A \\ [u]_b \\ [\beta u_n]_A \\ [\beta u_n]_B
\end{bmatrix},
$$

$$
M_1^A =
\begin{bmatrix}
1 & 0 & 0 & 0 & 0 & 0 & 0 & 0 \\
0 & 1 & 0 & 0 & 0 & 0 & 0 & 0 \\
0 & 0 & 1 & 0 & 0 & 0 & 0 & 0 \\
0 & 0 & 0 & 0 & 1 & 0 & 0 & 0 \\
0 & 0 & 0 & a & 0 & 0 & -1 & 0 \\
0 & 0 & 0 & b & 0 & -1 & 0 & 0 \\
0 & 0 & 0 & \dfrac{(1+a)(1+b)}{4} & 0 & -\dfrac{1}{2} & -\dfrac{1}{2} & -\dfrac{1}{4} \\
0 & 0 & 0 & \dfrac{\varepsilon\beta^{\mathcal{U}} n_\xi(1+a)}{2h_1} + \dfrac{\varepsilon\beta^{\mathcal{U}} n_\eta(1+b)}{2h_2} & 0 & \dfrac{\varepsilon\beta^{\mathcal{B}} n_\xi}{\tilde{b}h_1} & \dfrac{\varepsilon\beta^{\mathcal{B}} n_\eta}{\tilde{a}h_2} & \dfrac{\varepsilon\beta^{\mathcal{B}} n_\xi}{2\tilde{b}h_1} + \dfrac{\varepsilon\beta^{\mathcal{B}} n_\eta}{2\tilde{a}h_2}
\end{bmatrix},
$$

$$
M_0^B =
\begin{bmatrix}
1 & 0 & 0 & 0 & 0 & 0 & 0 & 0 \\
-s^{\mathcal{U}} & s^{\mathcal{U}} & 0 & 0 & s^{\mathcal{U}} & 0 & s^{\mathcal{U}}q^{\mathcal{U}} & 0 \\
-1 & 0 & 0 & 1 & 0 & 0 & 0 & 0 \\
s^{\mathcal{U}} & -s^{\mathcal{U}} & s^{\mathcal{U}} & -s^{\mathcal{U}} & -s^{\mathcal{U}} & s^{\mathcal{U}} & -s^{\mathcal{U}}q^{\mathcal{U}} & s^{\mathcal{U}}q^{\mathcal{U}} \\
0 & 0 & 1 & 0 & 0 & 0 & 0 & 0 \\
0 & 0 & s^{\mathcal{B}} & -s^{\mathcal{B}} & 0 & s^{\mathcal{B}} & 0 & -s^{\mathcal{B}}q^{\mathcal{B}} \\
0 & 1 & -1 & 0 & 0 & 0 & 0 & 0 \\
-s^{\mathcal{B}} & s^{\mathcal{B}} & -s^{\mathcal{B}} & s^{\mathcal{B}} & s^{\mathcal{B}} & -s^{\mathcal{B}} & -s^{\mathcal{B}}q^{\mathcal{B}} & s^{\mathcal{B}}q^{\mathcal{B}}
\end{bmatrix},
$$

$$
\begin{aligned}
s^{\mathcal{U}} &= \frac{\beta^{\mathcal{U}}}{\xi_C \beta^{\mathcal{B}} + \tilde{\xi}_C \beta^{\mathcal{U}}}, \\
s^{\mathcal{B}} &= -\frac{\beta^{\mathcal{B}}}{\xi_C \beta^{\mathcal{B}} + \tilde{\xi}_C \beta^{\mathcal{U}}}, \\
q^{\mathcal{U}} &= \frac{\xi_C h_1}{\beta^{\mathcal{B}}}, \\
q^{\mathcal{B}} &= \frac{\xi_C h_1}{\beta^{\mathcal{U}}},
\end{aligned}
$$

$$
M_1^B =
\begin{bmatrix}
1 & 0 & 0 & 0 & 0 & 0 & 0 & 0 \\
0 & 0 & 1 & 0 & 0 & 0 & 0 & 0 \\
0 & 0 & 0 & 0 & 1 & 0 & 0 & 0 \\
0 & 0 & 0 & 0 & 0 & 0 & 1 & 0 \\
0 & a & 0 & 0 & 0 & -\tilde{a} & 0 & -\tilde{a} \\
0 & b & 0 & b & 0 & -\tilde{b} & 0 & 0 \\
0 & 0 & 0 & 1/4 & 0 & 0 & 0 & -1/4 \\
0 & \dfrac{\beta^{\mathcal{U}} n_\xi}{h_1} & 0 & \dfrac{\beta^{\mathcal{U}} n_\xi}{2h_1} + \dfrac{\beta^{\mathcal{U}} n_\eta(a+b)}{2h_2} & 0 & \dfrac{\beta^{\mathcal{B}} n_\xi}{h_1} & \dfrac{\beta^{\mathcal{B}} n_\xi}{2h_1} + \dfrac{\beta^{\mathcal{B}} n_\eta(\tilde{a}+\tilde{b})}{2h_2} & 0
\end{bmatrix}.
$$

Remark 1. In M_1^A, B_1^A, M_1^B, and B_1^B: n_ξ and n_η are computed at point C.

MIC(0) DD Preconditioning of FEM Elasticity Systems on Unstructured Tetrahedral Grids

Nikola Kosturski

Institute for Parallel Processing, Bulgarian Academy of Sciences
Acad. G. Bonchev, Bl. 25A, 1113 Sofia, Bulgaria
nkosturski@gmail.com

Abstract. In this study, the topics of grid generation and FEM applications are studied together following their natural synergy. We consider the following three grid generators: NETGEN, TetGen and Gmsh. The qualitative analysis is based on the range of the dihedral angles of the triangulation of a given domain. After that, the performance of two displacement decomposition (DD) preconditioners that exploit modified incomplete Cholesky factorization MIC(0) is studied in the case of FEM matrices arising from the discretization of the three-dimensional equations of elasticity on unstructured tetrahedral grids.

Keywords: finite element method, preconditioned conjugate gradient method, MIC(0), displacement decomposition.

1 Introduction

Mesh generation techniques are now widely employed in various scientific and engineering fields that make use of physical models based on partial differential equations. While there are a lot of works devoted to finite element methods (FEM) and their applications, it appears that the issues of meshing technologies in this context are less investigated. Thus, in the best cases, this aspect is briefly mentioned as a technical point that is possibly non-trivial.

In this paper we consider the problem of linear elasticity with isotropic materials. Let $\Omega \subset \mathbb{R}^3$ be a bounded domain with boundary $\Gamma = \partial\Omega$ and $\mathbf{u} = (u_1, u_2, u_3)$ the *displacement* in Ω. The components of the *small strain tensor* are

$$\varepsilon_{ij}(\mathbf{u}) = \frac{1}{2}\left(\frac{\partial u_i}{\partial x_j} + \frac{\partial u_j}{\partial x_i}\right), \quad 1 \leq i, j \leq 3$$

and the components of the *Cauchy stress tensor* are

$$\tau_{ij} = \sum_{k,l=1}^{3} c_{ijkl}\varepsilon_{kl}(\mathbf{u}), \quad 1 \leq i, j \leq 3,$$

where the coefficients c_{ijkl} describe the behavior of the material. In the case of isotropic material the only non-zero coefficients are

$$c_{iiii} = \lambda + 2\mu, \quad c_{iijj} = \lambda, \quad c_{ijij} = c_{ijji} = \mu.$$

I. Lirkov, S. Margenov, and J. Waśniewski (Eds.): LSSC 2007, LNCS 4818, pp. 688–695, 2008.

Now, we can introduce the *Lamé's* system of linear elasticity (see, e.g., [2])

$$(\lambda + \mu) \sum_{k=1}^{3} \frac{\partial^2 u_k}{\partial x_k \partial x_i} + \mu \sum_{k=1}^{3} \frac{\partial^2 u_i}{\partial x_k^2} + F_i = 0, \quad 1 \le i \le 3 \tag{1}$$

equipped with boundary conditions

$$u_i(\mathbf{x}) = g_i(\mathbf{x}), \quad \mathbf{x} \in \Gamma_D \subset \partial\Omega \ ,$$

$$\sum_{j=1}^{3} \tau_{ij}(\mathbf{x}) n_j(\mathbf{x}) = h_i(\mathbf{x}), \quad \mathbf{x} \in \Gamma_N \subset \partial\Omega \ ,$$

where $n_j(\mathbf{x})$ denotes the components of the outward unit normal vector $\mathbf{n}$ onto the boundary $\mathbf{x} \in \Gamma_N$. The finite element method (FEM) is applied for discretization of (1) where linear finite elements on a triangulation $\mathcal{T}$ are used. The preconditioned conjugate gradient (PCG) [1] method will be used for the solution of the arising linear algebraic system $K\mathbf{u_h} = \mathbf{f_h}$.

2   MIC(0) DD Preconditioning

We first recall some known facts about the modified incomplete Cholesky factorization MIC(0), see, e.g. [4,5]. Let $A = (a_{ij})$ be a symmetric $n \times n$ matrix and let

$$A = D - L - L^T \ ,$$

where D is the diagonal and $-L$ is the strictly lower triangular part of A. Then we consider the factorization

$$C_{\text{MIC(0)}} = (X - L)X^{-1}(X - L)^T \ ,$$

where $X = \text{diag}(x_1, \ldots, x_n)$ is a diagonal matrix, such that the row sums of $C_{\text{MIC(0)}}$ and A are equal

$$C_{\text{MIC(0)}}\mathbf{e} = A\mathbf{e}, \quad \mathbf{e} = (1, \ldots, 1) \in \mathbb{R}^n \ .$$

Theorem 1. *Let $A = (a_{ij})$ be a symmetric $n \times n$ matrix and let*

$$L \ge 0$$
$$A\mathbf{e} \ge 0$$
$$A\mathbf{e} + L^T\mathbf{e} > 0 \quad \text{where} \quad \mathbf{e} = (1, \ldots, 1)^T \ .$$

Then there exists a stable MIC(0) factorization of A, defined by the diagonal matrix $X = \text{diag}(x_1, \ldots, x_n)$, where

$$x_i = a_{ii} - \sum_{k=1}^{i-1} \frac{a_{ik}}{x_k} \sum_{j=k+1}^{n} a_{kj} > 0 \ .$$

It is known, that due to the positive offdiagonal entries of the coupled stiffness matrix K, the MIC(0) factorization is not directly applicable to precondition the FEM elasticity system. Here we consider a composed algorithm based on a separable displacement three-by-three block representation

$$\begin{bmatrix} K_{11} & K_{12} & K_{13} \\ K_{21} & K_{22} & K_{23} \\ K_{31} & K_{32} & K_{33} \end{bmatrix} \mathbf{u_h} = \mathbf{f_h} .$$

In this setting, the stiffness matrix K is spectrally equivalent to the block-diagonal approximations C_{SDC} and C_{ISO}

$$C_{\text{SDC}} = \begin{bmatrix} K_{11} & & \\ & K_{22} & \\ & & K_{33} \end{bmatrix}, \quad C_{\text{ISO}} = \begin{bmatrix} A & & \\ & A & \\ & & A \end{bmatrix}, \quad (2)$$

where $A = \dfrac{1}{3}(K_{11} + K_{22} + K_{33})$. The theoretical background of this displacement decomposition (DD) step is provided by the second Korn's inequality [2]. Now the MIC(0) factorization is applied to the blocks of (2). In what follows, the related preconditioners will be referred to as $C_{\text{SDC-MIC(0)}}$ and $C_{\text{ISO-MIC(0)}}$, cf. [2,4,6].

3   Diagonal Compensation

The blocks K_{11}, K_{22}, K_{33} and A correspond to a certain FEM elliptic problem on the triangulation $\mathcal{T}$. Here, we will restrict our analysis to the case of isotropic DD, i.e., we will consider the piece-wise Laplacian matrix

$$A = \sum_{e \in \mathcal{T}} A_e$$

where the summation sign stands for the standard FEM assembling procedure. In the presence of positive offdiagonal entries in the matrix, the conditions of Theorem 1 are not met. To meet these conditions we use *diagonal compensation* to substitute the matrix A by a proper M-matrix $\bar{A}$. After that the MIC(0) factorization is applied to $\bar{A}$. The procedure consists of replacing the positive offdiagonal entries in A with 0 in $\bar{A}$ and adding them to the diagonal, so that $A\mathbf{e} = \bar{A}\mathbf{e}$.

The following important geometric interpretation of the current element stiffness matrix holds (see, e.g., in [7])

$$A_e = \frac{P_e}{6} \begin{bmatrix} \displaystyle\sum_{1 \neq i < j} \ell_{ij} \cot \theta_{ij} & -\ell_{34} \cot \theta_{34} & -\ell_{24} \cot \theta_{24} & -\ell_{23} \cot \theta_{23} \\[2ex] -\ell_{34} \cot \theta_{34} & \displaystyle\sum_{2 \neq i < j \neq 2} \ell_{ij} \cot \theta_{ij} & -\ell_{14} \cot \theta_{14} & -\ell_{13} \cot \theta_{13} \\[2ex] -\ell_{24} \cot \theta_{24} & -\ell_{14} \cot \theta_{14} & \displaystyle\sum_{3 \neq i < j \neq 3} \ell_{ij} \cot \theta_{ij} & -\ell_{12} \cot \theta_{12} \\[2ex] -\ell_{23} \cot \theta_{23} & -\ell_{13} \cot \theta_{13} & -\ell_{12} \cot \theta_{12} & \displaystyle\sum_{i < j \neq 4} \ell_{ij} \cot \theta_{ij} \end{bmatrix},$$

where P_e is some constant, depending on the material coefficents, ℓ_{ij} denotes the length of the edge connecting vertices v_i and v_j of the tetrahedron e and θ_{ij} denotes the dihedral angle at that edge. This interpretation shows that each positive offdiagonal entry in the element stiffness matrix corresponds to an obtuse dihedral angle in the tetrahedron e. Also a positive entry tends to infinity when the dihedral angle tends to $180°$. In the presence of very large dihedral angles, the relative condition number $\kappa(\bar{A}^{-1}A)$ may become very large. Since the MIC(0) factorization is applied to the auxilary matrix $\bar{A}$, the performance of the preconditioner strongly depends on this relative condifion number. In the two-dimensional case an uniform estimate of the condition number, depending only on the minimal angle was derived (see [6]). In the three-dimensional case, however, it is much harder to obtain an uniform estimate, since the element matrices depend not only on the shape of the elements, but also on elements sizes.

4   Comparison of Mesh Generators

In this section we compare the following three mesh generators:

- NETGEN v.4.4 (http://www.hpfem.jku.at/netgen/);
- Tetgen v.1.4.1 (http://tetgen.berlios.de/);
- Gmsh v.2.0.0 (http://geuz.org/gmsh/).

In the previous section we have seen the impact of very large dihedral angles on the preconditioning. Very small and very large angles also affect the accuracy of the FEM approximation as well as the condition number of the related stiffness matrix.

The domain we chose for this comparison is

$$\Omega = \{(x, y, z) \mid 0.1 \le x^2 + y^2 + z^2 \le 1, x, y, z \ge 0\} \ . \tag{3}$$

Different parameters of the grid generators may affect the quality of the resulting meshes. Some generated meshes are shown in Fig. 1 and the minimal and

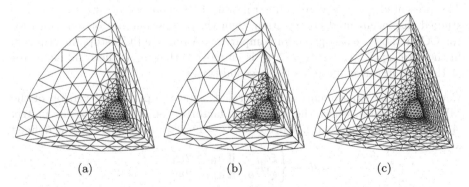

| (a) | (b) | (c) |

Fig. 1. Meshes, generated by: (a) NETGEN; (b) TetGen; (c) Gmsh

Table 1. Resulting Mesh Properties

Generator	Parameters	Min Angle	Max Angle	Elements	Nodes
NETGEN	grading = 1	14.3553 °	151.997 °	436	189
NETGEN	grading = 0.5	19.3608 °	142.821 °	650	245
NETGEN	grading = 0.2	26.1134 °	135.173 °	1882	504
TetGen	ratio = 2	5.06703 °	166.432 °	474	197
TetGen	ratio = 1.5	6.26918 °	169.619 °	714	251
TetGen	ratio = 1.2	6.12442 °	168.717 °	1484	417
Gmsh	$h = 0.05, H = 0.5$	13.3345 °	143.297 °	1192	344
Gmsh	$h = 0.03, H = 0.3$	20.9614 °	144.173 °	1553	436
Gmsh	$h = 0.015, H = 0.15$	18.7442 °	137.373 °	3718	940

maximal angles and numbers of nodes and elements for the three considered mesh generators with various values of the parameters are given in Table 1.

The mesh quality in NETGEN highly depends on the *mesh-size grading* parameter. Decreasing the value of this parameter leads to a mesh with better dihedral angles at the expense of larger number of elements and nodes. In Tet-Gen, the mesh element quality criterion is based on the *minimum radius-edge ratio*, which limits the ratio between the radius of the circumsphere of the tetrahedron and the shortest edge length. It seems, however, that this parameter does not directly reflect on the dihedral angles. With all tested values the resulting meshes contained both very small and very large dihedral angles. For Gmsh, the parameters h and H correspond to the *characteristic lengths*, assigned respectively to the vertices on the inner and the outer spherical boundary of the domain.

The results show that NETGEN generally achieved better dihedral angles than TetGen. Gmsh achieved similar dihedral angles, but with considerably larger number of elements/nodes than NETGEN.

5   Numerical Experiments

The presented numerical test illustrate the PCG convergence rate of the two studied displacement decomposition algorithms. The number of iterations for the CG method are also given for comparison. A relative PCG stopping criterion in the form $\mathbf{r}_k^T C^{-1} \mathbf{r}_k \leq \varepsilon^2 \mathbf{r}_0^T C^{-1} \mathbf{r}_0$ is employed. Here $\mathbf{r}_k$ is the residual vector at the k-th iteration and C is the preconditioner.

Remark 1. The experiments are performed using the perturbed version of the MIC(0) algorithm, where the incomplete factorization is applied to the matrix $\tilde{A} = A + \tilde{D}$. The diagonal perturbation $\tilde{D} = \tilde{D}(\xi) = \text{diag}(\tilde{d}_1, \ldots, \tilde{d}_n)$ is defined as follows:

$$\tilde{d}_i = \begin{cases} \xi a_{ii} & \text{if } a_{ii} \geq 2w_i \\ \xi^{1/2} a_{ii} & \text{if } a_{ii} < 2w_i \end{cases},$$

where $0 < \xi < 1$ is a constant and $w_i = -\sum_{j>i} a_{ij}$.

Table 2. Model Problem in the Unit Cube, $\varepsilon = 10^{-6}$

Mesh	Elements	Nodes	CG	ISO-MIC(0)	SDC-MIC(0)
1	384	125	26	13	13
2	3 072	729	53	17	15
3	24 576	4 913	110	26	22
4	196 608	35 937	192	38	33
5	1 572 864	274 625	459	53	51

Remark 2. A generalized coordinate-wise ordering is used to ensure the conditions for a stable MIC(0) factorization.

Remark 3. Uniform refinement of the meshes in not used in the experiments, since it does not preserve the dihedral angles. For example let us consider the platonic tetrahedron (with dihedral angles $\approx 70.5288\,°$). After splitting it in 8 new tetrahedrons we obtain a mesh with dihedral angles ranging from $54.7356\,°$ to $109.471\,°$. Four of the new tetrahedrons are similar to the original one, and all the other four have one obtuse dihedral angle. The numbers of elements in the experiments with unstructured meshes, thus do not increase exactly 8 times.

5.1 Model Problem in the Unit Cube

We first consider a model pure displacement problem in the unit cube $\Omega = [0, 1]^3$ and $\Gamma_D = \partial\Omega$. The material is homogeneous with $\lambda = 1$ and $\mu = 1.5$, and the right-hand side corresponds to the given solution $u_1 = x^3 + \sin(y + z)$, $u_2 = y^3 + z^2 - \sin(x - z)$, $u_3 = x^2 + z^3 + \sin(x - y)$. An uniform initial (coarsest) triangulation with a mesh size $h = 1/4$ is used. The resulting convergence rates are given in Table 2.

5.2 Model Problem in a Curvilinear Domain

We consider the same model problem, but on the domain (3) (see Fig. 1(a)). The resulting convergence rates are given in Table 3. NETGEN is used to generate the meshes for this experiment.

Table 3. Model Problem in the Curvilinear Domain, $\varepsilon = 10^{-6}$

Mesh	Elements	Nodes	CG	ISO-MIC(0)	SDC-MIC(0)
1	1 882	504	54	16	16
2	13 953	3 022	117	17	16
3	107 530	20 589	291	23	21
4	843 040	150 934	715	31	31

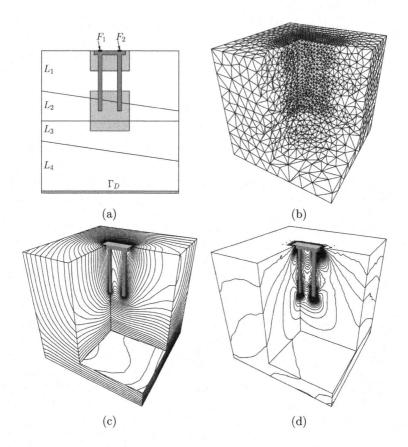

Fig. 2. Pile Foundation System: (a) Geometry; (b) A Mesh with Local Refinement; (c) Vertical Displacements; (d) Vertical Stresses

5.3   Computer Simulation of a Pile Foundation System

We consider the simulation of a foundation system in multi-layer soil media. The system consists of two piles with a linking plate. Fig. 2 (a) shows the geometry of Ω and the related weak soil layers. The generator used here is NET-GEN. Meshes are locally refined in areas with expected concentration of stresses, see Fig. 2 (b). The material characteristics of the concrete (piles) are $\lambda_p = 7666.67\,\text{MPa}$, $\mu_p = 11500\,\text{MPa}$. The related parameters for the soil layers are as follows: $\lambda_{L_1} = 28.58\,\text{MPa}$, $\mu_{L_1} = 7.14\,\text{MPa}$, $\lambda_{L_2} = 9.51\,\text{MPa}$, $\mu_{L_2} = 4.07\,\text{MPa}$, $\lambda_{L_3} = 2.8\,\text{MPa}$, $\mu_{L_3} = 2.8\,\text{MPa}$, $\lambda_{L_4} = 1.28\,\text{MPa}$, $\mu_{L_4} = 1.92\,\text{MPa}$. The forces, acting on the top cross-sections of the piles are $F_1 = (150\,\text{kN}, 2000\,\text{kN}, 0)$ and $F_2 = (150\,\text{kN}, 4000\,\text{kN}, 0)$. Dirichlet boundary conditions are applied on the bottom side. Fig. 2 (c) and (d) show contour plots of the solution. Table 4 contains the PCG convergence rate for Jacobi (the diagonal of the original matrix is used as a preconditioner) and the two MIC(0) DD preconditioners.

Table 4. Pile Foundation System, $\varepsilon = 10^{-6}$

Mesh	Elements	Nodes	Jaccobi	ISO-MIC(0)	SDC-MIC(0)
1	24 232	4 389	942	376	307
2	136 955	24 190	1680	564	505
3	859 895	149 111	3150	783	668
4	6 137 972	1 052 306	5416	972	929

5.4 Concluding Remarks

The rigorous theory of MIC(0) preconditioning is applicable to the first test problem only. For a structured grid with a mesh size h and smoothly varying material coefficients, the estimate $\kappa(C_h^{-1}A_h) = O(h^{-1}) = O(N^{1/3})$ holds, where C_h is the SDC-MIC(0) or ISO-MIC(0) preconditioner. The number of PCG iterations in this case is $n_{it} = O(N^{1/6})$. The reported number of iterations fully confirm this estimate. Moreover, we observe the same asymptotics of the PCG iterations for the next two problems, which is not supported by the theory up to now. As we see, the considered algorithms have a stable behaviour for unstructured meshes in a curvilinear domain (see Fig. 1(a)). The robustness in the case of local refinement and strong jumps of the coefficients is well illustrated by the last test problem.

Acknowledgment. The author gratefully acknowledges the support provided via EC INCO Grant BIS-21++ 016639/2005.

References

1. Axelsson, O.: Iterative solution methods, Cambridge University Press, Cambridge, MA (1994)
2. Axelsson, O., Gustafsson, I.: Iterative methods for the Navier equations of elasticity. Comp. Meth. Appl. Mech. Engin. 15, 241–258 (1978)
3. Axelsson, O., Margenov, S.: An optimal order multilevel preconditioner with respect to problem and discretization parameters. In: Minev, Wong, Lin (eds.) Advances in Computations, Theory and Practice, vol. 7, pp. 2–18. Nova Science, New York (2001)
4. Blaheta, R.: Displacement Decomposition — incomplete factorization preconditioning techniques for linear elasticity problems. Numer. Lin. Alg. Appl. 1, 107–126 (1994)
5. Gustafsson, I.: An incomplete factorization preconditioning method based on modification of element matrices. BIT 36(1), 86–100 (1996)
6. Kosturski, N., Margenov, S.: Comparative Analysis of Mesh Generators and MIC(0) Preconditioning of FEM Elasticity Systems. In: Boyanov, T., et al. (eds.) NMA 2006. LNCS, vol. 4310, pp. 74–81. Springer, Heidelberg (2007)
7. Shewchuk, J.: What Is a Good Linear Finite Element? — Interpolation, Conditioning, Anisotropy, and Quality Measures. In: Eleventh International Meshing Roundtable, pp. 115–126 (2002)

Parallelizations of the Error Correcting Code Problem

C. León, S. Martín, G. Miranda, C. Rodríguez, and J. Rodríguez

Dpto. de Estadística, I.O. y Computación, Universidad de La Laguna,
E-38271 La Laguna, Canary Islands, Spain
(cleon,samartin,gmiranda,casiano,jrpedri)@ull.es

Abstract. In this work, an exact approach to solve the Error Correcting Code problem is presented. For the implementation, the Branch and Bound skeleton of the MaLLBa library has been applied. This tool provides a hierarchy of C++ classes which must be adapted to the specific requirements of the problem. Then, it generates two parallel solvers: one based on the message passing paradigm and other designed on the basis of a shared memory model. For both parallel proposals the sequential algorithm follows the same principles. Note that with a single and simple specification of the problem, the tool gives the user two different parallel approaches. Computational results obtained with the OpenMP and MPI tools are shown.

1 Introduction

In the transmission of a message in binary code, interferences may appear. Interferences may cause to receive a message different from the one originally sent. If the error is detected, one possible solution is to request to the emitter the retransmission of the complete data block. However, there are many applications where the data retransmission is not possible or is not convenient in efficiency terms. In these cases, the message must be corrected by the receiver. For these particular situations *Error Correcting Codes* [6,8] are used. There are many types of error correcting codes: block (linear or cyclic) codes, convolutional codes, etc. This work is focused on block codes.

Due to the algorithmic complexity of the problem, most of the related works in the literature propose heuristics approaches [2,4]. In this work, an approach based on exact techniques will be exposed. A C++ tool for the implementation of problems by using Branch and Bound strategies has been used. This tool is provided by the MaLLBa skeleton library [1]. The way how MaLLBa::BnB skeleton [5,7] has been applied to solve the Error Correcting Code problem is shown.

The article contents are organized in the following way: section 2 gives a brief definition of the problem. The exact algorithm principles and its implementation are explained in section 3. Section 4 is devoted to expose the definition of the problem through the MaLLBa Branch and Bound skeleton. Also, the operation mode of the generated OpenMP and MPI parallelizations are described.

I. Lirkov, S. Margenov, and J. Waśniewski (Eds.): LSSC 2007, LNCS 4818, pp. 696–704, 2008.

Computational results are shown in section 5. Finally, the conclusions are given in section 6.

2   Error Correcting Code Problem

Let $A = \{a_1, a_2, ..., a_r\}$ be a set of r elements. Such set is called the *alphabet of the code* and its elements are known as the *symbols of the code*. A $r - ary$ block code C over an alphabet A is a not empty subset of A^n, where A^n is the set of all the words of fixed length n over A. The elements in C are denoted *codewords* and n represents the *length of the code*. The number of M words in the code C (that is the cardinality of the subset) is called *the size of the code*. The messages to be transmitted consist of sequences of such M codewords. The *Hamming distance* between two sequences v_i and v_j is the number of different code symbols between them. The *minimum distance of the code* C is denoted as $d(C)$ and it is defined as the minimum Hamming distance between all the different codewords:

$$d(C) = min\{d(c_i, c_j)|c_i, c_j \in C, c_i \neq c_j\}.$$

A code with M words of length n over an alphabet of r symbols and a minimum distance d_{min} is designated as a r-ary (n, M, d_{min}) code. The minimum distance d_{min} is related to the capacity of the code C to detect and correct errors [6]. A code C is able to detect v errors if and only if $d_{min} \geq v + 1$. A code C is able to correct e errors if and only if $d_{min} \geq 2e + 1$. Then, if a code C has a minimum distance of d_{min}, C will be able to detect $d_{min} - 1$ errors and correct $\lfloor (d_{min} - 1)/2 \rfloor$ errors in any word of the code. The error correcting is based on the *maximum likeliness principle*. This principle establishes that: when the receiver gets a codeword W' that is not included in the code C, the criterion to follow is to select as the correct codeword W the *nearest* codeword to W'. The *nearest* codeword to W' is the word of the code with the minimum distance to W':

$$d(W, W') < d(Y, W'), \forall Y \in C; Y \neq W.$$

When designing an error correcting code the objectives are:

- **Minimize n.** Find the codewords with minimum length in order to decrease the time invested in the message transmission.
- **Maximize d_{min}.** The Hamming distance between the codewords must be maximum to guarantee a high level of correction at the receiver. If the codewords are very different one from each other, it would be very unlikely to appear so many errors to transform one codeword into a different one.
- **Maximize M.** Maximize the number of codewords in the code (the final objective is to be as near as possible to A^n).

Even though, these objectives are incompatible, so, what it is usually done is to optimize one of the parameters (n, M or d_{min}) giving a specific fixed value for the other two. The most common approach for the problem is to maximize d_{min} for given n and M.

The problem solved here is posed as follows: starting with fixed values for the parameters M and n, it is necessary to get (from all the possible codes that can be generated) the code with the maximum minimum distance. The total number of possible codes with M codewords of n bits is equal to $\binom{2^n}{M}$. Depending on the problem parameters (M and n), this value could be very high. In such cases, the approach to the problem could be almost unfeasible in terms of computational resources. The execution time grows exponentially with the increase of any of the parameters M or n.

3   Exact Algorithm

The approach followed to get the code of M words of n bits with the maximum minimum distance is based on the *Subset Building Algorithm* [3]. This exact algorithm allows to generate all the possible codes with M words of length n. From all the possible codes, the one with the best minimum distance will be chosen as the final solution. Note that several codes with the same best distance can be obtained.

Taking into account some properties of the algorithm, several particularities have been applied to the implementation in order to reduce the total search space. One way to reduce the computational effort needed by the exhaustive search consists in forcing the implementation to generate only the subsets containing the word "00...00" (of n bits). As a result, the depth of the search tree is reduced (now only $M - 1$ words have to be selected). Another improvement introduced to the original algorithm consists in generating only non-equivalent codeword subsets.

The problem search space can be represented as a general search tree. The algorithm applied is similar to an exhaustive tree search strategy. By this reason, the principles of a Branch and Bound technique are introduced in order to avoid exploring branches that will never get to an optimal solution. When the algorithm gets a solution code with a certain minimum distance, it will be set as the current best solution. In the future, branches with current minimum distance lower or equal to the current best distance, will be bound. Moreover, the expected code must be able to correct a certain number of errors, so that, all the branches representing codes that break this constraint will be pruned.

Initially, an ad-hoc C++ implementation of the algorithm described in the previous paragraph was developed. This sequential approximation was not enough efficient to afford some of the problem big instances. But an exact solution to the problem is needed in order to verify the quality of the non-exact approximations. So, to improve the efficiency of the exact algorithm some kind of parallel techniques might be applied.

4   Problem Implementation with **MaLLBa**

In order to simplify the development of a solution to the problem and also with the aim of obtaining more efficient schemes, MaLLBa skeleton library [1] has been used. MaLLBa library consists of a set of algorithmic skeletons for solving combinatorial optimization problems. They provide an important advantage in comparison to a direct implementation of the algorithm from the beginning, not only in terms of code reuse but also in methodology and concept clarity. In this case, MaLLBa::BnB skeleton [5,7] has been applied to solve the Error Correcting Code problem.

The process of building all the possible codeword sets proposed in section 3 can be represented by using a tree structure. A tree node would represent a possible set of codewords (code). At every iteration, one node of the tree is chosen and branched. Branching a node consists in building all the new possible codeword sets by adding a new codeword to the current set. Nodes representing a possible solution code are identified to finally choose the best one. Note that the implementation follows a scheme very similar to a general Branch and Bound strategy. By this reason, instead of directly implement the algorithm from the beginning, MaLLBa::BnB skeleton will be tested. MaLLBa::BnB [5] implements a Branch and Bound technique over the problem search space. It needs some functions to calculate upper and lower bounds of each subproblem, in order to avoid exploring the hole search space. It explores the tree space, branching each subproblem and pruning the worse branches. When the exploration finishes, the solver returns the best solution found.

In general, the software that supplies skeletons presents declarations of empty classes. The user must fill these empty classes to adapt the given scheme for the resolution of a particular problem. In particular, MaLLBa::BnB requires to the user the specification of three classes: *Problem* stores the characteristics of the problem to solve, *Solution* defines how to represent the solutions and *Sub-Problem* represents a node in the tree or search space. This last class defines the search for a particular problem and it must contain a field of type *Solution* in which store the (partial) solution. The methods to define for this class are: *initSubProblem(pbm, subpbms)* creates the initial subproblem or subproblems from the original problem, *lower_bound(pbm)* calculates the subproblem accumulated cost, *upper_bound(pbm, sol)* calculates the subproblem estimated total cost, *branch(pbm, subpbms)* generates a set of new subproblems from the current one. The structure of the skeleton required classes for the implementation of the Error Correcting Code problem is shown in Figure 1.

The skeleton provides to the user two classes: *Setup* is used to configure all the search parameters and skeleton properties and *Solver* implements the strategy to do (a Branch and Bound in this case). Usually, each skeleton provides several solvers. Some of them are sequential and other are parallel. MaLLBa::BnB provides one sequential solver and two parallel solvers. The user can modify certain characteristics of the search by using the provided configuration class *Setup* and depending on the definition done for some methods of the *SubProblem* class.

```
class Problem {
    unsigned int M;          // Number of codewords in the code
    static int n;            // Codewords length (in bits)
    unsigned int numError;   // Minimum number of errors to correct
    int bestD;               // Current higher minimum distance
    unsigned int maxSet;     // Maximum number of obtainable codewords
    ...
}

class Solution {
    set<int> cw;             // Set of codewords in the code
    ...
}

class SubProblem {
    int d;                   // Minimum distance of the subproblem code
    Solution sol;            // Solution represented by the subproblem
    ...
}
```

Fig. 1. MaLLBa Classes Definition

One of the advantages of using MaLLBa skeletons consists in the fact that with only one definition of the corresponding classes and methods, the user gets several implementations: some sequential and other parallel. In particular, MaLLBa::BnB provides one sequential solver and two parallel ones. Besides, the user can tune the type of search to do by the skeleton. Giving certain values to the configuration parameters, the user can obtain a simple Branch and Bound algorithm or any type of search algorithm (i.e. an A* search).

MaLLBa::BnB uses a structure to store the nodes that are pending to be analyzed. First, all the initial subproblems are inserted into the list of pending nodes. At each step of the process, the first node in the list is removed and branched: all the new subproblems generated from the current one are inserted into the list of nodes. The type of insertions to do into the list of nodes depends on the type of search algorithm that is being implemented. The process ends when the list of nodes is empty. From all the suitable solutions reached during the search, the best one will be selected.

In order to improve the efficiency of the application, the two parallel approaches provided by the tool have been tested. The first parallel solver is based on the message passing paradigm and the second one relies on a shared memory scheme. The message passing parallelization has been implemented with MPI [10] while the shared memory approach uses OpenMP [9].

4.1 MPI Solver

This parallel design is based on a master-slave paradigm. The master sends to every slave a node from which generate new subproblems. Each initial node represents the first codeword from which begin to generate possible words subsets. Each slave generates all the possible subsets beginning from the codeword sent by the master. That is, each slave explores all the search space created from the initial received node. Between all the codeword subsets reached, the slave sends to the master the best solution found. The master receives all the solutions reached by each slave. Among them, the master will select the best one (the code with the higher minimum distance). In this implementation, the master distributes the work load between the slaves but it does not participate in the nodes branching.

In order to maintain always the best current solution updated, some extra communications are necessary between the master and the threads. Each time a slave reaches a better current solution, it must notify it to the master. Then, the master will transmit the value of the new higher minimum distance to the others slaves. Having the best current solution always updated allows the slaves to avoid exploring unnecessary areas.

4.2   OpenMP Solver

In this parallel implementation, the data structures to manage the nodes are stored in shared memory. The parallelization scheme is also based on a master-slave model. During the search, several slave threads can branch different subproblems at the same time. Meanwhile, at each step, the master thread is in charge of removing one node from the list of pending nodes and insert its corresponding new subproblems (previously created by a slave thread) into the same list. If the master removes an unbranched node, it will have to do the branching of the subproblem. Thus, the master removes nodes from the list and inserts their new subproblems while the slaves are branching the pending nodes in the list. The problem of this scheme lies in the fact that the master is modifying the list of nodes at the same time that the slave threads are accessing it to look for job to do. By this reason, special synchronization mechanisms between threads had to be designed. Synchronization tasks require a considerable computational effort, that is why it can seriously affect to the algorithm behaviour. Anyway, the user does not have to worry about these features, just try what the library gives. That is one of the great advantages of using the proposed tool.

The value of the current higher minimum distance is 0 updated by all the threads. That allows the threads to avoid exploring branches that will never get to a solution better than the current one.

5   Computational Results

For the computational study, several instances of the Error Correcting Code problem have been selected. For these instances, various values for n and M have been defined. The minimum number of errors to correct in all the instances is one. Thus, the efficiency of the implementations for solving instances with

Table 1. Sequential Results

Problem	C++ Ad Hoc Implementation		MaLLBa::BnB Implementation	
	Time	Nodes	Time	Nodes
n = 9 M = 17	206.326	255	223.79	221
n = 10 M = 12	25.034	511	24.614	469
n = 13 M = 5	4330,789	4095	4238.540	4022
n = 8 M = 18	2035.620	127	2332.458	100

Table 2. Parallel OpenMP Results

Problem	2 threads		4 threads		8 threads		16 threads	
	Time	Nodes	Time	Nodes	Time	Nodes	Time	Nodes
n = 9 M = 17	244.32	221	743.55	221	13.75	221	13.65	221
n = 10 M = 12	25.726	469	23.255	469	23.942	469	33.801	469
n = 13 M = 5	4279.372	4022	1717.342	4022	954.083	4022	627.311	4022
n = 8 M = 18	5149.484	100	9403.328	100	8605.960	100	733.830	100

Table 3. Parallel MPI Results

Problem	2 processors		4 processors		8 processors		16 processors	
	Time	Nodes	Time	Nodes	Time	Nodes	Time	Nodes
n = 9 M = 17	205.039	255	203.398	255	209.971	255	198.532	255
n = 10 M = 12	24.040	511	8.378	511	3.954	511	2.198	511
n = 13 M = 5	4167.337	4095	1394.192	4095	602.862	4095	283.006	4095
n = 8 M = 18	2034.787	127	1819.053	127	1785.241	127	1702.547	127

different computational efforts will be analyzed. The experiments have been run over an Origin 3800.

Table 1 shows the execution times (in seconds) and the number of computed nodes, that is, the number of branched nodes, for the sequential C++ ad hoc implementation and for the sequential MaLLBa::BnB solver. The obtained results are very similar, so the skeleton does not introduce too much overhead to the algorithm. Tables 2 and 3 show the execution times and the number of computed nodes for the parallel OpenMP and MPI implementations.

Figure 2 represents the speedups gotten with the parallel solvers. MPI speedup have a better behaviour than the ones obtained for the OpenMP version. Results for the MPI implementation do not always improve the sequential times but the behaviour of this version is more stable and predictable. For the OpenMP implementation, some strange behaviours could appear. Results for the problem instance "$n=9$, $M=17$" presents a case of superlinearity. This is due to an algorithmic speedup. It is important to take into account that a Branch and Bound strategy has been implemented and, for this particular instance, one of the threads found the solution immediately, making possible to avoid exploring the whole search space. In the sequential case, most of the tree branches would have to be explored before finding the problem solution.

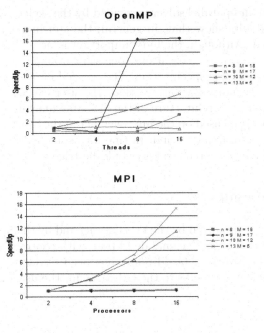

Fig. 2. Speedup

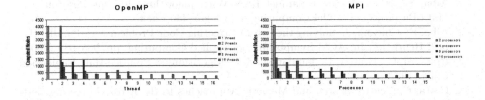

Fig. 3. Computed Nodes

Figure 3 shows the number of computed nodes per processor for the OpenMP and MPI implementations for the problem instance "$n=13$, $M=5$". In both cases, the load distribution between the processors is quite fair. In the MPI implementation, the load is distributed among the slaves but in the OpenMP case the master can also collaborate with its slaves.

6   Conclusions

Through this work, an exact algorithm for the Error Correcting Code problem has been presented. The algorithm follows a scheme very similar to a Branch and Bound strategy. For this reason, the algorithm has been implemented using the MaLLBa::BnB library. The flexibility, efficiency and simplicity of MaLLBa tools have been proved.

The sequential and parallel solvers provided by the skeleton have been tested. Computational results show that, in general, the MPI implementation presents a better behaviour. Although, the most important issue is that the user directly obtains three different implementations (one sequential and two parallel) from a single and simple problem specification. The parallel implementations are based on very different memory schemes, so that, the user can choose between them considering the parallel architecture of the available machines.

Currently, work is focused on the improvement of the exact algorithm through the introduction of an upper bound. In this way, some extra branches could be pruned and a first-best search strategy could be tried.

Acknowledgements

This work has been supported by the EC (FEDER) and by the Spanish Ministry of Education inside the 'Plan Nacional de I+D+i' with contract number TIN2005-08818-C04-04. The work of G. Miranda has been developed under the grant FPU-AP2004-2290. Also thanks to the CIEMAT for the usage of their computer systems.

References

1. Alba, E., et al.: Efficient parallel LAN/WAN algorithms for optimization: the MaLLBa project. Parallel Computing 32(5), 415–440 (2006)
2. Alba, E., Chicano, J.: Solving the Error Correcting Code Problem with Parallel Hybrid Heuristics. In: ACM Symposium on Applied Computing, pp. 985–989 (2004)
3. Brassard, G., Bratley, P.: Fundamentals of Algorithmics. Prentice-Hall, Englewood Cliffs (1997)
4. Cotta, C.: Scatter Search and Memetic Approaches to the Error Correcting Code Problem. In: Gottlieb, J., Raidl, G.R. (eds.) EvoCOP 2004. LNCS, vol. 3004, pp. 51–61. Springer, Heidelberg (2004)
5. González, J.R., León, C., Rodríguez, C.: An Asynchronous Branch-and-Bound Skeleton for Heterogeneous Clusters. In: Kranzlmüller, D., Kacsuk, P., Dongarra, J. (eds.) EuroPVM/MPI 2004. LNCS, vol. 3241, pp. 191–198. Springer, Heidelberg (2004)
6. Hill, R.: A First Course in Coding Theory. In: Oxford applied mathematics and computing science series, Oxford University Press, Oxford (1986)
7. Miranda, G., León, C.: An OpenMP skeleton for the A* heuristic search. In: Yang, L.T., et al. (eds.) HPCC 2005. LNCS, vol. 3726, pp. 717–722. Springer, Heidelberg (2005)
8. Morelos-Zaragoza, R.: The art of error correcting coding. Wiley, Chichester (2002)
9. OpenMP Architecture Review Board: OpenMP C and C++ Application Program Interface. Version 1.0 (1998), http://www.openmp.org
10. Snir, M., et al.: MPI: The Complete Reference. MIT Press, Cambridge (1996)

Benchmarking Performance Analysis of Parallel Solver for 3D Elasticity Problems

Ivan Lirkov[1], Yavor Vutov[1], Marcin Paprzycki[2], and Maria Ganzha[3]

[1] Institute for Parallel Processing, Bulgarian Academy of Sciences,
Acad. G. Bonchev, Bl. 25A, 1113 Sofia, Bulgaria
ivan@parallel.bas.bg, yavor@parallel.bas.bg
http://parallel.bas.bg/~ivan/, http://parallel.bas.bg/~yavor/
[2] Institute of Computer Science, Warsaw School of Social Psychology, ul.
Chodakowska 19/31, 03–815 Warszawa, Poland
marcin.paprzycki@swps.edu.pl
http://mpaprzycki.swps.edu.pl
[3] Systems Research Institute, Polish Academy of Science,
ul. Newelska 6, 01-447 Warszawa, Poland
maria.ganzha@ibspan.waw.pl
http://www.ganzha.euh-e.edu.pl

Abstract. In this paper we consider numerical solution of 3D linear elasticity equations described by a coupled system of second order elliptic partial differential equations. This system is discretized by trilinear parallelepipedal finite elements. Preconditioned Conjugate Gradient iterative method is used for solving large-scale linear algebraic systems arising after the Finite Element Method (FEM) discretization of the problem. The displacement decomposition technique is applied at the first step to construct a preconditioner using the decoupled block diagonal part of the original matrix. Then circulant block factorization is used to precondition thus obtained block diagonal matrix. Since both preconditioning techniques, displacement decomposition and circulant block factorization, are highly parallelizable, a portable parallel FEM code utilizing MPI for communication is implemented. Results of numerical tests performed on a number of modern parallel computers using real life engineering problems from the geosciences (geomechanics in particular) are reported and discussed.

1 Introduction

Our work concerns development and implementation of efficient parallel algorithms for solving elasticity problems arising in geosciences. Typical application problems include simulations of foundations of engineering constructions (which transfer and distribute the total loading into the bed of soil) and multilayer media with strongly varying material characteristics. Here, the spatial framework of the construction produces a complex stressed-strained state in the active

I. Lirkov, S. Margenov, and J. Waśniewski (Eds.): LSSC 2007, LNCS 4818, pp. 705–712, 2008.

interaction zones. The modern design of cost-efficient construction with a sufficient guaranteed reliability requires determining parameters of this stressed-strained state.

These engineering problems are described mathematically by a system of three-dimensional nonlinear partial differential equations. A finite element (or finite difference) discretization reduces the partial differential equation problem to a system of linear equations $K\mathbf{x} = \mathbf{f}$, where the stiffness matrix K is large, sparse and symmetric positive definite. The Conjugate Gradient (CG) type methods are recognized as the most cost-effective way to solve problems of this type [1]. To accelerate the iteration convergence a preconditioner M is combined with the CG algorithm. To make a reliable prediction of the construction safety, which is sensitive to soil deformations, a very accurate model is required. In the real-life applications, the linear system can be very large, containing up to several millions of unknowns. Hence, these problems have to be solved by robust and efficient parallel iterative methods on powerful multiprocessor computers.

Note that the numerical solution of linear systems is a fundamental operation in computer modeling of elasticity problems. Specifically, solving these linear systems is usually very time-consuming (requiring up to 90% of the total solution time). Hence, developing fast solvers for linear equations is essential. Furthermore, such algorithms can significantly speed up the simulation processes of real application problems. Due to the size of the system, an efficient iterative solver should not only have a fast convergence rate but also high parallel efficiency. Moreover, the resulting program has to be efficiently implementable on modern shared-memory, distributed memory, and shared-distributed memory parallel computers.

2   Elasticity Problems

For simplicity, in this work we focus our attention on 3D linear elasticity problems following two basic assumptions: (1) displacements are small, and (2) material properties are isotropic. A precise mathematical formulation of the considered problem is described in [5]; the 3D elasticity problem in the stressed-strained state can be described by a coupled system of three differential equations. This system of three linear differential equations is often referred to as Lamé equations.

We restrict our considerations to the case when the computational domain Ω is a rectangular parallelogram $\Omega = [0, x_1^{max}] \times [0, x_2^{max}] \times [0, x_3^{max}]$, where the boundary conditions on each wall of Ω are of fixed type.

Benchmark problems from [4] are used in numerical tests reported here. The engineering problems are as follows: a) single pile in a homogeneous sandy clay soil (see Fig. 1(a)) and b) two piles in an inhomogeneous sandy clay soil (Fig. 1(b)). In the solution process, uniform grid is used with n_1, n_2 and n_3 grid points along the coordinate directions.

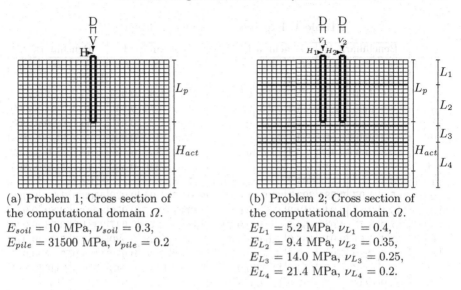

(a) Problem 1; Cross section of
the computational domain Ω.
$E_{soil} = 10$ MPa, $\nu_{soil} = 0.3$,
$E_{pile} = 31500$ MPa, $\nu_{pile} = 0.2$

(b) Problem 2; Cross section of
the computational domain Ω.
$E_{L_1} = 5.2$ MPa, $\nu_{L_1} = 0.4$,
$E_{L_2} = 9.4$ MPa, $\nu_{L_2} = 0.35$,
$E_{L_3} = 14.0$ MPa, $\nu_{L_3} = 0.25$,
$E_{L_4} = 21.4$ MPa, $\nu_{L_4} = 0.2$.

Fig. 1. Benchmark problems

3 Displacement Decomposition Circulant Block Factorization Preconditioner

There exists a substantial body of work dealing with preconditioning of iterative solution methods for elasticity systems discretized using the Finite Element Method. For instance, in [2] Axelsson and Gustafson construct their preconditioners based on the point-ILU (Incomplete LU) factorization of the displacement decoupled block-diagonal part of the original matrix. This approach is known as displacement decomposition (see, e.g., [3]). In [6] circulant block-factorization is used for preconditioning of the obtained block-diagonal matrix and a displacement decomposition circulant block factorization preconditioner is constructed. The estimate of the condition number of the proposed preconditioner shows that DD CBF solver is asymptotically as fast as preconditioners based on the point-ILU factorization [5,6]. Moreover DD CBF solver has a good parallel efficiency (see, e.g., [5,6]).

4 Benchmarking Performance Analysis

To solve the above described problems, a portable parallel FEM code was designed and implemented in C, while the parallelization has been facilitated using the MPI library [7,8]. The parallel code has been tested on cluster computers located in the National Energy Research Scientific Computing Center (NERSC), Oklahoma Supercomputing Center (OSCER), and in Bologna, Italy (CINECA). In our experiments, times have been collected using the MPI provided timer and report the best results from multiple runs. We report the elapsed time T_p in seconds on p processors, the speed-up $S_p = T_1/T_p$, and the parallel efficiency

Table 1. Experimental results on Jacquard

		Benchmark 1			Benchmark 2				Benchmark 1			Benchmark 2		
p	n	T_p	S_p	E_p	T_p	S_p	E_p	n	T_p	S_p	E_p	T_p	S_p	E_p
1	32	12.5			50.4			64	812.9			1277.8		
2		6.6	1.90	0.950	25.7	1.96	0.981		416.8	1.95	0.975	675.2	1.89	0.946
4		3.5	3.55	0.886	14.3	3.53	0.883		217.2	3.74	0.936	351.3	3.64	0.909
8		1.8	6.81	0.852	7.4	6.78	0.848		111.7	7.28	0.910	185.1	6.90	0.863
16		1.2	10.64	0.665	4.7	10.82	0.676		56.4	14.42	0.901	92.7	13.79	0.862
32		0.8	15.93	0.498	3.1	16.28	0.509		35.2	23.11	0.722	57.7	22.16	0.692
64									25.4	32.01	0.500	43.1	29.68	0.464
1	48	326.5			608.7			96	5259.8			8702.3		
2		165.9	1.97	0.984	303.1	2.01	1.004		2704.7	1.94	0.972	4503.9	1.93	0.966
3		115.1	2.84	0.946	212.2	2.87	0.956		1833.1	2.87	0.956	3083.6	2.82	0.941
4		87.0	3.75	0.939	158.3	3.85	0.961		1388.3	3.79	0.947	2331.9	3.73	0.933
6		59.3	5.51	0.918	107.9	5.64	0.940		952.7	5.52	0.920	1588.3	5.48	0.913
8		44.2	7.39	0.924	80.5	7.56	0.945		714.8	7.36	0.920	1188.9	7.32	0.915
12		30.1	10.85	0.904	54.9	11.09	0.924		480.3	10.95	0.913	796.3	10.93	0.911
16		25.9	12.62	0.789	47.0	12.96	0.810		358.1	14.69	0.918	590.5	14.74	0.921
24		17.7	18.47	0.769	32.1	18.94	0.789		240.1	21.91	0.913	399.9	21.76	0.907
32									182.8	28.77	0.899	299.8	29.03	0.907
48		12.5	26.21	0.546	23.6	25.80	0.537		177.2	29.69	0.618	293.0	29.70	0.619
96									140.0	37.58	0.391	231.5	37.58	0.392

$E_p = S_p/p$. For the benchmark problems described in Section 2, we used discretization with $n_1 = n_2 = n_3 = n$ where $n = 32, 48, 64$, and 96, while sizes of discrete problems were $3n^3$.

In Table 1 we present results of experiments performed on Jacquard (see http://www.nersc.gov/nusers/resources/jacquard/). It is a 712-CPU (356 dual-processor nodes) Opteron Linux cluster. Each processor runs at 2.2 GHz, and has a theoretical peak performance of 4.4 GFlop/s. Processors on each node share 6 GB of memory. The nodes are interconnected with a high-speed Infini-Band network. Shared file storage is provided by a GPFS file system. We have used the ACML Optimized Math Library and compiled the code using "mpicc -Ofast $ACML" command. The "-Ofast" option is a generic option leading to vendor suggested aggressive optimization.

As expected, parallel efficiency improves with the size of the discrete problems. For the largest problems in this set of experiments ($n = 96$), parallel efficiency is above 90% on up to 32 processors which confirms our general expectations that the proposed approach parallelizes very well.

Table 2 shows execution time on Topdawg. It is Dell Pentium4 Xeon64 Linux cluster (see http://www.oscer.ou.edu/resources.php). It has 512 dual-processor nodes. Each processor runs at 3.2 GHz and has a theoretical peak performance of 6.4 GFlop/s. Processors within each node share 4 GB of memory, while nodes are interconnected with a high-speed InfiniBand network. We have used Intel C compiler and compiled the code with the following options: "-O3 -parallel -ipo -tpp7 -xP" (collection of options for aggressive optimization suggested by Henry Neeman of OSCER).

Table 2. Experimental results on Topdawg

p	n	Benchmark 1 T_p	S_p	E_p	Benchmark 2 T_p	S_p	E_p	n	Benchmark 1 T_p	S_p	E_p	Benchmark 2 T_p	S_p	E_p
1	32	10.0			38.0			64	536.0			852.0		
2		6.0	1.67	0.83	22.0	1.73	0.86		359.0	1.49	0.75	592.0	1.44	0.72
4		3.1	3.23	0.81	11.0	3.45	0.86		180.0	2.98	0.74	293.0	2.91	0.73
8		1.8	5.56	0.69	6.3	6.03	0.75		82.0	6.54	0.82	236.0	3.61	0.45
16		1.2	8.47	0.53	3.9	9.84	0.62		44.0	12.18	0.76	71.0	12.00	0.75
32		1.1	9.09	0.28	3.5	10.86	0.34		24.0	22.33	0.70	39.0	21.85	0.68
64									18.0	29.78	0.47	29.0	29.38	0.46
1	48	244.0			444.0			96	4074.0			6766.0		
2		146.0	1.67	0.84	267.0	1.66	0.83		2353.0	1.73	0.87	3817.0	1.77	0.89
3		96.0	2.54	0.85	177.0	2.51	0.84		1538.0	2.65	0.88	2557.0	2.65	0.88
4		71.0	3.44	0.86	131.0	3.39	0.85		1207.0	3.38	0.84	1996.0	3.39	0.85
6		46.0	5.30	0.88	84.0	5.29	0.88		805.0	5.06	0.84	1344.0	5.03	0.84
8		34.0	7.18	0.90	62.0	7.16	0.90		602.0	6.77	0.85	999.0	6.77	0.85
12		24.0	10.17	0.85	41.0	10.83	0.90		406.0	10.03	0.84	675.0	10.02	0.84
16		19.0	12.84	0.80	34.0	13.06	0.82		307.0	13.27	0.83	509.0	13.29	0.83
24		13.0	18.77	0.78	24.0	18.50	0.77		207.0	19.68	0.82	343.0	19.73	0.82
32									158.0	25.78	0.81	262.0	25.82	0.81
48		9.7	25.15	0.52	18.0	24.67	0.51		115.0	35.43	0.74	190.0	35.61	0.74
96									70.0	58.20	0.61	115.0	58.83	0.61

The execution time on Topdawg is substantially smaller than that on Jac-quard (in computations that are primarily floating point arithmetic, Xeon64 processors running at 3.2 GHz are more efficient than Opteron processors at 2.2 GHz; which can be also seen comparing their theoretical peak performance). The communication time on both clusters is approximately the same (they both use InfiniBand network) and this is one of the reasons for higher parallel efficiency of Jacquard (slower processors combined with equally fast network). Again, parallel efficiency increases with the size of the discrete problems and for the largest problems reaches 60% on 96 processors.

Table 3 contains execution times collected on an IBM Linux Cluster 1350 made of 512 2-way IBM X335 nodes. Each computing node contains 2 Xeon Pentium IV processors running at 3 GHz and 2 GB of RAM. Nodes are interconnected via a Myrinet network with a maximum bandwidth of 256 Mb/s. We have used IBM Visual Age compiler and a "-O3" option.

The execution time on one processor is larger than the results from earlier mentioned computer systems. While the run-time on IBM Linux cluster is much longer than on Jacquard and Topdawg, its parallel efficiency is higher — it is higher than 50% for full set of experiments reported here. This indicates that the decrease in processor speed offsets the slower interconnection network.

Finally, Table 4 reports execution times collected on an IBM SP Cluster 1600 made of 64 nodes p5-575 (see http://www.ibm.com/servers/eserver/pseries/library/sp_books/). A p5-575 node contains 8 IBM Power5 proces-sors running at 1.9 GHz and has 16 GB of RAM. Nodes are interconnected

Table 3. Experimental results for the IBM Linux Cluster

p	n	Benchmark 1			Benchmark 2			n	Benchmark 1			Benchmark 2		
		T_p	S_p	E_p	T_p	S_p	E_p		T_p	S_p	E_p	T_p	S_p	E_p
1	32	22.7			90.3			64	1384.1			2232.3		
2		12.5	1.81	0.906	49.5	1.82	0.911		730.2	1.90	0.948	1195.9	1.87	0.933
4		6.5	3.50	0.876	25.7	3.51	0.877		393.3	3.52	0.880	633.5	3.52	0.881
8		3.4	6.75	0.843	13.2	6.84	0.855		208.8	6.63	0.829	339.6	6.57	0.822
16		1.9	12.03	0.752	7.3	12.33	0.771		99.0	13.99	0.874	164.9	13.54	0.846
32		1.4	16.02	0.501	5.6	16.21	0.507		54.1	25.59	0.800	86.5	25.80	0.806
64									33.6	41.20	0.644	54.5	40.96	0.640
1	48	600.6			1104.2			96	10080.4			17648.8		
2		323.6	1.86	0.928	594.3	1.86	0.929		5401.3	1.87	0.933	8953.1	1.97	0.986
3		220.2	2.73	0.909	399.0	2.77	0.922		3654.3	2.76	0.919	6061.5	2.91	0.971
4		168.4	3.57	0.892	311.0	3.55	0.888		2794.2	3.61	0.902	4633.8	3.81	0.952
6		115.8	5.19	0.864	214.2	5.15	0.859		1900.2	5.30	0.884	3158.8	5.59	0.931
8		84.7	7.09	0.887	155.6	7.10	0.887		1454.9	6.93	0.866	2415.8	7.31	0.913
12		57.5	10.44	0.870	105.1	10.50	0.875		972.7	10.36	0.864	1604.4	11.00	0.917
16		43.7	13.75	0.860	80.2	13.78	0.861		754.2	13.37	0.835	1249.0	14.13	0.883
24		30.2	19.86	0.827	55.5	19.91	0.830		477.2	21.13	0.880	793.5	22.24	0.927
32									355.7	28.34	0.886	589.3	29.95	0.936
48		18.9	31.72	0.661	35.1	31.46	0.655		248.0	40.65	0.847	411.8	42.86	0.893
96									151.3	66.64	0.694	251.8	70.08	0.730

Table 4. Experimental results for the IBM SP cluster

p	n	Benchmark 1			Benchmark 2			n	Benchmark 1			Benchmark 2		
		T_p	S_p	E_p	T_p	S_p	E_p		T_p	S_p	E_p	T_p	S_p	E_p
1	32	21.8			86.8			64	1257.8			2056.8		
2		10.7	2.03	1.015	43.1	2.01	1.007		670.2	1.88	0.938	989.9	2.08	1.039
4		5.4	4.03	1.006	21.1	4.11	1.027		313.0	4.02	1.005	527.4	3.90	0.975
8		2.7	8.04	1.005	10.6	8.22	1.027		152.1	8.27	1.034	252.4	8.15	1.019
16		1.5	15.01	0.938	5.9	14.70	0.919		76.6	16.43	1.027	126.2	16.29	1.018
32		1.0	21.68	0.677	3.1	28.18	0.881		39.3	31.98	0.999	65.1	31.58	0.987
64									21.0	60.01	0.938	34.2	60.19	0.940
1	48	541.7			993.6			96	9100.5			12338.7		
2		278.2	1.95	0.974	500.7	1.98	0.992		4501.5	2.02	1.011	6771.2	1.82	0.911
3		182.4	2.97	0.990	337.5	2.94	0.981		3001.4	3.03	1.011	3988.2	3.09	1.031
4		137.9	3.93	0.982	252.7	3.93	0.983		2313.5	3.93	0.983	2982.7	4.14	1.034
6		90.1	6.01	1.002	159.3	6.24	1.039		1477.4	6.16	1.027	1961.8	6.29	1.048
8		67.3	8.05	1.006	122.9	8.08	1.010		1095.2	8.31	1.039	1473.9	8.37	1.046
12		45.1	12.00	1.000	82.4	12.06	1.005		740.3	12.29	1.024	1016.5	12.14	1.012
16		34.2	15.84	0.990	58.6	16.96	1.060		560.5	16.24	1.015	774.3	15.94	0.996
24		24.2	22.35	0.931	43.8	22.67	0.945		382.2	23.81	0.992	512.6	24.07	1.003
32									283.6	32.09	1.003	383.5	32.17	1.005
48		12.4	43.78	0.912	21.2	46.79	0.975		193.8	46.96	0.978	258.0	47.82	0.996
96									100.9	90.20	0.940	146.5	84.21	0.877

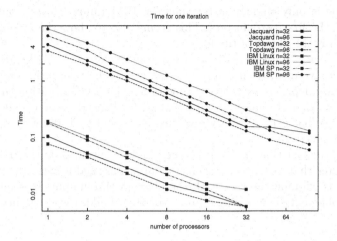

Fig. 2. Time for one iteration on the parallel computer systems

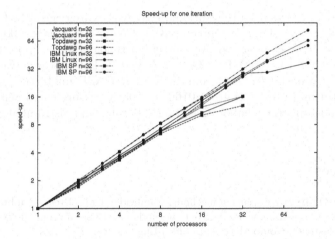

Fig. 3. Speed-up for one iteration on the parallel computer systems

with a pair of connections to the Federation High Performance Switch (HPS). The HPS interconnect is capable of a unidirectional bandwidth of up to 2 Gb/s. We have used the IBM Visual Age compiler and compiled the code using "-O4 -qipa=inline" options. One can see that for relatively large problems the speed-up is close to the theoretical limit — the number of processors. This result was expected because communications between processors is not only very fast, but also its start-up time is faster than in the case of other machines. Interestingly, a super-linear speed-up is observed in some cases. The main reasons for this fact can be related to splitting the entire problem into subproblems which helps memory management in the case of 8 processor nodes; in particular it allows for better usage of cache memories of individual parallel processors. Interestingly, this machine is only slightly faster that the IBM Linux Cluster, but remains slower than the first two clusters. This seems also to show the age of this machine, which is the oldest of the four.

A comparison of parallel performance of the developed C+MPI code obtained on all four above mentioned computer systems can be seen in Figures 2 and 3. In Figure 2 we depict the execution time of a single PCG iteration ($n = 32, 96$) of our code, while in Figure 3 we represent parallel speed-up of a single iteration ($n = 32, 96$). What is particularly revealing is the fact that all four systems have very similar speed-up.

However, the fact that the largest speed-up was obtained on the IBM SP machine indicates that as far as large clusters are concerned it is till the processing power that is winning the race with the network throughput. It is much easier to solve problems fast on a single processor than build a well-balanced parallel computer.

Acknowledgments

Computer time grants from the National Energy Research Scientific Computing Center and the Oklahoma Supercomputing Center are kindly acknowledged. The parallel numerical tests on two clusters in Bologna were supported via the EC Project HPC-EUROPA RII3-CT-2003-506079. This research was partially supported by grant I-1402/2004 from the Bulgarian NSF and by the project BIS-21++ funded by FP6 INCO grant 016639/2005. Work presented here is a part of the Poland-Bulgaria collaborative grant: "Parallel and distributed computing practices".

References

1. Axelsson, O.: Iterative solution methods. Cambridge Univ. Press, Cambridge (1994)
2. Axelsson, O., Gustafsson, I.: Iterative methods for the solution of the Navier equations of elasticity. Comp.Meth.Appl.Mech.Eng. 15, 241–258 (1978)
3. Blaheta, R.: Displacement decomposition-incomplete factorization preconditioning techniques for linear elasticity problems. Num. Lin. Alg. Appl. 1, 107–128 (1994)
4. Georgiev, A., Baltov, A., Margenov, S.: Hipergeos benchmark problems related to bridge engineering applications, REPORT HG CP 94–0820–MOST–4
5. Lirkov, I.: MPI solver for 3D elasticity problems. Math. and computers in simulation 61(3-6), 509–516 (2003)
6. Lirkov, I., Margenov, S.: MPI parallel implementation of CBF preconditioning for 3D elasticity problems. Math. and computers in simulation 50(1-4), 247–254 (1999)
7. Snir, M., et al.: MPI: The Complete Reference. Scientific and engineering computation series. The MIT Press, Cambridge (1997) (Second printing)
8. Walker, D., Dongara, J.: MPI: a standard Message Passing Interface. Supercomputer 63, 56–68 (1996)

Re-engineering Technology and Software Tools for Distributed Computations Using Local Area Network

A.P. Sapozhnikov, A.A. Sapozhnikov, and T.F. Sapozhnikova

Joint Institute for Nuclear Research,
Dubna, Russia

Abstract. A new technology is proposed for integration the old standalone Fortran — written computational programs into more large distributed computer systems. This is a re-engineering technology, because the main developer's tool becames a F2F program for source modules converting. This converter provides the maintenance of all rules, needed for transformed program, created initially for monoprocessor computer systems, into Computational Server, working in distributed network area. The principles of F2F converter and details of communications between Client and Server are discussed.

Starting from the early 90-s of XX century the object-oriented programming technologies, based on C++ and Pascal languages, have been intensively progressed. Now it is already impossible to imagine a serious application, working in an old MS-DOS style, i.e. with no mouse and modern windowed graphics. But the large amount of numeric programs, fundamental for modern applied systems, comes into PC-world from an antique epoch of mainframes. Actually it means that these programs were Fortran-written.

The static nature of Fortran does not allow it to be the general tool for object-oriented programming. Moreover, the current generation of programmers prefers C++ and Delphi, i.e. more dynamic languages. On the other hand, all the difficulties in numeric algorithms programming does not depend upon a language used, but only depend on understanding the algorithms themselves. As for a serious numeric program, to redesign it from Fortran to C++, one needs to know all delicate details of it, i.e. to be its author, but the author is most likely to be very far from active programming now! Using of language converters like F2C [1] cannot resolve the problem because:

- the program is still remaining a "black box" as it was before;
- there are no converters from Fortran to C++ or Pascal.

That's why we can suppose that the large numeric programs made in the old days using Fortran, will be forced to stay in their Fortran-incarnation in the foreseeable future. Some of these programs, being very popular in the past, still stay indispensable at present, because they personify the unique experience of

I. Lirkov, S. Margenov, and J. Waśniewski (Eds.): LSSC 2007, LNCS 4818, pp. 713–720, 2008.

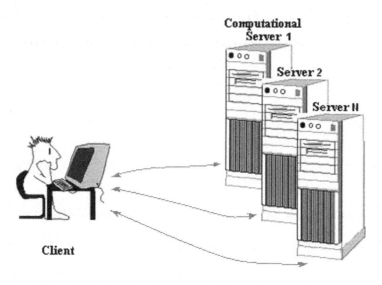

Fig. 1. The distributed system for large computational job

outstanding specialists. Below we intend to discuss namely about such kind of programs. Let us call them for short "unique programs".

Another particular feature of the present-day programming is the tendency to distribute calculation between several computers in the computer networks. In general they may be even of different types. The existing technologies of parallelization, such as OpenMP and MPI [2], are oriented to decomposition of the whole computational job onto a number of smaller processes. As a rule, this decomposition is performed manually while software development or modernization. At the same time, there is no technology to integrate the large standalone made computational blocks into huge-scale distributed systems.

Here we are proposing a new technology for integration of unique programs onto more large distributed systems. This is a re-engineering technology, because the main developer's tool becames a program for converting of standalone-developed Fortran-written programs onto integrated parts of large distributed computer systems. The general architecture of the proposed distributed computational system also seems to be quite non-traditional (see Fig.1). User's workstation contains a single Client-process, one of whose aims is to distribute jobs among a number of independent Computational Servers. Each of them is a separate process, which executes a corresponding program and works, in a general case, on separate computer. All the requests on input-output operations, needed for each Computational Server, are re-addressed to Client and interpreted by it.

The key idea is the automation of building the Computational Server from standalone unique programs, without any manual transformation of their source modules. Exactly this idea we call here "re-engeneering". This offers an opportunity to integrate well tested during long time old programs, made by bygone generation of experienced programmers, into large-scale modern present-day systems of data processing, including advanced visualization facilities, databases and others mechanisms for human-computer communications.

Our approach to integration is not a general solution, we do not pretend to universal usage, as the Service-Oriented Architecture (SOA) does. Our main aim is to propose the technology and software tools namely for the unique programs, mentioned above, without any modification of their source codes. Moreover, the terms "Client" and "Server" themselves are here rather relative, to not invent a new substances. They designate only two processes, which communicate with each other in Local Area Network or even inside the single computer.

The Server's aim is to solve a specific computational job, ordered by Client. While executing this job, Server can consume the external data from standalone-made files or directly from Client. Server can produce output data for Client. All needed input-output operations are ordered in Server's side using traditional Fortran operators Read/Write, but will be interpreted on Client's side.

User's workstation executes a Client-program, distributing the whole task between a number of independent Computational Servers. Besides launching Servers and processing of their input-output requests, Client is the single point for communication with User. This way offers an opportunity for full separation of two stages: solving of computational problem and interpretation of results. Therefore, for each of these stages we can use both the most appropriate tools and the different and independent groups of developers.

A special F2F (Fortran-To-Fortran) converter has been created for transformation of old unique programs into Computational Server, mentioned above. The converted program works in own separate address-space, maybe in remote computer. Particularly, this increases the reliability of the whole system, because possible errors on Client-side cannot influence errors on Server-side. F2F organizes all needed environment for interaction with Client. F2F automatically substitutes all Input/Output operators in Fortran source modules for calling special subroutines, realizing a Client-Server interaction protocol. The auxiliary information is built into these subroutines: current running Input/Output operator, source line number, name of current data file and current processed record number. It gives an opportunity to get a detailed information about possible I/O-Errors, that's why the converted program becomes even better than the original one. Besides, F2F substitutes all operators OPEN of opening data files for calling special subroutine, which can search specified files following various rules. Further, F2F builds into all I/O operators catching possible errors for preventing the Server against unexpected "hanging" in it. Another quite attractive F2F feature is the ability to built into converted program the debugging information about subroutines calling. It is useful while looking for errors in computational program.

This is an example of how F2F works:

```
Subroutine S                    !here is fragment of the original:
c=sin(a)+cos(b)                 !some computations
```

```
Open(unit=Lun,file='myfile.dat',status='old')
Write(1),a,b,c                !binary output to file
Write(*,*),a,b,c              !consoleoutput
Read(*,'(i5)',err=9)n    !console input with own error catching
Stop
End


Subroutine S                  !The result of converting:
Use ComFort !here is all environment and interface with Client
c=sin(a)+cos(b)    !of course all computations remains intact !
mess='subroutine S,line14:Open(unit=Lun,file="myfile.dat",...'
Call Open_File(Lun,'myfile.dat','old','formatted')
if(fo_error.ne.0) goto 123
mess = 'subroutine S, line 15: Write(1),a,b,c'
Write(1, err=123) a,b,c       !catching of possible I/O errors
mess ='subroutine S, line 16: Write(*,*),a,b,c'
Write (iobuf,err=123)a,b,c !format transformation
Call InterfaceIO (jwrite)     !then  output re-addressing
mess ='subroutine S, line 17: Read(*,"(i5)",err=9),n'
Call InterfaceIO(jread)       !input re-addressing
Read(iobuf,'(i5)',err=9) n    !then format transformation
Call Instead_Of_Stop          !normal finish of Server
123 Call IOError(mess)        !abnormal finish of Server
End
```

The aim of the F2F converter is to transform a program, destined for interaction directly with human, into a program for interaction with the Client-process while working in a distributed computer system. Actually, F2F is a translator from Fortran to Fortran. Namely, F2F is the basic re-engeneering tool in our technology (Fig.2).

Though F2F has to perform rather a deep syntactical analysis of the program being converted, it works more quickly than a native Fortran-compiler. Initially, we supposed to exploit F2F only in a semi-automatic mode, because Fortran has a very complicated syntax and has no grammar at all. In some cases the human-help assumed. However, we had managed all the syntactical problems. Now F2F does not need any manual revision for resulted Fortran-code. Capacity F2F for work was practically tested on a lot of real numeric programs with huge size.

1   Communications Interface between Client and Computational Server

A simple symmetric model is proposed for communications between Client-program and any of its Servers. The single basic unit in this model is a line of text, i.e. a sequence of chars with a special char at the end. Analyzing the input line, both sides can monosemantically interpret what to do.

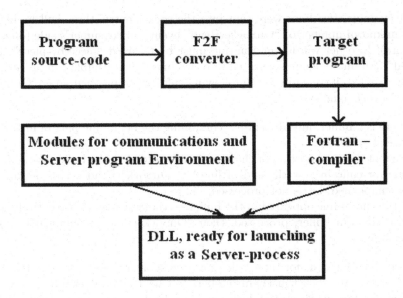

Fig. 2. Building an Computational Server from standalone unique program

For example, requests from Client to Server:

```
Execute  MyFavoriteJob.dll        - run specified program
Take Line  <arbitrary line>        - answer for text input
Give Time                          - request of elapsed time
Suspend
Resume
Stop
```

The similar requests from Server to Client:

```
Take Line  <arbitrary line >       - text output
Give Line                          - waiting for text input
```

The only structural requirement to these communications is to be non-synchronous. This allows for both sides to stay permanently active, not being in waiting status without special necessity. To meet this requirement, there is enough to use classical socket mechanism, which appeared for the first time in OS UNIX and then was successfully transferred into Windows.

Moreover, having chosen Delphi as the main tool for programming in both sides, the whole Client and Server's communication level, we used TServerSocket and TClientSocket classes from rich Delphi Component Palette. These classes were developed by Borland Inc. specially for using in inter-computer communications based on socket mechanism. They ensure required non-synchronosity, because they can process specific event "OnSocketRead".

While working in a network area, the most serious problem is a non-predicted size of information, being received by consumer in the single input. Therefore, if, for example, two lines were sent into the socket entry, there will be before-hand unknown number of "OnSocketRead" events on consumer's side (one, two or more)! In other words, information must be received "piece-by-piece", then "glued" and finally again separated into two lines! This trouble can be very elegantly avoided, if we will use the following rules while programming "OnSocketRead" event handler:

- there is a buffer, initially empty, containing the beginning part of line, being already received;
- current received portion of chars is entirely added to the buffer. If the buffer doesn't contain now the end of line — "OnSocketRead" error declared as not happened, i.e. is not processed;
- if the end of line appeared in the buffer — the first line will be entirely withdrawed from buffer, processed, then "OnSocketRead" event will be called recursively.

Corresponding code for these rules got even shorter than their verbal description! Usage of the socket mechanism offers to Client and Server to be absolutely independent upon their mutual location — in the same computer or in the different computers in a network. As for Server — there is up to it, who is connecting with it. As for Client, being forced to specify the address or name of Server, there is enough to specify the keyword "localhost" to designate the Server, located in the same computer.

The described above interface is the second basis of the proposed technology for the distributed system building. We ought to specify, that this technology does not oblige to build the Server part of system namely from the standalone Fortran-written programs. If the Computational Server is developed "from zero", the developer can choose:

1. to use his own tools observing non-burdensome requirements of our interface;
2. to use good old Fortran with habitual operators for Input/Output, and then convert his program via F2F.

This approach particularly allows one to involve into Computational Servers development the experienced specialists on numeric methods, not wishing to exceed the framework of their habitual Fortran, while the young developers can concentrate on the Client development with usage of more contemporary technologies.

Below we want to illustrate the proposed methodics of automated building the remote Computational Server from standalone Fortran-written numeric program.

MINUIT [3] — the program for general kind functions minimizing, was created by F. James (CERN) about 40 years ago, but still stays the most popular tool while solving wide class of problems. It is a typical "unique program" in our terminology. The Fortran-written source module (about 7500 lines), getting directly from the author, without any manual changes was processed by our F2F-converter. After that all communications between MINUIT and outer

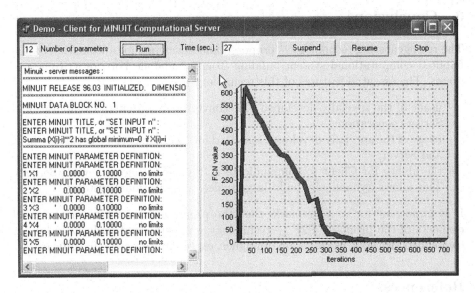

Fig. 3. The tiny Client for MINUIT program, performing visualization of convergence to the minimum of function

world got switched to rather ordinary (120 lines) Client-program. This Client was written using Delphi. It engaged in an interpretation of all I/O requests from MINUIT and in visualization of a convergence to the minimum. Of course, that visualization was not foreseen in original MINUIT.

Thus this example (Fig.3) demonstrates the following advantages of F2F-technology:

- the ability to integrate old Fortran-written programs with modern graphical applications;
- the simplicity in buildings a Client-applications, based on proposed standards;
- the appearance of new, attractive properties in old computational programs.

2   Further Perspectives

The idea of automatic source modules converting seems to be extremely productive. Particularly, many source transformations needed to integrate with MPI-package, may be successfully performed by our F2F converter. For example, all MPI-programs must perform Input/Output operations not for all processes, as MPI-paradigm SPMD requires, but as the rule — by the single master-process [2]. All needed checkings, broadcastings, initial and final MPI-operations can be automatically added by F2F-converter while porting the program under MPI-package. Such automation of routined works while parallelization of unique programs, initially developed for execution on single processor, can be an important direction for the F2F-technology evolution.

3   Conclusions

The re-engeneering technology is proposed and developed for automation of distributed computational system building from standalone Fortran-written programs. The essential features of this technology are:

- usage of the specially developed F2F-converter to automate Computational Servers generation;
- usage of the specially developed standard for communications between Client and Server;
- the ability for Client to communicate at the same time to several Servers. This allows dynamically build large-scaled computational systems.

It is important for us to remark, that the programmatic achievement for all elements of the proposed F2F — technology was performed exclusively by our own efforts, without usage of any foreigner, especially commercial, software.

References

1. Feldman, S.I., Weinberger, P.J.: A Portable Fortran 77 Compiler, UNIX Time Sharing System Programmer's Manual, 10th edn., AT&T, Bell Laboratories, vol. 2 (1990)
2. Snir, M., et al.: MPI: The complete Reference. MIT Press, Cambridge (1997)
3. CERN Program Library Long Writeup D506. James. F. MINUIT. CERN, Geneva, Switzerland

On Single Precision Preconditioners for Krylov Subspace Iterative Methods

Hiroto Tadano[1,3] and Tetsuya Sakurai[2,3]

[1] Graduate School of Informatics, Kyoto University, Kyoto 606-8501, Japan
[2] Department of Computer Science, University of Tsukuba, Tsukuba 305-8573, Japan
[3] Core Research for Evolutional Science and Technology,
Japan Science and Technology Agency, Japan

Abstract. Large sparse linear systems $Ax = b$ arise in many scientific applications. Krylov subspace iterative methods are often used for solving such linear systems. Preconditioning techniques are efficient to reduce the number of iterations of Krylov subspace methods. The coefficient matrix of the linear system is transformed into MA or AM in the left or right preconditioning, where M is a preconditioning matrix. In this paper, we analyze the influence of perturbation in the computation of preconditioning of Krylov subspace methods. We show that the perturbation of preconditioner does not affect the accuracy of the approximate solution when the right preconditioning is used. Some numerical experiments illustrate the influence of preconditioners with single precision arithmetic.

1 Introduction

Large and sparse linear systems

$$Ax = b, \tag{1}$$

where A is a nonsingular and non-Hermitian $n \times n$ matrix, appear in many scientific fields. Since almost all of the computational time is spent to solve linear systems (1) of these applications, fast solvers are desired.

It is known that Krylov subspace iterative methods are efficient [2,7,9] for solving such linear systems. Preconditioning techniques are often used to improve the convergence rate of Krylov subspace methods. Several preconditioning techniques such as the incomplete LU factorization (ILU) preconditioner [7] and the sparse approximate inverse (SAI) preconditioner [4] have been proposed [3]. It has been verified by many numerical experiments that these preconditioners are effective. However, the computational cost of the preconditioning part is sometimes large. Thus, it is valuable to reduce the computational time of this part.

Recently, the calculation techniques using the Cell processor and the graphics processing unit (GPU) have attracted attention for wide area of scientific computing. These processors provide high performance in single precision arithmetic.

I. Lirkov, S. Margenov, and J. Waśniewski (Eds.): LSSC 2007, LNCS 4818, pp. 721–728, 2008.

The purpose of this study is to consider the influence of the preconditioners performed with single precision arithmetic.

This paper is organized as follows. In the next section, we describe left and right preconditioners. In addition, the polynomial preconditioner and the variable preconditioner are also described. In Section 3, we analyze the influence of the perturbation which occurs in the preconditioning part. Some numerical experiments illustrate the influence of the single precision preconditioners in Section 4. Finally, we present concluding remarks and future works in Section 5.

2   Preconditioners for Krylov Subspace Iterative Methods

In this section, we briefly describe preconditioners of Krylov subspace iterative methods which we will use. The rate of convergence of Krylov subspace methods depends on the coefficient matrix A. It is known that the residual of these methods converges in a few iterations when the coefficient matrix is close to the identity matrix. In order to improve the convergence rate, the coefficient matrix is transformed to MA or AM, where M is an $n{\times}n$ preconditioning matrix. They are called left preconditioning and right preconditioning, respectively.

The linear systems (1) are transformed to

$$MA\boldsymbol{x} = M\boldsymbol{b}$$

by the left preconditioning. In a similar way, we can obtain the right preconditioned linear systems

$$AM\boldsymbol{y} = \boldsymbol{b}, \quad \boldsymbol{y} = M^{-1}\boldsymbol{x}. \tag{2}$$

The algorithms of the BiCGSTAB method with the left and right preconditioners are shown in Fig. 1. Here, ε is a small constant for stopping criterion.

2.1   Polynomial Preconditioners

In the polynomial preconditioners, the preconditioning matrix M is defined by a polynomial of A. As one of these polynomials, the Neumann series expansion [7] is known.

We assume that $\|I - A\| < 1$, where I is the $n{\times}n$ identity matrix. The Neumann series expansion of A^{-1} is represented as follows:

$$A^{-1} = (I - (I - A))^{-1} = \sum_{j=0}^{\infty}(I - A)^j.$$

The truncated polynomial of degree m

$$A^{-1} \approx M = \sum_{j=0}^{m}(I - A)^j,$$

has been used as the preconditioning matrix M. Since the product of the matrix M by a vector is computed by the product of the coefficient matrix A by the vector, M is not computed explicitly.

x_0 is an initial guess,
compute $r_0 = M(b - Ax_0)$,
set $p_0 = r_0$,
choose r_0^* such that $(r_0^*, r_0) \neq 0$,
for $k = 0, 1, \ldots,$ until $\|r_k\|_2 \leq \varepsilon\|b\|_2$ do:
begin

$$u_k = Ap_k,$$
$$v_k = Mu_k,$$
$$\alpha_k = \frac{(r_0^*, r_k)}{(r_0^*, v_k)},$$
$$t_k = r_k - \alpha_k v_k,$$
$$s_k = At_k,$$
$$q_k = Ms_k,$$
$$\zeta_k = \frac{(q_k, t_k)}{(q_k, q_k)},$$
$$x_{k+1} = x_k + \alpha_k p_k + \zeta_k t_k,$$
$$r_{k+1} = t_k - \zeta_k q_k,$$
$$\beta_k = \frac{\alpha_k}{\zeta_k} \cdot \frac{(r_0^*, r_{k+1})}{(r_0^*, r_k)},$$
$$p_{k+1} = r_{k+1} + \beta_k(p_k - \zeta_k v_k),$$
end

(a) Left preconditioning.

x_0 is an initial guess,
compute $r_0 = b - Ax_0$,
set $p_0 = r_0$,
choose r_0^* such that $(r_0^*, r_0) \neq 0$,
for $k = 0, 1, \ldots,$ until $\|r_k\|_2 \leq \varepsilon\|b\|_2$ do:
begin

$$u_k = Mp_k,$$
$$v_k = Au_k,$$
$$\alpha_k = \frac{(r_0^*, r_k)}{(r_0^*, v_k)},$$
$$t_k = r_k - \alpha_k v_k,$$
$$s_k = Mt_k,$$
$$q_k = As_k,$$
$$\zeta_k = \frac{(q_k, t_k)}{(q_k, q_k)},$$
$$x_{k+1} = x_k + \alpha_k u_k + \zeta_k s_k,$$
$$r_{k+1} = t_k - \zeta_k q_k,$$
$$\beta_k = \frac{\alpha_k}{\zeta_k} \cdot \frac{(r_0^*, r_{k+1})}{(r_0^*, r_k)},$$
$$p_{k+1} = r_{k+1} + \beta_k(p_k - \zeta_k v_k),$$
end

(b) Right preconditioning.

Fig. 1. The preconditioned BiCGSTAB method

2.2 Variable Preconditioner

As one of the flexible preconditioners, the variable preconditioning [1] has been proposed by Abe et al. in 2001. In this preconditioner, the linear systems of the form $Az = w$ are solved approximately instead of computing the product of M by a vector. As solutions of these linear systems, various iterative methods (e.g., stationary iterative methods, Krylov subspace methods) can be applied. This part is called the inner-loop.

The iteration of the inner-loop is stopped by one of the following conditions:

$$\|z^{(\ell)} - z^{(\ell-1)}\|_\infty / \|z^{(\ell)}\|_\infty \leq \gamma, \tag{3}$$

$$\|w - Az^{(\ell)}\|_2 / \|w\|_2 \leq \gamma, \tag{4}$$

where $z^{(\ell)}$ denotes the ℓ-th approximate solution of $Az = w$, and γ is a positive number. The condition (3) is employed when stationary iterative methods are used as the solutions of the inner-loop. When the linear systems $Az = w$ are solved by Krylov subspace methods, condition (4) is used. In addition, the iteration of the inner-loop is stopped when the number of iteration exceeds $N_{\max}$.

3   The Influence of the Perturbation in the Preconditioning Part

In this section, we describe the influence of the perturbation in the preconditioning part. First, we describe the case of left preconditioner. We assume that the first $(k-1)$ steps of the left preconditioned BiCGSTAB have been already computed without the perturbation. Moreover, we assume that the perturbation occurs in the preconditioning parts of the kth step of BiCGSTAB. Let $\tilde{v}_k$ and $\tilde{q}_k$ be

$$\tilde{v}_k = v_k + \delta v_k, \quad \tilde{q}_k = q_k + \delta q_k,$$

respectively. The vectors δv_k and δq_k denote the perturbation of v_k and q_k. Throughout this section, the scalars and vectors which include the perturbation are denoted with the symbol ~. The $(k+1)$-th approximate solution $\tilde{x}_{k+1}$ is computed by

$$\tilde{x}_{k+1} = x_k + \tilde{\alpha}_k p_k + \tilde{\zeta}_k \tilde{t}_k.$$

On the other hand, the $(k+1)$-th residual $\tilde{r}_{k+1}$ is obtained as follows:

$$\begin{aligned}
\tilde{r}_{k+1} &= \tilde{t}_k - \tilde{\zeta}_k \tilde{q}_k \\
&= r_k - \tilde{\alpha}_k (MAp_k + \delta v_k) - \tilde{\zeta}_k (MA\tilde{t}_k + \delta q_k) \\
&= M[b - A(x_k + \tilde{\alpha}_k p_k + \tilde{\zeta}_k \tilde{t}_k)] - \tilde{\alpha}_k \delta v_k - \tilde{\zeta}_k \delta q_k \\
&= M(b - A\tilde{x}_{k+1}) - \tilde{\alpha}_k \delta v_k - \tilde{\zeta}_k \delta q_k.
\end{aligned}$$

Thus, the relation $\tilde{r}_{k+1} = M(b - A\tilde{x}_{k+1})$ between the approximate solution and the residual no longer holds.

Next, we describe the case of right preconditioner. In a similar way, we assume that the first $(k-1)$ steps of the right preconditioned BiCGSTAB have been already computed without perturbation. In addition, we assume that the perturbation occurs in the preconditioning parts of the k-th step of BiCGSTAB. Let $\tilde{u}_k$ and $\tilde{s}_k$ be

$$\tilde{u}_k = u_k + \delta u_k, \quad \tilde{s}_k = s_k + \delta s_k.$$

The vectors δu_k and δs_k denote the perturbation of u_k and s_k. The $(k+1)$-th approximate solution $\tilde{x}_{k+1}$ is computed by

$$\tilde{x}_{k+1} = x_k + \tilde{\alpha}_k \tilde{u}_k + \tilde{\zeta}_k \tilde{s}_k.$$

The $(k+1)$-th residual $\tilde{r}_{k+1}$ is denoted as follows:

$$\begin{aligned}
\tilde{r}_{k+1} &= \tilde{t}_k - \tilde{\zeta}_k \tilde{q}_k \\
&= r_k - \tilde{\alpha}_k A\tilde{u}_k - \tilde{\zeta}_k A\tilde{s}_k \\
&= b - A(x_k + \tilde{\alpha}_k \tilde{u}_k + \tilde{\zeta}_k \tilde{s}_k) \\
&= b - A\tilde{x}_{k+1}.
\end{aligned}$$

As a consequence, the relation $\tilde{r}_{k+1} = b - A\tilde{x}_{k+1}$ between the approximate solution and the residual holds. This implies that the preconditioning can be performed with single precision arithmetic.

From (2), the $(k+1)$-th approximate solution x_{k+1} of the right preconditioned BiCGSTAB can be computed as follows:

$$\begin{aligned}
y_{k+1} &= y_k + \alpha_k p_k + \zeta_k t_k, \\
x_{k+1} &= M y_{k+1}.
\end{aligned}$$

However, the accuracy of the approximate solution deteriorates if these equations are used.

4   Numerical Experiments

In this section, we present several numerical experiments to investigate the influence of single precision preconditioner. These examples carried out on Dell Precision Workstation 470 (CPU: Intel Xeon 3.2 GHz, RAM: 2.0 GBytes, OS: Red Hat Enterprise Linux WS 4, Compiler: Intel Fortran ver. 9.1, Compile option: `-O3 -xP`). We used the preconditioned BiCGSTAB method [8] for solving linear systems. The polynomial preconditioner and the variable preconditioner were used. In the inner-loop of the variable preconditioner, the Jacobi method was used. These preconditioners were performed with single precision arithmetic. Other parts were performed with double precision arithmetic. Iteration was started with $x_0 = 0$ and $r_0^* = r_0$. The iteration of the BiCGSTAB were stopped when $\|r_k\|_2/\|b\|_2 \leq 10^{-15}$. The degree m of polynomial preconditioner was 5. The parameters γ and $N_{\max}$ of the variable preconditioner were 10^{-1} and 5, respectively.

In numerical experiments, we consider the linear system derived from the Quantum Chromodynamics (QCD). As a coefficient matrix A of the linear system, we used `conf5.4-0018x8-1000` [6]. The size n of A is $49,152$, and the right-hand vector b was given by $b = [5.4, 5.4, \ldots, 5.4]^T$. This linear system has a coefficient matrix A of the form

$$A = I_n - \kappa D, \quad D = \begin{bmatrix} D_{00} & D_{01} \\ D_{10} & D_{11} \end{bmatrix},$$

where I_n is the $n \times n$ identity matrix and κ is a scalar. In the following examples, we used $\kappa = 0.182$. The nonzero structure of A is shown in Fig. 2(a).

We can obtain the reordered coefficient matrix $\tilde{A}$ of the form

$$\tilde{A} = I_n - \kappa \tilde{D}, \quad \tilde{D} = \begin{bmatrix} O & \tilde{D}_{01} \\ \tilde{D}_{10} & O \end{bmatrix},$$

by using the Red-Black reordering. In Fig. 2(b), the nonzero structure of $\tilde{A}$ is shown. In addition, the SSOR preconditioner [5] was used for reducing the number of iterations. The coefficient matrix $\hat{A}$ after applying the SSOR preconditioner is represented by

$$\hat{A} = (I_n + L)^{-1}\tilde{A}(I_n + U)^{-1} = (I_n - L)\tilde{A}(I_n - U) = I_n - \kappa^2 \hat{D},$$

where, the matrices L, U, and $\hat{D}$ are denoted by

$$L = \begin{bmatrix} O & O \\ -\kappa \tilde{D}_{10} & O \end{bmatrix}, \quad U = \begin{bmatrix} O & -\kappa \tilde{D}_{01} \\ O & O \end{bmatrix}, \quad \hat{D} = \begin{bmatrix} O & O \\ O & \tilde{D}_{10}\tilde{D}_{01} \end{bmatrix},$$

respectively. Hence, the coefficient matrix $\hat{A}$ and the vectors $\hat{x}$, $\hat{b}$ of the preconditioned linear system $\hat{A}\hat{x} = \hat{b}$ are represented by

$$\hat{A} = \begin{bmatrix} I_{n/2} & O \\ O & I_{n/2} - \kappa^2 \tilde{D}_{10}\tilde{D}_{01} \end{bmatrix}, \quad \hat{x} = \begin{bmatrix} \hat{x}^{(0)} \\ \hat{x}^{(1)} \end{bmatrix}, \quad \hat{b} = \begin{bmatrix} \hat{b}^{(0)} \\ \hat{b}^{(1)} \end{bmatrix}. \tag{5}$$

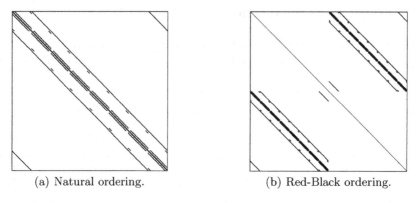

(a) Natural ordering.   (b) Red-Black ordering.

Fig. 2. Nonzero structures of the coefficient matrix

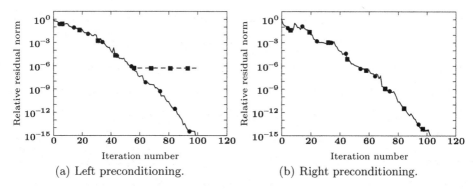

(a) Left preconditioning.   (b) Right preconditioning.

Fig. 3. Relative residual norm histories for the BiCGSTAB with the polynomial preconditioner. ●: computed relative residual norm, ■: true relative residual norm.

From (5), the upper part $\hat{\boldsymbol{x}}^{(0)}$ of the solution $\hat{\boldsymbol{x}}$ is given by $\hat{\boldsymbol{b}}^{(0)}$. Therefore, we solved the reduced linear system

$$(I_{n/2} - \kappa^2 \tilde{D}_{10}\tilde{D}_{01})\hat{\boldsymbol{x}}^{(1)} = \hat{\boldsymbol{b}}^{(1)}$$

in numerical experiments.

The relative residual norm histories of the BiCGSTAB with the polynomial preconditioner are shown in Fig. 3. The horizontal axis and the vertical axis denote the iteration number and the relative residual norm, respectively. The symbol ● in graphs denotes the relative residual norm $\|\boldsymbol{r}_k\|_2/\|\hat{\boldsymbol{b}}^{(1)}\|_2$ computed by the recurrence relation of $\boldsymbol{r}_k$ in Fig. 1. This value is called the computed relative residual norm. The symbol ■ denotes the true relative residual norm (TRR). The TRR was computed by $\|M[\hat{\boldsymbol{b}}^{(1)} - (I_{n/2} - \kappa^2 \tilde{D}_{10}\tilde{D}_{01})\hat{\boldsymbol{x}}_k^{(1)}]\|_2/\|\hat{\boldsymbol{b}}^{(1)}\|_2$ in the BiCGSTAB with left preconditioning. In the BiCGSTAB with right preconditioning, the TRR was computed by $\|\hat{\boldsymbol{b}}^{(1)} - (I_{n/2} - \kappa^2 \tilde{D}_{10}\tilde{D}_{01})\hat{\boldsymbol{x}}_k^{(1)}\|_2/\|\hat{\boldsymbol{b}}^{(1)}\|_2$.

The computed relative residual norm converged when using the left preconditioning of single precision. However, the true relative residual norm stagnated

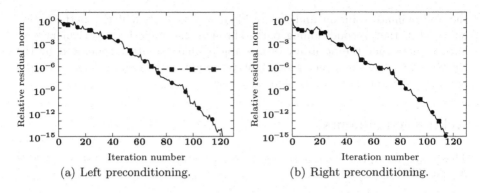

(a) Left preconditioning.   (b) Right preconditioning.

Fig. 4. Relative residual norm histories for the BiCGSTAB with the variable preconditioner. ●: computed relative residual norm, ■: true relative residual norm.

Table 1. Comparison of CPU Time [s]

(a) Polynomial preconditioner.

	Single precision	Double precision
Left preconditioning	12.55	14.10
Right preconditioning	12.91	14.59

(b) Variable preconditioner.

	Single precision	Double precision
Left preconditioning	16.06	17.86
Right preconditioning	15.18	17.55

around 10^{-6}. On the other hand, in the right preconditioning, the true relative residual norm reached 10^{-15} in spite of using the single precision preconditioner.

In Fig. 4, we show the relative residual histories of the BiCGSTAB with the variable preconditioner. In the left preconditioned BiCGSTAB, the accuracy of the approximate solution stagnated. The true relative residual norm of the right preconditioned BiCGSTAB satisfied the stopping criterion.

The CPU time of the single precision preconditioner and the double precision preconditioner are tabulated in Table 1. By using the single precision preconditioner, the CPU time decreased by about 1.5 seconds. We can expect to reduce the computation time of single precision arithmetic by using the inline assembler.

5   Conclusions

In this paper, we have considered the influence of the perturbation occurred in the preconditioning part of the Krylov subspace methods. In the left preconditioned BiCGSTAB, the relation between the approximate solution and the residual lost when the perturbation occurred in the preconditioning part. On the other hand, in the right hand preconditioned BiCGSTAB, the relation between the approximate solution and the residual was held even if the perturbation occurred in the preconditioning part. Through the numerical experiments, we have verified that the accuracy of the approximate solution generated by the BiCGSTAB with single precision right preconditioner is equivalent to results obtained with the double precision arithmetic.

Acknowledgements

This work was supported in part by CREST of the Japan Science and Technology Agency (JST).

References

1. Abe, K., et al.: A SOR-base variable preconditioned GCR method (in Japanese). Trans. JSIAM 11, 157–170 (2001)
2. Barrett, R., et al.: Templates for the solution of linear systems: Building blocks for iterative methods, 2nd edn. SIAM, Philadelphia (1994)
3. Benzi, M.: Preconditioning techniques for large linear systems: A survey. J. Comput. Phys. 182, 418–477 (2002)
4. Benzi, M., Tůma, M.: A sparse approximate inverse preconditioner for nonsymmetric linear systems. SIAM J. Sci. Comput. 19, 968–994 (1998)
5. Frommer, A., Lippert, T., Schilling, K.: Scalable parallel SSOR preconditioning for lattice computations in gauge theories. In: European Conference on Parallel Processing, pp. 742–749 (1997)
6. Matrix Market, http://math.nist.gov/MatrixMarket/
7. Saad, Y.: Iterative methods for sparse linear systems, 2nd edn. SIAM, Philadelphia (2003)
8. Van der Vorst, H.A.: Bi-CGSTAB: A fast and smoothly converging variant of Bi-CG for the solution of nonsymmetric linear systems. SIAM J. Sci. Stat. Comput. 13, 631–644 (1992)
9. Van der Vorst, H.A.: Iterative Krylov methods for large linear systems. Cambridge University Press, Cambridge (2003)

A Parallel Algorithm for Multiple-Precision Division by a Single-Precision Integer

Daisuke Takahashi

Graduate School of Systems and Information Engineering, University of Tsukuba
1-1-1 Tennodai, Tsukuba, Ibaraki 305-8573, Japan
daisuke@cs.tsukuba.ac.jp

Abstract. We present a parallel algorithm for multiple-precision division by a single-precision integer. This short division includes a first-order recurrence. Although the first-order recurrence cannot be parallelized easily, we can apply the parallel cyclic reduction method. The experimental results of multiple-precision parallel division by a single-precision integer on a 32-node Intel Xeon 3 GHz PC cluster are reported.

1 Introduction

Many multiple-precision division algorithms have been thoroughly studied [10, 12,13,15,16]. Knuth [11] described classical algorithms for n-digit division. These methods require $O(n^2)$ operations.

Division of two n-digit numbers can be performed by using the Newton iteration [13, 1, 10]. This scheme requires $O(M(n))$ operations, where $M(n)$ is the number of operations used to multiply two n-digit numbers.

Multiple-precision multiplication of n-digit numbers requires $M(n) = O(n^2)$ operations using an ordinary multiplication algorithm [11]. Karatsuba's algorithm [9] reduces the number of operations to $M(n) = O(n^{\log_2 3})$. It is known that multiplication of n-bit numbers can be performed in $M(n) = O(n \log n \log \log n)$ bit operations by using the Schönhage-Strassen algorithm [14] which is based on the fast Fourier transform (FFT) [4].

Parallel implementation of multiple-precision division of two n-digit numbers has been proposed [18]. This scheme requires $O(M(n)/P)$ operations on a parallel computer with P processors [18].

On the other hand, multiple-precision division by a single-precision integer is often used in multiple-precision arithmetic because it is much faster than the division of two multiple-precision numbers.

Several multiple-precision arithmetic packages [1, 2, 3, 7, 17] include a routine for multiple-precision division by a single-precision integer. We call such a routine short division.

Parallel implementation of the multiple-precision arithmetic on a shared memory machine have been presented by Weber [19]. Weber modified the MPFUN multiple-precision arithmetic package [1] to run in parallel on a shared memory multiprocessor. Fagin also implemented the multiple-precision addition [5]

I. Lirkov, S. Margenov, and J. Waśniewski (Eds.): LSSC 2007, LNCS 4818, pp. 729–736, 2008.

and multiplication [6] on the Connection Machine CM-2. However, a parallel algorithm for short division has not yet been presented.

In this paper, a parallel algorithm for multiple-precision division by a single-precision integer is presented.

We implemented the parallel algorithm for multiple-precision division by a single-precision integer on a 32-node Intel Xeon PC cluster, and the experimental results are reported herein.

Section 2 describes the algorithm for multiple-precision division by a single-precision integer. Section 3 describes the parallelization of multiple-precision by a single-precision integer. Section 4 gives experimental results. In section 5, we provide some concluding remarks.

2 Multiple-Precision Division by a Single-Precision Integer

In this paper, we discuss multiple-precision arithmetic with radix-b for the division of an n-digit dividend by an $O(1)$-digit divisor, which gives an n-digit quotient and an $O(1)$-digit remainder. For simplicity, we assume that we are working with a nonnegative integer.

Let us define an n-digit dividend $u = \sum_{i=0}^{n-1} u_i b^i$ and an $O(1)$-digit divisor v in radix-b notation, where $0 \leq u_i < b$ and $1 \leq v < b$.

The quotient q can be expressed as follows:

$$q = \lfloor u/v \rfloor = \sum_{i=0}^{n-1} q_i b^i, \tag{1}$$

where $0 \leq q_i < b$.

The remainder r and the partial remainder r_i can be expressed as follows:

$$r = u - vq, \tag{2}$$
$$r_i = br_{i+1} + u_i - vq_i, \qquad i = n-1, n-2, \cdots, 0, \tag{3}$$

where $0 \leq r_i < v$ and we assume $r_n = 0$.

Then, the partial quotient q_i and the partial remainder r_i can be expressed as follows:

$$q_i = \lfloor (br_{i+1} + u_i)/v \rfloor, \qquad i = n-1, n-2, \cdots, 0, \tag{4}$$
$$r_i = (br_{i+1} + u_i) \bmod v, \qquad i = n-1, n-2, \cdots, 0. \tag{5}$$

We note that (5) includes the first-order recurrence. Then, the remainder r is given by $r = r_0$.

The first-order recurrence of (5) can be evaluated sequentially by the definition of the recurrence with the following FORTRAN 77 code:

```
    r(n)=0
    do 10 i=n-1,0,-1
      r(i)=mod(b*r(i+1)+u(i),v)
 10 continue
```

where r and u have been declared as arrays. The arithmetic operation count of this algorithm is clearly $O(n)$.

3   Parallelization of Multiple-Precision Division by a Single-Precision Integer

Although the first-order recurrence cannot be parallelized easily, we can apply the parallel cyclic reduction method [8] to (5).

The first-order recurrence of (5) for the two successive terms can be written as follows:

$$r_i = (br_{i+1} + u_i) \bmod v, \tag{6}$$

$$r_{i+1} = (br_{i+2} + u_{i+1}) \bmod v. \tag{7}$$

Substituting (7) into (6), we obtain

$$r_i = (b^2 r_{i+2} + bu_{i+1} + u_i) \bmod v$$
$$= (b^{(1)} r_{i+2} + u_i^{(1)}) \bmod v, \tag{8}$$

where

$$b^{(1)} = b^2 \bmod v, \tag{9}$$

$$u_i^{(1)} = (bu_{i+1} + u_i) \bmod v. \tag{10}$$

By repeated application of the above procedure, we obtain

$$r_i = (b^{(k)} r_{i+2^k} + u_i^{(k)}) \bmod v \quad \begin{cases} k = 0, 1, \cdots, \lceil \log_2 n \rceil \\ i = 0, 1, \cdots, n - 1, \end{cases} \tag{11}$$

where

$$b^{(k)} = (b^{(k-1)})^2 \bmod v, \tag{12}$$

$$u_i^{(k)} = (b^{(k-1)} u_{i+2^{k-1}}^{(k-1)} + u_i^{(k-1)}) \bmod v, \tag{13}$$

and initially

$$b^{(0)} = b \bmod v, \tag{14}$$

$$u_i^{(0)} = u_i \bmod v. \tag{15}$$

We assume r_i and u_i are zero in (5) when $i \geq n$. Moreover, when $k = \lceil \log_2 n \rceil$, the subscript of $r_{i+2^k} = r_{i+2^{\lceil \log_2 n \rceil}}$ in (11) is outside the defined range $0 \leq i \leq n-1$. Therefore, all references to $r_{i+2^k} = r_{i+2^{\lceil \log_2 n \rceil}}$ are also zero in (11).

Finally, the solution to the recurrence is given by

$$r_i = u_i^{(\lceil \log_2 n \rceil)} \bmod v. \tag{16}$$

Fig. 1 shows the communication diagram for the evaluation of r_i on a parallel computer with 8 processors.

The parallel algorithm for the first-order recurrence can be implemented in a parallel form of Fortran 90, as follows:

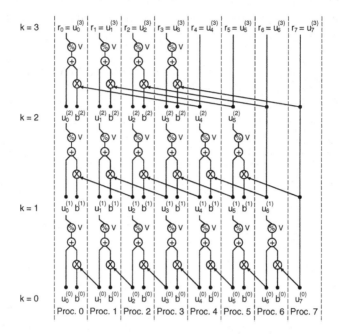

$k = 3$

$k = 2$

$k = 1$

$k = 0$

Proc. 0 | Proc. 1 | Proc. 2 | Proc. 3 | Proc. 4 | Proc. 5 | Proc. 6 | Proc. 7

Fig. 1. The communication diagram for (13)

```
bmod=mod(b,v)
r(0:n-1)=mod(u(0:n-1),v)
do k=1,ceiling(log2(n))
  if (bmod .eq. 0) exit
  r(0:n-1)=mod(bmod*eoshift(r(0:n-1),2**(k-1))+r(0:n-1),v)
  bmod=mod(bmod**2,v)
end do
```

where r and u have been declared as arrays.

When $b^{(j)} \bmod v$ $(j = 0, 1, \cdots, \lceil \log_2 n \rceil - 1)$ is zero, all references to $b^{(k)} \bmod v$ $(k = j + 1, j + 2, \cdots, \lceil \log_2 n \rceil)$ are also zero in (12), and $r_i = u_i^{(j)} \bmod v$ in (11). Therefore, the do loop of the above program can be interrupted when $b^{(k)} \bmod v = 0$.

In particular, when a radix b is a multiple of a divisor v, $b^{(0)} \bmod v = 0$. In this case, the arithmetic operation count of this algorithm is $O(n/P)$ on a parallel computer with P processors. On the other hand, when a radix b is not a multiple of a divisor v, we can compute the first-order recurrence in serial on an intraprocessor and also compute the first-order recurrence by using the parallel cyclic reduction method on an interprocessor. Therefore, the upper bound of the arithmetic operation count of this algorithm is $O((n/P) \log P)$.

Finally, we can obtain the quotient q in parallel by the following Fortran 90 code:

```
q(0:n-1)=floor((b*eoshift(r(0:n-1),1)+u(0:n-1))/v)
```

where q, r and u have been declared as arrays.

Then, the remainder r is given by r(0).

4   Experimental Results

In order to evaluate the parallel algorithm for multiple-precision division by a single-precision integer, the decimal digit n and the number of processors P were varied. We averaged the elapsed times obtained from 10 executions of the multiple-precision parallel division by single-precision integers ($\pi/2$ and $\pi/3$). We note that the respective values of n-digit π were prepared in advance. The selection of these values has no particular significance here, but was convenient to establish definite test cases, the results of which were used as randomized test data.

A 32-node Intel Xeon PC cluster (Irwindale 3 GHz, 12 K uops L1 instruction cache, 16 KB L1 data cache, 2 MB L2 cache, 1 GB DDR2-400 SDRAM main memory per node, Intel E7520, Linux 2.6.19-1.2911.6.4.fc6) was used. The nodes on the PC cluster are interconnected through a 1000Base-T Gigabit Ethernet switch. Open MPI 1.1.4 was used as a communication library. All routines were written in FORTRAN 77 with MPI.

The compiler used was g77 version 3.4.6. The compiler options used were specified as "g77 -O3 -fomit-frame-pointer." All programs were run in 64-bit mode. The radix of the multiple-precision number is 10^8. The multiple-precision number is stored with block distribution in the array of 32-bit integers.

Table 1 shows the average execution times of multiple-precision parallel division by a single-precision integer, $\pi/2$. The column with the n heading shows the decimal digits. The next six columns contain the average elapsed time in seconds.

Fig. 2 shows the speedup of multiple-precision parallel division by a single-precision integer ($\pi/2$) relative to the one-processor execution time. For $n = 2^{26}$ and $P = 32$, the speedup is approximately 31.9 times. The arithmetic operation count is $O(n/P)$ in the division of $\pi/2$, because the radix ($= 10^8$) is a multiple of the divisor ($= 2$). Moreover, interprocessor communications for the partial

Table 1. Execution time of multiple-precision parallel division by a single-precision integer ($\pi/2$) (in seconds), $n =$ number of decimal digits

n	$P = 1$	$P = 2$	$P = 4$	$P = 8$	$P = 16$	$P = 32$
2^{16}	0.00281	0.00176	0.00105	0.00065	0.00046	0.00039
2^{18}	0.01137	0.00603	0.00302	0.00166	0.00100	0.00066
2^{20}	0.04521	0.02298	0.01161	0.00601	0.00315	0.00175
2^{22}	0.18109	0.09087	0.04558	0.02294	0.01158	0.00601
2^{24}	0.72608	0.36308	0.18168	0.09107	0.04563	0.02299
2^{26}	2.90455	1.45305	0.72716	0.36293	0.18208	0.09107

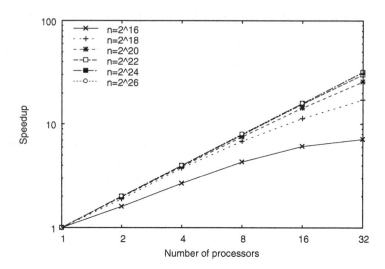

Fig. 2. Speedup of multiple-precision parallel division by a single-precision integer $(\pi/2)$, n = number of decimal digits

Table 2. Execution time of multiple-precision parallel division by a single-precision integer $(\pi/3)$ (in seconds), n = number of decimal digits

n	$P = 1$	$P = 2$	$P = 4$	$P = 8$	$P = 16$	$P = 32$
2^{16}	0.00284	0.00301	0.00254	0.00200	0.00216	0.00262
2^{18}	0.01132	0.00973	0.00707	0.00546	0.00399	0.00340
2^{20}	0.04535	0.03701	0.02589	0.01712	0.01118	0.00740
2^{22}	0.18112	0.14773	0.10149	0.06525	0.04064	0.02487
2^{24}	0.72807	0.59019	0.40603	0.25957	0.15827	0.09380
2^{26}	2.91205	2.34418	1.63208	1.03915	0.63284	0.37224

remainder occurs just twice. These are two reasons why the speedup of the computation of $\pi/2$ is nearly a linear speedup for larger digits.

Table 2 shows the average execution times of multiple-precision parallel division by a single-precision integer $(\pi/3)$. The column with the n heading shows the decimal digits. The next six columns contain the average elapsed time in seconds.

Fig. 3 shows the speedup of multiple-precision parallel division by a single-precision integer $(\pi/3)$ relative to the one-processor execution time. For $n = 2^{16}$, the speedup is under 1.0 on 2 processors, as shown in Fig. 3. Its speedup loss comes mainly from communication overhead. For $n = 2^{26}$ and $P = 32$, the speedup is approximately 7.8 times. The arithmetic operation count is proportional to $(n/P) \log P$ in the division of $\pi/3$, because the radix $(= 10^8)$ is not a multiple of the divisor $(= 3)$. Therefore, the speedup ratio is limited to $O(P/\log P)$. This is the reason why the computation of $\pi/3$ shows poor speedup.

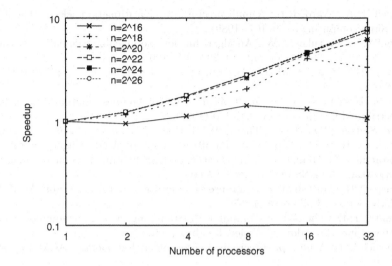

Fig. 3. Speedup of multiple-precision parallel division by a single-precision integer $(\pi/3)$, n = number of decimal digits

The computation of $\pi/2$ is up to approximately 4.1 times faster than that of $\pi/3$ when $n = 2^{30}$ and $P = 32$. This is mainly because the computation of $\pi/2$ has less arithmetic operations $O(n/P)$ compared with the computation of $\pi/3$.

5   Conclusion

This paper has presented a parallel algorithm for multiple-precision division by a single-precision integer. This short division can be derived from the first-order recurrence, which can be parallelized by the parallel cyclic reduction method.

In particular, when a radix b is a multiple of a divisor v, the arithmetic operation of the multiple-precision parallel division of an n-digit number by a single-precision integer is $O(n/P)$ on a parallel computer with P processors. On the other hand, when a radix b is not a multiple of a divisor v, the upper bound of the arithmetic operation of this algorithm is $O((n/P)\log P)$.

We implemented the parallel algorithm for multiple-precision division by a single-precision integer on a 32-node Intel Xeon PC cluster. The experimental results show that speedups of 31.9 times and 7.8 times for $n = 2^{26}$ decimal digits $\pi/2$ and $\pi/3$, respectively, on a 32-node Intel Xeon PC cluster.

References

1. Bailey, D.H.: Algorithm 719: Multiprecision translation and execution of FOR-TRAN programs. ACM Trans. Math. Softw. 19, 288–319 (1993)
2. Brent, R.P.: A Fortran multiple-precision arithmetic package. ACM Trans. Math. Softw. 4, 57–70 (1978)

3. Buell, D.A., Ward, R.L.: A multiprecise integer arithmetic package. The Journal of Supercomputing 3, 89–107 (1989)
4. Cooley, J.W., Tukey, J.W.: An algorithm for the machine calculation of complex Fourier series. Math. Comput. 19, 297–301 (1965)
5. Fagin, B.S.: Fast addition of large integers. IEEE Trans. Comput. 41, 1069–1077 (1992)
6. Fagin, B.S.: Large integer multiplication on hypercubes. Journal of Parallel and Distributed Computing 14, 426–430 (1992)
7. Free Software Foundation: The GNU MP Home Page, http://www.swox.com/gmp/
8. Hockney, R.W., Jesshope, C.R.: Parallel Computers. Adam Hilger, Bristol (1981)
9. Karatsuba, A., Ofman, Y.: Multiplication of multidigit numbers on automata. Doklady Akad. Nauk SSSR 145, 293–294 (1962)
10. Karp, A.H., Markstein, P.: High-precision division and square root. ACM Trans. Math. Softw. 23, 561–589 (1997)
11. Knuth, D.E.: The Art of Computer Programming, Vol. 2: Seminumerical Algorithms, 3rd edn. Addison-Wesley, Reading (1997)
12. Mifsud, C.J.: A multiple-precision division algorithm. Comm. ACM 13, 666–668 (1970)
13. Rabinowitz, P.: Multiple-precision division. Comm. ACM 4, 98 (1961)
14. Schönhage, A., Strassen, V.: Schnelle Multiplikation grosser Zahlen. Computing (Arch. Elektron. Rechnen) 7, 281–292 (1971)
15. Stein, M.L.: Divide-and-correct methods for multiple precision division. Comm. ACM 7, 472–474 (1964)
16. Smith, D.M.: A multiple-precision division algorithm. Math. Comput. 65, 157–163 (1996)
17. Smith, D.M.: Algorithm 693: A FORTRAN package for floating-point multiple-precision arithmetic. ACM Trans. Math. Softw. 17, 273–283 (1991)
18. Takahashi, D.: Implementation of multiple-precision parallel division and square root on distributed-memory parallel computers. In: Proc. 2000 International Conference on Parallel Processing (ICPP 2000) Workshops, pp. 229–235 (2000)
19. Weber, K.: An experiment in high-precision arithmetic on shared memory multiprocessors. SIGSAM Bulletin 24, 22–40 (1990)

Improving Triangular Preconditioner Updates for Nonsymmetric Linear Systems

Jurjen Duintjer Tebbens and Miroslav Tůma

Institute of Computer Science, Czech Academy of Sciences, Pod Vodárenskou věží 2, 18207 Praha 8, Czech Republic
{tebbens,tuma}@cs.cas.cz

Abstract. We present an extension of an update technique for preconditioners for sequences of non-symmetric linear systems that was proposed in [5]. In addition, we describe an idea to improve the implementation of the update technique. We demonstrate the superiority of the new approaches in numerical experiments with a model problem.

1 Introduction

Sequences of linear systems with large and sparse matrices arise in many applications like computational fluid dynamics, structural mechanics, numerical optimization as well as in solving non-PDE problems. In many cases, one or more systems of *nonlinear* equations are solved by a Newton or Broyden-type method [6], and each nonlinear equation leads to a sequence of linear systems. The solution of sequences of linear systems is the main bottleneck in many of the above mentioned applications. For example, some solvers need strong preconditioners to be efficient and computing preconditioners for individual systems separately may be very expensive.

In recent years, a few attempts to update preconditioners for sequences of large sparse systems have been made. If a sequence of linear systems arises from a quasi-Newton method, straightforward approximate small rank updates can be useful (this has been done in the SPD case in [9,3]). For shifted SPD linear systems, an update technique was proposed in [8] and a different one can be found in [2]. The latter technique, based on approximate diagonal updates, has been extended to sequences of parametric complex symmetric linear systems (see [4]). This technique, in turn, was generalized to approximate (possibly permuted) triangular updates for nonsymmetric sequences [5]. In addition, recycling of Krylov subspaces by using adaptive information generated during previous runs has been used to update both preconditioners and Krylov subspace iterations (see [7,10,1]).

In this paper we address two ways to improve the triangular updates of preconditioners for nonsymmetric sequences of linear systems from [5]. It was discussed in [5] that triangular updates may be particularly beneficial under three types of circumstances: first, if preconditioner recomputation is for some reason expensive (e.g. in parallel computations, matrix-free environment); second,

I. Lirkov, S. Margenov, and J. Waśniewski (Eds.): LSSC 2007, LNCS 4818, pp. 737–744, 2008.

if recomputed preconditioners suffer from instability and the updates relate to a more stable reference factorization; third, if the update is dominant, at least structurally, that is, if it covers a significant part of the difference between the current and reference matrix. Our first contribution is motivated by the third case. The updates from [5] exhibit this property whenever a permutation can be found such that one triangular part of the permuted difference matrix clearly dominates the other part. Experiments given there show that this is the case in many types of applications where the permutation may not even be needed. Nevertheless, these techniques neglect one of the two triangular parts of the (permuted) difference matrix and possibly useful information contained in this part is lost. We will describe here how both triangular parts can be taken into account by considering a simple but effective extension of the original technique. We compare the new idea with the original strategy and experiments demonstrate its improved power on a model problem. Our second contribution is of a more technical nature. We present a different implementation for the triangular updates. The important time savings with respect to the strategy used in [5] confirm what we only assumed there, and reveal more about the potential of the preconditioner updates.

In the next section we address the first improvement and Section 3 describes the second one. Numerical experiments are presented in Section 4. We denote by $\| \cdot \|$ an arbitrary, unspecified matrix norm.

2   Gauss-Seidel Type Updates

We consider a system $Ax = b$ with a factorized preconditioner $M = LDU$ and let $A^+x^+ = b^+$ be a system of the same dimension, and denote the difference matrix $A - A^+$ by B. We search for an updated preconditioner M^+ for $A^+x^+ = b^+$. We have $\|A - M\| = \|A^+ - (M - B)\|$, hence the norm of the difference $A^+ - M^+$ with $M^+ \equiv M - B$, called the *accuracy* of M^+ (with respect to A^+), is the same as that of M with respect to A. If $M^+ = M - B$ is the preconditioner, we need to solve systems with $M - B$ as system matrix in every iteration of the iterative solver. Clearly, for general B the preconditioner $M^+ = M - B$ cannot be used in practice since the systems are too expensive to solve. Instead, we will consider cheap approximations of $M - B$. If $M - B$ is nonsingular, we approximate it by a product of factors which are easier to invert. The approximation consists of two steps. First, we approximate $M - B$ as

$$M - B = L(DU - L^{-1}B) \approx L(DU - B), \tag{1}$$

or by

$$M - B = (LD - BU^{-1})U \approx (LD - B)U. \tag{2}$$

The choice between (1) and (2) is based on the distance of L and U to identity. If $\|I - L\| < \|I - U\|$ then we will base our updates on (1) and in the following C will denote the matrix $DU - B$. If on the other hand $\|I - L\| > \|I - U\|$, then we will use (2) and we define C by $C \equiv LD - B$. Our implementation chooses the appropriate strategy adaptively.

Our next goal is to find an approximation of C that can be used as a preconditioner. We split C as $C = L_C + D_C + U_C$, where L_C, D_C, U_C denote the strict
lower triangular, the main diagonal and the strict upper triangular part of C,
respectively. In [5], C was approximated by a single triangular factor $L_C + D_C$ or
$U_C + D_C$. In this paper we propose a Gauss-Seidel type of approach that takes
into account both triangular parts of C. We will use the classical symmetric
Gauss-Seidel approximation $C \approx (L_C + D_C)D_C^{-1}(U_C + D_C)$. Putting the two
approximation steps together, we obtain an updated preconditioner of the form

$$M^+ = L(L_C + D_C)D_C^{-1}(U_C + D_C), \quad C = DU - B, \tag{3}$$

when $\|I - L\| < \|I - U\|$, and otherwise we use

$$M^+ = (L_C + D_C)D_C^{-1}(U_C + D_C)U, \quad C = LD - B. \tag{4}$$

These updates can be cheaply obtained. C is a difference of two sparse matrices, the splitting of C is trivial. The updated preconditioner has one additional
factor compared with the original factorization LDU, hence its application is
a little more expensive. In cases where the sparsity patterns of B and L or U
differ significantly, the solves with the updated factors are also more expensive
than with the original factors. The choice between (3) and (4) can be based on
comparing the Frobenius norms of $I - L$ and $I - U$, which is very cheap with
sparse factors. As for storage costs, the original factorization and the reference
matrix A must be available when applying updates of this form.

We showed in [5] that the accuracy of the updates introduced there increases
with a factor L (or U) closer to identity and with a smaller error $\|C - D_C - U_C\|$ (or $\|C - D_C - L_C\|$). Also stability increases with these properties. Our
theoretical results explained why the updates are often more powerful than old
factorizations and may even be, in favorable cases, more powerful than newly
computed factorizations. For the updates introduced in this paper, similar results
hold. We will now concentrate on comparison of the new Gauss-Seidel updates
with the original technique. The updates from [5] can be written as

$$M^+ = L (D_C + U_C), \quad C = DU - B, \tag{5}$$

and

$$M^+ = (L_C + D_C) U, \quad C = LD - B. \tag{6}$$

Intuitively it is clear that the Gauss-Seidel type updates may be expected to be
more powerful than (5) and (6) if their approximation of C is stronger than with
one triangular part only. Let $E = A - LDU$ denote the accuracy error of the
preconditioner for A and let $G = C - (L_C + D_C)D_C^{-1}(U_C + D_C) = L_C D_C^{-1} U_C$
be the approximation error of C for the Gauss-Seidel update. We split B as $B = L_B + D_B + U_B$, where L_B, D_B, U_B denote the strict lower triangular, the main
diagonal and the strict upper triangular part of B, respectively. The accuracy
of (6) can then be written as

$$\|A^+ - (L_C + D_C)U\| = \|A - LDU - B + (L_B + D_B)U\| = \|E - B(I - U) - U_B U\|,$$

where we used that $L_C + D_C = LD - L_B - D_B$. Similarly, the accuracy of (4) is

$$\|A^+ - (L_C + D_C)(D_C^{-1}U_C + I)U\| =$$
$$\|E - U_B - (L_B + D_B)(I - U) - (L_C D_C^{-1} + I)U_C U\| =$$
$$\|E - U_B - (L_B + D_B)(I - U) + U_B U - G \cdot U\| =$$
$$\|E - B(I - U) - G \cdot U\|.$$

Hence the accuracies of (4) and (6) are of the form $\|X - GU\|$ and $\|X - U_B U\|$, where $X = E - B(I - U)$. As U_B is nothing but the error to approximate C according to (6), here we see the effect of the approximation errors G and U_B with the two update techniques. The updates (5) and (3) share a similar relation.

Now let us denote by *striu* the strict upper triangular part and by *tril* the lower triangular part of a matrix (including the main diagonal). In the Frobenius norm, denoted by $\|\cdot\|_F$, positive influence of the Gauss-Seidel technique can be expressed as follows. Assume that

$$\|tril(E - B(I - U) - G \cdot U)\|_F^2 + \tag{7}$$
$$\|striu(E - B + (B - G) \cdot U)\|_F^2 \leq \tag{8}$$
$$\|tril\,(E - B(I - U))\|_F^2 + \tag{9}$$
$$\|striu\,(E - B + (L_B + D_B)U)\|_F^2 , \tag{10}$$

then

$$\|A^+ - (L_C + D_C)(D_C^{-1}U_C + I)U\|_F^2 =$$
$$\|tril(E - B(I - U) - G \cdot U)\|_F^2 + \|striu(E - B + (B - G) \cdot U)\|_F^2 \leq$$
$$\|tril\,(E - B(I - U))\|_F^2 + \|striu\,(E - B + (L_B + D_B)U)\|_F^2 =$$
$$\|A^+ - (D_C + L_C)\,U\|_F^2.$$

The last equality follows from $striu(BU) = striu((U_B + L_B + D_B)U) = striu(U_B U) + striu((L_B + D_B)U) = U_B U + striu((L_B + D_B)U)$. Thus superiority of the Gauss-Seidel type update (6) may be expected if the contribution of $-GU$ to the lower triangular part of X reduces the entries of this part and if the contribution of $(B - G)U$ to the strict upper triangular part of X reduces the entries more than the contribution of $(L_B + D_B)U$ to this part. We will confirm exactly this behavior in the experiments in Section 4.

3   Alternative Implementation

The update (5) (or (6)) in [5] was implemented with two backward (or forward) solves. In the upper triangular case (5), the solve step with $D_C + U_C$, which is equal to $DU - (D_B + U_B)$, used separate loops with DU and with $D_B + U_B$ in [5]. These loops were tied together by scaling with the sum of the diagonal entries of DU and B. In detail: Let the entries of DU and B be denoted by $(du)_{ij}$ and b_{ij}, respectively, and consider a linear system $(DU - D_B - U_B)\,z = y$. Then for $i = n, n-1, \ldots, 1$ the subsequent cycles

$$z_i = y_i - \sum_{j>i}(du)_{ij}z_j, \qquad z_i = z_i - \sum_{j>i}b_{ij}z_j, \qquad (11)$$

were used, followed by putting

$$z_i = \frac{z_i}{(du)_{ii} + b_{ii}}. \qquad (12)$$

A first advantage of this implementation is that the solution process is straightforward. The sparsity patterns of DU and $D_B + U_B$, which are immediately available, do not need to be further processed. Another advantage of this implementation is that the difference matrix B may be sparsified in a different way for different matrices of a sequence. It was mentioned repeatedly in [5], however, that merging the two matrices DU and $D_B + U_B$ may yield better timings. Here we present results of experiments with merged factors which formed the sum $DU - D_B - U_B$, or its lower triangular counterpart, explicitly. This sum needs to be formed only once at the beginning of the solve process of the linear system, that is in our case, before the preconditioned iterations start. Every time the preconditioner is applied, the backward solve step with the merged factors may be significantly cheaper than with (11)–(12) if the sparsity patterns of DU and $D_B + U_B$ are close enough. In our experiments we confirm this.

4   Numerical Experiments

Our model problem is a two-dimensional nonlinear convection-diffusion model problem. It has the form (see, e.g. [6])

$$- \Delta u + Ru \left(\frac{\partial u}{\partial x} + \frac{\partial u}{\partial y}\right) = 2000x(1-x)y(1-y), \qquad (13)$$

on the unit square, discretized by 5-point finite differences on a uniform grid. The initial approximation is the discretization of $u_0(x, y) = 0$. In contrast with [5] we use here $R = 100$ and different grid sizes. We solve the resulting linear systems with the BiCGSTAB [11] iterative method with right preconditioning. Iterations were stopped when the Euclidean norm of the residual was decreased by seven orders of magnitude. Other stopping criteria yield qualitatively the same results.

In Table 1 we consider a 70×70 grid, yielding a sequence of 13 matrices of dimension 4900 with 24220 nonzeros each. We precondition with ILU(0), which has the same sparsity pattern as the matrix it preconditions. This experiment was performed in Matlab 7.1. We display the number of BiCGSTAB iterations for the individual systems and the overall time to solve the whole sequence. The first column determines the matrix of the sequence which is preconditioned. The second column gives the results when ILU(0) is recomputed for every system of the sequence. In the third column ILU(0) is computed only for the first system and reused (frozen) for the whole sequence. In the remaining columns this first factorization is updated. 'Triang' stays for the triangular updates from [5], that

Table 1. Nonlinear convection-diffusion model problem with n=4900, ILU(0)

ILU(0), psize ≈ 24000				
Matrix	Recomp	Freeze	Triang	GS
$A^{(0)}$	40	40	40	40
$A^{(1)}$	25	37	37	27
$A^{(2)}$	24	41	27	27
$A^{(3)}$	20	48	26	19
$A^{(4)}$	17	56	30	21
$A^{(5)}$	16	85	32	25
$A^{(6)}$	15	97	35	29
$A^{(7)}$	14	106	43	31
$A^{(8)}$	13	97	44	40
$A^{(9)}$	13	108	45	38
$A^{(10)}$	13	94	50	44
$A^{(11)}$	15	104	45	35
$A^{(12)}$	13	156	49	42
overall time	13 s	13 s	7.5 s	6.5 s

Table 2. Nonlinear convection-diffusion problem with n=4900: Accuracies and values (7)–(10)

i	$\|A^{(i)} - M_{GS}^{(i)}\|_F^2$	$\|A^{(i)} - M_{TR}^{(i)}\|_F^2$	value of (7)	value of (9)	value of (8)	value of (10)
1	852	857	*	*	*	*
2	938	1785	377	679	560	1105
3	1102	2506	373	843	729	1663
4	1252	3033	383	957	869	2076
5	1581	3975	432	1155	1149	2820
6	1844	4699	496	1303	1388	3395
7	2316	5590	610	1484	1706	4106
8	2731	6326	738	1631	1993	4695
9	2736	6372	735	1642	2002	4731
10	2760	6413	742	1650	2018	4763
11	2760	6415	742	1650	2018	4765
12	2760	6415	742	1650	2018	4765

is for adaptive choice between (5) and (6). The last column presents results for the Gauss-Seidel (GS) updates (3) and (4). The abbreviation 'psize' gives the average number of nonzeros of the preconditioners.

As expected from [5], freezing yields much higher iteration counts than any updated preconditioning. On the other hand, recomputation gives low iteration counts but it is time inefficient. The new GS strategy from Section 2 improves the power of the original triangular update. Table 2 displays the accuracies of (4) (here denoted by M_{GS}) and (6) (denoted by M_{TR}) in the Frobenius norm and the values of (7-10). These values reflect the efficiencies of the two updates

Table 3. Nonlinear convection-diffusion model problem with n=49729, ILUT(0.2/5)

ILUT(0.2/5), psize ≈ 475000, ptime ≈ 0.05				
Matrix	Recomp	Freeze	Triang	GS
$A^{(0)}$	113/2.02	113/2.02	113/2.02/2.02	113/2.02
$A^{(1)}$	119/2.06	112/1.94	104/1.95/1.81	122/2.26
$A^{(2)}$	111/1.94	111/1.95	104/1.91/1.78	100/1.84
$A^{(3)}$	94/1.66	115/2.00	92/1.64/1.45	96/1.77
$A^{(4)}$	85/1.44	116/2.00	92/1.77/1.55	90/1.67
$A^{(5)}$	81/1.45	138/2.44	93/1.73/1.47	83/1.55
$A^{(6)}$	72/1.28	158/2.75	101/1.89/1.63	85/1.59
$A^{(7)}$	72/1.28	163/2.86	101/1.91/1.59	92/1.69
$A^{(8)}$	78/1.36	161/2.84	94/1.77/1.53	82/1.48
$A^{(9)}$	72/1.23	159/2.72	92/1.72/1.73	80/1.55
$A^{(10)}$	73/1.27	153/2.66	97/1.91/1.61	82/1.48

and confirm the remarks made after (7-10). Note that the first update in this sequence is based on (3), resp. (5) and thus the values (7-10) do not apply here.

In Table 3 we use the grid size 223 and obtain a sequence of 11 linear systems with matrices of dimension 49729 and with 247753 nonzeros. The preconditioner is ILUT(0.2,5), that is incomplete LU decomposition with drop tolerance 0.2 and number of additional nonzeros per row 5. This experiment was implemented in Fortran 90 in order to show improvements in timings for the alternative implementation strategy discussed in Section 3. The columns contain the BiCGSTAB iteration counts, followed by the time to solve the linear system, including the time to compute the (updated or new) factorization. In the column 'Triang' the last number corresponds to the implementation with merged factors as explained above and 'ptime' denotes the average time to recompute preconditioners.

The benefit of merging is considerable. Still, even with this improved implementation, the Gauss-Seidel type of updates happens to be faster than the standard triangular updates for several systems of the sequence. As for the BiCGSTAB iteration counts, for the majority of the linear systems Gauss-Seidel updates are more efficient. We have included in this table the results based on recomputation as well. In contrast to the results of the previous example, the decomposition routines are very efficient and exhibit typically in-cache behaviour. Then they often provide the best overall timings. This does not need to be the case in other environments like matrix-free or parallel implementations, or in cases where preconditioners are computed directly on grids.

5   Conclusion

In this paper we considered new ways for improving triangular updates of factorized preconditioners introduced in [5]. We proposed a Gauss-Seidel type of approach to replace the triangular strategy, and we introduced a more efficient implementation of adaptive triangular updates. We showed on a model nonlinear problem that both techniques may be beneficial. As a logical consequence, it seems worth to combine the two improvements by adapting the new implementation strategy for Gauss-Seidel updates. We expect this to yield even more efficient updates. For conciseness, we did not present some promising results with the Gauss-Seidel approach generalized by adding a relaxation parameter.

Acknowledgement

This work was supported by the project 1ET400300415 within the National Program of Research "Information Society". The work of the first author is also supported by project number KJB100300703 of the Grant Agency of the Academy of Sciences of the Czech Republic.

References

1. Baglama, J., et al.: Adaptively preconditioned GMRES algorithms. SIAM J. Sci. Comput. 20, 243–269 (1998)
2. Benzi, M., Bertaccini, D.: Approximate inverse preconditioning for shifted linear systems. BIT 43, 231–244 (2003)
3. Bergamaschi, L., et al.: Quasi-Newton Preconditioners for the Inexact Newton Method. ETNA 23, 76–87 (2006)
4. Bertaccini, D.: Efficient preconditioning for sequences of parametric complex symmetric linear systems. ETNA 18, 49–64 (2004)
5. Duintjer Tebbens, J., Tůma, M.: Preconditioner updates for solving sequences of large and sparse nonsymmetric linear systems. SIAM J. Sci. Comput. 29, 1918–1941 (2007)
6. Kelley, C.T.: Iterative methods for linear and nonlinear equations. SIAM, Philadelphia (1995)
7. Loghin, D., Ruiz, D., Touhami, A.: Adaptive preconditioners for nonlinear systems of equations. J. Comput. Appl. Math. 189, 326–374 (2006)
8. Meurant, G.: On the incomplete Cholesky decomposition of a class of perturbed matrices. SIAM J. Sci. Comput. 23, 419–429 (2001)
9. Morales, J.L., Nocedal, J.: Automatic preconditioning by limited-memory quasi-Newton updates. SIAM J. Opt. 10, 1079–1096 (2000)
10. Parks, M.L., et al.: Recycling Krylov subspaces for sequences of linear systems. SIAM J. Sci. Comput. 28, 1651–1674 (2006)
11. van der Vorst, H.A.: Bi-CGSTAB: A fast and smoothly converging variant of Bi-CG for the solution of non-symmetric linear systems. SIAM J. Sci. Stat. Comput. 12, 631–644 (1992)

Parallel DD-MIC(0) Preconditioning of Nonconforming Rotated Trilinear FEM Elasticity Systems

Yavor Vutov

Institute for Parallel Processing, Bulgarian Academy of Sciences
Acad. G. Bonchev, Bl. 25A, 1113 Sofia, Bulgaria
yavor.vutov@gmail.com

Abstract. A new parallel preconditioning algorithm for 3D noncon-forming FEM elasticity systems is presented. The preconditioner is con-structed in two steps. First, displacement decomposition of the stiffness matrix is used. Then MIC(0) factorization is applied to a proper auxiliary M-matrix to get an approximate factorization of the obtained block-diagonal matrix. The auxiliary matrix has a special block structure — its diagonal blocks are diagonal matrices themselves. This allows the solution of the preconditioning system to be performed efficiently in par-allel. Estimates for the parallel times, speedups and efficiencies are de-rived. The performed parallel tests are in total agreement with them. The robustness of the proposed algorithm is confirmed by the presented experiments solving problems with strong coefficient jumps.

Keywords: nonconforming finite element method, preconditioned con-jugate gradient method, MIC(0), parallel algorithms.

1 Introduction

We consider the weak formulation of the linear elasticity problem in the form: find $\mathbf{u} \in [H_E^1(\Omega)]^3 = \{\mathbf{v} \in [H^1(\Omega)]^3 : \mathbf{v}_{\Gamma_D} = \mathbf{u}_S\}$ such that

$$\int_\Omega [2\mu\varepsilon(\mathbf{u}) : \varepsilon(\mathbf{v}) + \lambda \operatorname{div} \mathbf{u} \operatorname{div} \mathbf{v}]d\Omega = \int_\Omega \mathbf{f}^t\mathbf{v}d\Omega + \int_{\Gamma_N} \mathbf{g}^t\mathbf{v}d\Gamma, \qquad (1)$$

$\forall \mathbf{v} \in [H_0^1(\Omega)]^3 = \{\mathbf{v} = [H^1(\Omega)]^3 : \mathbf{v}_{\Gamma_D} = 0\}$, with the positive constants λ and μ of Lamé, the symmetric strains $\varepsilon(\mathbf{u}) := 0.5(\nabla\mathbf{u} + (\nabla\mathbf{u})^t)$, the volume forces $\mathbf{f}$, and the boundary tractions $\mathbf{g}$, $\Gamma_N \cup \Gamma_D = \partial\Omega$. Nonconforming rotated trilinear elements of Rannacher-Turek [6] are used for the discretization of (1).

To obtain a stable saddle-point system one usually uses a mixed formulation for $\mathbf{u}$ and $\operatorname{div} \mathbf{u}$. By the choice of non-continuous finite elements for the dual variable, it can be eliminated at the macroelement level, and we get a symmetric positive definite finite element system in displacement variables. This approach is known as *reduced and selective integration* (RSI) technique, see [5].

Let $\Omega^H = w_1^H \times w_2^H \times w_3^H$ be a regular coarser decomposition of the domain $\Omega \subset \mathbb{R}^3$ into hexahedrons, and let the finer decomposition $\Omega^h = w_1^h \times w_2^h \times w_3^h$

I. Lirkov, S. Margenov, and J. Waśniewski (Eds.): LSSC 2007, LNCS 4818, pp. 745–752, 2008.

be obtained by a regular refinement of each macro element $E \in \Omega^H$ into eight similar hexahedrons. The cube $\hat{e} = [-1,1]^3$ is used as a reference element in the parametric definition of the rotated trilinear elements. For each $e \in \Omega^h$, let $\psi_e : \hat{e} \to e$ be the trilinear 1–1 transformation. Then the nodal basis functions are defined by the relations $\{\phi_i\}_{i=1}^6 = \{\hat{\phi}_i \circ \psi_e^{-1}\}_{i=1}^6$, where $\hat{\phi}_i \in span\{1, \xi_j, \xi_j^2 - \xi_{j+1}^2, j = 1, 2, 3\}$. Mid-point (MP) and integral mid-value (MV) interpolation conditions can be used for determining the reference element basis functions $\{\hat{\phi}_i\}_{i=1}^6$. This leads to two different finite element spaces V^h, referred as Algorithm MP and Algorithm MV.

The RSI finite element method (FEM) discretization reads as follows: find $\mathbf{u}^h \in V_E^h$ such that

$$\sum_{e \in \Omega^h} \int_e \left[2\mu \varepsilon^*(\mathbf{u}^h) : \varepsilon^*(\mathbf{v}^h) + \lambda \operatorname{div} \mathbf{u}^h \operatorname{div} \mathbf{v}^h \right] de = \int_\Omega \mathbf{f}^t \mathbf{v}^h d\Omega + \int_{\Gamma_N} \mathbf{g}^t \mathbf{v}^h d\Gamma,$$

(2)

$\forall \mathbf{v}^h \in V_0^h$, where $\varepsilon^*(\mathbf{u}) := \nabla \mathbf{u} - 0.5 I_L^{Q^H} [\nabla \mathbf{u} - (\nabla \mathbf{u})^t]$, V_0^h is the FEM space, satisfying (in nodalwise sense) homogeneous boundary conditions on Γ_D, the operator $I_L^{Q^H}$ denotes the L^2–orthogonal projection onto Q^H, the space of piecewise constant functions on the coarser decomposition Ω^H of Ω. Then a standard computational procedure leads to a system of linear equations

$$\begin{bmatrix} K_{11} & K_{12} & K_{13} \\ K_{21} & K_{22} & K_{23} \\ K_{31} & K_{32} & K_{33} \end{bmatrix} \begin{bmatrix} \mathbf{u}_h^1 \\ \mathbf{u}_h^2 \\ \mathbf{u}_h^3 \end{bmatrix} = \begin{bmatrix} \mathbf{f}_h^1 \\ \mathbf{f}_h^2 \\ \mathbf{f}_h^3 \end{bmatrix}.$$

(3)

Here the stiffness matrix K is written in block form corresponding to a separate displacements components ordering of the vector of nodal unknowns. Since K is sparse, symmetric and positive definite, we use the preconditioned conjugate gradient (PCG) method to solve the system (3). PCG is known to be the best solution method for such systems [2].

2   DD MIC(0) Preconditioning

Let us first recall some well known facts about the modified incomplete factorization MIC(0). Let us split the real $N \times N$ matrix $A = (a_{ij})$ in the form

$$A = D - L - L^T,$$

where D is the diagonal and $(-L)$ is the strictly lower triangular part of A. Then we consider the approximate factorization of A which has the following form:

$$C_{MIC(0)}(A) = (X - L)X^{-1}(X - L)^T,$$

where $X = \operatorname{diag}(x_1, \ldots, x_N)$ is a diagonal matrix determined such that A and $C_{MIC(0)}$ have equal row sums. For the purpose of preconditioning we restrict ourselves to the case when $X > 0$, i.e., when $C_{MIC(0)}$ is positive definite. In this case, the MIC(0) factorization is called *stable*. Concerning the stability of the MIC(0) factorization, we have the following theorem [4].

Theorem 1. *Let $A = (a_{ij})$ be a symmetric real $N \times N$ matrix and let $A = D - L - L^T$ be a splitting of A. Let us assume that (in an elementwise sense)*

$$L \geq 0, \quad Ae \geq 0, \quad Ae + L^T e > 0, \qquad e = (1, \cdots, 1)^T \in \mathbb{R}^N,$$

i.e., that A is a weakly diagonally dominant matrix with nonpositive offdiagonal entries and that $A + L^T = D - L$ is strictly diagonally dominant. Then the relation

$$x_i = a_{ii} - \sum_{k=1}^{i-1} \frac{a_{ik}}{x_k} \sum_{j=k+1}^{N} a_{kj} > 0 \tag{4}$$

holds and the diagonal matrix $X = diag(x_1, \cdots, x_N)$ defines a stable MIC(0) factorization of A.

Remark 1. The numerical tests presented in this work are performed using the perturbed version of MIC(0) algorithm, where the incomplete factorization is applied to the matrix $\tilde{A} = A + \tilde{D}$. The diagonal perturbation $\tilde{D} = \tilde{D}(\xi) = diag(\tilde{d}_1, \ldots \tilde{d}_N)$ is defined as follows: $\tilde{d}_i = \xi a_{ii}$ if $a_{ii} \geq 2w_i$, and $\tilde{d}_i = \xi^{1/2} a_{ii}$ otherwise, where $0 < \xi < 1$ is a constant and $w_i = -\sum_{j>i} a_{ij}$.

We use PCG with an isotropic displacement decomposition (DD) MIC(0) factorization preconditioner in the form:

$$C_{DDMIC(0)}(K) = \begin{bmatrix} C_{MIC(0)}(B) & & \\ & C_{MIC(0)}(B) & \\ & & C_{MIC(0)}(B) \end{bmatrix}$$

Matrix B is a modification of the stiffness matrix A corresponding to the bilinear form

$$a(u^h, v^h) = \sum_{e \in \Omega^h} \int_e E \left(\sum_{i=1}^{3} \frac{\partial u^h}{\partial x_i} \frac{\partial v^h}{\partial x_i} \right) de.$$

Here E is the modulus of elasticity. Such DD preconditioning for the coupled matrix K is theoretically motivated by the Korn's inequality which holds for the RSI FEM discretization under consideration [3]. The auxiliary matrix B is constructed element-by-element: Following the standard FEM assembling procedure we write A in the form $A = \sum_{e \in \Omega^h} L_e^T A_e L_e$, where L_e stands for the restriction mapping of the global vector of unknowns to the local one corresponding to the current element e and $A_e = \{a_{ij}\}_{i,j=1}^{6}$ is the element stiffness matrix The local node numbering and connectivity pattern is displayed in Fig. 1 (a). Now we will introduce the structure of two variants for local approximations B_e. They will later be referred to as Variant B1 and Variant B2.

<table>
<tr><td align="center">**Variant B1**</td><td align="center">**Variant B2**</td></tr>
</table>

$$B_e = \begin{bmatrix} b_{11} & & a_{13} & a_{14} & a_{15} & a_{16} \\ & b_{22} & a_{23} & a_{24} & a_{25} & a_{26} \\ a_{31} & a_{32} & b_{33} & & a_{35} & a_{36} \\ a_{41} & a_{42} & & b_{44} & a_{45} & a_{46} \\ a_{51} & a_{52} & a_{53} & a_{54} & b_{55} & \\ a_{61} & a_{62} & a_{63} & a_{64} & & b_{66} \end{bmatrix} \qquad B_e = \begin{bmatrix} b_{11} & & a_{13} & a_{14} & a_{15} & a_{16} \\ & b_{22} & a_{23} & a_{24} & a_{25} & a_{26} \\ a_{31} & a_{32} & b_{33} & & & \\ a_{41} & a_{42} & & b_{44} & & \\ a_{51} & a_{52} & & & b_{55} & \\ a_{61} & a_{62} & & & & b_{66} \end{bmatrix}$$

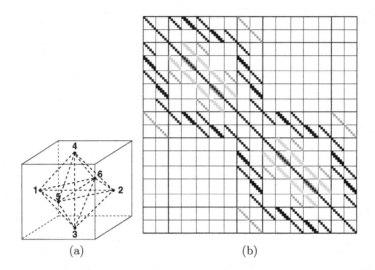

Fig. 1. (a) Local node numbering and connectivity pattern; (b) Sparsity pattern of the matrices A and B (both variants) for a division of Ω into 2x2x6 hexahedrons. Non-zero elements are drawn with boxes: ■ non-zero in A and B (both variants), ▨ non-zero in A and B Variant B1, ■ non-zero only in A. With thicker lines are bordered blocks in the matrix B Variant B2.

The matrices B_e are symmetric and positive semidefinite, with nonpositive off-diagonal entries, such that $B_e\mathbf{e} = A_e\mathbf{e}$, $\mathbf{e}^T = (1,1,1,1,1,1)$. Then we construct the global matrix $B = \sum_{e \in \Omega_h} \lambda_e^{(1)} L_e^T B_e L_e$, where $\{\lambda_e^{(i)}\}_{i=1}^5$ are the nontrivial eigenvalues of $B_e^{-1}A_e$ in ascending order. The matrix B is a M-matrix and has a special block structure with diagonal blocks being diagonal matrices, see Fig. 1(b). These blocks correspond to nodal lines and plains for variants B1 and B2, respectively. Lexicographic node numbering is used. This allows a stable MIC(0) factorization and efficient parallel implementation. It is important, that A and B are spectrally equivalent, and the relative condition number $\kappa(B^{-1}A)$ is uniformly bounded [1].

3   Parallel Algorithm

3.1   Description

The PCG algorithm is used for the solution of the linear system (3). Let us assume that the parallelogram domain Ω is decomposed into $n \times n \times n$ equal nonconforming hexahedral elements. The size of the resulting nonconforming FEM system is $N = 9n^2(n+1)$. To handle the systems with the preconditioner one has to solve three times systems $\tilde{L}\mathbf{y} \equiv (X - L)\mathbf{y} = \mathbf{v}$, $X^{-1}\mathbf{z} = \mathbf{y}$ and $\tilde{L}^T\mathbf{w} = \mathbf{z}$, where L is the strictly lower triangular part of the matrix B. The triangular systems are solved using standard forward or backward recurrences.

This can be done in $k_{B1} = 2n^2 + 2n$ and $k_{B2} = 2n + 1$ stages for variants B1 and B2, respectively. Within stage i the block $\mathbf{y}_i$ is computed. Since the blocks $\tilde{L}_{ii}$, are diagonal, the computations of each component of $\mathbf{y}_i$ can be performed in parallel. Let the $p \leq n/2$ processors be denoted by P_1, P_2, ..., P_p. We distribute the entries of the vectors corresponding to each diagonal block of B among the processors. Each processor P_j receives a strip of the computational domain. These strips have almost equal size. Elements of all vectors and rows of all matrices that participate in the PCG algorithm are distributed in the same manner. Thus the processor P_j takes care of the local computations on the j-th strip.

3.2   Parallel Times

On each iteration in the PCG algorithm one matrix vector multiplication $K\mathbf{x}$, one solution of the preconditioner system $C\mathbf{x} = \mathbf{y}$, two inner products and three linked triads of the form $\mathbf{x} = \mathbf{y} + \alpha\mathbf{z}$ are computed. The matrix vector multiplication can be performed on the macroelement level. In the case of rectangular brick mesh, the number of non-zero elements in the macroelement stiffness matrix is 1740. The number of operations on each PCG iteration is:

$$\mathcal{N}^{it} = \mathcal{N}(K\mathbf{x}) + \mathcal{N}(C^{-1}\mathbf{x}) + 2\mathcal{N}(<.,.>) + 3\mathcal{N}(\mathbf{x} = \mathbf{y} + \alpha\mathbf{z})$$
$$\mathcal{N}^{it} \approx 24N + \mathcal{N}(C^{-1}\mathbf{x}) + 2N + 3N, \; \mathcal{N}^{it}_{B1} \approx 40N, \; \mathcal{N}^{it}_{B2} \approx 38N$$

An operation is assumed to consist of one addition and one multiplication. Estimations of the parallel execution times are derived with the following assumptions: a) executing M arithmetical operations on one processor lasts $T = Mt_a$, b) the time to transfer M data items between two neighboring processors can be approximated by $T^{comm} = t_s + Mt_c$, where t_s is the startup time and t_c is the incremental time for each of the M elements to be transferred, and c) send and receive operations between each pair of neighboring processors can be done in parallel. We get the following expressions for the communication times:

$$T^{comm}(K\mathbf{x}) \approx 2t_s + \frac{4}{3}N^{2/3}t_c,$$

$$T^{comm}(C_{B1}^{-1}\mathbf{x}) \approx \frac{2}{3}N^{2/3}t_s + \frac{8}{3}N^{2/3}t_c, \quad T^{comm}(C_{B2}^{-1}\mathbf{x}) \approx \frac{2}{9}N^{1/3}t_s + \frac{8}{3}N^{2/3}t_c.$$

Two communication steps for the matrix vector multiplication are performed to avoid duplication of the computations or extra logic. For the solution of the triangular systems, after each nodal column (variant B1) or each nodal plain (variant B2) of unknowns is commputed some vector components must be exchanged.

The three systems of the preconditioner (one for each displacement) are solved simultaneously. Thus no extra communication steps for different displacements are required. The above communications are completely local and do not depend on the number of processors. The inner product needs one broadcasting and one gathering global communication but they do not contribute to the leading terms of the total parallel time. The parallel properties of the algorithm do not depend on the number of iterations, so it is enough to evaluate the parallel time per iteration, and use it in the speedup and efficiency analysis. As the computations are almost equally distributed among the processors, assuming there is no overlapping of the communications and computations one can write for the total time per iteration on p processors the following estimates:

$$T_{B1}^{it}(p) = \frac{40N}{p}t_a + \frac{2}{3}N^{2/3}t_s + 4N^{2/3}t_c, \quad T_{B2}^{it}(p) = \frac{38N}{p}t_a + \frac{2}{9}N^{1/3}t_s + 4N^{2/3}t_c$$

The relative speedup $S(p) = T(1)/T(p)$ and efficiency $E(p) = S(p)/p$, will grow with n in both variants up to their theoretical limits $S(p) = p$ and $E(p) = 1$. Since on a real computer $t_s \gg t_c$ and $t_s \gg t_a$ we can expect good efficiencies only when $n \gg p\,t_s/t_a$. The efficiency of Variant B2 is expected to be much better than the one of Variant B1, because about $3n$ times fewer messages are sent.

4   Benchmarking

4.1   Convergence Tests

The presented numerical tests illustrate the PCG convergence rate of the studied displacement decomposition algorithms when the size of the discrete problem and the coefficient jumps are varied. The computational domain is $\Omega = [0,1]^3$ where homogeneous Dirichlet boundary conditions are assumed at the bottom face. An uniform mesh is used. The number of intervals in each of the coordinate directions for the finer grid is n.

A relative stopping criterion $(C^{-1}\mathbf{r}^i, \mathbf{r}^i)/(C^{-1}\mathbf{r}^0, \mathbf{r}^0) < \varepsilon^2$ is used in the PCG algorithm, where $\mathbf{r}^i$ stands for the residual at the i-th iteration step, and $\varepsilon = 10^{-6}$. The interaction between a soil media and a foundation element with varying elasticity modulus is considered. The foundation domain is $\Omega_f = [3/8, 5/8] \times [3/8, 5/8] \times [1/2, 1]$. The mechanical characteristics are $E_s = 10$ MPa, $\nu_s = 0.2$ and $E_f = 10^J$ MPa, $\nu_f = 0.2$ for the soil and foundation respectively. Experiments with $J = 0, 1, 2, 3$ are performed. The force acting on the top of the foundation is 1 MN. In Tables 1 and 2 the number of iterations are collected for both variants B1 and B2 for Algorithms MP and MV respectively. In Table 1 also is added Variant B0 corresponding to the application of the MIC(0)

Table 1. Algorithm MP, number of iterations

		J												
		0			1			2			3			
n	N	B0	B1	B2	B0	B1	B2	B0	B1	B2	B0	B1	B2	
32	304 128	161	147	113	186	173	130	227	253	189	361	343	247	
64	2 396 160	264	223	162	284	262	186	428	391	271	565	523	357	
128	19 021 824	367	331	230	424	389	264	638	581	385	843	780	509	
256	151 584 768		486	327		570	377		852	542		1 148	725	

Table 2. Algorithm MV, number of iterations

n	N	\multicolumn{2}{c}{0}		\multicolumn{2}{c}{1}		\multicolumn{2}{c}{2}		\multicolumn{2}{c}{3}	
		B1	B2	B1	B2	B1	B2	B1	B2
32	304 128	173	255	197	280	313	348	405	411
64	2 396 160	295	648	310	744	486	904	630	1069
128	19 021 824	471	916	536	1 053	778	1 281	1 013	1 517
256	151 584 768	730	1 282	857	1 486	1 198	1 813	1 600	2 154

factorization directly to the matrix A. Note that this is possible only for the Algorithm MP (because of the positive offdiagonal entries in A in algorithm MV) and only in a sequential program. One can clearly see the robustness of the proposed preconditioners. The number of iterations is of order $O(n^{1/2}) = O(N^{1/6})$. It is remarkable that for Algorithm MP, the number of iterations for Variants B2 is less than that number for Variant B1, and it is even less than the number of iterations obtained without the modification of the matrix A.

4.2 Parallel Tests

Here we present execution times, speedups, and efficiencies from experiments performed on three parallel computing platforms, referred to further as C1, C2, and C3. Platform C1 is an "IBM SP Cluster 1600" consisting of 64 p5-575 nodes interconnected with a pair of connections to the Federation HPS (High Performance Switch). Each p5-575 node contains 8 Power5 SMP processors at 1.9 GHz and 16 GB of RAM. The network bandwidth is 16 Gb/s. Platform C2 is an IBM Linux Cluster 1350, made of 512 dual-core IBM X335 nodes. Each node contains 2 Xeon Pentium IV processors and 2 GB of RAM. Nodes are interconnected with an 1 Gb Myrinet network. Platform C3 is a "Cray XD1" cabinet, fully equipped with 72 2-way nodes, totaling in 144 AMD Opteron processors at 2.4 GHz. Each node has 4 GB of memory. The CPUs are interconnected with the Cray RaidArray network with a bandwidth of 5.6 Gb/s.

Since the parallel properties of the algorithm do not depend on the discretization type and the number of iterations, experiments only for Algorithm MP and for the case with the strongest coefficient jumps are performed. In Table 3 sequential execution times $T(p)$ are shown in seconds. The relative speedups $S(p)$ and efficiencies $E(p)$ for various values of n and number of processors p are collected in Table 4. Results for both variants B1 and B2 are included. For a fixed number of processors the speedup and efficiency grow with the problem size. Conversely for fixed n, the efficiency decrease with the number of processors. This is true for all platforms and confirms our analysis.

For Variant B1, reasonable efficiencies are obtained, only when n/p is sufficiently large. And again, as we expected, for a given p and n Variant B2 performs far better even for smaller ratios n/p. It is clearly seen, how reducing the number of communication steps in the solution of the preconditioner improves the parallel performance.

Table 3. Sequential times

	Variant B1			Variant B2		
n	C1	C2	C3	C1	C2	C3
32	52.18	30.87	29.47	28.16	18.61	21.18
64	578.4	336.8	347.6	336.1	228.4	224.2
128	6596	3793	3556	3887	2556	2610

Table 4. Parallel speedups and efficiencies

		Variant B1						Variant B2					
		C1		C2		C3		C1		C2		C3	
n	p	$S(p)$	$E(p)$	$S(p)$	$E(p)$	$S(p)$	$E(p)$	$S(p)$	$E(p)$	$S(p)$	$E(p)$	$S(p)$	$E(p)$
32	2	1.49	0.74	1.31	0.66	1.77	0.88	1.93	0.96	1.33	0.66	1.97	0.99
	4	1.83	0.45	1.49	0.37	2.40	0.60	3.53	0.88	2.08	0.51	3.25	0.81
	8	2.11	0.26	1.22	0.15	3.34	0.42	5.78	0.72	3.07	0.38	5.20	0.65
	16	1.61	0.10	0.92	0.06	3.22	0.20	9.45	0.59	3.93	0.25	7.63	0.48
64	2	1.68	0.84	1.38	0.69	2.02	1.01	2.02	1.01	1.35	0.68	1.77	0.88
	4	2.46	0.61	1.98	0.49	3.17	0.79	3.92	0.98	2.49	0.62	3.50	0.87
	8	3.27	0.41	1.93	0.24	4.26	0.53	7.38	0.92	4.21	0.52	5.91	0.73
	16	3.78	0.23	2.06	0.13	6.03	0.38	12.83	0.81	6.53	0.40	8.64	0.54
128	2	1.82	0.91	1.51	0.76	1.56	0.78	2.00	1.00	1.49	0.74	1.93	0.96
	4	2.96	0.74	2.40	0.60	2.73	0.68	3.90	0.98	2.54	0.63	3.72	0.93
	8	4.50	0.56	2.70	0.34	5.34	0.67	7.33	0.92	4.59	0.57	7.30	0.91
	16	5.83	0.36	3.64	0.23	7.64	0.48	12.73	0.80	7.51	0.47	12.21	0.76

Acknowledgments

The numerical tests are supported via EC Project HPC-EUROPA RII3-CT-2003-506079. The author also gratefully acknowledges the support provided via EC INCO Grant BIS-21++ 016639/2005.

References

1. Arbenz, P., Margenov, S., Vutov, Y.: Parallel MIC(0) Preconditioning of 3D Elliptic Problems Discretized by Rannacher–Turek Finite Elements. Comput. Math. Appl (to appear)
2. Axelsson, O.: Iterative Solution Methods. Cambridge University Press, Cambridge (1994)
3. Axelsson, O., Gustafsson, I.: Iterative methods for the Navier equation of elasticity. Comp. Math. Appl. Mech. Engin. 15, 241–258 (1978)
4. Blaheta, R.: Displacement Decomposition — incomplete factorization preconditioning techniques for linear elasticity problems. Numer. Lin. Alg. Appl. 1, 107–126 (1994)
5. Malkus, D., Hughes, T.: Mixed finite element methods. Reduced and selective integration techniques: an uniform concepts. Comp. Meth. Appl. Mech. Eng. 15, 63–81 (1978)
6. Rannacher, R., Turek, S.: Simple nonconforming quadrilateral Stokes Element. Numer. Methods Partial Differential Equations 8(2), 97–112 (1992)

Author Index

Author Index

Lecture Notes in Computer Science

Sublibrary 1: Theoretical Computer Science and General Issues

For information about Vols. 1– 4647
please contact your bookseller or Springer